AF132775

Optical Properties of Excited States in Solids

NATO ASI Series

Advanced Science Institutes Series

A series presenting the results of activities sponsored by the NATO Science Committee, which aims at the dissemination of advanced scientific and technological knowledge, with a view to strengthening links between scientific communities.

The series is published by an international board of publishers in conjunction with the NATO Scientific Affairs Division

A	**Life Sciences**	Plenum Publishing Corporation
B	**Physics**	New York and London
C	**Mathematical and Physical Sciences**	Kluwer Academic Publishers
D	**Behavioral and Social Sciences**	Dordrecht, Boston, and London
E	**Applied Sciences**	
F	**Computer and Systems Sciences**	Springer-Verlag
G	**Ecological Sciences**	Berlin, Heidelberg, New York, London,
H	**Cell Biology**	Paris, Tokyo, Hong Kong, and Barcelona
I	**Global Environmental Change**	

Recent Volumes in this Series

Volume 294—Single Charge Tunneling: Coulomb Blockade Phenomena in Nanostructures
edited by Hermann Grabert and Michel H. Devoret

Volume 295—New Symmetry Principles in Quantum Field Theory
edited by J. Fröhlich, G.'t Hooft, A. Jaffe, G. Mack, P. K. Mitter, and R. Stora

Volume 296—Recombination of Atomic Ions
edited by W. G. Graham, W. Fritsch, Y. Hahn, and J. A. Tanis

Volume 297—Ordered and Turbulent Patterns in Taylor–Couette Flow
edited by C. David Andereck and F. Hayot

Volume 298—Chaotic Dynamics: Theory and Practice
edited by T. Bountis

Volume 299—Time-Dependent Quantum Molecular Dynamics
edited by J. Broeckhove and L. Lathouwers

Volume 300—Equilibrium Structure and Properties of Surfaces and Interfaces
edited by A. Gonis and G. M. Stocks

Volume 301—Optical Properties of Excited States in Solids
edited by Baldassare Di Bartolo
Assistant Editor: Clyfe Beckwith

Series B: Physics

Optical Properties of Excited States in Solids

Edited by

Baldassare Di Bartolo

Boston College
Chestnut Hill, Massachusetts

Assistant Editor

Clyfe Beckwith

Boston College
Chestnut Hill, Massachusetts

Springer Science+Business Media, LLC

Proceedings of an International School of Atomic and Molecular Spectroscopy Tenth Course and NATO Advanced Study Institute on Optical Properties of Excited States in Solids, held June 16–30, 1991, in Erice, Italy

NATO-PCO-DATA BASE

The electronic index to the NATO ASI Series provides full bibliographical references (with keywords and/or abstracts) to more than 30,000 contributions from international scientists published in all sections of the NATO ASI Series. Access to the NATO-PCO-DATA BASE is possible in two ways:

—via online FILE 128 (NATO-PCO-DATA BASE) hosted by ESRIN, Via Galileo Galilei, I-00044 Frascati, Italy

Library of Congress Cataloging-in-Publication Data

Optical properties of excited states in solids / edited by Baldassare Di Bartolo.
p. cm. -- (NATO ASI series. Series B, Physics ; v. 301)
"Published in cooperation with NATO Scientific Affairs Division."
"Proceedings of an International School of Atomic and Molecular Spectroscopy, tenth course, and NATO Advanced Study Institute on Optical Properties of Excited States in Solids, held June 16-30, 1991, in Erice, Italy"--T.p. verso.
Includes bibliographical references and index.
ISBN 978-0-306-44316-9 ISBN 978-1-4615-3044-2 (eBook)
DOI 10.1007/978-1-4615-3044-2
1. Solids--Optical properties--Congresses. 2. Exciton theory--Congresses. 3. Excited state chemistry--Congresses. 4. Molecular spectroscopy--Congresses. I. Di Bartolo, Baldassare. II. North Atlantic Treaty Organization. Scientific Affairs Division. III. International School of Atomic and Molecular Spectroscopy. IV. NATO Advanced Study Institute on Optical Properties of Excited States in Solids (1992 : Erice, Italy) V. Series.
QC176.8.060618 1992
530.4'12--dc20 92-30374
CIP

ISBN 978-0-306-44316-9

Friendship is the comfort, the inexpressible
comfort of feeling safe with a person,
having neither to weigh thoughts nor measure words,
but pouring all right out just as they are,
chaff and grain together,
certain that a faithful friendly hand
will take and sift them, keep what is worth keeping,
and with a breath of comfort, blow the rest away.

—George Eliot

PREFACE

This book presents an account of the course "Optical Properties of Excited States in Solids" held in Erice, Italy, from June 16 to 30, 1991. This meeting was organized by the International School of Atomic and Molecular Spectroscopy of the "Ettore Majorana" Centre for Scientific Culture.

The purpose of this course was to present physical models, mathematical formalisms and experimental techniques relevant to the optical properties of excited states in solids.

Some active physical species, such as ions or radicals, could survive indefinitely if they were completely isolated in space. Other active species, such as excited molecular and solid-state systems, are inherently unstable, even in isolation, due to the spontaneous mechanisms that may convert their excitation energies into radiation or heat. Physical parameters that may be used to characterize these excited systems are the localization or delocalization, and the coherence or incoherence, of their state excitations.

In solids the excited states, whether they are localized (as for impurities in insulators) or delocalized (as they may occur in semiconductors), are relevant in several regards. Their de-excitation is extremely sensitive to the nature of the excitations of the systems, and a study of the de-excitation processes can yield a variety of information. For example, the excited states may represent the initial condition of the onset of such processes as Stokes-shifted emission, hot luminescence, symmetry-dependent Jahn-Teller and scattering processes, tunneling processes, energy transfer to like and unlike centers, superradiance, coherent radiation, and excited state absorption.

Because of the inherent instability of excited electronic states, the physical phenomena related to them are of transient nature. This fact poses particular problems to the theoretical treatments and the experimental techniques. New ultra-fast detection devices have allowed measurement of the rapid processes that follow an electronic excitation. New developments have also been made in the techniques of coherent spectroscopy. The course reviewed the theoretical and experimental advances in the field and examined the role of the optical properties of excited solids.

Each lecturer developed a coherent section of the program starting at a somewhat fundamental level and ultimately reaching the frontier of knowledge in the field. The sequence of the lectures was in accordance with the logical development of the subject of the meeting. The formal lectures were complemented by seminars and discussions on specific topics and applications and by problem sessions.

The course was addressed to workers in the field of solid-state physics from universities and industries. Senior scientists in related fields were encouraged to participate.

A total of eighty-five participants came from fifty-five laboratories and twenty-four different countries (Belgium, Canada, Egypt, Finland, France, Germany, Greece, India, Ireland, Israel, Italy, Japan, the Netherlands, P.R. of China, Poland, Portugal, the Soviet Union, Spain, Sweden, Switzerland, Taiwan, Turkey, the United Kingdom, and the United States).

The secretaries of the course were Clyfe Beckwith and John Di Bartolo.

Thirty lectures divided into nine series were given. In addition, fifteen (one-, two-, or three-hour) "long seminars" and thirty-two "short seminars" were presented. Several problem sessions were organized: the purpose of these sessions was to create a closer interaction between participants and lecturers. During a problem session, a lecturer did not produce any new material, but explained in greater detail some points of the theory already presented, answered more questions, presented examples, etc.

Two round-table discussions were held. The first round-table discussion took place after three and a half days of lectures and had as subjects an evaluation of the work done in the first days of the course and the consideration of suggestions and proposals regarding the format of the lectures, the level of presentations, etc. This round-table discussion favorably affected the progress of the course. The second round-table discussion was held at the end of the course: the attendees had the opportunity of evaluating the work done during the entire meeting and to discuss various proposals for the next course of the International School of Atomic and Molecular Spectroscopy.

During the final session, Professor Imbusch presented a summary of the meeting and identified the major themes which appeared in various degrees throughout the school.

I wish to express my sincere gratitude to Drs. Gabriele and Savalli, to Ms. Zaini, to Mr. Pilarski, and to all the personnel of the "Ettore Majorana" Centre, who contributed so much to create a congenial atmosphere for our meeting. I wish to acknowledge the sponsorship of the meeting by the NATO Scientific Affairs Division, the ENEA Organization, Boston College, the European Physical Society, the Italian Ministry of Public Education, the Italian Ministry of Scientific and Technological Research, and the Sicilian Regional Government.

I would like to thank the members of the Organizing Committee (Dr. Baldacchini, Professor Reuss, and Dr. Auzel) and Professor Imbusch, the secretaries of the course (Clyfe Beckwith and John Di Bartolo), Brian Bowlby, Xuesheng Chen, Giovanni Galfano, and Daniel Di Bartolo.

I would also like to acknowledge the many friends from Trapani and Erice who were close to me during the time of the course, especially Michele Strazzera, Pino Luppino, Benedetto Fontana, Professore Adragna, Signori Piacentino, Signori Tartamella and my former schoolmates of the Liceo Scientifico of Trapani.

Every Erice meeting gives me the opportunity to return to Sicily and to my hometown Trapani. The memories of my early youth come back to my mind with renewed intensity and I savor them with the same relish with which I drink the strong Marsala wine. To direct a course at the Majorana Centre adds to my feelings because in Erice I spent several pleasant summers with my parents and relatives.

During the preparation of this book I have been making plans for the next meeting of the International School of Spectroscopy in 1993 and I am looking forward to it.

To encounter in my native land so many fine people who participate in my courses, to be able to share the Erice experience with them and to call them friends gives me the opportunity to tie in my thoughts the present with the past and to project towards the future a serene hope.

B. Di Bartolo
Director of the International School of Atomic and Molecular Spectroscopy of the "Ettore Majorana" Centre

Erice, June 30, 1991

CONTENTS

LONG SEMINARS

SPECIAL TOPICS

SHORT SEMINARS

THE NATURE OF THE ELECTRONIC EXCITED STATES OF MOLECULAR SYSTEMS

B. Di Bartolo

Department of Physics
Boston College
Chestnut Hill, MA 02167
USA

ABSTRACT

The purpose of this article is to present the background material necessary to deal with the properties of molecular species in excited electronic states.

In the first section the basic chemical and spectroscopic properties of excited molecular species are briefly examined.

The second section treats the interaction of radiation with atoms and molecules which are considered simply as two-level systems. The concept of transition rate is introduced. In addition, the optical Bloch equations, the Rabi oscillations, and the mechanisms responsible for the broadening of spectral lines are dealt with.

The third section presents a treatment of the density matrix formulation, first in a general context, and then in regard to a two level system. The formalism is applied to the case of an ensemble of radiating systems and the properties of the density matrix to be used are examined.

The fourth section uses the gyroscopic model in order to illustrate the rotating wave approximation. The model is made more relevant by adding damping and losses to it. The experimental conditions leading to the formation of photon echoes are reviewed and interpreted in terms of the gyroscopic model.

The fifth section examines the properties of two-level systems, when excited coherently and incoherently. First the sources responsible for such excitations are examined; then the attention is focussed on the two characteristic times, the longitudinal relaxation time T_1 and the transversal relaxation time T_2, and on the mechanisms responsible for them. The effects of such mechanisms on the widths of spectral lines are examined. Methods for the determination of T_1 and T_2 are also considered.

Optical Properties of Excited States in Solids
Edited by B. Di Bartolo, Plenum Press, New York, 1992

I. SPECTROSCOPY AND CHEMISTRY OF EXCITED MOLECULAR SPECIES

I.A. General Considerations

In order to explain the physical properties of matter, physicists have attempted to relate these properties to the elementary constituents of the physical bodies. Since matter, in its gaseous, liquid or solid form, is made up of atoms and molecules, the key to the understanding of the physical properties of matter is the understanding of atomic and molecular structure and processes.

Various atomic models were already in existence by the middle of the past century: Dalton's model in which atoms had hooks and eyes and could be joined together in chemical compounds, or the model used in kinetic theory in which the atoms were thought to be smooth and tiny balls continually moving and colliding. Each model had some range of applications and validity; however, at the end of the past century no model existed that could explain all the features of the atom. Bunsen and Kirchoff, by using the light emitted by atoms for chemical analysis, showed that each atom had its own characteristic spectrum and made physicists realize that the understanding of atomic structure had to be based on the interpretation of atomic spectra. Subsequently the work of Planck and Bohr established the model of the quantum atom with its discrete energy levels. In the hydrogen atom, the simplest atomic system, the fundamental energy level lies at -13.6 electron volts and the next highest level at -3.4 electron volts, so that the energy gap between the first two energy levels is 10.2 electron volts; in addition the energy gap between adjacent levels becomes smaller and smaller going up in energy and disappears completely for energies above the ionization level. It turns out that all atoms and molecules have a large energy gap between their lowest and their next highest electronic energy levels: the reason for the generality of this property will become evident a little later. This large energy gap is the cause of the relative stability of atomic and molecular species. These species are continuously colliding with other molecules or atoms which perform endless thermal motion. However, the energy that can be transferred to a molecule in such a collision is much smaller than the electronic energy gap; therefore the large gap assures that no collision will result in an excitation of the molecule that could be the initial step of an undesired transmutation. In the long complex molecules which embody our genetic code, the large gap assures their stability and the proper functioning of our body chemistry.

A breakthrough in the understanding of atoms was produced by Louis de Broglie, who, considering the dual (wave-particle) nature of light, argued that particles, and in particular electrons, could also present a dual nature and act as waves in some circumstances. The discreteness of the energy levels and of the electron orbits in atoms was, in light of this concept, explainable as due to the discrete "resonances" that occur when an integral number of wavelenghts fits the length of the orbit. In the simplest case of the H atom, while the de Broglie wavelength grows proportionally to the principal quantum number n, the radius and consequently the perimeter of the orbits grow proportional to n^2; in other words, the first orbit corresponds to one wavelength, the second to two wavelengths, the thousandth to a thousand wavelengths, etc. The fact that the length of the orbit becomes increasingly larger than the wavelength is responsible for the "crowding" of the energy levels near the ionization limit.

A similar crowding of resonances takes place in other physical systems. I recall from my days as a microwave engineer the fun I had in designing high-Q microwave resonant cavities (called echo-boxes) for radar-testing. The high Q required dimensions for the cavity considerably larger than the microwave wavelength; and here the trouble started, for the longitudinal dimension of the cavity had to be varied for tuning purposes. The resonant modes of the cavity were so closely spaced that a slight change in the volume of the cavity would unravel a complex maze of spurious resonances.

I.B. The Nature of the Electronic Excited State

The electronic excitation of molecular species has effects on their chemical and spectroscopic properties. The electronic excitation may result in an electron being raised to a higher and less bound energy state or in an electron actually leaving the molecule which remains as a positive ion.

An excited species may then undergo one of two possible processes:

1) It may return to its original deexcited form losing its energy by some physical process (for example, by radiationless decay). In the case of an ion it may return to its atomic form by capturing an electron.

2) It may react with a neighboring molecule. The chemical reactivity of an excited species depends essentially on the time involved in the process 1) above.

Ionic species (and, for that matter, radicals) are not inherently unstable, in the sense that their mean life depends on the possibility of reacting with nearby molecules or ions; in most cases they would last indefinitely if completely isolated. As expected, they have a longer life in the gaseous than in the condensed phase where the probability of recombination is much higher and may reduce the actual lifetime to the time it takes to undergo a single collision.

In contrast to this, excited molecular species are unstable, even in isolation. This is due to several "spontaneous" deexcitation mechanisms that are always present: the radiative decay [1] by which the excited molecule reemits a photon and returns to its ground state configuration and the radiationless decay by which the molecule uses its electronic excitation energy to enhance the thermal vibrations of its atoms, transforming in effect its electronic energy into heat energy. The times involved in such processes vary greatly from molecule to molecule but are not greatly different for molecules in the condensed or gaseous phase.

The characteristic times for the radiative processes are 10^{-8} - 10^{-9} sec, but for some (metastable) states may be as long as 10^{-3} sec. (Radiationless processes are effective when they take place in shorter times.) These times are longer than collision times in, say, solutions; therefore there may be enough time for the excited molecule to encounter another molecular species before losing its excitation energy.

I.C. The Excited State and Chemical Reactions

A chemical reaction is an event in which bonds between atoms are broken or rearranged, as the result of the nuclei moving to positions of minimum energy in the potential set up by an electron charge distribution different from the one of the isolated atoms. The very first act of a chemical reaction may be thought of as consisting of the rearrangement of the electron charge distribution while the system preserves its geometrical structure; the setting up of this new electron pattern creates a new potential in which the nuclei move to seek positions of minimum energy. The rearrangement of the electron charge takes place because the electron orbits are modified by the "mixing" with (i.e., the partial use of) orbits not normally used in the reacting molecules. This mixing of normal with excited orbital states is the very first event of a chemical reaction and precedes the formation of reaction intermediates; the primary act of a reaction is therefore electronic in nature [2]. Since in most cases the reactant species are stable by themselves, the perturbation of the electron charge distribution results in an increase of the potential energy of the system. Consequently a small change in the positions of the atoms will simply be followed by the movement of the reaction back in the direction of the reactants; if the change exceeds some critical value, the atoms will continue to seek their new positions and the reaction will proceed in the direction of the products. In effect, the system must go over a "hump" for the reaction to ensue; the reduction of the height of this hump, i.e. the reduction of the amount of the energy that must be given to the system in order to undergo the decisive initial rearrangement, is the task of catalytic agents.

In the simple case of a reaction in which two linear molecules are involved the path of the reaction can be followed by plotting the potential energy of the system versus the relative distance of the two molecules. In such a diagram, reactants and products reside in stable equilibrium positions corresponding to minima in the potential curve; the "activated complex", an intermediate arrangement of atoms, is a metastable equilibrium point at the top of the potential energy barrier. For more complex molecules it may be impossible to visualize the reaction path by plotting the potential curve versus a single coordinate; since several nuclei may be simultaneously changing their positions, a multi-dimensional plot may be in principle necessary to represent the evolution of the system.

One question now arises: how do we provide the system with energy sufficient to go over the hump, i.e., how do we get the reaction started? The necessary energy may be provided thermally by heating the system, or by an exothermic reaction occurring in the vicinity. The energy may also be provided by radiation if the system absorbs light. In the first two cases the reaction takes place while the molecules, which are initially in their ground electronic states, remain in these states; the rearrangement of the electron charge takes place by the partial use of excited electronic states, but the process enfolds continuously without any electron jump from an orbit to another orbit of higher energy.

On the other hand, when abosrption of radiation brings the system to an excited electronic state, the ensuing charge distribution change may be such as to modify drastically the potential energy pattern, therefore altering the evolution of the reaction. It should not be surmised that electronic excitation necessarily accelerates the course of a reaction; however, it may happen that the changes of the potential energy are such as to remove completely the hump and to make the reaction proceed much faster than it would in the absence of electronic excitation.

I.D. The Spectroscopy of the Excited State

In order to deal with the effects of the electronic excitation of molecular species on the spectral properties of these systems we shall cite the following topics:

1) Excited state absorption, observed if the number of excited species is large enough,
2) coherence and incoherence of excitation, and
3) pathways and patterns of decay from the excited states.

This article will deal extensively with the last two topics. We shall treat now briefly the first subject.

The most basic spectroscopic technique consists in sending a beam of light containing a wide spectrum of wavelengths through the material to be examined and in comparing the spectral distribution of the output beam with that of the input beam. This comparison uncovers the wavelengths at which the material absorbs and therefore it may be used to establish the energy level scheme of the system. Absorption spectroscopy is the best spectroscopic tool for the identification of molecular species. When these species are in excited electronic (or for that matter, vibrational) states, some complications may arise.

The absorption spectra of molecular species are normally related to transitions that initiate in the ground state and end in an excited state. The basic assumption in absorption spectroscopy is that, before the absorption process, a negligible number of molecules are in excited states. Situations may arise in which the number of excited molecules is not negligible; such a situation may be created by "pumping" enough light into the molecule's absorption bands. From these bands the excited molecules may decay by some rapid (radiationless) processes to a metastable state. When equilibrium between the radiation and the molecular system is reached, a relevant number of molecules may actually find themselves in the metastable state. In these conditions the system behaves differently, with respect to an incoming beam of light, than an "unpumped" system, because an absorption process may take place from the metastable level to higher levels.

Excited-state absorption has been observed in molecular systems. The lowest triplet levels of organic molecules have very long lifetimes and can be populated by intersystem crossing from upper singlet states; absorption spectra corresponding to transitions from the triplet levels upwards have been observed in many aromatic molecules in fluid solvents by Porter and Windsor [3]. The use of laser excitation has made it possible to observe transitions also from upper singlet levels, which have much shorter lifetime; the use of the laser has allowed the observation of absorption spectra arising from the lowest excited singlet states of several aromatic molecules [4].

Simple dihalide molecules such as Cl_2 and Br_2 also present absorption arising from the excited state corresponding to their normal abosrption bands [5].

Excited-state absorption experiments have been done in such solids as $Al_2O_3{:}Cr^{3+}$ (ruby) [6,7], in glasses doped with Gd [8], U, and Er [9] and in $SrF_2{:}Sm^{2+}$ [10]. Sometimes a greater cross section can be expected for transitions starting from an excited state than from the ground state; this may be the case of Gd^{3+}, in which the ground state is ^8S and the excited states are sextets. Excited-state absorption transitions from the resonant level ($^6P_{7/2}$) to other upper sextets respect the spin selection rule $\Delta S = 0$, contrary to what happens in ground state absorption transitions.

It is clear, therefore, that excited excited-state absorption may represent an important tool for the evaluation of energy levels that are weakly connected to the ground state and that can be seen only with a high concentration of absorbing centers in conventional absorption.

When a system is being pumped for the purpose of achieving laser oscillations, excited-state absorption may alter the kinetics of the excitation process and may affect the efficiency of the laser system [11,12].

II. INTERACTION OF RADIATION WITH ATOMS AND MOLECULES

II.A. Two-Level Systems

Let us consider a system with a time-independent Hamiltonian H_o. The time-dependent Schroedinger equation gives

$$H_o\psi = i\hbar \frac{\partial\psi}{\partial t} \tag{2.1}$$

If the system is in a stationary state labeled i

$$\psi(t) = \psi_i(t) = e^{-i(E_i/\hbar)t}\, \psi_i(0) \tag{2.2}$$

where the energy values are given by

$$H_o\, \psi_i(0) = E_i\, \psi_i(0) \tag{2.3}$$

We shall assume that the wavefunctions $\psi_i(t)$ are orthonormal.

Let us now suppose that the system is subjected to a time-dependent perturbation represented by $H'(t)$. The system will be represented by a wavefunction $\psi(t)$ such that

$$H\psi(t) = (H_o + H')\, \psi(t) = i\hbar \frac{\partial\psi(t)}{\partial t} \tag{2.4}$$

We can expand $\psi(t)$ in terms of the complete set $\psi_i(t)$

$$\psi(t) = \sum_i c_i(t)\, \psi_i(t) \tag{2.5}$$

If $H' = 0$, the coefficients c_i's are time-independent. Replacing Eq. (2.5) in Eq. (2.4),

$$\left(H_o + H'\right) \sum_i c_i(t)\, \psi_i(t) = i\hbar \left[\sum_i c_i(t) \frac{\partial\psi_i(t)}{\partial t} + \sum_i \frac{\partial c_i(t)}{\partial t}\, \psi_i(t)\right] \tag{2.6}$$

Then

$$\sum_i c_i(t)\, H'\, \psi_i(t) = i\hbar \sum_i \frac{\partial c_i(t)}{\partial t}\, \psi_i(t) \tag{2.7}$$

where we have taken advantage of eq. (2.2) and (2.3). Multiplying by $\psi_k^*(t)$ and integrating over all space we obtain

$$i\hbar \frac{\partial c_k(t)}{\partial t} = \sum_i c_i(t) < \psi_k(t)|H'|\psi_i(t) > = \sum_i c_i(t)\, M_{ki}\, e^{i\omega_{ki}t} \tag{2.8}$$

where

$$\omega_{ki} = \frac{E_k - E_i}{\hbar}\ ;\ M_{ki} = <\psi_i(0)|H'|\psi_i(0)> \tag{2.9}$$

We shall now make the following simplifying assumptions:

(a) The system has only two energy levels, say 1 and 2, and
(b) the diagonal matrix elements of H' are zero.

The coupled equations (2.8) become

$$\begin{cases} i\dot{c}_2(t) = c_1 V_{21}\, e^{i\omega_o t} \\ i\dot{c}_1(t) = c_2 V_{12}\, e^{-i\omega_o t} \end{cases} \tag{2.10}$$

where

$$\omega_o = \frac{E_2 - E_1}{\hbar} \; ; \quad V_{ij} = \frac{M_{ij}}{\hbar} \tag{2.11}$$

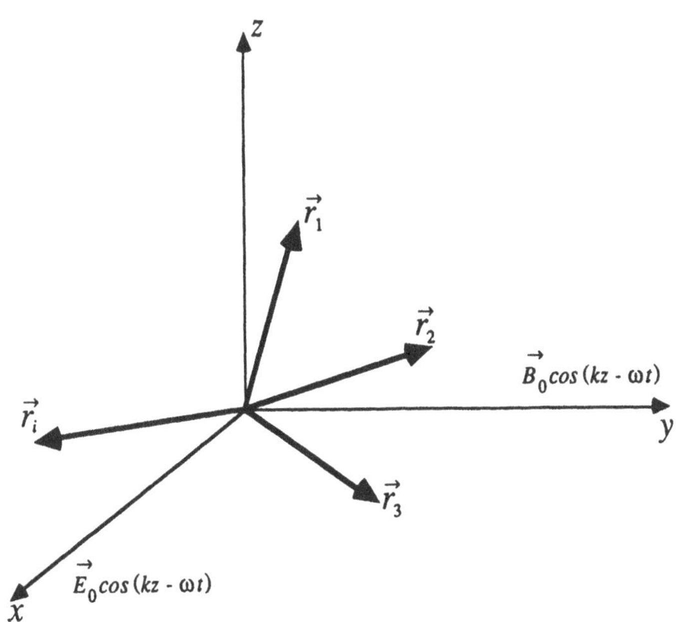

Figure 1. An Atom Interacting with Radiation.

II.B. The Hamiltonian of the Interaction with Radiation

Let a polarized electromagnetic wave interact with an atom. Let the nucleus of the atom be at the origin of a system of coordinates as in Fig. 1 and let the electrons have coordinates $\vec{r}_i$; let also be x the direction of polarization of the $\vec{E}$ field. The size of an atom is of the order of the Bohr radius

$$a_o = \frac{4\pi\epsilon_o \hbar^2}{me^2} = 5 \times 10^{-9} \text{ cm} << \lambda \text{ radiation} \tag{2.12}$$

Then

$$k\ a_o << 1 \tag{2.13}$$

and the field acting on the atom is given by

$$\vec{E} = \vec{E}_o \cos \omega t \tag{2.14}$$

with no variation of $\vec{E}$ across the atom. The total electric dipole moment is given by

$$e\vec{p} = e \sum_{i=1}^{Z} \vec{r}_i \tag{2.15}$$

where Z = number of electrons in the atom. The interaction Hamiltonian can be written as follows

$$H' = -e\vec{p} \cdot \vec{E}_o \cos \omega t \tag{2.16}$$

$\vec{p}$, and therefore H', are odd operators: then

$$V_{11} = V_{22} = 0 \tag{2.17}$$

But

$$V_{12} = \frac{1}{\hbar} <\psi_1(0)|H'|\psi_1(0)> = \frac{1}{\hbar} <\psi_1(0)|-e\vec{p}\cdot\vec{E}_o \cos \omega t|\psi_2(0)> =$$

$$= - \frac{eE_o}{\hbar} \cos \omega t <\psi_1(0)|p^x|\psi_2(0)> = - \frac{eE_o\ p_{12}^x}{\hbar} \cos \omega t =$$

$$= V \cos \omega t \tag{2.18}$$

where

$$V = - \frac{eE_o\ p_{12}^x}{\hbar} = - \frac{eE_o}{\hbar} <\psi_1(0)|p^x|\psi_2(0)> \tag{2.19}$$

and

$$p^x = \sum_{i=1}^{Z} e\ x_i \tag{2.20}$$

<u>Example</u>: Atom of Hydrogen [13]

Let $|\psi_1> = 1s$ state

$|\psi_2> = 2p_x$ state

We can write:

$$\psi_1 = \frac{1}{\sqrt{\pi\ a_o^3}}\ e^{-r/a_o}$$

$$\psi_2 = \frac{1}{\sqrt{2^5 \pi a_o^5}} e^{-r/2a_o} x$$

The energy of the various levels is

$$E_n = - \frac{me^4}{32\pi^2 \varepsilon_o^2 \hbar^2 n^2}$$

Then

$$V = - \frac{eE_o}{\hbar} \langle\psi_1|x|\psi_o\rangle = 5.93 \times 10^4 E_o \sec^{-1}$$

But

$$\hbar\omega_o = E_2 - E_1 = - \frac{me^4}{32\pi^2 \varepsilon_o^2 \hbar^2} \left(\frac{1}{4} - 1\right) = \hbar \, (1.63 \times 10^{16})$$

In order to make $V \approx \omega_o$ we need a field strength of $\sim 3{\times}10^{11}$ V/m. For light beams produced by conventional (non-laser) sources

$$V << \omega_o$$

The upper limit for conventional spectroscopic sources is represented by the field $E_o = 10^3$ V/m produced by a mercury lamp with its emission line at 2537 A.

II.C. Transition Rates

Using the expression (2.18) for the matrix element V_{12}, we can rewrite the eqs. (2.10) as follows:

$$\begin{cases} V \cos \omega t \, e^{-i\omega_o t} c_2 = i\dot{c}_1 \\ V^* \cos \omega t \, e^{i\omega_o t} c_1 = i\dot{c}_2 \end{cases} \tag{2.21}$$

$|c_2(t)|^2$ represents the probability of finding the system in its upper state ψ_2 at time t, and $|c_2(t)|^2/t$ the rate at which this probability increases.

We set the initial conditions:

$$\begin{cases} c_1(0) = 1 \\ c_2(0) = 0 \end{cases} \tag{2.22}$$

and use the approximation

$$c_1(t) \approx 1 \tag{2.23}$$

We obtain

$$c_2(t) = \frac{V^*}{2}\left[\frac{1 - e^{i(\omega_o+\omega)t}}{\omega_o + \omega} + \frac{1 - e^{i(\omega_o-\omega)t}}{\omega_o - \omega}\right] \tag{2.24}$$

We use an additional approximation by neglecting the first term in the square brackets in (2.24). We shall have other occasions to discuss at length this approximation, called, for reasons which will become clear later, the rotating wave approximations. For the moment it will suffice to say that when $\omega \approx \omega_o$ the second term in [ ] is much larger than the first and there is a reason for such an approximation. Then

$$c_2(t) = \frac{V^*}{2}\ \frac{1 - e^{i(\omega_o-\omega)t}}{\omega_o - \omega} \tag{2.25}$$

$$|c_2(t)|^2 = \frac{|V|^2}{4(\omega_o-\omega)^2}\left(1 - e^{i(\omega_o-\omega)t}\right)\left(1 - e^{-i(\omega_o-\omega)t}\right) =$$

$$= |V|^2\ \frac{\sin^2 \frac{\omega_o-\omega}{2}\ t}{(\omega_o-\omega)^2} \tag{2.26}$$

When $\omega = \omega_o$

$$|c_2(t)|^2 = \frac{1}{4}\ |V|^2\ t^2 \tag{2.27}$$

namely the probability of excitation is proportional to t^2. For $\omega \neq \omega_o$ such probability has an oscillatory behavior. Fig. 2 shows the dependence of the probability of excitation at a certain time t on the angular frequency ω.

Integrating over a range $\Delta\omega$ we obtain

$$|c_2(t)|^2 = \int_{\omega_o - \frac{\Delta\omega}{2}}^{\omega_o + \frac{\Delta\omega}{2}} |V|^2\ \frac{\sin^2\left[(\omega_o-\omega)t/2\right]}{(\omega_o-\omega)^2}\ d\omega$$

$$= \frac{2e^2|p^x_{12}|^2\ W(\omega_o)}{\varepsilon_o\hbar^2}\int_{\omega_o - \frac{\Delta\omega}{2}}^{\omega_o + \frac{\Delta\omega}{2}} \frac{\sin^2\left[(\omega_o-\omega)t/2\right]}{(\omega_o-\omega)^2}\ d\omega\ q \tag{2.28}$$

where we have made use of the relations

$$V = \frac{e\ E_o\ p^x_{12}}{\hbar} \tag{2.29}$$

$$\frac{1}{2}\ \varepsilon_o\ E_o^2 = W(\omega)d\omega \tag{2.30}$$

$W(\omega)d\omega$ being the energy density of the wave for frequencies in $(\omega,\ \omega+d\omega)$.

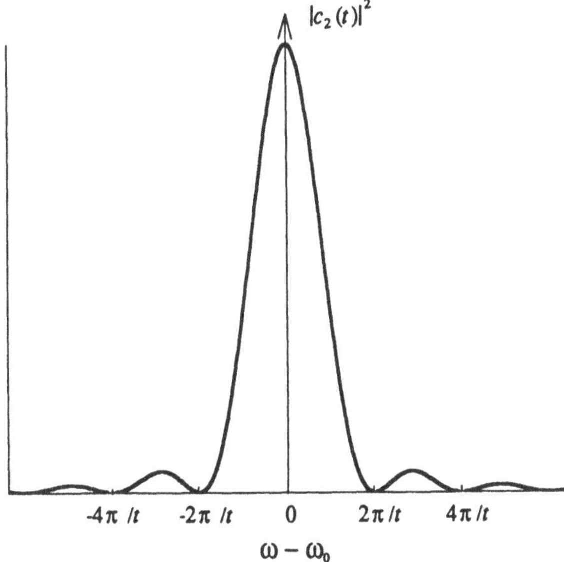

Figure 2. The Probability of Finding the System in State ψ_2 at Time t.

The value of the integral is $\frac{t^2}{4}\Delta\omega$ for $t\,\Delta\omega << 1$ and $\frac{1}{2}\pi t$ for $t\Delta\omega >> 1$. The latter condition, which we shall hold true, leads to a transition probability $|c_2(t)|^2$ proportional to t:

$$|c_2(t)|^2 = \frac{2e^2|p_{12}^x|^2\, W(\omega_o)}{\varepsilon_o \hbar^2}\,\frac{1}{2}\,\pi\, t = \frac{\pi\, e^2\, |p_{12}^x|^2\, W(\omega_o)t}{\varepsilon_o \hbar^2} \tag{2.31}$$

The linear approximation expressed above breaks down for t long enough to make $|c_2(t)|^2 > 1$, contrary to the normalization condition. However, as long as $t >> \tau_R$ = radiative lifetime, the linear dependence of the transition probability on time is valid.

II.D. Optical Bloch Equations

In the presence of a perturbation H'(t) the wavefunction of a two-level system is expressed by

$$\psi(\vec{r},t) = c_1(t)\,\psi_1(\vec{r},t) + c_2(t)\,\psi_2(\vec{r},t) \tag{2.32}$$

where $c_1(t)$ and $c_2(t)$ are such that

$$|c_1(t)|^2 + |c_2(t)|^2 = 1 \tag{2.33}$$

and are solutions of

$$\begin{cases} V \cos \omega t \; e^{-i\omega_o t} \; c_2 = i\dot{c}_1 \\ V^* \cos \omega t \; e^{i\omega_o t} \; c_1 = i\dot{c}_2 \end{cases} \tag{2.34}$$

We now look for more general solutions of these equations, by doing the following:

1) We use the rotating wave approximation,
2) we retain terms in all orders in V, and
3) we assume the electromagnetic radiation interacting with the two-level system monochromatic.

We define an <u>atomic density matrix</u> as follows:

$$\begin{cases} D_{11} = c_1 c_1^* = |c_1|^2 = \frac{N_1}{N} \\ D_{22} = c_2 \, c_2^* = |c_2|^2 = \frac{N_2}{N} \\ D_{12} = c_1 \, c_2^* \\ D_{21} = c_2 \, c_1^* \end{cases} \tag{2.35}$$

where N_1, N_2 and N are the populations in level 1, level 2 and total, respectively. Note that

$$\begin{cases} D_{11} + D_{22} = 1 \\ D_{12} = D_{21}^* \end{cases} \tag{2.36}$$

Then

$$\dot{D}_{22} = c_2^* \, \dot{c}_2 + \dot{c}_2 \, c_2^* = c_2^* \frac{V^*}{i} \cos \omega t \; e^{i\omega_o t} \; c_1 + \text{c.c.}$$

$$= -i \cos \omega t \left[V^* \, e^{i\omega_o t} \, D_{12} - V \, e^{-i\omega_o t} D_{21} \right] = - \dot{D}_{11} \tag{2.37}$$

and

$$\dot{D}_{12} = c_1 \dot{c}_2^* + c_2^* \dot{c}_1 = c_1 \left(\frac{V^*}{i} \cos \omega t \, e^{i\omega_o t} c_1 \right)^* + c_2^* \left(\frac{V}{i} \cos \omega t \, e^{-i\omega_o t} c_2 \right)$$

$$= iV \cos \omega t \, e^{-i\omega_o t} \left(D_{11} - D_{22} \right) = \dot{D}_{21}^* \tag{2.38}$$

We now use the rotating wave approximation; accordingly we consider the terms oscillating with frequence $\omega_o + \omega$ neglibible with respect to the terms oscillating at frequency $\omega_o - \omega$. We write

$$\begin{cases} \dot{D}_{22} = - \dot{D}_{11} = - \frac{i}{2} V^* e^{i(\omega_o-\omega)t} D_{12} + \frac{i}{2} V e^{-i(\omega_o-\omega)t} D_{21} \\ \dot{D}_{12} = \dot{D}_{21}^* = \frac{i}{2} V e^{-i(\omega_o-\omega)t} (D_{11} - D_{22}) \end{cases} \tag{2.39}$$

These equations are called the Bloch Optical equations [14].

II.E. Rabi Oscillations

The optical Bloch equations can be written as follows:

$$\begin{cases} \dot{D}_{11} = \frac{i}{2} V^* e^{i(\omega_o-\omega)t} D_{12} - \frac{i}{2} V e^{-i(\omega_o-\omega)t} D_{21} \\ \dot{D}_{22} = - \frac{i}{2} V^* e^{i(\omega_o-\omega)t} D_{12} + \frac{i}{2} V e^{-i(\omega_o-\omega)t} D_{21} \\ \dot{D}_{12} = \frac{i}{2} V e^{-i(\omega_o-\omega)t} D_{11} - \frac{i}{2} V e^{-i(\omega_o-\omega)t} D_{22} \\ \dot{D}_{21} = - \frac{i}{2} V^* e^{i(\omega_o-\omega)t} D_{11} + \frac{i}{2} V^* e^{i(\omega_o-\omega)t} D_{22} \end{cases} \tag{2.40}$$

We shall introduce the following trial solutions

$$\begin{cases} D_{11} = D_{11}^o e^{\alpha t} \\ D_{22} = D_{22}^o e^{\alpha t} \\ D_{12} = D_{12}^o e^{-i(\omega_o-\omega)t} e^{\alpha t} \\ D_{21} = D_{21}^o e^{i(\omega_o-\omega)t} e^{\alpha t} \end{cases} \tag{2.41}$$

with the quantities D_{ij}^o and α independent of time.

Using the relations (2.41) in the equation (2.40) we obtain a system of homogeneous equations in the unknown quantities D^o_{ij}:

$$\begin{pmatrix} -\alpha & 0 & \frac{i}{2}V^* & -\frac{i}{2}V \\ 0 & -\alpha & -\frac{i}{2}V^* & \frac{i}{2}V \\ \frac{i}{2}V & -\frac{i}{2}V & i(\omega_o-\omega)-\alpha & 0 \\ -\frac{i}{2}V^* & \frac{i}{2}V^* & 0 & -i(\omega_o-\omega)-\alpha \end{pmatrix} \begin{pmatrix} D^0_{11} \\ D^0_{22} \\ D^0_{12} \\ D^0_{21} \end{pmatrix} = 0 \tag{2.42}$$

These equations admit solutions if the determinant of the coefficients is equal to zero:

$$\alpha^2\left[\alpha^2 + (\omega_o-\omega)^2 + |V|^2\right] = 0 \tag{2.43}$$

The possible values of α are

$$\begin{cases} \alpha_1 = 0 \\ \alpha_2 = i\Omega \\ \alpha_3 = -i\Omega \end{cases} \tag{2.44}$$

where

$$\Omega = \sqrt{(\omega_o-\omega)^2 + |V|^2} \tag{2.45}$$

We note here that the dependence of Ω on $|V|^2$ indicates a change in the frequency of oscillations of the "coupled" system which consist of (atom and light beam): this effect is called the dynamic Stark effect.

The most general solution is then

$$D_{ij} = D^{(1)}_{ij} + D^{(2)}_{ij}\, e^{i\Omega t} + D^{(3)}_{ij}\, e^{-j\Omega t} \tag{2.46}$$

Additional oscillatory exponentials are present in the off-diagonal elements. By using the initial conditions and the optical Bloch equations we can obtain the constant coefficients.

Example

$$\begin{aligned} c_1(0) = 1 &\longrightarrow D_{22}(0) = 0 \\ c_2(0) = 0 &\longrightarrow D_{12}(0) = 0 \end{aligned} \tag{2.47}$$

The density matrix elements are

$$\begin{cases} D_{22} = \dfrac{|V|^2}{\Omega^2} \sin^2 \frac{1}{2}\,\Omega t \\[2ex] D_{11} = 1 - D_{22} = 1 - \dfrac{|V|^2}{\Omega^2} \sin^2 \Omega t \\[2ex] D_{12} = e^{-i(\omega_o-\omega)t} \dfrac{V}{\Omega^2} \sin \frac{1}{2}\,\Omega t \left[- (\omega_o-\omega) \sin(\frac{1}{2}\Omega t) + i\,\Omega \cos (\frac{1}{2}\Omega t)\right] \\[2ex] D_{21} = e^{i(\omega_o-\omega)t} \dfrac{V^*}{\Omega^2} \sin \frac{1}{2}\,\Omega t \left[-(\omega_o-\omega) \sin (\frac{1}{2}\,\Omega t) - i\,\Omega \cos (\frac{1}{2}\,\Omega t)\right] \end{cases} \tag{2.48}$$

For zero detuning ($\omega = \omega_o$): $\Omega = |V|$ and

$$\begin{cases} D_{22} = \sin^2 \frac{1}{2}\,|V|t \\ D_{12} = i\,\dfrac{V}{|V|} \sin (\frac{1}{2}|V|t) \cos (\frac{1}{2}|V|t) \end{cases} \tag{2.49}$$

The behavior of the quantity D_{22} for different values of detuning

$$d = \frac{\omega_o-\omega}{|V|} \tag{2.50}$$

is represented in Fig. 3. We shall make the following observations:

1) For zero detuning ($\omega = \omega_o$) the atom oscillates between ground and excited states: these oscillations are called <u>Rabi oscillations</u> and their frequency $|V|$ is called <u>Rabi frequency</u>. Solutions for the similar problem of a spin system in an oscillatory magnetic field were obtained by Rabi [15]. When

$$\frac{|V|}{2}t = \frac{eE_o|p_{12}^x|}{2\hbar}t = \pi \tag{2.51}$$

$D_{22} = 1$, all the atoms are in the upper state, i.e. the population is completely <u>inverted</u>. When

$$\frac{|V|}{2}t = \frac{eE_o|p_{12}^x|}{2\hbar}t = 2\pi \tag{2.52}$$

the original situation, with all the atoms in the ground state, is restored.

2) Since the rotating wave approximation was originally used in deriving the optical Bloch equations the solutions (2.48) are valid only if $(\omega_o - \omega) << (\omega_o + \omega)$.

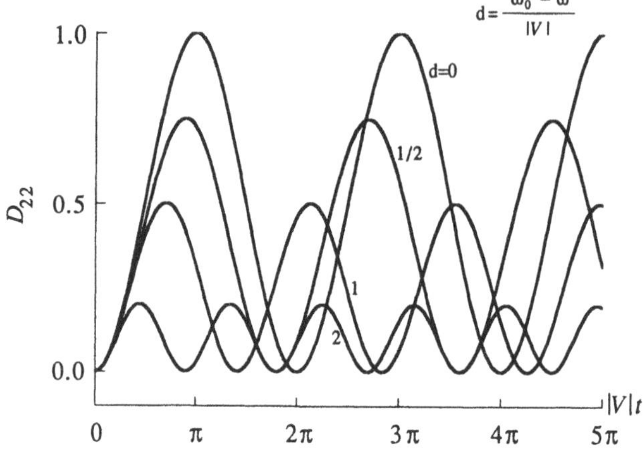

Figure 3. Probability of Excitation as Function of Time.

3) The solutions for D_{22} and D_{12} refer to monochromatic radiation. In effect the oscillations can be seen experimentally when the frequency spread of the electromagnetic radiation is much smaller than the linewidth of the transition.

4) We have to dedicate some attention to the fact that the processes that broaden the linewidth of the transition introduce modifications in the optical Bloch equations. We shall return on this last point.

II.F. Broadening of Spectral Lines

1. Definition of Susceptibility. Consider a gas of atoms and apply to it an electric field

$$E(t) = \int_{-\infty}^{+\infty} E(\omega)\, e^{-i\omega t}\, d\omega \tag{2.53}$$

where

$$E(\omega) = \frac{1}{2\pi} \int_{-\infty}^{+\infty} E(t)\, e^{i\omega t}\, dt \tag{2.54}$$

Since E(t) is real

$$E\,(-\omega) = E^*(\omega) \tag{2.55}$$

Let P(t) = atomic polarization:

$$P(t) = \int_{-\infty}^{+\infty} P(\omega)\, e^{i\omega t}\, d\omega \tag{2.56}$$

where

$$P(\omega) = \varepsilon_o\, \chi(\omega)\, E(\omega) \tag{2.57}$$

and $\chi(\omega)$ = electric susceptibility.

Then

$$P(t) = \varepsilon_o \int_{-\infty}^{+\infty} \chi(\omega)\, E(\omega)\, e^{-i\omega t}\, d\omega \tag{2.58}$$

A "real" stimulus E(t) must produce a "real" response P(t): then

$$\chi(-\omega) = \chi^*(\omega) \tag{2.59}$$

Therefore, if we write in general

$$\chi(\omega) = \chi'(\omega) + i\chi''(\omega) \tag{2.60}$$

(2.59) gives

$$\begin{cases} \chi'(-\omega) = \chi'(\omega) \\ \chi''(-\omega) = -\chi''(\omega) \end{cases} \tag{2.61}$$

The above relations are called crossing relations for the real and imaginary parts of the susceptibility.

2. Electric Dipole Moment. Consider a gas of Z-electron, 2-level atoms, and apply to it an electric field along the x direction:

$$E(t) = \frac{1}{2} E_o (e^{-i\omega t} + e^{i\omega t}) \tag{2.62}$$

A polarization will set in:

$$P(t) = \frac{1}{2} \varepsilon_o E_o \left[\chi(\omega)\, e^{-i\omega t} + \chi(-\omega)\, e^{i\omega t} \right] \tag{2.63}$$

The electric dipole moment of one atom in the x direction will be

$$d(t) = - \langle \psi(t) | e\, p^x | \psi(t) \rangle \tag{2.64}$$

where

$$ep^x = e \sum_{i=1}^{Z} x_i \tag{2.65}$$

But

$$\psi(t) = c_1(t)\, \psi_1(\vec{r}, t) + c_2(t)\, \psi_2(\vec{r}, t)$$

$$= c_1(t)\, e^{-i(E_1/\hbar)t}\, \psi_1(\vec{r}) + c_2(t)\, e^{-i(E_2/\hbar)t}\, \psi_2(\vec{r}) \tag{2.66}$$

Then

$$d(t) = -\langle c_1\, e^{-i(E_1/\hbar)t}\, \psi_1(\vec{r}) + c_2\, e^{-i(E_1/\hbar)t}\, \psi_2(\vec{r}) | ep^x | c_1\, e^{-i(E_1/\hbar)t}\, \psi_1(\vec{r})$$

$$+ c_2\, e^{-i(E_2/\hbar)t}\, \psi_2(\vec{r}) \rangle = -e \left[c_1^* c_2\, p_{12}^x\, e^{-i\omega_o t} + c_1 c_2^*\, p_{21}^x\, e^{i\omega_o t} \right] \tag{2.67}$$

where c_1 and c_2 are solutions of the equations

$$\begin{cases} V \cos \omega t \; e^{-i\omega_o t} \; c_2 = i\dot{c}_1 \\ V^* \cos \omega t \; e^{i\omega_o t} \; c_1 = i\dot{c}_2 \end{cases} \tag{2.68}$$

In the absence of electromagnetic field $V = 0$ and $\dot{c}_1 = \dot{c}_2 = 0$: the probabilities of occupancy of the two levels are constant and no "spontaneous" emission is possible. To remedy the situation we introduce an additional damping term in the second equation (2.68) and write

$$i\dot{c}_2 = V^* \cos \omega t \; e^{i\omega_o t} \; c_1 - i\gamma c_2 \tag{2.69}$$

In absence of a perturbing field

$$\dot{c}_2 = -\gamma c_2 \tag{2.70}$$

and

$$c_2(t) = c_2(0) \; e^{-\gamma t} \tag{2.71}$$

For a gas of atoms with a population $N_2(0)$ at time $t = 0$ in the upper level:

$$N_2(t) = N_2(0) \; e^{-2\gamma t} \tag{2.72}$$

We identify the quantity 2γ with the Einstein's A coefficient:

$$2\gamma = A_{21} \tag{2.73}$$

We now take eq. (2.69) and set in it $c_1 = 1$: we then have

$$i\dot{c}_2 = V^* \cos \omega t \; e^{i\omega_o t} - i\gamma c_2 \tag{2.74}$$

and

$$c_2(t) = -\frac{1}{2} V^* \left[\frac{e^{i(\omega_o+\omega)t}}{\omega_o+\omega-i\gamma} + \frac{e^{i(\omega_o-\omega)t}}{\omega_o-\omega-i\gamma} \right] \tag{2.75}$$

$|c_2(t)|^2$ is of the order $|V|^2$; this makes $|c_1(t)|^2$ differing from 1 by a term of the order $|V|^2$: then

$$d(t) \simeq -e\left[c_2 \, p^x_{12} \, e^{-\omega_o t} + c^*_2 \, p^x_{21} \, e^{i\omega_o t}\right]$$

$$= \frac{e^2 |p^x_{12}|^2 E_o}{2\hbar} \left[\frac{e^{i\omega t}}{\omega_o+\omega-i\gamma} + \frac{e^{-i\omega t}}{\omega_o-\omega-i\gamma} + \frac{e^{-i\omega t}}{\omega_o+\omega+i\gamma} + \frac{e^{i\omega t}}{\omega_o-\omega+i\gamma} \right] \tag{2.76}$$

We average over the random orientations of the atoms by introducing a factor of $\frac{1}{3}$ and writing

$$d(t) = \frac{e^2|p_{12}|^2 E_o}{6\hbar}\left[\frac{e^{i\omega t}}{\omega_o+\omega-i\gamma} + \frac{e^{-i\omega t}}{\omega_o-\omega-i\gamma} + \frac{e^{-i\omega t}}{\omega_o+\omega+i\gamma} + \frac{e^{i\omega t}}{\omega_o-\omega+i\gamma}\right] \qquad (2.77)$$

where p_{12} = magnitude of $\vec{p}_{12}$. Then the polarization is given by

$$P(t) = \frac{N}{V}\, d(t) = \frac{Ne^2|p_{12}|^2 E_o}{6\hbar V}\left[\frac{e^{i\omega t}}{\omega_o+\omega-i\gamma} + \frac{e^{-i\omega t}}{\omega_o-\omega-i\gamma} + \frac{e^{-i\omega t}}{\omega_o+\omega+i\gamma} + \frac{e^{i\omega t}}{\omega_o-\omega+i\gamma}\right]$$

$$= \frac{1}{2}\,\varepsilon_o\, E_o\left[\chi(\omega)e^{-i\omega t} + \chi(-\omega)e^{-i\omega t}\right] \qquad (2.78)$$

Then

$$\chi(\omega) = \frac{Ne^2|p_{12}|^2}{3\varepsilon_o\hbar V}\left(\frac{1}{\omega_o-\omega-i\gamma} + \frac{1}{\omega_o+\omega+i\gamma}\right) \qquad (2.79)$$

and

$$\chi(-\omega) = \frac{Ne^2|p_{12}|^2}{3\varepsilon_o\hbar V}\left(\frac{1}{\omega_o+\omega-i\gamma} + \frac{1}{\omega_o-\omega+i\gamma}\right) \qquad (2.80)$$

We note that

$$\chi(-\omega) = \chi^*(\omega) \qquad (2.81)$$

3. Radiative Broadening. We shall first seek a relation between the atomic absorption coefficient and the imaginary part of the susceptibility.

A gas of atoms can be considered as a dialectic medium in which the polarization is related to the electric field through the electric susceptibility χ:

$$\vec{P} = \varepsilon_o\,\chi\,\vec{E} \qquad (2.82)$$

The susceptibility is frequency-dependent. The dielectric constant K is related to the susceptibility as follows:

$$K = \left(\frac{kc}{\omega}\right)^2 = 1 + \chi = n^2 \qquad (2.83)$$

where $k = \frac{2\pi}{\lambda}$ = magnitude of the wave vector and n = index of refraction. If we let

$$n = n_r + in_i \qquad (2.84)$$

we obtain

$$k = \frac{\omega}{c}\,n_r + i\,\frac{\omega}{c}\,n_i \qquad (2.85)$$

We can also write

$$(n_r + in_i)^2 = n_r^2 - n_i^2 + 2i\,n_r\,n_i = \left(\frac{kc}{\omega}\right)^2 = 1 + \chi' + i\chi'' \qquad (2.86)$$

and

$$\begin{cases} n_r^2 - n_i^2 = 1 + \chi' \\ 2n_r\,n_i = \chi'' \end{cases} \qquad (2.87)$$

A traveling wave moving in, say, the z-direction would have time and space dependences given by

$$e^{i(kz-\omega t)} = e^{i(\frac{\omega}{c} n_r + i\frac{\omega}{c} n_i)z - i\omega t} = e^{i\omega(\frac{n_r}{c} z - t)} e^{-\frac{\omega}{c} n_i z} \tag{2.88}$$

The intensity of the wave will go down as

$$e^{-\alpha z} = e^{-\frac{2\omega n_i}{c} z} \tag{2.89}$$

where

$$\alpha = \text{absorption coefficient} = \frac{2\omega n_i}{c} = \frac{\omega\chi''}{n_r c} \tag{2.90}$$

We now do the following:

a) we consider the gas dilute and set $n_r = 1$, and

b) we use the rotating wave approximation:

$$\chi(\omega) = \chi'(\omega) + i\chi''(\omega) = \frac{Ne^2|p_{12}|^2}{3\varepsilon_o\hbar V} \frac{1}{\omega_o-\omega-i\gamma}$$

$$= \frac{Ne^2|p_{12}|^2}{3\varepsilon_o\hbar V} \frac{\omega_o-\omega}{(\omega_o-\omega)^2 + \gamma^2} + i \frac{Ne^2|p_{12}|^2\pi}{3\varepsilon_o\hbar V} \frac{\gamma/\pi}{(\omega_o-\omega)^2 + \gamma^2} \tag{2.91}$$

Then

$$\chi''(\omega) = \frac{Ne^2|p_{12}|^2\pi}{3\varepsilon_o\hbar V} \frac{\gamma/\pi}{(\omega_o-\omega)^2 + \gamma^2} \tag{2.92}$$

and

$$\alpha(\omega) = \frac{\pi\, Ne^2|p_{12}|^2\omega_o}{3\varepsilon_o c\hbar V} f_L(\omega) \tag{2.93}$$

where

$$f_L(\omega) = \text{Lorentzian lineshape} = \frac{\gamma/\pi}{(\omega_o-\omega)^2 + \gamma^2} \tag{2.94}$$

The function $f_L(\omega)$ is represented in Fig. 4.

4. Power Broadening. We have obtained the following result for the susceptibility correct to second order in p_{12}:

$$\chi(\omega) = \frac{Ne^2|p_{12}|^2}{3E_o\hbar V} \left(\frac{1}{\omega_o-\omega-i\gamma} + \frac{1}{\omega_o+\omega+i\gamma} \right) \tag{2.95}$$

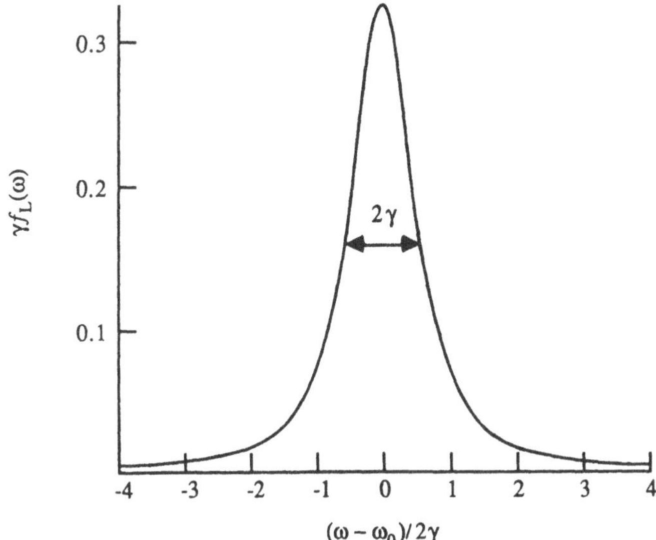

Figure 4. Shape of a Lorentzian Line

This result is consistent with a linear response of atoms to the electric field of the light beam. If we want to include higher terms in $\chi(\omega)$ we have to use the optical Bloch equations:

$$\begin{cases} \dot{D}_{22} = -\dot{D}_{11} = -\frac{i}{2}V^* e^{i(\omega_o-\omega)t} D_{12} + \frac{i}{2} V e^{-i(\omega_o-\omega)t} D_{21} \\ \\ \dot{D}_{12} = \dot{D}^*_{21} = \frac{i}{2}V e^{-i(\omega_o-\omega)t} (D_{11} - D_{22}) \end{cases} \qquad (2.96)$$

but we have to include spontaneous emission. We do so by considering the two equations

$$\begin{aligned} \dot{c}_2 &= - i V^* \cos \omega t \, e^{i\omega_o t} c_1 - \gamma c_2 \\ \dot{c}_1 &= - i V \cos \omega t \, e^{-i\omega_o t} c_2 \end{aligned} \qquad (2.97)$$

Then

$$\dot{D}_{22} = c_2^* \frac{dc_2}{dt} + c_2 \frac{dc_2^*}{dt}$$

$$= c_2^* \left(- iV^* \cos \omega t \, e^{i\omega_o t} c_1 - \gamma c_2\right) + c_2 \left(i V \cos \omega t \, e^{-i\omega_o t} c_1^* - \gamma c_2^*\right)$$

$$= - iV^* \cos \omega t \, e^{i\omega_o t} D_{22} - \gamma D_{22} + iV \cos \omega t \, e^{-i\omega_o t} D_{21} - \gamma D_{22}$$

$$\underset{=}{\text{RWA}} - \frac{i}{2} V^* e^{i(\omega_o-\omega)t} D_{12} + \frac{i}{2} V e^{-i(\omega_o-\omega)t} D_{21} - 2\gamma D_{22}$$

$$\dot{D}_{12} = c_1 \frac{dc_2^*}{dt} + c_2^* \frac{dc_1}{dt} = i V \cos \omega t \, e^{-i\omega_o t} (D_{11} - D_{22}) - \gamma D_{12}$$

$$\underset{=}{\text{RWA}} \frac{i}{2} Ve^{-i(\omega_o-\omega)t} (D_{11}- D_{22}) - \gamma D_{12}$$

where RWA indicates the use of the rotating wave approximation. In summary

$$\begin{cases} \dot{D}_{22} = - \dot{D}_{11} - \frac{i}{2} V^* e^{i(\omega_o-\omega)t} D_{12} + \frac{i}{2} V e^{-i(\omega_o-\omega)t} D_{21} - 2\gamma D_{22} \\ \dot{D}_{12} = \dot{D}_{21}^* = \frac{i}{2} V e^{-i(\omega_o-\omega)t} (D_{11} - D_{22}) - \gamma D_{12} \end{cases} \tag{2.98}$$

The solutions of these equations are no longer purely oscillatory; after a certain time we have a steady state situation

$$\begin{cases} D_{22} = \dfrac{|V|^2/4}{(\omega_o-\omega)^2 + \gamma^2 + |V|^2/2} \\ \\ D_{12} = - e^{-i(\omega_o-\omega)t} \dfrac{\frac{|V|}{2} (\omega_o-\omega-i\gamma)}{(\omega_o-\omega)^2 + \gamma^2 + |V|^2/2} \end{cases} \tag{2.99}$$

The atomic dipole moment is given, according to eq. (2.67) by

$$d(t) = -e \left[D_{21} p_{12}^x e^{-i\omega_o t} + D_{12} p_{21}^x e^{i\omega_o t}\right]$$

$$= e \left[e^{i(\omega_o-\omega)t} \frac{\frac{V^*}{2} (\omega_o - \omega + i\gamma) p_{12}^x}{(\omega_o-\omega)^2 + \gamma^2 + |V|^2/ 2} e^{i\omega_o t} + \text{c.c.}\right]$$

$$= e \left[e^{i\omega t} \frac{\frac{V^*}{2} (\omega_o - \omega + i\gamma) p_{12}^x}{(\omega_o-\omega)^2 + \gamma^2 + |V|^2/ 2} + \text{c.c.}\right] \tag{2.100}$$

Then

$$P(t) = \frac{N}{V} d(t) = e^{-i\omega t} \frac{N}{V} \frac{e^2 |p^x_{12}|^2 E_o}{2\hbar} \frac{\omega_o - \omega + i\gamma}{(\omega_o - \omega)^2 + \gamma^2 + |V|^2/2}$$

$$+ e^{i\omega t} \frac{N}{V} \frac{e^2 |p^x_{12}|^2 E_o}{2\hbar} \frac{\omega_o - \omega - i\gamma}{(\omega_o - \omega)^2 + \gamma^2 + |V|^2/2}$$

$$= \frac{1}{2} \varepsilon_o E_o \chi(\omega) e^{-i\omega t} + \frac{1}{2} \varepsilon_o E_o \chi(-\omega) e^{i\omega t} \quad (2.101)$$

Then

$$\chi(\omega) = \frac{Ne^2 |p_{12}|^2}{3\varepsilon_o \hbar V} \frac{(\omega_o - \omega) + i\gamma}{(\omega_o - \omega)^2 + \gamma^2 + |V|^2/2} \quad (2.102)$$

This expression is not "complete" because we have used the RWA; the complete expression can be written as follows:

$$\chi(\omega) \propto \left[\frac{\omega_o - \omega + i\gamma}{(\omega_o - \omega)^2 + \gamma^2 + |V|^2/2} + \frac{\omega_o + \omega - i\gamma}{(\omega_o - \omega)^2 + \gamma^2 + |V|^2/2} \right] \quad (2.103)$$

If we use the complete expression we obtain

$$\chi(-\omega) = \chi^*(\omega) \quad (2.104)$$

a condition that is not verified for the simpler expression (2.102).

Now we can write

$$\chi''(\omega) = \frac{Ne^2 |p_{12}|^2}{3\varepsilon_o \hbar V} \frac{\gamma}{(\omega_o - \omega)^2 + \gamma^2 + |V|^2/2} \quad (2.105)$$

Because of the presence of the term $|V|^2/2$ in the denominator the susceptibility is dependent on the field strength, the rate of absorption of incident light is reduced and the linewidth of the atomic transition is given by

$$2 \sqrt{\gamma^2 + \frac{1}{2} |V|^2} \quad (2.106)$$

The additional contribution to the linewidth is called <u>power broadening</u> or <u>saturation broadening</u>.

5. <u>Damped Rabi Oscillations</u>. The diagonal elements of the atomic density matrix give us, when multiplied by the total atomic population N, the populations of atoms in states ψ_1 and ψ_2

$$\begin{cases} N_2 = ND_{22} = \dfrac{N |V|^2/4}{(\omega_o - \omega)^2 + \gamma^2 + |V|^2/2} \\ N_1 = N D_{11} = N - N_2 \end{cases} \quad (2.107)$$

These expressions apply in the steady state. In the limit of weak intensity of incident light the value of N_2 is proportional to the intensity of light. We note also that the steady state value above is independent of the initial conditions.

For studying the transients we need, however, the initial conditions. The equations to consider are eqs. (98); let us deal with the simple case in which $\omega_o = \omega$ (zero detuning) and

$$\begin{cases} D_{12}(0) = 0 & (c_2=0) \\ D_{12}(0) = 0 & (c_1 = 1) \end{cases} \tag{2.108}$$

R. A. Smith [16] has given and C. Yang [17] has worked out in great detail the solutions for D_{22} in three different situations for this case of zero detuning:

$|V| > \frac{1}{2}\gamma$

$$D_{22} = \frac{|V|^2/2}{2\gamma^2 + |V|^2}\left[1 - (\cos at + \frac{3\gamma}{2a}\sin at)\exp(-\frac{3}{2}\gamma t)\right] \tag{2.109}$$

where

$$a = \left(V^2 - \frac{\gamma^2}{4}\right)^{1/2} \tag{2.110}$$

$|V| = \frac{1}{2}\gamma$

$$D_{22} = \frac{1}{18}\left[1 - (\frac{3}{2}\gamma t + 1)\exp(-\frac{3}{2}\gamma t)\right] \tag{2.111}$$

$|V| < \frac{1}{2}\gamma$

$$D_{22} = \frac{|V|^2/2}{2\gamma^2 + |V|^2}\left[1 - (\cosh\frac{1}{2}a't + \frac{3\gamma}{a'}\sinh\frac{1}{2}a't)\exp(-\frac{3}{2}\gamma t)\right] \tag{2.112}$$

where

$$a' = \left(\gamma^2 - 4|V|^2\right)^{1/2} \tag{2.113}$$

For $\gamma = 0$, $a = |V| = \Omega$, and we recover the expression (2.49)

$$D_{22} = \frac{1}{2}(1 - \cos|V|t) = \sin^2\frac{1}{2}|V|t \tag{2.114}$$

In Fig. 5 we represent the variations of D_{22} as given by (2.109) for the case of zero detuning. We note the following:

1) The greater is γ the more damped are the oscillations. For $\eta = \frac{\gamma}{|V|} = \frac{1}{3}$ a single maximum remains.

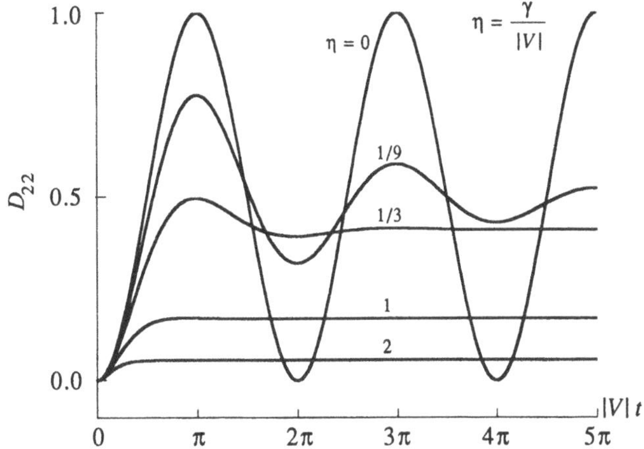

Figure 5. D_{22} for Zero Detuning and Various Values of $\eta = \frac{\gamma}{|V|}$.

2) The incident light must be of large enough intensity in order to make

$$\eta << \frac{1}{3}, \text{ i.e. } |V| >> 3\gamma$$

in order to generate significant oscillations in the populations of atoms in the two states.

3) The steady state value of D_{22}

$$\frac{|V|^2/2}{2\gamma^2 + |V|^2} = \frac{1}{2 + 4\eta^2} \qquad (2.115)$$

decreases with increasing η.

4) The oscillatory behavior of the populations is called <u>optical nutation</u>. The nutation frequency depends on $|V|$, detuning, and radiation damping.

5) The effect of a beam of light on atomic excitation can be summarized as follows:
 a) The atoms are initially in the ground state. The beam is attenuated as energy is transferred from the beam to the atoms.
 b) If the intensity of the beam is large enough so that

$$|V| > (\omega_o - \omega) \text{ and } |V| >> \gamma$$

 the excited state component of the wave function will exceed the ground state component and energy is transferred back from the atoms into the beam, increasing its initial intensity.

c) This "cycle" repeats itself producing optical nutation. The oscillations in the atomic populations are accompanied by oscillations in the intensity of the transmitted light. Experimental observations of this type were made by MacGillivray and co-workers [18] who observed the effects of optical nutation in the D_2 transition of sodium atoms. A detailed theory of experiments of this type is given in [19].

6. <u>Collision Broadening</u>. Consider a gas of molecules and single out a molecule with velocity $\vec{v}$. We call p(t) the probability that such a molecule does <u>not</u> collide in a time t. We call also wdt the probability that the molecule collides in a time in the interval (t, t+dt): we shall assume w = w(v) independent of past collisions, and then independent of time. We can write

$$p(t+dt) = p(t)(1-wdt) \tag{2.116}$$

or

$$p(t) + \frac{dp}{dt}dt = p(t) - p(t)\,wdt$$

$$\frac{1}{p}\frac{dp}{dt} = -w \tag{2.117}$$

Since v does not change in a time w^{-1} between collisions we can integrate and obtain

$$p(t) = e^{-wt} \tag{2.118}$$

The probability that a molecule, not having collided in time t, collides in (t, t+dt) is

$$\mathcal{P}(t)dt = p(t)w\,dt = e^{-wt}w\,dt \tag{2.119}$$

We verify that

$$\int_0^\infty \mathcal{P}(t)dt = \int_0^\infty e^{-wt}\,wdt = 1 \tag{2.120}$$

The mean time between collisions is

$$\tau = \bar{t} = \int_0^\infty p(t)t\,dt = \int_0^\infty e^{-wt}\,wtdt = \frac{1}{w} \tag{2.121}$$

Then we can write

$$p(t)\,dt = \frac{e^{-t/\tau}}{\tau}\,dt \tag{2.122}$$

Since w = w(v), τ = τ(v). We can then characterize a gas of atoms by the average collision time of the atoms traveling with the mean speed $\bar{v}$:

$$\tau = \tau(\bar{v}) \tag{2.123}$$

For a gas of N atoms of mass m and radius $\frac{d}{2}$ in a volume V we find [20]

$$\tau^{-1} = \frac{4d^2N}{V}\left(\frac{\pi kT}{m}\right)^{1/2} \tag{2.124}$$

In a gas of N_2 molecules at room temperature and atmospheric pressure, $\tau = 6 \times 10^{-10}$ sec.

Collisions among atoms or molecules can be elastic or inelastic. An elastic collision leaves the atom in the same quantum state, but changes the phase of the atomic wavefunction. An inelastic collision produces a change in the state of the colliding atom: inelastic collisions are taken into account in the Bloch equations by suitably increasing the rate of decay γ.

We shall consider only the elastic collisions because it is found experimentally that they represent the dominant line-broadening mechanism in a very large number of cases and conditions. The presence of collisions of this type changes the off-diagonal terms D_{12} and D_{21}, but does not change D_{11} and D_{22}:

$$\begin{cases} \dot{D}_{22} = -\frac{i}{2} V^* e^{i(\omega_o-\omega)t} D_{12} + \frac{i}{2} V e^{-i(\omega_0-\omega)t} D_{21} - 2\gamma D_{22} \\ \dot{D}_{12} = \frac{i}{2} V e^{-i(\omega_o-\omega)t} (D_{11} - D_{22}) - \gamma' D_{12} \end{cases} \tag{2.125}$$

The steady state solutions of these equations are

$$\begin{cases} D_{22} = \dfrac{(|V|^2/4)\, \gamma'/\gamma}{(\omega_o-\omega)^2 + \gamma'^2 + \frac{1}{2}\frac{\gamma'}{\gamma} |V|^2} \\ D_{12} = e^{-i(\omega_o-\omega)t} \dfrac{(V/2)(\omega_o-\omega-i\gamma')}{(\omega -\omega)^2 + \gamma'^2 + \frac{1}{2}\left(\frac{\gamma'}{\gamma}\right)|V|^2} \end{cases} \tag{2.126}$$

The susceptibility can be rederived as follows:

$$\chi(\omega) = \frac{Ne^2|p_{12}|^2}{3\varepsilon_o \hbar V} \frac{\omega_o - \omega + i\gamma'}{(\omega_o-\omega)^2 + \gamma'^2 + \frac{\gamma'}{\gamma}\frac{|V|^2}{2}} \tag{2.127}$$

The linewidth is now given by

$$2\left[\gamma' + \frac{1}{2}\left(\frac{\gamma'}{\gamma}\right) |V|^2\right]^{1/2} \tag{2.128}$$

where

$$\gamma' = \gamma + \gamma\ \mathrm{coll} = \gamma + \frac{1}{\tau} \tag{2.129}$$

and contains the effects of radiative broadening (γ), power broadening ($|V|$) and collision broadening (τ).

7. Doppler Broadening. A photon of energy $\hbar\omega$ carries a momentum

$$\hbar\vec{k} = \hbar \frac{\omega}{c} \frac{\vec{k}}{|\vec{k}|} \tag{2.130}$$

An atom, residing originally in the ground level, absorbs this photon and moves to an excited level. Let E_1 and E_2 be the energies of the ground and excited level of the atom, respectively. Let also $\vec{v}_1$ and $\vec{v}_2$ be the velocities of the atom before and after the absorption of the photon.

Conservation of momentum and conservation of energy are expressed by the relations

$$\begin{cases} m\,\vec{v}_1 + \hbar\vec{k} = m\vec{v}_2 \\ E_1 + \frac{1}{2}\,m\,v_1^2 + \hbar\omega = E_2 + \frac{1}{2}\,m\,v_2^2 \end{cases} \tag{2.131}$$

We shall call

$$\omega_o = \frac{E_2 - E_1}{\hbar} \tag{2.132}$$

and we shall take the direction in which the photon travels as the x-direction. The above relations give us

$$\omega_o = \omega - \frac{\omega v_{1x}}{c} - \frac{\hbar\omega^2}{2mc} \tag{2.133}$$

Typical values of some of the above quantities are

$$\frac{\omega\, v_{1x}}{c} \simeq 10^{-8}$$

$$\frac{\hbar\omega^2}{2mc} \simeq 10^{-9}$$

Then the third term in (2.130) is negligible with respect to the second term and we can write, dropping the subscript 1,

$$\omega = \frac{\omega_o}{1 - \frac{v_x}{c}} \simeq \omega_o\left(1 + \frac{v_x}{c}\right) \tag{2.134}$$

A photon can be absorbed by an atom which has an x-component of the velocity v_x if its frequency ω is related with the frequency ω_o of the atom by the relation (2.131).

The probability that an atom in a gas at temperature T has the x-component of its velocity in $(v_x,\ v_x+dv_x)$ is proportional to

$$\exp\left(-\frac{mv_x^2}{2kT}\right)\,dv_x = \exp\left(-\frac{mc^2(\omega-\omega_o)^2}{2\omega_o^2 kT}\right)dv_x \tag{2.135}$$

Let $g(\omega)$ be the profile of the spectral line. We can write

$$g(\omega)\;d\omega = g(v_x)\;dv_x \tag{2.136}$$

and, if we use (2.131),

$$g(\omega) = g(v_x)\,\frac{d\omega}{dv_x} = g(v_x)\,\frac{c}{\omega_o} \tag{2.137}$$

Then

$$\exp\left(-\frac{mv_x^2}{2kT}\right) dv_x = \frac{c}{\omega_o} \exp\left(-\frac{mc^2(\omega-\omega_o)^2}{2\omega_o^2 kT}\right) d\omega \tag{2.138}$$

The profile represented by (2.135) is called a Gaussian lineshape. The FWHM (full width at half maximum height) of this line is

$$2\Delta\omega = 2\omega_o \left(\frac{2kT \ln 2}{mc^2}\right)^{1/2} \tag{2.139}$$

The mean square spread is

$$\sigma = \left[\overline{(\omega-\omega_o)^2}\right]^{1/2} = \omega_o \left(\frac{kT}{mc^2}\right)^{1/2} = \frac{\Delta\omega}{\left(2 \ln 2\right)^{1/2}} \tag{2.140}$$

We can then write for a normalized Gaussian line:

$$f_G(\omega) = \frac{1}{\sqrt{2\pi\sigma^2}} \exp\left(-\frac{(\omega-\omega_o)^2}{2\sigma^2}\right) \tag{2.141}$$

8. Composite Lineshape. Both the collision broadening and the Doppler broadening are proportional to $T^{1/2}$ and $m^{-1/2}$ and may be changed by changing the temperature of the gas. But the collision broadening is also proportional to the density $\frac{N}{V}$ and, for this reason, is also called pressure broadening.

A "composite" lineshape may arise when two different mechanisms contribute to the broadening of a spectral line. If these two mechanisms produce, say, the profiles $F_1(\omega)$ and $F_2(\omega)$, then the profile of the composite line is the convolution of the two spectral functions:

$$f(\omega) = \int_{-\infty}^{+\infty} F_1(s)\, F_2(\omega-\omega_o-s)ds \tag{2.142}$$

where ω_o = common central frequency.

We can now make the following statements:

1) Any number of line-broadening mechanisms can be combined by repeated convolutions. The resultant lineshape is independent of the order in which the convolutions are performed.

2) For two Lorentzian broadening mechanisms giving widths $2\gamma_1$ and $2\gamma_2$,

$$2\gamma = 2\gamma_1 + 2\gamma_2 \tag{2.143}$$

3) For two Gaussian broadening mechanisms giving widths 2Δ, and $2\Delta_2$,

$$\Delta^2 = \Delta_1^2 + \Delta_2^2 \tag{2.144}$$

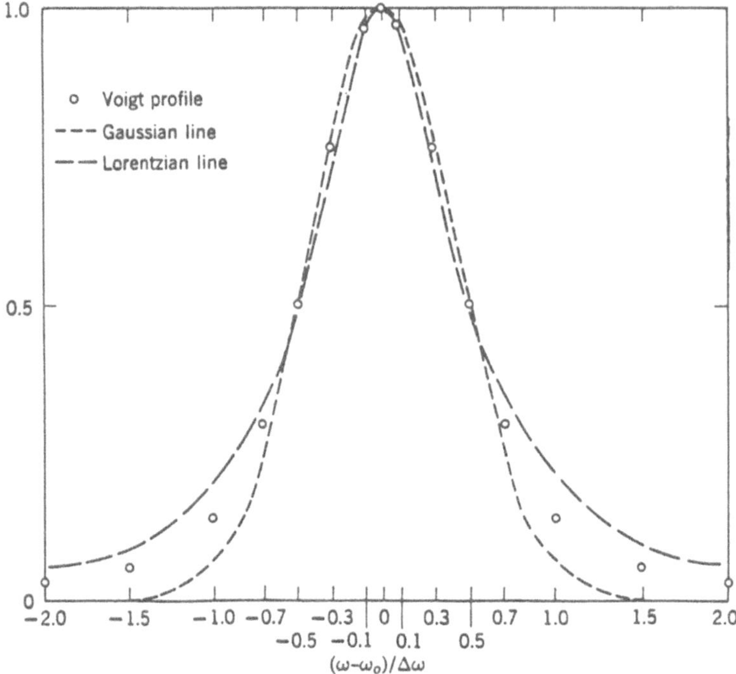

Figure 6. Lorentzian, Gaussian and Voigt Line Shapes

4) The convolution of a Lorentzian and a Gaussian profiles is a <u>Voigt profile</u> [21]. If $\Delta\nu_L$ and $\Delta\nu_G$ are the widths given by the Lorentzian and Gaussian mechanisms, respectively, the Voigt profile is given by

$$r_v(\nu) = \frac{2 \ln 2}{\pi\sqrt{\pi}} \frac{\Delta\nu_L}{\Delta\nu_G} \int_{-\infty}^{+\infty} \frac{e^{-y^2}}{a^2 + (\Omega - y)^2} dy \qquad (2.145)$$

where

$$\Omega = \frac{2\nu\sqrt{\ln 2}}{\Delta\nu_G} \qquad (2.146)$$

$$a = \frac{\Delta\nu_L}{\Delta\nu_G} \ln 2 \qquad (2.147)$$

Note that

$$r_v(\nu) \xrightarrow[\Delta\nu_L \to 0]{} \text{Gaussian lineshape}$$

$$r_v(\nu) \xrightarrow[\Delta\nu_G \to 0]{} \text{Lorentzian lineshape}$$

The Voigt profile is important because it occurs in the case when collision broadening and Doppler broadening are both significant and are affecting the spectral line independently. In Fig. 6 we report the three shapes of a Gaussian, a Lorentzian and a Voigt profiles with the same half width.

The mechanisms which broaden spectral atomic lines fall into two broad categories and situations:

1) Different atoms absorb or emit radiation at somewhat different ω's. The profile is in this case Gaussian and the broadening is called inhomogeneous. Examples: Doppler broadening in gases and broadening of lines of laser ions in solids at very low temperatures.

2) Each atom absorbs or emits in the same way, with no particular frequency associated to a particular atom. The width is related to the time interval Δt in which the atom is left undisturbed:

$$\Delta\omega \ \Delta t \geq 1$$

The profile in this case is Lorentzian, and the broadening is called homogeneous. Examples: radiative broadening, collision broadening, broadening of lines of laser ions at $T > 77K$.

We shall report now the example of Neon atoms at $p = 0.5$ torr and room temperature. The two pressure independent mechanisms give the following contributions to the linewidth:

Doppler broadening:	1.7 GHz
Radiative broadening:	20 MHz

The pressure-dependent mechanism gives:

Collision broadening:	0.64 MHz

At sufficiently high pressures, the collision broadening predominates over the Doppler broadening.

III. DENSITY MATRIX FORMULATION

III.A. Density Matrix

Consider a physical system with a wavefunction $\psi(\vec{r},t)$ which satisfies the equation

$$i\hbar \frac{\partial\psi}{\partial t} = H \psi \tag{3.1}$$

Let $\psi_n(\vec{r})$ be a complete set of orthonormal time-independent functions. In the Schroedinger representation we may write

$$\psi(\vec{r},t) = \sum_n a_n(t) \ \psi_n(\vec{r}) \tag{3.2}$$

Then (3.1) can be written as follows

$$i\hbar \sum_m \dot{a}_n \ \psi_n = \sum_n a_m \ H\psi_m \tag{3.3}$$

If we multiply by ψ_k^* on the left side of each term and integrate over space coordinates we obtain

$$i\hbar\,\dot{a}_k = \sum_n a_n H_{kn} \tag{3.4}$$

where

$$H_{kn} = \langle\psi_k|H|\psi_m\rangle = H^*_{nk} \tag{3.5}$$

Similarly

$$i\hbar\,\dot{a}_m = \sum_n a_n H_{mn} \tag{3.6}$$

Consider the following

$$\frac{d}{dt}(a_k a^*_m) = \dot{a}_k a^*_m + a_k \dot{a}^*_m = \frac{1}{i\hbar}\sum_n a_n a^*_m H_{kn} + a_k\left(\frac{1}{i\hbar}\sum_n a_n H_{mn}\right)^* =$$

$$= \frac{1}{i\hbar}\sum_n \left[H_{kn}(a_n a^*_m) - (a_k a^*_n)H_{nm}\right] \tag{3.7}$$

Let

$$\rho_{km} = a_k a^*_m \tag{3.8}$$

Then

$$\frac{d}{dt}\rho_{km} = \frac{i}{\hbar}\sum_n (\rho_{kn} H_{nm} - H_{kn}\rho_{nm}) \tag{3.9}$$

or

$$\dot{\rho} = \frac{i}{\hbar}[\rho, H] \tag{3.10}$$

If the ψ_n functions in (3.2) are N in number, this relation represents N^2 equations in the N^2 unknown matrix elements of ρ. The N^2 constants that appear in the solutions are fixed by the initial conditions which may be specified by giving the entire matrix ρ at time t=0.

Consider now an operator Ω. The expectation value of the observable it represents is given by

$$\langle\Omega\rangle = \langle\psi|\Omega|\psi\rangle = \langle \sum_m a_m \psi_m|\Omega|\sum_n a_n\psi_n\rangle$$

$$= \sum_m\sum_n a^*_m a_n \langle\psi_m|\Omega|\psi_n\rangle = \sum_m\sum_n \rho_{nm}\Omega_{mn} = \mathrm{Tr}(\rho\Omega) = \mathrm{Tr}(\Omega\rho) \tag{3.11}$$

If we take the trace of the matrices in (3.10) we can write

$$\sum_m \dot{\rho}_{mm} = \frac{i}{\hbar}\mathrm{Tr}(\rho H - H\rho) = \frac{i}{\hbar}\left[\mathrm{Tr}(\rho H) - \mathrm{Tr}(H\rho)\right] = 0 \tag{3.12}$$

Hence

$$\sum_m \dot{\rho}_{mm} = 0 \longrightarrow \sum_m \rho_{mm}(t) = \mathrm{const} \tag{3.13}$$

This constant is 1 if the wavefunction $\psi(\vec{r},t)$ is normalized. Once the initial state of ρ is specified, we may obtain the time dependence of the matrix elements of ρ. We may then calculate the product of ρ and any other operator Ω and the time dependence of the expectation value of the observable related to Ω.

III.B. Perturbed Hamiltonian

Let H_o be the unperturbed Hamiltonian of a conservative (t-independent) system, and let H′(t) be an external time-dependent perturbation. In a representation in which H_o is diagonal, H′ will not in general be so. The total Hamiltonian is

$$H = H_o + H'(t) \tag{3.14}$$

We may write

$$\dot{\rho} = \frac{i}{\hbar}\left[\rho, H\right] = \frac{i}{\hbar}\left[\rho, H_o\right] + \frac{i}{\hbar}\left[\rho, H'\right] \tag{3.15}$$

Then

$$\dot{\rho}_{mn} = \frac{i}{\hbar}\sum_k \left[\rho_{mk}\,(H_o)_{kn} - (H_o)_{mk}\,\rho_{kn}\right] + \frac{i}{\hbar}\sum_k \left(\rho_{mk}\,H'_{kn} - H'_{mk}\,\rho_{kn}\right) \tag{3.16}$$

But

$$\begin{aligned} (H_o)_{kn} &= E_n\,\delta_{kn} \\ (H_o)_{mk} &= E_m\,\delta_{mk} \end{aligned} \tag{3.17}$$

Then

$$\begin{aligned} \dot{\rho}_{mn} &= \frac{i}{\hbar}\left(\rho_{mn}\,E_n - \rho_{mn}\,E_m\right) + \frac{i}{\hbar}\sum_k \left(\rho_{mk}\,H'_{kn} - H'_{mk}\,\rho_{kn}\right) \\ &= \frac{i}{\hbar}\left(E_m - E_n\right)\rho_{mn} + \frac{i}{\hbar}\sum_k \left(\rho_{mk}\,H'_{kn} - H'_{mk}\,\rho_{kn}\right) \end{aligned} \tag{3.18}$$

and

$$\dot{\rho}_{mm} = \frac{i}{\hbar}\sum_k \left(\rho_{mk}\,H'_{km} - H'_{mk}\,\rho_{km}\right) = \frac{i}{\hbar}\sum_k \left(\rho_{mk}\,H'_{km} - \text{c.c.}\right) \tag{3.19}$$

III.C. Two-Level System

In the representation in which H_o is diagonal the various matrices will look as follows

$$H_o = \begin{pmatrix} E_a & 0 \\ 0 & E_b \end{pmatrix} \tag{3.20}$$

$$H'(t) = ex\,E(t) = eE(t)\begin{pmatrix} 0 & p_{ab} \\ p_{ba} & 0 \end{pmatrix} \tag{3.21}$$

where $p_{ab} = \langle a|x|b\rangle$ and where we have assumed, as in section II.B., that

$$p_{aa} = p_{bb} = 0 \tag{3.22}$$

Then

$$\dot{\rho}_{aa} = \frac{i}{\hbar}\left(\rho_{aa}\,H'_{aa} + \rho_{ab}\,H'_{ba} + \text{c.c.}\right) = \frac{i}{\hbar}\left(\rho_{ab}\,H'_{ba} - \text{c.c.}\right) \tag{3.23}$$

$$\dot{\rho}_{bb} = \frac{i}{\hbar}\left(\rho_{ba}\,H'_{ab} + \rho_{bb}\,H'_{bb} + \text{c.c.}\right) = \frac{i}{\hbar}\left(\rho_{ba}\,H'_{ab} - \text{c.c.}\right) \tag{3.24}$$

$$\begin{aligned} \dot{\rho}_{ab} &= -\frac{i}{\hbar}\left(E_a - E_b\right)\rho_{ab} + \frac{i}{\hbar}\left(\rho_{aa}\,H'_{ab} - H'_{ab}\,\rho_{bb}\right) \\ &= i\omega_o\,\rho_{ab} + \frac{i}{\hbar}\left(\rho_{aa} - \rho_{bb}\right)H'_{ab} \end{aligned} \tag{3.25}$$

where

$$\omega_o = \frac{E_b - E_a}{\hbar} \tag{3.26}$$

Summarizing

$$
\begin{cases}
\dot{\rho}_{aa} = \frac{i}{\hbar}\left(\rho_{ab}\,H'_{ba} - \rho_{ba}\,H'_{ab}\right) \\
\dot{\rho}_{bb} = -\frac{i}{\hbar}\left(\rho_{ab}\,H'_{ba} - \rho_{ba}\,H'_{ab}\right) = i\,\dot{\rho}_{aa} \\
\dot{\rho}_{ab} = i\omega_o\,\rho_{ab} + \frac{i}{\hbar}\left(\rho_{aa} - \rho_{bb}\right) H'_{ab} \\
\dot{\rho}_{ba} = \dot{\rho}_{ab} = i\,-\omega_o\,\rho_{ba} - \frac{i}{\hbar}\left(\rho_{aa} - \rho_{bb}\right) H'_{ba}
\end{cases}
\tag{3.27}
$$

The expectation value of the operator ex is

$$
\langle ex\rangle = \mathrm{Tr}\,(ex\,\rho) \tag{3.28}
$$

The relevant matrix is

$$
ex\,\rho = \begin{pmatrix} 0 & p_{ab} \\ p_{ba} & 0 \end{pmatrix}\begin{pmatrix} \rho_{aa} & \rho_{ab} \\ \rho_{ba} & \rho_{bb} \end{pmatrix} = \begin{pmatrix} p_{ab}\,\rho_{ba} & p_{ab}\,\rho_{bb} \\ p_{ba}\,\rho_{aa} & p_{ba}\,\rho_{ab} \end{pmatrix} \tag{3.29}
$$

Then

$$
\mathrm{Tr}\,(ex\,\rho) = p_{ab}\,\rho_{ba} + p_{ba}\,\rho_{ab} \tag{3.30}
$$

and

$$
\langle ex\rangle = \mathrm{Re}\,(2p_{ab}\,\rho_{ba}) \tag{3.31}
$$

III.D. Density Matrix of an Ensemble of Radiating Systems

1. Ensemble of Quantum Systems. Measurements of physical parameters are usually performed on quantum systems in equilibrium whose states are controlled by a Boltzmann distribution over the available stationary energy levels. This distribution is characterized by temperature. An ensemble of systems which are so distributed is called a canonical ensemble.

Averages of physical parameters for such an ensemble must include: 1) quantum averaging, i.e. calculation of the expectation value: $\langle\Omega\rangle$, and 2) classical averaging, i.e. ensemble averaging: $\bar{\Omega}$.
A combination of the two averaging processes results in the notation $\langle\bar{\Omega}\rangle$.
In general, $\langle\bar{\Omega}\rangle \neq \langle\Omega\rangle$.

2. Perturbations. The perturbations that may influence the systems of the ensemble are called coherent if they affect every system in the same way, and incoherent if they affect the systems differently.

Examples of coherent perturbations are the oscillating electric and magnetic fields produced at radio frequencies by oscillators, at microwave frequencies by klystrons and masers, and at optical and infrared frequencies by lasers.

As for the incoherent perturbations the oscillating fields may be produced by such sources as globars, heated filaments or electric discharges. Relaxation processes related to the so-called T_1 and T_2 time constants, to be introduced later, represent also incoherent perturbations always present in spectroscopic measurements. Spontaneous and nonradiative deexcitation processes almost always act as incoherent perturbations. On account of the often pervasive incoherent pertubations no two systems of the ensemble feel each other or the thermal bath with which they are in contact in the same way.

3. Ensemble-Quantum Averages. We shall focus our attention on quantum systems whose wavefunctions do not have much spatial overlap. These systems are coupled together for two reasons:
1) they may be affected by the presence of the same radiation field, and
2) they are all relaxing to the same thermal bath.

The ensemble-quantum average of the physical parameter represented by the operator Ω is given by the ensemble average of the possible expectation values of Ω (each one of these expectation values corresponding to a system of the ensemble):

$$\langle\bar{\Omega}\rangle = \frac{1}{N} \sum_{n=1}^{N} \langle\Omega\rangle_n \tag{3.32}$$

where N = number of systems in the ensemble,

$$\langle\Omega\rangle_n = \lim_{\nu\to\infty} \frac{1}{\nu} \sum_{i=1}^{\nu} \omega_{ni} \tag{3.33}$$

and ω_{ni} = result of a measurement on a system which is in state n.

4. The Density Matrix. We shall assume that the systems of the ensemble have little wavefunction overlap and interact with one another weakly. In addition:
1) any perturbation applied to the systems is coherent,
2) at the time the perturbation is switched on, all the systems of the ensemble are in identical states, and
3) the experiments do not last longer than a time $\tau \ll T_1, T_2$, i.e. relaxation effects are not important.

Under these restrictive conditions we may define a matrix ρ for which two relations are of relevance:

Equation of motion for ρ: $$\dot{\rho} = \frac{i}{\hbar} [\rho, H] \tag{3.34}$$

Average value of Ω: $$\langle\Omega\rangle = \mathrm{Tr}(\Omega\rho) = \langle\bar{\Omega}\rangle \tag{3.35}$$

If the systems of the ensemble have wavefunctions with little overlap, but do not follow the three additional requirements referred to above, then we have to describe the ensemble in different terms:

1) Let the ensemble consist of N identical systems and let the nth system be represented by a wavefunction

$$\psi_n(q_n, t) = a_{n1}\psi_1(q) + a_{n2}\psi_2(q) + \ldots\ldots \tag{3.36}$$

2) We define an operator $\rho^{(n)}$ as follows

$$\rho_{ij}^{(n)} = a_{ni}\, a_{nj}^{*} \tag{3.37}$$

3) The expectation value of a physical parameter represented by the operator Ω is

$$\langle\Omega\rangle_n = \mathrm{Tr}\left[\rho^{(n)}\, \Omega\right] \tag{3.38}$$

4) The ensemble average of Ω is

$$<\bar{\Omega}> = \frac{1}{N}\sum_{n=1}^{N} <\Omega>_n = \frac{1}{N}\sum_{n=1}^{N} \mathrm{Tr}\left[\rho^{(n)}\Omega\right] \quad (3.39)$$

5) We now utilize the following properties of matrices

$$AB + AC = A(B+C) \quad (3.40)$$

$$\mathrm{Tr}(AB) + \mathrm{Tr}(AC) = \mathrm{Tr}\left[A(B+C)\right] \quad (3.41)$$

Then

$$<\bar{\Omega}> = \frac{1}{N}\sum_{n=1}^{N} \mathrm{Tr}\left[\rho^{(n)}\Omega\right] = \mathrm{Tr}\left[\left(\frac{1}{N}\sum_{n=1}^{N} \rho^{(n)}\right)\Omega\right] = \mathrm{Tr}\left[\bar{\rho}\ \Omega\right] \quad (3.42)$$

where the matrix $\bar{\rho}$ represents the density operator

$$\bar{\rho} = \frac{1}{N}\sum_{n=1}^{N} \rho^{(n)} \quad (3.43)$$

The matrix of $\bar{\rho}$ is called the density matrix of the ensemble.

6) The equation of motion of the density operator is

$$i\hbar \frac{\partial\bar{\rho}}{\partial t} = [H,\bar{\rho}] \quad (3.44)$$

7) Prescriptions on how to calculate $<\bar{\Omega}>$:

a) Choose an initial state for the ensemble by choosing the elements of $\bar{\rho}$ at time t=0.

b) Solve the equation of motion for $\bar{\rho}$ given by (3.44) and find $\bar{\rho}$ at the time t of the measurement.

c) Calculate the time-independent matrix of Ω using the same eigenfunctions.

d) Multiply the two matrices and sum the diagonal elements. The result is the experimental value of the parameter Ω, averaged over the systems of the ensemble.

5. Properties of the Density Matrix. We can now investigate the properties of the ensemble density matrix $\bar{\rho}$:

1) Since the trace of each $\rho^{(n)}$ matrix is 1, the trace of $\bar{\rho}$ is also 1, if the ensemble is closed.

2) The total Hamiltonian of a system in the ensemble contains three terms:

H_o: time independent, very large, establishes the stationary states of a system,

H': time dependent, a perturbation, induces transitions among eigenstates of H_o, and

H_R: provides for the exchange of energy among the systems of the ensemble, and between the systems and the heath bath.

3) The most convenient choice of basis functions is the set of eigenfunctions of H_o.

4) Since all the $\rho^{(n)}$ are Hermitian, $\bar{\rho}$ is Hermitian.

5) The off diagonal elements of a matrix $\rho^{(n)}$ can be expressed as follows:

$$\rho_{ij}^{(n)} = |\rho_{ij}^{(n)}| \exp\left(i\phi_{ij}^{(n)}\right) \tag{3.45}$$

Then

$$\bar{\rho}_{ij} = \frac{1}{N} \sum_{n=1}^{N} |\rho_{ij}^{(n)}| \exp\left(i\phi_{ij}^{(n)}\right) \tag{3.46}$$

6) We can state the Principle of Equal a Priori Probabilities as follows: "All configurations of any system, or ensemble of systems, which have the same energy are equally probable at equilibrium".

A consequence of this principle is the Hypothesis of Random Phases: "In an ensemble at equilibrium all the off-diagonal elements of the ensemble density matrix must be zero, because of the destructive interferences among the corresponding elements of the various individual $\rho^{(n)}$ matrices".

7) The diagonal elements of $\bar{\rho}$ do not vanish when the ensemble is in thermal equilibrium. They are given by

$$\bar{\rho}_{ii}^{e} = \frac{1}{N} \sum_{n=1}^{N} \rho_{ii}^{(n)} = \frac{1}{N} \sum_{n=1}^{N} |a_i^{(n)}|^2 \tag{3.47}$$

where the superscript e indicates equilibrium and where

$$|a_i^{(n)}|^2 = \frac{e^{-E_i/kT}}{Z} = \frac{e^{-E_i/kT}}{\sum_{i=1}^{M} e^{-E_i/kT}} = \text{fraction of systems with energy } E_i \tag{3.48}$$

with

$$Z = \sum_{i=1}^{M} e^{-E_i/kT} = \text{partition sum.} \tag{3.49}$$

Then

$$\overline{\rho^{e}} = \frac{1}{Z} \begin{pmatrix} e^{-E_1/kT} & & & & 0 \\ & e^{-E_2/kT} & & & \\ & & e^{-E_3/kT} & & \\ & & & \ddots & \\ 0 & & & & e^{-E_M/kT} \end{pmatrix} \tag{3.50}$$

8) If a perturbation H' which, by acting on the systems, has driven the ensemble away from equilibrium, is removed, each element of $\bar{\rho}$ begins to decay back to its equilibrium value according to

$$\frac{d\bar{\rho}_{ij}}{dt} = - \frac{\bar{\rho}_{ij} - \bar{\rho}_{ij}^{e}}{T_{ij}} \tag{3.51}$$

The effect of the Hamiltonian H_R is felt through these relaxation mechanisms.

9) Not all the T_{ij}'s are independent of each other:

$$T_{ij} = T_{ji}$$

because $\bar{\rho}_{ij} = \bar{\rho}^*_{ji}$. Also there are conditions on the T's for the diagonal elements because, in any case,

$$\sum_{i=1}^{M} \bar{\rho}_{ii} = 1 \tag{3.52}$$

10) For two-level systems (M = 2) there are two relaxation times: T_1 = relaxation time for the diagonal elements, called longitudinal relaxation time, and T_2 = relaxation time for the off-diagonal elements, called transversal relaxation time.

We can now write the following equations that represent the time evolution of the elements of $\bar{\rho}$:

$$\left\{\begin{aligned}
\dot{\bar{\rho}}_{11} &= \frac{1}{i\hbar}\left[(H_o + H'), \bar{\rho}\right]_{11} - \frac{\bar{\rho}_{11} - \bar{\rho}^e_{11}}{T_1} \\
\dot{\bar{\rho}}_{22} &= \frac{1}{i\hbar}\left[(H_o + H'), \bar{\rho}\right]_{22} - \frac{\bar{\rho}_{22} - \bar{\rho}^e_{22}}{T_1} \\
\dot{\bar{\rho}}_{12} &= \frac{1}{i\hbar}\left[(H_o + H'), \bar{\rho}\right]_{12} - \frac{\bar{\rho}_{12}}{T_2} \\
\dot{\bar{\rho}}_{21} &= \frac{1}{i\hbar}\left[(H_o + H'), \bar{\rho}\right]_{21} - \frac{\bar{\rho}_{21}}{T_2}
\end{aligned}\right. \tag{3.53}$$

6. Equation of Motion of the Ensemble Density Matrix Operator. In order to deal with such an equation we like to go through the following calculations:

$$\left[(H_o + H'), \bar{\rho}\right] = \left[H_o, \bar{\rho}\right] + \left[H', \bar{\rho}\right] \tag{3.54}$$

$$\left[H_o, \bar{\rho}\right] = H_o\bar{\rho} - \bar{\rho}H_o$$

$$\left(H_o\bar{\rho}\right)_{pq} = \sum_{r=1}^{M} (H_o)_{pr}\, \bar{\rho}_{rq} = (H_o)_{pp}\, \bar{\rho}_{pq} = E_p\, \bar{\rho}_{pq} \tag{3.55}$$

$$\left(\bar{\rho}H_o\right)_{pq} = \sum_{r=1}^{M} \bar{\rho}_{pr} \, (H_o)_{rq} = \bar{\rho}_{pq} \, (H_o)_{qq} = E_q \, \bar{\rho}_{pq}$$

$$\left[H_o, \bar{\rho}\right]_{pq} = \left(E_p - E_q\right) \bar{\rho}_{pq}$$

$$\left[H', \bar{\rho}\right] = H'\bar{\rho} - \bar{\rho}H'$$

$$\left(H'\bar{\rho}\right)_{pq} = \sum_{r=1}^{M} (H')_{pr} \, \bar{\rho}_{rq}$$

$$\left(\bar{\rho}, H'\right)_{pq} = \sum_{r=1}^{M} \bar{\rho}_{pr} \, (H')_{rq} \qquad (3.56)$$

$$\left[H', \rho\right] = \sum_{r=1}^{M} \left[(H')_{pr} \, \bar{\rho}_{rq} - \bar{\rho}_{pr} \, (H')_{rq}\right]$$

We apply now these finding to the two-level system case:

<u>M = 2</u>

$$\left(H'\bar{\rho}\right)_{11} = H'_{12}\bar{\rho}_{21} \qquad \left(\bar{\rho}H'\right)_{11} = \bar{\rho}_{12}H'_{21}$$

$$\left(H'\bar{\rho}\right)_{12} = H'_{12}\bar{\rho}_{22} \qquad \left(\bar{\rho}H'\right)_{12} = \bar{\rho}_{11}H'_{12}$$

$$\left(H'\rho\right)_{21} = H'_{21}\bar{\rho}_{11} \qquad \left(\bar{\rho}H'\right)_{21} = \bar{\rho}_{22}H'_{21}$$

$$\left(H'\rho\right)_{22} = H'_{21}\bar{\rho}_{12} \qquad \left(\bar{\rho}H'\right)_{22} = \bar{\rho}_{21}H'_{12}$$

$$\left[H', \bar{\rho}\right]_{11} = \left(H'\rho\right)_{11} - \left(\rho H'\right)_{11} = H'_{12} \, \bar{\rho}_{21} - H'_{21} \, \bar{\rho}_{12}$$
$$= 2 \text{ Im} \left(H'_{12} \, \bar{\rho}_{21}\right)$$

$$\left[H', \bar{\rho}\right]_{22} = \left(H'\rho\right)_{22} - \left(\rho H'\right)_{22} = H'_{21} \, \bar{\rho}_{12} - H'_{12} \, \bar{\rho}_{21}$$
$$= -2 \text{ Im} \left(H'_{12} \, \bar{\rho}_{21}\right)$$

$$\left[H', \bar{\rho}\right]_{12} = \left(H'\bar{\rho}\right)_{12} - \left(\bar{\rho}H'\right)_{12} = H'_{12}\left(\bar{\rho}_{22} - \bar{\rho}_{11}\right)$$

$$\left[H', \bar{\rho}\right]_{21} = \left(H'\rho\right)_{21} - \left(\bar{\rho}H'\right)_{21} = H'_{21}\left(\bar{\rho}_{11} - \bar{\rho}_{22}\right)$$

We can then write:

$$
\left\{
\begin{aligned}
\dot{\bar{\rho}}_{11} &= \frac{1}{i\hbar}\left(H'_{12}\,\bar{\rho}_{21} - H'_{21}\,\bar{\rho}_{12}\right) - \frac{\bar{\rho}_{11} - \bar{\rho}^{e}_{11}}{T_1} \\
\dot{\bar{\rho}}_{22} &= \frac{1}{i\hbar}\left(H'_{21}\,\bar{\rho}_{12} - H'_{12}\,\bar{\rho}_{21}\right) - \frac{\bar{\rho}_{22} - \bar{\rho}^{e}_{22}}{T_1} \\
\dot{\bar{\rho}}_{12} &= \frac{1}{i\hbar}\left(E_1 - E_2\right)\bar{\rho}_{12} + \frac{H'_{12}\left(\bar{\rho}_{22} - \bar{\rho}_{11}\right)}{i\hbar} - \frac{\bar{\rho}_{12}}{T_2} = \\
&= i\omega_o\,\bar{\rho}_{12} + \frac{H'_{12}}{i\hbar}\left(\bar{\rho}_{22} - \bar{\rho}_{11}\right) - \frac{\bar{\rho}_{12}}{T_2} \\
\dot{\bar{\rho}}_{21} &= -\,i\omega_o\,\bar{\rho}_{21} + \frac{H'_{21}}{i\hbar}\left(\bar{\rho}_{11} - \bar{\rho}_{22}\right) - \frac{\bar{\rho}_{21}}{T_2}
\end{aligned}
\right. \tag{3.57}
$$

IV. ROTATING WAVE APPROXIMATION

IV.A. The Gyroscopic Model

We shall consider again a two-level system with a Hamiltonian H_o and the wavefunctions of the two eigenstates:

$$
\left\{
\begin{aligned}
\psi_1(\vec{r},t) &= \psi_1(\vec{r})\ e^{-i(E_1/\hbar)t} \\
\psi_2(\vec{r},t) &= \psi_2(\vec{r})\ e^{-i(E_2/\hbar)t}
\end{aligned}
\right. \tag{4.1}
$$

where

$$
H_o\ \psi_i(\vec{r}) = E_i\ \psi_i(\vec{r}) \tag{4.2}
$$

If we introduce a perturbation $H'(t)$, the Hamiltonian of the system is

$$
H = H_o + H'(t) \tag{4.3}
$$

and the general state of the system is represented by

$$
\psi(\vec{r},t) = c_1(t)\ \psi_1(\vec{r},t) + c_2(t)\ \psi_2(\vec{r},t) \tag{4.4}
$$

where

$$
\left\{
\begin{aligned}
i\hbar\dot{c}_1 &= H'_{12}\ c_2\ e^{-i(E_2-E_1)t/\hbar} \\
i\hbar\dot{c}_2 &= H'_{21}\ c_1\ e^{i(E_2-E_1)t/\hbar}
\end{aligned}
\right. \tag{4.5}
$$

If we set

$$\begin{cases} a(t) = c_1(t)\, e^{-iE_1 t/\hbar} \\ b(t) = c_2(t)\, e^{-iE_2 t/\hbar} \end{cases} \tag{4.6}$$

the eqs. (4.5) become

$$\begin{cases} i\,\hbar\,\dot{a} = E_1 a + H'_{12} b \\ i\,\hbar\,\dot{b} = E_2 b + H'_{21} a \end{cases} \tag{4.7}$$

and (4.4) becomes

$$\psi(\vec{r},t) = a(t)\ \psi_1(\vec{r}) + b(t)\ \psi_2(\vec{r}) \tag{4.8}$$

At this point we define, with Feyman, Vernon and Hellwarth [22], a vector $\vec{R}$, which has the following real components in an abstract space with axes 1,2 and 3:

$$\begin{cases} R_1 = ab^* + a^*b \\ R_2 = i(ab^* - a^*b) \\ R_3 = aa^* - bb^* \end{cases} \tag{4.9}$$

We note that, given the coefficients a and b, $\vec{R}$ is determined. Viceversa if $\vec{R}$ is given a and b will be determined, apart from a physically insignificant phase factor. We can then say that $\vec{R}$ describes completely the physical state of the system.

We define also a vector $\vec{\omega}(t)$ which, in the same system of axes, has the components

$$\begin{cases} \omega_1 = (H'_{12} + H'_{21})/\hbar \\ \omega_2 = i\ (H'_{12} - H'_{21})/\hbar \\ \omega_3 = (E_1 - E_2)/\hbar = -\omega_o \end{cases} \tag{4.10}$$

We note that these components are real. $\vec{\omega}$ describes the perturbation; it is determined for a certain system when the perturbation is known.

The two vectors $\vec{R}(t)$ and $\vec{\omega}(t)$ are connected by the following relation

$$\frac{d\vec{R}}{dt} = \vec{\omega} \times \vec{R} \tag{4.11}$$

which is formally identical to the Bloch equations. We shall now prove the relation above.

We shall begin by considering

$$\frac{d\vec{R}}{dt} = \vec{i}\,\frac{dR_1}{dt} + \vec{j}\,\frac{dR_2}{dt} + \vec{k}\,\frac{dR_3}{dt} \tag{4.12}$$

The first component is

$$\begin{aligned}
\frac{dR_1}{dt} &= \frac{d}{dt}\,(ab^* + a^*b) = \dot{a}b^* + a\dot{b}^* + \dot{a}^*b + a^*\dot{b} \\
&= \frac{1}{i\hbar}\left(E_1a + H'_{12}b\right)b^* + a\left[-\frac{1}{i\hbar}\left(H'_{12}a + E_2b\right)\right] \\
&\qquad - \frac{1}{i\hbar}\left(E_1a^* + H'_{21}b^*\right)b + a\left[\frac{1}{i\hbar}\left(H'_{21}a + E_2b\right)\right] \\
&= \frac{1}{i\hbar}\Big(E_1ab^* + H'_{12}bb^* - H'_{12}aa^* - E_2ab^* \\
&\qquad - E_1a^*b - H'_{21}b^*b + H'_{21}aa^* + E_2a^*b\Big]
\end{aligned} \tag{4.13}$$

As for the cross-product appearing in (4.11):

$$\vec{\omega} \times \vec{R} = \vec{i}\,(\omega_2R_3 - \omega_3R_2) + \vec{j}(\omega_3R_1 - \omega_1R_3) + \vec{k}(\omega_1R_2 - \omega_2R_1) \tag{4.14}$$

The first component is

$$\begin{aligned}
\omega_2R_3 - \omega_3R_2 &= \\
&= \frac{i}{\hbar}\left(H'_{12} - H'_{21}\right)\left(aa^* - bb^*\right) - i\left(\frac{E_1-E_2}{\hbar}\right)\left(ab^* - a^*b\right) \\
&= \frac{1}{i\hbar}\Big[- H'_{12}aa^* + H'_{12}bb^* + H'_{21}aa^* - H'_{21}bb^* \\
&\qquad + E_1ab^* - E_1a^*b - E_2ab^* + E_2a^*b\Big]
\end{aligned} \tag{4.15}$$

The two first components (4.13) and (4.15) agree. The same can be found for the second and third components. With this (4.11) is proved.

The relation (4.11) lends itself to a pictorial presentation, where $\vec{R}$ precesses around the vector $\vec{\omega}$ as in Fig. 7. Because of (4.11) we can write

$$dR = \omega\, R \sin\theta\, dt \tag{4.16}$$

$$d\alpha = \frac{dR}{R\sin\theta} = \omega\, dt \tag{4.17}$$

$$\frac{d\alpha}{dt} = \omega \tag{4.18}$$

The vector $\vec{R}$ precesses around $\vec{\omega}$ and the angular precession velocity is ω. Also, since

$$\vec{R} \cdot \frac{d\vec{R}}{dt} = \frac{1}{2}\,\frac{d}{dt}(R^2) = \vec{R} \cdot \vec{\omega} \times \vec{R} = 0 \tag{4.19}$$

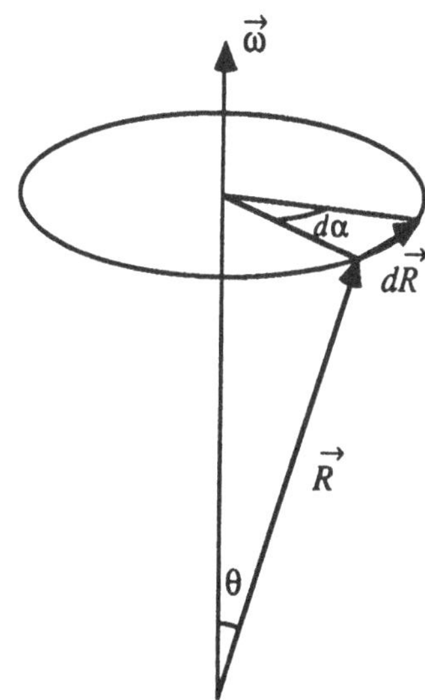

Figure 7. Precession of $\vec{R}$ around $\vec{\omega}$

R = const, i.e. the magnitude of vector $\vec{R}$ does not change in time. If we impose the usual condition of normalization

$$c_1c_1^* + c_2c_2^* = aa^* + bb^* = 1 \tag{4.20}$$

we find that

$$R = 1 \tag{4.21}$$

IV.B. The Case of a Linearly Polarized Field

Let us recall the form of the interaction. The total electric dipole moment is $e\vec{p}$, where

$$\vec{p} = \sum_{i=1}^{Z} \vec{r}_i \tag{4.22}$$

The interaction Hamiltonian is

$$H' = - e\vec{p} \cdot \vec{E}(t) \tag{4.23}$$

Taking advantage of the arbitrary value of the phase of the eigenfunction $\psi_1(\vec{r})$ or $\psi_2(\vec{r})$ we can make the following matrix element real:

$$e\vec{p}_{12} = \langle\psi_1|e\vec{p}|\psi_2\rangle = \vec{\mu} \tag{4.24}$$

Then

$$H'_{12} = H'_{21} = - \mu E(t) \tag{4.25}$$

where μ = projection of $\vec{\mu}$ in the direction of $\vec{E}$. In this case the three components of $\vec{\omega}$ are given by

$$\begin{cases} \omega_1 = (H'_{12} + H'_{21})/\hbar = -\dfrac{2\mu}{\hbar} E(t) \\ \omega_2 = i(H'_{12} - H'_{21}) = 0 \\ \omega_3 = -\omega_o \end{cases} \tag{4.26}$$

As for the components of $\vec{R}$:

$$R_3 = aa^* - bb^* = c_1c_1^* - c_2c_2^* = \rho_{11} - \rho_{22} \tag{4.27}$$

= difference between probabilities of occupancy of the two levels.

We note that

$$\begin{aligned} \langle e\vec{p}\rangle &= \langle\psi|e\vec{p}|\psi\rangle = \langle a\psi_1 + b\psi_2|e\vec{p}|a\psi_1 + b\psi_2\rangle \\ &= a^*b\,\langle\psi_1|e\vec{p}|\psi_2\rangle + ab^*\,\langle\psi_2|e\vec{p}|\psi_1\rangle = (a^*b + ab^*)\,\vec{\mu} = \vec{\mu}\,R_1 \end{aligned} \tag{4.28}$$

Then R_1, when multiplied by $\vec{\mu}$, represents the expectation value of the electric dipole moment. R_2 has no physical meaning in this case of linearly polarized electric field.

Considering now in the present case the equation relating the two vectors $\vec{R}$ and $\vec{\omega}$ we can write

$$\frac{d\vec{R}}{dt} = \vec{\omega} \times \vec{R} = \begin{pmatrix} \vec{i} & \vec{j} & \vec{k} \\ -\dfrac{2\mu}{\hbar}E & 0 & -\omega_o \\ R_1 & R_2 & R_3 \end{pmatrix} \tag{4.29}$$

or

$$\begin{cases} \dot{R}_1 = \omega_o R_2 \\ \dot{R}_2 = -\omega_o R_1 + \dfrac{2\mu}{\hbar} R_3 E \\ \dot{R}_3 = -\dfrac{2\mu}{\hbar} E \end{cases} \tag{4.30}$$

R_2, a parameter without meaning in the present case, can be eliminated from the three equations above; then we write

$$\begin{cases} \dot{R}_3 = -\dfrac{2\mu}{\hbar\omega_o} E\dot{R}_1 \\ \ddot{R}_1 + \omega_o^2 R_1 = \dfrac{2\mu\omega_o}{\hbar} ER_3 \end{cases} \tag{4.31}$$

These equations can be solved once we know the function E(t) and the initial conditions.

Example

$$\begin{cases} E = E_o \cos \omega t \quad (E_o = \text{const}) \\ R_3(0) = |a|^2 - |b|^2 = 1 \end{cases} \tag{4.32}$$

We put $R_3 = 1$ in the second of eqs. (4.31) and write

$$\ddot{R}_1 + \omega_o^2 R_1 = \frac{2\mu\omega_o}{\hbar} E_o \cos \omega t \tag{4.33}$$

This is the equation of a harmonic oscillator without damping driven by a sinusoidal field. We note that the quantity $R_1(t)$ produces beats with the field E (via the term $E\dot{R}_1$), and a variation of R_3 leading to absorption. We can also see the mechanism responsible for absorption as due to the fact that the field generates a dipole which then interacts with the same field.

IV.C. Introduction of Damping and Losses

In order to take losses into account we add "ad hoc" terms to the eqs. (4.31):

$$\begin{cases} \dot{R}_3 + \dfrac{R_3 - 1}{T_1} = -\dfrac{2\mu}{\hbar\omega_o} E\dot{R}_1 \\ \\ \ddot{R}_1 + \omega_o^2 R_1 + \dfrac{2}{T_2} \dot{R}_1 = \dfrac{2\mu\omega_o}{\hbar} ER_3 \end{cases} \tag{4.34}$$

If we consider the first of the two equations above and put E = 0, we get

$$\dot{R}_3 + \frac{R_3 - 1}{T_1} = 0 \tag{4.35}$$

Taking $R_3(0) = 0$ and $R_3(\infty) = 1$, we find

$$R_3(t) = 1 - e^{-t/T_1} \tag{4.36}$$

T_1 is the time constant with which R_3 relaxes to its equilibrium value. As for the second equation the term added to it is a simple "damping" term accounting for the line width of the spectral line.

IV.D. Homogeneous and Inhomogenous Broadening

1. Homogeneous Broadening. In this case the eqs. (4.34) apply:

$$\begin{cases} \dot{R}_3 + \dfrac{R_3-1}{T_1} = -\dfrac{2\mu}{\hbar\omega_o} E\dot{R}_1 \\ \ddot{R}_1 + \dfrac{2}{T_2}\dot{R}_1 + \omega_o^2 R_1 = \dfrac{2\mu\omega_o}{\hbar} ER_3 \end{cases} \tag{4.37}$$

Let

$n_t = \dfrac{N}{V}$ = number of atoms per unit volume, and

$n = n_t R_3$ = population difference between lower level and upper level per unit volume.

Then

$$R_3 = \frac{n}{n_t} \tag{4.38}$$

and

$$P = n_t \langle\mu\rangle = n_t \mu R_1 = \text{polarization} \tag{4.39}$$

The eqs. (4.37) can now be written

$$\begin{cases} \dot{n} + \dfrac{n-n_t}{T_1} = -\dfrac{2}{\hbar\omega_o} E\dot{P} \\ \ddot{P} + \dfrac{2}{T_2}\dot{P} + \omega_o^2 P = \dfrac{2\omega_o\mu^2}{\hbar} E\, n \end{cases} \tag{4.40}$$

2. Inhomogeneous Broadening. In this case

$$\begin{cases} n = n_t \displaystyle\int_0^\infty g^* R_3' \, d\omega_o' \\ P = n_t \mu \displaystyle\int_0^\infty g^* R_1' \, d\omega_o' \end{cases} \tag{4.41}$$

where $n_t g^* d\omega_o'$ = number of atoms per unit volume with transition frequency in $(\omega_o', \omega_o' + d\omega_o')$. Also

$$\begin{cases} \dot{R}_3' + \dfrac{R_3'-1}{T_1} = -\dfrac{2\mu}{\hbar\omega_o'} E\dot{R}_1' \\ \ddot{R}_1' + \dfrac{2}{T_2}\dot{R}_1' + \omega_o'^2 R_1' = \dfrac{2\mu\omega_o'}{\hbar} ER_3' \end{cases} \tag{4.42}$$

IV.E. The Rotating Wave Approximation

Let us take another look at the equation of motion of the vector $\vec{R}$:

$$\frac{d\vec{R}}{dt} = \vec{\omega} \times \vec{R} \tag{4.43}$$

where

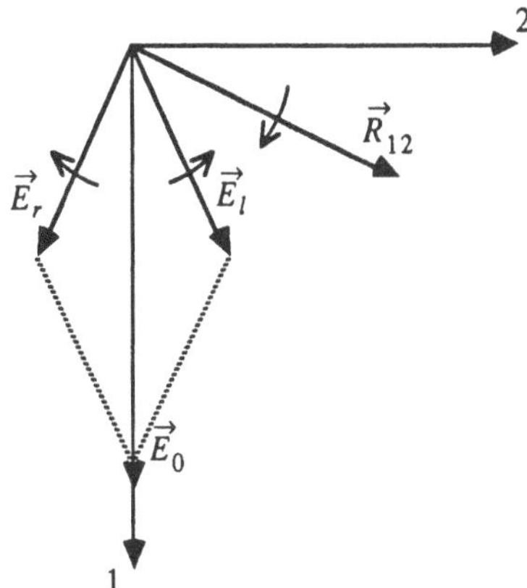

Figure 8. Motion of the Projection of $\vec{R}$ and of an Applied Field $\vec{E}$ in the Plane (1,2).

$$\vec{\omega} \equiv \left(-\frac{2\mu}{\hbar}E(t),\ 0,\ -\omega_o\right) \tag{4.44}$$

$$\vec{R} \equiv \left(a^*b + a^*b,\ i(ab^* - a^*b),\ aa^* - bb^*\right) \tag{4.45}$$

If $\vec{E} = 0$, $\vec{R}$ precesses around the axis 3 with a precession velocity ω_o. The projection of the vector $\vec{R}$ on the (1,2) plane is called R_{12} and rotates as indicated in Fig. 8 in the right circular sense.

An applied field

$$\vec{E} = \vec{E}_o(t) \cos\left[\omega t + \phi(t)\right] \tag{4.46}$$

lines along the direction of axis 1, and can be decomposed into two components E_ℓ and E_r, both of magnitude $E_o/2$, and rotating in the left circular and right circular sense, respectively.

Looking at eqs. (4.37) we notice that the term containing the product $E_\ell \dot{R}_1$ oscillates at a frequency 2ω and does not involve any net energy exchange between the electromagnetic field and the atomic system when averaged over time. This is the basis of the Rotating Wave Approximation which consists in neglecting the effect of the E_ℓ component.

In the Rotating Wave Approximation we find it convenient to use a new set of axes:

III ≡ 3 (this axis coincides with axis 3)

I, II: these two axes rotate in the right circular sense, with angular velocity $\omega + \dot{\phi}$, with I coinciding in direction with E_r.

The following relations (see Fig. 9) allow us to go from the coordinates of $\vec{R}$ in (1,2,3) to the coordinates of $\vec{R}$ in (I,II,III):

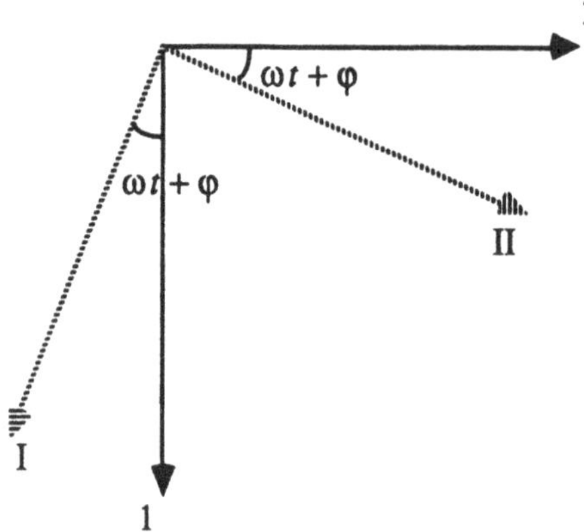

Fig. 9. The Rotation of System (I,II,III) with Respect to System (1,2,3).

$$\begin{cases} R_1 = R_I \cos(\omega t + \phi) + R_{II} \sin(\omega t + \phi) \\ R_2 = - R_I \sin(\omega t + \phi) + R_{II} \cos(\omega t + \phi) \\ R_3 = R_{III} \end{cases} \tag{4.47}$$

In the new system of axes:

$R_{III} = R_3$ = difference between probabilities of occupation of the lower and upper levels,

μR_I = component of the electric dipole μR_1 in phase with the field,

and

μR_{II} = component of the electric dipole μR_1 out of phase (90°) with the field.

In order to describe the motion of $\vec{R}$, as seen by an observer at rest in the rotating frame (I,II,III), we start from the relations

$$\begin{cases} \dot{R}_1 = \omega_o R_2 - \dfrac{R_1}{T_2} \\ \dot{R}_2 = -\omega_o R_1 + \dfrac{2\mu}{\hbar} ER_3 - \dfrac{R_2}{T_2} \\ \dot{R}_3 = - \dfrac{2\mu}{\hbar} ER_2 - \dfrac{R_3 - 1}{T_1} \end{cases} \tag{4.48}$$

We then derive the eqs. (4.47) with respect to time

$$\begin{cases} \dot{R}_1 = R_I \cos\beta - \dot{\beta} R_I \sin\beta + \dot{R}_{II} \sin\beta + \dot{\beta} R_{II} \cos\beta \\ \dot{R}_2 = -R_I \sin\beta - \dot{\beta} R_I \cos\beta + \dot{R}_{II} \cos\beta - \dot{\beta} R_{II} \sin\beta \\ \dot{R}_3 = \dot{R}_{III} \end{cases} \tag{4.49}$$

where $\beta = \omega t + \phi$. Then, using (4.48) and (4.49) we obtain the following two relations:

$$\begin{cases} \dot{R}_I \cos\beta - \dot{\beta} R_I \sin\beta + \dot{R}_{II} \sin\beta + \dot{\beta} R_{II} \cos\beta \\ \quad = \omega_o\left(-R_I \sin\beta + R_{II} \cos\beta\right) - \frac{1}{T_2}\left(R_I \cos\beta + R_{II} \sin\beta\right) \\ -\dot{R}_I \sin\beta - \dot{\beta} R_I \cos\beta + \dot{R}_{II} \cos\beta - \dot{\beta} R_{II} \sin\beta \\ \quad = -\omega_o\left(R_1 \cos\beta + R_{II} \sin\beta\right) + \frac{2\mu}{\hbar} ER_{III} - \frac{1}{T_2}\left(-R_I \sin\beta + R_{II} \cos\beta\right) \end{cases} \tag{4.50}$$

Multiplying the first of these two relations by $\cos\beta$ and the second by $\sin\beta$ and subtracting we obtain

$$\dot{R}_I + (\Delta\omega + \dot{\phi}) R_{II} - \frac{R_I}{T_2} = 0 \tag{4.51}$$

where we have set

$$\cos\beta \sin\beta = \langle\cos\beta \sin\beta\rangle = 0$$

$$\sin^2\beta = \langle\sin^2\beta\rangle = 0$$

and

$$\Delta\omega = \omega - \omega_o .$$

Analogously multiplying the first of (4.50) by $\sin\beta$ and the second by $\cos\beta$ and adding we find

$$R_{II} - (\Delta\omega + \dot{\phi}) R_I = \frac{\mu E_o}{\hbar} R_{III} - \frac{R_{II}}{T_2} \tag{4.52}$$

where we have set the terms in $\sin\beta \cos\beta$ equal to zero and

$$\sin^2\beta = \cos^2\beta = \frac{1}{2}$$

Finally the third of (4.50) gives

$$\dot{R}_{III} = -\frac{\mu E_o}{\hbar} R_{II} - \frac{R_{III} - 1}{T_1} \tag{4.53}$$

Putting together these results we may write

$$\begin{cases} \dot{R}_I = -\left(\Delta\omega + \dot{\phi}\right)R_{II} - \dfrac{R_1}{T_2} \\ \dot{R}_{II} = \left(\Delta\omega + \dot{\phi}\right)R_I + \dfrac{\mu E_o}{\hbar} R_{III} - \dfrac{R_I}{T_2} \\ \dot{R}_{III} = -\dfrac{\mu E_o}{\hbar} R_{II} - \dfrac{R_{III}-1}{T_1} \end{cases} \tag{4.54}$$

In order to give these equations an easy interpretation we wish to derive them from the equation of motion of vector $\vec{R}$:

$$\frac{d\vec{R}}{dt} = \vec{\omega} \times \vec{R} \tag{4.55}$$

We define

$$\left(\frac{d\vec{R}}{dt}\right)^* = \text{time variation of } \vec{R} \text{ as measured by an observer at rest in (I,II,III)} \tag{4.56}$$

Now

$$\frac{d\vec{R}}{dt} = \left(\frac{d\vec{R}}{dt}\right)^* + \vec{\omega}' \times \vec{R} \tag{4.57}$$

where

$$\vec{\omega}' \equiv \left(0,\ 0,\ -(\omega + \dot{\phi})\right) \tag{4.58}$$

is a vector which lies along the axis III ≡ 3 and describes the rotation of (I,II,III) with respect to (1,2,3). From (4.55) and (4.57) we obtain

$$\left(\frac{d\vec{R}}{dt}\right)^* = \left(\vec{\omega} - \vec{\omega}'\right) \times \vec{R} = \vec{\Omega} \times \vec{R} \tag{4.59}$$

where

$$\vec{\Omega} = \vec{\omega} - \vec{\omega}' \tag{4.60}$$

The different vectors appearing in (4.57) and (4.59) are

$$\begin{cases} \dfrac{d\vec{R}}{dt} \equiv \left(\dot{R}_1, \dot{R}_2, \dot{R}_3\right) \\ \left(\dfrac{dR}{dt}\right)^* = \left(\dot{R}_I, \dot{R}_{II}, \dot{R}_{III}\right) \\ \vec{\omega} = \left(-\dfrac{2\mu}{\hbar} E_r, 0, -\omega_o\right) \qquad \left(E_r = \dfrac{E_o}{2}\right) \\ \vec{\omega}' \equiv \left(0, 0, -(\omega + \dot{\phi})\right) \\ \vec{\Omega} \equiv \left(-\dfrac{\mu E_o}{\hbar}, 0, \Delta\omega + \dot{\phi}\right) \qquad \left(\Delta\omega = \omega - \omega_o\right) \end{cases} \tag{4.61}$$

Then

$$\left(\frac{dR}{dt}\right)^* = \begin{vmatrix} \hat{I} & \hat{II} & \hat{III} \\ -\dfrac{\mu E_o}{\hbar} & 0 & \Delta\omega+\dot{\phi} \\ R_I & R_{II} & R_{III} \end{vmatrix} \tag{4.62}$$

and

$$\begin{cases} \dot{R}_I = -(\Delta\omega + \dot{\phi})R_{II} \\ \dot{R}_{II} = (\Delta\omega + \dot{\phi})R_I + \dfrac{\mu E_o}{\hbar} R_{III} \\ \dot{R}_{III} = -\dfrac{\mu E_o}{\hbar} R_{II} \end{cases} \tag{4.63}$$

These equations are the same as the equations (4.54) except for the absence of the damping terms in them.

Comparing the two equations of motion for $\vec{R}$

$$\frac{d\vec{R}}{dt} = \vec{\omega} \times \vec{R} \tag{4.64}$$

$$\left(\frac{d\vec{R}}{dt}\right)^* = \vec{\Omega} \times \vec{R} \tag{4.65}$$

we note that in the first equation $\vec{\omega}$ has a rapid variation, whereas in the second equation $\vec{\Omega}$ is constant if E_o and ϕ are constant or slowly varying in time if, as it is often the case

$$\frac{dE_o}{dt} << \omega E_o \tag{4.66}$$

$$\frac{d\phi}{dt} << \omega \tag{4.67}$$

IV.F. Photon Echoes [23]

The phenomenon of photon echoes furnishes the occasion for an application of the theory presented in the previous section IV.E. We shall consider a two-level system whose spectral line is predominantly inhomogeneous. We shall make the following additional assumptions:

1) $$\Delta\omega_o = \frac{2}{T_2} = \text{homogeneous width} << \Delta\omega_o^* = \frac{2}{T_2^*} = \text{inhomogeneous width} \tag{4.68}$$

as for ruby (or generally for any laser crystal) at very low T (T < 77K).

2) For any ω_o' (frequency of resonance of a particular atom), the light pulses are such that (assuming $\dot{\phi} \simeq 0$):

$$\Omega_I >> \Omega_{II} = \omega - \omega_o' \longrightarrow \frac{\mu E_o}{\hbar} >> \Delta\omega_o^* \tag{4.69}$$

3) The light pulses are π or $\frac{\pi}{2}$ pulses:

$$\frac{\mu}{\hbar}\int_o^{t_p} E_o dt = \pi \text{ or } \frac{\pi}{2} \tag{4.70}$$

Then

$$\frac{\mu E_o t_p}{\hbar} \simeq \pi \longrightarrow t_p \simeq \frac{1}{\mu E_o/\hbar} \tag{4.71}$$

But

$$\frac{\mu E_o}{\hbar} >> \Delta\omega_o^* = \frac{1}{T_2^*} \tag{4.72}$$

Then

$$t_p \simeq \frac{1}{\mu E_o/\hbar} << \frac{1}{\Delta\omega_o^*} = T_2^* \tag{4.73}$$

Then

$$t_p << T_2^* << T_2 \tag{4.74}$$

4) For the sake of simplicity in our treatment we shall neglect the effect of damping (T_1, $T_2 \longrightarrow \infty$).

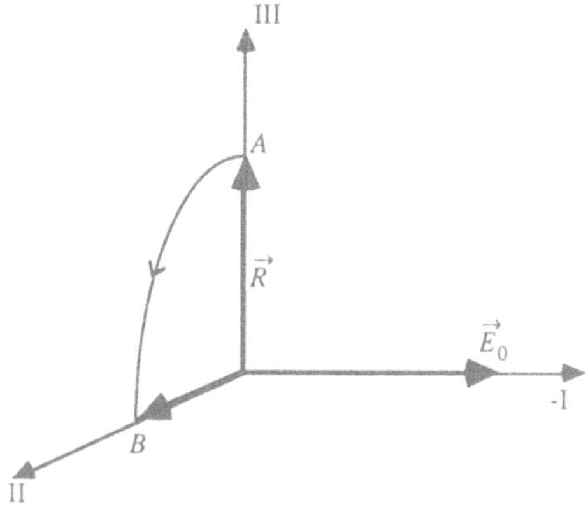

Figure 10. Application of a $\frac{\pi}{2}$ Pulse.

We are now ready for the action which enfolds as follows:

1) The situation of the material is reflected in the initial conditions (point A, Fig. 10)

$$\begin{cases} R_{III} = |a|^2 = 1 \\ R_I = R_{II} = 0 \end{cases}$$

2) We send a $\frac{\pi}{2}$ pulse into the material. Since $\Omega_I >> \Omega_{III}$ the vector $\vec{R}$ precesses around $\vec{E}_o$ until it arrives to point B (Fig. 10).

3) At this time the pulse ends and $E_o = 0$. The individual atoms begin to precess around their respective vectors

$$\Omega_{III} = \omega - \omega_o'$$

where ω_o' = frequency of resonance of a particular atom.

As an example we represent in Fig. 11 the precession of three atoms whose frequencies of resonance are

1. $\omega_o = \omega$

2. $\omega_o < \omega$

3. $\omega_o > \omega$

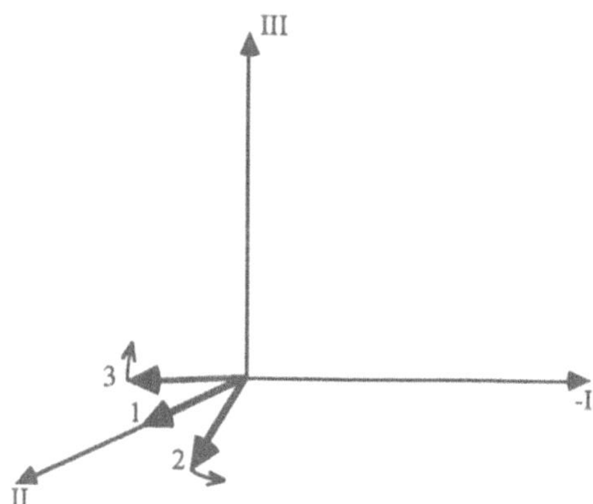

Figure 11. Precession of Individual Atoms around Ω_{III}.

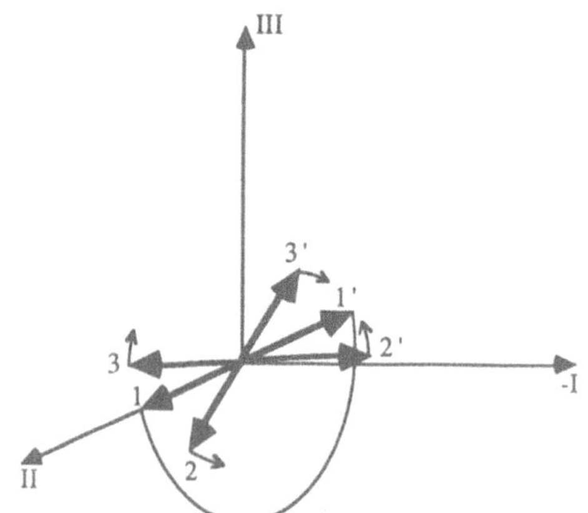

Figure 12. The Application of a π Pulse.

4) At a time τ after the end of the $\frac{\pi}{2}$ pulse, the three vectors (Fig. 11) will be out of phase. At this time we send a π pulse (Fig. 12).

5) The three vectors precess around the axis -I until they arrive at positions 1′, 2′ and 3′. E_o at the end of the pulse is zero.

6) The vectors resume their precession about their respective vectors $\Omega_{III} = \omega - \omega_o'$. 2′ and 3′ get in phase again with 1′ after a time τ equal to the time between the two pulses.

7) All the atoms behave indeed in the same manner; therefore, at time τ after the π pulse, all the vectors corresponding to the various atoms are again in phase and aligned along the axis II.

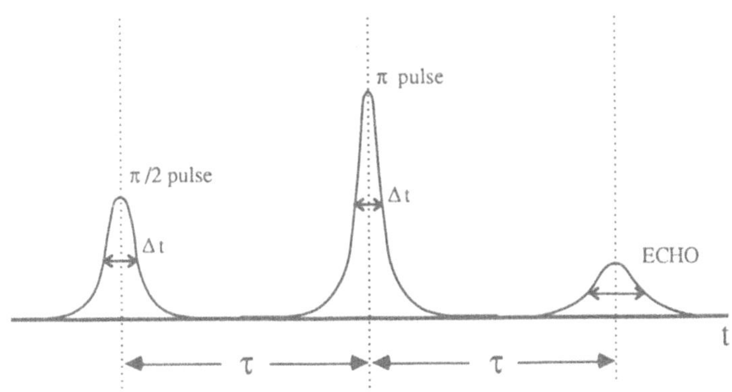

Figure 13. $\frac{\pi}{2}$ Pulse, π Pulse and Echo.

8) The result of the rephasing of the individual atom-vectors is a large electric dipole, 90° out of phase with respect to the field. This electric dipole produces a light pulse called photon echo.

9) The photon echo is emitted in the same direction as the $\frac{\pi}{2}$ and π pulses and is delayed by a time τ equal to the time separation between the $\frac{\pi}{2}$ and π pulses (see Fig. 13).

10) The duration of the photon echo pulse is equal to the time it takes the vectors 1, 2, 3 to dephase from each other, i.e., T_2^*. Since $t_p <<$ T_2^*, the photon echo pulse is much larger than the $\frac{\pi}{2}$ and π pulses.

At this point I would like to direct the attention of the reader to Fig. 14 which reproduces the front cover of the November 1953 issue of Physics Today. The figure gives an excellent illustration of the photon echo phenomenon.

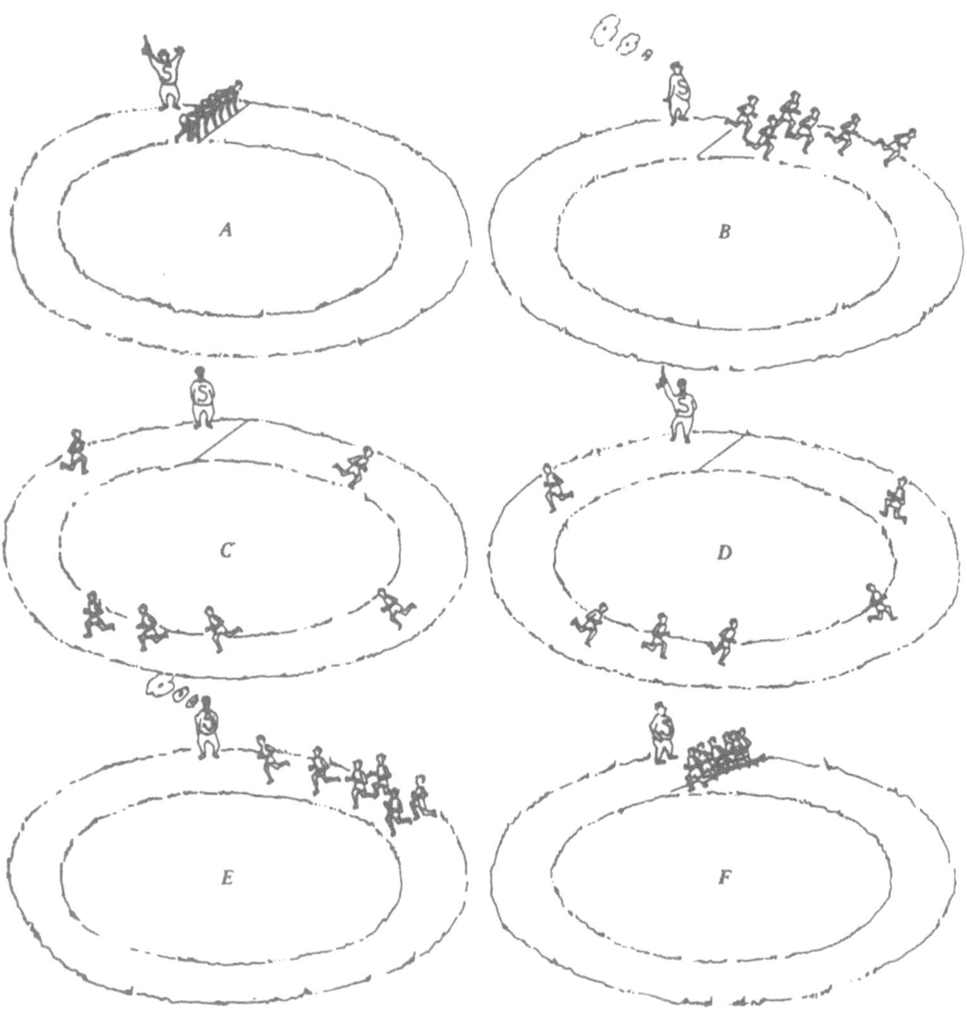

Figure 14. Dephasing and the Effect of Reversal on a Race Track (from the front cover of the November 1953 issue of Physics Today; reproduced by permission).

V. COHERENCE AND INCOHERENCE

V.A. Coherent and Incoherent Sources

A two-level atomic system emitting radiation at the resonance frequency can be viewed as a radiating dipole oscillating at the beat frequency between the two stationary atomic states. The radiation emitted by the atom has the characteristic dipole pattern which spans all directions, except the direction of the dipole.

We shall compare for a while emitting atoms to radiating dipoles and begin by considering two dipoles driven by the same transmitter. If the electric fields due to the two dipoles are $\vec{E}_1$ and $\vec{E}_2$, the total field is

$$\vec{E} = \vec{E}_1 + \vec{E}_2 \tag{5.1}$$

and the corresponding Poynting vector

$$S = |\vec{S}| \propto |\vec{E}_1 + \vec{E}_2|^2 = |E_1|^2 + |E_2|^2 + 2|E_1||E_2|\cos\theta_{12} \tag{5.2}$$

where $|E|$ = real magnitude of the complex field $\vec{E}$. The phase angle θ_{12} depends on:

a) the orientations of the dipoles,
b) the distance between the dipoles,
c) the location of the point of observation, and
d) the phases of the oscillations of the two dipoles.

Assume $|E_1| = |E_2| = |E|$; then

$$S \propto 2|E|^2 (1 + \cos\theta_{12}) \tag{5.3}$$

Note the following:

if $\theta_{12} = 180^\circ$ (destructive interference), $S = 0$;

if $\theta_{12} = 90^\circ$, $S = 2S_1 = 2S_2$, i.e. intensity = 2× intensity of single dipole emission;

if $\theta_{12} = 0^\circ$ (constructive interference), $S = 4S_1 = 4S_2$, i.e. intensity = 4× intensity of single dipole emission.

Consider now the case of N dipoles:

$$S_N \propto \left| \sum_{i=1}^{N} \vec{E}_i \cdot \sum_{j=1}^{N} \vec{E}_j \right|^2 = \sum_{n=1}^{N} |E_n|^2 + \sum_{\substack{i=1 \\ i\neq j}}^{N} \sum_{j=1}^{N} |E_i|\,|E_j| \cos\theta_{ij} \tag{5.4}$$

If all $|E_n|$ are equal to $|E|$:

$$S_N \propto |E|^2 \left(N + \sum_{\substack{i=1 \\ i\neq j}}^{N} \sum_{j=1}^{N} \cos\theta_{ij} \right) \tag{5.5}$$

Example

$$\begin{cases} N = 4 \\ \theta_{ij} = 0 \text{ if } i + j \text{ even} \\ \theta_{ij} = 180^\circ \text{ if } i + j \text{ odd} \end{cases} \tag{5.6}$$

In this case

$$\begin{aligned} \sum_i \sum_{\substack{j \\ i \neq j}} \cos\theta_{ij} &= \cos\theta_{12} + \cos{}_{13} + \cos\theta_{14} \\ &\quad + \cos\theta_{21} + \cos\theta_{23} + \cos\theta_{24} \\ &\quad + \cos\theta_{31} + \cos\theta_{32} + \cos\theta_{34} \\ &\quad + \cos\theta_{41} + \cos\theta_{42} + \cos\theta_{43} \\ &= -1 + 1 - 1 \\ &\quad - 1 - 1 + 1 \\ &\quad + 1 - 1 - 1 \\ &\quad - 1 + 1 - 1 = -4 \end{aligned}$$

Then

$$N + \sum_i \sum_{\substack{j \\ i \neq j}} \cos\theta_{ij} = 4 - 4 = 0 \tag{5.7}$$

We can see that in this case, and in the case of N even we have complete destructive interference. If N is odd and the phase angles have the same relations (5.6), then we have "almost" complete destructive interference.

Assume now that we arrange the physical conditions in such a way that $\theta_{ij} = 0$ for all i's and j's; then

$$S_N \propto |E|^2 \left[N + N(N-1)\right] = N^2 \, |E|^2 \tag{5.8}$$

The phase angles θ_{ij} depend on the position of the point of observation; therefore the intensity of the radiation coming out of the array of N dipoles can vary in different directions from zero (destructive interference) to a value $\propto N^2|E|^2$ (constructive interference).

If the phase θ_{ij} are random, as it may be the case when the N dipoles are disconnected from their transmitter,

$$\sum_i \sum_{\substack{j \\ i \neq j}} \cos\theta_{ij} \propto \langle\cos\theta\rangle = 0 \tag{5.9}$$

and

$$S_N \propto N \, |E|^2 \tag{5.10}$$

The randomization of the phase angles produces a loss of "coherence" and may be observed at a certain point as a decrease of intensity from $\alpha\ N^2|E|^2$ to $\alpha\ N|E|^2$.

We can now make the following observations:

1) We can associate an atomic density matrix $\rho^{(n)}$ to each atom-dipole and an ensemble density matrix $\bar{\rho}$ to the array of dipoles. If the dipoles are driven by the same transmitter or by transmitters which are locked in phases, then the off-diagonal elements of the individual matrices $\rho^{(n)}$ will also be synchronized, and the off-diagonal terms of $\bar{\rho}$ will be non-zero.

2) When such phase-locking is present, the dipole moments of the atoms emit coherent radiation. This radiation will be highly directional and very intense in certain directions (N^2 times as intense as that of the individual atom).

3) If the N dipoles are driven by transmitters having random phases the corresponding off diagonal terms of the ensemble density matrix $\bar{\rho}$ will vanish and no net macroscopic (i.e. experimentally observable) oscillating polarization will be present.

4) This is the case with light sources in emission spectroscopy which generally produce an incoherent mixture of electromagnetic waves. If the light from these sources is monochromatized by passing it through a monochromator or a filter, the outcome is a radiation analogous to that produced by N transmitters operating with random phases at the same frequency. Under these conditions there cannot be any measurable macroscopic polarization in the ensemble of dipoles which is a poor mimic of the individual quantum system making up the ensemble. The radiation emitted is isotropic and has an intensity proportional to the number N of radiators.

V.B. Creating Coherence

Light sources employed in spectroscopic studies are often of the incoherent (non-laser) type. In these systems the random phases of the individual atomic systems cancel the off-diagonal elements of the matrix $\bar{\rho}$ and wipe out all the details of the absorption and emission processes. A detailed knowledge of these processes by which radiation interacts with the atoms is not necessary to conduct experiments in "linear" spectroscopy.

If, on the other hand, one is interested in the details of the spectroscopic transitions that are not measurable in linear spectroscopy, then one has to induce phase correlation among the wavefunctions of the quantum systems which make up the ensemble. Such correlations are generally produced by exciting the systems with coherent radiation.

Assume that a perturbation H′ represented by a coherent radiation field has been applied for a while to the ensemble of systems (each with an unperturbed Hamiltonian H_o) and has then been turned off. H′ has driven the ensemble away from equilibrium; in its absence the ensemble, being in contact with the heat reservoir, goes back to its equilibrium situation which consists of a Boltzmann distribution over the eigenstates of H_o.

The relaxation of the ensemble to equilibrium is represented by the equations of motion (3.57), in which the matrix elements of the perturbation H′ are set equal to zero and the only remaining terms are the relaxation terms. For this reason our attention turns at this time to the phenomenological constants T_1 and T_2.

1. T_1, the Longitudinal Relaxation Time. Given an ensemble of two-level systems we define as effective temperature of the ensemble the quantity

$$T_s = \frac{E_2 - E_1}{k\ln(\bar{\rho}_{22}/\bar{\rho}_{11})} \tag{5.11}$$

The temperature of the bath is

$$T = \frac{E_2 - E_1}{k\ln(\bar{\rho}^e_{22}/\bar{\rho}^e_{11})} \tag{5.12}$$

The quantity $(T_s - T)$ is a measure of the energy received by the ensemble due to the exciting perturbation H'. The phenomenological constant T_1 characterizes the rate of flow of energy from the ensemble to the heat bath and vice versa: it represents indeed the 1/e cooling (or heating) time necessary to bring $T_s \rightarrow T$.

If the ensemble is excited by such random agents as incoherent radiation, the off diagonal elements of $\bar{\rho}$ are zero and the only relaxation process possible is the one associated with T_1. Of course, even if we know the value of this constant, we may not know what particular process or combination of processes is responsible for the relaxation to equilibrium.

A possible mechanism for getting rid of the excess energy is the one due to spontaneous emission of radiation, otherwise called luminescence. This emission is incoherent and is proportional to the number N of quantum systems in the ensemble. The excess population in the excited state decays exponentially with the rate $1/T_1$ and the intensity of the radiation emitted decays with the same rate. For this reason in such case the parameter T_1 is also called luminescence lifetime.

Spontaneous emission is not the only T_1 process present in spectroscopic samples. Other processes called radiationless or nonradiative may convert the excess energy directly into heat.

Luminescent emission is generally associated with large energy gaps (few eV) and with electronic transitions. In luminescent materials incoherent spontaneous emission may be the dominant decay process in correspondence to these transitions, a desirable feature in phosphors or light emitters. On the other hand, the fact that many transitions, allowed by selection rules, are not accompanied by luminescence emission indicates that nonradiative processes are predominant in many instances. It should also be noted that, even when luminescence emission is present, nonradiative processes may actually compete with the radiative ones by providing the excited material with an alternate path of decay to equilibrium.

At radio frequencies, in NMR samples, nonradiative processes dominate completely the T_1 relaxation, because the spontaneous emission process is extremely slow ($\sim 10^{15}$ years) on account of the very small energy splitting of the levels involved.

2. T_2, the Transversal Relaxation Time. At the end of the T_1 processthe proper exchange of energy between the ensemble and the thermal bath has taken place and the ensemble is in thermal equilibrium. For

temperatures much lower than $\hbar\omega/k$ every system of the ensemble is in the ground state: i.e.

$$\begin{cases} c_{1n} = 1 \\ c_{2n} = 0 \end{cases} \tag{5.13}$$

In these circumstances the off diagonal elements $c^*_{1n}c_{2n}$ and $c_{1n}c^*_{2n}$ of each density matrix $\rho^{(n)}$ are zero, and, of course, the off diagonal elements of $\bar{\rho}$ are zero. Therefore T_1 processes are also T_2 processes, because the T_1 relaxation mechanism can make all the $\bar{\rho}_{ik}(i \neq k)$ elements go to zero. If the T_2 relaxation takes place via the T_1 processes,

$$T_1 = T_2 \tag{5.14}$$

The T_1 processes are sufficient, but not necessary to produce the T_2 relaxation. In general,

$$T_2 < T_1 \tag{5.15}$$

because other processes may occur that cause the off-diagonal elements of the density matrix $\bar{\rho}$ to decay to zero faster than the diagonal elements to their Boltzmann equilibrium values.

We shall consider now the T_2 processes point by point:

1) A perturbation H′ has excited the ensemble coherently and has been turned off.

2) Each system at this time has an excess of energy and will decrease it be emitting radiation.

3) Before the systems eliminate their excess energy other processes will cause them to scramble their phases ($T_2 < T_1$).

4) The synchronization forced on the systems by H′ disappears gradually, the phases become unlocked, the radiation becomes incoherent and isotropic and its intensity (in regions where constructive interference had made it large) is drastically reduced.

5) Finally the T_1 processes end all emission.

An effective visualization of the interplay of T_1 and T_2 processes is given in the following illustration taken from the book by J.D. Macomber cited in the Acknowledgements, and reproduced by permission.

We like to add the following relevant points:

6) For ensembles of quantum systems in materials of macroscopic sizes the difference in intensity between the initial coherent emission and the following incoherent one can be very large. Ensembles of quantum systems with a high degree of phase correlation are called <u>superradiant</u>. The process of decay of the coherent (superradiant) radiation to situations of lower and lower coherence in a time T_2 is called <u>free induction decay</u> by NMR workers.

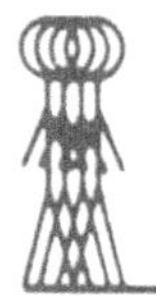

Coherently driven soldiers.

Drill sargeant counts cadence.

(a)

Shoot the sargeant!

No immediate breakup of marching formation because each soldier has memory of sargeant's cadence.

(b)

After time $t = T_2$, complete disarray of formation, but march continues.

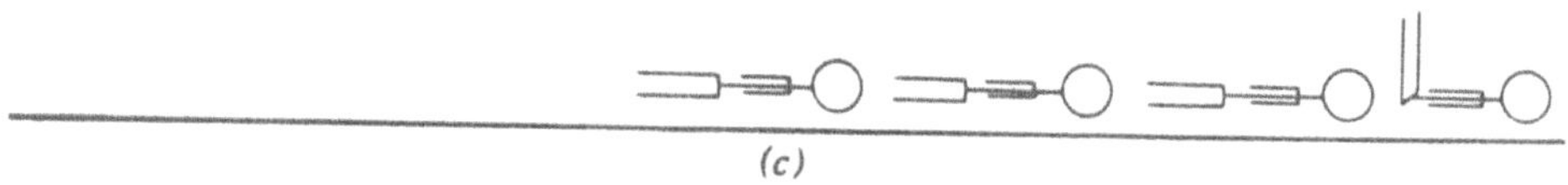

(c)

At time $t = T_1$ soldiers cease marching and fall by exhaustion.

7) We have already seen that in NMR nonradiative processes dominate the T_1 relaxation. However, in NMR samples (of course, coherently excited) the excess energy can be eliminated quickly by coherent superradiant emission and the free induction decay can represent a mechanism that competes effectively with nonradiative processes in bringing about the T_1 relaxation. This feature of the decay path is called radiation damping.

V.C. Relaxation Processes and Effects on Spectral Lines

1. T_1 Processes. The description of the effects of the interaction of radiation with an ensemble of quantum systems has to include in the equations of motion the relaxation terms if the time of observation may be larger than the two characteristic relaxation times T_1 and T_2. We shall consider first the T_1 processes and remind ourselves that $1/T_1$ is the rate at which the population difference between the two quantum states involved in the transition returns to its equilibrium value via the transfer of energy from the ensemble to the heat bath.

Assume that the ensemble is a dilute gas in a container. For this material the heat bath is represented by the walls of the container which have a high heat capacity. At low temperatures the atoms of the gas move slowly and collide seldom with each other. If they are excited their main relaxation mechanism is the spontaneous emission of radiation. This is, as we know, a random process; therefore we cannot predict exactly the time that will elapse between the excitation of a particular atomic system and its spontaneous emission. However, looking at the overall output from the sample, we can expect the emission to start immediately after excitation and to last for a time $T_1 = 1/A$ (A = Einstein coefficient of spontaneous emission).

The radiation output from such (incoherently excited) a material will be an exponentially decaying signal with a carrier frequency ω_0 and lasting for a time T_1. The Fourier transform of such a signal has a Lorentzian shape whose width is $2/T_1 = 2A$. If no other mechanism is present $T_2 = T_1$; however it may happened that the excess energy finds its way to the thermal bath via some degrees of freedom other than the ones involved in the transition under consideration, before spontaneous emission has a chance to occur.

If a dilute gas is excited coherently, the spontaneous emission following the end of the excitation will also be coherent. All the induced dipole moments will oscillate in phase with amplitudes proportional to the off diagonal elements of the density matrix. These elements will decay exponentially with the time constant T_1 which is the characteristic time in which the populations of the two levels will relax to equilibrium: after a time T_1 the coherence of the quantum systems is simply destroyed, because the systems do not radiate any longer. Again, in this case, the T_1 processes are also T_2 processes.

2. T_2 Processes. If we increase the density of a gas by compressing it isothermally, we decrease the collision time which may go down to the value of the longitudinal relaxation time T_1. Since most collisions are elastic, the increased density will not result in the shortening of the time constant T_1; the time constant T_2 will, however, be affected.

A radiating molecule A produces a decaying signal which has definite frequency and phase. If such a molecule, while radiating, collides with a molecule B and transfers its remaining energy to it, B will begin radiating with a new phase. The time required by the system to lose its excess energy is still T_1, but the radiation has been partitioned into trains of coherent oscillations which last each a time T_2. The resulting spectral line has a Lorentzian shape with width $1/T_2 > 1/T_1$, with T_2 = mean time between collisions. Actually even an encounter in which no transfer of energy takes place will result in the dephasing of the wavefunctions of the colliding molecules, and in a T_2 event.

In effect two types of encounters take place in a gas of identical molecules: collisions in which energy is exchanged, and collisions in which energy is not exchanged. Both processes are random and represent mechanisms for T_2 relaxation: however we cannot distinguish them.

If an ensemble of quantum systems is excited coherently, the following free induction decay is accompanied by the randomization of the phases of the elementary dipoles by either type of T_2 relaxation processes.

If an ensemble of quantum systems is excited incoherently and then emits spontaneous (luminescence) radiation, the spectral line of the transition is affected by the T_2 relaxation processes brought about by the collisions. This broadening effect is called pressure broadening and is homogeneous; the line in this case is Lorentzian in shape with a width equal to $2/T_2$.

V.D. Determination of the Relaxation Constants

1. Determination of T_1. It is possible to determine the relaxation constant T_1 by applying to the material a series of π, $\pi/2$ pulses, which are all very short in comparison to T_1 or T_2. We proceed as follows (see Fig. 15):

1) We begin with the ensemble at equilibrium and $R_{III} = M$ ($M \leq 1$; point A).

2) We send a π pulse and make the R_{III} component equal to $-M$ (point B).

3) The ensemble, being in contact with the heat reservoir, then relaxes: the vector $\vec{R}$ contracts along the negative part of axis III and returns to its equilibrium value in a time T_1.

4) At time t′ (point C) the relaxation and the contraction of R_{III} have proceeded for a while. The extent of the relaxation process is given by the length of the R_{III} component at C which can be measured by applying a $\pi/2$ pulse that brings the $\vec{R}$ vector from C to D.

5) $\vec{R}$ at D generates a signal that then decays exponentially with the time constant T_2.

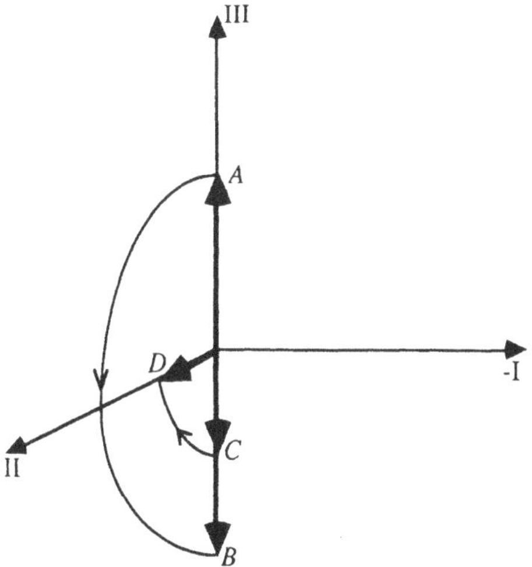

Figure 15. Sequence of Pulses for the Measurement of T_1.

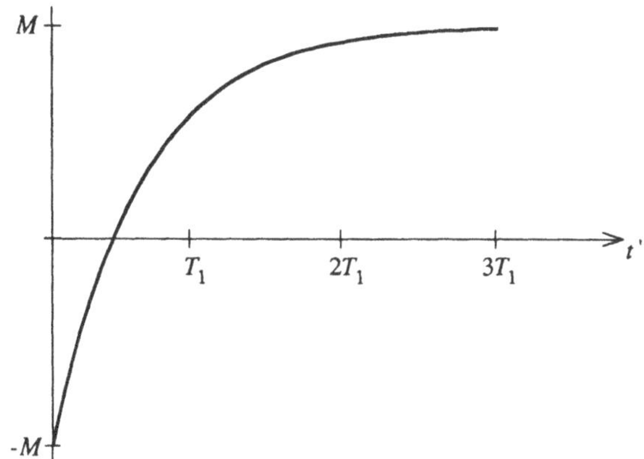

Figure 16. Amplitude of the Free Induction Signal After a Sequence of π, $\pi/2$ Pulses.

6) At this point the experiments stops for a time longer than T_1 in order to allow the ensemble to return to thermal equilibrium and the R_{III} component of $\vec{R}$ to relax to its equilibrium value M. The experiment is then repeated using a different time t'.

7) The initial emplitude of the free induction signal is then plotted versus t', the time separation of the two pulses. This plot allows one to trace the entire T_1 decay process (see Fig. 16).

2. Determination of T_2. The measurement of T_2 is performed by using a procedure designed by Carr and Purcell [24], which is similar to the one used to measure T_1.

It is indeed possible to determine the relaxation constant T_2 by applying to the material a series of $\pi/2$, π, pulses, which are all very short in comparison to T_1 or T_2. A sequence of pulses $\pi/2(0)$, $\pi(\tau)$, $\pi(3\tau)$, $\pi(5\tau)$, $\pi(7\tau)$, etc. is used to excite the ensemble. Successive echoes are produced at times 2τ, 4τ, 6τ, etc., with amplitudes M $\exp(-2\tau/T_2)$, M $\exp(-4\tau/T_2)$, M $\exp(-6\tau/T_2)$, etc. The plot of these amplitudes allows one to trace the entire T_2 decay process.

V.E. Coherent Spectroscopic Effects

It may be worthwile at this point to examine briefly the effects of coherence on the outcome of some spectroscopic measurements.

1. Optical Nutation. When a perturbation H' is applied to an ensemble of quantum systems as a step function the response of the ensemble is called the transient nutation effect, or the Torrey effect [25] (named after the discoverer of nutation in nuclear spin systems). We have already considered this effect in subsection II.F.5, but we want to view it now in the gyroscopic model.

In such a model the nutation effect can be seen as a wobble, superimposed on the fast precessional motion, and damped by the relaxation mechanisms (for this reason the effect is called "transient").

If T_1, $T_2 \gg 1/|V|$ the applied field oscillates several times before the damping becomes relevant. The first part of each oscillation of the field corresponds to an increase of the energy of the ensemble and a decrease of the energy of the field (absorption); the second part corresponds to a flow of energy from the ensemble into the radiation field (stimulated emission).

The precession of the vector $\vec{R}$ about the vector $\vec{\omega}$ in the fixed system of axes (1,2,3) is represented in Fig. 17.

2. Free Induction Decay. Superradiant ensembles consist of quantum systems with great phase correlation. The decay of the superradiant emission takes place in a time T_2 and is called free induction decay. This effect was discovered in spin systems by E.L. Hahn [26]. Brewer and Shoemaker discovered later the quantum "optical" free induction decay [27].

3. Self-Induced Transparency. A pulse of light can pass through a material without experiencing any attenuation, if the material is transparent at the light frequency. Transparency of an opaque material can be produced under certain circumstances by the light pulse itself and, for this reason, is called self-induced transparency. This effect was predicted and observed by McCall and Hahn at an optical frequency

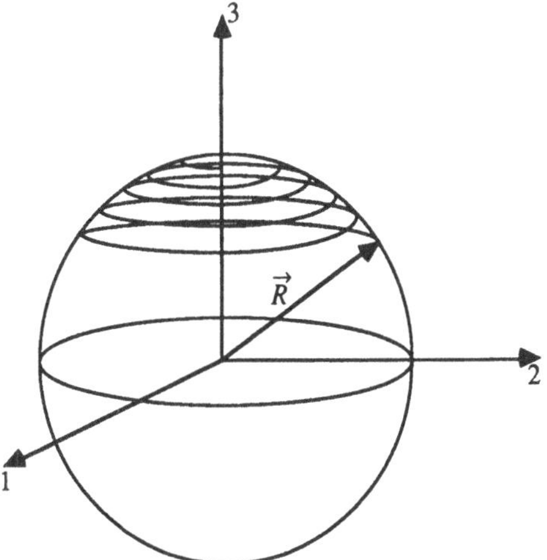

Figure 17. Precession of the Vector $\vec{R}$ about the Vector $\vec{\omega}$ as Viewed in the Fixed Frame (1,2,3).

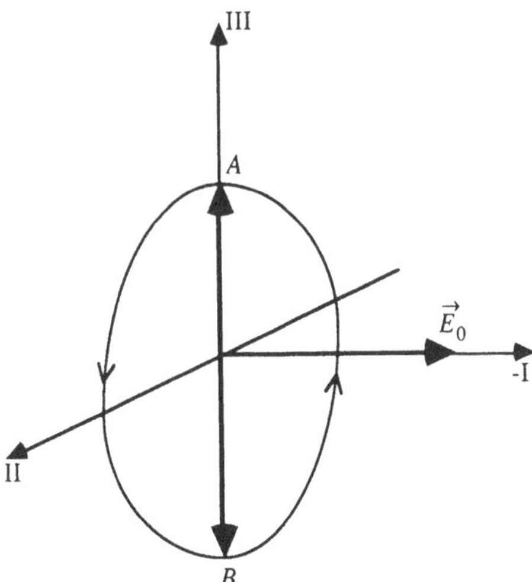

Figure 18. Self-Transparency Due to a 2π Pulse.

that matched a resonant transition in ruby [28]. The process unfolds as follows:

1) One starts from the condition $R_{III} = 1$ (point A in Fig. 18). All the population is in the ground state.

2) One then applies a 2π pulse, i.e. a pulse for which

$$I = \frac{\mu}{\hbar} \int_{-\infty}^{+\infty} E_o dt = 2\pi \tag{5.16}$$

The first part of the pulse, up to such a time t_π that

$$\frac{\mu}{\hbar} \int_{-\infty}^{t_\pi} E_o dt = \pi \tag{5.17}$$

produces a precession of $\vec{R}$ about the axis -I which brings it from point A to point B. This first half cycle corresponds to the absorption of field energy by the material.

3) The vector $\vec{R}$ continues its precession around the axis -I and returns to A at the end of the pulse. The energy abosrbed from the leading edge of the pulse is given back to its trailing edge; the energy is thus returned to the original beam.

The following considerations can be made:

1) Self-induced transparency can in principle be produced with 2π, 4π, 6π, etc. pulses in materials undergoing a transition in any region of the electromagnetic spectrum. However, only for optical transitions the wavelength is much smaller than the dimension of the sample and the echoes of the exciting pulses actually propagate, making this effect more than a simple process consisting of the absorption and subsequent reemission of some field energy.

2) The pulse of radiation propagates with velocity smaller than c because part of the energy spends some time in the material and during this time does not propagate. The larger is the magnitude of the absorption at the wavelength of the transition, the larger is the amount of the energy taken up by the material. One can thus achieve a greater delay in the propagation by increasing the concentration of absorbing species. One can also achieve such a delay by increasing the pulse width, but one has to be careful not to allow the energy to stay with one absorbing center a time longer than T_2. Velocities of propagation as low as $10^{-3}c$ have been observed experimentally.

3) If the pulse length is such that the quantity I of (5.16) is less than 2π, the pulse is completely absorbed. If I is more than 2π but less than 4π, the pulse is attenuated until it reaches the value 2π, and then propagates without losses. If I is greater than 4π the pulse is broken into a sequence of 2π pulses and the rest of the energy is absorbed.

The self-induced transparency effect is illustrated by Fig. 19, a schematic illustration of the mechanical analog presented in the August 1967 issue of Physics Today. A rolling ball, representing the light pulse, strikes the first pendulum ball of a row of such pendula which represent the material sample. The pendulum turns 360° and strikes the rolling ball on the backside restoring its energy. The ball moves then in the original direction and repeats the cycle at the location of the next pendulum, and so on.

V.F. Coherence of Radiating Systems

If the phase constants of the off-diagonal elements of the individual density matrices $\rho^{(n)}$ of the systems of an ensemble are distributed at random, the off-diagonal elements of the ensemble density matrix $\bar{\rho}$ are zero. If, on the other hand, the phase constants are all the same, all the wavefunctions of the systems of the ensemble are oscillating in synchronism in a way similar to that of coherent electromagnetic waves.

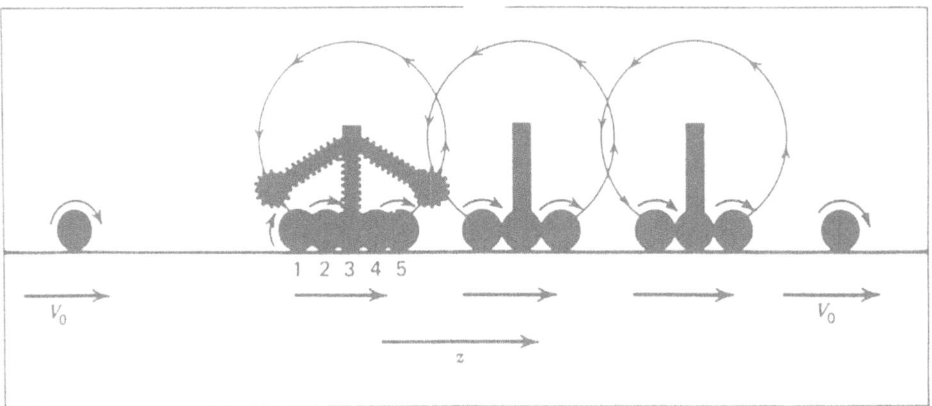

Figure 19. Mechanical Analog to the Self-Transparency Effect (from the August 1967 issue of Physics Today; reproduced by permission).

We know from statistical mechanics that it is impossible to deduce all the properties of the individual systems of an ensemble in thermal equilibrium from the ensemble-averaged behavior of the macrosystem. Indeed the individual systems present in general a greater variety of properties than the macrosystem: "the individual system can do everything the ensemble does and more".

But, if we are able to render the phases of the individual systems of the ensemble more and more similar, the ensemble density matrix $\bar{\rho}$ approaches in all its elements the typical matrix $\rho^{(n)}$. When complete coherence among the wavefunctions of the systems is achieved, the behavior of the macrosystem is a perfect reflection of the behavior of the individual systems. This means that in the presence of coherence the quantum behavior of individual atoms and molecules can be "seen" in the macroscopic world, namely it can be in effect evaluated by actual measurements.

ACKNOWLEDGEMENTS

The author wishes to thank his graduate students Clyfe Beckwith, Brian Bowlby, Gang Lei, and Chunlai Yang for helpful criticism of this paper.

He wishes also to acknowledge the benefit of reading or consulting the following books which he would like to recommend to the readers of this article for further readings:

1. R. Loudon, The Quantum Theory of Light, Clarendon Press, Oxford, 1983.
2. J.D. Macomber, The Dynamics of Spectroscopic Transitions, John Wiley & Sons, New York, 1976.
3. L. Allen and J. Eberly, Optical Resonance and Two-Level Atoms, John Wiley & Sons, New York, 1975.
4. O. Svelto, Principles of Lasers, Plenum Press, New York and London, 1976.
5. J. I. Steinfeld, Molecules and Radiation, The MIT Press, Cambridge, Massachusetts, 1985.
6. B.W. Shore, The Theory of Coherent Atomic Excitation, John Wiley & Sons, New York, 1990.

Some of the books cited above are general references for the various sections of this article: book #1 for section II, books #2 and 3 for sections III and V, and book #4 for section IV.

In addition, the author would like to acknowledge Physics Today for their permission to reproduce two figures which appeared in past issues of this journal as Figs 14 and 19 of this article, and John Wiley & Sons for their permission to reproduce as an illustration for subsection V.B.1 of this article a figure appearing in the book by J. D. Macomber cited above.

REFERENCES

1. B. Di Bartolo, "Interaction of Radiation with Atoms and Molecules", in Spectroscopy of the Excited State, B. Di Bartolo, editor, Plenum Press, New York, 1976, p.1.
2. C. Reid, Excited States in Chemistry and Biology, Butterworths, London, 1957.
3. G. Porter and M. Windsor, Proc. Roy. Soc. A245, 238 (1958).
4. J.R. Novak and M.W. Windsor, Proc. Roy. Soc. A308, 95 (1968).
5. A.G. Briggs and R.G.W. Norrish, Proc. Roy. Soc. A276, 51 (1963).
6. F. Gires and G. Mayer, Comptes Rendues 254, 659 (1962); also in Quantum Electronics Proceedings of the Third International Congress, P. Grivet and N. Bloembergen, editors, Columbia, New York, 1964, p.841.
7. C.S. Naiman, B. Di Bartolo and A. Linz, in Physics of Quantum Electronics, P.L. Kelley, B. Lax and P.E. Tannenwald, editors, McGraw-Hill, New York, 1966, p. 315.
8. U.S. Naval Research Laboratory; Memorandum Report 1483.
9. C.C. Robinson, J. Opt. Soc. Am. 55, 1576 (1965); also J. Opt. Soc. Am. 57, 4 (1967).
10. J. W. Huang and H.W. Moos, Bull. Am. Phys. Soc. 12, 1068 (1967).
11. M.D. Galanin, V.N. Smorchkov, and Z.A. Chizhikova, Opt. Spectr. 19, 168 (1965).
12. R.L. Greene, J.L. Emmett, and A.L. Schawlow, Appl. Opt. 5, 350 (1966).
13. L. Pauling and E.B. Wilson, Introduction to Quantum Mechanics, McGraw Hill, New York, 1935, pp.133-136.
14. F. Bloch, Phys. Rev. 70, 460 (1946).

15. I.I. Rabi, Phys. Rev. 51, 652 (1937).
16. R.A. Smith, Proc. Roy. Soc. A362, 1 (1978).
17. C. Yang, Boston College, Department of Physics, Spectroscopy Laboratory Internal Report, 1992.
18. W.R. MacGillivray, D.T. Pegg, and M.C. Standage, Optics Commun. 25, 355 (1978).
19. R.G. Brewer in Nonlinear Spectroscopy, N. Bloembergen, editor, North-Holland, Amsterdam (1977), p.87.
20. F. Reif, Fundamentals of Statistical and Thermal Physics, McGraw Hill, New York 1965, p.470.
21. D.W. Posener, Austral. J. Phys. 12, 184 (1959).
22. R.P. Feynman, F.L. Vernon, Jr., and R.W. Hellworth, J. Appl. Phys. 28, 49 (1957).
23. I.D. Abella, N.A. Kurnit, and S.R. Hartmann, Phys. Rev. 141, 391 (1966).
24. H.Y. Carr and E.M. Purcell, Physical Review 94, 630 (1954).
25. H.C. Torrey, Phys. Rev. 76, 1059 (1949).
26. E.L. Hahn, Phys. Rev. 77, 297 (1950).
27. R.G. Brewer and R.L. Shoemaker, Phys. Rev. Lett. 27, 631 (1971).
28. S.L. McCall and E.L. Hahn, Phys. Rev. Lett. 18, 908 (1967); also Phys. Rev. 183, 457 (1969).

PROPERTIES OF THE EXCITED STATES OF COMPLEX MOLECULES: WHAT CAN WE LEARN FOR SOLIDS

J. Reuss

Fysisch Laboratorium
Katholieke Universiteit Nijmegen
Toernooiveld, 6525 ED Nijmegen
The Netherlands

ABSTRACT

The general idea is to go from the study of simple small molecules to that of big ones where the occurrence of many resonances and the presence of high denisty states leads finally to the limit of infrared spectroscopy. Simple molecules to be discussed are NH_3 and C_2H_4; bigger ones are (highly excited) SF_6 and butyne.

Raman Spectroscopy is another means to explore vibrational and sometimes vibrational-rotational structures of molecules. After discussion of simple examples we shall proceed to the case of molecules with internal motions and describe effects of tunneling and vibrational motions changing to nearly free rotations of high excitation.

Clusters form the transition from single molecules to little pieces of solids. For increasing cluster size the spectra become denser and finally show mainly gross features due to non-equivalent positions of constituents $(NH_3)_2$ and $(CH_3OH)_2$ and Davidov splittings ($(SF_6)_n$ and $(SiF_4)_n$). For smaller clusters sizes (especially dimers) narrow spectral features sometimes testify the long-levity of vibrationally excited complexes ($(C_2H_4)_2$).

I. INTRODUCTION

In the following some lines of research are discussed, where spectroscopic methods are applied to molecules in the gasphase. In this case, interaction between molecules is normally considered as perturbing leading to line broadening or quenching, that is to a loss of information, regarding a molecule as an isolated entity. This is quite different from the case of solid state spectroscopy. The solid state owes its entire existence to the presence of interaction between neighbors.

To be sure, even an isolated molecule is a thing of beauty, very much alive. There are internal motions, isomerisms, dissociation upon high excitation, changes of behaviour in strong electromagnetic fields, tunnelling between equivalent structures -- phenomena that keep one busy for quite some time if one wants to clear up and rationalize the behaviour of our molecular "masterpieces of art". In spite of their isolation, complex molecules can teach some lessons to solids state people. Evidently, if one does not understand the reaction of a single molecule to strong laser fields, the explanation of what a solid does in such a case is still far away. We will discuss at some length a phenomenon coined rapid adiabatic passage (RAP). This passage means a change of state of a quantum mechanical system where a transition probability of one can be achieved and e.g. inversion can be produced without restricting the duration of the interaction to a well defined time interval like in the case of π-pulse excitation. RAP is rather insensitive to the intensity of the interaction, too, and can easily be realised in a set-up where a molecular beam is crossed by a laser beam. RAP forms the extreme alternative to Rabi oscillations, where a system undergoes a periodic change of state, as long as it remains in a laser field.

One encounters an ideal situation to study the interaction between two neighbors if a dimer is looked at by spectrocopic means. These dimers are formed by supersonic expansion of molecular beams. They are cooled down by collisions in the initial part of the beam to travel further downstream, without experiencing further collisions. Many hydrogen-bonded systems (e.g. $(H_2O)_2$) have been investigated by spectroscopic means and structural information as well as lifetimes of excited species have been obtained.

High excitation of a molecular system means that the density of states increases very much, especially if low-lying modes like internal rotations are present. The problem of level ordering in case of interacting internal rotors will be discussed, on the one hand to show the beauty of this type of analysis and the depth of our understanding of such a micro cosmos of possible flections and motions, but on the other hand also the limit of our systematic knowledge of parameters like barrier heights (to be penetrated by internal motions), if we compare different molecular systems.

There is a limit to spectroscopy. If the density of states becomes so high that even the most refined techniques are unable to resolve spectral structure one really starts to suffer severely from congestion. We shall discuss how this condition is approached by observations of dark states coupled to and doubling (or even multiplexing) the number of observed bright states, until finally an undigestible spectral soup is produced where spectroscopists start to give up and run away, normally.

II. IR-SPECTROSCOPY OF CLUSTERS

Clusters are often said to form the intermediates between molecules and small pieces of solids. Sometimes this reasoning goes so far as to justify funding of cluster research as an important issue to get information on solids when these solids more or less consist of surfaces.

The spectroscopy of clusters is difficult since one often deals with complex systems, with a high density of states and of a floppy nature. What is encountered in molecules as an exception, internal motion like the (hindered) rotation of the two methyl groups in dimethylamine ($(CH_3)_2NH$) and tunneling of the H-atom of the same molecule between two equivalent positions, one encounters as a rule for Van der Waals complexes, e.g. for NH_3-Ar[II.1], $(NH_3)_2$[II.2], $(H_2O)_2$[II.3], $(HF)_2$[II.4], etc.

The structure of e.g. dimers is determined by the intermolecular forces between its constituents. This is different from the conventional way of thinking with respect to molecules. The reason is that in a dimer more often than not we deal with two slightly perturbed entities which retain their molecular properties whatever their orientation in the complex. In a molecule all constituting elements are normally heavily influenced by the formation of chemical bonds. For instance, in the salt molecule NaCl,Na(Cl) resembles little the original alkali (halogen) atom having gained (lost) much of an electron by the bond formation.

For dimers, the above reasoning can be turned around. Since the intermolecular potential is responsible for the structure and the Van der Waals vibrational modes of the system, much can be learned from the IR-spectroscopy of dimers with respect e.g. to the angle-dependence of the intermolecular potential. This information is very precise -- but also limited. For instance, dimer spectroscopy explores the intermolecular potential for distances very near to the equilibrium position of the constituents and neither for long distances of dispersion forces nor for very short distances of mainly repulsive forces.

A similar statement holds for the angular dependent part. Only those angular regions are explored where the dimer structure yields a high probability, i.e. a large local value of the squared wave function describing the angular distribution. Other regions remain "invisible". They must be investigated by studying (inelastic) collisions.

If we compare a solid and the interaction amongst its constituents with what we have said about clusters and molecules, one discovers similarities with both. For instance a piece of metal resembles a large molecule in that the atoms have changed their properties considerably because they have all given off one electron to produce the common stock of binding electrons which can move rather freely through the entire piece. Common electrons are also encountered in molecules like H_2 and C_6H_6. Likewise ionic crystals like kitchen salt have their counterparts in molecules like NaCl. However, if one considers molecular crystals like solid H_2O, solid NH_3 or NH_3 in an argon matrix the molecular units are seen to remain intact and in some cases even to rotate (in some hindered fashion) like what we have encountered in floppy dimers.

There are two features which illuminate the similarities between molecular solids and molecular clusters. First, site-effects are observed both in solids and in cluster spectra. For example $(NH_3)_2$[II.5] has such a structure that the two ammonia constituents occupy non-equivalent positions (if we disregard tunnelling for the moment, which interchanges the role of the two partners). Another example is formed by the H-F...H-F dimer, where clearly only one molecule provides the H-atom for the H-bonding [II.4]. Excitation both of the umbrella-mode in case of $(NH_3)_2$ and of the H-F stretch in case of $(HF)_2$ leads to two different excitation peaks, with rather different lifetime-determined linewidth for $(HF)_2$. Second, there are sometimes different isomeric forms observed e.g. for $(HCN)_3$ and N_2O-HF [II.6], which correspond to different crystal structures in a solid. However, whereas different structures are assumed by a solid through rather well-defined phase-transitions, the isomeric clusters co-exist in spite of their different energy content.

In dimers like $(SF_6)_2$ and $(SiF_4)_2$, a simple spectral structure is observed, which is caused by resonant dipole-dipole forces [II.7]. The excitation of the dimer (or large cluster) does not take place in a localized manner. Instead, two (or more) constituents divide the one-photon excitation among themselves in a symmetric or asymmetric fashion, in such a way that the degeneracy is lifted mainly by the interaction between the two transition dipole moments ($\sim \mu^2/R^3$, where R is the dimer distance). In case of $(SF_6)_2$, this interaction leads to a splitting of 20 cm^{-1}. In solids, similar effects have been given the name Davydov-Effect. The situation may become more complicated since one constituent can be surrounded by more than one equivalent neighbor and the width of RI-absorption lines may be so large as to render the Davydov-Effect sometimes unobservable.

III. RAP AND RABI

III.A. RAP and Rabi, Classically

Consider the coupled classical oscillators, e.g. each one an undamped LC-circuit with some mutual magnetic induction, the strength determined by the constant K. If one puts $LC = \omega^{-2}$ and $L_oC_o = \omega_o^{-2}$ one can describe the time dependence of the charge Q and Q_o, the current Q and Q_o and its derivative $\ddot{Q}$ and $\ddot{Q}_o$ by

$$\ddot{Q} + \omega^2 Q + K\ddot{Q}_o = 0 \qquad (1)$$

$$\ddot{Q}_o + \omega_o^2 Q_o + K\ddot{Q} = 0 \qquad (2)$$

The solution is assumed to be of the form $Q = A \cdot e^{i\tilde{\lambda}t}$ and $Q_o = A_o \cdot e^{i\tilde{\lambda}t}$, with

$$\tilde{\lambda}^2 = \frac{\omega^2 + \omega_o^2}{2(1 - K^2)} \pm \frac{|\omega^2 + \omega_o^2}{2(1 - K^2)} \pm \frac{1}{2(1 - K^2)} \left\{ \sqrt{(\omega^2+\omega_o^2)^2 + 4K\omega^2\omega_o^2} - |\omega^2+\omega_o^2| \right\} \qquad (3)$$

First, the case of resonance is discussed, $\omega = \omega_o$. One finds

$$\tilde{\lambda}^2 = \frac{1}{1 - K^2}\,\omega_o^2 \pm \frac{K}{1 - K^2}\,\omega_o^2 = \frac{\omega_o^2}{1 \pm K} \qquad (4)$$

There are two solutions with different frequencies, which can be called sationary since the amplitudes A and A_o remain constant as does the energy for each oscillator. Other solutions are obtained by superposition of the two stationary ones. Especially amplitudes and phases can be chosen so that for instance all energy stays in one oscillator for t = 0. As time develops, the energy is transferred to the other oscillator, at $t = (\Delta\tilde{\lambda})^{-1} \cdot \pi$, and back, at $t = (\Delta\tilde{\lambda})^{-1} \cdot 2\pi$, etc. This is precisely the classical analogon with quantum mechanical Rabi-oscillations, where the splittings, due to the dynamic Stark effect, $\hbar\Omega_{Rabi}$, determines the energy exchange between a molecule and an optical mode in the same way as the splitting $\Delta\tilde{\lambda}$ determines the energy exchange between the two oscillators, which would be degenerate if they were not coupled through the coefficient K.

For originally nearly degenerate oscillators, the energy transfer is incomplete and the exchange frequency increases, in both cases, classically and quantummechanically.

Next adiabatic passage is considered. To this end one start with ω_o for $t < 0$, $\omega = \omega_o$ for $t = 0$ and $\omega > \omega_o$ for $t > 0$. This change (e.g. by variation of the capacity C) has to take place very slowly, so that the solution of our differential equation changes "adiabatically" from that belonging to a constant value of C at a certain instant to that for a neighbouring C-value. For $t \ll 0$, all energy is assumed to reside in a single oscillator. Since $\omega \ll \omega_o$ in the remote past

$$\tilde{\lambda}_1 \approx \omega_o \text{ and } \tilde{\lambda}_2 \approx \omega \qquad (5)$$

Therefore, the assumed initial state is stationary, as discussed above.

Since $\tilde{\lambda}_1$ and $\tilde{\lambda}_2$ never cross, one always has to follow the unique path containing the initial $\tilde{\lambda}_i$-value, for t << 0. If we consider $\tilde{\lambda}_2 = \omega$, for the initial value. With time, $\tilde{\lambda}$ increases and assumes the value $\frac{\omega_o^2}{1+K}$ at t = 0. In the remote future, t >> 0, where $\omega >> \omega_o$, $\tilde{\lambda}^2$ approaches ω_o and the energy resides in the oscillator that initially has been at rest. This then is the perfect RAP process, however, classically.

This "gedanken experiment" resembles -- in all modesty -- the famous undamped pendulum, where the length of the suspension was slowly shortened and then again lengthened to its original value. Both classically and quantummechanically the motion passed adiabatically (without quantum jumps) to arrive finally again at its original form. This adiabatic passage is said to have formed the subject of profound discussions amongst the founders of wuantummechanics. The common aspect is that here also no level crossing occurs, so that the system remains in its state though the state is changing slowly in time.

III.B. RAP Quantum Mechanically, for Two-, Three- and Four-Level Systems

1. Avoided crossing in a two-level system. The principle of avoided crossing will be the red thread through the following. First, a two-level system in a single mode nearly resonant laser field (frequency ω) is discussed, with n photons initially present and the molecular system in its lower state $|1>$ [III.1]. Without coupling between the molecule and the laser field, two nearly degenerate states exist, i.e. $|n>|1>$ and $|n-1>|2>$ with $E_2 - E_1 \approx \hbar\omega$. To be definite we assume $\hbar\omega < E_2 - E_1$ for t < 0, $\hbar\omega = E_2 - E_1$ for t = 0 and $\hbar\omega > E_2 - E_1$ for t > 0. With coupling dynamics Starkshifts appear, yielding a splitting by $\hbar\Omega_{Rabi} = \mu_{12}\varepsilon$ at t = 0, ε being the amplitude of the electric field and μ_{12} the transition dipole moment for the two considered levels.

At t < 0, the state $|n>|1>$ has the lower energy value $n\hbar\omega+E_1$; its value increases in time since ω is increasing, but the two levels never cross. At t=0, the energy-value is $n(E_2 - E_1) + E_1 - \hbar\Omega_{Rabi}/2$ for the lower state, and $(n-1)(E_2 - E_1)+E_2 + \hbar\Omega_{Rabi}/2$ for the upper one. For t>0 and assumed adiabatic passage, the system stays on its lower level and approaches $(n-1)\hbar\omega + E_2$, for t>0 and $\hbar\omega > E_2 - E_1$. In this discussion we have assumed two isolated but interacting entities, the molecular system and the (excited) optical mode. Spontaneous decay has been neglected, i.e. the interaction of the molecule with the infinite number of non-excited optical modes (i.e their zero-point fields) is discarded, a procedure allowed if n >> 1 and the considered passage times shorter than actual spontaneous decay times. The latter condition is easily met especially with IR-lasers. Experimentally, the two-level effects have been studied by the group of Bordé [III.2] and at Nijmegen [III.3 and III.4], extensively. A numberical study is found in [III.14].

2. Three-level system and one color. Next, we consider a three-level system, a single mode optical field with frequency ω and the resonance condition $2\hbar\omega = E_3 - E_1$, for $t = 0$; the intermediate level is assumed to be slightly detuned. Here, we have three nearly degenerate levels $|n\rangle|1\rangle, |n-1\rangle|2\rangle, |n-2\rangle|3\rangle$. To be definite we assume (see Fig.1)

$$\begin{aligned} &\text{for } t < 0 \;\; 2\hbar\omega < E_3 - E_1,\; E_2 - E_1 < \hbar\omega \\ &\text{for } t = 0 \;\; 2\hbar\omega = E_3 - E_1,\; E_2 - E_1 < \hbar\omega \\ &\text{for } t > 0 \;\; 2\hbar\omega > E_3 - E_1,\; E_2 - E_1 < \hbar\omega \end{aligned} \tag{6}$$

Precise resonance would occur for $t = 0$ between the two dressed states $|n\rangle|1\rangle$ and $|n-2\rangle|3\rangle$ were it not for the coupling. This time the coupling is indirect; $|3\rangle$ is coupled to $|1\rangle$ via $|2\rangle$ and the splitting at $t = 0$ amounts to $(\mu_{12}^2 + \mu_{23}^2)\varepsilon^2/[E_2 - E_1 - \hbar\omega]$; the denominator represents the detuning of the intermediate level. One starts, for $t < 0$, on level $|1\rangle$, in the dressed state $|n\rangle|1\rangle$.

Adiabaticity demands that level $|3\rangle$ is reached, in state $|n-2\rangle|3\rangle$, for $t > 0$, see Fig. III.1.

Inversion can thus be produced by changing the laser frequency ω, for two- and three-level systems. In the latter case, the indirect coupling is weaker than in the first case and thus larger laser power is required. For experiments, see [III.4].

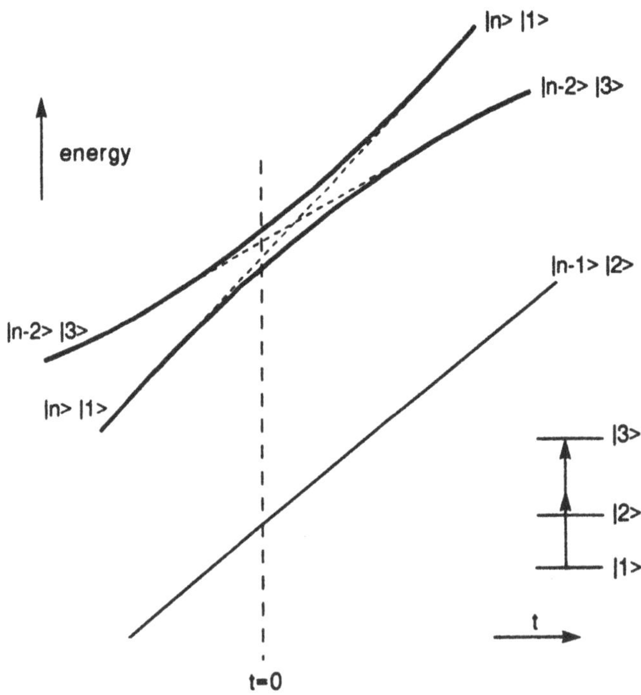

Figure 1. Three level system in a one-color laser field, with chirping and detuned intermediate levels $|2\rangle$. The heavy lines include the field-molecule interaction.

3. <u>Three-level system and two colors.</u> A three-level-system can also be addressed by two colors, i.e. laser frequencies ω_1 and ω_2; in this case there is no need for a variation of the laser frequencies to obtain inversion. As before, one has three, this time really degenerate levels $|n_1 > |n_2 > |1 >, |n_1 - 1 >|n_2 >|2 >$ and $|n_1 - 1 >|n_2 - 1 >|3 >$ (degenerate in the dressed atom picture, neglecting the dynamic Stark-effect). The molecular level $|1 >$ is coupled to $|2 >$ by lasermode "1" with ω_1 containing initially n_1 photons. Level $|2 >$ is coupled to $|3$

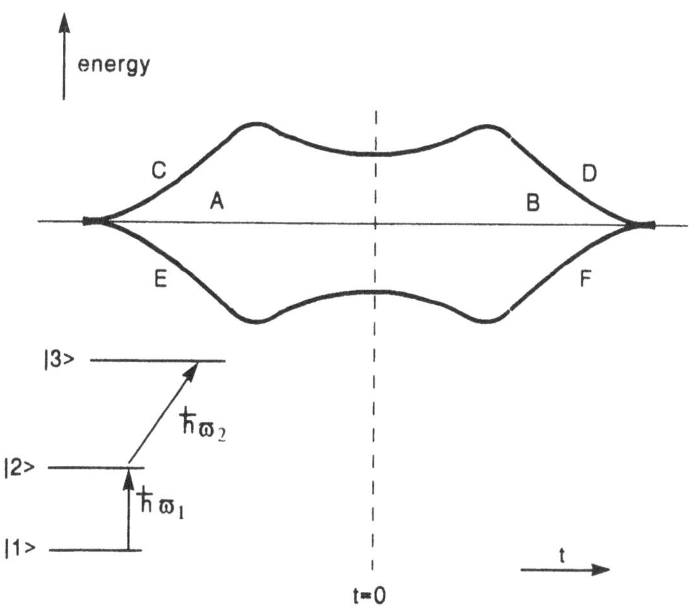

Figure 2. Three-level system in a resonant two-color laser field, with $A=|n_1 >|n_2 > |1 >$ (initially populated) and $B = |n_1 - 1 >|3 >$ (finally populated). Laser mode "2" overlaps but precedes laser mode "12". C and E are mixtures of states $|n_1 - 1 >|n_2 >|2 >$ and $|n_1 >|n_2 - 1 >|3 >$; D and F of $|n_1 >|n_2 >|1 >$ and $|n_1 - 1 >|n_2 >|2 >$. Note that a new resonant two-photon transition is involved.

$>$ by lasermode "2" with ω_2 containing initially n_2 photons. To obtain inversion, i.e. total population transfer from $|1 >$ to $|3 >$, lasermode "2" has to interact with the molecule before lasermode "1" does (naturally with some temporal overlap). In that way, adiabaticity demands that the molecule remains on the same level, as shown in Fig.2. However, the character of the state belonging to this level changes. For t < 0 the two dressed states containing |2 > and |3 > repel each other due to the dynamic Stark-effect. Whereas for t = 0 all three levels contain |1 > and |2 > and |3 >, for t > 0 the sole presence of lasermode "1" produces the repelling of the two states containing |1 > and |2 >; thus the intermediate level contains totally |3 >. The soundness of this reasoning has been demonstrated experimentally, [III.5 and III.6 and III.7 and III.8].

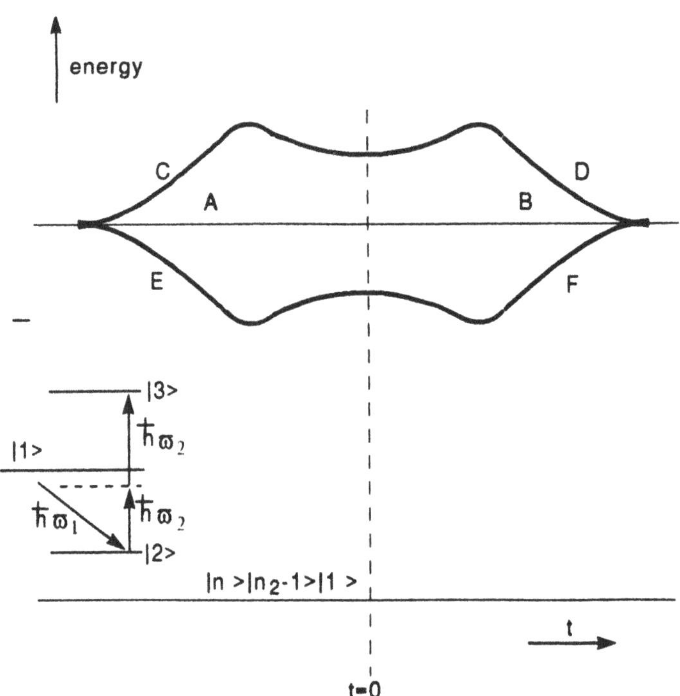

Figure 3. Three level system in a resonant two-color laser field, with A = $|n_1 - 1>|n_2>|1>$ and B = $|n_1 - 1>|n_2 - 1>|3>$. Laser mode "2" overlaps but precedes laser mode "12". C and E are mixtures of states $|n_1>|n_2>|2>$ and $|n_1>|n_2 - 1>|3>$; D and F of $|n_1>|n_2>|2>$ and $|n_1 - 1>|n_2>|1>$. Note that a new resonant two-photon transition is involved.

4. <u>Three-level system and two colors, modified</u>. The next example forms a modification of the previous one; again we look at a three-level system but now molecular levels |2 > and |3 > are coupled by a two-photons resonance (lasermode "2") involving the slightly detuned molecular level |1 >. Level |1 > is the starting level which is coupled resonantly to |2 > by lasermode "1" (Fig.3). In the dressed-level picture we deal with four nearly (sic!) degenerate states $|n_1 - 1>|n_2>|1>$; $|n_1>|n_2>|2>$; $|n_1>|n_2 - 1>|1>$; $|n_1 - 1>|n_2 - 1>|3>$. Of

these 4 states one possesses a slightly detuned energy, $|n_1>|n_2 - 1>|1>$, whereas the other three would be exactly on resonance were there no coupling and dynamic Stark-effect. The experimental trick again is the time order of interactions; first, mode "2" must be switched on to produce repelling of the two levels that contain states $|2>$ and $|3>$ and thereafter -- with overlap -- mode "1" interacts and causes the repelling between the levels containing $|1>$ and $|2>$. The always slightly detuned level $|n_1>|n_2 - 1>|1>$ fulfills the part of spectator state, important mediator for the indirect interaction between $|2>$ and $|3>$, but never attaining any sizeable population. This scheme of excitation is tried out experimentally, at this moment.

5. Four-Level System and One Color. One-frequency three-photon transitions in a four-level system occur, in general, if $3h\nu = E_4 - E_1$ allowing even for a detuning of the intermediate levels, $E_2 - E_1 - h\nu = \Delta_2 \neq 0$ and $E_3 - E_4 + h\nu = \Delta_3 \neq 0$. With O_{ij} as off-diagonal element between the dressed states $|i>$ and $|j>$, i,j = 1,...,4 and

$$
\begin{aligned}
|1> &= |1,n>,\\
|2> &= |2,n-1>,\\
|3> &= |3,n-2>,\\
|4> &= |4,n-3>,
\end{aligned}
$$

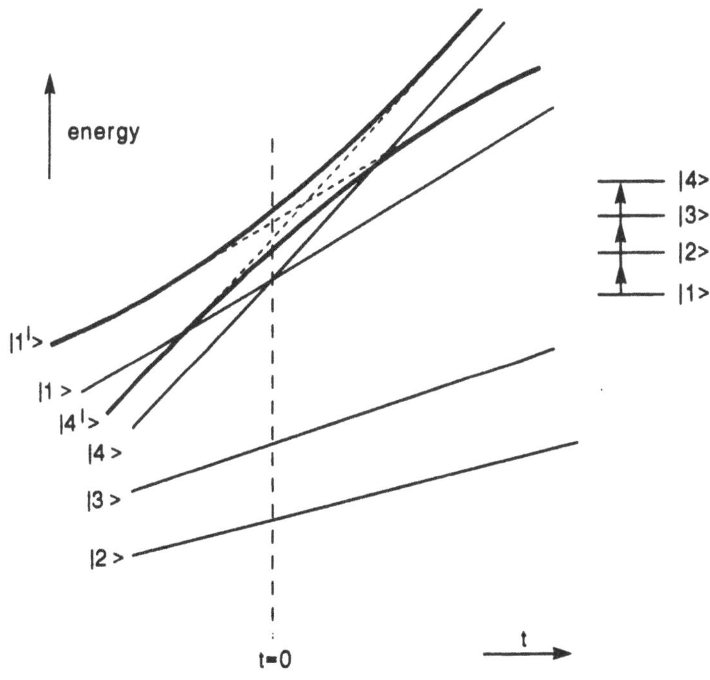

Figure 4. Four-level system with chirped one-color laser field. The levels $|1>,\ldots,|4>$ correspond to the dressed atom picture, neglecting the molecule-field interaction. Levels $|1'>$ and $|4'>$ show the dynamic Stark effect due to the interaction between $|1>$ and $|2>$, and $|3>$ and $|4>$. The heavy lines show the avoided crossings as discussed in the text.

one finds for the Hamiltonian matrix

$$\begin{pmatrix} 0 & O_{12} & 0 & 0 \\ O_{12} & \Delta_2 & O_{23} & 0 \\ 0 & O_{23} & \Delta_3 & O_{34} \\ 0 & 0 & O_{34} & 0 \end{pmatrix} \tag{7}$$

In order to make the problem tractable for discussion we introduce some assumptions concerning the magnitude of the detunings i = 2,3 and the (real) off-diagonal-elements; these latter ones obey $O_{ij} = (1/2)\hbar\Omega_{ij}\varepsilon$ i.e. they depend on the amplitude ε of the laserfield.

In our case it is assumed that the detunings Δ_i are large both with respect to the off-diagonal-elements O_{ij} and with respect to the dynamic Stark shifts of the energy of the two quasi-degenerate dressed states |1> and |4>. First then we diagonalize first the upper two and the lower two blocks of the Hamiltonian and obtain for the interaction between level |1> and |4>

$$\begin{pmatrix} -\dfrac{O_{23}^2}{\Delta_2} & O_{23}\cdot\dfrac{O_{12}}{\Delta_2}\cdot\dfrac{O_{34}}{\Delta_3} \\ O_{23}\cdot\dfrac{O_{12}}{\Delta_2}\cdot\dfrac{O_{34}}{\Delta_3} & -\dfrac{O_{34}^2}{\Delta_3} \end{pmatrix} \tag{8}$$

This Hamiltonian yields a splitting of the quasi-degenerate levels by about twice

$$O_{23}\cdot\frac{O_{12}}{\Delta_2}\cdot\frac{O_{34}}{\Delta_3} = \frac{1}{8}\hbar\Omega_{Rabi,23}\cdot\frac{\hbar\Omega_{Rabi,12}}{\Delta_2}\cdot\frac{\hbar\Omega_{Rabi,34}}{\Delta_3} \tag{9}$$

Multilevel systems have been studied experimentally by Liedenbaum et al [III.9]. Further information on multiphoton transition of SF_6 is contained in [III.10-13].

IV. RAMAN SCATTERING

IV.A. Historical Note

Chandrasekhara Raman (1888-1970) got the Nobel Prize in 1930 for his discovery that scattered light contained a weak component that was frequency shifted (blue and red, i.e. antistokes and stokes) with respect to the incoming light. Shortly afterwards Landsberg and Mandelstam of Russia observed a similar effect in light scattered by liquids and solids.

The Raman-effect in gases reflects the energy-levels of molecules that can be (de)-excited by the photons leading to an energy (increase) decrease. Szilard explained this effect classically, e.g. by a modulation of the $\bar{\alpha}$ - polarizability of scattering CH_4 molecules due to the all-symmetric CH-stretch-vibration. The molecules act as an antenna emitting with the thus modulated intensity, proportional to

$$2 \cos \omega t \cdot \cos\omega_0 t = \cos\left((\omega_0 + \omega)t\right) + \cos\left((\omega_0 - \omega)t\right).$$

IV.B. The Physics Behind the Raman-Effect

The spontaneous Raman-effect has some features in common with what we have discussed above in connection with three-level two-frequency two-photon transitions (double resonance).

In both cases, $|1\rangle$ and $|3\rangle$ represent real levels of a molecule belonging e.g. to different rovibrational states. The intermediate level $|2\rangle$ is, however, resonantly addressed by the double resonance technique, where as in the case of spontaneous Raman effect, this level (and all the other possible intermediate ones) are far off resonance.

Therefore, one customarily talks about a virtual (i.e. non-existent) level to which the molecule is excited and from which it decays to level $|3\rangle$. Here, the second basic difference comes into the picture.

Whereas in Λ-type double resonance schemes the second step is induced by a second radiation source, with $E_2 - E_3 = h\nu_2$, this step is a decay process, for Raman scattering, i.e. a consequence of the interaction of the molecule with the zero-point energy in the appropriate electromagnetic vacuum-modes which may acccept a photon of energy $h\nu_2$.

As a consequence of both these differences, Raman signals are extremely weak and obtainable from any non-resonant laser-radiation. This, in a nutshell are the weak and strong points of the Raman technique.

IV.C. Application of Raman Scattering

Because all molecules and radicals scatter non-rsonanatly incoming light wave-length-shifted in a characteric way, Raman scattering is an ideal probing technique for dynamic processes like combustion in flames, for detecting the composition of gases from stacks in power stations etc. Pulsed lasers with their high energy density make up for the weak Raman scattering cross section. Box-car techniques allow signal-sampling exclusively during the presence of the laser pulse. Turning of the laser (note that slightly but sufficiently tunable excimer-lasers are on the market) provides the possibility to avoid largely fluorescence of the present molecules. Finally, the use of UV-radiation renders the contribution from dust particles (Mie scattering, proportional to the light frequency) practically negligible as compared to Raman- and Raleigh-scattering. For many applications, one needs to get information from an extended sample, especially when turbulent flow fields produce large fluctuations in signals from point-like samples. It is easy to observe line-wise Raman scattered light from a light-sheat, dispersed by a monochromator, and to displace the sample line from pulse to pulse, in a high repition rate regime. 2-D imaging and computer sampling permit reconstruction of flow-fields (temperature and gas, composition) with an unprecedented density of information.

IV.D. Raman Overtone Studies

In Table IV.1 some simple data on the barrier potential of molecules with one or two methylic internal rotors are collected. There are two main experimental methods to obtain this information; first microwave-data on rotational transitions revealing splitting due to tunnelling through the barrier and, second, overtone Raman spectra where transitions between different torsional states are probed, with levels even above the torsional barrier. It is this second method which is of our concern here.

Raman scattering yeilds weak signals; overtone scattering means that one has to be content with very weak signals. In order to overcome this handicap, one works with an intra-cavity set up; the practical gain amounts to about a factor 100, [IV.12,13]. This extra-effort seems justified since it yields information on vibrationally highly excited systems. We concentrate here on molecules with internal torsional motion. Often, the excitation of a single quantum step requires an energy transfer of about 200-300 cm^{-1}. Highly excited does not mean a big amount of energy residing in a molecule, on an absolute scale, whatever that may be. It means, however, some or many photons put into certain vibrational modes what consequently leads to a rather high density of states around the final level of excitation. Molecules with internal torsion play an important role in the technique of production of FIR-radiation by IR-laser pumping. For instance, methanol absorbes CO_2-laser radiation and experiences a decay-cascade through collisions, where a high density of intermediate levels helps to ease collisional de-excitation with possible inversion occurring for two "neighboring" levels of the cascade ladder. Methanol has this two internal groups - CH_3 and OH - which are able to rotate or vibrate torsionally with respect to each other. A list of assigned transitions in the range of the CO_2-laser can be found in [IV.1].

Similar molecules are discussed here, based upon experimental information gathered in Table IV.1. Basically, a methyl-group experiences a torsion-potential which approximately can be described by $V_3 \cos 3\alpha$, α being the torsional angle. For bound (vibrational) states low in the wells, a small splitting yeilds a two-fold degenerate E-state and a non-degenerate A-state, both possessing one or two or more torsional quanta absorbed. For states near the potential barrier or even above, state-mixing occurs so that the number of torsional quanta is no longer very meaningful.

For two methyl-groups, the number of torsional quanta can be shared so that torsional polyads with a fixed number of torsional quanta can be distinguished -- in the limit of low excitation, at least. Here, then, finer effects come into play; mutual interaction between the two methyl-groups produce extra splittings. For propane, all the constants have been determined which determine the rotor-frame and rotor-rotor interactions [IV.2]. There is an enormous discrepancy with the results from a recent microwave-investigation [IV.3], the latter one yeilding a 20% lower barrier parameter V_3. This clearly is not caused by bad measurements or erroneous analysis of data. The source of the discrepancy might be that the analysis of microwave data for other than the most simple systems is highly complicated. Inclusion of rotor-rotor terms in the hamiltonian would make the calculation rather untractable. Consequently, microwave data are usually analyzed with truncated effective Hamiltonians [IV.4-6]. Satisfyingly, the results for simple systems (C_2H_6 and CH_3CD_3) are in fair agreement [IV.7-11].

IV.E. Problems with the MW- and Raman-Method, Alternative Methods

Inspection of Table IV.1 [IV.12,13] demonstrates that there are problems e.g. for propane. In our opinion their origin can be traced back easily. MW-spectroscopy yields extremely precise data, but mainly concerning the torsional ground -- and first excited state. Though there are considerably many data on rotational transitions, they all are concerned with states far below the barrier and transitions within a torsional state. These transitions can be described with a few effective parameters like the barrier height V_3 and the rotational constant F for the internal top. Inclusion of more parameters only demonstrates the strong correlations between these parameters (for those low lying states) without yielding new insight. The Raman technique suffers from the opposite. Few transitions can be measured, and the experimental resolution is 10^4 or even more times worse as compared to the MW-data. But one measures transitions between torsional states (mainly $\Delta v = 2$, but also $\Delta v = 1$ and $\Delta v = 4$) and the upper state may be even above the barrier. Therefore, in spite of the catastrophically bad resolution, the high states become very informative with respect to tunnelling since the tunnelling splitting increases dramatically for the higher torsional states and becomes easily observable.

There is a great danger for this Raman approach. One deals with very weak transitions and thus every series of hot bands or strong transitions of low concentration impurities might be mistaken for torsional transitions.

The lesson seems simple: both methods should only be used together so that a common potential explains both sets of data. In the past, this has never been practiced. Instead, more often than not, the mutual disliking of the other method led to big barriers with low tunnelling = exchange probability. In addition, MW-measurements require permanent dipole moments so that molecules with a center of inversion can not be treated by MW-absorption where as Raman spectroscopy is on its best for these symmetric molecules since the selection rules forbid many otherwise perhaps confusing transitions.

So there is a need for alternative methods. First, it is nowadays possible to measure sensitivily very weak infrared transitions, with high resolution, utilizing long absorption paths. $\Delta v = 1$ torsional transition becomes weakly allowed for molecules like C_2H_6 and C_2D_6[IV.10; IV.11], by precarious distorsion effects. In recent time this method has emerged to be a valuable alternative for simple molecules.

Another method runs under the name of curve-crossing spectroscopy; its application is limited to simple molecules with permanent dipole moment -- but where it is applicable it yields many data with high resolution also on torsional transitions and accordingly must be regarded as a major asset of experimentalists in this field.

Finally and even more recently high resolution electronic transitions (by LIF) became a valuable tool to observe torsional properties extensively, not only for the ground state but also of the excited fluorescing electronic state. At this moment, the molecule acetone is under study [IV.14].

IV.F. The Lesson for Solid State Physicists

These floppy molecules which possess internal degrees of freedom have as their counterparts molecules emobdied in a solid state matrix e.g. of noble gas atoms. Not necessarily has one to think here of molecules like C_2H_6 and question the possible quenching of their internal rotation. Already a simple rotor like H_2 or NH experiences the crystal field and has equivalent positions; between these tunnelling may take place resulting in a splitting of the rotational levels [IV.15].

Table IV.1. Comparison of torsional barrier heights determined from MW-data and IR/Raman-data (values in cm^{-1})

'ONE-CH_3-TOP' MOLECULES	'TWO-CH_3-TOP' MOLECULES
ETHANE (CH_3)CH_3 $V_3 = 1009$ Raman[a] $V_3 = 1012.0$ IR[b]	PROPANE (CH_3)CH_2(CH_3) $V_3 = 1353$ Raman[c] $F = 5.72$ $V_3 = 1108$ MW[d] $F = 6.13$
ETHANE (CD_3)CD_3 $V_3 = 990$ Raman[e] $F = 1.3416$ $V_3 = 989$ IR[f] $F = 1.3416$	
METHYLAMINE (CH_3)NH_2 $V_3 = 684.71$ MW[g] $F = 7.1$	DIMETHYLAMINE (CH_3)NH(CH_3) $V_3 = 1350$ Raman[h] $F = 6.7$
ACETALDEHYDE (CH_3)HC=O $V_3 = 406$ MW[i] $F = 7.5$	ACETONE $(CH_3)_2C{=}O$ $V_3 = 279.4$ IR[j] $F = 5.73$ $V_3 = 266.3$ MW[k] $F = 5.59$

a) R. Fantoni, K. van Helvoort, W. Knippers and J. Reuss, Chem. Phys. 110 (1986) 1; *b*) N. Moazzen-Ahmadi, H.P. Gush, M. Halpern, H. Jagannath, A. Leung and I. Ozier, J. Chem. Phys. 88 (1988) 563; *c*) R. Engeln, J. Reuss, D. Consalvo, J. v. Bladel, A. v.d. Avoird and V. Pavlov-Verevkin, Chem. Phys. 144 (1990) 81; *d*) G. Bestmann, W. Lalowski and H. Dreizler, Z. Naturf. 40a (1985) 271; *e*) J.M. Fernandez-Sanchez, A.G. Valdenebro and S. Montero, J. Chem. Phys. 91 (1989) 3327; *f*) N. Moazzen-Ahmadi, A.R.W. McKellar, J.W.C. Johns and I. Ozier, to be published; *g*) N. Ohashi, S. Tsunek, K. Takagi and J.T. Hougen, J. Mol. Spectr. 137 (1989) 33; *h*) R. Engeln, J. Reuss, D. Consalvo, J. v. Bladel and A. v.d, Avoird, Chem. Phys. Lett. 170 (1990) 206; *i*) W. Liang, J.G. Baker, E. Herbst, R.A. Booker and F.C. De Lucia, J. Mol. Spect. 120 (1986) 298; *j*) P. Groner, G.A. Guirgis and J.R. Durig, J. Chem. Phys. 86 (1987) 565; *k*) J.M. Vacherand, B.P. van Eijck, J. Burie and J. Demaison, J. Mol. Spectr. 118 (1986) 355.

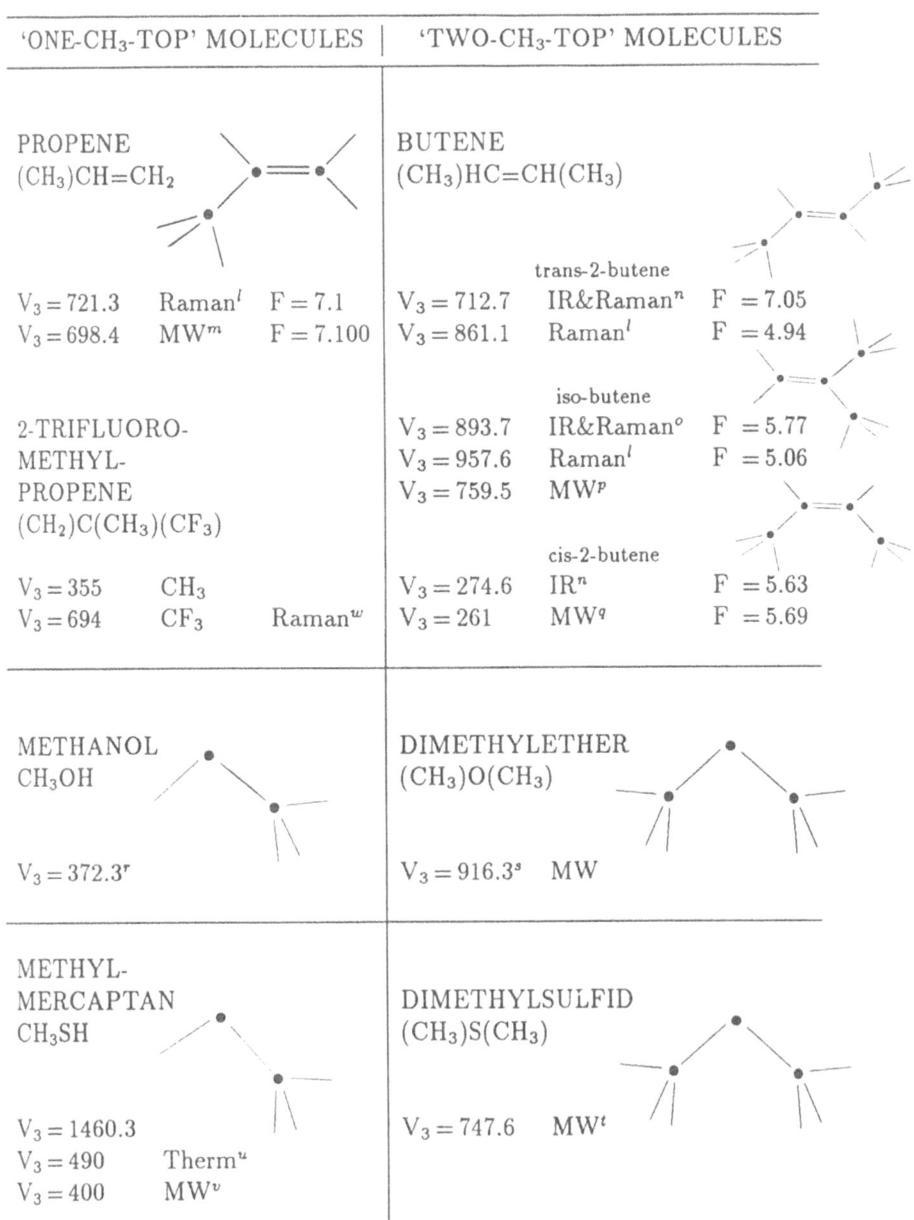

'ONE-CH_3-TOP' MOLECULES	'TWO-CH_3-TOP' MOLECULES
PROPENE $(CH_3)CH{=}CH_2$ $V_3 = 721.3$ Raman[l] $F = 7.1$ $V_3 = 698.4$ MW[m] $F = 7.100$	BUTENE $(CH_3)HC{=}CH(CH_3)$ trans-2-butene $V_3 = 712.7$ IR&Raman[n] $F = 7.05$ $V_3 = 861.1$ Raman[l] $F = 4.94$ iso-butene $V_3 = 893.7$ IR&Raman[o] $F = 5.77$ $V_3 = 957.6$ Raman[l] $F = 5.06$ $V_3 = 759.5$ MW[p] cis-2-butene $V_3 = 274.6$ IR[n] $F = 5.63$ $V_3 = 261$ MW[q] $F = 5.69$
2-TRIFLUORO-METHYL-PROPENE $(CH_2)C(CH_3)(CF_3)$ $V_3 = 355$ CH_3 $V_3 = 694$ CF_3 Raman[w]	
METHANOL CH_3OH $V_3 = 372.3$[r]	DIMETHYLETHER $(CH_3)O(CH_3)$ $V_3 = 916.3$[s] MW
METHYL-MERCAPTAN CH_3SH $V_3 = 1460.3$ $V_3 = 490$ Therm[u] $V_3 = 400$ MW[v]	DIMETHYLSULFID $(CH_3)S(CH_3)$ $V_3 = 747.6$ MW[t]

l) our result; *m*) Eizi Hirota, J. Chem. Phys. 45 (1966) 1984; *n*) J.R. Durig, S.D. Hudson and W.J. Natter, J. Chem. Phys. 70 (1979) 5747; *o*) J.R. Durig, W.J. Natter and P. Groner, J. Chem. Phys. 67 (1977) 4948; *p*) J. Demaison and H.D. Rudolph, J. Mol. Struct. 24 (1975) 325; *q*) S. Kondo, Y. Sakurai, E. Hirota and Y. Morino, J. Mol. Spectr. 34 (1970) 231, recently confirmed by the group of Prof. Dreizler (private communications); *r*) R.M. Lees, J. Chem. Phys. 59 (1973) 2690; *s*) H. Lutz, H. Dreizler, Z. Naturf. 33a (1978) 1498; *t*) J. Demoison, B.T. Tan, V. Typke, H.D. Rudolph, J. Mol. Spec. 86 (1981) 406; *u*) G. Herzberg, Molecular Spectra and Molecular Structure (1945) van Nostrand N.Y.; *v*) C.H. Townes and A.L. Schawlow, Microwace Spectroscopy (1975) Dover N.Y.; *w*) G.A. Guirgis, Y.S. Li, J.R. Durig, Raman Spectroscopy, Eds. Lascombe& Huong (1982) Wiley& Sons.

V. GAS PHASE SPECTROSCOPY - TOWARDS CONGESTION

V.A. Introduction

Resolving power -- that has been a strong driving force in spectroscopy ever since its beginnings, the promised land being where one understands molecular (quantum-) mechanics and hopefully fundamental laws in finest details -- perhaps once for good. Electronic transition- bands have been measured with rotational resolution, e.g. for molecules as large as benzene [V.1], pyrazine [V.2], pyrimidine [V.3], dimethyl-naphtaline [V.4] and acetone [V.5].

However, this route may come to a dead end before the promised land comes in sight. For the solid state physicist this seems trivial; a solid, i.e. a clump of material has so many narrowly spaced transitions that resolving single ones is out of the question. Still, he considers spectroscopy a very valid tool to interrogate nature.

The molecular physicist experiences the limit of high resolution spectroscopy as a disgrace; it occurs in two forms. First, he might still be able to measure many sharp transitions -- but he no longer can make (much) sense out of it. This limit leads to chaotic behaviour of molecular mechanics; a new category of scientic questioning is found, like the investigation of distributions of transition strengths and transition spacings.

Second, the path to progress may be blocked by congestion; the density of (high lying) levels becomes so high that continuous and broad spectra are obtained. This is the common situation in molecular physics, in the past normally overcome by the invention of finer spectroscopic methods. But here, too, doomes a final limit; if the distance to neighboring levels becomes smaller than the width of life-time-limited transitions, a dead end is reached certainly. The life-time limit, in our modern understanding, might be formed by intra-molecular energy redistribution (IVR) rates or (pre-)dissociation or spontaneous emission or whatever you like -- it blocks the path to further progress.

In the following we will discuss examples of (mainly) ir-spectroscopy which bear on this issue.

V.B. A Fermi-Resonance

Historically, the first resonance of this type was found for the CO_2-molecule where double excitation of the bending mode, $2v_3$, leads to a level of nearly the same excitation energy as the symmetric stretch, v_2. The $2v_3^0$ - level (with zero vibrational angular momentum) belongs to the same symmetry as the v_2-level; thus both levels may interact with each other.

$$H = \begin{pmatrix} E(2v_3^0) & I_{12} \\ I_{12} & E(1v_2) \end{pmatrix} \qquad (10)$$

Diagonalization yields the in reality measured levels -- and their eigen functions being superpositions of $|2v_3^0>$ and $|v_2>$. Because $|2v_3^0>$ belongs to an ir-active state and $|v_2>$ does not, the observed experimental transition strength yeilds the amount of mutual admixture.

It is clear, that all J-levels possess a rather similar admixture, since the off-diagonal elements I_{12} are present for all pairs of vibrational levels with equal J-value. Other interaction terms (notably the Coriolis-interaction) are J-dependent and lead to splittings especially for high J-values or to local effects when two interacting J-levels of different vibrational states come near to each other.

Naturally, there may be more than two vibrational states which interact with each other, leading to multiple Fermi-resonances.

V.C. Rovibrational Spectra of Symmetric Tops

The molecule $HC \equiv C-CH_3$ is an example of a symmetrical top, with the principle axis of smallest inertial moment along the HC≡C-C - axis (prolate top). The molecules possess a permanent electrical dipole-moment along the symmetry axis and consequently obeys the selection rule, for rotational transitions

$$\Delta J = \pm 1,0, \quad \Delta K = 0 \tag{11}$$

The quantum numbers J and K ≥ 0 determine the total rotational angular momentum and its projection on the symmetry axis. The rotational energy is given by

$$E_{Rot} = BJ(J + 1) + (A - B)K^2, \quad A > B \tag{12}$$

Here, centrigugal effects are neglected.

If such a symmetric top molecule is vibrationally excited, the transition dipole moment lies parallel to the symmetry axis, for ‖-transitions, or perpendicular, for ⊥-transitions. The selection rules are, for ‖-transitions

$$v' \neq v'', \ |J' - J''| = 1,0, \ |K' - K''| = 0 \tag{13}$$

and for ⊥-transitions

$$v' \neq v'', \ |J' - J''| = 1,0, \ |K' - K''| = 1 \tag{14}$$

The rotational constants normally change (slightly) upon the transition

B″ and A″ for the ground state
B′ and A″ for the excited state

In case of excitation of a degenerate vibrational mode, an extra term must be added to the energy equation (e.g.), $\mp 2ApK\ell$. The projection of the vibrational angular momentum on the symmetry axis determining...

Typical spectra can be found in the unequalled book of G. Herzber, [V.6].

V.D. The Case of $HC \equiv C-CH_3$, a FERMI-Resonance

McIlroy and Nesbitt [V.7] performed a detailed study of the rovibrational spectrum of propyne, $HC=C-CH_3$, around 3000 cm^{-1}, with a difference frequency infrared laser spectrometer with $10^{-1} cm^{-1}$ resolution. Propyne was expanded through a pulsed slit nozzle (40mm × 57μm, 19 Hz rep rate, 500 μs open time), in a He-propyne mixture with about 1% propyne in 1000 torr He. A rotational temperature of T_R=4,5K has been achieved.

The tunability of the laser radiation at the appropriate frequency has been obtained by mixing the light of a single mode Ar^+ laser to that from a Ar^+ laser pumped ring dye laser (R6G), on a Li NbO_3-crystal. The measurements consist in the determination of direct absorption (by the molecular jet) of the radiation from the source continuously tunable from 2.2 to 4.2 μm.

The vibrational frequencies, ω_e, and the diagonal anharmonicities, $\omega_e x$, are shown in Table V [V.7], together with the description of the mode and the shift of the rotational constant, B_v, as consequence of the vibrations. Note that the vibrational energy is

$$E = w_e(v + 1/2) - x_e w_e(v + 1/2)^2 \tag{15}$$

We concentrate on v_1 and v_2, both of A_1-symmetry, i.e. ‖-transitions. For v_1, the acetylinic CH-stretch, a $P_0(1)$ and a $P_i(2)$, i = 0, 1 transition are observed resulting in well resolved narrow lines. The subscripts indicate the K-value; only K = 0 and 1 are populated, at the low T_R-values of the experiment. Note, K = 0 and 1 levels are about equally populated due to their different nuclear spin symmetries, E and A_1, which exclude relaxation during the slit expansion.

$P_0(1)$ means explicitely $(n_1''=0,\ J''=1,\ K''=0) \rightarrow (n_1'=1,\ J'=0,\ K''=0)$; it is evident that the $P_1(1)$-transition does not exist since $|K''-K'|=0$ is observed.

In addition to these P-transitions a Q-branch is found, Q(1) - Q(6), all belonging to K = 1 because ΔJ = 0 transitions are forbidden for K = 0.

These observations result in the determination of the following parameters, for the v_1-stretch, v_0 = 3335.0594(3) cm^{-1}, B' = 0.2843846(66) cm^{-1}, A' = 5.31714(32) cm^{-1}.

In contrast to the acetylinic stretch, the methylinic on, v_2, is less isolated from the rest of molecular motions so that if any than couplings should manifest themselves for this vibrational excitation; and indeed they do. The experimental and simulated spectra are displayed in Fig.4 and show "too many lines". As is clear from the indicated K-values (remember ΔK=0) there is one line too many for all observed K=0 transitions. This fact is attributed to the interaction with a neighbouring state, with K=0, through a constant off-diagonal matrix-element $\langle v_{cs}|\hat{H}|v_2\rangle$. The analysis led to the following parameters

$$\begin{aligned} &v_2 \\ &v_0 = 2940.9995(4)\text{cm}^{-1},\ B' = 0.2851437(56)\text{cm}^{-1},\ A' = 5.3637(14)\text{cm}^{-1} \\ &v_{cs} \\ &v_0 = 2940.8331(9)\text{cm}^{-1},\ B' = 0.2857792(94)\text{cm}^{-1},\ \langle v_{cs}|\hat{H}|v_2\rangle = 0.09641(38)\text{cm}^{-1} \end{aligned} \tag{16}$$

McIlroy and Nesbitt discuss that there is only one candidate for this coupled state, i.e. the combination $v_5 + v_8 + 3v_{10}$. And with this a beautiful example is given where state mixing occurs, by a fortuitous coincidence, already for low excitations and densities of vibrational states of about 1/cm^{-1}. By the way, replacement of a ^{12}C atom by a ^{13}C

atom leads to a completely regular spectrum with no coupled states at all.

The same group [V.8] has performed measurements on the acethylinic CH stretch of 1-butyne, $HC\equiv C-CH_2-CH_3$, a larger molecule with an additional torsional degree of freedom ($V_3=|14|cm^{-1}$, for the CH_3-torsion [V.9]) leading to a density of states of about $90/cm^{-1}$, at about 3330 cm^{-1} of excitation. Instead of a regular spectrum, e.g. the P(1) transition was found to be split into at least 7 transitions, demonstrating the presence of more than one coupled state. With 1-butyne and even more with 1-pentyne, $HC\equiv C-CH_2-CH_2-CH_3$, we enter the field of heavily perturbed systems where statistical analysis of line spectra and IVR-considerations have been applied. The molecule trifluyoropropyne $HC\equiv C-CF_3$, has been investigated by Lehmann at al [IV.10]; it has a higher density of states than propyne. Anharmonic splittings were found, too.

V.E. Towards Congestion

As evident from above, congestion and non-regular behaviour is conditioned by a large density of states; this situation occurs for large molecules with their many degrees of freedom as well as for smaller systems if they become vibrationally highly excited. A simple molecule like CH_3OH is already heavily perturbed and the transitions can be assigned only with fatigue, at an excitation-level of about 1000 cm^{-1} [V.11]. Debet of this situation is the low lying torsional mode which produces many near coincidences higher up on the vibrational ladder. A molecule like SF_6, on the other hand, appears entirely congested at about 5000 cm^{-1} [V.12]; however, still very narrow inversion spikes (by rapid actiabatic passage) can be produced showing the absence of fast IVR-processes at this level of excitation. For SF_6-dimers and NH_3-tetramers the density of levels also for the ground state has increased so far that no narrow inversion spikes can be observed anymore so here really the endpoint of useful high resolution, ir- spectroscopy has been reached [V.13]. The reason for this must be sought in the presence of 6 Van der Waals degrees of freedom, in a dimer, with excitation levels of about 50^{-1} only.

REFERENCES

[II.1] D.H. Gwo, M. Havenith, K.L. Busarow, R.C. Cohen, C.H. Schmuttenmaer and R.J. Saykally, Mol. Phys.
R.C. Cohen, K.L. Busarow, Y.T. Lee and R.J. Saykally, J. Chem. Phys. 92(1990) 169.
E. Zwart, H. Linnartz, W.L. Meerts, G.T. Fraser, D.D. Nelson Jr. and W. Klemperer, "Microwave and Submillimeter Spectroscopy of $Ar-NH_3$ States Correlating with Ar + NH_3(j=1, -k- = 1), (to be published).

[II.2] L.H. Coudert and J.T. Hongen, J. Mol. Spectr. 139 (1990) 259.
M. Havenith, R.C. Cohen, K.L. Busarov, D.H. Gwo, Y.T. Lee and R.T. Saykally, J. Chem. Phys., (accepted for publication).

[II.3] E. Zwart, J.J. ter Meulen, W.L. Meerts and L.H. Coudert, The submillimeter rotation-tunnelling spectrum of the water dimer. To be published.

[II.4] S.P. Belov, E.N. Karyakin, I.N. Kozin, A.F. Krupnov, O.L. Polyansky, M. Yu. Tretyakov, N.F. Zobov, R.D. Suenram and W.J. Lafferty, J. Mol. Spectr. 141 (1990) 204.

[II.5] D.D. Nelson Jr., W. Klemperer, G.T. Fraser, F.J. Lovas and R.D. Suenram, J. Chem. Phys. 87 (1987) 6364.

[II.6] K.W. Jucks and R.E. Miller, Near infrared spectroscopic observation of the linear and cyclic isomers of the hydrogen cyanide trimer, J. Chem., Phys. 131 (1989) 403.

[II.7] B. Heymen, A. Bizzarri, S. Stolte and J. Reuss, Ir double resonance experiments on SF_6 and SiF_4 clusters, Chem. Phys. 132 (1990) 331.

[III.1] L. Allen and J.H. Eberly, Optical resonance and two-level atoms, Dover 1975, New York 1987.

[III.2] C.J. Borde, Developments riants en spectroscopie infrasrouge a ultra-haute resolution, Revue de Cethedec - Ondes et Signal NS83-1 (1983).

[III.3] C. Liedenbaum, S. Stolte and J. Reuss, Inversion produced and reversed by adiabatic passage, Phys. Rep. 178 (1989) 1.

[III.4] A. Linskens, S. te Lintel Hekkert and J. Reuss, One and two photon spectra of SF_6 molecular beam measurements, Infr. Phys. (in print).

[III.5] Huo-Zhong He, A. Kuhn, S. Schieman and K. Bergmann, Population transfer by stimulated Raman scattering with delayed pulses and by stimulated mission pumping method: a comparative study, J. Opt. Soc. Am. B7 (1990).

[III.6] N. Dam, L. Oudejans and J. Reuss, Relaxation rates of ethylene obtained from their effect on coherent transitions, Chem. Phys. 140 (1990) 217.

[III.7] U. Gaubatz, P. Rudecki, S. Schieman and K. Bergmann, Population transfer between molecular vibrational evels by stimulated Ramman scattering with partially overlapping laser fields: a new concept and experimental results, J. Chem. Phys. 92 (1990) 5363.

[III.8] J. Reuss and N. Dam, Multiphoton ir spectroscopy, Plenum Press 1990 (Nato-ASI series), W. Demtroder (ed.).

[III.9] C. Liedenbaum, S. Stolte and J. Reuss, Multiphoton excitation of molecules by single mode cw lasers, Infrared Phys. 29 (1989) 397.

[III.10] S.S. Aliempiev, G.S. Baranow, D.K. Bronnikov, A.E. Varfolomeev, I.I. Zasavitskii, S.M. Nikiforov, B.G. Sartakov and A.P. Shotov, Experimental Study and Mathematical modelling of high resolution spectra of excited vibrational levels of SF_6 molecules, Opt. Spectr. (USSR) 67 (1989) 625.

[III.11] C.W. Patterson, F. Herlemont. M. Azizi and J. Lemaire, J. Mol. Spec. 108 (1984) 31.

[III.12] M. Bobin, Dijon, private communication.

[III.13] C.W. Patterson, Los Alamos, private communication.

[III.14] B. Wichman, C. Liedenbaum and J. Reuss, A numerical investigation of occurrence conditions and line broadening effects for a rapid adiabatic passage process, Appl. Phys. B 51 (1990) 358.

[IV.1] G. Moruzzi, F. Strumia, P. Carnesecchi, R.M. Lees, I Mukhopadhyay and J.W.C. Johns, Fourier Spectrum of CH_3OH between 950 and 1100 cm^{-1}, Infrared Phys. 29 (1989) 583.

[IV.2] R. Engeln, J. Reuss, D. Consalvo, J.W.I. van Bladel, A. van der Avoird and V. Pavlov-Verevkin, Torsional motion of the CH_3 groups of propane studied by Raman overtone spectroscopy, Chem. Phys. 144 (1990) 81.

[IV.3] G. Bestmann, W. Lalowski and H. Dreizler, Determinations of a high barrier hindering internal rotation from the ground state spectrum: The methylbarrier of propane, Z. Naturforsch. 40a (1985) 271.

[IV.4] J.D.R. Herschbach, J. Chem. Phys. 31 (1959) 91.

[IV.5] Molecular Spectroscopy: Modern Research (Ed. K.N. Rao), Academic Press 1972, Rotational Spectra of Molecules with two internal degrees of freedom, H. Dreizler, p. 59-71.

[IV.6] P. Grouw, J.F. Sullivan and J.R. Durig, in Vibrational spectra and structure, vol. 9, ed. J.R. Durig (Elsevier, Amsterdam, 1981 ch.6).

[IV.7] R. Fantoni, K. van Helvoort, W. Knippers and J. Reuss, Direct observation of torsional levels in Raman spectra of C_2H_6, Chem. Phys. 110 (1986)1.

[IV.8] K. van Helvoort, R. Fantoni, W.L. Meerts and J. Reuss, Internal Rotation in CH_3CD_3: Raman spectroscopy of torsional overtones, Chem. Phys. Lett. 128 (1986) 494.

[IV.9] K. van Helvoort, W. Knippers, R. Fantoni and S. Stolte, The Raman spectrum of ethane from 600 to 6500 cm^{-1} stokes shifts, Chem. Phys. 111 (1987) 445.

[IV.10] N. Moazzen-Ahmadi, H.P. Gush, M. Halpern, H. Jagannath, A. Leung and I. Ozier, The torsional spectrum of CH_3CH_3, J. Chem. Phys. 88 (1988) 563.

[IV.11] N. Moazzen-Ahmadi, A.R.W. McKellar, J.W.C. Johns and I. Ozier, The torsional spectrum of CD_3CD_3, J. Chem. Phys. 94 (1991) 2387.

[IV.12] R. Engeln and J. Reuss, Methylic torsion in $C_2H_{4-n}(CH_3)_n$, n=1,2, Chem. Phys. submitted.

[IV.13] R. Engeln, J. Reuss, D. Consalvo, J.W.I. van Bladel and A. van der Avoird, Internal motion of two-top molecules: propane and dimethylamine, Chem. Phys. Lett. 170 (1990) 206.

[IV.14] H. Zuckermann, M. Drabbels, J. Heinze and W.L. Meerts, private communications.

[IV.15] C. Blindauer, N. van Riesenbeck, K. Seranski. W. Winter, A.C. Becker and V. Shuhrath, Rotational structure and splitting phemomena in electronic spectra of matrix-isolated NH, Che. Phys. 150 (1990) 93.

[V.1] E. Riedle, H.J. Neusser, E.W. Schlag and S.H. Lin, Intramolecular vibrational relaxation of benzene, J. Phys, Chem. 88 (1984) 198.

[V.2] W. Siebrand, W.L. Meerts and D.W. Pratt, Analysis and deconvolution of some $J' \neq 0$ rovibronic transitions in the high resolution $S_1 \leftarrow S_0$ fluorescence excitation spectrum of pyrazine, J. Chem. Phys. 90 (1989) 1313.
P. Uijt de Haag, J. Heinze and W.L. Meerts, Rotational assignments in the S_1 ($^1B_{3u}$ state of pyrazine by UV-UV pump-probe laser spectroscopy, Chem. Phys. Lett. 177 (1991) 357-360.

[V.3] P. Uijt de Haag and W.L. Meerts, Vibrational and rotational effects on the intersystem crossing in pyrazine and pyrimidine, Chem. Phys. 156 (1991) 197-207.

[V.4] M. Ebben, R. Spooren, J.J. ter Meulen and W.L. Meerts, A LIF monitor for potato-sprout inhibitors, J. Phys. D: Appl. Phys. 22 (1989) 1549-1551.
P. Uijt de Haag, R. spooren, M. Ebben and W.L. Meerts, Internal rotation in 1,4-dimenthylnaphthalene studied by high resolution laser spectroscopy, Mol. Phys. 69 No. 2(1990) 165-280.

[V.5] H. Zuckermann, M. Drabbels, J. Heinze and W.L. Meerts, private communication.

[V.6] G. Herzberg, Infrared and Raman Spectra, v. Nostrand Company, Princeton (1964) 400-440.

[V.7] A. McIlroy and D.J. Nesbitt, High resolution, slitjet infrared spectroscopy of hydrocarbons: Quantum state specific mode mixing in CH stretch-excited propyne, J. Chem. Phys. 91 (1989) 104.

[V.8] A. McIlroy and D.J. Nesbitt, vibrational work mixing in terminal acetylenes: High-resolution infrared laser study of isolate J. states, J. Chem. Phys. 92 (1990) 2229.

[V.9] B.M. Landsberg and R.D. Suenram, 1-butyne microwave spectrum, barrier to internal rotation, and molecular dipole moment, J. Mol. Spectr. 98 (1983) 210.

[V.10] K. Klehmann, B.H. Pate and G. Scoles, Eigenstate-resolved Y_1 spectrum of CF_3CCH: an harmonic ouplings to the bath, J. Chem. Soc. Faraday Trans. 86 (1990) 2071.

[V.11] G. Moruzzi, F. Strumia, P. Carnesecchi, R.M. Lees, I. Mukhopad hyay and J.W.C. Johns, Fourier spectrum of CH_3OH between 950 and 1100 cm_1, Infrared Phys. 29 (1989) 583.

[V.12] C. Liedenbaum, S. Stolte and J. Reuss, Inversion produced and reversed by rapid adiabatic passage, Phys. Rep. 178 (1989) 1.

[V.13] B. Heymen, A. Bizzarri, S. Stolte nad J. Reuss, IR excitation and dissociation of $(NH_3)_n$, n=2,3,4,5, and Ar-NH_3, Chem. Phys. 132 (1990) 331.

RATES OF PROCESSES INVOLVING EXCITED STATES

A.M. Stoneham

AEA Industrial Technology
Harwell Laboratory
Didcot, Oxon OX11, ORA, UK

ABSTRACT

I review the dynamics of excited states, concentrating on two features. First, when is the ionic motion significantly different from the simple oscillator description commonly adopted? Examples include "breathers" in conducting polymers and the consequences of dephasing in conventional host solids; the recognition of entropy and of the differences between constant pressure and constant volume behaviour also emerge. Secondly, how might one exploit the newer electronic structure methods within the context of non-radiative transitions, so as to go beyond common and naive assumptions?

I. INTRODUCTION

The recovery of a solid after excitation can be by various routes. There may be radiative transitions, or there may be heat generated; there may even be other channels. The exchange of energy between electronic excitation and vibrational energy is a good example; the exchange of energy between some correlated vibrational motion into incoherent thermal vibrations -- heat -- is another. The situations showing non-radiative transitions can vary enormously, from electrode processes through suppression of lasing to the efficient use of solar energy in photosynthesis.

Traditionally, the theory of non-radiative transitions has been based on gross approximations. It is true that even the simplest theory is often complicated algebraically, and it is true that the early theories were developed at a time when neighter electronic structure nor dynamics of solids were very well developed; these same approximations are repeatedly restated, wtih far less excuse. It is prudent to look again at the ideas and the methods of the field: can one take advantage of the new self-consistent approaches to electronic properties? Is anything qualitatively wrong with the simple models? I shall discuss some of the new approaches and phenomena, and give some indication of how electronic and ionic dynamics can be brought to bear on the behaviour of excited states.

II. TYPES OF NON-RADIATIVE TRANSITION

II.A. Background [1-5]: Types of Transition

There are four main types of transition. First there are "cooling" transitions, [3,4] in which vibrational excitation (perhaps the motion of a particular atom following collision with a projectile or an electronic transition) is transferred to other vibrations. Only a single adiabatic energy surface is involved; correlated local motion is degraded into heat. Secondly, there are processes associated with special, transient, configurations [5,6], as in transitions via intersections of energy surfaces such as thermal diffusion or the motion of small polarons. Here thermal vibration allows an ionic rearrangement, possibly ionic diffusion, but perhaps only electronic relocation, as in the motion of small polarons. Thirdly, one has energy transfer between electronic excitation and ionic motion. The commonest example is non-radiative recovery after optical excitation [1,2], but recombination-enhanced diffusion and electronically-stimulated defeat production are among the varied cases where energy is tranferred from electrons to ions. However, the transfer of energy from ions to electrons can also be significant in cases of high excitation [7], not least because it damps the motion of ions. Finally, there are the Auger processes, where electronic energy is redistributed over the electrons in one way or another.

I shall only deal with cases involving lattice vibrations, i.e. the first three classes of process. However, I should remark that phonon-assisted Auger transitions can have interesting features. Indeed, one of the rare self-consistent calculations of rates [8] is of just such a process, where the exchange of energy between the electron and hole excitations in an exciton (effectively [e*h] converting to [eh*]) is a key step in the production of a vacancy-interstitial pair after bank-gap excitation.

II.B. Are Steps Independent? Hierarchical Processes

A further point to note concerns the independence of successive steps in rate processes: can one really presume (as in mucc of statistical physics) that each step is independent, e.g. that a heat bath recovers its equilibrium between steps? Can one really write rate equations, rather than using the full equations of motion for the density matrix? In part, these points are addressed in the classic paper of Chandrasekhar [9] and Tolmans' book [10]. A brief review of issues regarding rate equations is given by Stoneham ([1] section 14.5).

The classic case of Debye relaxation arguably describes independent, uncorrelated events with constant probability per unit time. This is a special case of ideas introduced in different ways by Boltzmann and Markov and others. Yet there are well-characterised experiments (Jonscher [11]) which show so-called "stretched exponential" behaviour (see also [12, 13, 14]). In such cases the explanation can be "hierarchical relaxation" in which the occurrence of a process at one level requires certain conditions at a higher level. There are thus at least two types of possible correlation: those between successive events (e.g. in diffusion when jump rates are of the order of vibrational frequencies) and those controlled by other "layers" of events (e.g. when diffusion of one slow-moving species limits the diffusion of another).

II.C. Simple Models

It helps to start by reviewing some of those simplifications which are commonly made. Most of these are inessential conveniences, but their effect can be to hide real and significant features.

The first approximation is that there is a way to separate the electronic and ionic calculations. For cooling transitions, the adiabatic approximation suffices, and often the system has no electronic degeneracy, when the Born-Oppenheimer approximation will be enough. For the transfer of electronic energy to vibration, the factoring into electronic and ion components is more subtle, and the nature of the transition needs defining. The basic process is one in which an electronic state is changed from i to f and energy is redistributed over phonons, changing their occupancy from {n} to {n'}. The transition probability for specific n, n' is then given by

$$w\left(i,f;\ \{n\},\ \{n'\}\right) = M(i,f)^2 \times \text{(factor for energy conservation)}$$

with M(i,f) a matrix element which may depend on the positions of ions. The energy conservation factor is essentially the "line shape function" of optical spectra. What is observed experimentally is, of course, usually an average (normally thermal) over the initial phonon occupancies {n} and a sum over final state occupancies {n'}, i.e.

$$W(i,f) = \sum{}' < w\left(i,f;\ \{n\},\ \{n'\}\right) > .$$

A further simplification -- rarely dangerous -- is to separate phonon modes into two classes: those which drive the transition (and so enter into M(i,f)) are promoting modes; those which merely appear to conserve energy are accepting modes. However, the second simplification -- which can be very bad -- is that the matrix element is independent of ion coordinates. This, the so-called Condon approximation (from the theory of optical transitions, where it works well) can lead to major errors. These errors (casually accepted in dozens of papers) force us to examine just how best to exploit recent developments in electronic structure theory (see later, § 6).

The second group of simplifications concerns lattice vibrations. The first standard assumption is that the vibrations are harmonic. This is rarely serious, and can often be corrected. Less satisfactory is the second simplification to say a single phonon frequency suffices. Conceptually, this is the root of the Configuration Coordinate Diagram, where even stronger assumptions are common, notably that this frequency does not depend on electronic state. As a concept, as a tool to guide through the complexities of coupled electron-phonon systems, the Configuration Coordinate diagram is indispensible; problems come when more subtle questions are answered, as I shall note later. Two clear instances are (a) dephasing, where modes with two different frequencies will get "out of step" after a few vibrations, and (b) the distinction between modes (dynamically independent combinations of displacements and momenta in the harmonic approximation) and coordinates (e.g. reaction coordinates or other symmetrised combinations of displacements).

A third type of simplification is the common belief that thermodynamics is not needed, and that constant volume calculations are sufficient. This is especially unreliable for diffusion processes, where it might be taken to imply limits on "attempt frequencies" (or, more strictly, on prefactors). Experiments are, however, normally constant pressure experiments; experimental prefactors often imply impossibly large "attempt frequencies". If data suggest a prefactor far in excess of any vibrational frequency, these data carry a message: if you have not misinterpreted (e.g. from a local fit over a very narrow temperature range) there is a large "entropy" (not always technically an entropy), which may come from degeneracy, from soft vibrational modes, or -- more commonly -- from simpler features. The prefactor can be altered by orders of magnitude [6] from simple effects like the change of activation internal energy through thermal expansion. This contribution can also lead to the Meyer-Neldel rule: large prefactors are found with large activation energies (for large barriers fall more rapidly as thermal expansion occurs [6]).

Related to this, but another class of error, has been the historical view that the best calculation is that which gives the fastest transition rate. It is often possible to put bounds on rates [3], and these are often much slower, since they recognise that a lattice reorganization will often be bounded by a vibrational frequency (but see §5 in relation to prefactors) and electronic reorganization by a plasma frequency or an electronic scattering time.

III. COOLING TRANSITIONS I: GENERAL IDEAS

III.A. Modes and Coordinates [4]

The vibrations of any harmonic system (even if amorphous, macroscopic like a collection of continuum elastic components, finite like a molecule or cluster, or even full of defects) can be expressed in terms of dynamically-independent normal modes, i.e. the vibrations can be regarded as those of a set of suitably-chosen independent oscillators. Each of the normal modes Q is a linear combination of the individual ionic displacements u:

$$Q = A.u$$

However, as aready noted, any particular combination of displacements (e.g. the radial displacements of the neighbours of an impurity) does not constitute a normal mode. Such reaction coordinates (or configuration coordinates) q are linear combinations of the Q

$$q = B.Q = B.A.u,$$

and will usually contain modes with a range of frequencies. In consequence, motion in one of these coordinates will fade into noise as dephasing occurs, with important consequences for enhanced diffusion and other phenomena.

III.B. Cooling of Simple Systems

Even these simple ideas have profound consequences. First, if a normal mode Q is excited in a perfectly harmonic lattice, it will stay excited: it will not convert that excitation to heat (transfer energy to other modes). Transfer of energy between truly normal modes needs anharmonicity. Secondly, if the lattice were still harmonic but Q were only an approximate normal mode, there would be terms in the energy proportional to QX or QX″ (if X, X″ indicate sensibly-exact modes other than Q) and these can give an apparent ransfer of energy. This transfer is real enough if we are concerned with the coherent motion in some reaction coordinate; we should not forget, however, that we can always go to a normal mode representation in which there is only dephasing, not energy transfer from one mode to another. Thirdly, therefore, assume there is weak anharmonicity. The leading new terms of interest are ones proportional to QXX". A simple calculation shows the rate at which oscillator Q makes a transition from occupancy n to occupancy (n-1) is proporational to n:

The cooling rate between adjacent vibrational levels of a sensibly-normal mode is proportional to its degree of excitation.

This rule, whilst neither fully general nor rigorous, has useful and major implications which seem to have wider qualitative validity.

III.C. Configuration Coordinate Diagrams

Configuration coordinate diagrams are standard in analyzing optical transitions. As a pedagogical tool, they are excellent. How far can one take them? For some applications, they are far more general than they might appear, e.g. in predicting bounds on mean square displacement or mean square momenta at one temperature from data at another ([1] section 10.8.3). For other applications, the distinction between a coordinate (and that is what it is called) and a mode is crucial. More generally, one must ask how the coordinate is defined, and which range of frequencies is involved. Here one might choose differently for themodynamic properties rather than optic properties, or generalised to depend on electronic state (which is often talked about, but less often adopted). A most useful discussion of realistic energy surfaces and reaction paths is given by Muller [15] which should dispel ideas that a single coordinate, a harmonic system, and a single frequency are typical and realistic.

IV. COOLING TRANSITIONS II: REALISTIC MODEL SYSTEMS

I now consider two cases where detailed dynamic behavior has been modelled. They show marked differences from the simplest assumptions.

IV.A. Radiation- or Recombination-Enhanced Motion: Phonon Kick

This case concerns what happens when a particular atom in a crystal is suddenly given extra energy. The source of energy might be radioactive decay (e.g. in the Mössbauer effect, or β-decay) or collision with a projectile from a particle beam, or it could come from a localized electronic transition which changes the forces on the ion. The net effect is an impulse on the ion, and an enhance probability of some other transition (like a diffusion jump) when the motion along a reaction coordinate exceeds some threshold. The main features can be seen from a simple calculation [16] in which an interestitial in a cubic crystal (the structure and interatomic forces actually mimic Fe, but that is not critical).

A typical response to a "kicked" atom is shown in fig (1). It is immediately clear that the local motion is damped: there is an initial excursion, a swing in the opposite direction, and so on, until the enhanced local motion merges into the thermal motion. This damping would be absent in models which use only one or two frequencies, where the peak amplitude would recur regularly (or at least frequently, if there are fetures which introduce chaos). If a reaction occurs at some critical coordinate (e.g. where two energy surfaces intersect) then there will be

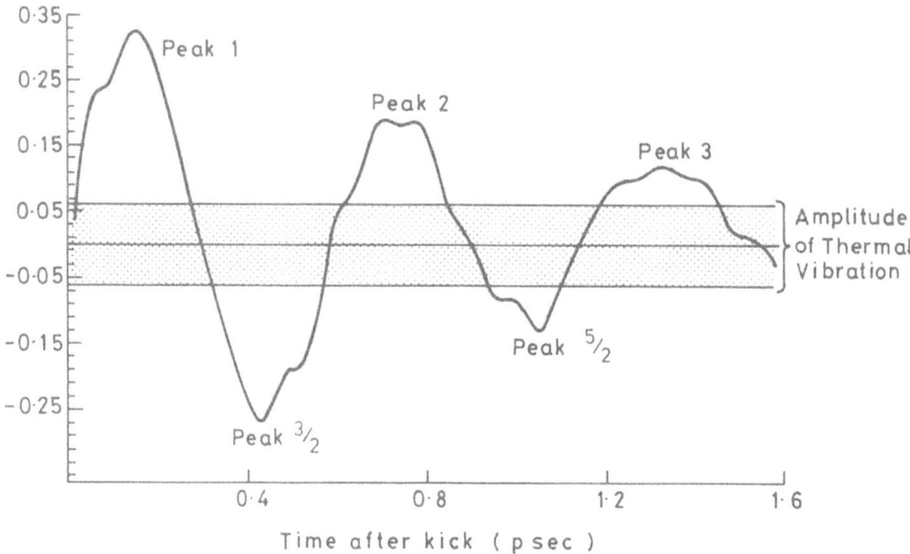

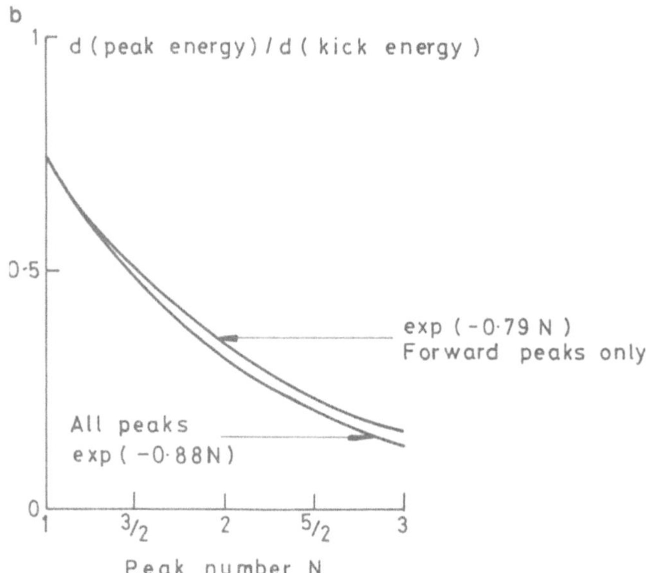

Fig.1. Dephasing of a multimode reaction coordinate shown by molecular dynamics. In (a) the forward and backward peaks are compared with thermal vibration amplitudes; possible critical coordinates would correspond to choices of ordinate; in (b) the damping shows an approximately exponential fall-off largely due to de-phasing (after [22]). For a simple Debye model, fall-off as exp. (-1.03N) would be expected.

a finite number of opportunities for a transition above the purely thermal transitions. This can be seen by simply drawing lines corresponding to different vaues of displacement along the chosen reaction coordinate. The choice of different critical displacements, incidentally, means that the same molecular dynamics runs can be used in several ways for, in effect, there are three relevant energies defining two key dimensionless parameters: the strain energy needed to achieve the critical displacement, the energy of the kick, and the thermal energy. Once can also choose different directions of critical displacement, for the kick (which will be determined by the relationships between the two energy surfaces) may not have a simple crystallographic axis.

What emerges from these calculations? First, dephasing is important. Suppose the forward maximum displacements occur at times separated by period τ, related to the average frequency $\langle\Omega\rangle$ by $\tau = 2\pi/\langle\Omega\rangle$.

Dephasing will occur because of a spread $\delta\Omega$ (which we shall take as $\sqrt{[\langle\Omega^2\rangle - \langle\Omega\rangle^2]}$) in the frequencies of the modes contributing to this reaction coordinate, and the highest and lowest frequencies will be out of phase by π after a dephasing time related to Ω. Putting all this together, for a Debye spectrum, we expect the peak displacements to fall off with peak number as exp(-1.03N). The molecular dynamics "experiments" give exp(-0.79N) for forward peaks, and exp(-0.88N) for backward peaks. Clearly dephasing is critical, and one- or two-frequency models grossly mislead (there is no damping for one mode, and the Poincaré cycle is short for two modes). Secondly, one might guess that a kick giving velocity v at angle θ to the "ideal" direction would be as effective as a kick with velocity v cos(θ) along the ideal direction. This is close to what happens, though anharmonicity makes the fall-off with θ somewhat faster for the actual system treated. Thirdly, the maximum displacement corresponds to a static strain energy which is roughly linear in the kinetic energy given in the kick. Finally, to identify a point which will arise later, one should be able to distinguish between rate processes going as $\exp[-(E - \epsilon'')/kT]$ and those going as $\exp[-E/(\epsilon + kT)]$, since these depend on factors like the rate of energy transfer between modes. Unfortunately, statistics were not good enough to be decisive (note the two expressions are the same to lowest order in ϵ when $\epsilon'' = E\epsilon/kT$).

In unpublished work [17] similar calculations were begun for the excited state of the F centre in NaCl. Here the damping proved relatively weak, i.e. the amplitude fell off as exp $(-t/\tau)$ with τ of about 20 periods of the dominant period of the breathing distortion. This makes modelling difficult, as many phonon periods must be modelled. There are practical bounds from computer limitations (whether speed or merely budget) and possibly from chaos, where I mean specifically that minor changes in initial conditions may lead to a dramatically different evolution. In this particular case there were further complications because of repeated level crossings; this is returned to in section 8.

IV.B. Other Aspects of Large Vibrational Excitation

Bickham and Sievers [18] have suggested that transitions at defects might generate displacements of sufficient amplitude to produce intrinsic localised modes. There were no signs of this in the molecular dynamics of the system just described, but the system we turn to next (breathers in t-PA) certainly shows analogous behavior.

Another complication we should note concerns whether or not a highly-excited system does return to equilibrium and (analogous to the picture above) how one may distinguish between genuine relaxation in some sense and dephasing. This relates too to the comments on chaos above. These issues arise especially in the dynamics of small molecules or in their representations as coupled Morse oscillators (e.g. [19]). Here use is made of the concept of the Kolmogorov entropy, K, and of Sinai's interpretation of 1/K as a relaxation time for highly-excited, almost completely irregular systems. In effect, molecules behave like sets of separate oscillators at low energies, but their intrinsic complexity leads to efficient energy transfer between degrees of freedom at higher excitations.

IV.C. Amplitude Breathers in Trans- and Cis- Polyacetylene (t-PA, c-PA)

We can assess the dynamical behavior in t-PA and c-PA, and especially the existence of "breathers" [20] following the creation of an electron-hole pair. This example shows that, when there is strong electron-phonon coupling, the way the excited state evolves may be very different from one structure to another, and also different from simpler models. The breather, a spatially-localised, charge-neutral oscillation, emerges as the solitons generated by photo-excitation move apart. This dynamical defect, has been postulated as the species responsible for the 1.35eV peak in the photoinduced absorption spectrum of t-PA. An isotope effect is predicted. Consistent with this, Vardeney et al [21] find the breather peak at 1.39 eV in hydrogenated t-PA $(CH)_n$ and 1.35 eV in deuterated t-PA $(CD)_n$. The possible existence of breathers in c-PA has not been studied before, partly for lack of a satisfactory model. The dynamical nature of our code, coupled with its flexibility and speed, makes it ideally suited to this class of problem.

Photo-excitation of t-PA produces an electron-hole pair which evolves rapidly into a charged soliton-antisoliton pair. The "mid-gap" photo-induced absorption at about 0.5 eV is similar to the 0.6-0.7 eV absorption associated with charged solitons created in t-PA by chemical doping. However, the photo-induced absorption contains an extra transition at 1.35 eV [21-23] apparently associated with a so-called "amplitude breather" [20], a spatially-localised charge-neutral oscillation emerging as the photo-excited solitons move apart. Later work pointed to several types of breather. Excitation of an electron at the top of the valence band to the bottom of the conduction band requires the band gap energy 2G); the separated soliton and antisoliton have rest energy of about 0.65G each, and move apart with about 0.2G kinetic energy. Within SSH theory the "central breather" has two electronic states within the gap; the observed 1.39 eV (t-PA; 1.35 eV in deuterated t-PA [22]) transition is consistent with excitation from the doubly-occupied lower state to the empty upper one.

Self-consistent molecular dynamics [24] based on the lowest spin triplet state of t-$C_{48}H_{50}$, shows clear evidence for the localised central breather in t-PA. After electron-hole pair creation in a uniformly-dimerised chain, vibration of the central atoms is enhanced. The energy in the breather is slowly dissipated; no electronic transition is involved, the decay being mainly into phonons, though here a part comes from the finite timestep of our calculation. The motion has a period 12 fs (corresponding to 2800 cm-1), close to the estimated

Raman-active LO mode (2660 cm-1) in agreement with previous theory. The periodic breather motion is still clear at the longest times we considered, nearly 40 timesteps (fig. 2). In contrast, in c-PA there is no stable localised oscillation: the electron-hole pair rapidly becomes an exciton plus longitudinal optic vibration of the whole chain (fig. 3), implying the absence of a breather absorption band in this material.

V. MORE THAN ONE ENERGY SURFACE

V.A. The Landau-Stückelberg-Zener Ideas

Many calculations invoke the idea of quasi-classical motion over an energy surface, and combine it with a "branching" when two surfaces cross. Usually the branching ratio is given by the formula given by Landau, Stückelberg and Zener in their (separate) papers. The forula has wide applicability; however, its application to solid-state problems is not simple. Qualitatively, there is no problem: at an intersection of two energy surfaces, the system will follow the adiabatic energy surface if the velocity is slow enough; at high velocities, the passage is "ballistic". The critical velocity is defined in terms of the relative gradients of the energy surfaces and the matrix elements linking them. There are several difficulties and open questions, even for molecules. What matters here is that for solids there are extra problems: some are technical (e.g. whether the matrix element really is independent of the coordinate, or whether the velocity over the energy surface is constant), but others are deeper. In condensed matter, there are many extra degrees of freedom (notably vibrational); further, the separated states between which the Landau et al branching occurs are not clearly defined. Fortunately, the LSZ approach tends to give predictions similar to those from even simpler, but more appropriate, arguments.

V.B. Branching Between Configuration Coordinate Surfaces [25]

The Stoneham-Bartram approach provides an especially simple model which applies in the case of weak damping, i.e. when transitions between vibrational levels occur slower than the period of vibration. The prediction makes use of the result outlined earlier, namely that for a single energy surface, the rate of transition between adjacent energy levels is proportional to the energy with which the oscillator is excited. Thus the transition from n to n-1 phonons is proportional to the energy $n\hbar\Omega$. Suppose a defect is excited optically to a vibronic state above the crossover between the configuration coordinate curves of ground and excited states. Energy transfer to other degrees of freedom will occur until the system reaches those states with energies close to the crossover. If there is sufficient admixture (i.e. a large enough range of states depending on the matrix element linking the electronic components) then the system will be in a state which involves a substantial component of both electronic states. From such an admixed state, the system may evolve as the ground electronic state or as the excited electronic state. The branching between them will be largely determined by the vibrational excitation. Whilst presented in different ways, the qualitative features of this Bartram-Stoneham approach and the Landau-Zener approach are very similar.

V.C. Luminescence or Not?

Can one predict from absorption data whether luminescence will occur? In a useful range of cases, the answer is yes. Early suggestions by Seitz proposed that the excited system could simply "roll down" the

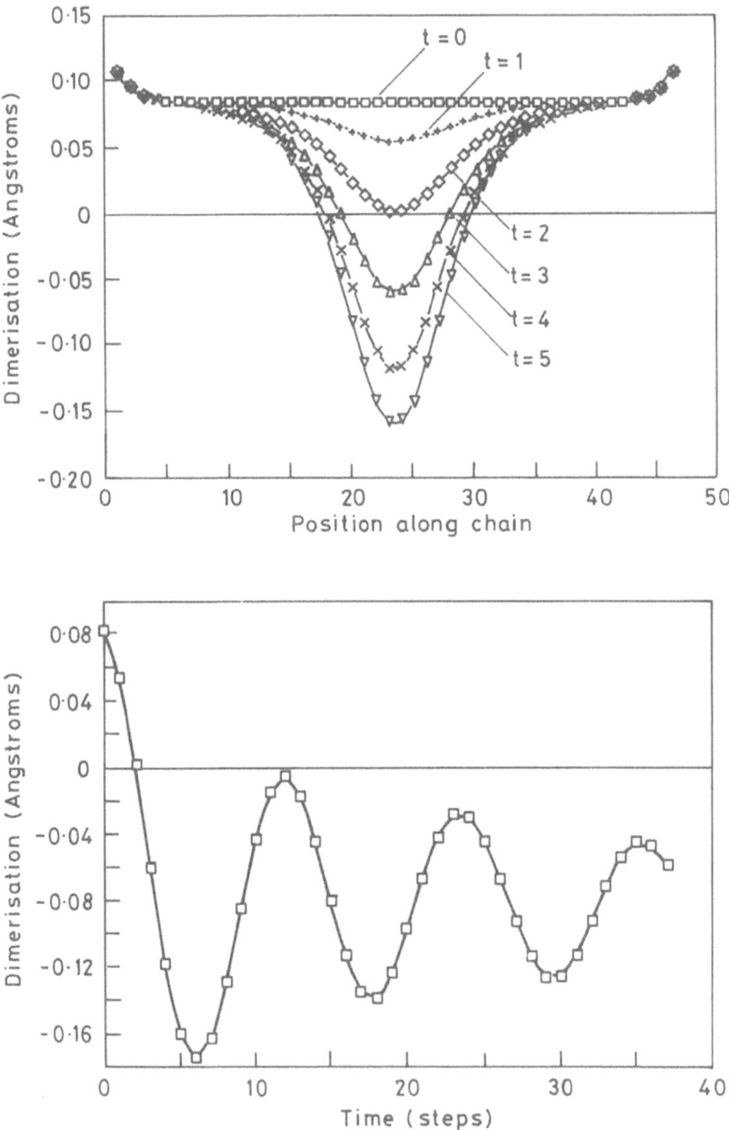

Fig.2. Breather formation in t-PA following injection of an electron-hole pair into a uniformily-dimerised chain. The dimerisation which develops corresponds to a bound soliton-antisoliton pair with an additional local oscillation of the bonding pattern at the centre of the defect. The lower figure shows the time-dependence for motion at the chain centre, (the timestep is about 1 fs), showing weakly-damped oscillation.

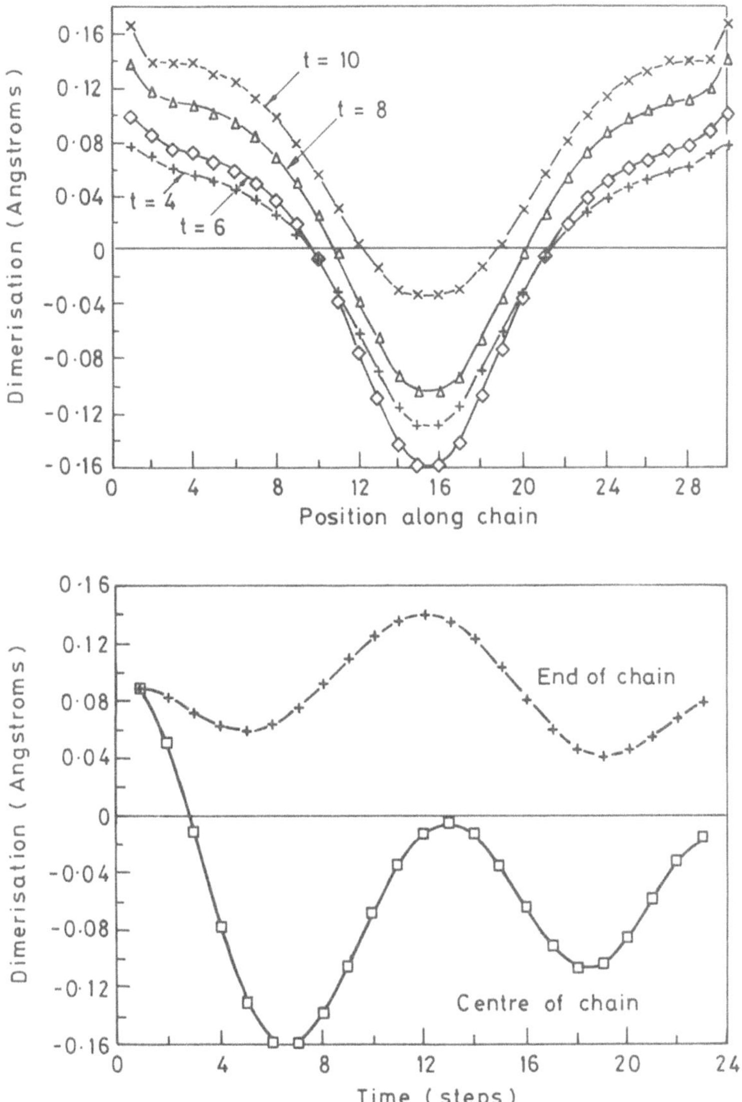

Fig.3. Lack of breather formation in c-PA following injection of an electron-hole pair in a uniformily dimerised chain. The extra oscillation superimposed represents excitation of the LO phonon mode, and involves motion of the whole chain.

configuration coordinate curve, leading to a condition which does not work well (in essence, it is not wrong; rather it is inappropriate in that it applies to a different regime of damping of motion in the configuration coordinate). Dexter, Klick and Russell suggested a far better model [25]. They argued that luminescence would be suppressed whenever the Franck-Condon final state for optical excitation was higher in energy than the crossover between ground and excited configuration coordinate curves. Bartram and Stoneham [26] took this further, partly by developing the theory to show quantitatively what should be the efficiency, partly by showing a better way of analysing data so as to cover non-luminescing cases, and partly by an extensive comparison with experiment.

The general picture (fig.4) is extremely good and, as such, has encouraged experiments to challenge it. I will not discuss the details here, except to note that what this complex of ideas does is predict the branching ratio for a system with two or more intersecting energy levels in the weak damping limit (several oscillations per phonon emitted). It does not exclude other routes to some particular excited state, and it is such alternate channels which seem to give rise to the confusion in some discussions ([27] discusses some possibilities). The prediction of branching rates is also applicable, in principle, to bistable systems and the population of their possible forms [28].

Further variants are possible; one extension which may prove useful is to allow a further level close to the excited state; this might be a free carrier state, or the 2s state for an F centre. If one chooses the extra state to be a conduction band state, then the energy splitting of the excited states can be related (for a particular value of configuration coordinate, at least; if all states have the same force constants, the result is easier) to the ground-state to excited state transition energy. Thus no new parameters are needed, but a richer variety of behavior -- including ionisation -- can be identified.

V.D. Luminescence of Polyacetylene Chains versus Length

In the polydiacetylenes, there is a systematic dependence of properties on the number of units in the chain. Each unit comprises 4 carbons and associated radicals on the end C of each unit. There are several bonding patterns, but all show a common trend: the zero phonon energy falls as the chain becomes longer. There is a second feature; the longer chains do not luminesce [29]. Why might that be? The first explanation is that we have another illustration of the DKR mechanism. If the relaxation energies are roughly constant (and this does seem to be so from the observed Stokes shifts for shorter chains), then there should be no luminescence when the zero phonon energy falls below a threshold. With the data available, the threshold is 1.5-1.6eV. The observed threshold is about 2.1eV. There is no simple way to reconcile the numbers, and it would seem the explanation is far simpler: optical transitions become more probable at higher energies (there is a higher photon density of states factor) whereas non-radiative transitions become less efficient, simply because more phonons are needed to carry away the energy. The polydicetylenes may simply be showing a crossover between the two mechanisms.

V.E. Luminescence Quenching in Cis-Polyacetylene (c-PA)

The ideas which allow one to understand whether F-centres luminesce or not also offer a possible explanation of the behavior of c-PA, and point

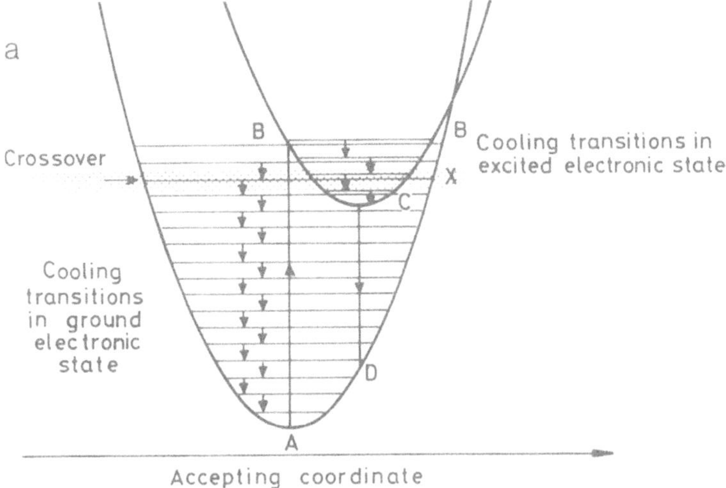

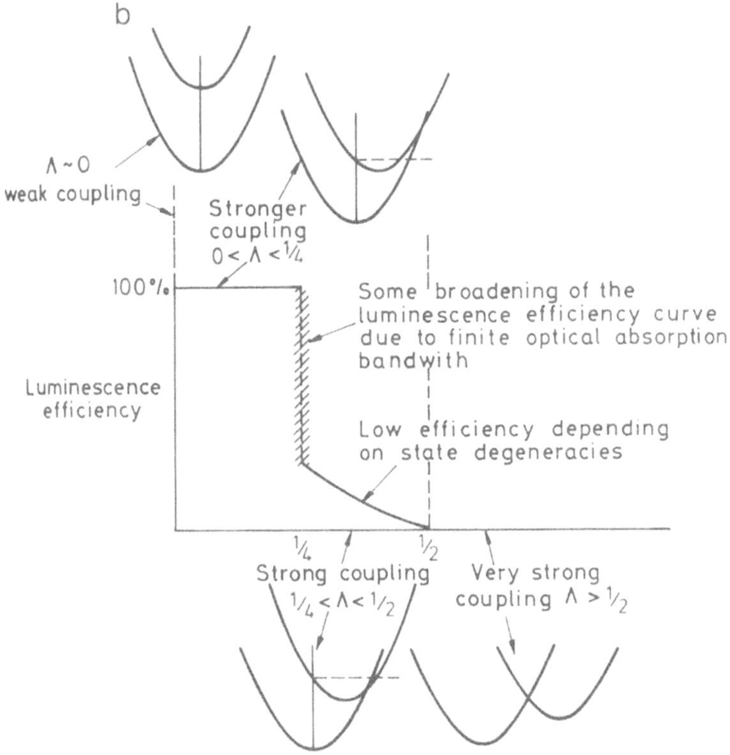

Fig. 4. (a) Cooling transitions in the ground and excited electronic state following a mixture in the crossover region.
(b) Luminescence efficiency as a function of coupling strength in the theory of Bartram and Stoneham.

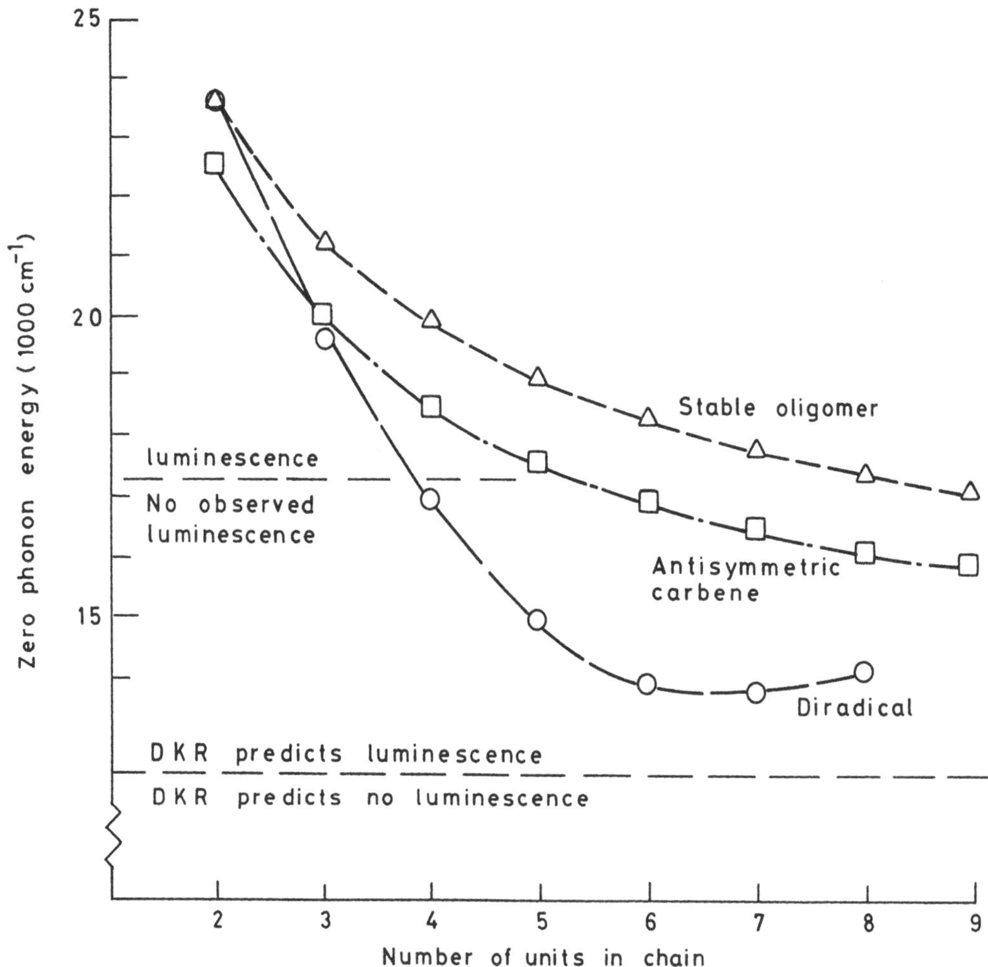

Fig.5. Luminescence and its absence for polydiacetylenes. Data are from ref.29, and give zero phonon line energies for chains of different length (the stable oligomer and the antisymmetric carbene differ only in their termination, so their difference gives a measure of end effects). The DKR line is a prediction by A.M. Stoneham using optical data from ref.29. The discrepancy between this line and the experimental line is significant!

to weaknesses in the standard Su-Schrieffer-Heeger (SSH) model. This important model has several key limitations; it is usually not self-consistent, and it also displays charge-conjugation symmetry, and this later simplification affects selection rules.

Neither approximation is made in the self-consistent molecular dynamics developed by Wallace. Another problem with the SSH model for c-PA seems to stem from the method by which it has been adjusted to cope with the special features of t-PA.

Experimentally, the situation is this [30]. There is a broad luminescence peak at 1.9 eV, with a low quantum efficiency (less than 10-5) and a short lifetime, of order 9ps; this luminescence requires an excitation energy of 2.05 eV. The inefficiency of luminescence is not consistent with the charge conjugation symmetry of SSH theory, which forbids non-radiative decay of a photo excited state to the ground state. There is also photo-induced absorption, with two components close to 1.6 eV, one with a lifetime of miliseconds, the other (perhaps twice as intense) decaying in picoseconds. Theoretically, we know that the qualitative behavior could be rationalised by a large relaxation, so that one could have both the fast but weak luminescence and the possibility of an excited state surviving to give the photoexcited states. The SSH model, in fact, predicts 100% luminescence with the Dexter-Klick-Russell criterion, irrespective of its selection rules, since it gives only a modest relaxation. Self-consistent molecular dynamics [31] does give a much larger relaxation energy, indeed large enough to place the excited state minimum outisde the ground state configuration coordinate curve. This allows (cf fig.4 in the very strong coupling limit) general agreement with experiment, though the predictions are still of limited accuracy.

VI. THE DIFFUSION PHENOMENON: INCOHERENT CHANGES OF CONFIGURATION

The oldest (and in some ways the simplest) non-radiative transition is diffusion. This shows at least one essential extension of the ideas already given, namely the need to go to constant pressure rather than constant volume predictions, and the imprtance of entropy. However, there are useful parallels among the several classes of non-radiative transitions. One is the idea of a critical coordinate: for a non-radiative transition, one is concerned with an intersection of two energy surfaces; for diffusion, there is a saddle point on one energy surface. Even at this stage, one can see there are bounds: the fastest vibration will limit the number of times the critical coordinate will be achieved (whether or not there is a transition brings in other factors too).

In the more complex cases of excited states -- or even ground states -- there are often two or more local energy minima. These may correspond to different orientations of an asymmetric defect, or to different sites for a diffusing particle, or to qualitatively-different states (e.g. self-trapped and not self-trapped excitons, or other bistable systems). These two minima can be on a single-valued energy surface (e.g. the reorientation of off-centre Li in KCl) or on two separate energy surfaces which intersect (as for a reorienting Jahn-Teller system). How does the system make a transition from one to the other? What difference does it make when there is an excess of vibrational energy? Does it matter how the system is prepared, e.g. whether the initial state is reached optically or thermally?

Broadly, there are three classes of mechanism [6], counting only those which are incoherent (i.e. where there are jumps uncorrelated in time, and where the probability is negligible for a return jump to precisely the same initial state -- vibrational as well as site and electronic state). First, there is classical diffusion, in which transitions occur when there is a thermal fluctuation which provides ionic positions and momenta appropriate for the "saddle point" of the jump. Informally, one says that the reaction path is one for which the saddle point has lowest static energy, with all atomic positions relaxed. Secondly, there are the classes of quantum diffusion and small polaron behavior (fig. 6), where again there is a thermal fluctuation, but where the key state is that of lowest energy such that a light particle (electron, proton, muon, etc) an tunnel from one site to the other without exchanging energy with the rest of the lattice. Whilst tunnelling is, of course, a quantum concept, this quantum description (that the moving particle should find initial and final sites equivalent) is also implied classically, since energy transfer between particles of very different masses is inefficient (indeed, molecular dynamics modelling of the motion of H in Si shows many features of quantum diffusion even with purely classical dynamics). The third class of motion involves excitation to above any relevant barriers, i.e. to a state in which coherent motion is possible. This can be electronic (e.g. the optical excitation of the self-trapped hole in alkali halides), or thermal (as for molecular oxygen in the alkali halides, or possibly H in liquid Fe).

What is important here is the way that the many phonon modes contribute to the transition rate, and the frequent failure of naive descriptions like that of the "approach frequency". In classical diffusion, we can consider two important theories. One class, which I shall call "thermodynamic" (Reaction Rate theory, or Vineyard theory in its most useful form) asks "What is the probability that, for a system in thermal equilibrium, the momenta and positions of ions will correspond to the diffusing particle at the saddle point and moving in the right direction?" The other class, which I shall call "first passage" (Dynamical theory) asks "How long does it take for a reaction coordinate Q which is a linear combination of independent oscillators, to exceed a critical value Q° (with dQ/dt of the right sign) for the first time?" The thermodynamic and first-passage theories are not equivalent, and indeed they are best applied in different circumstances. The first, thermodynamic, description is often best for standard diffusion processes in solids; as I shall show, it allows a simple explanation of "approach frequencies" which far exceed all vibrational frequencies of the solid. The second, "first passage", description is more appropriate to molecular systems, and gives a natural limit to this highest fequency. However, one should realize that even simple molecular systems show extensive mixing and complexity if excited to close to thier dissociation limit, i.e. the choice depends both on the system and the initial conditions.

The large amount of diffusion data points directly to a need to consider entropy terms. Indeed, any system which shows a frequency prefactor larger than the highest phonon frequencies is posing a question about interpretation. I will simply note here that many problems vanish when calculations are constant pressure ones. At its simplest, thermal expansion causes barriers to diffusion to fall, and this (as simple manipulation of the Arrhenius expression shows) enhances the apparent frequency prefactor. Factors as large as 500-1000 are noted [6]. Moreover, since large barriers tend to fall faster with expansion, one can understand the Meyer-Neldel rule relating prefactors and activation energies.

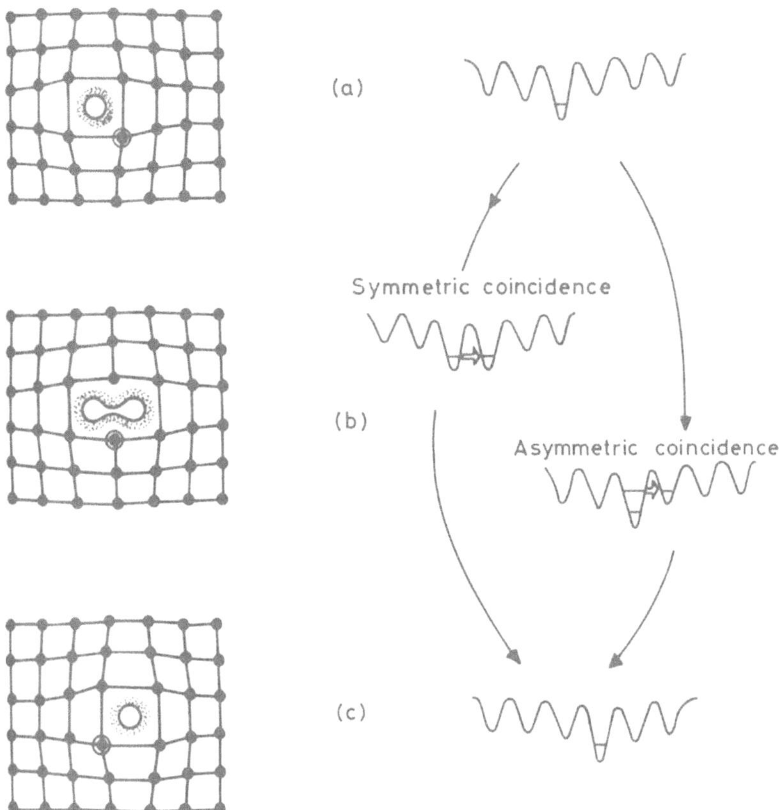

Fig.6. Schematic diagram of the reorientation of a trapped small polaron or of the quantum diffusion of trapped hydrogen. Note the role of self-trapping, the symmetric and asymmetric coincidence channels, and the distinct differences from classical behaviour.

VII. THE ELECTRONIC PROBLEM

So far I have concentrated on qualitative features and on relatively simple models. Sometimes the simple models give useful quantitative predictions (e.g. whether or not luminescence will occur). These models also imply caution: the mechanism may not be the "obvious" one; configuration coordinate diagrams can miss critical aspects; one-frequency models may prove qualitatively misleading. Future progress in the vibrational aspects of non-radiative transitions and of excited state dynamics needs accurate dynamic models, not just the simplistic ones we have all used. However, it is clear that many features are determined largely by the energy surfaces alone, including some branching ratios. I move now to absolute rates and to the electronic problem -- how are the electronic matrix elements to be calculated -- and to discuss how the latest electronic structure codes might be used. Such discussion must include bounds on non-radiative rates.

It is surprising how little has been done to apply good quantum methods to the electronic aspects of non-radiative transitions. Almost all calculations (and there have been very many) factor the problem into part concerned with accepting modes and a part which is essentially dismissed, i.e. the electronic matrix element regarded as a parameter, at best fixed from a continuum approach or empirically adjusted. I can recall only one calculation using self-consistent wavefunctions in estimating the matrix element (8). Reasons for the gap may be understood from the following discussion; however, it should also be clear there is a route to using the best modern methods (27, 31) and it is to be hoped that they will be demonstrated.

We wish to calculate rates of non-radiative transitions between adiabatic (Born-Oppenheimer) states. These transitions are induced by the non-adiabaticity operator. We may assume that the term linear in the promoting coordinates provides the key matrix elements; we cannot, of course, assume that we can concentrate on the linear terms alone of the accepting coordinates, so there is good reason for keeping them formally distinct.

The transition probability of interest is an average (usually thermal) over initial states and a sum over final states basic transition probabilities between specific vibronic states. These contain a matrix element and a factor conserving total energy. It is the matrix element within the vibrational parts of the vibronic states which concerns us here. The operator has two factors: one is the commutator of nuclear kinetic energy and the promoting coordinate, and comes directly from the non-adiabaticity operator; the other (R(f,i;Q)) is the matrix element within the electronic sttes of the promoting mode momentum. The key points which will emerge are these. First, we can rewirte the matrix element R in several ways which are formally identical, yet which are quite different with common approximations. Secondly, by the choice of a suitable form, we can see routes to use the better electronic structure calculations. Thirdly, we can see shy many early calculations obtained very bad answers. Fourthly, we can see several traps for future attempts, and hope fervently that they will be avoided.

The problem with R(Q) is its very rapid dependence on Q, so that vibrational averages (like the use of a selected value of Q only) can be very bad. The dependence is most rapid near avoided crossings of configuration coordinate curves. The solution is to transform to make most of this dependence on Q explicit. This will be demonstrated. I note in passing a subsidiary problem. We may need matrix elements of d/dQ and, in practice, this would normally involve numerical differentiation. Unfortunately, we cannot concentrate on valence electrons alone -- the transitions occur just because the electrons do not follow the nuclei and the core electrons adiabatically. Thus the displacements dQ which are small enough to keep even core overlaps sizeable will be very small, and this will lead to problems in numerical differentiation.

Whilst one could work with R directly, the averaging is difficult; a Monte-Carlo integration might be needed, and certainly the approach would be expensive as well as prone to inaccuracy. If we chose to use R evaluated at just a single value of Q, we should have -- in effect -- the so-called Condon approximation, known to give poor values for non-radiative transitions. However, one can remove the major part of the Q dependence explicitly, giving a new factor S(f,i;Q) which is much better-behaved. An integral over Q will still be needed, but it is much less troublesome. A third for can be found too, using closure. Here an approximation allows major simplification: we neglect one term because energy conservation suggests it will be small. This leads to a matrix element in which the operator is S(Q)Q, again insensitive to Q, again involving a Q integral one would rather avoid. However, we should realize that the integrand is now sufficiently smooth that we can approximate by taking a specific value for Q, and using S evaluated for just that value. What we have regained is something close to the so-called static approximation. What we have gained in addition is an expression in which we can use state-of-the-art Hartree Fock codes.

The way the wavefunction is involved means we cannot use density functional methods so easily. Nor can we use pseudopotentials or valence electrons only without ammendments. In part this is a practical problem, for the displacements which allow accurate numberical evaluation of the matrix elements give changes in core overlap which are not small. However, it is possible -- but not proven -- that, as in parallel cases (e.g. hyperfine interactions) a simple fix can be obtained.

VIII. COUPLED ELECTRONIC AND VIBRATIONS

I turn next to a further complication. How is one to treat a system in which there are repeated level-crossings? An example was mentioned above: when an F-centre is excited, the 2s and 2p states are nearly degenerate; the loss of energy from the coherent motion can be slow; there will therefore be many transitions between the 2s and 2p state. How do we characterise the evolution of the system? After all, we may want to predict more than the coarse branching ratio derived above, which establishes whether the relaxed excited state is reached or not; we may hope to predict how luminescence, for instance.

Let us assume that we can calculate the adiabatic energy surface for any ionic configuration. Let us assume too that the time-steps in our molecular dynamics are close enough and our knowledge of other electronic states (specifically the one which intersects the current state) sufficient to use LZS theory or similar to estimate the probability of a non-adiabatic transition in any particular crossing encounter. This would normally involve recognizing when the two states are close enough to mix significantly (cf. section 4), and deriving information equivalent to (i) the mixing matrix element and (ii) the rate of approach of the non-interacting energy surfaces. The encounter can be defined either by an energy criterion or by some other (e.g. dephasing time relative to the time of crossing of the non-interacting surfaces). This level of LSZ transition probability is easily done in analytical models, but is extremely difficult to achieve with any sophistication (e.g. in the better atomic collision theories, it is necessary to include effects of the mixing of states on the particle trajectories).

However, given success to this extent, in any molecular dynamics run, there would be a series of crossings i = 1,...,N, with a probability between 0 and 1 of non-adiabatic transition associated with each. How does one combine these probabilities?

One approach [32] is to sum the probabiities over transits, then dividing by the total time of the molecular dynamics run, and normalising this rate to a separately-determined rate constant. Their approach has the merit of giving an answer in all cases, and that answer is one which agrees with simpler limits. Yet it can hardly be satisfactory, as it implies a systematic build-up of probability, rather than options for back transitions and modified evolution. One of the simplest ways beyond this [17] would seem the following. Starting from particular initial conditions, let the system evolve. allow it to progress adiabatically through some number N of crossings, obtaining probabilities for each. Repeat the run from the same initial conditions, but now accept a non-adiabatic transition at one of the crossings. The evolution will change, of course, because the forces on the ions have altered. Continue this approach until al the options for transitions or not have been explored within some chosen time from this initial state. Repeat again for other initial conditions. The results may then be weighted and used to predict observables.

Anyone but a computer manufacturer will be disturbed by the approach; moreover, even the enthusiast will recognize that the number of crossings N (which need not be constant -- it is the run time which is more important) must be small. It is clear, therefore, that such massive projects are best employed as ways to verify simpler methods, rather than for general application.

IX. EVEN MORE COMPLEX SYSTEMS

My final remarks concern the future of this field. Which problems remain? Where can one expect advances? The gaps are very obvious. Yet the last decade has brought several major improvements. The route to calculating electronic matrix elements is identified; approaches to accurate ion dynamics are recognized; even dynamics with self-consistent electronic structure is commonplace. Classical diffusion (and even quantum diffusion with classical potentials) are established as reliable for rates as well as activation energies. Enhanced diffusion is another matter, with the early momentum dissipated as practical ways were found to avoid the technological problems it caused. Absolute rates for non-radiative transfer of electronic energy to heat are still appallingly rare at any level of sophistication. Yet there are even harder problems waiting for attention. The initial stages of photosynthesis are an example; the reactions occurring at receptors are another and affect the processes of life.

REFERENCES

1. A.M. Stoneham, Theory of Defects in Solids. Oxford University Press (1985).
2. R. Englman, Non-radiative Decay of Ions and Molecules in Solids, North Holland (1979).

3. A.M. Stoneham, Rep. Prog. Phys. 44, 1251 (1981).
4. W. Hayes and A.M. Stoneham, Defects and Defect Processes in Non-Metallic Solids. John Wiley (1985).
5. A.M. Stoneham, Rev. Sol. St. Sci. 4, 161 (1989).
6. A.M. Stoneham, Physica Sripta T25, 17 (1989).
7. A.M. Stoneham, Nucl. Instr. Meth. B48, 389 (1989).
8. N. Itoh, A.M. Stoneham and A.H. Harker, J. Phys. Soc. Japan. 49, 1364 (1979).
9. S. Chandrasekhar, Rev. Mod. Phys. 15, 1 (1943).
10. R.C. Tolman, Principles of Statistical Mechanics, Oxford University Press (1938).
11. A. Jonscher, Dielectric Relaxation in Solids, Chelsea Dielectrics Press (1983).
12. M. Schlesinger, Ann. Rev. Phys. Chem. 39, 269 (1988).
13. R.G. Palmer, D.L. Stein, E. Abrahams, and P.W. Anderson, Phys. Rev. Lett. 53, 958 (1984).
14. D. Walton, Phys. Rev. Lett. 65, 1599 (1990).
15. K. Müller, Angewandte Chemie, 19, 1-13 (1980).
16. P.M. Masri and A.M. Stoneham, J. Electronic Mat. 14, 205 (1984).
17. Catherine Ingham, M.J. Gillan and A.M. Stoneham, unpublished (1985).
18. S.R. Bickham and A.J. Sievers, Phys. Rev. B. (1990).
19. I. Hamilton, D. Carter, and P. Brummer, J. Phys. Chem. 86, 2124 (1982); Ya G Sinai, Acta. Phys. Austr. Suppl. X, 575 (1973).
20. A.R. Bishop, D.K. Campbell, P.S. Lomdahl, B. Horovitz and S.R. Philpott, Phys. Rev. Lett. 52, 671 (1984).
21. Z. Vardeny, J. Strait, D. Moser, T.C. Chung and A. Heeger, Phys. Rev. Lett. 49, 161 (1982).
22. C.V. Shank, R. Yen, R.L. Fork, J. Orenstein and G.L. Baker, Phys. Rev. Lett. 49, 1660 (1982).
23. C.L. Wang and F. Martino, Phys. Rev. B34, 5540 (1986).
24. D.S. Wallace, D. Phil. Thesis, Clarendon Laboratory, Oxford (Harwell Report AERE TP-1331) (1989); D.S. Wallace, Synthetic Metals 28, D457 (1989); D.S. Wallace, A.M. Stoneham, W. Hayes, A.J. Fisher, A.H. Harker and A. Testa, J. Phys. Cond. Mat. 3 (in press) (1991).
25. D.L. Dexter, C.C. Klick and G.A. Russell, Phys. Rev. 100, 603 (1956).
26. A.M. Stoneham and R.H. Bartram, Sol. St. Electronics 21, 1325 (1978); R.H. Bartram and A.M. Stoneham, Sol. St. Comm. 17, 1593 (1975); Semic & Insulators 5, 297 (1983).
27. R.H. Bartram, J. Phys. Chem. Sol. 51, 641 (1990).
28. A. Chantre, Appl. Phys. A48, 3-9 (1989).
29. H. Sixl, Polydiacetylene (edited D. Bloor), NATO Advanced Science Series, Martinus Nijhoff (see especially p.51 for data) (1984); theory quoted is unpublished work by A.M. Stoneham (1986).
30. W. Hayes, C.N. Ironside, J.F. Ryan, R.P. Steele and R.A. Taylor, J. Phys. C16, L729 (1983). The theory is discussed in ref. 24.
31. R.H. Bartram and A.M. Stoneham, J. Phys. C18, L549 (1985).
32. J.W. Halley and J. Hautman, Phys. Rev. B38, 1170 (1988).

EXCITED STATES IN SEMICONDUCTORS

C. Klingshirn

Department of Physics of the University
D–6750 Kaiserslautern, Germany

ABSTRACT

In the first section of this contribution we introduce the main elementary excitations in semiconductors namely phonons, plasmons and excitons and discuss their interaction with the electromagnetic radiation field in the weak and strong coupling limit leading to the concept of polaritons. In the second section we describe the optical properties of semiconductors starting with bulk material and proceeding to systems of reduced dimensionality. After a short introduction to the concepts of nonlinear optics in section III, we follow in section IV and V two different aspects of excited state spectroscopy in semiconductors: First we consider the fate of a single electronic excitation i.e. of an exciton, introducing the concepts of phase – and intraband relaxation and of interband recombination times and illustrating them by various examples. Then we consider what happens if we add to one exciton a second or many resulting in biexcitons, scattering processes and finally in the transition to an electron – hole plasma.

In a final section we come back to the excitonic polariton, introduce the concepts of spatial dispersion and present various techniques of $\vec{K}$–space spectroscopy which allow to measure directly its dispersion relation.

I. ELEMENTARY EXCITATIONS IN SEMICONDUCTORS

Basically it is a straight forward task to write down the Hamiltonian of a semiconductor (SC) consisting of the kinetic energy terms of N electrons with mass m_0 and of N' ion cores with M_j and the Coulomb–interaction between them

$$H_{tot} = -\sum_{i=1}^{N} \frac{\hbar^2}{2m_0} \vec{\nabla}^2_{\vec{r}_i} - \sum_{j=1}^{N'} \frac{\hbar^2}{2M_j} \vec{\nabla}^2_{\vec{R}_j} + \frac{1}{4\pi\epsilon_0} \left[\sum_{i>i'} \frac{e^2}{|\vec{r}_i - \vec{r}_{i'}|} + \sum_{j>j'} \frac{Q_j Q_{j'}}{|\vec{R}_j - \vec{R}_{j'}|} + \sum_{i,j} \frac{Q_j e}{|\vec{r}_i - \vec{R}_j|} \right] \quad (1)$$

Unfortunately the sums run over several 10^{22} particles per cm^{-3} and therefore it is impossible to find the eigenfunctions ϕ (r_i, R_j) and eigenenergies of (1) directly. Instead, scientists have developped in the last few decades a different approach to treat solids involving the concept of elementary excitations and of quasiparticles which we shall outline below.

I.A. Phonons, Plasmons, Excitons

We first note that $M_j >> m_0$. Since the Coulomb–forces acting on electrons and ions are comparable, we can state that the motion of ions is much slower than the motion of the electrons. This means, that electrons can follow easily the motion of the ions but not vice versa. This leads to the adiabatic or Born – Oppenheimer approximation (see eg.[1]) and allows us to treat first the motion of the atoms or of the ions located in the IR frequency regime.

The effective interaction potential as a function of the interatomic distance (or of the lattice constant a) looks for all types of binding (ionic, covalent, metallic, etc) rather similar and is shown in Fig 1. It consists of a Coulomb–type attractive part and a short range repulsive part caused by exchange interaction i.e. by Pauli's principle. The mathematical description is e.g. by Born – Mayer or Lenard – Jones type expressions. What is important for us here is the harmonic approximation which can be fitted to the potential minimum and which allows to use a simple model of masses and springs to describe the motion of the atoms or ions. In Fig 2 we introduce therefore in analogy to many textbooks on solid state physics (see e.g. [2]) the one–dimensional linear chain model, using for the moment equal atomic masses M and spring constants D.

We consider longitudinal waves i.e. an elongation in the direction of propagation, describe the displacement of atom n by $u_n(t)$ and get the classical equation of motion.

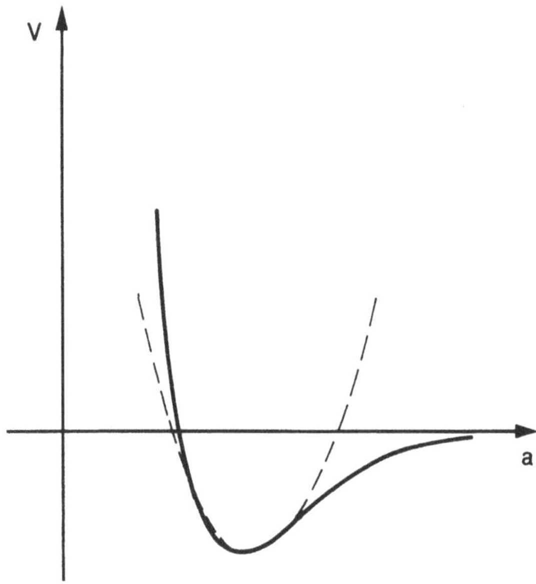

Fig 1 Schematic drawing of the crystal potential V as a function of the interatomic distance a (———); harmonic approximation (- - - - -).

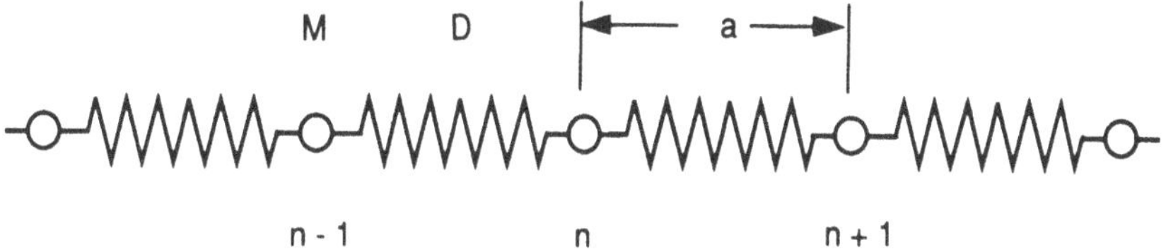

Fig 2 The linear chain model for the description of phonons.

$$M\,\ddot{u}_n(t) = D\,[(u_{n+1}(t) - u_n(t)) - (u_n(t) - u_{n-1}(t))] =$$

$$D\,[u_{n+1} - 2\,u_n + u_{n-1}] \tag{2}$$

With the ansatz

$$u_n = u_{n,0} \exp\,[i(\omega t - Kna)]$$

$$u_{n\pm1} = u_{n\pm1,0} \exp\,[i(\omega t - K(n\pm1)a)] \tag{3}$$

where $K = 2\pi\lambda^{-1}$ is the wave – "vector" and noting that

$$u_{n,0} = u_{n\pm1,0} \tag{4}$$

we get the dispersion relation $\omega(K)$ for the vibrations of our one–dimensional chain

$$\omega = \left(\frac{4\,D}{M}\right)^{1/2} \left|\sin \frac{Ka}{2}\right| \tag{5}$$

which is depicted in Fig 3. Evidently the dispersion relation is periodic in K space and we can restrict ourselves to the region

$$-\frac{\pi}{a} \leq K \leq \frac{\pi}{a} \tag{6}$$

which is the first Brillouin zone.

Actually $\pm\frac{\pi}{a}$ correspond to the largest physically meaningful K–value i.e. to the shortest wavelength. As shown in Fig 4 shorter wavelength have no real significance.

The next step is usually to go to the diatomic chain in order to introduce the concept of acoustic and optic lattice vibrations [2].

The corresponding equations of motion get slightly more complex than the ones in (2) and lead to a secular equation.

We want to use here another approach assuming that we "paint" the atoms in Fig 2 alternatingly red and green without changing their physical properties. As a consequence we get a new lattice constant.

$$a' = 2a \tag{7}$$

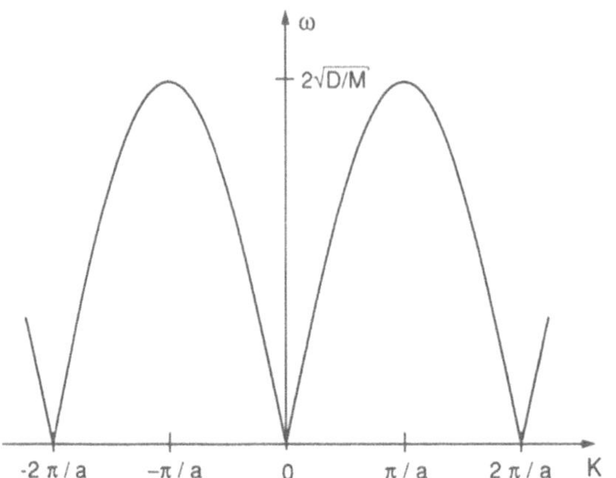

Fig 3 The dispersion of longitudinal oscillations of the linear chain of Fig 2.

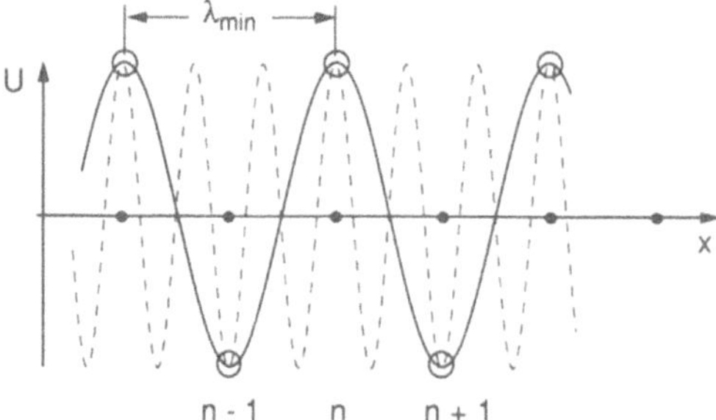

Fig 4 The elongation u of the linear chain for the shortest physically meaningful wavelength.

and a first Brillouin zone between $-\frac{\pi}{a'} \le K \le \frac{\pi}{a'}$, which is just one half of the original one (Fig 5). Again we expect, that everything of physical relevance happens in the first Brillouin zone. As we shall see in a moment, it is indeed possible, to shift the outer parts of the dispersion relation with the help of reciprocal lattice vectors into the first Brillouin zone as indicated by a dash–dotted line.

If we finally allow that the red and green atoms are physically different, having e.g. different masses M and M', the degeneracy at the boarder of the Brillouin zone is lifted (dotted lines) and we get two branches. The lower one, which starts a K = 0 and $\omega = 0$ is known as acoustic branch (Fig 6a), because it describes the propagation of sound waves. The upper branch is called optic branch. If the atoms forming the lattice carry some electric charge (i.e. if we have a certain contribution of ionic binding) then the optic modes may be connected with an electric dipole moment and couple to the IR light field as shown in Fig 6b.

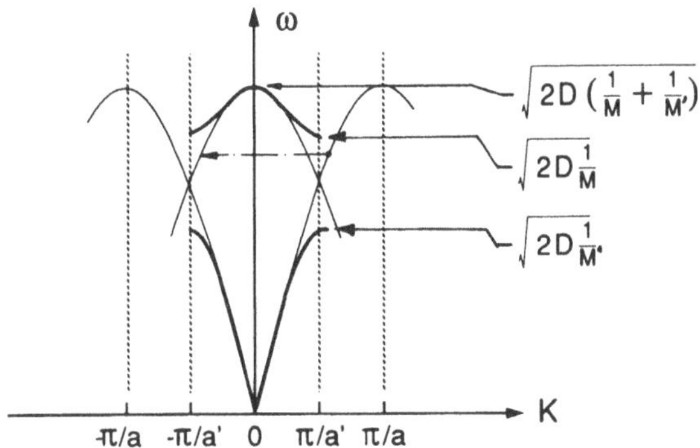

Fig 5 The derivation of the dispersion for the diatomic chain (heavy solid lines) from the one of the one–atomic chain (light solid lines).

At this point it seems necessary to say a few words about real and reciprocal space. We shall come back with some new insight to the problem of lattice vibrations in a few moments.

A spatially periodic arrangement of lattice points can be described in three dimensions by three noncomplanar basic or primitive translation vectors $\vec{a}_1$, $\vec{a}_2$, $\vec{a}_3$. All lattice points can be reached by vectors $\vec{R}$

$$\vec{R} = n_1\vec{a}_1 + n_2\vec{a}_2 + n_2\vec{a}_3;\ n_i = 0, \pm1, \pm2... \tag{8}$$

In other words a translation $\vec{R}$ transforms the crystal into itself.

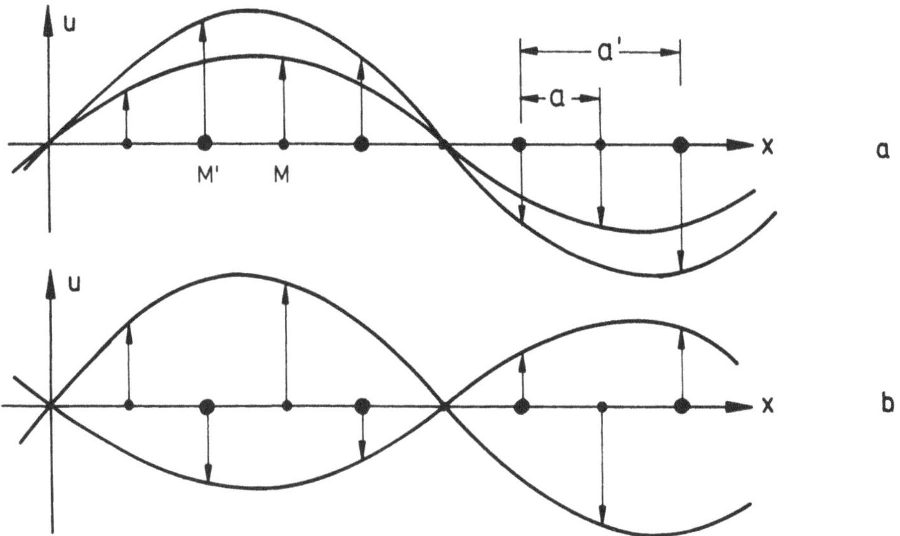

Fig 6 Schematic representation of an acoustic (a) and an optic wave (b).

The $\vec{a}_i$ span a parallelelepiped which is usually called unit cell. The whole space can be completely filled by placing identical unit cells next to each other as shown for simplicity for a twodimensional system in Fig 7a. The choice of the primitive unit cell is in principle not unique for a given arrangement of lattice points, but by agreement there is one generally accepted choice for every lattice. In addition to the $\vec{a}_i$ we must know, where the atoms sit in the unit cell and this is described by the basis. In our example of Fig 7 there is one atom at (0,0) and the other at ($1/3\ \vec{a}_1$, $1/3\ \vec{a}_2$). If the $\vec{a}_i$ and the basis are known, we know everything about the arrangement of the atoms in a perfectly periodic lattice.

Similarly to the $\vec{a}_i$ in real space, we can define a reciprocal space, which is introduced by basic translation vectors $\vec{b}_i$ defined by

$$\vec{b}_1 = 2\pi \frac{\vec{a}_2 \times \vec{a}_3}{\vec{a}_1(\vec{a}_2 \times \vec{a}_3)} \quad \text{and c.p.} \tag{9}$$

and with vectors of this reciprocal lattice $\vec{G}$ defined in analog to (8)

$$\vec{G} = m_1\vec{b}_1 + m_2\vec{b}_2 + m_3\vec{b}_3 \text{ with } m_i = 0, \pm 1, \pm 2 \ldots \tag{10}$$

This reciprocal space is the adequate one to describe processes which involve wavevectors $\vec{K}$.

There are many important properties of the reciprocal space and relations to the real space, which can be found in textbooks on solid state physics [2]. We shall not go into details here but state only without proof, that all wavevectors in a periodic lattice are only defined modulo integer mupltiples of the $\vec{b}_i$:

$$\vec{K} \mathrel{\hat{=}} \vec{K} + \vec{G} \tag{11}$$

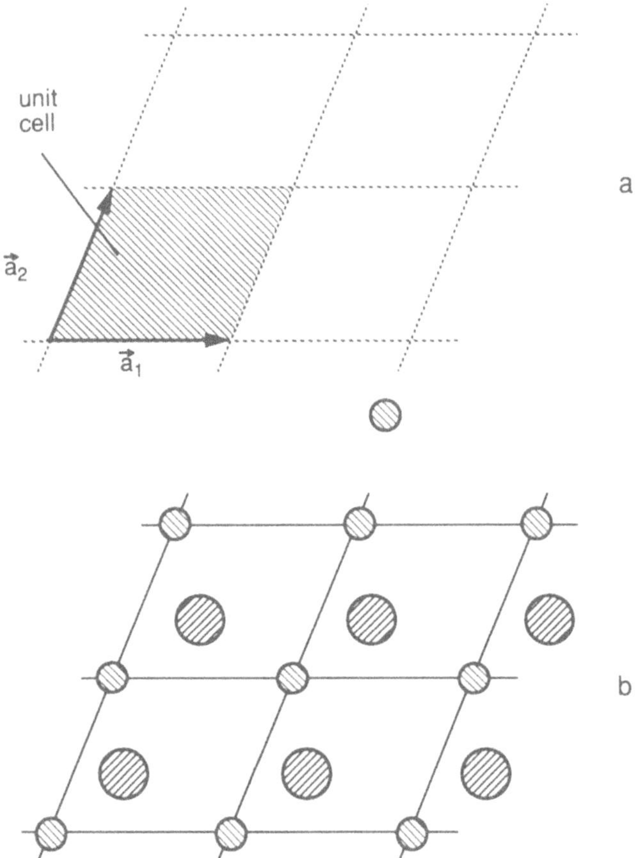

Fig 7 The elementary translation vectors $\vec{a}_1$ and $\vec{a}_2$ and the unit–cell for a twodimensional periodic lattice (a) and a basis consisting of two atoms at (0,0) and at $(\frac{1}{3}\vec{a}_1, \frac{1}{3}\vec{a}_2)$(b).

This statement can be made clear by inspecting Noether's theorem, which says that a conservation law follows from every invariance of the Hamiltonian.

Examples are e.g. invariance against infinitesmal translations in time or space

$$H(t) = H(t + dt); \Rightarrow \text{conservation of energy} \tag{12a}$$

$$H(x) = H(x + dx); \Rightarrow \text{conservation of x–component of momentum } p_x \tag{12b}$$

in a periodic lattice (12b) is no longer valid but replaced with (8) by

$$H(\vec{r}) = H(\vec{r} + \vec{R}) \tag{13}$$

and as a consequence the wavevector $\vec{K}$ or the momentum $\hbar\vec{K}$ is conserved only up to some $\vec{G}$ vectors. All dispersion relations are therefore periodic in $\vec{K}$

$$\omega(\vec{K}) = \omega(\vec{K} + \vec{G})$$

and branches outside the first Brillouin zone can be shifted into it as we did in connection with Fig 5 and vice versa. This is one of the reasons why the quantity $\hbar\vec{K}$ in a periodic lattice is sometimes called quasi–momentum.

If we identify the origin of the reciprocal space with one of its lattice points, the first Brillouin zone can be defined as the part of the reciprocal space, which is closer to or equally far away from the origine than to any other point of the reciprocal space. For a simple cubic lattice this definition results in

$$-\frac{\pi}{a} \leq K_i \leq \frac{\pi}{a}$$

This result has already been used e.g. in Fig. 3 and 5 or in (6).

It is easy to show the connection between (12b) and (13). In the limes a => 0 (13) coincides with (12b). On the other hand the first Brillouin zone (6) covers then the whole reciprocal space and the use of $\vec{G}$ vectors becomes obsolete.

Now let us return to the lattice vibrations. We get for every wave–vector apart from the longitudinal mode two transversal ones, which may be degenerate or not, depending e.g. on the crystal symmetry. Usually the transverse branches are situated in frequency below the longitudinal ones.

The transition to three dimensions has some influence on the density of states (see below), but for lattice vibration not much on the dispersion relation. We get for every $\vec{K}$ vector always three acoustic branches and 3s – 3 optical ones, where s is the number of atoms in the basis.

In Fig 8 we give the situation for a crystal of rather low symmetry with two atoms in the unit cell.

We now add in a rather intuitive way what quantum mechanics says about lattice vibrations introducing with this example the concept of quasi–particles. Eq (2) is very similar to the equation of motion of a harmonic oscillator except for the "non–diagonal" elements u_{n-1}. Indeed linear combinations of the u_n can be found, so–called normal coordinates, in which the equation of motion becomes identical with the one of a harmonic oscillator. Quantization results then in energy units $\hbar\omega_i(\vec{K})$ where the index i denotes the various branches e.g. in Fig 8.

The units or quanta of energy $\hbar\omega$ are generally called quasiparticles or quanta of the elementary excitations. In the special case of lattice vibrations these quanta are called phonons.

The dispersion–relation e.g. of Fig 8 is not influenced by a linear transformation of the coordinates nor by quantization and is therefore still valid. The total energy in the phonon system is given by

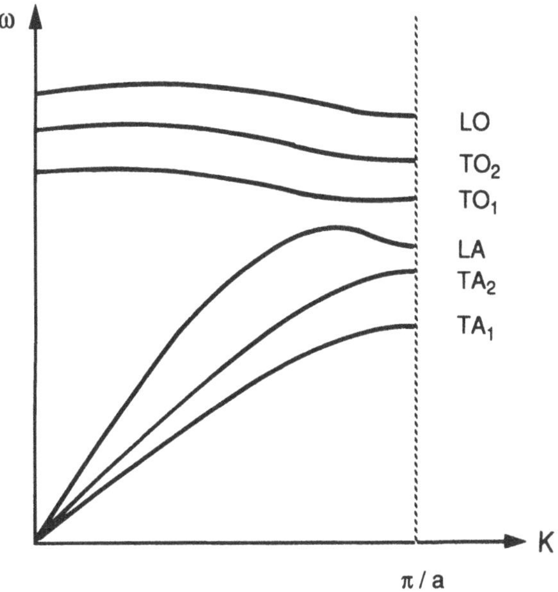

Fig 8 Schematic drawing of the phonon dispersion of a crystal with two atoms per unit cell and low symmetry so that all degenacies are lifted.

$$E = \sum_{\vec{K},i} (n_{\vec{K},i} + \tfrac{1}{2})\hbar\omega_i(\vec{K}) \tag{14}$$

where $n_{\vec{K},i}$ is the occupation number of the state with $\vec{K}$ on branch i. The term 1/2 in (14) accounts for the zero–point energy and is neglected in the future. In the picture of second quantization it is possible to introduce creation and annihilation operators $B^+_{\vec{K},i}$ and $B_{\vec{K},i}$, respectively. They obey the commutation relations for Bosons, and the $n_{\vec{K},i}$ are the eigenvalues of the number operator

$$B^+_{\vec{K},i}\, B_{\vec{K},i} \tag{15}$$

The dispersion of phonons can be measured by various techniques among which inelastic neutron scattering proved to be most powerful [2].

Incident monochromatic low energy neutrons with energy E_i and momentum $\vec{K}_i$ are scattered from the solid into final states E_f and $\vec{K}_f$. Energy and momentum conservation allow to deduce the properties of the created (or annihilated) phonons according to

$$\hbar\omega_{phonon} = E_i - E_f \tag{16a}$$

$$\vec{K}_{phonon} = \vec{K}_i - \vec{K}_f + \vec{G} \tag{16b}$$

Dispersion curves of phonons in real semiconductors are found in many testbooks on solid–state or semiconductor physics [2, 3] or in the comprehensive collection of semiconductor data [4].

The next quasiparticle which we want to introduce is the plasmon, the quanta of the collective oscillation of the electrons (or holes) of a partly filled band. From the definition it is immediately clear, that plasmons are of especial importance for metals and indeed they are responsible for the high reflectivity of metals in the IR and large parts of the visible spectrum. In semiconductors, partly filled bands occur in heavily doped materials or in highly excited ones where an electron–hole plasma (see section V) has been formed.

We proceed as in the case of phonons and start with the classical equation of motion. In Fig 9 we displace the electron cloud of density n with respect to a positive (ionic) background by an amount z. Similarly as for a capacitor we get surface charges ρ_s and an electric field E_z given by

$$\rho_s = \pm nez \tag{17a}$$

$$E_z = \frac{ne}{\epsilon_b \epsilon_0} z; \ \epsilon_b = \text{background dielectric constant} \tag{17b}$$

Fig 9 Schematic drawing of the displacement of the electron cloud (- - - - -) with respect to the positively charged background of the lattice to deduce the equation of motion for plasmons.

This leads for the electrons, which we describe by an effective mass m (see below), to an equation of motion

$$m\ddot{z} = \frac{e^2 n}{\epsilon_b \epsilon_0} z \tag{18}$$

This is already the equation of motion for a harmonic oscillator the eigenfrequency of which is known as plasma frequency ω_{pl} [2, 3]

$$\omega_{pl} = \left(\frac{e^2 n}{m \epsilon_b \epsilon_0}\right)^{1/2} = \omega_L ; \omega_T = 0 \tag{19}$$

This is a longitudinal oscillation. The transverse eigenfrequency ω_T of the electron gas is zero since an electron gas has, like other gases, no shear–stifness.

The quanta of this oscillation are the next type of quasi–particles which e meet in this contribution. They are again Bosons and are called plasmons. For finite $\vec{K}$–values the plasma dispersion shows in three dimensional systems a weak parabolic dispersion (Fig 10). The plasma – frequency is situated for usual carrier densities in doped semiconductors ($10^{15} \lesssim n \lesssim 10^{19}$ cm^{-3}) in the IR in contrast to metals, where ω_{pl} is situated due to the $n^{1/2}$ dependence in the visible or UV part of the spectrum. Apart from the collective excitation there is also a whole range of single particle excitations, shown in Fig 10 by the shaded area and explained in Fig 11.

It should be mentioned that the constant electric field of eq (17b) is valid only for three–dimensional systems. A displacement of the electrons in quasi two – or one–dimensional systems leads to charged lines or points and to an electric field which decays for long wavelengths as r^{-1} or r^{-2}, respectively. As a consequence the plasmon dispersion starts in these cases for K = 0 at $\omega = 0$ [5].

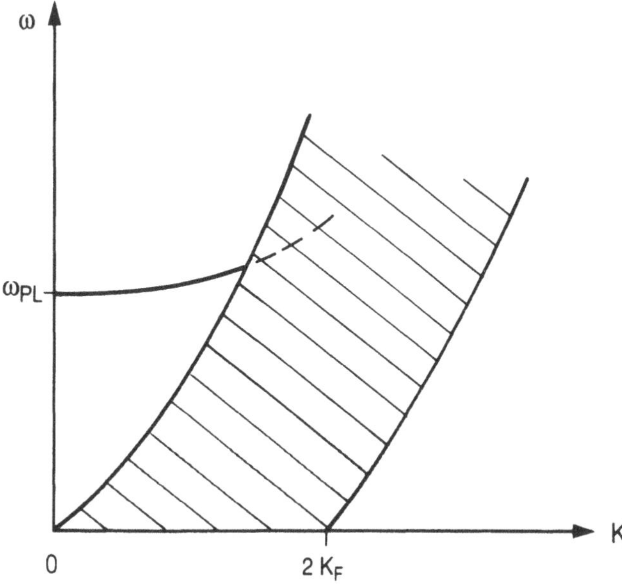

Fig 10 Schematic drawing of the plasmon dispersion in a three–dimensional metal (——) and the range of one (two–) particle excitations (shaded area). K_F = wave–vector at the Fermi energy.

Before we can proceed to the third example of elementary excitations in SC namely to the excitons, we have to spend some time on the concept of crystal electrons and holes as a further example of quasiparticles, since excitons are composed particles consisting of an electron and a hole. Again one cannot use the general Hamiltonian (1) but introduce some simplifications. The ion cores are assumed to be in their equilibrium positions. Consequently they will produce a spatially periodic contribution to the potential seen by the electrons. Motions of the ion cores will be treated later in perturbation as electron – or exciton – phonon interaction. The more important approximation is however the one – electron approximation. One assumes, that all electron – electron interactions, which appear e.g. in a Hartree – Fock approach for a many electron system form also a periodic potential, leading to a Schrödinger equation for one particle [1, 2, 6]

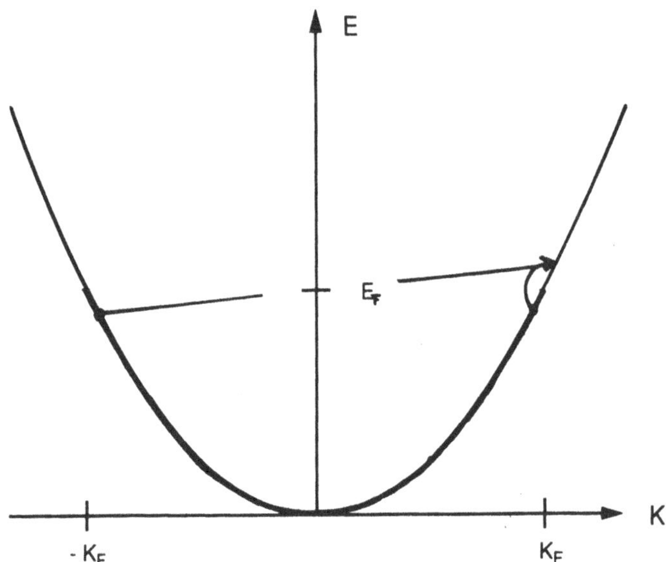

Fig 11 Occupied (▬▬) and empty (——) states in a degenerate electron gas at low temperature and some one (two–) particle excitations with small and large wave–vector.

$$H = -\frac{\hbar^2}{2m_0}\vec{\nabla}^2 + V(\vec{r})$$

$$\text{with } V(\vec{r}) = V(\vec{r} + \vec{R}) \tag{20}$$

One calculates the solutions of (20) $\varphi(\vec{r})$ and fills the states with the number of electrons which are in the crystal according to Fermi – Dirac statistics.

It is well known, that the eigenfunctions of a periodic potential are Bloch – waves [1 – 3, 6] of the form

$$\varphi_{\vec{k},i}(\vec{r}) = e^{i\vec{k}\vec{r}}\, u_{\vec{k},i}(\vec{r})$$

$$\text{with } u_{\vec{k},i}(\vec{r}) = u_{\vec{k},i}(\vec{r} + \vec{R}) \tag{21}$$

i.e. a plane wave term modulating a lattice periodic function $u_{\vec{k},i}(\vec{r})$ which remembers the parent atomic orbitals, in the more ionic SC or the hybrid orbitals (e.g. sp^3) for the more covalent SC. The index i labels the various bands and $\vec{k}$ is obviously the wave vector. Since Bloch – waves are no eigenfunctions of the momentum operator $\hbar i^{-1} \vec{\nabla}$, there is another reason to call the quantity $\hbar\vec{k}$ quasi – momentum.

The statements which we gave for the $\vec{K}$–vector in a periodic system above are also valid for the electron states. Consequently one has for the eigenenergies $E_i(\vec{k})$ of the Bloch – states.

$$E_i(\vec{k}) = E_i(\vec{k} + \vec{G}) \tag{22}$$

In connection with Fig 12 we want to discuss the concept of the bandstructure in some more detail. The dashed line in Fig 12a is the dispersion of non – relativistic free electrons with a mass m_0. If we introduce e.g. in the frame of nearly free electron approach a weak periodic potential, the dispersion relation will not change much (solid line in Fig 12a). Only for some special wave vectors, and these are just the boundaries of the various Brillouin zones, all back–scattered waves interfere constructively for an incident wave resulting finally in a standing wave.

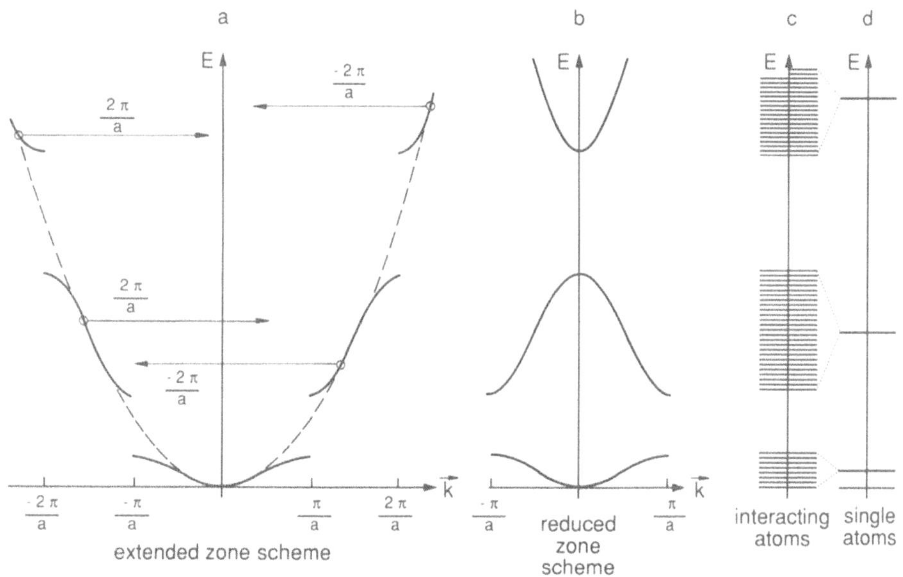

Fig 12 The deduction of the band–structure in the reduced zone scheme, starting from free electrons (l.h.s.) and from atomic orbitals (r.h.s.). From [7].

Such a standing wave can have for the same $\vec{k}$–value i.e. for the same kinetic energy its maxima of $|\varphi^*\varphi|$ either at the positions of the periodic potential minima or between them, resulting in two different values for the potential energy and thus for the total one $E_i(k_{x,z,y} = \pm n\pi/a)$. Since the group – velocity

$$v_g = \frac{d\omega}{dk} \tag{23}$$

of wave – pakets formed essentially from standing waves is zero, we expect a vanishing slope of the dispersion relation at the boundaries of the various Brillouin zones.

Using eq (22) we can either go from the extended zone scheme of Fig 12a to a periodic one, similarly to Fig 3 or we can shift everything into the first Brillouin zone, leading to the generally used reduced zone scheme of Fig 12b.

The other approach to the band – structure known e.g. as linear combination of atomic orbitals (LCAO) or tight binding approximation starts from the atomic levels of the atoms which constitute the crystal (Fig 12d). Moving the atoms closer together to form the solid, the atomic terms split into bands due to the interaction (overlap – integrals) between the wave functions. This effect is quite analogous to the situation of coupled identical pendula in classical mechanics. If we have N pendula, which are not coupled, each one oscillates with its eigenfrequency ω_0. If we couple two pendula there is a splitting into two frequencies, if we couple three of them we get three frequencies and so on, and if we couple 10^{23} of them we get 10^{23} frequencies. The same happens for the atoms Fig 12c. Since the width of the resulting bands is for SC typically of the order of 10 eV, there is no possibility to see the 10^{23} energy levels per cm^{-3} of SC seperately but we get a continous band. Applying finally symmetry arguments i.e. the periodicity of the lattice we come from Fig 12c to 12b.

As already mentioned, we have to distribute the electrons now on the bands. If we end up at T = 0K with a partly filled band we have a metal , if we get only completely filled (valence–) bands separated by a forbidden region (energy gap E_g) from the empty (conduction–) bands we have an insulator for $E_g \geq 4eV$ and a semiconductor (SC) for

$$0 < E_g \leq 4eV \tag{24}$$

Materials with $E_g \approx 0$ are called semimetals, those with $E_g \lesssim 0.5eV$ narrow gap semiconductors.

In the following we shall only speak about SC.

The band – structure for real SC is somewhat more complicated as in Fig 12b. We give in Fig 13 schematically the bandstructure of SC crystallizing in the most frequently observed structures of diamond (point – group O_h; examples C, Si, Ge) Zinkblende (Td; GaAs; Inb, GaP, ZnS, ZnSe, ZnTe, CdTe, CuCl, CuBr) and Wurtzite (C_{6v}; ZnO, ZnS, CdS, CdSe, GaN). We want to outline now some of the basic features of the bandstructures of SC.

If the global extrema of the upmost valence band and of the lowest conduction band are situated at the same point of the Brillouin zone (usually at k = 0, the so – called Γ – point) we speak about direct gap SC, because the transition between the band extrema is "directly" possible with photons, which have a negligible momentum ($k \approx 10^5\ cm^{-1}$) compared to the boundary of the first Brillouin zone ($k \approx 10^8\ cm^{-1}$).

In the opposite case, e.g. when the valence band maximum is still situated at the Γ point but the conduction band minimum occurs e.g. in Δ – direction (Fig 13a) or near the L – point (dashed curve in Fig 13b) one has an indirect gap, since a momentum conserving phonon is usually necessary in addition to the photon to excite an electron from the maximum of the valenceband to the minimum of the conduction band across the forbidden gap.

Examples for indirect SC are the group IV materials C, Si, Ge, some of the III–V SC like GaP or AlAs or some I–VII SC like AgBr and for direct ones some III–V SC like GaAs or InP, the II–VI SC (ZnO, ZnS, ZnSe, ZnTe, CdS, CdSe, CdTe) or the copper–halides again as representatives of the I–VII SC. In the following we shall concentrate on direct gap SC if not explicitely stated otherwise.

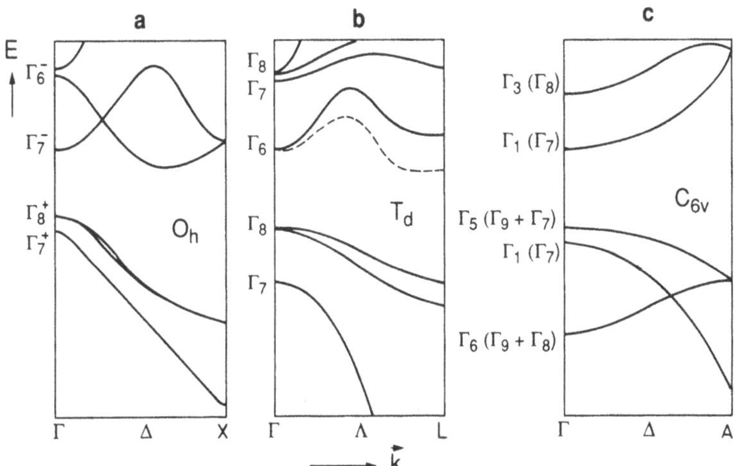

Fig 13 Schematics of the bandstructure of semiconductors crystallizing in O_h, T_d and C_{6v} symmetries, including spin in (a) and (b) and in (c) for the irreducible representation given in brackets. From [7].

The next concept which we want to introduce is the "effective mass" concept. We have learned with eq (21), that the eigenstates of a periodic potential are Bloch – waves. To investigate transport properties, one would intuitively prefer to consider the electrons as small particles. This can be done by describing electrons as wave – pakets formed by the superposition of Bloch – waves $\varphi_{k,i}(r)$ of a certain region around $\vec{k}_0$.

$$\Psi(\vec{r}) = \sum_{\vec{k}-\vec{k}_0} a_{\vec{k}}\, \varphi_{\vec{k},i}(\vec{r}) \tag{25}$$

These wave pakets are known e.g. as Wannier – functions.

Such a wavepaket moves under the influence of an external field ($\vec{E}$, $\vec{B}$ etc) as if it has an effective mass m given by

$$m^{-1} = h^{-2} \frac{\partial^2 E}{\partial k^2} \tag{26a}$$

or more precisely, if we are dealing with SC of lower symmetry

$$(m^{-1})_{i,j} = h^{-2} \frac{\partial^2 E}{\partial k_i \partial k_j} \tag{26b}$$

This means, that an electron in a crystal behaves like a particle with mass inversely proportional to the curvature of the band.

Inspection of the band structures of Fig 13 or in [4] shows, that the dispersion E(k) has in the vicinity of the extrema a parabolic shape i.e. a constant curvature. This finding is used in the "effective mass approximation" where one considers the crystal electrons simply as free particles however with an effective mass, which is basically the result of the fact, that the electrons do not move in free space but in the periodic potential of a crystal.

Typical values of the effective masses around the conduction – band minima and the valence band maxima, respectively, are

$$0.2m_0 \gtrsim m_e \gtrsim 0.05m_0 \tag{27a}$$

and

$$3m_0 \gtrsim |m_h| \gtrsim 0.6m_0 \tag{27b}$$

We shall use in the following the effective mass approximation.

The next important concept, which we introduce is the one of electrons and holes. We stated earlier, that a semiconductor has a T = 0K a completely empty conduction band and a completely filled valence band. We can bring some electrons into the conduction band e.g. by optical or thermal excitation of electrons from the valence band to the conduction band or by doping with donors according to

$$D^{\circ} \leftrightarrow D^{+} + e_c \tag{28a}$$

with a ionisation energy of the donor–electron of usually a few meV, and describe these electrons as effective mass particles with charge $e = -1.6\ 10^{-19}$ As as mentioned above.

If we remove an electron from the otherwise filled valenceband again by excitation into the conduction band or by doping with acceptors

$$A^0 + e_v \leftrightarrow A^- \tag{28b}$$

we can in principle consider the $(10^{23} - 1)$ cm^{-3} electrons in the valence band. Obviously it is simplier to consider the one unoccupied state. Indeed this state can be described as a quasiparticle called hole or defect electron with the properties given in Table 1.

Table 1. Properties of the hole compared to the ones of the electron removed from the valence band.

	property of hole		property of missing electron in the valence band
momentum	$\vec{k}_h$	=	$-\vec{k}_{me}$
charge	$q_h = \|e\|$	=	$-e$
spin	s_h	=	$-s_{me}$
effective mass	m_h	=	$-m_{me}$

Corresponding to table 1, holes have a positive charge and effective mass. In the following we use the electron and hole concept throughout, i.e. with a free "electron" we always mean an electron in the conduction band with effective mass m_e and charge e and with a free "hole" we always mean a quasiparticle in the valence band.

Equation (28b) will be therefore always used in the form

$$A^\circ \longleftrightarrow A^- + h \tag{28c}$$

After this short introduction to the one particle states and the corresponding band–structure of semiconductors we proceed to optical excitations of the electronic system, which we discuss in connection with Fig 14.

In Fig 14a we show the band–structure of a direct gap semiconductor in the vicinity of the Γ point. It becomes immediately clear that all optical excitations are two particles excitations, i.e. if we excite an electron from the valence– to the conduction band, we create necessarily simultaneously a hole in the valence band. In a recombination process an electron and a hole are simultaneously annihilated.

If the electron and hole would be non – interacting particles, one would expect from Fig 14a for a three – dimensional semiconductor, in the case of a k–independent band–to–band transition matrix element, an absorption spectrum which follows a square–root dependence as shown by the dashed line in Fig 14c

$$\alpha(\hbar\omega) = \alpha_0(\hbar\omega - Eg)^{1/2} \text{ for } \hbar\omega \geq Eg$$

$$\alpha = 0 \text{ for } \hbar\omega < Eg \tag{29}$$

reflecting the square–root dependence of the density of states resulting from the effective mass approximation for a three – dimensional semiconductor (see below).

Indeed, a behaviour according to (29) has never been observed, essentially because electrons and holes are interacting with each other via their Coulomb – interaction potential.

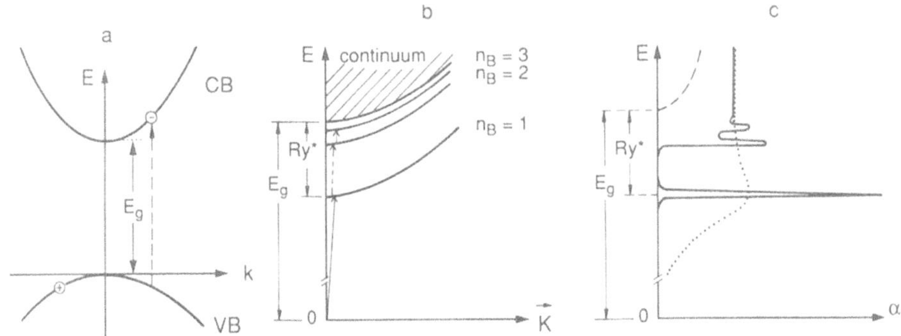

Fig 14 The conduction– and valencebands with an electron–hole pair (a) the dispersion of the series of exciton resonances (b) and the absorption spectrum resulting from coupling of excitons with $\vec{K} \approx 0$ to the electro–magnetic field (c) for small (——) and strong (••••••) damping. The dashed line gives the square–root absorption expected without Coulomb interaction between electron and hole.

The problem of a positive and a negative charge has been solved in quantum – mechanics for the H atom or the positronium long time ago. In the effective mass approximation we can adopt the result with some modifications. We get a hydrogen – like series of bound electron – hole pair states known as excitons, followed by the ionisation continum which corresponds to the band–to–band transitions.

The dispersion relation of excitons $E_x(\vec{K}, n_B)$ for a simple direct gap SC with nondegenerate bands reads in the three–dimensional case

$$E_x(\vec{K}, n_B) = E_g - Ry^* \frac{1}{n_B^2} + \frac{\hbar^2 \vec{K}^2}{2M} \tag{30}$$

with $\vec{k}_e + \vec{k}_h = \vec{K}$

$\vec{K}$	momentum of center off–mass motion
$M = m_e + m_n$	translational mass
n_B	main quantum number

$$Ry^* = 13{,}6\ \text{eV}\ \frac{\mu}{m_0}\ \epsilon^{-2} \qquad \text{modified Rydberg energy of the exciton} \tag{31}$$

Depending on the reduced mass

$$\mu = \frac{m_e m_h}{m_e + m_h} \tag{32}$$

and the dielectric constant, Ry* ranges

$$5\text{meV} \lesssim Ry^* \lesssim 200\ \text{meV} \tag{33}$$

for so called Wannier excitons, i.e. excitons with Bohr radius a given by

$$a_B = a_0 \cdot \epsilon \cdot \frac{m_0}{\mu} \tag{34a}$$

$$1.0\text{nm} \lesssim a_B \lesssim 20\text{nm} \tag{34b}$$

which is much larger than the lattice constant a.

The wavefunction of the exciton can be written schematically as

$$\phi = \Omega^{-1/2} e^{i \vec{K}\vec{R}}\ \phi_{n_B,l,m}(\vec{r}_e - \vec{r}_h)\ \varphi(\vec{r}_e)\ \varphi(\vec{r}_h) \tag{35}$$

$\phi_{n_B,l,m}$ is the Hydrogen – atom like envelope function and $\varphi(\vec{r}_{e,h})$ are Wannier functions (see (25)) describing wave – pakets. $\Omega^{-1/2}$ is the normalization factor.

Complications to the above equations induced e.g. by more complicated band – structures, a frequency dependence of ϵ etc. are beyond the scope of this paper. The reader is referred for details to [5–9] and references therein.

For a strictly two – dimensional system, i.e. one in which electron and hole can move only in a quasi – twodimensional plane but interact still with a three – dimensional r^{-1} Coulomb potential one finds (36) instead of (30)

$$E_x(\vec{K}, n_B) = E_g + E_q - Ry^*(n_B - \tfrac{1}{2})^{-2} + \frac{\hbar^2 \vec{K}^2}{2M} \tag{36}$$

The realization of such systems will be explained below. E_q is the sum of the quantization energies of both carriers.

For strictly one– and zero dimensional systems, the Coulomb – attraction diverges, so that expressions analogous to (30) or (36) cannot be given and numerical calculations for finite dimensions are necessary.

The excitons are the most important elementary excitations of the electronic system of SC. They behave at low densities as Bosons, but show with increasing density some deviations [5, 9], which are one of the reasons why a clear – cut evidence for a Bose – condensation of excitons has not yet been observed [10].

Before we got into details of the optical properties of the various excited states in SC, we want to address a big difference in the spectroscopy of excitons and of atoms:

In atomic spectroscopy like the one of a H–atom and including the spectroscopy of atoms in some matrices, which form the main topic of this series of schools, the atom is present before the light comes in. Consequently one observes e.g. transitions from the 1s to the 2p state (dashed arrow in Fig 14b) or, in excited state spectroscopy from 2p to higher levels. In the spectroscopy of excitons, the H–like quasiparticle, one usually creates the exciton by absorption of light (solid arrow in Fig 14b), i.e. one goes from a state without exciton (sometimes called "vacuum" state) to a state with one exciton. If the band–to–band transition is dipole allowed, one reaches with one photon transition usually the exciton states with $\vec{K} = 0$ and S type envelope function. In the schematic drawing of Fig 14c one sees for a sample of high quality at low temperature the transitions to 1, 2 and 3 excitons, followed by transitions in the ionization continuum using the assumption that $K_{photon} \approx 0$ and the weak coupling approach (see below). The absorption coefficient in the ionization continuum is strongly enhanced compared to the simple band–to–band transition model (dotted line) in a three – dimensional system by the so–called Sommerfeld – factor [8, 11], which has a square–root singularity for $E \to E_g$, $E > E_g$ and which describes the fact, that the probability to find electron and hole in the same unit cell is enhanced by their Coulomb attraction even in the ionization continuum. In a two–dimensional system, the Sommerfeld enhancement is less pronounced and varies only by a factor of two between $E = E_g$ and $E \to \infty$. In a quasi one–dimensional system the "enhancement" – factor is even below one above E_g [11].

For higher temperatures and/or for samples of lower quality the exciton resonances are broadened (dotted line in Fig 14c) but the Coulomb attraction is still present.

Indeed one observes very often an exponential tail of the absorption coefficient below the lowest free exciton known as Urbach or Urbach–Martienssen rule [12]

$$\alpha(\hbar\omega, T) = \alpha_0 \exp\left[\frac{\sigma(T).(\hbar\omega - E_0)}{k_B T}\right] \tag{37}$$

where σ, E_0 and α_0 are material parameters. E_0 coincides with the energy of the lowest free exciton at $T = 0K$ and σ is an only weakly temperature dependent quantity.

The Urbach tail is due to interaction with thermally excited phonons. Detailed theoretical appoaches are given in [13].

I.B. Coupling to the Radiation Field

After introducing the phonons, plasmons and excitons as the most important quasiparticles or elementary excitations necessary for the understanding of the optical properties of semiconductors, we start now to describe in a more rigorous way the interaction of the radiation field with these excitations.

We start with the semiclassical treatment of radiation [14]. The momentum in the Hamiltonian H describing the SC, or the relevant part of it has to be changed in the following way

$$\vec{p} \rightarrow \vec{p} - e\vec{A} \tag{38}$$

where $\vec{A}$ is the vector potential with

$$\vec{E} = \dot{\vec{A}} - \text{grad } \varphi \tag{39}$$

$$\vec{B} = \text{rot } \vec{A} \tag{40}$$

and $\vec{E}$ the electric field strength and $\vec{B}$ the magnetic flux density.

If we use the so–called Coulomb–gauge for $\vec{A}$, since $\vec{A}$ is not uniquely defined by (40)

$$\vec{\nabla} \cdot \vec{A} = 0 \tag{41}$$

φ is the electrostatic potential.

With the replacement (38) and $\vec{p} = -\frac{\hbar}{i}\vec{\nabla}$ we get two new terms in the Hamiltonian apart from those describing the semiconductor (H_{sc}) and the radiation field without matter (H_{rad}):

$$H \rightarrow H_{sc} + H_{rad} - \frac{e}{m}\vec{A}\frac{\hbar}{i}\vec{\nabla} + \frac{e^2}{2m}\vec{A}^2$$

$$= H_{sc} + H_{rad} + H_1 + H_2 \tag{42}$$

H_1 and H_2 are now considered as pertubations of first and second order. H_1 can be further modified to give the well–known dipole approximation by noting that $K_{photon} \approx 0$ i.e.

$$\vec{A} = \vec{A}_0 e^{i\vec{K}\vec{r}} e^{-i\omega t} \approx \vec{A}_0 e^{-i\omega t} \tag{43}$$

and by replacing the transition matrix element element $<\vec{p}>_{if}$ which comes from H_1 and appears e.g. in Fermi's golden rule by

$$<\vec{p}>_{if} \Rightarrow -im\omega<\vec{r}>_{if} \tag{44}$$

where $e<\vec{r}>$ is just the dipole operator.

In this approach, which is also known as "weak coupling" case,we would state e.g. that the transition probability w from the crystal ground state $|i>$ to an exciton state $|f>$ is proportional to

$$w_{if} \sim \vec{A}^2_0 \, |<f|\vec{r}|i>|^2 \tag{45}$$

This approach is good for all diluted systems like gases, atoms in a matrix or in semiconductors for transitions with small oscillator–strength (e.g. forbidden ones).

For transitions with "strong" coupling one has to go a step further to get a quantitative description of the optical properties.

I.C. Polaritons

In order to introduce the polariton–concept first in an intuitive way, we try to imagine, what is propagating in a semiconductor, if a light field in the frequency range of one of the elementary excitations impinges on a sample. The considerations are similar for phonons, excitons or plasmons, but we consider in the moment the exciton case.

An incoming photon creates with a certain probability (see eq(45)) an exciton. This exciton decays, emitting a photon, which creates an exciton etc. We show this process schematically by the diagram in Fig 15 where photons create electron–hole pairs and vice versa. The vertical lines in the bubbles symbolize the Coulomb interaction leading to the formation of excitons from electron–hole pairs as outlined above.

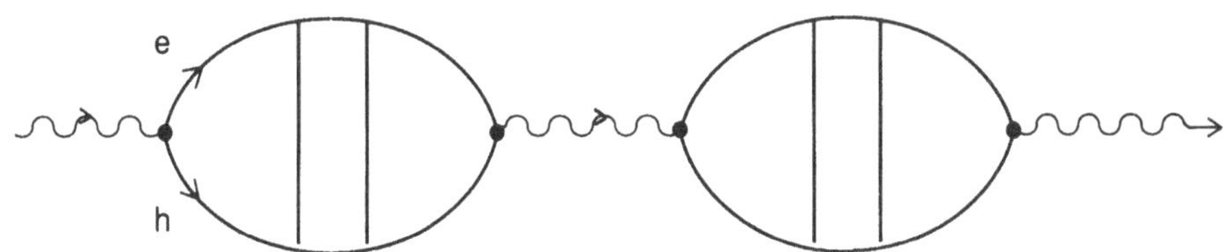

Fig 15 The concept of the exciton–polariton illustrated in a graph.

What is evidently propagating in matter is a mixture of an electromagnetic wave and a polarization wave described e.g. by the virtual or coherent excitation and annihilation excitons in our present example or of phonons or plasmons. This mixture can be quantized. The quanta are a new type of quasiparticles called polaritons. We distinguish between exciton–polaritons, phonon–polaritons and plasmon polaritons if the polarization wave accompaning the electron–magnetic one is predominantly exciton–, phonon– or plasmon–like, respectively.

For those, who do not like this intuitive introduction of the polariton concept, we give in the following an outline of its derivation in the picture of second quantization, using again the exciton–polariton and following the presentation in [5, 9].

We define a hole operator β which acts only in the valence band. The annihilation of an electron there corresponds to the creation of a hole, with momentum –k and spin $-s_e$

$$\beta^{+}_{-\vec{k},-\vec{s}} = a_{v,\vec{k},\vec{s}} \quad (46a)$$

the creation operator for an electron is correspondingly

$$\alpha^{+}_{\vec{k},\vec{s}} = a^{+}_{c\,\vec{k},\vec{s}} \quad (46b)$$

The quantity

$$\alpha_{\vec{k}}^{+}\,\beta_{-\vec{k}'}^{+} \tag{46c}$$

is the creation operator of an electron–hole pair. For convenience we drop the spin so we are not bothered in the moment with the difference between singulet excitons with anti–parallel electron and hole spins which can couple strongly to the radiation field and triplet excitons with parallel spins which have only very weak oszillator strength.

The exciton operator can be written as

$$B^{+}_{\nu,\vec{K}} = \sum_{\vec{k},\vec{k}'} \delta(\vec{K}-(\vec{k}-\vec{k}'))\,\psi_{\nu}\left(\frac{\vec{k}+\vec{k}'}{2}\right)\alpha^{+}_{\vec{k}}\beta^{+}_{-k'} \tag{47}$$

where the index ν stands for n_B, l, m in eq (35) and the ψ are connected with the Fourier transform of the envelope function $\phi_{n,l,m}$.

The Hamiltonian for the interacting exciton ($B^+ B$) and photon ($c^+ c$) system can be written as

$$H = \sum_{\vec{K}} \left\{ \sum_{\nu} E_{\nu,\vec{K}} B^{+}_{\nu\vec{K}} B_{\nu\vec{K}} + h\omega_{\vec{K}} c^{+}_{\vec{K}} c_{\vec{K}} \right.$$

$$\left. ih \sum_{\nu} g_{\nu\vec{K}} (B^{+}_{\nu\vec{K}} c_{\vec{K}} - \text{h.c.}) \right\} \tag{48}$$

where the coupling coefficients $g_{\nu\vec{K}}$ contain the matrix elements like in (45).

The polariton–concept is now reached if we do not treat the third term on the r.h.s. of (48) as a perturbation, but if we diagonalize the whole Hamiltonian by a suitable linear combination leading to polariton operators $p_{\vec{K}}$

$$p_{\vec{K}} = u_{\vec{K}} B_{\vec{K}} - v_{\vec{K}} c_{\vec{K}} \tag{49a}$$

with

$$|u_{\vec{K}}|^2 + |v_{\vec{K}}|^2 = 1 \tag{49b}$$

The dielectric function $\epsilon(\omega,\vec{K})$ resulting from the above procedure for a single resonance is the same as obtained for a classical Lorentz–oscillator [15]

$$\epsilon(\omega,\vec{K}) = \epsilon_b\left(1 + \frac{f}{\omega_0^2(\vec{K}) - \omega^2 - i\omega\gamma}\right) \tag{50}$$

with ω_0 transversal frequency of the exciton, phonon or plasmon, ϵ_b background dielectric constant describing the contribution of resonances with higher eigenfrequencies, f oscillator–strength proportional to $|\langle f|H|i\rangle|^2$ in eq (45) and a phenomenologically introduced damping term γ.

What is new in comparison to the classical model is the dependence of ω_0 (and eventually also of f and γ) on $\vec{K}$, making ϵ a function of the two independent variables ω and $\vec{K}$. The $\vec{K}$ – dependence is known as "spatial dispersion". We come back to this aspect in section VI.

The dispersion relation of the polaritons can be found from the polariton equation

$$\epsilon(\omega,\vec{K}) = \frac{c^2\vec{K}^2}{\omega^2} \tag{51}$$

and (50).

In Fig 16 we give the dispersion curves of phonon–, exciton – and plasmon polaritons. For simplicity we show only the Re $\{\vec{K}\}$ and consider the case without damping i.e. $\gamma = 0$ in eq (50). For the more complicated situation, where finite damping i.e. $\gamma \neq 0$ are considered see e.g. [16]. The polariton dispersion results from the one of photons and of the elementary excitations (phonon, exciton or plasmon) by applying the non–crossing or level–repulsion rule. The corresponding dispersion relations are shown by thin solid lines. They are shown only for clarity but do not correspond to any state in matter. The oscillator strength f in eq (50) which describes the coupling is directly proportional to the longitudinal transverse splitting. The longitudinal eigenmodes usually do not couple to the radiation field.

The parts of the lower and upper polariton branches (LPB and UPB, respectively) which have a constant slope are called photon–like, though they are also not pure electromagenetic waves as follows directly from the fact, that the refractive indices deduced from the slopes are

$$n_{LPB} = \left[\epsilon_b(1 + \frac{f}{\omega_0^2})\right]^{1/2} \tag{52a}$$

$$n_{UPB} = \epsilon_b^{1/2} \tag{52b}$$

and deviate usually from one in contrast to what is expected for a pure electro–magnetic wave. The other parts of the polariton dispersion are called phonon–, exciton– or plasmon–like, respectively.

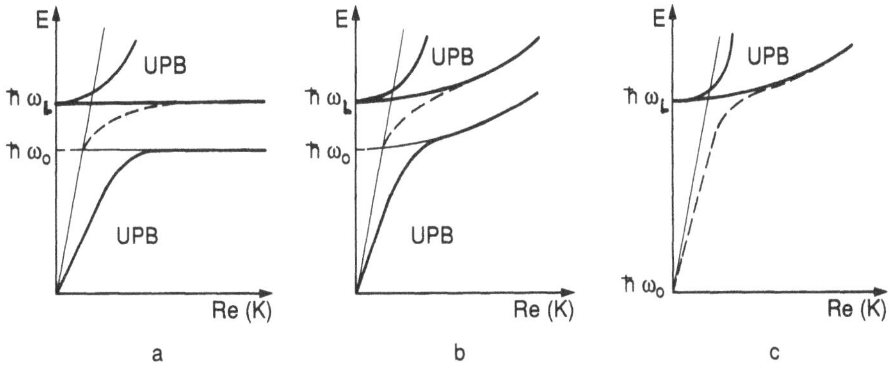

Fig 16 The dispersion relation of the phonon– (a), the exciton– (b) and the plasmon– polariton (c) for a single resonance.
The heavy lines give the dispersion relation of the transversal lower and upper polariton branches (LPB and UPB, respectively) and of the longitudinal branch. The thin solid lines give for comparison the dispersion relation of photons in vacuum and of the excitation in the SC without coupling to the radiation field. The dashed lines indicate the dispersion of surface polaritons. Zero damping has been assumed in all three cases and only the real part of K is shown.

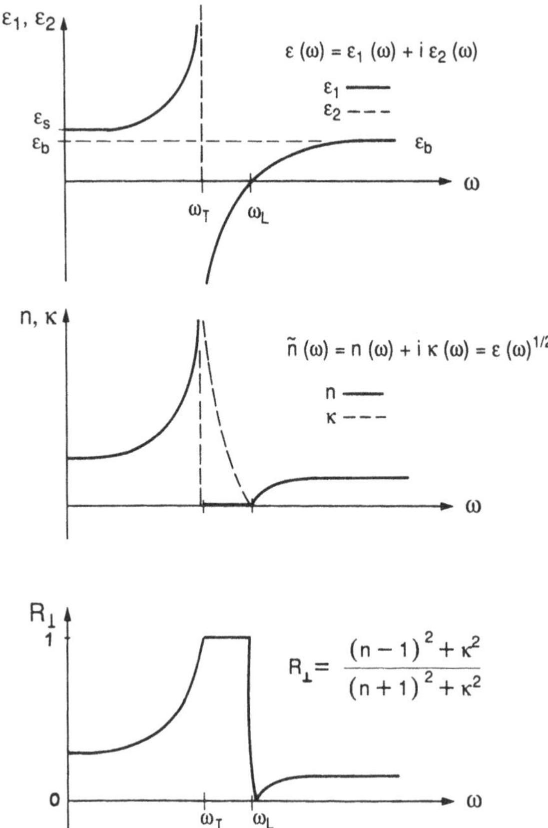

Fig 17 The complex dielectric function $\epsilon(\omega)$, the complex index of refraction $n(\omega)$ and the reflection spectrum in the vicinity of a resonance for zero damping ($\gamma = 0$) and without spatial dispersion (schematic).

Finally we give the dispersion of surface polaritons. These are elementary excitations which can propagate only along the interface between two different media (one is usually chosen to be vacuum). Their eigenfrequencies are situated between the longitudinal and the transverse ones and their dispersion relations exclude the decay into the bulk of the SC and into vacuum, due to the conservation of the component of $\vec{K}$ parallel to the interface. For more information on surface polaritons see e.g. [17] and references therein.

To conclude this introductory section we give in Fig 17 the complex dielectric function $\epsilon(\omega) = \epsilon_1(\omega) + i\epsilon_2(\omega)$ neglecting damping ($\gamma = 0$) and spatial dispersion ($\omega_0 \neq \omega_0(\vec{K})$), the complex index of refraction $\tilde{n}(\omega)$

$$\tilde{n}(\omega) = \sqrt{\epsilon(\omega)} \tag{53}$$

and the reflection spectrum for normal incidence

$$R_{\perp} = \frac{(n-1)^2+\kappa^2}{(n+1)^2+\kappa^2} \tag{54}$$

which shows for the above assumptions nicely the so–called Reststrahlenbande or stop–band between $\hbar\omega_0$ and $\hbar\omega_L$. It should be mentioned, that the results of Fig 17 follow already from the classical oscillator model [15].

The "recipie" for the translation between Fig 17b and Fig 16a is to exchange ordinate and abszissa, to replace

$$\omega <=> \hbar\omega \tag{55a}$$

and

$$\mathrm{Re}\,\{\vec{K}\} <=> 2\,\frac{\pi\omega}{c}\,\mathrm{Re}\,\{\tilde{n}(\omega)\} = 2\,\frac{\pi\omega}{c}\,n(\omega) \tag{55b}$$

In section VI we shall come back to the exciton polariton and methods to measure its dispersion relation.

II. LINEAR OPTICAL PROPERTIES OF SEMICONDUCTORS

After the first introductory part, in which we outlined the main elementary excitations in semiconductors and their coupling to the radiation field, we present in this section examples, starting with bulk – or three dimensional semiconductors.

II.A. Bulk material

The first example which we consider is the reflection spectrum of CdS (Fig 18) in the spectral region around the optical phonon resonances i.e. in the IR. The reflectivity in the Reststrahlenbande reaches values close to unity and can be nicely fitted e.g. by the model of Fig 17c if a small damping is introduced. Effects of spatial dispersion are usually negligible for optical phonons due to the almost horizontal dispersion relation in the region of K up to 10^5 cm^{-1}.

The dispersion relation of the phonon–polariton can be nicely measured in Raman–scattering [19]. The wave–vector of the phonon polariton $\vec{K}_{pp}$ is given by (56). In a backward geometry momentum conservation gives for the created phonon polariton (pp)

$$\vec{K}_{pp} = \vec{K}_i - \vec{K}_{out} \tag{56}$$

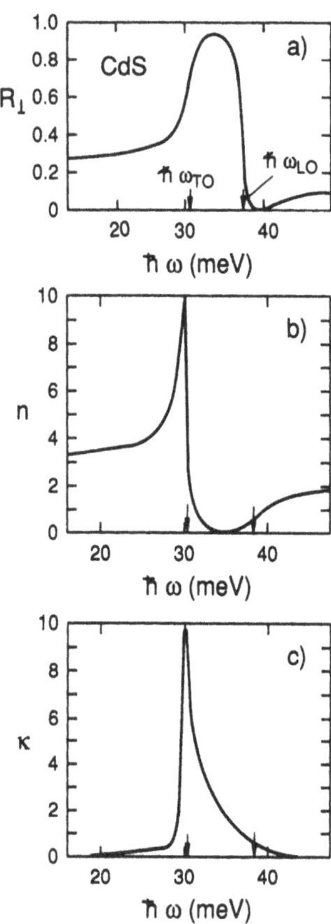

Fig 18 Spectra of reflection, and of the real and imaginary parts of the complex index of refraction for CdS around the optical phonon resonances. According to [4, 18] with some modifications.

For visible light ($\lambda = 0.5\mu m$) and a refractive index of 2, $\vec{K}_{ph} \approx 5 \cdot 10^5\ cm^{-1}$ and this is clearly on the phonon–like part of the polariton dispersion. In a forward geometry, however, $\vec{K}$ can be tuned from zero to larger values, changing the angle in the inset of Fig 19. The Raman shift gives the energy and with (56) the wave vector of the phonon polariton. Beautiful experiments have been performed in [19, 20] based on this concept to measure the dispersion of phonon polaritons. We give in Fig 19 an example for GaP.

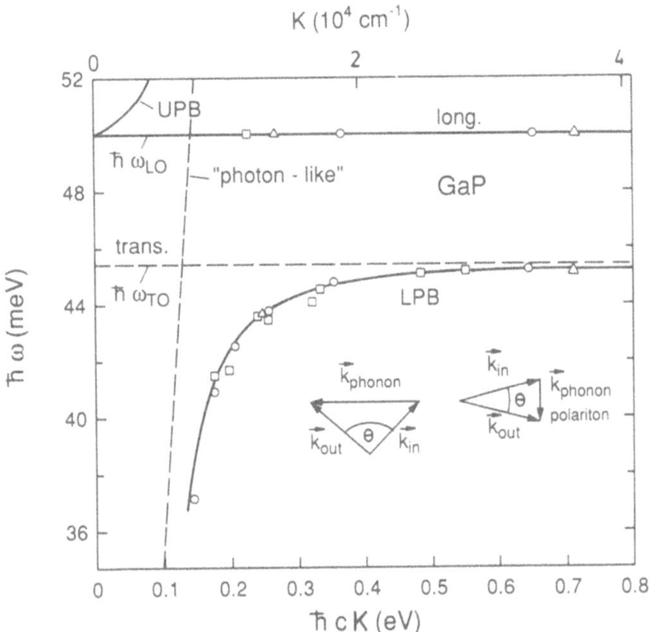

Fig 19 Dispersion of the phonon–polariton in GaP determined by angle resolved Raman scattering. According to [20].

Our next example are plasmon polaritons. As already mentioned, plasmons and polaritons exist only in strongly doped or highly excited SC. In Fig 20 from [21] we show reflection spectra of InSb samples with various n – doping levels. Since the transverse eigenfrequency at K = 0 is zero for plasmons as mentioned above, the Reststrahlenbande extends from zero frequency up to ω_{pl}. The reflection minimum coincides according to (54) with $n(\omega) = 1$ and is therefore situated only slightly above the longitudinal eigenfrequency. The shift of the reflection minimum demonstrates nicely the $n^{1/2}$ dependence of ω_{pl} in eq (19).

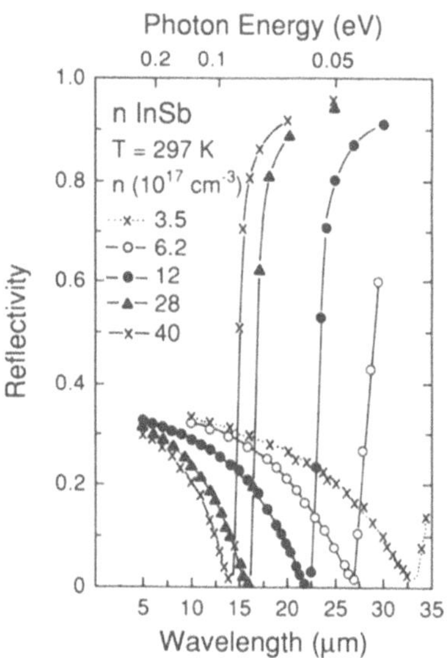

Fig 20 Reflection spectra in the IR around the plasma resonance of various n–doped InSb samples. According to [21].

Our last example for a three-dimensional system concerns the exciton polariton in CdS. In Fig 21a we give the bandstructure in the vicinity of the Γ point (compare also Fig 13c). The conduction band comes mainly from the 5s states of Cd^{++}. It is isotropic and parabolic around the Γ point. The valence band comes mainly from the 3p states of S^{--} and is split into three subbands by spin – orbit coupling and the hexagonal crystal field. These bands have symmetries Γ_9, Γ_7 and Γ_7 and are usually labelled A, B and C valenceband from higher to lower energies. The band-to-band transitions from the Γ_7 valence bands to the Γ_7 conduction band are dipole-allowed for both polarizations $\vec{E} \perp \vec{c}$ and $\vec{E} \parallel \vec{c}$ and from the Γ_9 band only for $\vec{E} \perp \vec{c}$ [22]. The valence-bands show some anisotropy concerning their effective masses for the directions parallel and perpendicular to the crystall-graphic $\vec{c}$ axis. In addition, Γ_7 bands may have from group – theoretical considerations a $\vec{k}$-linear term separating the two spin orientations for $\vec{k}_{\perp c} \neq 0$. Three exciton-series result from this bandstructure, known as A, B and C excitons. In Fig 21b we show the dispersion relation of the n = 1 $A\Gamma_5$ and $B\Gamma_5$ exciton polaritons for the orientation $\vec{E}\perp\vec{c}$ and $\vec{K}\perp\vec{c}$ (solid lines), the exciton states which couple to the light field (dashed lines) and the longitudinal and the $A\Gamma_6$ triplett excitons, which do not strongly couple to the radiation field (dash-dotted lines). The $\vec{k}$-linear term mixes the $B\Gamma_1$ which is usually dipole-allowed only for $\vec{E}\|\vec{c}$ and the triplet $B\Gamma_2$ state with the $B\Gamma_5$ level for $\vec{K} \neq 0$ giving them some oscillator strength. This results in the additional polariton branch.

Fig 21c finally gives the reflection spectrum in the same energy range. We see clearly, that the A excitons show up only for $\vec{E}\perp\vec{c}$ in contrast to the B-excitons. The oscillator strength decreases with n_B^{-3} making states $n_B > 3$ hardly visible in reflection spectra.

The fact that R does not reach 1 but only values around 0.4 in the reststrahlenbande is not an indication of considerable damping, but of the spatial dispersion $\omega_0(\vec{K})$ which gives one or even more propagating modes for all frequencies including the range between $\hbar\omega_o$ and $\hbar\omega_L$ as can be seen from Fig 21b. More details concerning the quantitative evaluation of reflection spectra of excitonic polaritons are given e.g. in [8, 21] or in section VI.1.

II.B. Reduced dimensionalities

A topic which attracts in the last decade increasing interest is the investigation of systems of reduced dimensionality like quasi-twodimensional quantum wells (QW), quasi-onedimensional quantum-wires and quasi-zerodimensional quantum dots (QD).

We start to outline first some general features concerning especially the density of states and give then examples for excitons in quantum wells and quantum dots.

Since all the elementary excitations considered so far have plane wave character, we outline the density of states for plane waves in a d-dimensional cube of length L.

The usual approach for normalization of a plane wave in a box assume either standing waves with nodes at the boundary equally spaced for every direction

$$K_{i,n+1} - K_{i,n} = \frac{\pi}{L};\ i = x, y, z;\ n_i = 1, 2, 3... \qquad (57a)$$

or periodic boundary conditions resulting in

$$K_{i,n+1} - K_{i,n} = \frac{2\pi}{L};\ i = x,y,z;\ n_i = \pm 1, \pm 2, \pm 3...; \qquad (57b)$$

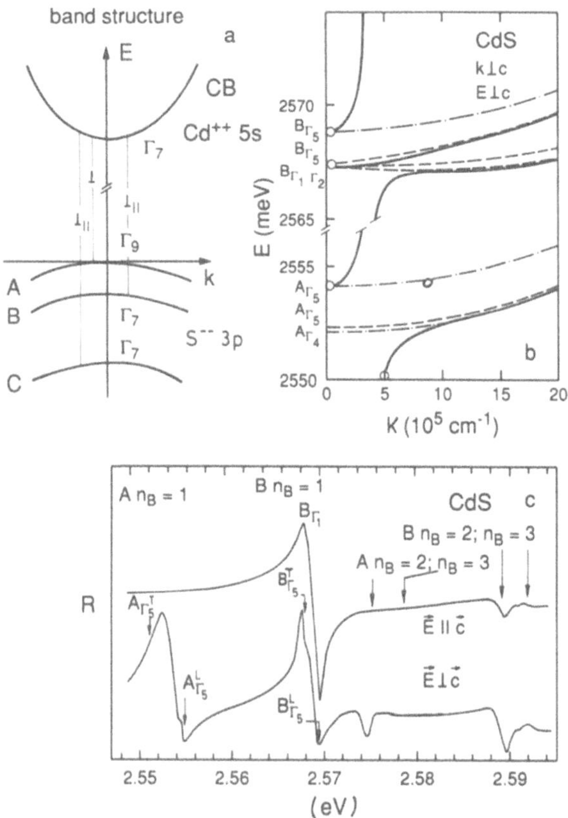

Fig 21 The band structure of CdS (schematic) (a), the calculated polariton dispersion in the n = 1 resonances of the A and B exciton series, together with experimental data (b) and the reflection spectra for the orientations E ⊥ c and E ∥ c in CdS. According to [22].

In both cases one finds, that the volume per state in $\vec{K}$–space V_k is proportional to

$$V_k \sim \left(\frac{\pi}{L}\right)^d \tag{58}$$

where d is the dimensionality of the system.

Considering the density of states D in $\vec{K}$–space in polar coordinates we get

$$D(K)dK \sim K^{d-1}dK \tag{59}$$

For many purposes, e.g. in Fermies golden rule [14], one needs the density of states on the energy axis.

While eqs (57) to (59) are generally valid for all plane waves independent if they describe phonons, electrons or excitons, the special dispersion relation enters in the density of states as a function of energy according to

$$D(E)\, dE = D(\vec{K}(E))\, \frac{dK}{dE}\, dE \tag{60}$$

We use here and in the following as already mentioned the effective mass approximation i.e. a parabolic dispersion relation

$$E(k) = E_0 + \frac{\hbar^2 \vec{K}^2}{2m} \tag{61}$$

and get

$$D(E) \sim (E - E_0)^{\frac{d}{2} - 1} = \begin{cases} (E - E_0)^{1/2} & d = 3 \quad (62a) \\ (E - E_0)^0 & d = 2 \quad (62b) \\ (E - E_0)^{-1/2} & d = 1 \quad (62c) \end{cases}$$

for $E \geq E_0$

We illustrate these dependences in Fig 22 for d = 3, 2, 1 and 0. In the latter case we get basically a set of δ – functions.

The $E^{-1/2}$ dependence of the one dimensional density of states is also known for Landau–levels in magnetic fields [2], where the motion of charged particles is quantized in the direction perpendicular to $\vec{B}$ and free only for $\vec{K} \parallel \vec{B}$. The difference is only, that the Landau levels are equally spaced with

$$\hbar\omega_{c,n} = \hbar\, \frac{eB}{m}\, (n_L + \frac{1}{2}) \tag{63}$$

while the energies in quantum – wires follow rather a n^2 law.

Disorder which occurs e.g. in alloy semiconductors or amorphous materials leads often to localized states forming an exponential tail in the density of states, separated by a more or less sharp mobility edge from the extended states. We indicate this tail schematically in Fig 22 for d = 3. More details of this topic are found e.g. in [23] and references therein.

Now we want to discuss how systems of reduced dimensionality can be realized.

Usually one sandwiches a material with low energy gap between a material with a larger one.

In Fig 23 we give the lattice constants and the energy gaps of various semiconductors. The lines connecting them represent alloy semiconductors e.g. of the type $Ga_{1-y}In_yAs$. Solid lines indicate direct gap materials, dashed lines indirect ones.

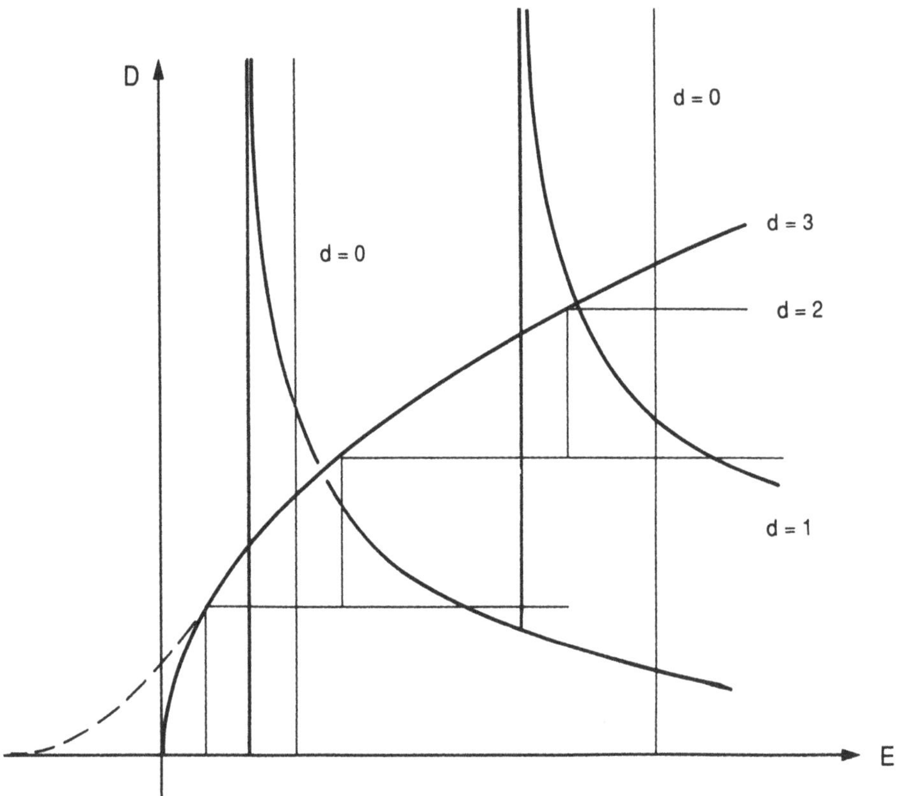

Fig 22 The density of states as a function of energy for effective mass particles in 3, 2, 1 and 0–dimensional systems. The dashed lines give the exponential tail of localized states appearing in (weakly) disordered systems.

Usually one prefers to grow hetero–interfaces between materials of approximately equal lattice constants like the system $GaAs/Al_{1-y}Ga_yAs$ and $InP/Al_{1-y}In_yAs/Ga_{1-y}In_yAs$. In the other case one gets stained layers which have with increasing thickness a tendency to form unwanted dislocations to relax the lattice misfit (e.g. ZnS/ZnSe).

The next crucial point is the relative alignment of the bands. We consider in the following only type I superlattices, where electrons and holes are confined in the same material.

In Fig 24a we show the first two quantized states in z–direction. A percularity of many cubic (O_h and T_d) SC is the fourfold degenerate upper valence–band state of symmetry $\Gamma_8^{(+)}$ which splits for $\vec{k} \neq 0$ into a heavy and light hole band of different curvature. Consequently one gets in a QW two series of quantized hole states the heavy (hh) and light (lh) holes, respectively. In the x–y plane the carriers and the excitons move as free quasiparticles.

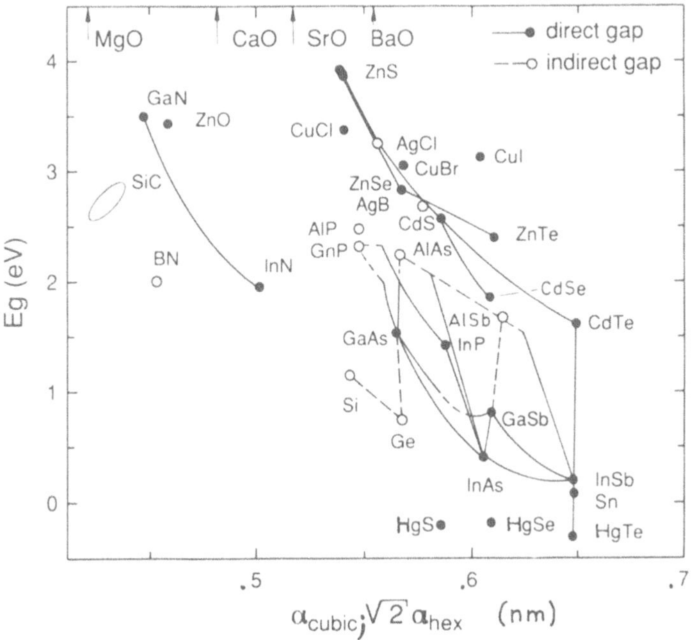

Fig 23 The energy gap of various semiconductors as a function of the lattice constant. According to data from [4].

In Fig 24b we see the resulting absorption spectrum. The plateaux correspond to the band–to–band transitions with the simple selection rule

$$\Delta n_z = n_z^e - n_z^n = 0 \tag{64}$$

reflecting the 2d constant density of states in eq (62b), since the Sommerfeld – enhancement is not very pronounced in two dimensions. Aditionally we see for every transition the exciton – resonances. Usually, one observes only the $n_B = 1$ states, since the separation of the $n_B \geq 2$ states from the continuum is relatively smaller for 2d than for 3d systems, eqs (30, 36). Additionally there are broadening processes, which are typical for quantum wells, especially well width fluctuations.

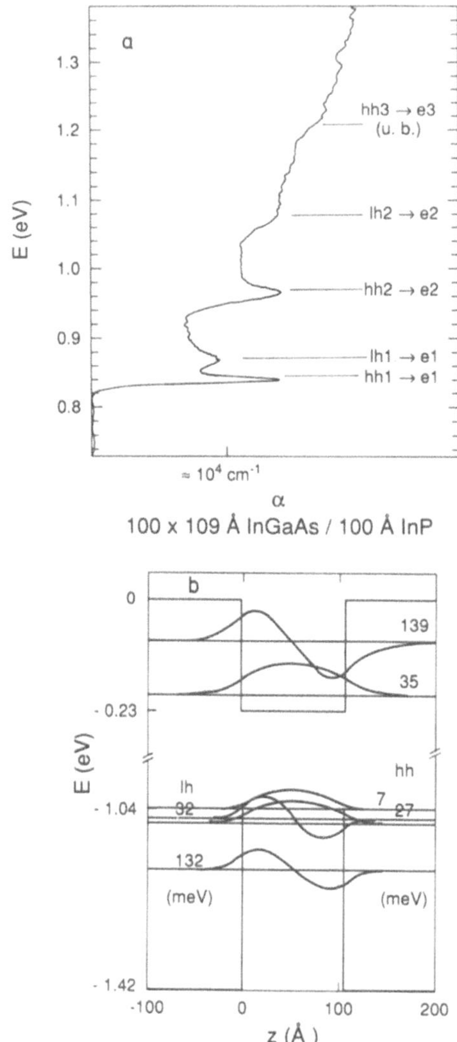

Fig 24 The first two quantized eigenstates calculated for electrons, heavy– and light holes in InGaAs/InP (a) and the observed absorption spectrum (b). According to [24, 25].

The quantization energy $E_Q(n_z)$ for an effective mass particle is for the idealized situation of infinitely high barriers given by

$$E_Q(n_z) = \frac{\hbar^2 \pi^2}{l_z{}^2 m} n_z{}^2 \tag{65}$$

The relative variation ΔE_Q for $n_z = 1$ grows then inversely proportional to l_z

$$\frac{\Delta E_Q}{E_Q} = 2 \frac{\Delta l_z}{l_z} \tag{66}$$

For typical values of $l_z = 10$ nm and $\Delta l_z = 0.3$ nm and $m_e = 0.1$ m, we get

$$E_Q \approx 35 \text{ meV and } \Delta E_Q \approx 2 \text{ meV} \tag{67}$$

Connected with the finite barrier height in real systems there is only a finite number of quantized states, e.g. for the electrons in Fig 24a only two. Consequently the optical properties go for higher energies gradually over to a three – dimensional behaviour. For the same reason the limit $l_z \Rightarrow 0$ leads to the optical properties of the barrier material.

More details about quantum wells are found e.g. in some recent reviews [26] and references therein.

Quantum wires can be fabricated e.g. by microstructuring of quantum wells. This technique has still some problems e.g. the different dimensions in the two quantization directions ($l_z \approx 10$ nm, $l_y \approx 150$ nm). We do not go into this field and refer the reader e.g. to [26, 27].

Similar arguments hold partly for quantum–dots produced by microstructuring. Beautiful experiments can be nevertheless performed in them concerning e.g. one electron states in magnetic fields [27].

Alternative ways to produce quantum dots are their growth by a diffusion limited process during annealing of semiconductor doped glasses or by chemical precipitation [28].

Though the dots produced by these techniques are not necessarily spherical, they have roughly equal diameters in all three dimensions.

Systems which are widely investigated in this field are micro crystals (sometimes also called nano–crystals or mesoscopic systems) of Cu – halides and especially the II–VI compounds $CdS_{1-x}Se_x$ (including $x = 0$ and $x = 1$) and of CdTe.

In Fig 25 we show the absorption spectra of $CdS_{1-x}Se_x$ QD produced from the same glass melt, by annealing for the same time but at different temperatures. The average radius of the dots R increases both with annealing time and temperature.

The blue shift with decreasing radius is obvious.

In simpliest approximation (i.e.) using effective mass particles in an infinite, sherical well and neglecting the Coulomb–energy one finds for the lowest electron hole pair state in a QD an energetic shift ΔE_{QD} [28a].

$$\Delta E_{QD} = C \, Ry^* \, (a_B \, \pi/\bar{R})^2 \tag{68}$$

where C is from theoretical consideration 0.67 while a fit to experimental data reveals values ranging from .3 to 10 [30]. The origine of this discrepancy is not fully

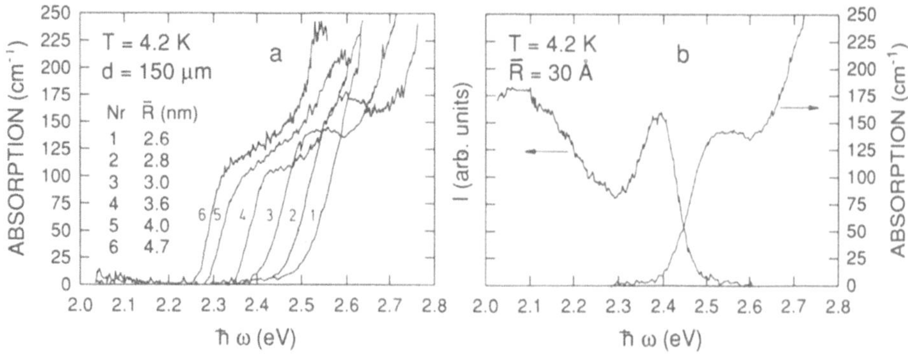

Fig 25 Absorption spectra of $CdS_{1-x}Se_x$ doped glasses with different crystallite radii (a) the comparison of absorption and emission of one sample (b). From [29].

understood and might be connected with uncertainties of the determination of $\bar{R}$ by X–ray scattering and with deviation from a precisely sherical shape. Though first and partly second quantized levels can be easily seen, the structures of Fig 25 deviate significantly from a series of δ – functions expected from the density of states in Fig 21. There are various arguments, which can explain this discrepancy:

Due to the growth process there is always a certain distribution of R values around $\bar{R}$. Theoretical considerations [31] indicate that minimal values of $\Delta R/\bar{R}$ around .1 have to be expected. Via (68) every R value leads to a different energetic position. For details see [32]. This phenomen can, however, not explain the Stokes shift between emission and absorption seen in Fig 25b around 2.5 eV (The broad emission peaking at 2.1 eV is connected with some deep centers). Such shifts are well known in the spectroscopy of ions in a polar matrix and can be explained by a coupling to the optical phonons. Recent investigations [28h, 30, 33] indicate that the Huang–Rhy factor describing this coupling [34] is around 1 to 2 for the systems under consideration here, while it is usually below 1 for bulk semiconductors. A third contribution to the broadening, which also enhances to the Stokes shift is the participation of impurity – or interface – states into which the electron–hole pairs may relax after excitation [30, 35] eventually also combined with a lattice relaxation. Lively research is presently going on in various groups to clarify these points [28k].

With this example we finish the presentation of the linear optical properties of semiconductors and proceed to effects which involve optical nonlinearities.

III. NONLINEAR OPTICS

In preceeding sections we considered the elementary excitations of SC and the linear optical properties connected with them. Linear optics means that the polarization of a medium is a linear response function of the incident electric field amplitude. Consequences are that the optical properties $\epsilon(\omega, \vec{K})$ or $\tilde{n}(\omega)$ are independent on the light intensity I (I is the energy flux density of a light beam or the Poynting vector $\vec{S} = \vec{E} \times \vec{H}$ averaged over several periodes) and that light beams can cross in matter without interaction. The term "nonlinear optics" comprises consequently all phenomena in which the optical properties change in a reversible way under illumination.

There are two concepts to describe optical nonlinearities. When the changes of the dielectric function or of the susceptibility χ

$$\chi = \epsilon - 1 \tag{69}$$

follow instantoneaously the incident fields $\vec{E}_i$, then the nonlinearities can be described by a dependence of χ on $\vec{E}_i$. Usually an expansion of this dependence in a power series is used

$$\frac{1}{\epsilon_0}\vec{p}_i = \chi_{ij}^{(1)}\vec{E}_j + \chi_{ijk}^{(2)}\vec{E}_j\vec{E}_k + \chi_{ijkl}^{(3)}\vec{E}_j\vec{E}_k\vec{E}_l + \ldots \tag{70}$$

Such an expansion is useful e.g. for coherent processes with short lifetime such as virtually excited states.

The first term on the r.h.s. of (70) describes linear optics as mentioned above, the second one effects like second harmonic, sum and difference frequency generation, since the product $\vec{E}_j\vec{E}_k$ contains terms like exp $\{i(\omega_j \pm \omega_k)t\}$ and the third one e.g. four wave mixing.

If the frequencies of the various $\vec{E}_i$ are equal an expansion of $\tilde{n}$ in a power series of the intensities is possible resulting in

$$\tilde{n} = \tilde{n}_0 + \tilde{n}_2 I + \ldots \tag{71}$$

The approach according to eqs (70) and (71) breaks down, if the optical nonlinearity is due to an incoherent population of some excited species (e.g. excitons or phonons) with a finite life time T_1. Changes of ϵ can persist for a time T_1 even after the laser i.e. the E_i in (70) have been switched off. A situation, which is clearly not covered by this approach. The appropriate way to describe such situations is to assume that χ depends on the density n_p of excited species present at a time t

$$\chi = \chi(\omega, n_p(t)) \tag{72}$$

n_p in turn is given by the integral over the generation rate G(t') in the past, weighted by some decay function e.g. an exponential

$$n_p(t) = \int_{-\infty}^{t} G(t')e^{-t/T_1}\,dt' \tag{73}$$

the generation rate G is given for one– and two– photon excitation by

$$G(t') = \alpha(\hbar\omega_{exc})I_{exc}(t') + \beta(\hbar\omega_{exc})I^2_{exc}(t') + \ldots \tag{74}$$

where the index exc refers to the exciting beam.

Eventually the quantities T_1, α and β in (73) and (74) depend themselves on n_p, leading them to a rather complex set of coupled integrodifferential equations. We shall see various examples of both cases in the following.

Consequences of the nonlinearities are among others interaction processes between light beams in matter. This phenomenon can be used in turn to measure the optical nonlinearities.

One method is the so–called pump–and–probe beam technique. One measures the transmission or reflection spectrum of a sample with a weak probe–beam, once without and once with additional pump beam. Changes in the transmission or reflection spectra with pump allow to investigate optical nonlinearites.

Another widely used technique is the spectroscopy with laser induced gratings (LIG). It has many variants. The basic idea is the following: two coherent beams 1 and 2 interfere in a certain spatial region. In the interference pattern one gets a periodic spatial modulation of the intensity. If a sample is placed in this area, one gets a periodic modulation of its properties if there is any optical nonlinearity, resulting in a phase and/or an amplitude grating depending whether the excitation–induced changes of the real and/or of the imaginary part of $\tilde{n}$ dominate.

Such a grating diffracts light, and we assume in the following, that we are in the regime of thin gratings (Raman – Nath regime) defined in the simpliest case by

$$2\pi\alpha\lambda/n_1 << \Lambda^2 \tag{75}$$

where d is the thickness of the sample, λ the wavelength of the light and Λ the grating period, which depends on λ and the angle Θ between beams 1 and 2 according to

$$\Lambda = \lambda(2 \sin \Theta/2)^{-1} \tag{76}$$

In the simpliest case, the two beams forming the grating are diffracted from it in forward or backward configuration. This self–diffraction gives information about the self–renormalization of the optical properties. Depending on the type of the grating and of the shape of the transmission – function there can be first or higher diffracted orders. The first orders go into the directions $2\vec{K}_1 - \vec{K}_2$ or $2\vec{K}_2 - \vec{K}_1$ and are also known as degenerate fow wave mixing (DFWM). If one reads the grating with a beam with different frequency ω_{probe} one can learn something about the change of e.g. $\tilde{n}$ at ω_{probe} induced by excitation at ω_{exc}.

If finally the beams 1 and 2 have different frequencies ω_{exc1} and ω_{exc2} but are still coherent, one gets a grating moving laterally with a velocity

$$v_{grating} = \frac{|\omega_1 - \omega_2|}{|\vec{K}_1 - \vec{K}_2|} \tag{77}$$

The orders diffracted from such a moving grating are Doppler shifted in frequency by integer multiples of $\omega_1 - \omega_2$. This phenomenon is known as non – degenerate four wave mixing (NDFWM). We shall get examples of the various above mentioned methods in the following.

To conclude this section, we give a first rather simple example for an optical nonlinearity, namely a collision broadening of the exciton resonances. In Fig 26 we show schematically the absorption spectrum of a SC in the exciton region at low temperature. The first three exciton resonances are clearly resolved. If we pump the sample and create excitons at high densities they will scatter with each other resulting in a collision broadening of the resonances and we can clearly see that there are spectral regions with excitation – induced increase or decrease (bleaching) of absorption. Furthermore a change of the absorption spectrum will result via Kramers – Kronig relations in a change of the real part of n. We shall get in the next section examples for such process.

More details about the theoretical description and experimental results of optical nonlinearities are found e.g. in [5, 7, 9, 22, 25, 26, 36].

IV. THE FATE OF AN OPTICAL EXCITATION IN A SEMICONDUCTOR

The first aspect of excited state spectroscopy, which we want to follow in some detail in this section, is the fate of an optically excited electron–hole pair, usually an exciton, in a SC.

As is presently still commonly done in this field, we use the weak coupling approach, i.e. we consider the electro–magnetic field of an incident laser–pulse $\vec{E}$ and the polarization $\vec{P}$ which it produces in the medium as two independent quantities, neglecting for the moment the polariton concept which describes a new entitity formed from $\vec{P}$ and $\vec{E}$, but we shall come back to this topic at the end of this section.

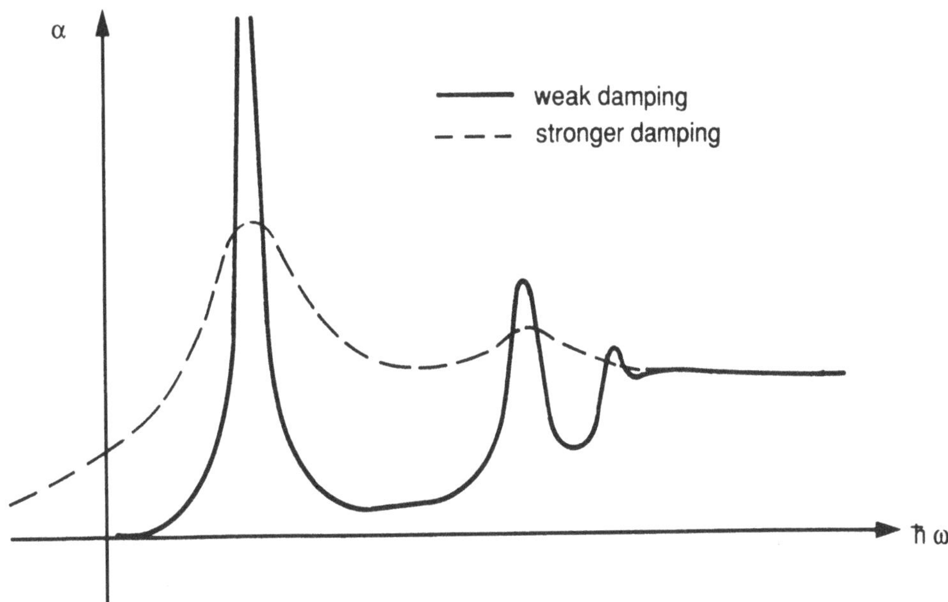

Fig 26 Schematic drawing of the changes of an excitonic absorption spectrum with increasing damping, caused e.g. by collision broadening.

First we define various time–constants with the help of Fig 27. We create with a short laser pulse an exciton, here in the continuum states.

The first scattering process destroys the coherence of the polarization wave with the exciting pulse. The characteristic time is usually called phase–relaxation time T_2 and we can write for the coherent part of the polarization $\vec{P}$

$$\vec{P} = \vec{P}_0 \exp(-t/T_2) \tag{78}$$

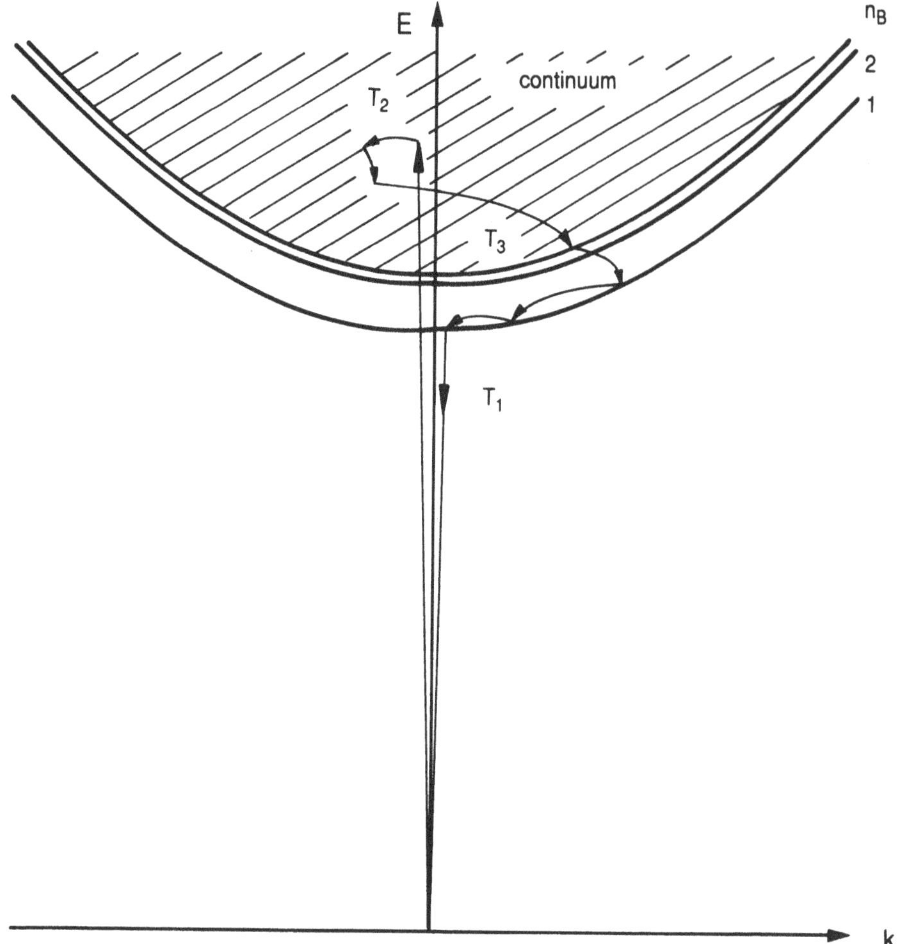

Fig 27 Schematic drawing of various relaxation and recomnination processes of an exciton after optical creation, introducing the phase–relaxation time T_2, the intraband relaxation time T_3 and the interband recombination time T_1.

The scattering process may or may not be connected with a change in energy or in the amount or the direction of $\vec{K}$. It is only necessary that the phase is disturbed.

A second group of processes is the energetic intrabandrelaxation e.g. by emission of optical and acoustic phonons. This relaxation is described sometimes by a time T_3. If one excites resonantly in the ground state, there is not much reason to introduce T_3, but still there are processes which destroy the phase.

Finally the population in the excited state decays by radiative or radiationless interband recombination. This decay is usually characterized by T_1.

T_1 and T_2 are also called longitudinal and transversal relaxation times, respectively. If the interband recombination is the first process which destroys the phase, we get an upper limit at T_2 by comparing eq. (78) with the decay of the population N

$$N = N_0 e^{-t/T_1} \quad (79)$$

and bearing in mind that the population is proportional to the amplitude squared

$$N = N_0 e^{-t/T_1} \sim P^2 \sim P_0 e^{\frac{t^2}{T_2}} \quad (80)$$

resulting in

$$T_2 \lesssim 2T_1 \quad (81)$$

T_2 is closely connected with the damping γ in eq (50) and often with the homogenous line – width ΔE_h

$$\gamma \cdot T_2 = 1; \ \Delta E_h = \hbar T_2^{-1} \quad (82)$$

In principle T_2 can therefore be determined from the homogenous width of a transition. However, most of the optical transitions in SC are inhomogenously broadened and γ or T_2 have to be deduced in a rather indirect way, e.g. from fitting of excitonic reflection spectra like the ones in Fig 21, or from spectral hole–burning experiments which are treated e.g. in [37g, 38].

More recently the spectroscopy in the ps regime allowed to measure T_2 directly in the time domain using four –wave mixing or photon–echo techniques.

The mathematical description of these processes via the Bloch equations is given in great detail in another contribution to this school [37g] or in [37]. So we restrict ourselves to an intuitive description, considering first homogenously broadened systems, where all oscillators have exactly the same eigenfrequency ω_0 and the finite width comes only from the finite phase–relaxation time T_2 and then inhomogenously broadened systems where the eigenfrequencies of the resonators cover a certain frequency range $\Delta\omega_{inh}$ around ω_0.

We start with the homogenous situation. An incident pulse with wavevector $\vec{K}_1$ (Fig 28b), produces a polarisation in the medium, the coherent part of which decays with T_2. During the decay the polarization radiates according to the well known law

$$I_{rad} \sim |\vec{P}|^2 \quad (83)$$

This radiation is known as free induction decay. Usually it has only a minor influence on T_2 which is generally limited by scattering processes.

After a delay τ a second pulse arrives which is coherent with the first one and has the same frequency ω and wavevector $\vec{K}_2$. It interferes with the coherent part of the polarization, left from the first pulse. This is described by the off–diagonal elements ρ_{12} in the Bloch equations using the density matrix formulation, which depend on the product of a dipole operator μ_{12} creating the polarization and $\vec{E}$. The interference produces a LIG which radiates its first order in the directions

$$2\vec{K}_2 - \vec{K}_1 \quad (84)$$

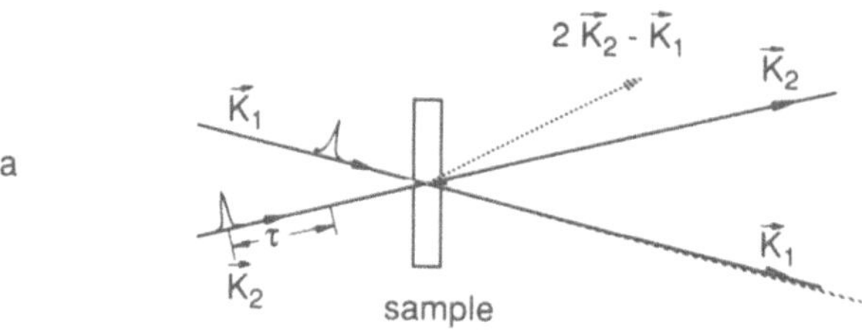

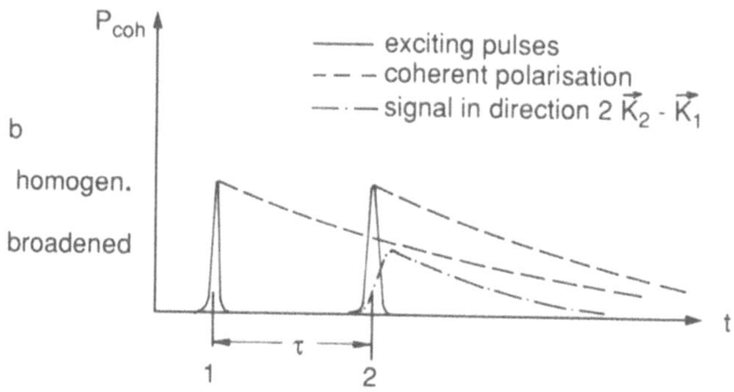

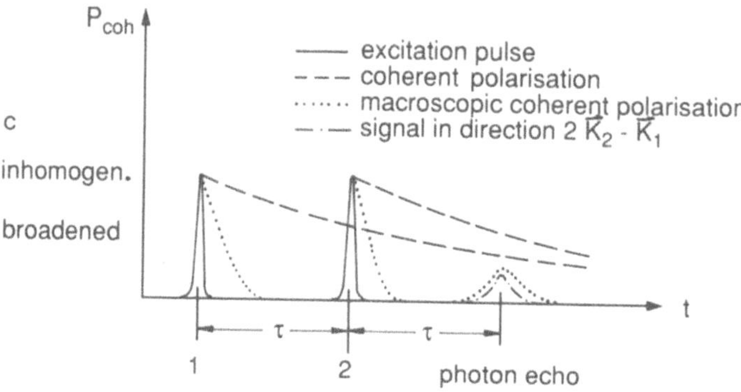

Fig 28 The setup for a degenerate four wave mixing experiment (a). The temporal evaluation of the coherent part of the polarization in a homogeneously broadened system (b) and in an inhomogeneous one (c), where the dotted line describes the dynamics of the macroscopic polarisation.

and $\vec{K}_1$. The latter signal is difficult to detect, so one usually observes the one given by eq (84). Since all oscillators have the same frequency, their coherent parts are all in phase and the diffracted intensity I_s starts to develop instantaneously with the arrival of the second pulse and decays according to

$$I_s^{hom} \sim \vec{P}^2 \sim \exp^2(-\tau/T_2) \sim \exp(2\tau/T_2) \tag{85}$$

The same relation holds for the time integrated signal. See Fig 28b.

For the inhomogenously broadened case of Fig 28c, the phase of every oscillator decays again with T_2. But since the oscillators have different eigenfrequencies, they run out of phase with respect to each other in a shorter time proportional to $(\Delta\omega_{inh})^{-1}$ resulting essentially in destructive interference of the radiation of every oscillator.

The emission due to the free induction decay is consequently also limited by this quantity (dotted line in Fig 28c). When the second pulse arrives after τ, a rephasing of the oscillators starts, and after another time τ they are in phase again radiating the so–called photon–echo. Its temporal width is limited by the free induction decay. The signal intensity is given in this case by

$$I_s^{inhom} \sim P^2(T = 2\tau) \sim \exp^2(-2\tau/T_2) = \exp(-4\tau/T_2) \tag{86}$$

The appearance of an echo as shown in Fig 29 is therefore a clear indication of an inhomogenous broadening. A pecularity at the description on this level is the fact, that no diffracted order occurs into the direction $2\vec{K}_2 - \vec{K}_1$ of eq (84) for negative delays i.e. if pulse 2 arrives before pulse 1.

Now we want to present examples of photon echo experiments for various excitations in semiconductors.

We start with free excitons in three–dimensional semiconductors. In Fig 30a we give the time–integrated photon echo signal measured in the $A\Gamma_5$ (n = 1) resonance. Since the sample is opaque in this range, the reflected signal has been used as shown in the inset. The signal increases with increasing incident intensity but the T_2 time which is deduced from the fit decreases with increasing exciton density. The same behaviour can be seen in Fig 30b, where the photon echo was measured at the transition energy between exciton and the biexciton.

A biexciton or excitonic molecule is another quasi–particle in a SC which consists of two excitons bound together with a binding energy E^b_{biex} in analogy to the H_2 or the positronium molecule. The dispersion relation of the biexciton is given by

$$E_{biex}(\vec{K}) = 2E_x(\vec{K}_x = 0, n_B = 1) - E^b_{biex} + \frac{\hbar^2\vec{K}^2}{4M} \tag{87}$$

The one photon transition from the crystal ground state to the biexciton is forbidden, but it has a high oscillator strength for two–photon transitions i.e. for

$$2\hbar\omega_{exc} = E_{biex} \tag{88}$$

or for two step transition converting an exciton with a photon into a biexciton

$$E_x + \hbar\omega = E_{biex} \tag{89}$$

This second possibility has been used in Fig 30b, where a prepulse with a temporal distance larger than T_2 produced a well defined density of excitons.

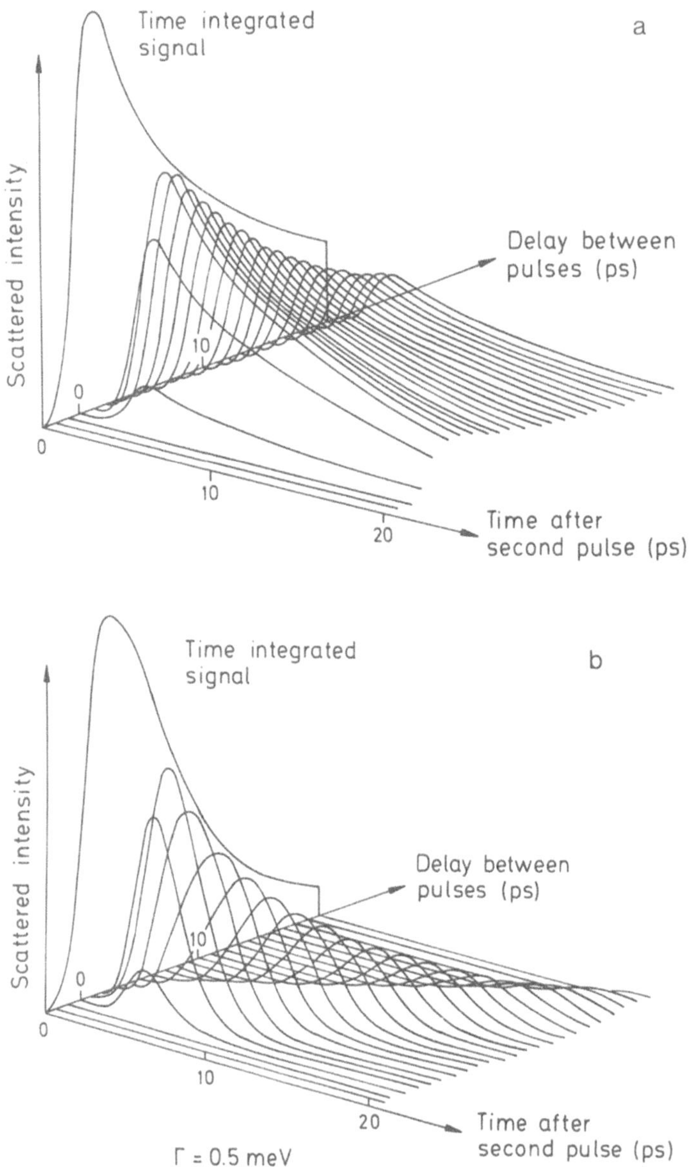

Fig 29 The calculated signal intensity of a two–beam DFWM experiment for a homogeneously (a) and an inhomogeneously broadened resonance (b). Parameters used in the calculation: $\tau_{pulse} = 1.4$ ps, $T_1 = T_2 = 20$ ps; $\Gamma_{inhom} = 0.45$ meV. From [39].

In Fig 30c one sees finally the dephasing rate T_2^{-1} as a function of the exciton density at low temperature. The increase of T_2^{-1} with increasing exciton density is just an example of the collision broadening shown schematically in Fig 26. The slope gives a scattering cross section for exciton–exciton scattering.

$$\sigma_{ex\text{-}ex} = 6.8\ \pi\ a_B^2 \tag{90}$$

i.e. slightly larger than the geometric value.

With increasing temperature, T_2 of the exciton decreases, too, due to scattering with phonons reaching at 300 K values of the order of several 100 fs.

The extrapolation of T_2 to zero temperature and density gives values of T_2 around 50 ps in high quality samples. The value is limited e.g. by scattering of free excitons with impurities or by capture into some traps and is much smaller than the life time T_1, which is about 1ns.

Photon echo measurements of the exciton resonance in GaAs by [40] gave at low temperature comparable values for T_2 around 15 ps.

Now we want to outline some other techniques to determine T_2. In Fig 31 from [41] we show the diffraction efficiency as a function of the inverse grating period Λ^{-1} (see eq (76)). Quasistationary excitation conditions with pulses of several ns duration were used. The excitation was slightly above the exciton resonance and the phase – grating was read by a beam in the transparent region below. The decay of the efficiency with increasing Λ^{-1} is due to diffusion. A fit gave as diffusion length l_D at T = 40 K of 0.78 μm and at 5 K of 1.4 μm. With a life – time T_1 of the excitons $\lesssim$ 1 ns one gets via

$$l_D = (DT_1)^{1/2} \tag{91}$$

a value of the diffusion constant D $\approx$ 20 $cm^2\ s^{-1}$. With the relations between the mobility μ, D and T_2

$$\mu = e\frac{T_2}{M} \text{ and } D = \mu\frac{k_BT}{e} \tag{92}$$

we get

$$D = \frac{k_BT}{M}T_2 \tag{93}$$

resulting in T_2^{cds} (T_L = 5 K) $\approx$ 30 ps.

Finally it is possible to deduce the damping γ from a detailed analysis of the excitonic reflection spectra. For high quality samples one finds at low temperature generally values < 1 meV resulting with (82) also in T_2 of the order of 10 ps.

To conclude we can state that T_2 values of ten or a few tens of ps seam typical for free excitons in high quality bulk semiconductors at low temperature and exciton density.

The life times T_1 are under the same conditions around 1 ns.

Now we consider, still in 3d SC, excitons which cannot move freely but are spatially localized by some effect. We start with bound exciton complexes. These are excitons which are bound at some point – defect like neutral acceptors or donors or ionized donors, called $A^\circ X$, $D^\circ X$ and D^+X, respectively [42].

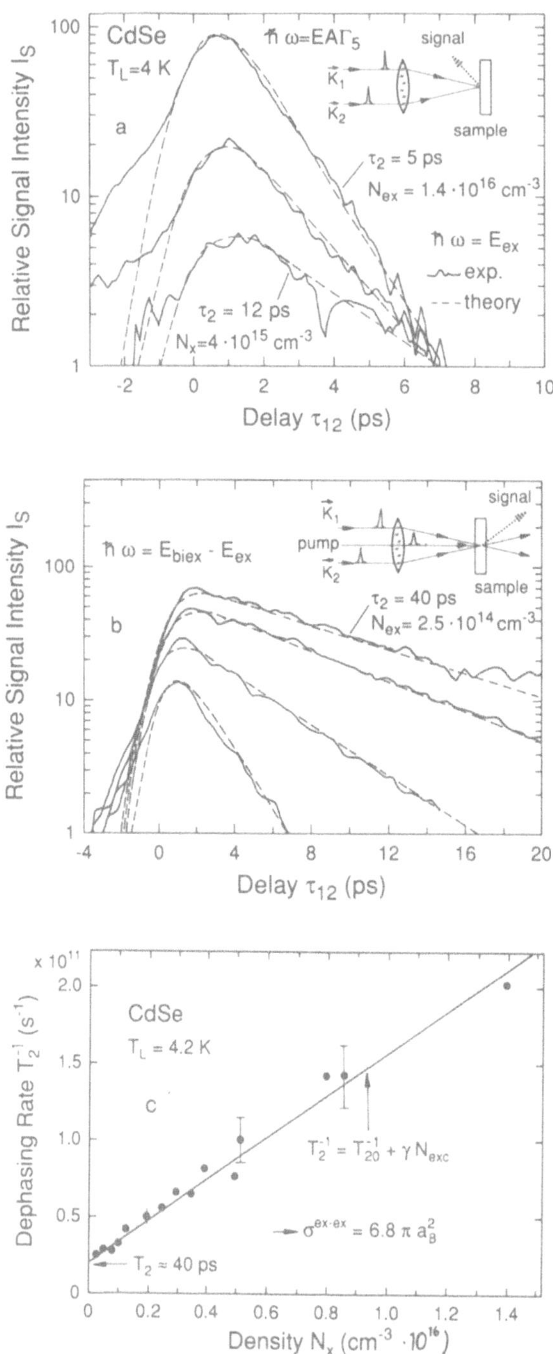

Fig 30 The photon–echo signal for various exciton densities in the n = 1 $A\Gamma_5$ exciton resonance of CdSe in reflection (a) in the exciton → biexciton transition in transmission (b) and the resulting dependence of the phasing rate T_2^{-1} as a function of exciton density (c). According to [39].

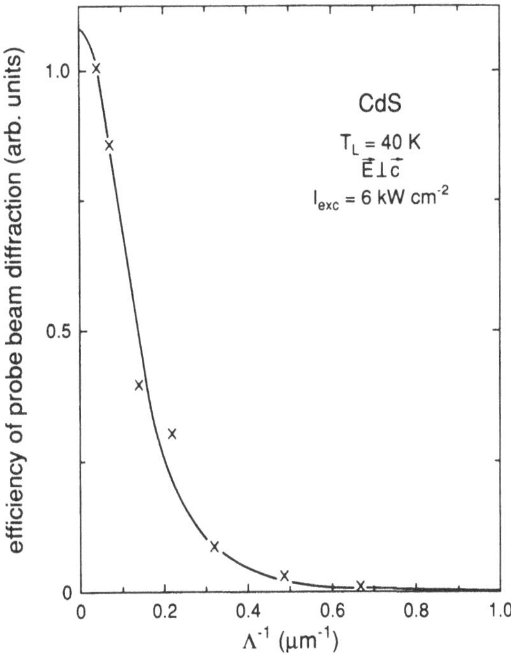

Fig 31 The decay of the diffraction efficiency of a probe beam as a function of the inverse grating constant Λ^{-1}. According to [41]. x experimental data, —— fit function for a diffusion length L = 0.78 μm.

In Fig 32a from [43] we show the time resolved photon – echo for the laser tuned in the resonance of the $A^\circ X$ complex in CdSe. Already the first inspection shows, that the decay times are much larger for this BEC than for the free exciton. Indeed a value of $T_2 = 600$ ps can be deduced from the data for T => 0. Since the life–time T_1 is of the same order of magnitude we come here close to the limit given by eq (81).

The temperature dependence of T_2^{-1} in Fig 32b can be fitted by an activation law

$$T_2^{-1} = \gamma_0 + \nu_0 \exp\{-Ea / k_B T\} \tag{94}$$

with $\gamma_0 = 1.6 \cdot 10^9\ s^{-1} = 600$ ps, $v_0 = 3 \cdot 10^{11}\ s^{-1}$ and $E_a = 5.6$ meV. The activation energy corresponds approximately to the binding energy of the exciton to the complex and so we can conclude, that the phase – destroying effect is with increasing temperature the thermal ionization of the exciton from the defect.

In CdS a T_2 value of 400 ps has been reported the D^+X complex [44] measured by the observation of quantum beats [45]. In quantum beat spectroscopy one excites two energetically close lying transitions with a short laser pulse of sufficient spectral width. The polarizations of the two resonances interfere with each other, resulting in a modulation of the free induction decay with the beat frequency. From the decay of the modulated amplitude the T_2 value can be deduc ed. In [44] the splitting and thus the beat frequency have been influenced by a magnetic field allowing thus in addition a precise determination of g – values.

The next example which we present concerns excitons localized by weak disorder in alloy semiconductors like $CdS_{1-x}Se_x$. The partial substitution of S by Se in the lattice leads to composition fluctuations and via the x dependence of Eg and consequently via eq. (30) of the exciton energies to an exponential tail of localized states as introduced in context with Fig 21. See also [46]. The existence of the localized states has been proven by site selective excitation [47]. In Fig 33 we show the spectral dependence of the phase relaxation time together with two luminescence spectra. Spectrum 1 is recorded under band–to–band excitation into extended states. The high energy peak corresponds to recombination from localized states. The transition region between localized and extended states coincides with its high energy edge [47]. The second peak is the first LO phonon replica. Under excitation around the mobility "edge" (spectrum 2) the LO phonon modes of the CdS and CdSe binding can be resolved. If spectrally narrow laser – pulses of a duration between 5 ps [49] and 10 ps [48] are used one finds T_2 values around the mobility – edge of 400 ps increasing up to values of 3 ns for deeply localized states. These values are of the same order of magnitude as the intraband relaxation times T_3 and the interband recombination times T_1 which can reach values from several hundred ps to some ns depending on energy and excitation conditions [50]. With increasing temperature the T_2 values get shorter due to increasing scattering with phonons [48].

An interesting topic can be treated in connection with Fig 34 from [48] where we give the temporal decay of the time–integrated photon–echo signal in $CdS_{1-x}Se_x$ for various laser pulse durations τ_p at low temperature (4.2 K $\lesssim$ T $\lesssim$ 10 K). In Fig 34b and d even the same sample has been used. Two effects are striking. The exponential tail from which we deduce T_2 is missing for $\tau_p = 120$ fs and gets increasingly longer for increasing τ_p. The ratio of the first initial spike, which coincides approximately with the auto–correlation function of the pulse and the longer decaying part is decreasing with increasing τ_p. The qualitative interpretation follows Fig 34e, where we show qualitatively the density of states of extended (E $\gtrsim$ ME) and localized (E < ME) exciton states and the spectral half – width of a short (120 fs) and a long (10 ps) laser pulse imposed by the uncertainty relation

$$\Delta E \cdot \tau_p \gtrsim \hbar \tag{95}$$

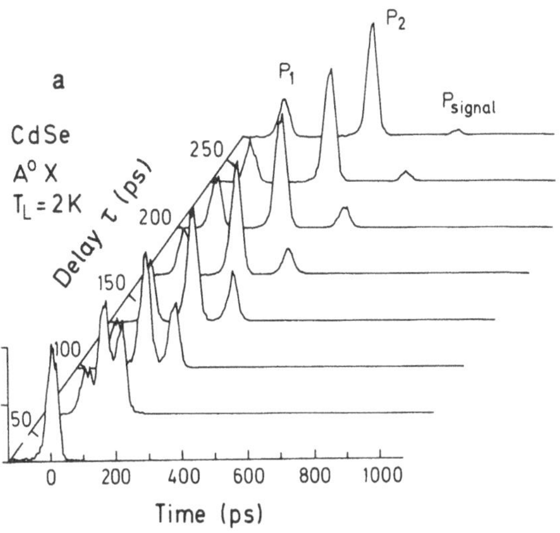

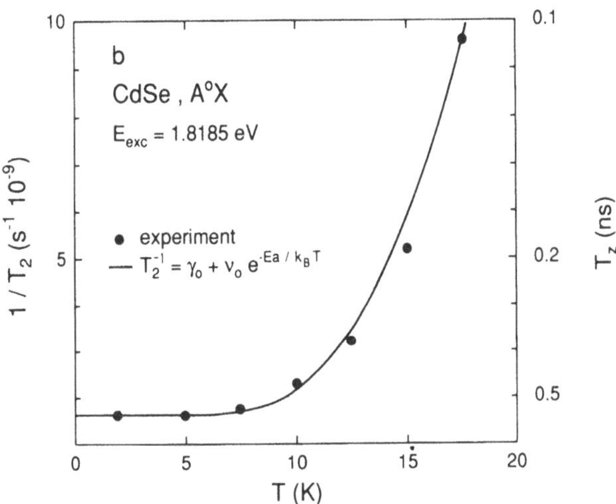

Fig 32 The time resolved photon echo in the A°X (I_1) bound exciton resonance of CdSe (a) and the temperature dependence of the dephasing rate T_2^{-1}(b). Fit with $\gamma_0 = 1.6\ 10^9\ s^{-1}$, $\upsilon_0 = 3 \cdot 10^{11}\ s^{-1}$ and Ea = 5.6 meV. According to [43].

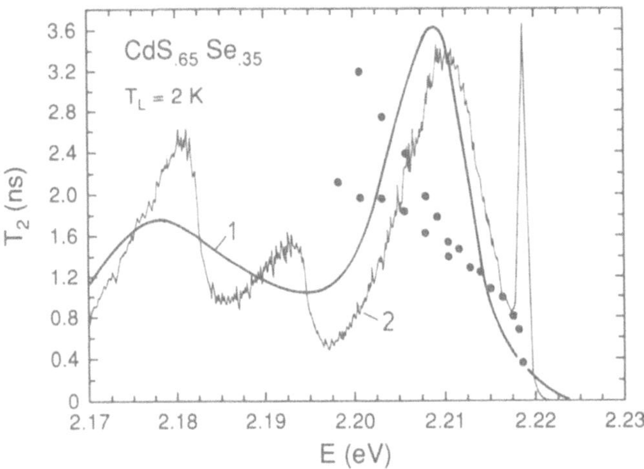

Fig 33 Luminescence spectra of CdS $_{.65}$ Se $_{.35}$ under resonant (2) and nonresonant (1) excitation and the phase relaxation times (·) measured at different spectral positions. According to [43b, 48].

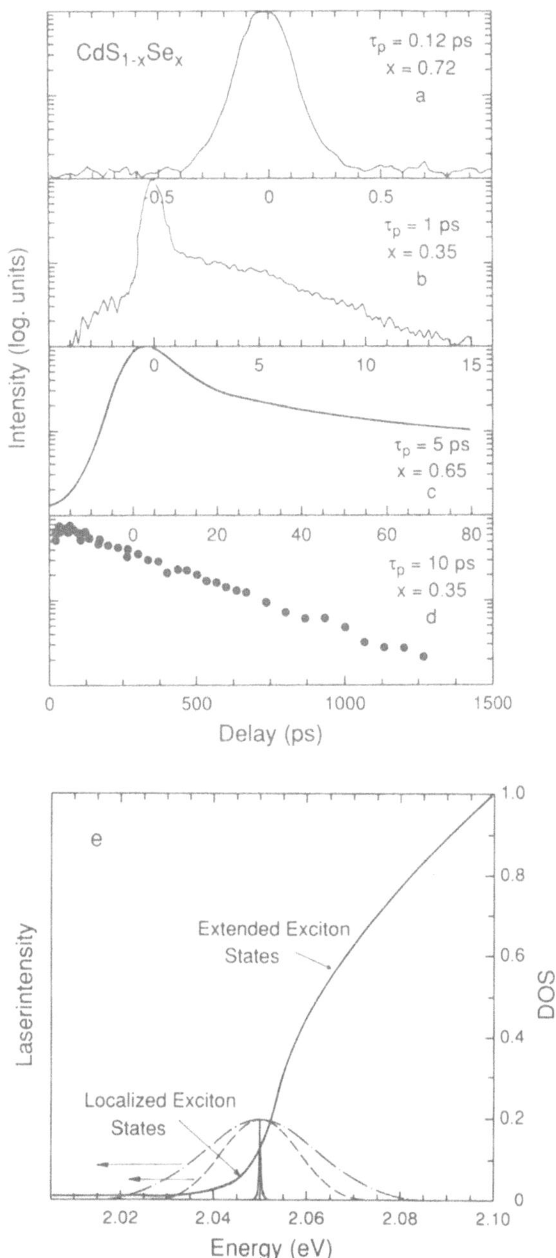

Fig 34 The dependence of the temporal evolution of the photon echo in $CdS_{1-x}Se_x$ for various durations of the exciting laser pulse τ_p. According to [43b+c].

A spectrally narrow pump – pulse tuned into the localized exciton states allows consequently to measure the T_2 times of these states. If the pulse – length is reduced, the high energy tail extends into the region of extended states. These states have in alloy SC a very short phase – relaxation time T_2 because of alloy scattering. This fast process explains the short spikes in Fig 34. Independent NDFWM experiments in $CdS_{1-x}Se_x$ showed indeed that T_2 is of the order of 100 fs [51]. Furthermore the extended states scatter also with the localized ones, reducing their observed T_2 values. These findings and interpretations solve a discrepancy outlined in [52]. Localized excitons do hardly interact which other [49] as has been found independently from the weak influence of the excitation intensity on the polarization memory of the luminescence of localized states [53] and the fact, that many particle effects, e.g. the formation of an electron – hole plasma (see below) are supressed at low temperatures for densities up to several 10^{18} cm $^{-3}$ i.e. when excitons populate only the localized states [54].

A strong dependence of the observed T_2 values on the spectral width of the pump laser can be expected not only for the alloy SC considered here but for all cases, where states with different T_2 values fall into the spectral width of the ps or sub ps laser.

From this experience we can deduce the warning, that the "quality" of physics or of the physicist does not necessarily increase with decreasing laser pulse duration τ_p, but that some considerations may be necessary to adopt τ_p to the physical question under consideration.

To conclude this subsection we present some examples of the fate of excitations in quasi–twodimensional QW.

In Fig 35a we show the time integrated luminescence spectrum of a GaAs/AlGaAs MQW sample. The emission is rather broad and consists of two peaks. This is characteristic for well – width fluctuations [57] via eqs (65, 66). The photo–luminescence excitation spectrum of the low energy peak is also shown in Fig 35a. It shows the $n_z = 1$ hh and lh exciton resonances, and the slight Stokes – shift between emission and absorption of about 3 meV also indicates some disorder like the well – width fluctuations mentioned above or the alloy disorder in the barriers.

In Fig 35b we give the temporal evolution of the luminescence if we excite in the second PLE – peak of Fig 35a. The emission shows only the low energy peak and decays with a time constant of a few hundred ps which is characteristic for this system [58].

Excitation is the hh resonance of the PLE spectrum of the higher peak around 1.56 eV (not shown in Fig 35a) gives a more complex behaviour. The resonantly excited high energy peak is first dominant, but also decays faster than the low energy one which has a delayed onset [55].

The reason for this complex behaviour is not necessarily a different oscillator strength of both transition, but the fact, that the high energy excitons sitting in spatial regions of narrower well width diffuse and feed the exciton states at places with large well width where they get trapped.

Now we investigate a sample, which has less pronounced well width fluctuations (Fig 36). The luminescence shows the $n_z = 1$ hh and lh emission. The shoulder on the low energy side of the hh peak can have various origins, namely again well–width fluctuations, recombination from a BEC e.g. a $D^\circ X$ or recombination of a biexciton leaving behind a free exciton and a photon. The rather broad peak around 1.53 eV definitely involves some impurties e.g. in a free to bound transition. The PLE spectrum shows the $n_z = 1$ hh and lh resonances and the $n_z = 2$hh. The high quality of the sample follows from the absence of a Stokes shift between emission and absorption in the PLE spectra and from the small FWHM of the peaks of 1.5 meV.

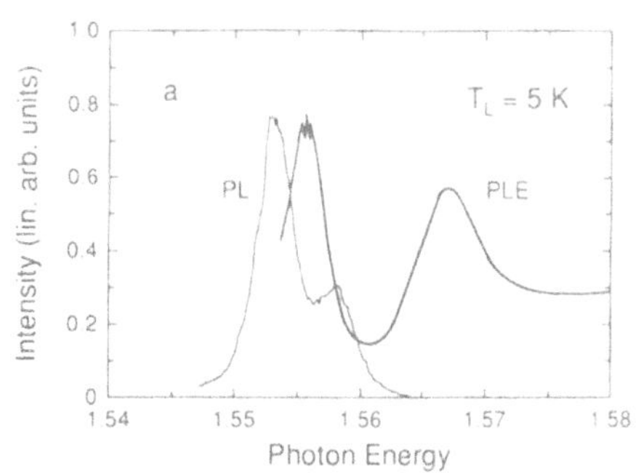

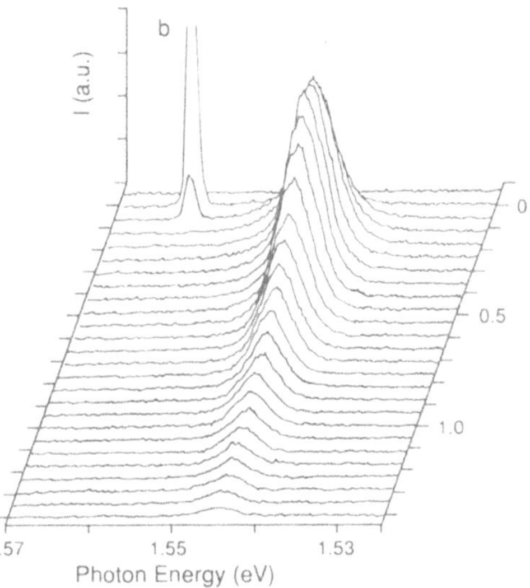

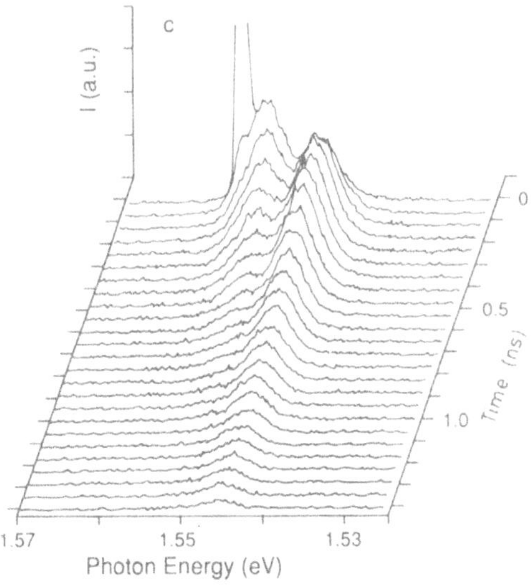

Fig 35 The photoluminescence (PL) and the ~ excitation spectrum (PLE) of an AlGaAs MQW sample 50 x (7nm GaAs/20nm Al $_3$Ga $_2$As) (a) and the luminescence dynamics for two different excitation energies (b) and (c). According to [55, 56].

In Fig 36b and c we show decay curves of the hh exciton emission for various excitation conditions. The most striking feature is the fact that a simple exponential decay is almost never observed.

For low, resonant excitation (curve 1 in Fig 36b) we observe a rapid decay of the luminescence, which is attributed in [55, 59] to a rapid capture of excitons into some deep centers. For increasing excitation this process saturates and we get for $t>200$ ps an exponential decay with a time constant of about 250 ps. We think that this value is eventually representative for the radiative and non – radiative decay of excitons in this type of MQW at low temperature. With increasing pump power a hump develops in the decay dynamics, which becomes even more pronounced under excitation with some excess energy (Fig 36c). We interpret this feature by a thermalization or T_3 process. Increasing I_{exc} or $\hbar\omega_{exc}$ leads to the creation of a hot gas of excitons with a certain average kinetic energy and with finite $\vec{K}_{\parallel}$-vectors. Due to the conservation of $\vec{K}_{\parallel}$ in the recombination process, only excitons with $\vec{K}_{\parallel}$ around zero can participate in the radiative recombination. The humps or the plateaux in curves 7 and 8 reflect according to [55] the cooling of the exciton gas. A modelling of the reaction kinetics along the ideas outlined above gave the curves in Fig 36d. The good agreement of curves 1, 2 and 7, 8 in Fig 36d with those in Fig 36b and respectively, give some support to this model. Indeed it has been found in [58] that the T_1 time increases for higher lattice temperatures T, which leads of course also to a rise of the exciton temperature.

The phase relaxation times T_2 of excitons in MQW have been investigated by the photon – echo technique in [40] and values around 20 ps have been obtained, which are slightly higher than in bulk GaAs. The T_2 values decrease again with increasing lattice temperature and increasing density of excitons and of free carriers. Since exciton – exciton interactions and exciton free carrier interaction are of dipole – dipole and of dipole – monopole type, respectively, it is understandable that free carriers are more efficient in reducing T_2 than free excitons [40].

We show here in Fig 37 more recent results, where T_2 is plotted versus T . The values are deduced directly from photon – echo experiments in transmission of samples from which the opaque GaAs substrate has been removed by selective etching and from the diffusion constant measured with ps LIG techniques. The data obtained with both techniques agree reasonably well.

It is presently an open question if localized states exhibit generally longer T_2 values at low T compared to free or extended ones in comparable materials or not. Theoretical considerations of this problem are under way [61]. Very recently quantum beats have been observed between the $n_z = 1$ hh and lh exciton and between the $n_z = 1$ hh exciton and the shoulder in Fig 36a [55].

To conclude this section we want to come back to a remark made at the beginning, where we stated, that phonon – echo experiments are generally treated in the weak coupling limit. A consequence of this approach is the fact, that a diffracted order in the direction $2\vec{K}_2 - \vec{K}_1$ (see Fig 28) is expected only if pulse 2 arrives after pulse 1 i.e. for $\tau > 0$.

In a more appropriate description in terms of polariton – polariton interaction such unsymmetries would partly disappear. A first approach in this direction has been made in [62], where a nonlinear polarisation – polarisation interaction has been considered, which describes e.g. the facts, that a diffracted order can be observed already for negative τ like in Fig 30a and even a minimum in the diffraction efficiency at $\tau = 0$ [62].

V. FROM ONE EXCITON TO THE ELECTRON – HOLE PLASMA

As indicated in the abstract, we want to follow now another aspect of excited state spectroscopy in semiconductors, namely the question, what happens, if we increase the density of electron – hole pairs in a SC.

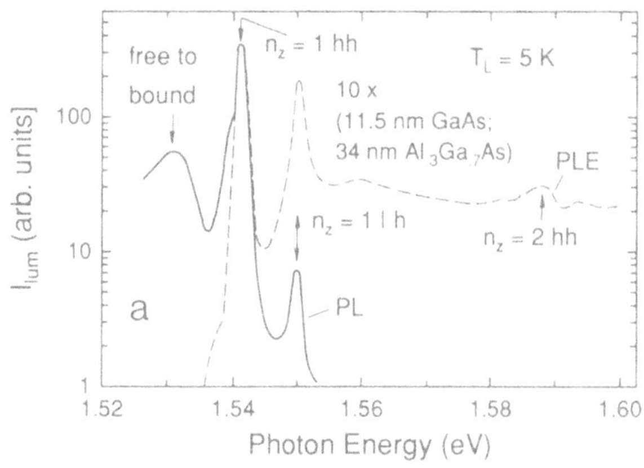

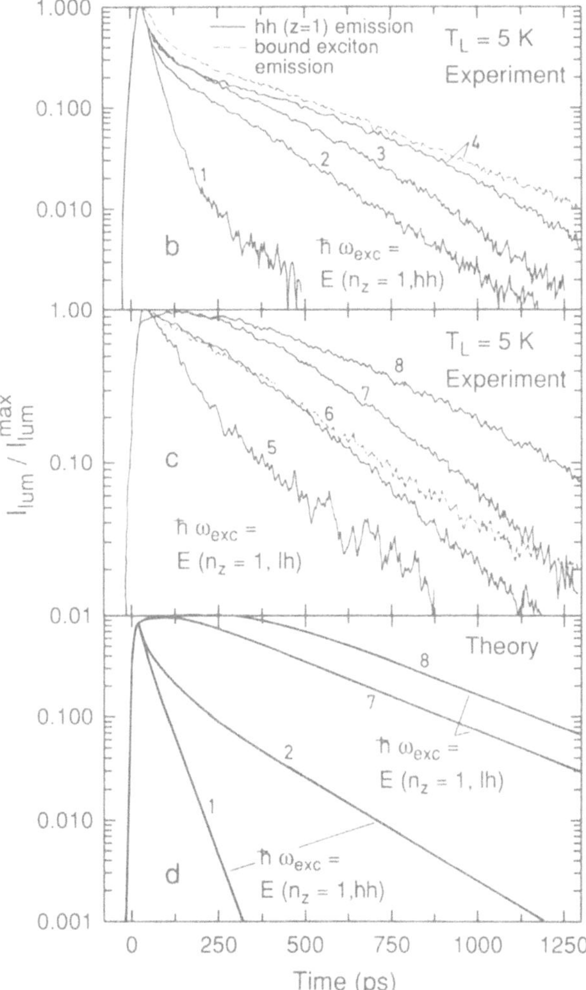

Fig 36 The PL and PLE spectra of a very high quality AlGaAs MQW sample (a) the measured $n_z = 1$, hh exciton luminescence kinetics for resonant excitation and increasing pump power from (1) to (4) (b) and for non–resonant excitation (5) to (8); calculations of the decay kinetics for the situation of curves (1), (2), (7) and (8). According to [55, 56, 59].

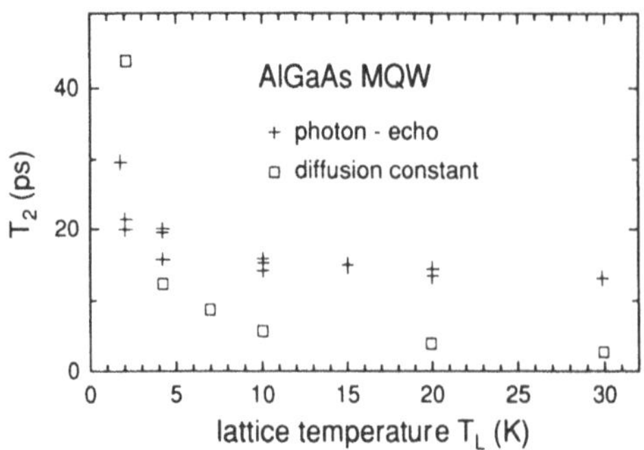

Fig 37 The T_2–time of the $n_z = 1$, hh exciton of a high quality AlGaAs sample deduced from photon–echo (+) and from diffusion experiments (□). According to [55].

In the last few decades a general scenario has been found for these so–called high excitation phenomena. At low light levels, the optical properties are determined by single excitons, which may be free or bound e.g. to some point defect. In an intermediate density regime there are elastic and inelastic scattering processes between excitons or at higher temperatures also between excitons and free carriers. At the highest densities a new collective phase of electrons and holes is formed, the electron–hole plasma [63].

One aspect of these scattering processes has been already presented in the proceeding section, namely a decrease of T_2 and a collision broadening of the exciton resonances with increasing density.

Another consequence especially of inelastic scattering processes is the appearance of new bands of emission, induced absorption or gain [5, 7, 9, 22, 25]. We give an example in Fig 38. In part a.) we show the collision process, schematically. Two exciton – like polaritons scatter via dipole – dipole interaction. One of them reaches a higher excited state ($n_B = 2, 3,...$) while the other one ends up under energy and momentum conservation on the photon – like part of the polariton dispersion, leaving the crystal which high probability as a luminescence photon.

Energy and momentum conservation for the process result in

$$E_x(\vec{K}_1, n_B = 1) + E_x(\vec{K}_2, n_B = 1)$$

$$= \hbar\omega + E_x(\vec{K}_1 + \vec{K}_2 - \vec{K}_f, n_B^f \geq 2) \tag{96}$$

where $\vec{K}_f$ is the momentum of the photon – like polariton $\hbar\omega$ and n_B^f the main quantum number of the exciton in the final state.

In simpliest description by reaction kinetics we expect that the emission increases quadritically with the exciton density and thus superlinearly with the excitation intensity I_{exc} in case of one photon excitation.

$$I_p \sim N_{ex}^2 \sim I_{exc}^2 \tag{97}$$

Such a superlinear increase occurs very often under high excitation conditions though generally with an exponent smaller than two. In Fig 38b we show an emission spectrum of ZnO where the P_2 and P_∞ bands can be clearly seen. The indices give the value of n^f_B in eq (96). The (thermal) distribution of the excitons in the initial state and the splitting of the exciton states with $n_B \geq 2$ caused e.g. by S and P type envelope functions contribute to the line – shape and the finite width of the emission bands.

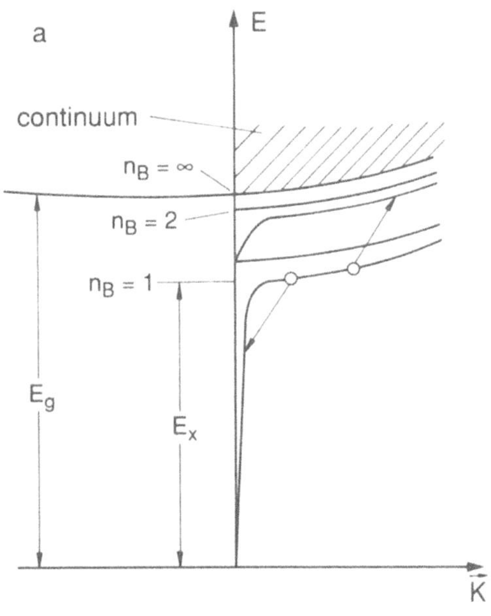

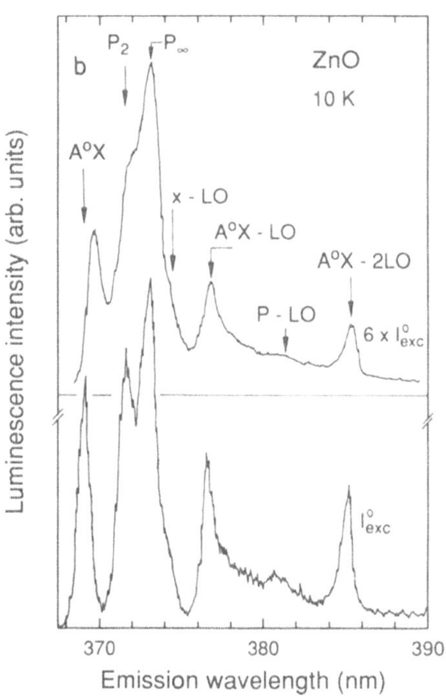

Fig 38 Schematic drawing of the inelastic exciton–exciton scattering (a), the resulting P_2 and P emission bands in ZnO (b). According to [22a, 64].

Similar scattering processes are also known between free and bound excitons and free carriers or optical or (in the case of bound excitons) also acoustic phonons. Details are given in [22, 25].

A second process which is characteristic for the intermediate density regime are transitions involving the excitonic molecule or biexciton, which we introduced in connection with eq (87).

As already mentioned, biexciton states have a high probababilily for creation by two photons, or by two steps. We illustrate this process now in some detail in the polariton picture (Fig 39) where we show by a solid line the real and imaginary parts of $\vec{K}$ around a $n_B = 1$ exciton resonance for the lower and upper polariton branches (compare with Fig 16 or 21). If we shine (laser–) light at an energy $\hbar\omega_{exc}$ on the sample, we populate the corresponding polariton state in the sample. A polariton at $\hbar\omega_{exc}$ can be converted into a biexciton by a second quantum of energy $\hbar\omega_{abs}$ given by

$$\hbar\omega_{abs} = E_{biex} - \hbar\omega_{exc} \tag{98}$$

resulting in an excitation – induced peak in the imaginary part at this energy and a dispersive singularity for the real part (dashed lines).

The oscillator – strength of the resonance at $\hbar\omega_{abs}$ increases with the population at $\hbar\omega_{exc}$.

We show in Fig 40 an example for the two – photon or more precisely two polariton absorption for CdS. The energy of the pump – beam is given on the rhs. The reabsorption dip, which swifts through the luminescence band of the CdS sample obeys eq (98). A variation of the scattering geometry allows to determine even the dispersion of the biexciton [22, 66]. The resonance shows also up in LIG spectroscopy [22, 67].

The radiative decay of a biexciton into a photon – like and an exciton – like polariton was the first indication of the existence of this quasi – particle [22, 36a].

The transition from the exciton – like part of the polariton – dispersion to the biexciton is shown by a dotted line in Fig 39. This was the resonance used in the photon – echo experiments in Fig 30b. We shall come back to the biexciton in the next chapter, where we discuss two – photon (or hyper –) Raman scattering involving virtually excited biexcitons as intermediate states. More details about the biexciton state and the nonlinearities connected with it are found e.g. in [22, 36a] and the references given therein.

The last example which we present here for the intermediate density regime is a coherent process, namely the so–called optical or ac – Stark effect [68]. It can occur in systems of various dimensions [70]. A simple explanation is in terms of level repulsion. We assume that the energy of the incident photons $\hbar\omega_p$ is chosen slightly below the exciton resonance. Then a state containing m photons and n = 0 excitons is close in energy with a state $|m-1, n=1\rangle$. Since both states are coupled via dipole interaction, we expect the common effect of level repulsion. The laser frequency is kept constant from outside and consequently the exciton resonance shifts to the blue by an amount δE given e.g. by [69–71]

$$\delta E = \frac{2|e\vec{r}_{cv}\vec{E}_p|^2 |\phi_{1s}(r_e - r_n = 0)|^2}{(E_{1s} - \hbar\omega_p)\, N_s p} \tag{99}$$

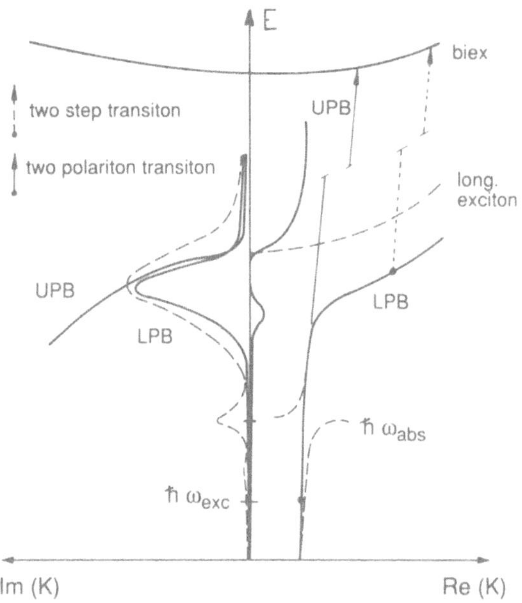

Fig 39 Schematic drawings of the two polariton and the two step transitions to the biexciton in the polariton dispersion.

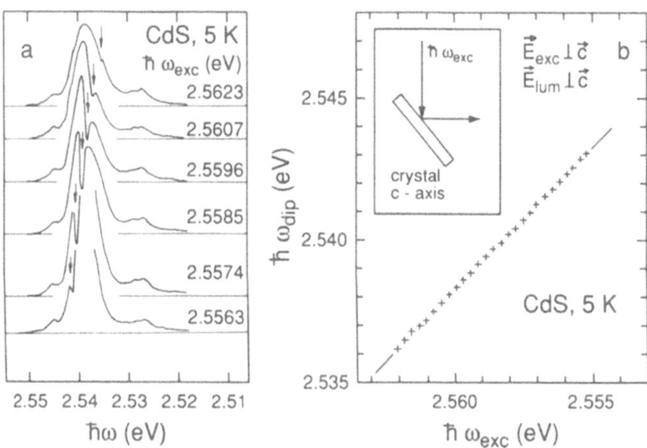

Fig 40 The (re–) absorption dip in the luminescence of CdS for variable exciting photon energies (a) and the relation between $\hbar\omega_{dip}$ and $\hbar\omega_{exc}$ (b). From [65].

where $e\vec{r}_{cv}$ is the band–to–band dipole–matrix element $\vec{E}_p$ the field amplitude of the pump laser, $|\phi_{1s}(r_e - r_h = 0|^2$ the enhamcement of the transition probability of the oscillator strength of the 1s exciton, $E_{1s} - h\omega_p$ the detuning and $N_s{}^p$ the density necessary to block the exciton resonance by phase–space filling (see below). The term $|e\,\vec{r}_{cv}\,\vec{E}_p|^2$ gives the so–called Rabifrequency, which describes in the weak coupling limit the frequency with which the excitation oscillates coherently between exciton and photon.

The name optical or ac–Stark effect obviously comes from the fact that it describes a shift of the exciton eigenfrequency caused by the electric field of the light. A perfect description, which explains not only the shift of the exciton resonance but also the appearance of a gain peak symmetrically on the low energy side of the pump laser, is again in terms of polariton – polariton interaction [69].

Since the ac – Stark effect is a coherent process, most of the successful attempts to observe it used ps laser – pulses with $\tau_p \lesssim T_2$ [70].

However, it has been shown in GaAs/AlGaAs MQW–structures that it is also possible under favourable conditions, to observe this effect by ns – pulses. Favourable conditions means in this case a rather low intensity I_p and a large detuning to avoid real excitation. Unfortunately this means with eq (99) also a small shift, so that sensitive differential transmission spectroscopy (DTS) has to be used.

In Fig 41 we show such a DTS spectrum together with the $n_z = 1$ hh exciton resonance for a GaAs/AlGaAs MQW system. The quantity

$$\sigma = \frac{\delta E \cdot \delta}{I_p} \text{ with } \delta = E_{1s} - \hbar\omega_p \tag{100}$$

deduced from the DTS spectrum is $\sigma_{exp} = 8.5 \cdot 10^{-8}\,(meV)^2\,cm^2\,W^{-1}$ and compares favourably with the theoretical one $\sigma_{theory} = 5.8 \cdot 10^{-8}\,(meV)^2\,cm^2\,W^{-1}$ [69–71] giving thus some support for the interpretation of the ns DTS spectrum in terms of ac–Stark effect.

If we pump the samples even harder, we leave the intermediate density regime and come in the high density regime to the electron – hole plasma.

Here the density of electron – hole pairs becomes comparable with the excitonic Bohr radius. As a consequence one can no longer speak about excitons or biexcitons as individual quasiparticles. They loose their identity and a new collective state is formed, the so–called electron – hole plasma (EHP). The transition from a low density gas of (neutral) excitons to the metal–like EHP is sometimes called a Mott–transition and is connected with significant changes of the electronic states and their optical properties which we want to outline below. For more details e.g. [5, 9, 22, 63, 36d] and references therein.

In Fig 42 we show the bandstructure of an EHP with a reduced gap $E_g'(n_p)$ where n_p is the electron–hole pair density of the – at low temperature – degenerate population of the bands with electrons and holes. The chemical potential of the EHP is indicated, which is defined as the energetic distance between the quasi Fermilevels describing the population of electrons and holes in their respective bands.

In Fig 42b we show various energies as a function of the electron–hole pair density n_p. The gap is a monotonously decreasing function of n_p due to exchange and correlation effects. In polar materials like the II–VI compounds, there is an enhancement of this decrease by the interaction of plamsons and optical phonons [22, 63]. The exciton resonances broaden and disappear at some density, when their binding energy tends to zero due to screening of the Coulomb – interaction and phase–space filling, i.e. occupation of the states which are necessary to form a Wannier exciton e.g. according to eqs. (35, 47). The latter contribution is especially important for quasi – two dimensional systems [72]. The decrease of the gap and the decrease of the excitonic binding energy almost compensate each other so that the absolute energy of the exciton does not change much [5, 9, 22, 63]. Additionally we show the chemical potential μ of the electron–hole pair system for various temperatures. Population inversion i.e.

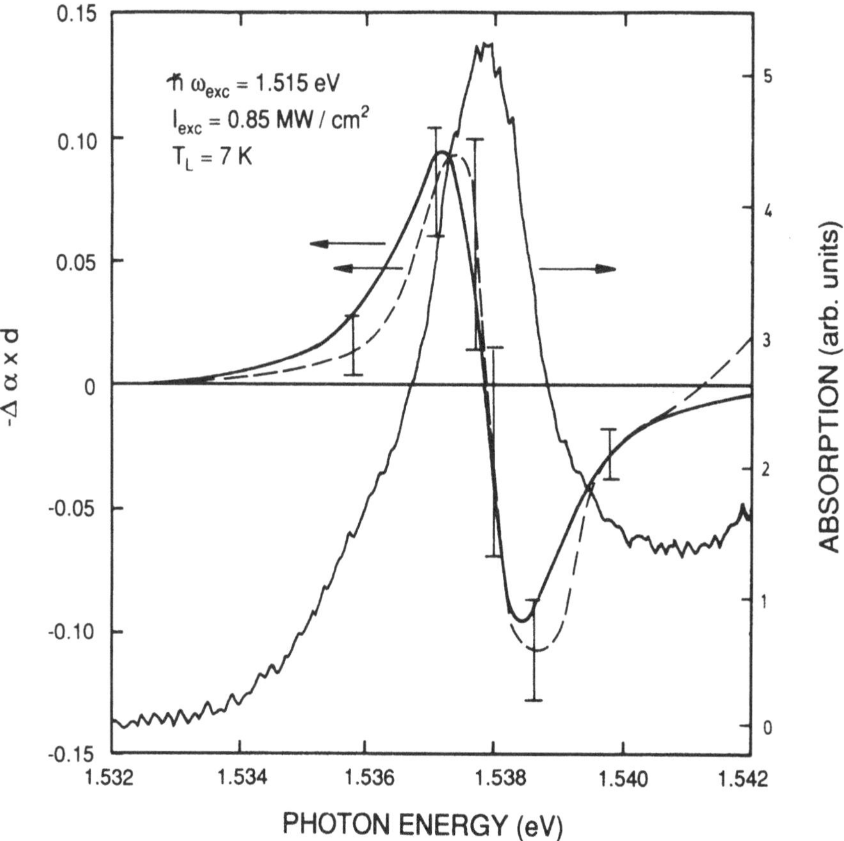

Fig 41 The n_z = 1 hh exciton absorption peak of an AlGaAs MQW structure (r.h.s. scale) and the differential transmission spectrum caused by below resonance excitation with a ns pulse. From [63a].

$$\mu(n_p, T_p) > E_g'(n_p) \tag{101}$$

occurs at densities increasing with temperature. This population inversion leads to optical amplification and is the basis for all laser–diodes [3]. Fig 42c shows the phase–diagramm which can be deduced from (b) assuming that quasi –equilibrium conditions are reached. It predicts a first order phase transition between a low density

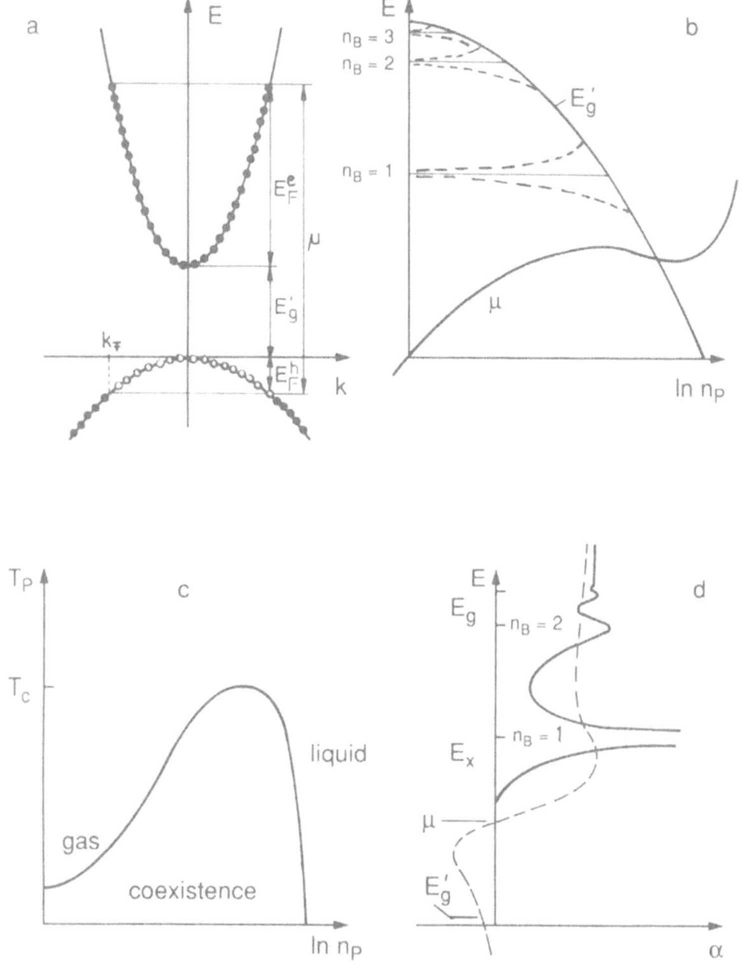

Fig 42 Schematic drawings of the population of conduction– and valence– bands in a degenerate electron–hole plasma (a) of the renormalized band gap Eg', the exciton resonances including damping and the chemical potential μ as a function of the plasma density n_P (b) and of the phase–diagram of the plasma in equilibrium (c) and of the absorption spectra without and with plasma (d). According to [7, 9, 22, 63].

gas of excitons and free carriers and a liquid like EHP state below a critical temperature T_c with a coexistence region, very similar to what is known for a real or van der Waals gas. Modifications of the phase – diagramm by "entropy ionisation" etc. are discussed e.g. in [36d]. In Fig 42d we illustrate finally the chances of the optical properties connected with the transition from a low density exciton gas to the EHP. We show schematically the exciton absorption spectrum and the one in presence of an EHP. The exciton resonances are gone as indicated above. For a degenerate situation there is optical amplification (gain) essentially between $E_g'(n_p)$ and μ (n_p, T_p) and absorption due to band to band transitions above. The quantitative interpretation of the gain spectra involves some additional considerations. Basically one starts with band–to–band transition and $\vec{k}$–conservation. This would give a square–root increase of the gain for $\hbar\omega \geq E_g'(n_p)$ and a transition to absorption at μ (n_p, T_p) given by the Fermifunctions of electrons and holes $f_{e,h}$

$$\alpha(\hbar\omega) \sim (\hbar\omega - E_g'(n_p))^{1/2} \cdot (1 - f_e - f_h) \text{ for } \hbar\omega \geq E_g'$$
$$\alpha(\hbar\omega) = 0 \text{ for } \hbar\omega < E_g' \qquad (102)$$

Two modifications are necessary [22a, 36d, 73] to compare the simple expression of (102) with experiment. There is an energy dependent (final state–) damping which shows a temperature dependent minimum around μ and which causes a tail of the gain–spectrum extending below $E_g'(n_p)$. Around μ (n_p, T_p) there is an also temperature dependent enhancement of the transition probability caused by the rest of the Coulomb interaction between electrons and holes and the fact, that occupied and empty states are available around $\hbar\omega = \mu$ (n_pT_p) allowing scattering between the carriers in the sense of an exciton – like orbit. Consequently this fact has been introduced as "excitonic enhancement" in [36d] and references therein for the EHP in bulk materials. It has been (re–)invented under names like Fermi – edge singularity for MQW [74] including the situation of one electron in a sea of holes and vice–versa, which occurs e.g. under high doping levels.

Now we want to give some examples for the properties of the EHP starting again with bulk materials. The first observation of an EHP was in Si (see [75] and references therein) and so we start here with examples for indirect gap materials. In Fig 43a we show the LA phonon assisted emission of Ge at low temperature and high excitation. One observes the emission of the free exciton and as a broad and relatively intense band the recombination from an EHP in a liquid like state. The analysis of this band gives the values of $E_g'(n_p)$, $\mu(n_p, T_p)$ etc.

From the onset of plasma emission with increasing excitation and the line–shape analysis the phase–diagramm of the EHP can be deduced. We give in Fig 43b and c examples for Ge and Si. Similar results have been found also in other indirect gap materials like GaP [77] or indirect $Al_{1-y}Ga_yAs$ [78]. In the latter compound, the density of localized states is smaller than e.g. in $CdS_{1-x}Se_x$ (see above) so that an EHP can be created under high excitation even at low temperatures.

Huge so–called γ–drops of EHP liquid can be formed applying inhomogenous stress to the samples [79], bound multi exciton complexes in Si or Ge can be considered as precurses or nucleation centers for the EHP [80] and the disappearance of the direct gap exciton in Ge simultaneously with the creation of a plasma has been reported in [81].

Now we return to the direct gap materials. The high optical gain, which is connected with a degenerate EHP in a direct gap SC makes a quantitative evaluation of luminescence spectra difficult, because they are often distorted by stimulated emission. Therefore, the pump–and–probe beam technique is the most appropriate one to get information about the EHP in III–V semiconductors like GaAs [82] or the II–VI compounds [22], for which we give examples in the following. A concise summary on the EHP in this group of SC will be published shortly by the author [63b]. The main effects observed in connection with the formation of an EHP are the

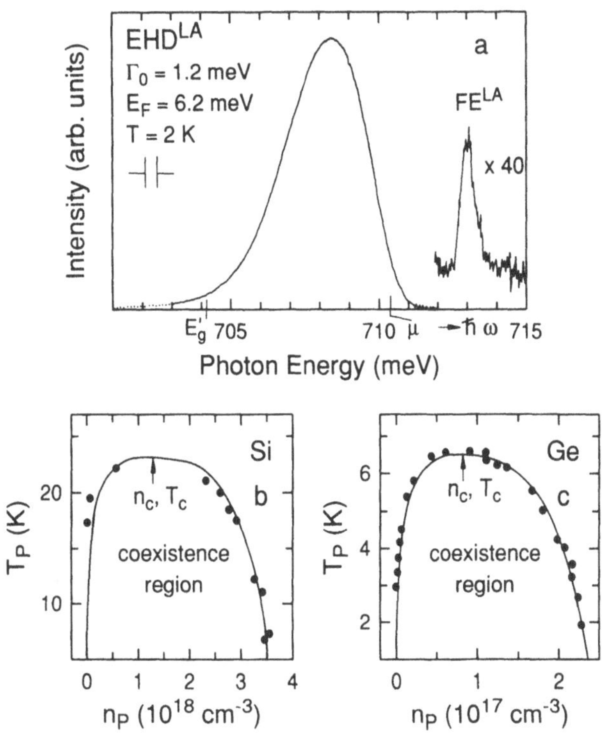

Fig 43 The LA–phonon assisted luminescence of free excitons (FE) and of the electron–hole liquid droplets (EHD) in Ge (a), the phase diagram of the EHD in Si (b) and Ge (c). According to [75, 76].

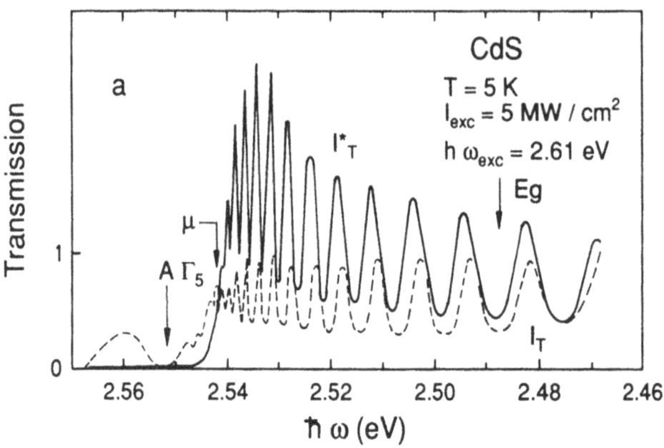

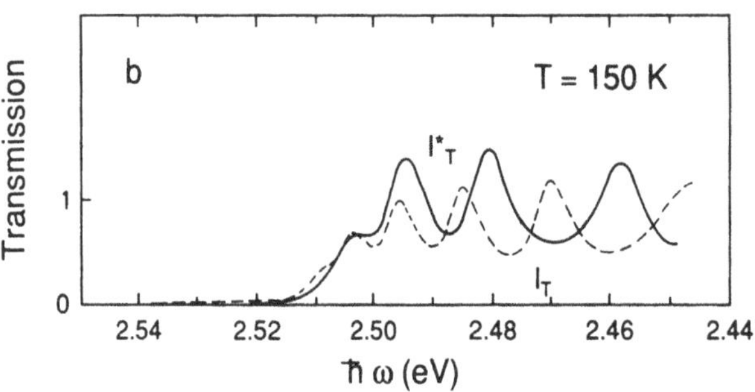

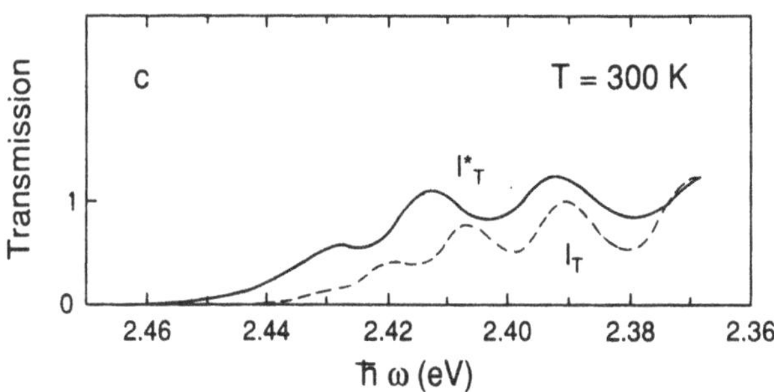

Fig 44 Transmission spectra of a CdS platelet of a few μm thickness without (————) and with (------) additional pump for various lattice temperatures. According to [83].

gradual disappearance of the excitonic features form the spectra of transmission and reflection with increasing I_{exc} and the appearance of bleaching, gain and/or induced absorption depending on the relative position of μ and the excitonic absorption tail and other parameters of the excitation conditions [73, 83].

In Fig 44 we give transmission spectra of CdS platelets without and with additional pump for various lattice temperatures. Without excitation one sees Fabry–Perot modes in the transparent part of the spectrum caused by the plan – parallel surfaces of the platelet type sample and the $A\Gamma_5$ exciton absorption at 5 K which broadends with increasing temperature and develops a tail to the low energy side, known as Urbach tail ([12, 13]) and eq (37)). With pump we see clearly the optical gain connected with electron–hole pair recombination in the EHP at T = 5 K and 150 K. At 300 K population inversion is no longer reached under these pump conditions. Additionally we see bleaching at 300 K or induced absorption at 5 K. For the latter temperature one can easily define the position of μ from the cross–over between gain and absorption at 2.54 eV and E_g' comes down from about 2.58 eV of the unexcited sample to values around 2.49 eV.

Furthermore one observes a blue shift of the Fabry–Perot modes at all temperatures indicating a decrease of the refractive index under excitation. The plasma density deduced from a fit of the gain spectra gives values for n_p up to $3 \cdot 10^{18} cm^{-3}$ under quasistationary excitation conditions. The fact that n_p increases with I_{exc} indicates, that the plasma does not reach a liquid like state in direct gap SC due to its short lifetime around 100 ps which does not allow the formation of a phase separation into liquid like droplets surrounded by an exciton gas [83, 84, 85].

The drift length of the EHP under inhomogenous excitation measured with spatially resolved pump–and–probe beam techniques or with LIG similar to Fig 31 gives values of (10 ± 5) μm for a degenerate and $\lesssim 1\mu$m for a nondegenerate EHP. Details concerning the above results are found e.g. in [63b, 73, 86] and references therein.

In Fig 45 we summarize results of the EHP in SC. The band–gap renormalization is plotted versus the plasma density n_p. Both quantities are normalized with the excitonic Rydberg energy Ry* (see eq (31)) and the excitonic Bohr radius a_B (see eq (34)). Experimental data for the less ionic bound group IV and III–V SC coincide nicely with the universal theories given, as shown e.g. in [63b, 82, 87]. The experimental points for various II–VI compounds given in Fig 45 follow the general trend of the theories, however, with an average shift of about 0.5 Ry*. This discrepancy has two origines. One is that the experimental values of the exciton binding energy i.e. the energetic distance between the polaron gap and the lowest free exciton do not coincide exactly with Ry* according to eq (31) as discussed e.g. in [22a, 63b, 89] in the context with a universal curve for the biexciton binding energy and that the band–gap renormalization in the more ionic materials involve an additional contribution by plasmon–phonon mixed states which are not contained in the universal theories. More detailed calculations e.g. for CdS coincide nicely with experiment [84, 85].

We proceed now again to quasi twodimensional SC. The EHP has been preferentially investigated in GaAs [90] and InP based MQW [72, 74, 91] partly also replacing high excitation by (modulation) doping. On the first glance the results look similar, i.e. one observes a disappearance of the exciton resonances, the appearance of optical gain, band–gap renormalizations etc.. In Fig 46 we see the band–gap renormalization Δ E_g and the chemical potential μ as a function the now two–dimensional density n_p. The experimental data coincide nicely with theory. A more detailed analysis, however, reveals some differences. The direct Coulomb screening is weaker and the influence of phase space filling is stronger in quasi–twodimensional systems than in three–dimensional ones [9, 36d, 74, 91–93].

The shift of the higher subbands $n_z \geq 2$ is lower than that of the fundamental one ($n_z = 1$) as long as the first ones are not populated [95]. A first explanation of these findings assumes that the inter–subband exchange interaction is weaker in quasi 2d systems [90c–93]. More recent calculations find a less pronounced influence of the dimensionality on the renormalization of higher subbands [94].

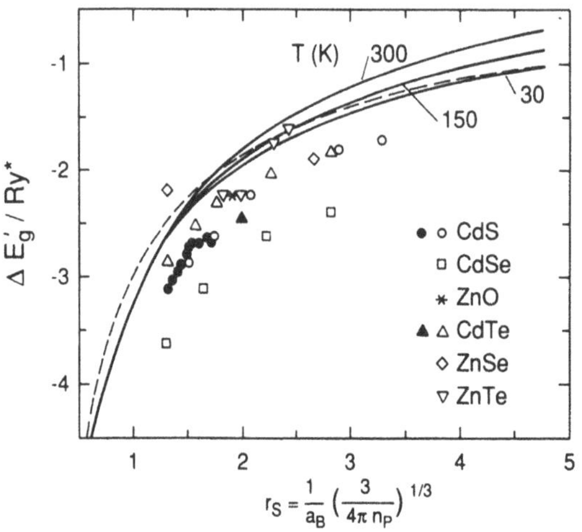

Fig 45 The band gap renormalization ΔEg as a function of the plasma density n_p in normalized units. Universal theories from [87] (- - - - -) and from [36d, 88](———); from [89].

As a last example of this section, we show an optical nonlinearity in $CdS_{1-x}Se_x$ quantum dots (Fig 47). Part a shows the absorption spectrum without excitation, which shows the absorption peaks of the transitions between the unresolved A and B and the spin–orbit splitt off C band, and the bleaching induced by excitation at $\hbar\omega_{ex}$ = 2.5 eV. Both bands are bleached because they involve the same final state in the conduction band. Tuning of $\hbar\omega_{exc}$ leads to a slight broadening of the DTS signals (Fig 47b) but not to a shift as expected for simple spectral hole burning. In contrast to a tentative explanation given in [28f] this feature remains in this sample also at much lower pump intensities [95]. Eventually the size of the QD plays an important role, too.

The dip in the DTS spectra, which extends partly even to negative value, i.e. to induced absorption, is interpreted by various authors as a transition from a one electron–hole pair state to a two electron–hole pair state in the dot [95, 96] in agreement with the interpretation of a luminescence feature occuring at high excitation in the same spectral range. It is a question of semantics if such a state is considered as a biexciton or as an EHP in a QD. First results about the dynamics of the optical nonlinearities and the photo–darkening observed in many glasses are found e.g. in [28] and references therein.

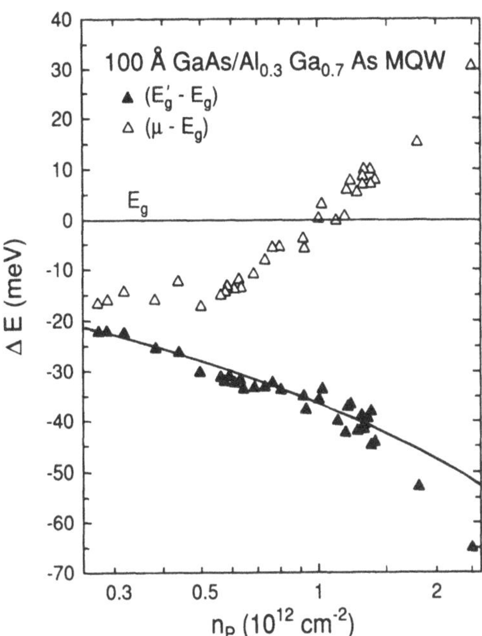

Fig 46 The renormalization of the forbidden gap (Δ) and the chemical potential (Δ) of an EHP in AlGaAs MQW compared to theory. From [92].

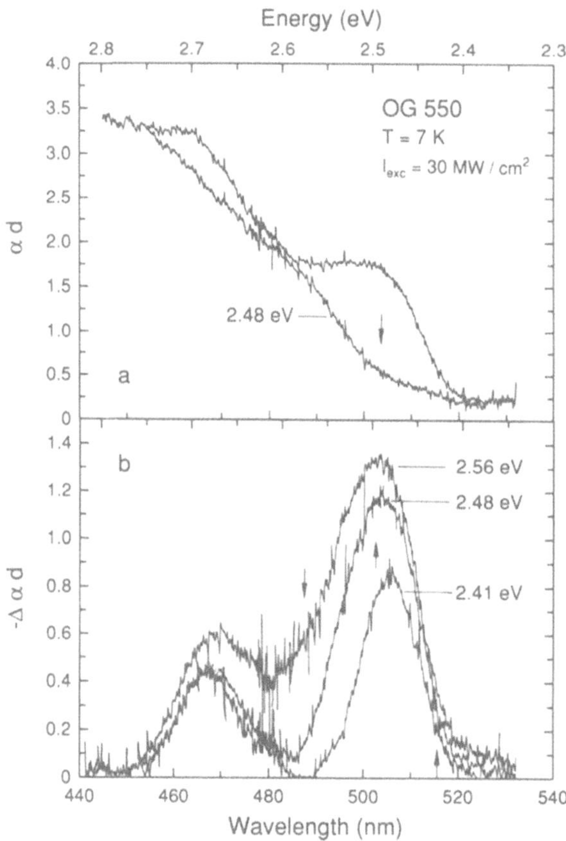

Fig 47 The bleaching of the absorption of $CdS_{1-x}Se_x$ microcrystallites in a glass matrix under high excitation (a) and the resulting differential transmission spectra for various values of $h\omega_{exc}$(b). According to [95].

For some time, people were very optimistic about the possible use of QD in optical data handling, because one expected a significant increase of the oscillator strength by the confinement and an increase of the resulting nonlinearites [97]. This hope has faded in the meantime because there are some unavoidable broadening processes of the resonances like size–distribution, phonon–coupling or surface states, which prevent working close to the resonance. Furthermore the volume fraction of the semiconductor in the organic or anorganic (glass) matrix is usually $\lesssim 10^{-3}$. In agreement with these considerations it has been found from LIG experiments, that the effective $\chi_{eff}{}^{3}$ for semiconductor doped glasses is several orders of magnitude smaller than e.g. of bulk CdS. Maximum values are $\chi_{eff}{}^{3} = 2 \cdot 10^{-9}$ and $9.5 \cdot 10^{-3}$, respectively [95c]. Finally semiconductors doped glasses have almost no electric properties, which are a necessary perequisit in their application as hybrid electro – optic devices which will pave the way of optics into data handling in the next decade.

On the other hand there are many unsolved problems in this field [28], which are very interesting from the point of fundamental research, e.g. the role of interface states [98] so that exciting basic research activities on QD can be expected in the future.

VI. THE EXCITON – POLARITON REVISITED

As already announced in section I.3 we come now back to the exciton–polariton. We elucidate first the concept of spatial dispersion and some of its consequences, present then a number of methods of $\vec{K}$ space spectroscopy which allow to measure the dispersion of exciton–polaritons directly and give then a short outlook.

VI.A. Spatial dispersion

As already mentioned earlier, the combination of the dielectric function $\epsilon\ (\omega, \vec{K})$ (50) and of the polariton equation (51) gives an implicite representation of the polariton dispersion.

The dependence of ω_0 on $\vec{K}$ is known as spatial dispersion. We repeat in Fig 48a the dispersion relation of the exciton polariton for $\gamma = 0$. For $\omega > \omega_L$ we get two propagating modes in the sample or even more, if the dispersion relation is more complex or if the longitudinal branch couples to the radiation field as it may occur for $\vec{K} \neq 0$ or for oblique incidence in uniaxial crystals.

We show in Fig 48b such a situation. The components of $\vec{K}$ parallel to the surface are conserved for the reflected and transmitted beams due to the translational invariance of the problem in this direction. Since $\vec{K}$ vectors of the modes in the samples are different, the beams propagate into different directions and this is one reason why the $\vec{K}$–dependence of ω_0 is known as "spatial" dispersion. Below ω_0 we have at least one propagating and one evanescent wave which gets, however, with finite damping γ also a small real part.

We have to consider at all frequencies more than one field amplitude in the medium (partly of propagating, partly of evanscent waves). The boundary conditions deduced from Maxwell's equations (e.g. steadyness of the tangential component of the electric field strength $\vec{E}$ and of the normal component of the dielectric displacement $\vec{D}$) account for a given incident beam only for one reflected and one transmitted beam. If there are more transmitted beams, additional boundary conditions (abc)

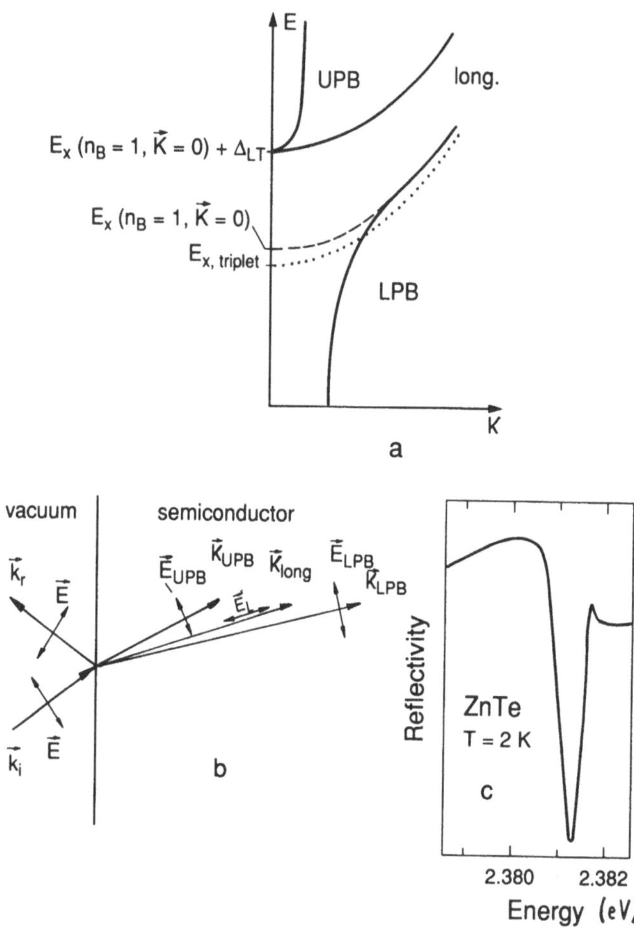

Fig 48 The dispersion of the exciton polariton without damping (a) the reflected and transmitted beams in the vicinity of the exciton resonance for $h\omega > h\omega$ (b) and a reflection spectrum of the $n = 1$ Γ_5 exciton in ZnTe according to [99] (c).

have to be introduced like the disappearance of the polarisation at the surface etc. Details of this problem are given e.g. in [100] and references therein.

Another consequence of the spatial dispersion is the absence of a strict stop–band between ω_0 and ω_{L0}. As one can see from Fig 48a there is at least one propagating mode for every frequency. As a consequence, the reflectivity does not reach unity between ω_0 and ω_{L0} even for vanishing damping in contrast to Fig 17. See e.g. the reflection spectrum of CdS in Fig 21c or the extreme case for ZnTe in Fig 48c, where mainly a reflection minimum is left around ω_{L0}, more precisely at the frequency where the refractive index n is one on the UPB (see eq (54)).

A fit of the reflection spectra allows to deduce the parameters of the resonance, however in a rather indirect way. Therefore various methods of $\vec{K}$–space spectroscopy have been developped to measure the dispersion–relation of the exciton polariton more directly. We shall outline several of them in the following.

VI.B. Methods of $\vec{K}$-space spectroscopy

The first method of $\vec{K}$ – or momentum space spectroscopy concerns the investigation of the Fabry–Perot modes, which occur e.g. in as – grown, thin, platelet–type samples. Transmission maxima occur in a Fabry–Perot if an integer number of half–waves fits into the resonator, i.e. for

$$K_m(\omega_m) = m\,\frac{\pi}{d};\; m = 1, 2, 3, \dots \tag{103}$$

where d is the geometric thickness of the sample. If ω_m and K_m (or $n(\omega_m)$) are precisely known for one m, then the whole dispersion curve can be reconstructed from (103). In Fig 49a we give an example for CuCl where the modes equidistantly spaced in $\vec{K}$ are indicated on the dispersion relation (from [101]) and in Fig 49b we show the measured and calculated transmission of a CdS platelet (from [102]). There we can see clearly the modes of the LPB above 4854 nm and the overlap of modes with narrow and wide spacing below, which indicated that part of the light propagates on the LPB and the other part on the UPB. Similar measurements exist also for CdSe [103].

The next method is resonant Brillouin–scattering. While Raman–scattering stands for a coherent process in which optical phonons (or phonon–polaritons [17–19]) are created or annihilated, Brillouin–scattering is a similar process involving acoustic phonons. Due to the linear dispersion relation of them, the momentum transfer is proportional to the shift in frequency. Fig 50a gives schematically the processes possible around the resonance in a backward configuration. Since the coupling of the acoustic phonons to the exciton polariton is very weak and the frequency shifts small ($\lesssim$ 1 meV), the observation of the Brillouin scattering needs some experimental skill.

Nevertheless it has been observed e.g. in CdS ([104] and references therein) and GaAs [105]. We give in Fig 50b data and fit curves for the $A\Gamma_5$ (n = 1) resonance in CdS from [104b]. The parameters given in the inset allow of course also the calculation of the dispersion relation. Fig 50c from [105] gives data for GaAs.

It is well known that diffraction from a prism can be used to determine the real part of $\tilde{n}$ and thus of $\vec{K}$. If sufficiently thin prism–shaped samples (d ≈ 1μm) are available it is possible to use this technique also in the resonance region, where one has high absorption. In Fig 51 we give two examples from [106, 107] for CdS. In Fig 51a one sees again the dichroism for the polarization $\vec{E} \perp \vec{c}$ (Γ_5) and $\vec{E} \parallel \vec{c}$ (Γ_1) already known from the reflection spectra of Fig 21b. In Fig 51b the dispersion of the LPB could be followed up to $n_1 \approx 25$ corresponding to $\epsilon \approx 600$ or $K \approx 3 \cdot 10^6$ cm^{-1}!

If one compares the time of flight of a ps–pulse through a sample with the propagation in vacuum, one can deduce the group velocity v_g. Since

$$v_g = \frac{d\omega}{dK} \tag{104}$$

this measurement gives the slope of the dispersion. In Fig 52 the values of v_g deduced from the calculated dispersion relation in CuCl are compared with experimental data [108]. The good agreement between experiment and theory prooves again the validity of the polariton concept. One can see from Fig 52 that v_g can be as low as $5 \cdot 10^{-5}$ c.

Similar measurement have been performed for GaAs [109] or CdS [110].

A nonlinear method of $\vec{K}$–space spectroscopy is two–photon– or hyper– Raman scattering. An incident laser creates in the sample two photon–like polaritons. This two polariton state decays under energy and momentum conservation into a photon–like polariton, which is observed as Raman emission, and another photon – or exciton like particle. The virtually excited intermediate biexciton state enhances the transition probability. An example for a scattering process is shown in Fig 53a.

In Fig 53b we show again the polariton dispersion of the lowest Γ_5 exciton polariton in CuCl together with data–points determined by hyper –Raman scattering and by two–photon absorption.

Hyper–Raman scattering has been observed in many SC. Examples and a rather detailed list of references are given in [22].

To conclude this subsection we present a method which allows to measure the dispersion of the surface polaritons shown in Fig 16 namely the attenuated (or frustrated) total reflection (ATR). A light beam is sent into a prism (Fig 54a) in a way that total internal reflection occurs on its base. An evanescent wave propagates under these conditions in the vacuum as indicated schematically. The frequency of this evanscent wave can be tuned trivially, its wave–vector $\vec{K}_{\parallel}$ by varying the angle of incidence. If a semiconductor is brought close enough (distance $< \lambda$) to the base of the prism, the evansecent wave can couple to the surface–polariton modes if $\hbar\omega$ and $K_{\parallel}$ coincide. In this case the total reflection is attenuated. This method can be used for surface phonon–, plasmon– and exciton polaritons [17–19]. We show in Fig 54b measured and calculated ATR spectra and in c the bulk and surface modes of the $C\Gamma_1$ ($n_B = 1$) resonance in ZnO. Some further aspects of this problem and other materials are discussed in [112, 113] and references therein.

VI.C. Outlook

The investigation of polaritons and $\vec{K}$–space spectroscopy was a very active field of semiconductor research in the seventies and the early eighties. Then MQW samples and other systems of reduced dimensionality came up and drew a large fraction of the interest in SC–optics on them. In the beginning the available samples were of rather poor quality and the resonances were consequently so broad, that the effects which are characteristic for the difference between the weak coupling and the strong coupling or polariton approach were completely hidden, with the consequence that only the simplier weak coupling approach was used. In the meantime, the quality of

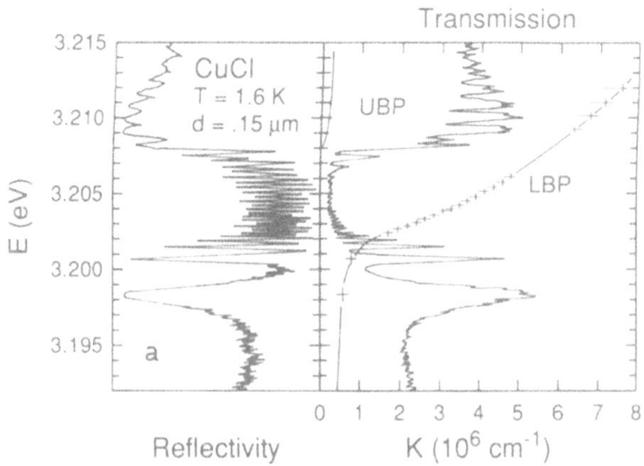

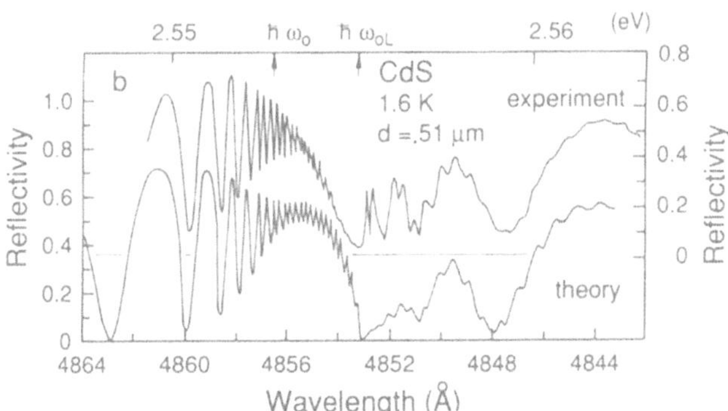

Fig 49 The transmission and reflection spectra of a thin CuCl film and the extrema indicated on the dispersion curve of the exciton polariton, according to [101] (a) and the measured and calculated reflection spectra of a CdS platelet type sample, according to [102] (b).

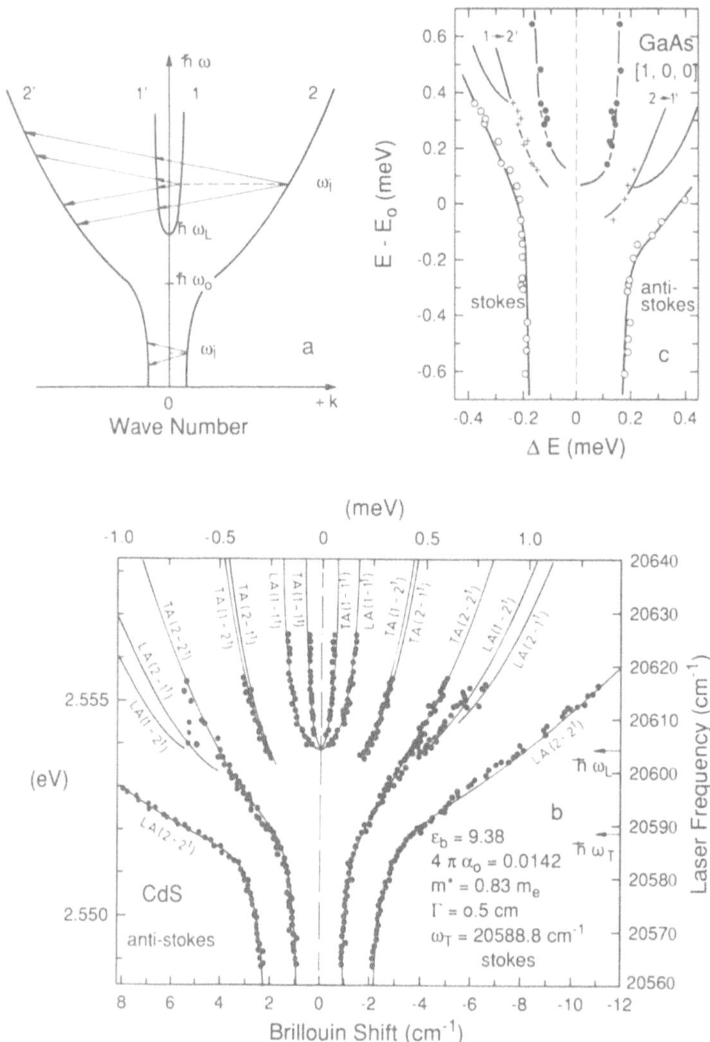

Fig 50 Schematic representation of Brillouin–scattering in a backward geometry in the vicinity of an exciton polariton resonance. In principle the corresponding set of scattering processes should appear twice, involving transverse (TA) and longitudinal (LA) acoustic phonons (a). Experimental results for CdS according to [107b] (b) and GaAs according to [105] (c).

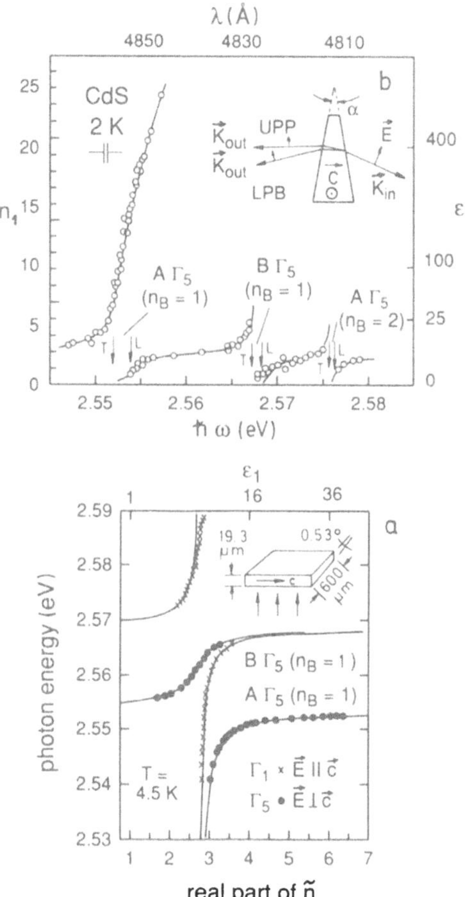

Fig 51 The dispersion of the real parts of the refractive index and of the dielectric function $n(h\omega)$ and $\epsilon_1(h\omega)$, respectively, in the vicinity of the A and B exciton resonances of CdS; (a) according to [106] and (b) according to [107].

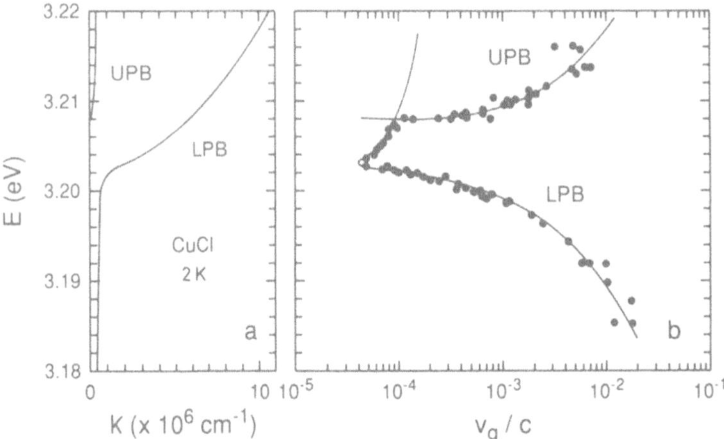

Fig 52 The dispersion of the n $= 1\Gamma_5$ exciton polariton resonance in CuCl (a) and the resulting group velocity (——) in (b) compared to data deduced from the propagation of a short pulse through the sample (·) in (b). According to [108].

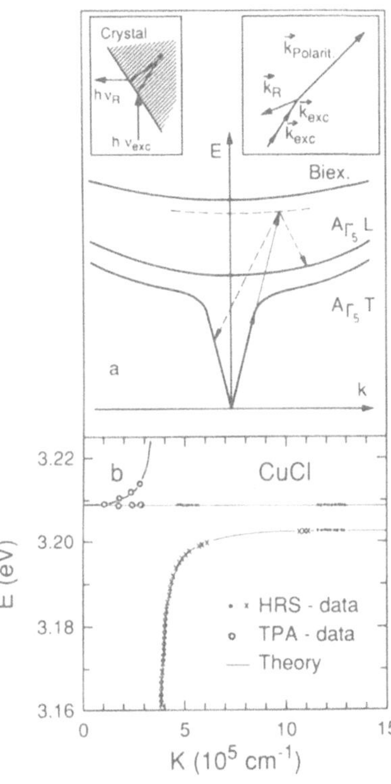

Fig 53 The schematics of two–photon or hyper Raman scattering (a) and the polariton dispersion of CuCl with data points from HRS and TPA according to [22b, 111] (b).

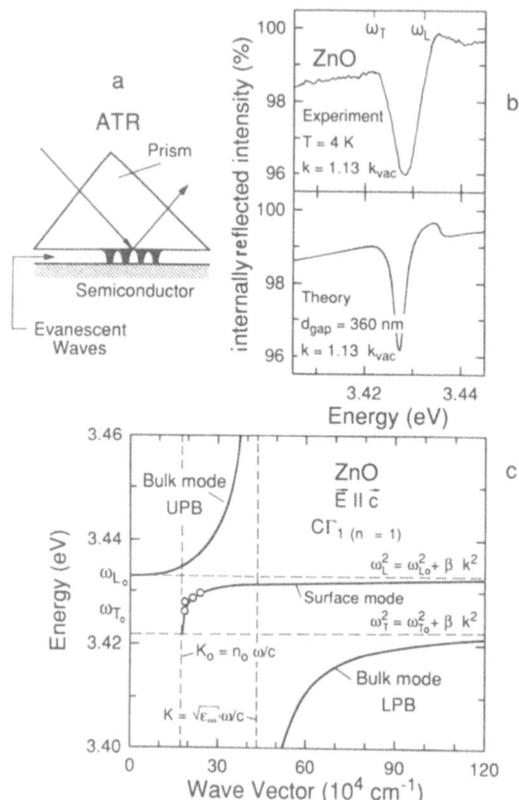

Fig 54 The basic concept of attennated total reflection (ATR) (a) a measured and a calculated ATR spectrum (b) and the bulk– and surface polariton dispersion in the $C\Gamma_1$ (n = 1) resonance of ZnO (c). According to [17a, 112].

the samples became better and a new generation of young scientists started their carreers. They begin gradually to ask themselves, what is really propagating in a MQW if they shine light on it or what is interacting in a photon–echo experiment or in the ac–Stark effect, and they will end to rediscover the polariton concept. Obviously every generation has to invent things for themselves, and the author feels, that this statement is true not only for exciton resonances or the interpretation of photon–echos.

VII. CONCLUSION

The aim of this contribution was to introduce the reader in an intuitive way to the fascinating scientific world of excited states spectroscopy in semiconductors. A more detailed book, which covers the areas mentioned here and some more, and which uses the same approach is under preparation [114]. Several topics which are directly connected with excited states of semiconductors have not been mentioned here at all like the application of the optical and electro–optical nonlinearities which are a necessary consequence of the excitation of a SC in optical bistability, in opto–electronic and optic data handling or in the realisation and analysis of various scenarios of nonlinear dynamics and of synergetics. We recommend some literature e.g. [115] and the reference given therein.

ACKNOWLEDGEMENTS

The author wants to thank his scientific teachers and many colleagues around the world for fruitful and stimulating discussions and his coworkers for their busy research activities in the lab. A large part of the work presented here has been supported by the Deutsche Forschungsgemeinschaft.

REFERENCES

[1] J.E. Bernard, D.E. Berry and F. Williams in "Energy Transfer Processes in Condensed Matter", B: Di Bartolo ed., NATO ASI series 114 1 (1984) Plenum Press

[2] Ch. Kittel, Introduction to Solid State Physics 6th Edition, John Wiley and Sons, New York (1986)
H. Ibach and H. Lüth, Festkörperphysik, 3rd Edition, Springer, Berlin (1990)
O. Madelung, Introduction to Solid State Theory, Springer Series in Solid–State Sciences 2, Springer, Berlin (1978)

[3] K. Seeger, Semiconductor Physics, 3rd Edition, Springer Series in Solid–State Sciences 40, Springer, Berlin (1985)
S.M. Sze, Physics of Semiconductor Devices, 2nd Edition, John Wiley and sons, New York (1981)

[4] Landolt–Börnstein, New Series, Group III, Vol 17a–g; O. Madelung, M. Schulz and H. Weiss eds., Springer, New York (1982)

[5] H. Haug and S.W. Koch, Quantum Theory of Optical and Electronic Properties of Semiconductors, World Scientific, Singapore (1990)

[6] Collective Excitations in Solids, B. Di Bartolo ed., NATO ASI Series B 88 (1983)

[7] C. Klingshirn, Topics in Applied Physics 65 201 (1989)

[8] a.) Excitons, K. Cho ed., Topics Curr. Phys. 14, Springer, Berlin (1979)
b.) Excitons, E.I. Rashba and M.D. Sturge eds. Modern Problems in Condensed Matter Sciences, 2, North Holland, Amsterdam (1982)
c.) B. Hönerlage, R. Lévy, J.B. Grun, C. Klingshirn and K. Bohnert, Phys. Rep. 124 161 (1985)

[9] H. Haug and S. Schmitt–Rink, Progr. Quant. Electr. 9, 3, (1984)
[10] H. Haug and H.H. Kranz, Z. Physik B 53 151 (1983)
J.P. Wolfe and A. Mysyrowicz, Scientific American, 250 Nr 3 p 70 (1984)
[11] Excitons in Confined Systems, R. Del. Sole, A. D'Andrea and A. Lapiccirella, M. Shinada, S. Sugano, J. Phys. Soc. Japan 21 1936 (1966)
U. Ekenberg and M. Altarelli, Phys. Rev. B 35 7585 (1987)
T. Ogawa and T. Takagahara, Phys. Rev. B in press (1991)
[12] F. Urbach, Phys. Rev. 92 1324 (1953)
W. Martienssen, J. Phys. Chem. Sol. 2257 (1957) and ibid. 8 294 (1959)
[13] J.D. Dow and D. Redfield, Phys. Rev. 85 94 (1972)
M. Sumi and Y. Toyozawa, J. Phys. Soc, Japan 31 342 (1971)
M. Schreiber and Y. Toyozawa, J. Phys. Soc. Japan 51 1528, 1537, 1544 (1982) and ibidem 52 318 (1983)
J.G. Liebler, S. Schmitt–Rink and H. Haug, J. Luminesc. 34 1 (1985)
[14] L.I. Schiff, Quantum Mechanics, Mc. .Graw Hill, New York (1955)
A. Messiah, Mécanique Quantique, Dunod, Paris (1964)
K. Gottfried, Quantum Mechanics, W.A. Benjamin, New York (1966)
[15] C. Klingshirn, NATO ASI series B 114 285 (1984)
[16] A. Stahl and Ch. Uihlein, Advances in Solid State Physics XIX 159 (1979)
M. Matsushita, J. Wickstead and H. Z. Cummins, Phys. Rev. B. 29 3362 (1984)
[17] J. Lagois and B. Fischer, Advances in Solid State Physics XVIII 197 (1978)
Surface Polaritons in Modern Problems in Condensed Matter Sciences, D.L. Mills and V.M. Agranovich eds., North Holland (1981)
[18] G. Borstel, H.J. Falge and A. Otto, Springer Tracts in Modern Physics 74 107 (1974)
Festkörperspektroskopie, H. Kuzmany, Springer, Berlin (1990)
[19] J. Brandmüller, R. Claus and L. Merten, Springer Tracts in Modern Physics 75 (1975)
R. Claus, phys. stat. sol. b 100 9 (1980)
[20] C.H. Henry and J.J. Hopfield, Phys. Rev. Lett. 15 964 (1965)
[21] G.W. Spiker and H.Y. Fan, Phys. Rev. 106 882 (1957)
[22] a.) C. Klingshirn and H. Haug, Phys. Reports 70 315 (1981)
b.) B. Hönerlage, R. Lévy, J.B. Grun, C. Klingshirn and K. Bohnert, Phys. Rep. 124 161 (1985)
[23] N.F. Mott and E.A. Davies, Electronic Processes of Non–Crystalline Solids, Clarendon press, Oxford (1971)
J.M. Ziman, Models of Disorder, Cambridge University Press (1979)
R. Zallen, Physics of Amorphous Solids, Wiley, New York (1983)
B.I. Shklovskii and A.L. Efros, Properties of Doped Semiconductor, Springer Series in Solid State Sciences 45 (1984)
Localization, Interaction and Transport Phenomena, B. Kramer, G. Bergmann and Y. Bruyseraede eds., Springer series in Solid State Sciences 61 (1984)
Localisation and Interaction, W.M. Finlayson ed.; 31 Scottish University, Summer School in Physics (1986)
Disordered Solids, B. Di Bartolo ed., Ettore Majorana Intern. Science Series, Physical Sciences 46 Plenum Press, New York (1989)
[24] D.S. Chemla, unpublished
[25] C. Klingshirn in Laser Spectroscopy of Solids II, W.M. Yen ed., Topics in Applied Physics 65 Springer, Berlin p 201 (1989)
[26] Semiconductor Quantum Wells and Superlattices, D.S. Chemla and A. Pinczuk eds., IEEE J. QE. 22 No. 9 (1986)
Physics and Applications of Quantum Wells and Superlattices, E.E. Mendez and K.V. Klitzing eds., NATO ASI Series B 175, Plenum, New York (1987)
Optical Switching in Low–Dimensional Solids, H. Haug and L. Banyai eds.; NATO ASI series B 194 (1989)
E.O. Göbel and K. Ploog, Progr. Quant. Electr. 14 289 (1990)
C.W.J. Beenabber and H. van Houten, Sol. State Phys. 44 1 (1991)
G. Bastard, J.A. Brun and R. Ferreira, ibid. p 229

[27] D. Heitmann, T. Demel, P. Grombow and K. Ploog, Advances in Solid State Physics 29 285 (1989)
U. Merkt, ibid. 30 77 (1990)

[28] a.) Al. L. Efros and A.L. Efros, Sov. Phys. Semicond. 16 772 (1982)
b.) L. Brus, IEEE J. QE. 22 1909 (1986)
c.) N.F. Borrelli, D.W. Hall, H.J. Holland and D.W. Smith, J. Appl. Phys. 61 5399 (1987)
d.) A.I. Ekimov and A.L. Efros, phys. stat. sol. b 150 627 (1988)
e.) N. Peyghambarian, B. Fluegel, D. Hulin, A. Migus, M. Joffre, A. Antonetti, S.W. Koch and M. Lindberg IEEE J. QE 25 2516 (1989)
f.) F. Henneberger, J. Puls, Ch. Spiegelberg, A. Schülzgen, H. Rossmann, V. Jungnickel and A.I. Ekimov, Semicond. Sci. and Technology 6 No 9A, p A41 (1991)
g.) V. Esch, K. Kong, B. Fluegel, Y.Z. Hu, G. Khitrova, H.M. Gibbs, S.W. Koch and N. Peyghambarian, J. Nonlin. Optics, in press
h.) A. Uhrig, L. Banyai, S. Gaponenko, A. Wörner, N. Neuroth and C. Klingshirn, Z. Phys. D 20 345 (1991)
i.) R. Reisfeld, Semiconductor Quantum Dots in Amorphous Materials, in the Proc. of this school
k.) II–VI '91, Tamano, Japan, Sept (1991) to be published in J. Crystal Growth
l.) L. Brus, Applied Optics in press; M.G. Bawendi, P.J. Carroll, W.L. Wilson and L. Brus, J. Chem Phys in press

[29] A. Uhrig, L. Banyai, Y.Z. Hu, S.W. Koch, C. Klingshirn and N. Neuroth, Z. Phys. B 81 385 (1990)

[30] A. Uhrig, A. Woerner, C. Banyai, S. Gaponenko, I. Lacis, N. Neuroth, B. Speit, K. Remnitz and C. Klingshirn in Ref [28k]

[31] I.M. Lifshitz and V.V. Slezov, Sov. Phys. JETP 35 331 (1959)

[32] S.W. Koch, Physikal. Blätter 46 167 (1990)

[33] M.C. Klein, F. Hache, D. Richard and C. Flytzanis, Phys. Rev. B 42 11123 (1990)
M.G. Bawendi, W.L. Wilson, L. Rothberg, P.J. Carroll, T.M. Jedju, M.L. Steigerwald and L.E. Brus, Phys. Rev. Lett. 65 1623 (1990)

[34] Spectroscopy of Solid State Laser Type Materials, B. Di Bartolo ed., Ettore Majorana Intern. Science Series, Physical Sciences 30 (1987)

[35] I. Rückmann, U. Woggon, J. Kornack, M. Müller, J. Cesnulevicius, J. Kolenda and M. Retranskas, SPIE 1513 (1991) in press

[36] a.) Excitonic Processes in Solids, M. Ueta, H. Kanzaki, K. Kobayashi, Y. Toyozawa and E. Hanamura, Springer Series in Solid–State Sciences 60 (1986)
b.) Laser–Induced Gratings, J.H. Eichler, P. Günter and D.W. Pohl, Springer Series in Optical Sciences 50 (1986)
c.) Optical Nonlinearities and Instabilities in Semiconductors, H. Haug ed., Academic Press, Boston (1988)
d.) R. Zimmermann, Many Particle Theory of Highly Excited Semiconductors, Teubner Texte zur Physik 18 Leipzig (1988)
e.) S. Schmitt–Rink, D.S. Chemla and D.A.B Miller, Advances in Physics 38 89 (1989)

[37] a.) L. Allen and J. Eberly, Optical Resonance and Two–Level Atoms, John Wiley and Sons, New York (1975)
b.) J.D. Macomber, The Dynamics of Spectroscopic Transitions, John Wiley and Sons, New York (1976)
c.) R.L. Shoemaker, Coherent Transient Infrared Spectroscopy, Plenum Press, New York (1978)
d.) Y.R. Shen, The Principles of Nonlinear Optics, J. Wiley and Sons, New York (1984)
e.) A.M. Weiner, S.D. Silvestri and E.P. Ippen, JOSA B2 654 (1985)
f.) M.D. Levenson and S.S. Kano, Introduction to Nonlinear Laser Spectroscopy, Academic Press Inc., London (1988)

[38] R. Macfarlane, this issue
[39] C. Dörnfeld, Ph. D. Thesis, Kaiserslautern (1990) and C. Dörnfeld and J.M. Hvam, IEEE J. QE 25 904 (1989)
[40] J. Kuhl, A. Honold, L. Schultheis and Ch.W. Tu, Advances in Solid State Physics 29 157 (1989)
[41] Ch. Weber, U. Becker, R. Renner and C. Klingshirn, Z. Physik B 72 379 (1988)
[42] P.J. Dean and D.C. Herbert in Topics in Current Physics 14, 55 K. Cho ed., Springer (1979)
[43] a.) H. Schwab, V.G. Lyssenko and J.M. Hvam, Phys. Rev. B 44 3999 (1991)
b.) H. Schwab, Ph. D. Thesis, Kaiserslautern (1992)
c.) H. Schwab, V.G. Lyssenko, J.M. Hvam, M. Urban and C. Klingshirn, Proc. II–VI '91, Tamano, Japan, J. Crystal Growth in press
[44] H. Stolz, V. Langer, E. Schreiber, S. Permogorov and W. von der Osten, Phys. Rev. Lett. 67 679 (1991)
[45] V. Langer, H. Stolz and W. von der Osten, Phys. Rev. Lett. 64 854 (1990)
[46] C. Klingshirn, in Disordered Solids, B. Di Bartolo ed., Ettore Majorana International Science Series, Physical Sciences 46 111, Plenum Press (1989)
[47] E. Cohen and M.D. Sturge, Phys. Rev. B 25 3828 (1982)
S.A. Permogorov, A. Reznitsky, P. Flügel, S. Verbin, G.O. Müller and M. Nikiforova, phys. stat. sol. b 113 589 (1982)
S. Permogorov and A. Reznitsky, J. Luminesc. (1992) in press
[48] H. Schwab, V. Lyssenko, J.M. Hvam and C. Klingshirn, Phys. Rev. B 44 3413 (1991)
[49] G. Noll, U. Siegner, S.G. Shevel and E.O. Göbel, Phys. Rev. Lett. 64 792 (1990)
[50] E. Cohen, Proc. 17th Intern. Conf. on the Physics of Semiconductors, J.D. Chadi and W.A. Harrison, eds., p. 1221, Springer (1984).
S.A. Permogorov, A.N. Reznitsky, S.Yu. Verbin and V.G. Lyssenko, JETP Lett. 37 463 (1983) and Sol. State. Commun. 47 5 (1983).
S. Permogorov, A. Reznitsky, S. Verbin, A. Naumov, W. von der Osten and H. Stolz, J. Physique C7, 173 (1985).
J.A. Kash, A. Ron and E. Cohen, Phys. Rev. B 28, 6147 (1983).
S.A. Permogorov, A.N. Reznitsky, S.Yu. Verbin and V.A. Bonch–Bruevich, JETP Lett. 38 25 (1983) and J. Lumin. 39 111 (1987).
J. Aaviksoo, J. Lippmaa, S. Permogorov, A. Reznitsky, P. Lavallard and C. Gourdon, JETP Lett. 45, 391 (1987).
S. Shevel, R. Fischer, E.O. Göbel, G. Noll, P. Thomas and C. Klingshirn, J. Lumin. 37 45 (1987).
H.E. Swoboda, F.A. Majumder, S. Shevel, R. Fischer, E.O. Göbel, G. Noll, P. Thomas, S. Reznitsky and S. Permogorov, J. Lumin. 38 79 (1987).
M. Urban, H. Schwab and C. Klingshirn, phys. stat. sol. b 166 423 (1991)
[51] C. Dörnfeld, R. Renner, H. Schwab, J.M. Hvam, G. Noll, E.O. Göbel and C. Klingshirn, Lasers '89, D.G. Harris and T.M. Shay eds., p. 782, STS Press (1990)
[52] C. Dörnfeld, G. Noll, H. Schwab, J.M. Hvam, Ch. Weber, R. Renner, E.O. Göbel, A. Reznitsky, V. Lyssenko, S.A. Pendjur, O.N. Talensky and C. Klingshirn, J. Crystal Growth 101 678 (1990)
G. Noll, H. Siegner, E.O. Göbel, H. Schwab, R. Renner and C. Klingshirn ibid. p. 731
[53] H. Schwab, V. Lyssenko, A. Reznitsky and C. Klingshirn, J. Lumin 48/49 661 (1991)
[54] F.A. Majumder, S. Shevel, V.G. Lyssenko, H.E. Swoboda and C. Klingshirn, Z. Physik B 66 409 (1987)
[55] D. Oberhauser, Ph. D. Thesis, Kaiserslautern (1992)
[56] D. Oberhauser, H. Kalt, H. Nickel, W. Schlapp and C. Klingshirn, DPC '91, Leiden, to be published in J. Lumin.

[57] J. Christen, Advances in Solid State Physics 30 239 (1990).
M.A. Herman, D. Bimberg and J. Christen, J. Appl. Phys. 70 R1 (1991).
D. Gammon, B.V. Shanabrook and D.S. Katzer, Phys. Rev. Lett. 67 1547 (1991)

[58] E.O. Göbel, H. Jung, J. Kuhl and K. Ploog, Phys. Rev. Lett. 51 1588 (1983).
J. Feldmann, G. Peter, E.O. Göbel, P. Dawson, K. Moore, C. Foxon and R.J. Elliot, ibid. 59 2337 (1987)

[59] D. Oberhauser, H. Kalt, C. Klingshirn, G. Weimann, W. Schlapp and H. Nickel, J. Lumin. 48/49 717 (1991)

[60] D. Oberhauser et al. to be published

[61] P. Thomas, Marburg, private communication

[62] M. Wegener, D.S. Chemla, S. Schmitt–Rink and W. Schäfer, Phys. Rev. A 42 5675 (1990).
K. Leo, M. Wegener, J. Shah, D.S. Chemla, E.O. Göbel, T.C. Damen, S. Schmitt–Rink and W. Schäfer, Phys. Rev. Lett. 65 1340 (1990)

[63] a.) C. Klingshirn, Semicond. Science and Technology 5, 457 and 1006 (1990)
b.) C. Klingshirn, Proc. II–VI '91, Tamano, Japan (1991), J. Crystal Growth, in press, and references therein

[64] J.M. Hvam, Sol. State Comm. 12 95 (1973)

[65] H. Schrey, V. Lyssenko and C. Klingshirn, Sol. State Commun. 32 897 (1979)

[66] V.G. Lyssenko, K. Kempf, K. Bohnert, G. Schmieder, C. Klingshirn and S. Schmitt–Rink, Sol. State Commun. 42 401 (1982)

[67] H. Kalt, V.G. Lyssenko, R. Renner and C. Klingshirn, JOSA B 2 1188 (1985).
H. Kalt, R. Renner and C. Klinghirn, IEEE J. QE. 22 1312 (1986).

[68] R. Zimmermann, Advances in Solid State Sciences 30 295 (1990) and references therein

[69] D. Fröhlich, A. Nöthe and K. Reimann, Phys. Rev. Lett. 55 1335 (1985)

[70] A. Mysyrowicz, D. Hulin, A. Antonetti, A. Migus, W.T. Masselink and H. Morkoc, Phys. Rev. Lett. 56 274 (1986).
W.H. Knox, D.S. Chemla, D.A.B. Miller, J.B. Stark and S. Schmitt–Rink, Phys. Rev. Lett. 62 1189 (1989).
B.D. Flügel, J.P. Sokoloff, F. Jarka, S.W. Koch, M. Lindberg, N. Peyghambarian, M. Joffre, D. Hulin, A. Migus, A. Antonetti, S. Ell, L. Banyai and H. Haug, phys. stat. sol. b 150 357 (1988).
N. Peyghambarian, S.W. Koch, M. Lindberg, B. Flügel and M. Joffre, Phys. Rev. Lett. 62 1185 (1989).
P.C. Becker, R.L. Forck, C.H. Brito Cruz, J.P. Gordon and C.V. Shank, Phys. Rev. Lett. 60 2462 (1988).
M. Joffre, D. Hulin, A. Migus and M. Combescot, Phys. Rev. Lett. 62 74 (1989)
B. Flügel, N. Peyghambarian, G. Olbright, M. Lindberg, S.W. Koch, M. Joffre, D. Hulin, A. Migus and A. Antonetti, Phys. Rev. Lett. 59 2588 (1987)
N. Peyghambarian and S.W. Koch, Rev. Phys. Appl. 22 1711 (1987)

[71] S. Schmitt–Rink, D.S. Chemla and H. Haug, Phys. Rev. B 37 94 (1988)

[72] D.S. Chemla, I. Bar–Joseph, J.M. Kuo, T.Y. Chang, C. Klingshirn, G. Livescu and D.A.B. Miller, IEEE JQE 24 1664 (1988)

[73] F.A. Majumder, H.–E. Swoboda, K. Kempf and C. Klingshirn, Phys. Rev. B 32 2407 (1985)

[74] S.Schmitt–Rink, C. Ell and H. Haug, Phys. Rev. B 33 1183 (1986).
G. Livescu, D.A.B. Miller, D.S. Chemla, M. Ramaswamy, T.Y. Chang, N. Sauer, A.C. Gossard and J.H. English, IEEE JQE 24 1677 (1988).
H. Kalt, K. Leo, R. Cingolani and K. Ploog, Phys. Rev. B 40 12017 (1989)

[75] T.M. Rice, Solid State Physics 32 1 (1977).
J.C. Hensel, T.G. Philips and G.A. Thomas, ibid. p 88.
V.B. Timofeev in ref [8b] p 349

[76] R.W. Martin and H.L. Störmer, Sol. State Commun. 22 523 (1977).
A. Forchel, B. Laurich, G. Moersch, W. Schmid and T.L. Reinecke, Phys. Rev. Lett 46 678 (1981) and Phys. Rev. B 25 2730 (1982).
G.A. Thomas, T.M. Rice and J.C. Hensel, Phys. Rev. Lett. 33 219 (1974)

[77] D. Bimberg, M.S. Skolnick and L.M. Sander, Phys. Rev. B 9 2231 (1979). R.Schwabe, F. Thuselt, H. Weinert and R. Bindemann, phys. stat. sol. b 95 571 (1979)

[78] H. Kalt, K. Reimann, W.W. Rühle, M. Rinker and E. Bauser, Phys. Rev. B 42 7058 (1990)

[79] P.L. Gourley and J.P. Wolfe, Phys. Rev. Lett. 40 526 (1978)

[80] G. Kirczenow, Can. J. Phys. 55 1787 (1977) and references therein

[81] H. Schweizer, W. Forchel, A. Haugleiter, S. Schmitt–Rink, J.P. Löwenau and H. Haug, Phys. Rev. Lett. 51 698 (1983)

[82] O. Hildebrandt, E.O. Göbel, K.M. Romaneck, H. Weber and G. Mahler, Phys. Rev. B 17 4775 (1978).
S. Tonaka, H. Kobayashi, H. Saito and S. Shionoya, J. Phys. Soc. Japan 49 1051 (1980).
M. Capizzi, A. Frova, S. Modesti, A. Selloni, J.L. Staehli and M. Guzzi, Helv. Phys. Acta 58 272 (1985)

[83] H.E. Swoboda, F.A. Majumder, V.G. Lyssenko, C. Klingshirn and L. Banyai, Z. Phys. B 70 341 (1988)

[84] K. Bohnert, M. Anselment, G. Kobbe, C. Klingshirn, H. Haug, S.W. Koch, S. Schmitt–Rink and F.F. Abraham, Z. Physik B 42 1 (1981)

[85] H. Yoshida and S. Shionaya, phys. stat. sol b 115 203 (1983).
Y. Unuma, Y. Abe, Y. Masumoto and S. Shionoya, ibid. 125 735 (1984)

[86] Ch. Weber, U. Becker, R. Renner and C. Klingshirn, Z. Physik B 72 379 (1988) and Appl. Phys. B 45 113 (1988)

[87] P. Vashista and R.K. Kalia, Phys. Rev. B 25 6492 (1982)

[88] R. Zimmermann, phys. stat. sol b 146 371 (1988)

[89] C. Klingshirn, Advances in Solid State Physics, 30 335 (1990)

[90] a.) G. Bongiovanni, J.L. Staehli and D. Martin, phys. stat. sol. b 150 685 (1988)
b.) G. Tränkle, H. Leier, A.Forchel, H. Haug, C. Ell and G. Weimann, Phys. Rev Lett. 58 49 (1987)
c.) K.–H. Schlaad, Ch. Weber, J. Cunningham, C.V. Hoof, G. Borghs, G. Weimann, W. Schlapp, H. Nickel and C. Klingshirn, Phys. Rev. B 43 4268 (1991)

[91] E. Lach, Ph. D. Thesis Stuttgart (1991)
E. Lach, V.D. Kulakovskii, A. Forchel, T.L. Reinecke, J. Straka, D. Grützmacher and G. Weimann
phys. stat. sol b 159, 125 (1990)
E. Lach, A. Forchel. D.A. Broido, T.L. Reinecke, G. Weimann and W. Schlapp Phys. Rev. B 42 5395 (1990)

[92] C. Weber, C. Klingshirn, D.S. Chemla, D.A.B. Miller, J.E. Cunningham and C. Ell, Phys. Rev. B 38 12748 (1988)

[93] C. Klingshirn, Ch. Weber, D.S. Chemla, D.A.B. Miller, J.E. Cunningham, C. Ell and H. Haug, NATO ASI series B 194 353 (1989)

[94] R. Zimmermann, to be published

[95] a.) A. Uhrig, L. Banyai, Y.Z. Hu, S.W. Koch, C. Klingshirn and N. Neuroth, Z. Phys. B 81 385 (1990)
b.) A. Uhrig, L. Banyai, S. Gaponenko, A. Wörner, N. Neuroth and C. Klingshirn Z. Physik D 20 345 (1991)
c.) A. Uhrig, A. Wörner, C. Klingshirn, L. Banyai, S. Gaponenko, I. Lacis, N. Neuroth, B. Speit and K. Remitz, Proc. II–VI '91, Tomano, Japan (1991), J. Crystal Growth, in press

[96] Y.Z. Hu, S.W. Koch, M. Lindberg, N. Peyghambarian, E.L. Pollock and F.F. Abraham, Phys. Rev. Lett. 64 1805 (1989).
V. Esch, K. Kang, B. Flügel, Y.Z. Hu, G. Khitrova, H.M. Gibbs, S.W. Koch and N. Peyghambarian, J. Nonlin. Optics in press

[97] S. Schmitt–Rink, D.A.B. Miller and D.S. Chemla, Phys. Rev. B 35 8113 (1987)

[98] G. Jungk and U. Woggon, Superlattices and Mikrostructures 9 314 (1991).
U. Woggon and I. Rückmann, ibid. p 245.

[99] G. Schmieder, Ph. D. Thesis, Karlsruhe (1981)
[100] A. Stahl and Ch. Uihlein, Advances in Solid State Physics 19 159 (1979). P. Halevi and R. Fuchs, J. Phys. C 17 3869, 3889 (1984)
[101] T. Mita and N. Nagasawa, Sol. State Commun. 44 1003 (1982)
[102] I.V. Makarenko, I.N. Uraltsev and V.A. Kiselev, phys. stat. sol. b 98 773 (1980)
[103] V.A. Kiselev, B.S. Razbirin and I.N. Uraltsev. Proc. 12th Intern. Conf. Phys. Semicond., M.H. Pilkuhn ed., p 996, Teubner, Stuttgart (1974)
[104] a.) G., Winterling, E.S. Koteles and M. Cardona, Phys. Rev. Lett. 39 1286 (1977), E. Koteles and G. Winterling ibid. 44 948 (1980)
b.) J. Wicksted, M. Matsushita, H.Z. Cummins, T. Shigenari and X.Z. Lu, Phys. Rev. B 29 3350 (1984).
M. Matsushita, J. Wicksted and H.Z. Cummins, Phys. Rev. B 29 3362 (1984).
c.) I. Broser and M. Rosenzweig, Sol. State Commun. 36 1027 (1980)
[105] R.G. Ulbrich and C. Weisbuch, Adances in Solid State Physics 28 217 (1978)
[106] I. Broser, R. Broser, E. Beckmann and E. Birkicht, Sol. State Commun. 39 1209 (1981)
[107] M.V. Lebedev, M.I. Strashnikova, V.B. Timofeev and V.V. Chernyi, JETP Lett. 39 366 (1984)
[108] Y. Masumoto, Y. Unuma, Y. Tanaka and S. Shionoya, J. Phys. Soc. Japan 47 1844 (1979)
[109] R.G. Ulbrich and G.W. Fehrenbach, Phys. Rev. Lett. 43 963 (1979)
[110] Y. Segawa, Y. Aoyagi and S. Namba, J. Phys. Soc. Japan 52 3664 (1983)
[111] D. Fröhlich, E. Mohler and P. Wiesner, Phys. Rev. Lett. 26 554 (1971)
[112] J. Lagois, Sol. State Commun. 39 563 (1981) and Phys. Rev. B 23 5511 (1981)
[113] M. Fukui, A. Kamada and O. Tada, J. Phys. Soc. Japan 50 866 (1981) and ibid. 53 1185 (1984)
[114] C. Klingshirn and H. Kalt, Semiconductor Optics in preparation
[115] Optical Computing, B.S. Wherrett and F.A.P. Tooley eds. Proc. 34. Scottish Universities Summer School in Physics, Edinburgh, Aug (1988).
"Nichtlineare Dynamik in Festkörpern", DPG School, Bad Honnef, Oct. (1989) to be published by Springer
C. Klingshirn in NATO ASI "Advances in Nonradiative Processes in Solids", B. Di Bartolo ed. series B 249 p 529, Plenum (1991)

ADVANCES IN THE CHARACTERIZATION OF EXCITED STATES OF LUMINESCENT IONS IN SOLIDS

G.F. Imbusch

Department of Physics
University College
Galway, Ireland

ABSTRACT

We discuss the spectroscopic studies of ions in solids which throw light on the properties of excited states which are reached by optical absorption from the ground state. The difference in the ion-lattice coupling between ground and excited states determines the bandshape of the radiative transitions and influences the process of nonradiative decay from the excited state. Absorption processes and excitation transfer out of the excited state also contribute to the reduction in quantum efficiency of the excited state.

I. INTRODUCTION

In these lectures we will be concerned with excited electronic states of ions in solids, in particular those states which can be reached from the ground state by an optical absorption transition. We will pay special attention to luminescent excited states.

The spectroscopic technique involves a transition between two states and gives information about the difference between them. Since in general our absorption and emission transitions occur between the ground state and some excited state, our spectroscopic technique in general gives information on how the excited state differs from the ground state. The electronic state of the ion is different after the transition, and the way in which the ion interacts with its environment may also be different. This latter difference determines the bandshape of the transition, and strongly influences the process of nonradiative relaxation out of the excited state

We will confine our attention to rare earth ions and transition metal ions.

Optical Properties of Excited States in Solids
Edited by B. Di Bartolo, Plenum Press, New York, 1992

II. RARE EARTH IONS

II.A Transitions within $4f^n$ states of rare earth ions

We first consider transitions within the $4f^n$ electronic configuration. For these states the coupling between the 4f electrons and the neighbouring ions is weak. If this coupling can be neglected, we then have free ion states |SLJ>, with a number of sharp energy levels and with sharp transitions between the levels [1]. To take into account the coupling between the rare earth ion and the surrounding ions, it is helpful to consider the coupling as consisting of two parts. First is the effect of the neighbouring ions when they are in their time-average positions. This produces the "static" electrostatic crystal field which causes a splitting of the free ion levels. Second is the effect of the vibrations of the ions about their time-average positions, the dynamic effect.

We consider first the effect of the static crystal field. Knowing the symmetry of this crystal field potential is very helpful, since it allows us to predict the number of multiplet levels into which each free ion |SLJ> level splits in the crystal field. Conversely, from the observed splitting of the levels we may be able to infer the symmetry of the crystal field potential. The crystal field can be expressed in terms of a small number of parameters which characterize its symmetry and strength, and well-established techniques have been developed to calculate crystal field splitting [1,2,3]. Since the crystal field multiplets have sharp energy levels, the radiative transitions between them should be sharp.

Next the effect of the vibrating environment is taken into account. This manifests itself as weak sidebands accompanying the sharp radiative transitions. These vibrational sidebands are transitions in which the creation or annihilation of a quantum of vibrational energy (a phonon) accompanies the change in the electronic state. These sidebands are usually termed vibronic transitions. The pure electronic transitions between the static crystal field levels, occurring without a change in vibrational energy, are termed zero-phonon transitions.

II.B Vibronic transitions

In general, the integrated strength of these vibronic transitions is weak in comparison with the strength of the accompanying zero-phonon line, and since the sidebands are very broad in comparison with zero-phonon lines, the sidebands can be difficult to study. Further, the abundance of sharp zero-phonon lines found in the spectra of many rare-earth ions can mask the vibronic transitions. Nevertheless, a large amount of data has been accumulated in the last few years on the intensities of vibronic transitions accompanying the $^6P_{7/2} \rightleftarrows {}^8S_{7/2}$ zero-phonon transitions on Gd^{3+} ions in a number of materials [4,5,6].

The overall strength of the vibronic sideband in these transitions can vary from less than 1% to more than 10% of the strength of the associated zero-phonon line. Fig. 1 shows the lowest crystal field $^6P_{7/2} \rightarrow {}^8S_{7/2}$ luminescence transition of Gd^{3+} in (a) LaB_3O_6:Gd, (b) GdB_3O_6, (c) a lanthanide borate glass [7].

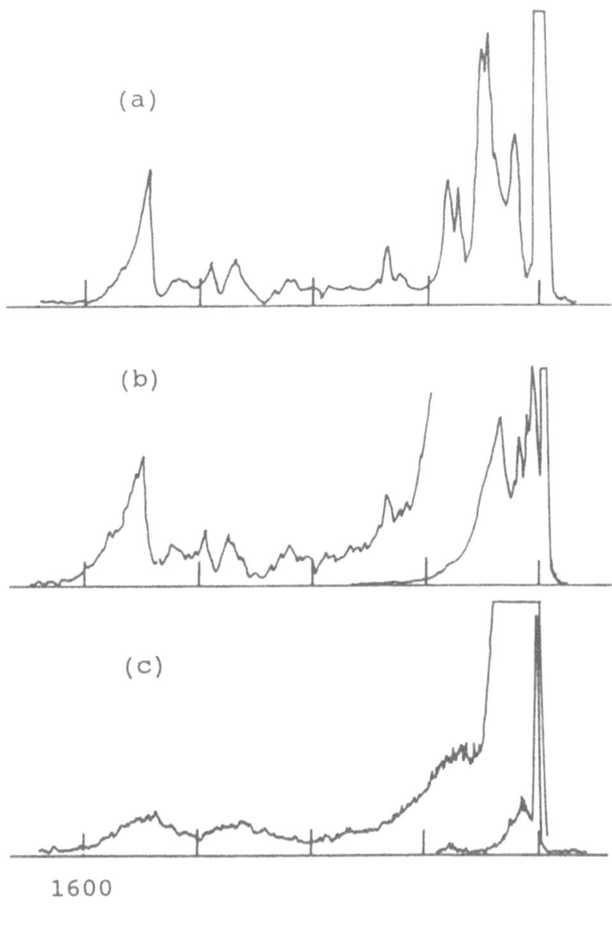

Fig.1 The $^6P_{7/2} \rightarrow {^8S_{7/2}}$ luminescence spectrum at 4.2 K showing the lowest energy zero-phonon line with accompanying sidebands of Gd^{3+} in (a) crystalline LaB_3O_6:Gd, (b) crystalline GdB_3O_6, and (c) Gd^{3+} in a lanthanide borate glass. In each case the zero-phonon line is seen at the zero of energy mark. In (b) and (c) the sideband region at low energy is masked by zero-phonon lines from a small number of Gd^{3+} ions in lower energy sites [7].

The spectra are taken at 4.2 K, and they are strongly amplified to show the sidebands clearly. We can think of the sidebands as occurring because the sharp transitions of the static crystal field are being <u>frequency modulated</u> by the lattice vibrations. Such FM sidebands, then, ought to reflect the nature and density of vibrational modes of the host material. From infrared spectroscopy we know that strong B-O vibrations of the BO_3 units occur at around 1400 cm^{-1}, vibrations of the BO_4 units occur at 900 - 1100 cm^{-1}, and borate bending vibrations occur at around 650 cm^{-1} [8]. Clearly, these vibrational modes show up in the vibronic spectrum. The low frequency sidebands below 400 cm^{-1} are ascribed to vibrations of the lattice as a whole rather than vibrations occurring within the borate units [8].

Obtaining a quantitative understanding of the intensity of vibronic transitions is a very difficult problem, but some progress has been made in recent years. The coupling of the 4f electrons with the crystalline environment is so weak that a simple frequency modulation of the sharp optical transitions cannot explain the complete vibronic intensity. We must seek an explanation by considering the dipole nature of the radiative transitions.

Transitions between pure $4f^n$ states take place between states of the same parity, and so are electric-dipole (ED) forbidden. Magnetic dipole (MD) transitions between pure $4f^n$ states are not forbidden, but allowed MD transitions are orders of magnitude weaker than allowed ED transitions [9]. The $^8S_{7/2} \rightleftarrows {}^6P_{7/2}$ zero-phonon transitions on the Gd^{3+} ions contain a large MD component. More often than not, radiative transitions between $4f^n$ electronic levels of rare earth ions in solids are found to be predominantly ED in nature. If the rare earth ion is in a site which lacks inversion symmetry, then the odd-parity component of static crystal field causes a small admixture of opposite-parity $4f^{n-1}5d$ states into the $4f^n$ states. Even though the amount of this mixing is small and has a negligible effect on the energy values, it can significantly affect the strength of the radiative transitions. This occurs because, although the admixture of a small amount of opposite-parity states allows only a weak ED process to occur, the ED process is some five orders of magnitude larger then the MD process, and this small ED process can dominate the transition. Even when the rare earth ion appears to occupy a site of inversion symmetry, the observed transitions are often found to have ED character, which may indicate a distortion of the site occupied by the rare earth ion. A general theory by Judd [10] and Ofelt [11], which takes into account the mixing of $4f^{n-1}5d$, $4f^{n-1}5g$ states with the $4f^n$ states by odd-parity static crystal field, gives a reasonably good explanation for the relative strengths of the ED transitions, although some modifications to this theory have been suggested recently [12].

In contrast to the zero-phonon transitions which can occur by MD and ED processes, vibronic transitions are believed to be predominantly ED in nature and are caused mainly by odd-parity vibrations which introduce an odd-parity dynamic crystal field potential and which, consequently, induces an ED transition on the rare earth ion. The specific process by which this can happen have been investigated by Faulkner and Richardson [13], Stavola et al. [14-16], Judd [17], Dexpert-Ghys and Auzel [18].

Judd considers two contributions. In the first, the nearby charged ligands polarize the f-electron system, and odd-parity vibrations of these ligands induce odd-parity dipole fields at the rare earth ion. In the second, the charge on the rare earth ion polarizes the nearby ligands, and their odd-parity vibrations induce electric dipole terms on rare-earth ion.

Judd [17] has given an expression for the vibronic oscillator strength, P_{vib}, arising from these two effects:

$$P_{\rm vib} = \frac{20\chi}{n}\ \frac{8\pi^2 m\nu}{3hg_{\rm j}}\ \frac{e^2}{R^6}\left[g + \frac{N\alpha}{R^3}\right]^2 \Xi^2 (U^{(2)})^2 (T^{(1)})^2 \qquad (1)$$

g_J is the degeneracy of the initial state, g is the charge on the nearby ligands, α is the polarizability of the ligands, N is the number of ligands, χ is the local field correction term, generally taken as $[(n^2 + 2)/3]^2$ where n is the refractive index. $(U^{(2)})$ is the reduced matrix element of the unit tensor operator $U^{(2)}$ linking the initial and final electronic states; values of this quantity have been calculated for the various rare earth transitions [2,19]. $T^{(1)}$ is an operator for the amplitude of the vibration, and $(T^{(1)})$ signifies its value where one phonon of frequency υ is created during the radiative process [20]. Ξ is an electric dipole-dipole matrix element, and the intermediate states in this double matrix element are the odd-parity electronic states $4f^{n-1}5d$, $4f^{n-1}5g$. These various terms have been evaluated [18,21,22], and the resulting vibronic transition probability would seem to be of the appropriate order of magnitude. It is doubtful if any greater degree of accuracy can be expected from such calculations. What we might hope to obtain are explanations for (i) the values of the sideband to zero-phonon line intensity ratio for the same zero-phonon transition on the same ion in different host materials, and (ii) the relative intensities of the sidebands accompanying different zero-phonon transitions on the same ion, when the $(U^{(2)})^2$ values for the different transitions can be compared. Blasse [6,23] lists the intensity ratio of sideband to zero-phonon line for the $^6P_{7/2} \rightarrow {}^8S_{7/2}$ transition of Gd^{3+} in a large number of compounds, and he seeks trends in the values of the intensity ratio which can be related to factors, such as the polarizability of neighbouring ligands, covalency, proximity to $4f^65d$ states, etc.

If the zero-phonon transition is an ED process of reasonable strength, then one should include the possibility of a sideband being caused by a frequency modulation of the zero-phonon transition by <u>even-parity</u> vibrations. Such a process could come about if, on being raised to the excited state, the coupling of the ion with its surrounding ligands changes by a significant amount, resulting in a contraction or expansion of the environment about the ion. This mechanism, which is much more common in transitions in which the electronic configuration changes between the ground and excited states (for example, $4f^n \rightleftarrows 4f^{n-1}5d$ transitions) will be described later. The appearance in a number of materials of sideband features of the $^8S_{7/2} \rightleftarrows {}^6P_{7/2}$ Gd^{3+} transitions due to coupling with the totally-symmetric vibrational mode have been pointed out [6]. These strongly suggest that such an FM effect is occurring in these materials. The need to take this FM sideband process into account has been stressed by Dexpert-Ghys and Auzel [18].

II.C $4f^n \rightleftarrows 4f^{n-1}5d$ transitions on rare earth ions in solids

A $4f^n \rightleftarrows 4f^{n-1}5d$ transition on a rare earth ion differs in two fundamental ways from a transition within the $4f^n$ configuration. First, a parity change occurs when an electron changes from 4f to 5d, and as a result an ED transition is no longer parity-forbidden. Next, a transition from 4f to 5d means that there is a change in the orbital nature of the electron state. Consequently, the coupling of the ion with the surrounding lattice changes, and this causes the transition to be very broad. Thus, in contrast to the sharp weak intra-$4f^n$ transitions, $4f^n \rightleftarrows 4f^{n-1}5d$ interconfigurational transitions are broad and strong.

Let us assume, initially, that the change in ion-lattice coupling takes the form of a totally-symmetric distortion in which the surrounding lattice expands or contracts about the ion, maintaining the overall site symmetry. This is the totally-symmetric stretch mode (or the breathing mode). If we further assume that the dynamic vibrations of the lattice are only breathing mode vibrations, then we can describe the effect on the ion-lattice system of the static and dynamic lattice variations by means of a single parameter, namely, the <u>configurational coordinate</u>, Q, which is a measure of the distance between the central ion and the first cage of adjacent lattice ions. If we now assume that the lattice restoring force is harmonic, then we have the simplest possible model to describe the ion-lattice system. The situation is sketched in Fig. 2; this is known as a <u>configurational coordinate diagram</u>.

The diagram plots the energy of the ion-plus-neighbouring-lattice system for the two electronic states (ground and excited states) as a function of Q. Q_o and Q_o' represent the equilibrium ion-lattice distances in the ground and excited electronic states, respectively. In the case where there is a significant change in the ion-lattice coupling when the ion goes from the ground electronic state to the excited electronic state, there will be a noticeable shift, $Q_o' - Q_o = \Delta Q$, between the two equilibrium positions. Given harmonic restoring forces, the potential energy of displacement in each electronic state is a parabola, and the lattice eigenstate is a harmonic oscillator eigenstate, χ_n, where n gives the number of quanta of vibrational energy associated with this lattice eigenstate. χ_n and χ'_m are the vibrational eigenstates associated with the ground and excited electronic states, respectively. If the parabolae in the ground and excited electronic states are identical - except for the horizontal shift ΔQ - then the vibrational quanta have the same energy in the two electronic states. We write this energy as $\hbar\omega$.

Using the configurational coordinate diagram we can now begin to understand the bandshapes of the optical transitions. We consider the absorption transition from the lowest vibrational eigenstate, χ_o, of the ground electronic eigenstate (g), such as we would have at low temperature. The transition can end up on any of the vibrational eigenstates, χ'_m, of the excited electronic state, e, and each such transition is governed by the dipole matrix element $\langle e,\chi'_m|D|g,\chi_o\rangle$, where D is the appropriate dipole operator. In the case where we are dealing with a change in electronic state, D is an electronic operator, so we can write the matrix element as

$$\langle e|D|g\rangle\langle\chi'_m|\chi_o\rangle \qquad (2)$$

that is, we can write it as a product of an electronic matrix element and a vibrational overlap integral.

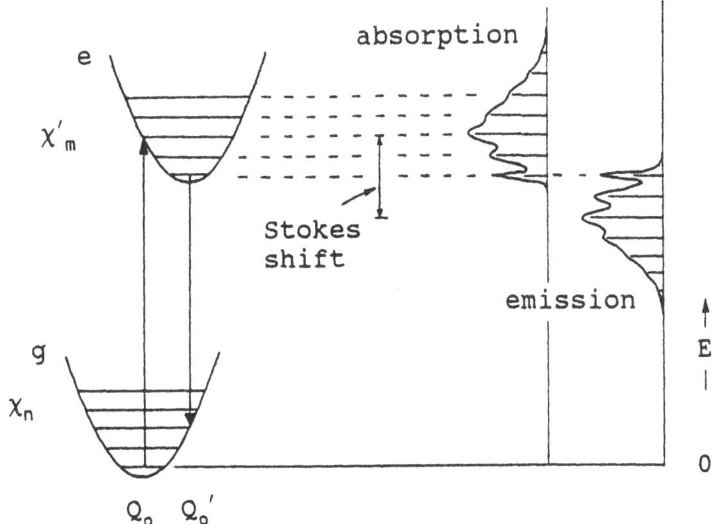

Fig. 2. Low temperature absorption and emission transitions between electronic states g and e. The zero-phonon line occurs at the same frequency in absorption and emission.

In the case where there is no difference in the interaction between the ion and the surrounding lattice in the two electronic states, then $\Delta Q = 0$, and the two sets of vibrational eigenstates are identical. Consequently, $\langle\chi_n|\chi'_m\rangle = \delta_{nm}$. Since the initial eigenstate is χ_0, the only final vibrational state with a non-zero transition probability state is the m = 0 state. We then have a single sharp transition with no change in vibrational state; phonons are neither created nor destroyed in the transitions. This is the <u>zero-phonon line</u>.

When $\Delta Q \neq 0$, the χ_n and χ_m' states are not identical, and the overlap integral $\langle\chi'_m|\chi_0\rangle$ is non-zero for a number of values of m. Thus, there are absorption transitions from state $|g,\chi_0\rangle$ to a number of vibrational levels of the excited state. The relative transition probability to the mth vibrational level [P(0,m)] is given by $|\langle\chi_0|\chi'_m\rangle|^2$, and this transition is higher in energy than the zero-phonon line by $m\hbar\omega$. The totality of these photon-plus-phonon transitions gives the overall transition its width. One of the transitions, from n= 0 to m = 0 is the zero-phonon line, but now it may be only a small sharp feature on the low energy side of the overall absorption transition.

The converse situation applies to the low temperature emission transition, going from $|e,\chi_0'\rangle$ to a number of $|g,\chi_n\rangle$ states. The zero phonon line, $|e,\chi'_0\rangle \rightarrow |g,\chi_0\rangle$, has the same frequency in emission and in absorption, but in emission the $|e,\chi_0'\rangle \rightarrow |g,\chi_n\rangle$ photon-plus-phonon transition has an energy which is $n\hbar\omega$ <u>below</u> the zero-phonon line. The zero-phonon line may now be only a weak sharp feature on the high energy side of the broad emission transition. The situation is sketched in Fig. 2.

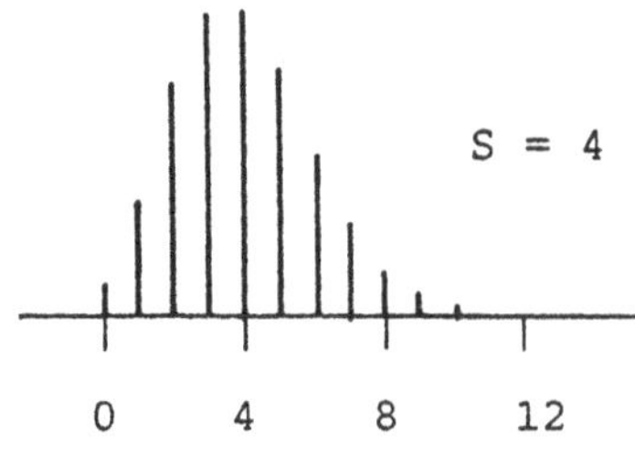

energy in units of $\hbar\omega$

Fig. 3. Relative intensities of the $0 \rightarrow m$ transitions for the case of S = 4.

An important parameter is the Stokes shift, the separation in energy between the absorption and emission peaks (Fig. 2). Clearly, the larger ΔQ the larger the Stokes shift. The size of the shift is given by $(2S-1)\hbar\omega$, where S is a dimensionless parameter (the Huang-Rhys parameter) which characterizes the difference in ion-lattice coupling between the two electronic states [9]. Based on the configurational coordinate model described above, we evaluate the relative intensity of the $|g,\chi_o\rangle \rightarrow |e,\chi'_m\rangle$ transition as

$$P(0,m) = |\langle\chi_m'|\chi_o\rangle|^2 = \exp(-S)S^m/m! \qquad (3)$$

Fig. 3 shows the relative intensities for the case of S = 4. We note that

$$\sum_m P(0,m) = 1 \qquad (4)$$

hence the fraction of the total transition strength contained in the zero-phonon line is exp(-S).

Although the model predicts that the absorption and emission transitions should take the form of sharp lines (corresponding to transitions to different vibrational levels of the final electronic state), we must allow for the existence of a band of vibrational modes rather than a single mode. This means that the transitions which involve a change in phonon state are usually broad and the overall shape can be a broad continuum, as predicted by the envelope of the sharp transitions in Fig. 3. This shape is Pekarian, a small distortion from Gaussian, in which there is a sharper fall off on the side containing the zero-phonon line, as is evident from the envelope of the transitions shown in Fig. 3. If a particular vibrational mode is dominant in the transition, the absorption and emission bands may contain a progression of peaks, corresponding to photon-plus-phonon transitions involving this mode, on top of the broad continuum, as drawn in the sketch in Fig. 2.

Although we have introduced the configurational coordinate model to describe the $4f^n \rightleftarrows 4f^{n-1}5d$ transition on rare earth ions, the model is of very general use, being applied successfully to transitions on colour centres and transition metal ions [9].

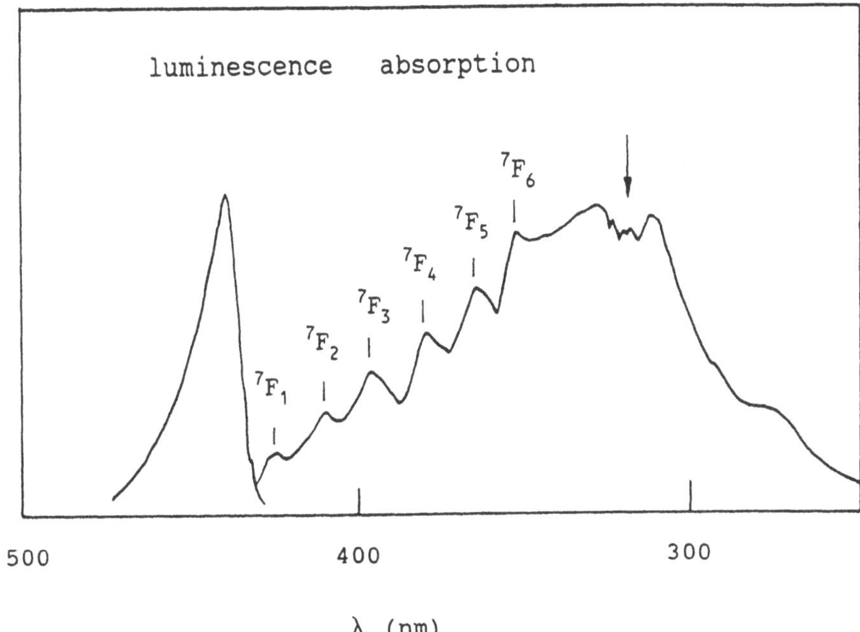

Fig. 4. Absorption and luminescence spectra of Eu in $Ba_5SiO_4Br_6$ at 4.2 K. The Eu^{2+} ions in dilute amounts substitute for Ba^{2+} [spectra redrawn from Ref. 24]

If the excited state undergoes a distortion other than of symmetry a_{1g}, say of e_g or t_{2g}, then an analogous model can be used, in which Q represents the magnitude of the distortion.

Despite the many simplifying assumptions inherent in the configurational coordinate model, it is surprisingly effective in predicting the shapes of broadband transitions on ions in solids. The closeness of the predicted and observed shapes should not, however, be taken as implying the validity of all the assumptions made in the model

As a first example of an interconfigurational transition on a rare earth ion, we consider the transitions on Eu^{2+} in $Ba_5SiO_4Br_6$ reported by Meijerink and Blasse [24]. The seven outer electrons of Eu^{2+} can have the $4f^7$ configuration, which is the same as that of the Gd^{3+} ion discussed earlier. The ground state is $^8S_{7/2}$, and the higher $4f^7$ levels are expected to be in very similar positions to that found for Gd^{3+}. Whereas in Gd^{3+} the $4f^65d$ states are over 50,000 cm^{-1} above the ground state, these $4f^65d$ states are much lower in Eu^{2+}, some being below the first excited state, ^6P, of the $4f^7$ configuration. Thus, the absorption in this material is characterized by the strong broad $4f^7(^8S_{7/2}) \rightarrow 4f^65d$ band. The lowest energy absorption spectrum is seen in Fig. 4. At low temperatures luminescence occurs from the lowest $4f^65d$ level to the $^8S_{7/2}$ final state. Because the ground state has essentially a single level, there is only a single zero-phonon line, and the broad luminescence band shown in Fig. 4 has the characteristic Pekarian shape predicted by the configurational coordinate model. A weak zero-phonon line is just detectable. The lifetime of the luminescence varies with temperature, but it is submicrosecond, consistent with the parity-allowed nature of the transition.

The shape of the absorption band seen in Fig. 4 is not a mirror image of the emission band; rather does it have a large amount of structure, which is indicative of splitting in the excited state, the terminal state of the absorption transition. The explanation for the series of peaks on the low energy side of the absorption band for this material was given by Meijerink and Blasse [24], following a model of Ryan et al. [25]. The excited state has the $4f^6 5d$ configuration, and the lowest state of the $4f^6$ configuration has the term ^7F_J, where J has values 0 → 6, and the different ^7F_J states cover a spread in energy of about 5000 cm^{-1}, with the $^7F_0$ state being lowest [$^7F_0$ is the ground term for the Eu^{3+} ion]. The peaks seen in the absorption spectrum correspond to the transitions to different ^7F_J levels of the excited state, as indicated in the figure. The sharp dips in the absorption spectrum at around 310 nm (indicated by the arrow in the figure) are also of interest; these coincide in energy with the positions expected for the weaker $4f^7\ ^8S_{7/2} \rightarrow 4f^7\ ^6I_J$ transitions. An analysis by Meijerink and Blasse [26] shows that these are Fano anti-resonances, due to an interference between the sharp $4f^7\ ^6I_J$ states and the broad $4f^6 5d$ continuum which occurs when these states have the same energy [27,28].

The next example we consider is the interesting spectrum of Yb^{2+} in SrF_2 [29]. Yb^{2+} has the filled-shell $4f^{14}$ ground state. The first excited state has the configuration $4f^{13}5d$, and this splits into two levels, at 3.2 eV (24,000 cm^{-1}) and 3.45 eV (27,750 cm^{-1}). Absorption into the lower level is forbidden, while absorption into the upper level is allowed. The absorption spectrum (A) is shown in Fig. 5. From comparison with the behaviour of Yb^{2+} in $SrCl_2$ and NaCl, one expects luminescence from these states with a Stokes shift to lower energy of around 1000 cm^{-1}. In $SrF_2:Yb^{2+}$, however, the luminescence is seen as a much broader band (curve F in Fig. 5), peaking at around 1.7 eV (13,700 cm^{-1}), shifted by around 10,000 cm^{-1} from the lowest absorption transition. Further, the decay rate, of the order of $10^3\ s^{-1}$, is much too small to be attributed to a 5d → 4f transition. Of particular interest in this material is the photoconductivity, which starts at just below 3 eV; the photoconductivity spectrum is shown by the broken curve (P) in Fig. 5.

McClure and Pedrini [29] offer an explanation for these spectra in terms of the existence of a trapped exciton, in which an electron is transferred from the Yb ion to the neighbouring lattice ions. In its lowest state, the electron and hole are bound to the Yb^{2+} ion; this can be called an <u>impurity-trapped exciton</u>.

The change from Yb^{2+} to Yb^{3+} should cause a sizeable reduction (~ .2 A) in the separation of the adjacent cage of F^- ions around the Yb ion in this exciton state, and this should lead to a very large Stokes shift. In the view of McClure and Pedrini, the absorption into this trapped exciton occurs where the first peak of the photoconductivity appears (3.2 eV), and this estimated absorption is shown by the chain curve (A′) in Fig. 5. The localized excited states of the Yb^{2+} ion in this interesting material are higher in energy than the ionization energy of the Yb^{2+} ion in the material.

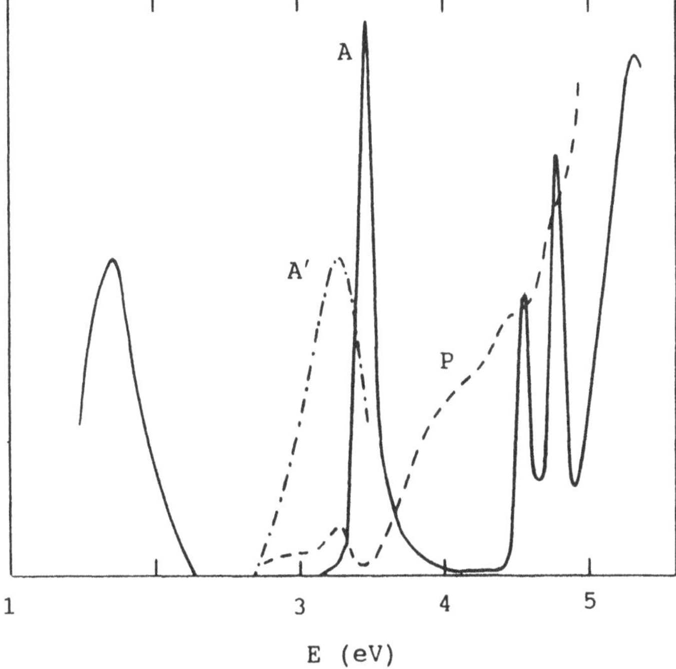

Fig. 5. The absorption (A), emission (F), and photoconductivity (P) spectra at low temperatures for $SrF_2:Yb^{2+}$. The curve A′ is the estimated trapped exciton absorption. [Spectra are redrawn from reference 29.]

II.D Two-photon spectroscopy of rare earth ions in solids

With the advent of intense tunable laser beams, spectroscopists now have the ability to carry out two-photon (TP) absorption studies, in which an ion is raised from the ground state to the excited state by the simultaneous absorption of two photons of light. Since this is a higher order radiative process, it has a much weaker transition probability than an allowed comparable one-photon (OP) process. The various higher order process which can lead to two-photon transitions between $4f^n$ states of a rare earth ion are discussed in a number of publications [30-34] and the present understanding of the subject has been reviewed by Downer [35]. The second order transition between the ground (g) and excited (e) states proceeds by the matrix element

$$\sum_i \frac{\langle e|E.D|i\rangle\langle i|E.D|g\rangle}{\Delta E_i} \qquad (5)$$

where i represents intermediate opposite-parity states, and ΔE_i is the energy separation between the $4f^n$ state in question and the opposite parity states, such as $4f^{n-1}5d$, $4f^{n-1}5g$. In general, terms higher than second order must be included in the two-photon process for rare earth ions [32]. In the above equation, D is the electric dipole operator, and E is the electric field of the radiation. The important point of this TP electric dipole process is that it is parity-allowed, states e and g can have the same parity, and so it is very suitable for transitions between $4f^n$ states. The allowed nature of the TP process between $4f^n$ states is in contrast to the OP ED process between $4f^n$ states, which in first order is parity-forbidden, and which depends for its strength on the mixing of opposite-parity states into the $4f^n$ states, the mixing being brought about by odd-parity crystal field components, either static or dynamic.

As a first example, we consider the transition between the ground $^8S_{7/2}$ state and the first excited $^6P_{7/2}$ state of the Gd^{3+} ion. Fig. 6 compares the OP and TP absorption spectra, taken at room temperature, on Gd^{3+} in $Cs_2NaGdCl_6$ [36]. The OP spectrum shows the zero-phonon transitions, the three lines arise because of crystal field splitting in the excited state. A Stokes sideband and a weak anti-Stokes sideband are seen. The relatively strong sideband reflects the fact that in this material the zero-phonon line is parity-forbidden and occurs only by MD process or by ED process on ions in distorted sites at which there is some odd-parity crystal field. The zero-phonon transition is thus weak. The sidebands are ED transitions induced by odd-parity vibrations, and due to the enhancement of the ED process, the sidebands are relatively

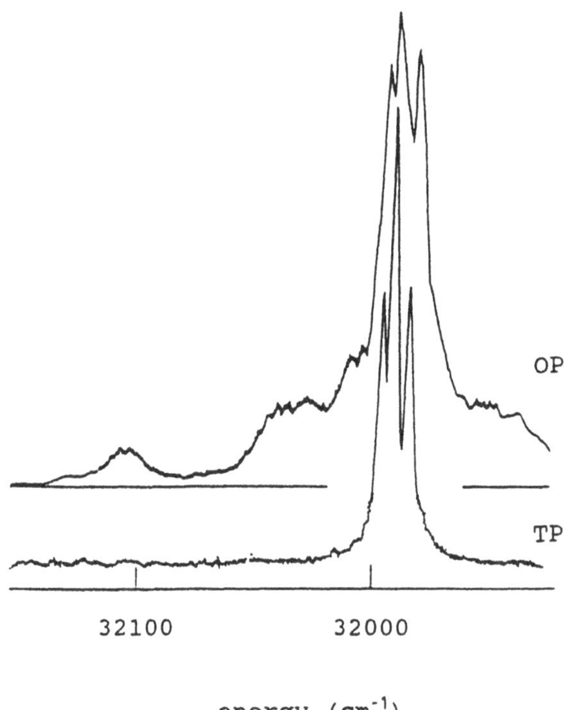

Fig. 6. Comparison of the OP and TP $^8S_{7/2} \rightarrow {}^6P_{7/2}$ absorption transitions on Gd^{3+} in $Cs_2NaGdCl_6$ at room temperature.

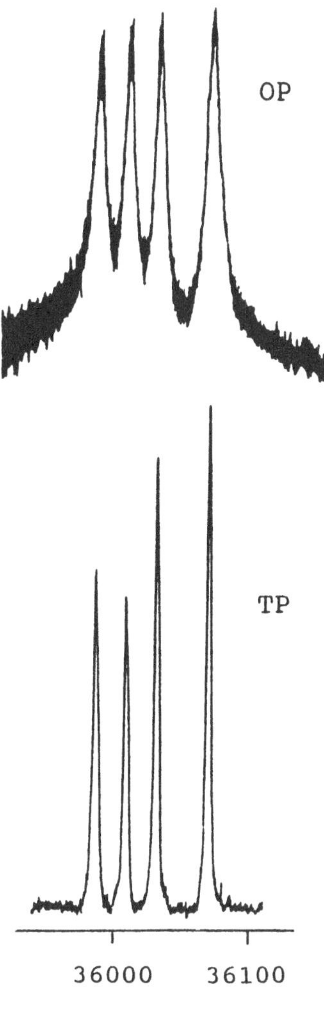

Fig. 7. A comparison between the OP and TP zero-phonon $^8S_{7/2} \rightarrow {}^6I_{7/2}$ absorption transitions on Gd^{3+} in GdB_3O_6 at room temperature.

strong. The TP spectrum shows a much sharper zero-phonon line but no detectable sideband features. The weakness of the sideband process in the TP spectrum is understandable since this process, like the TP zero-phonon process, is an even-parity transition, and no enhancement occurs through odd-parity vibrations. A sideband process in TP spectra will arise from a modulation of the electronic transition by even-parity modes, which, as we argued in section II.B, is a weaker process.

In a recent study of TP transitions in Gd^{3+} in $Cs_2NaGdCl_6$ [37], Bouazaoui et al. show TP spectra which appear to show weak sidebands accompanying the zero-phonon $^8S_{7/2} \rightarrow {}^6P_{7/2}$ and $^8S_{7/2} \rightarrow {}^6P_{5/2}$

lines. These indicate that modulation by even-parity modes, although weak, can be a detectable process, as pointed out in the case of OP transitions by Dexpert-Ghys and Auzel [18].

Figure 7 compares the room temperature zero-phonon OP and TP $^8S_{7/2} \rightarrow {}^6I_{7/2}$ transitions on Gd^{3+} in GdB_3O_6. The difference in the width of the transitions is striking. One possible explanation for this [36] is as follows. In the OP zero-phonon transition the strength is greatly enhanced by the presence of

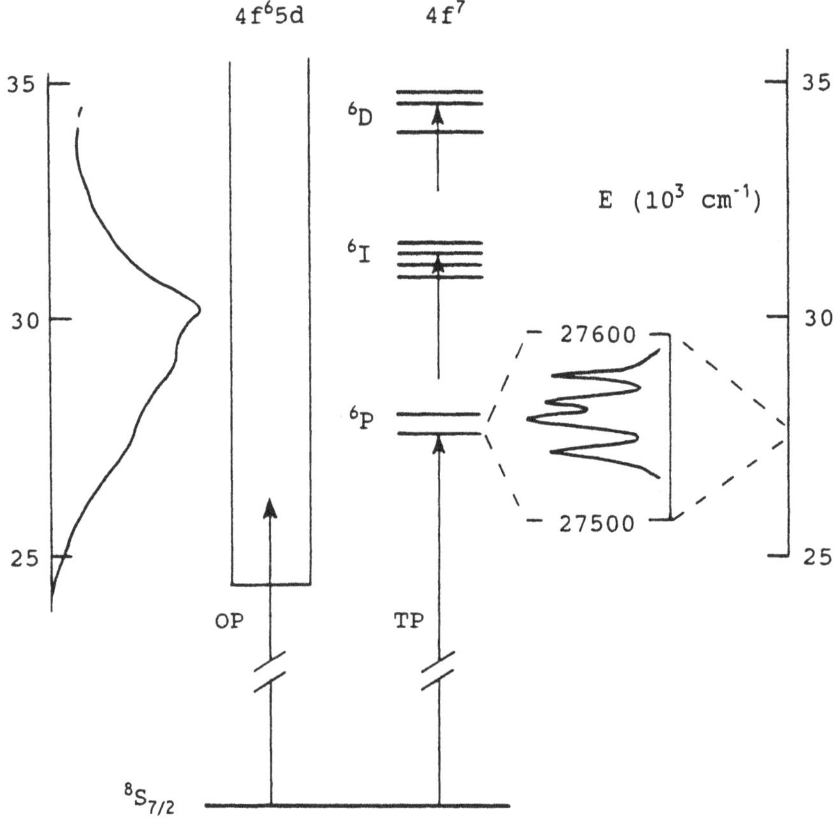

Fig. 8. OP and TP absorptions on Eu^{2+} in KCl [38]. The spectra are redrawn from this paper. The OP transitions involve the $4f^65d$ excited state, while the TP transitions involve the $4f^7$ excited state. Thus, one can discriminate between them.

static odd-parity crystal fields at the site of the optically-active ion; such fields will allow the transition to proceed by an ED process. This will tend to emphasize the transitions on the more distorted sites, since these are more likely to have odd-parity crystal field components and hence have a stronger ED transition. The emphasis on perturbed sites would be expected to lead to an increased inhomogeneous broadening. Zero-phonon TP transitions, on the other hand, since they are parity-allowed, will tend to have equal strengths for all the ions, whether perturbed or not. Thus, the inhomogeneous width of the TP transitions will reflect the true distribution of Gd^{3+} ions among the sites in the crystal.

If this explanation is valid, then sharper zero-phonon TP lines should be the general rule whenever OP zero-phonon lines are due to ED transitions induced by odd-parity static crystal field distortions from the regular site symmetry. Further, in such cases the radiative decay time of the excited state after TP excitation should be longer than that observed after OP excitation. To our knowledge, no such observations have been reported.

In the next example we compare the OP and TP absorption spectra of Eu^{2+} in KCl [38]. This ion has the same $4f^7$ configuration as Gd^{3+}, and the positions of the excited ^6P_J, ^6I_J, and ^6D_J states are shown in Fig. 8. For the divalent ion, the $4f^65d$ states are low in energy, stretching to below the ^6P_J state, as explained above. The parity-allowed $4f^7(^8S_{7/2}) \rightarrow 4f^65d$ transition dominates the OP absorption spectrum, and there is no evidence in the OP spectrum of, for example, the parity-forbidden $4f^7(^8S_{7/2}) \rightarrow 4f^7(^6P_{7/2})$ transition, in which the initial and final states have the same parity. Transitions between states of the same parity, however, dominate the TP absorption spectrum, and sharp spectral lines of the inter-configurational $4f^7$ transitions are found with splitting appropriate to the C_{4v} site symmetry. This study again illustrates the complementary nature of the OP and TP spectra. OP absorption transitions give information about the opposite-parity $4f^65d$ states, TP absorption transitions give information about the higher lying $4f^7$ states, which are overlapped by the strong broadband OP absorption transitions.

III. TRANSITION METAL IONS

Transition metal ions in solids lose their 4s electrons, and they possess an incomplete shell of 3d electrons on the outside of the ion. Radiative transitions occur within this $3d^n$ configuration. Being on the outside of the ion, the 3d electrons are much more sensitive to changes in the environment than are the 4f electrons on rare earth ions. For example, the same absorption transitions on the Cr^{3+} ion are responsible for the red colour of ruby and for the green colour of emerald. The colour difference comes about because of the difference in the energies of the $^4T_{2g}$ and $^4T_{1g}$ excited states, relative to the $^4A_{2g}$ ground state, in the two crystals and this is due to the different crystal field strengths in the two materials.

Transition metal ions in solids are often found in sites of six-fold coordination with a crystal field of predominantly octahedral symmetry. Let us consider a d electron in a static octahedral crystal field. In such a field the five-fold-degenerate electronic d orbital splits into a three-fold-degenerate t_2 orbital of lower energy and a two-fold-degenerate e orbital with higher energy. The simplest way of viewing a d^n-electron transition metal ion in an octahedral crystal field is to indicate how many of the n electrons are in t_2 orbitals and how many are in e orbitals. This gives the <u>strong field</u>

configuration. Electron-electron interaction and spin-orbit coupling cause a further splitting of the strong field configuration states. These calculations are given in reference [39]. In a static crystal field the energy levels of a d^n ion are sharp, hence transitions between the electronic levels in the static crystal field will be narrow.

Because of lattice vibrations, there are dynamic variations of the crystal field, these strongly modulate the radiative transitions between the electronic levels, and if the modulation is large enough, the transitions become very broad. To good approximation, the band shapes can be analyzed using the configurational coordinate diagram discussed previously.

III.A Inhomogeneous broadening and site distribution of ions in crystals

In an ideal crystal host, all the dopant ions would enter identical sites and the zero-phonon transitions would be extremely sharp, particularly at low temperatures. In real crystals, the presence of regions of different strain, disorder in the ionic arrangement, as well as the presence of nearby defects and impurity atoms all combine to produce a range of sites with slightly different crystal fields and a resultant broadening of the zero-phonon transitions. This is inhomogeneous broadening. The zero-phonon line in emission, for example, has a bandshape which is composed of a distribution of sharp line emissions from ions in distinct sites, and the inhomogeneous broadening can mask interesting structure in these spectral lines. In recent years a number of techniques involving narrow-band laser excitation, such as fluorescence line narrowing, coherent transient techniques, spectral hole burning, etc. have been employed to glean information hidden by the inhomogeneous broadening [40]. We will discuss the first of these: fluorescence line narrowing; the other techniques will be dealt with by Dr. Macfarlane.

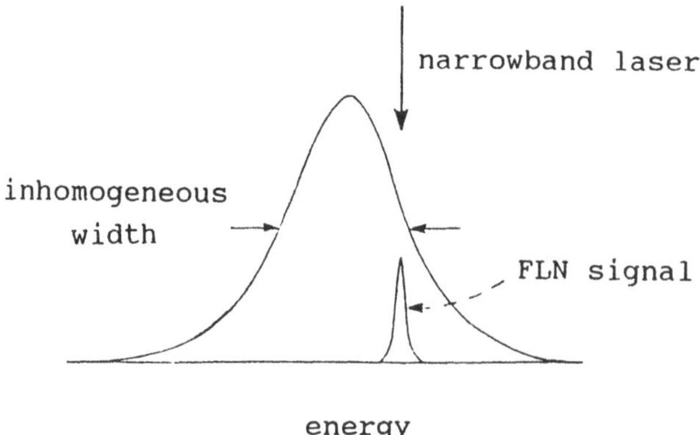

Fig. 9. When a narrowband laser excites a subset of ions within the inhomogeneously-broadened line, the resulting luminescence (the FLN signal) will have a width determined by the homogeneous broadening of the transition and by the laser bandwidth.

In the fluorescence line narrowing (FLN) experiment the sample is irradiated with a narrowband laser beam tuned to a frequency within the inhomogeneously-broadened absorption transition (Fig. 9). This excites a subset of the ions - those which are in resonance with the laser - and these excited ions in turn radiate with a very narrow bandwidth giving information about fine structure in the energy levels as well as giving information about excited state relaxation processes. An interesting recent example of such a study is the unravelling of the luminescence from the Cr^{3+} ions in distinct sites in GSGG: Cr^{3+}, a useful laser material, which exhibits a degree of disorder in the arrangement of the cations in the garnet structure [41-43].

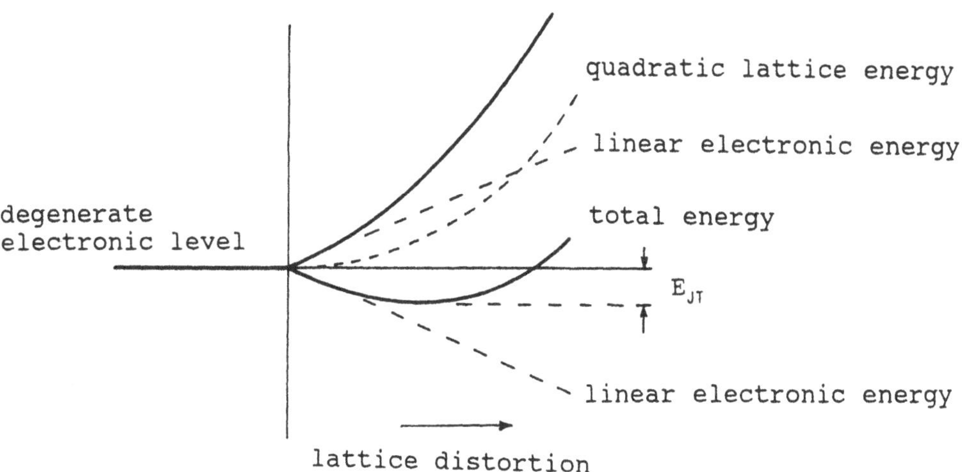

Fig. 10. A reduction in total energy can result from a reduction in site symmetry when the ion has an electronic degeneracy.

III.B Jahn-Teller Distortion

Studies of the excited states of transition metal ions (and colour centres) often show evidence of a Jahn-Teller distortion [44,45]. This distortion can come about as follows.

Consider a two-fold degenerate electronic state where the degeneracy can be removed by a lower symmetry distortion of the crystalline environment. We assume a linear splitting of the electronic energy level by the crystal field distortion; this is shown by the linear broken lines in Fig. 10. There is also a lattice distortion energy, which will increase quadratically with lattice distortion. This is shown as the broken parabolic curve in Fig. 10. Adding the two energy terms, linear electronic and quadratic lattice, we see that the total system can reach a lower energy state if it undergoes a lattice distortion, and the system may spontaneously distort to this lower energy configuration. The reduction in energy is written ΔE_{JT}. There can be similar possible distortions in other directions and corresponding reductions in energy. If ΔE_{JT} is large enough the system will stay in one of these lower-symmetry distorted states. This is the static Jahn-Teller effect. Further, because of the lattice distortion in one of the states, the transition involving that state will be broadened, as we saw in our discussion on the configurational coordinate model. Vibrations in this Jahn-Teller distortion mode can lead to a succession of peaks in the sideband.

If ΔE_{JT} is not too large, the system may be able to tunnel between the different distorted states; this is the dynamic Jahn-Teller effect [44-46].

III.C Excited States of Transition Metal Ions

We shall consider a number of recent examples of studies of transition metal ion spectra.

1. Ni^{2+} in MgO

Ni^{2+} is a much studied ion [47-51]. It has the $3d^8$ electronic configuration. When it substitutes for Mg^{2+} in MgO, it enters a site of perfect octahedral symmetry, and this high symmetry simplifies the subsequent analysis of the spectra [48]. The electronic energy level structure is shown in Fig. 11. On the left the electronic levels are shown in the octahedral crystal field, ignoring spin-orbit coupling. The states are classified by the spin multiplicity (2S+1) and by the appropriate irreducible representation of the octahedral group.

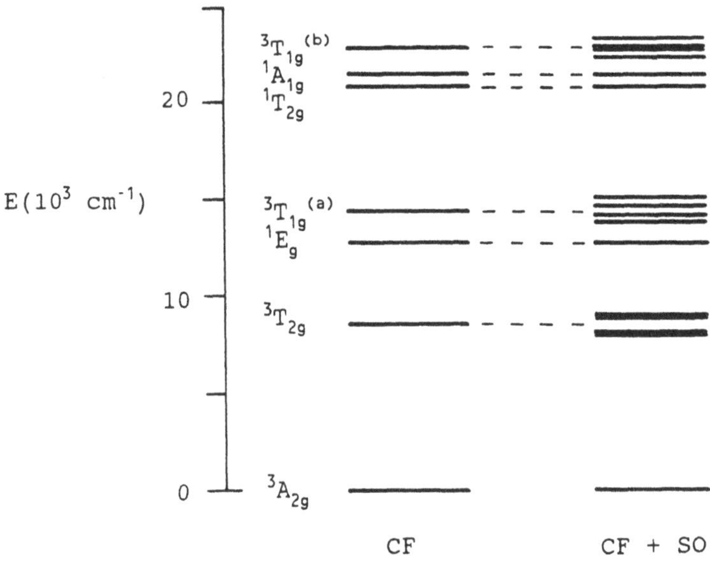

Fig. 11. The electronic energy level diagram for Ni^{2+} in an octahedral crystal field (CF) without taking spin-orbit coupling into account is shown on the left. When spin-orbit coupling (SO) is taken into account, the triplet states split, as shown on the right.

Spin-orbit coupling splits each of the triplet excited states, and this is represented on the right hand side of Fig. 11. The ground $^3A_{2g}$ state and the ^1E_g and $^1A_{1g}$ excited states have the $(t_2^6e^2)$ strong crystal field configuration, while the $^3T_{2g}$, $^3T_{1g}$ and $^1T_{2g}$ states have the $(t_2^5e^3)$ configurations. The t_2 and e crystal field orbitals interact differently with the crystal field. Since all the spin-allowed absorption transitions from the $(t_2^6e^2)^3A_{2g}$ ground state involve a change in strong crystal field configurations, $t_2^6e^2 \rightarrow t_2^5e^3$, we therefore expect these transitions to be strongly modulated by lattice vibrations and to be broad.

Since all the electronic states are of even parity, the zero-phonon transitions between them cannot occur by ED process. We next consider MD transitions. The MD operator is of type T_{1g}, which couples the $^3A_{2g}$ ground state only with the $^3T_{2g}$ excited state. The transition between $^3A_{2g}$ and $^1T_{2g}$ is spin forbidden. Thus, in the approximation of neglecting spin-orbit coupling, the only allowed zero-phonon transition is that between the $^3A_{2g}$ and $^3T_{2g}$ states, which is MD allowed. The vibrational spectrum of the MgO lattice shows a high density of optical modes at around 400 cm^{-1}, and these introduce crystal field distortions of T_{1u} and T_{2u} symmetry [49]. These mix odd-parity electronic states into the even-parity $(3d)^8$ states, and this may allow ED vibronic transitions to occur. There is also a peak in the density of modes at 260 cm^{-1}, but these are the longitudinal acoustical modes and they have little odd-parity character [49].

If we consider the dynamic T_{1u} distortions, and if we take the T_{1u} nature of the ED operator into account, we see that the effective dipole operator connecting the initial and final electronic states is of the form

$$T_{1u} \times T_{1u} = A_{1g} + E_g + T_{1g} + T_{2g} \tag{6}$$

which can connect the $^3A_{2g}$ ground state to the excited $^3T_{2g}$, $^3T_{1g}{}^{(a)}$, and $^3T_{1g}{}^{(b)}$ states by a vibronic process.

When spin-orbit coupling is taken into account it leads to a splitting into four of the $^3T_{2g}$, $^3T_{1g}{}^{(a)}$, and $^3T_{1g}{}^{(b)}$ states. The $^3T_{2g}$ state splits into two pairs of adjacent lines, these adjacent lines are unresolved in the figure. The $^3A_{2g}$ and $^1T_{2g}$ states are unsplit, but spin-orbit mixing of nearby $^3T_{1g}{}^{(a)}$ into $^1T_{2g}$ occurs, this relaxes the spin-forbidden nature of the $^3A_{2g} \rightarrow {}^1T_{2g}$ transition and allows a vibronic absorption process to occur.

We have seen from the selection rules that, except for the $^3A_{2g} \rightarrow {}^3T_{2g}$ transition, zero-phonon transitions are not allowed for the normal (one-photon) absorption of Ni^{2+} in MgO. Two-photon (TP) absorption spectroscopy, however, operates under a different set of selection rules, and two-photon spectroscopy of Ni^{2+} in MgO been studied by Moncorgé and Benyattou [50] and by Campochiaro et al. [51]. In two-photon spectroscopy the transition occurs by two simultaneous electric dipole processes whose transition operator is of the type $T_{1u} \times T_{1u}$, which, as noted above, allows transitions from $^3A_{2g}$ to the $^3T_{2g}$, $^3T_{1g}{}^{(a)}$ and $^3T_{1g}{}^{(b)}$ states, and these TP transitions occur as <u>zero phonon</u>

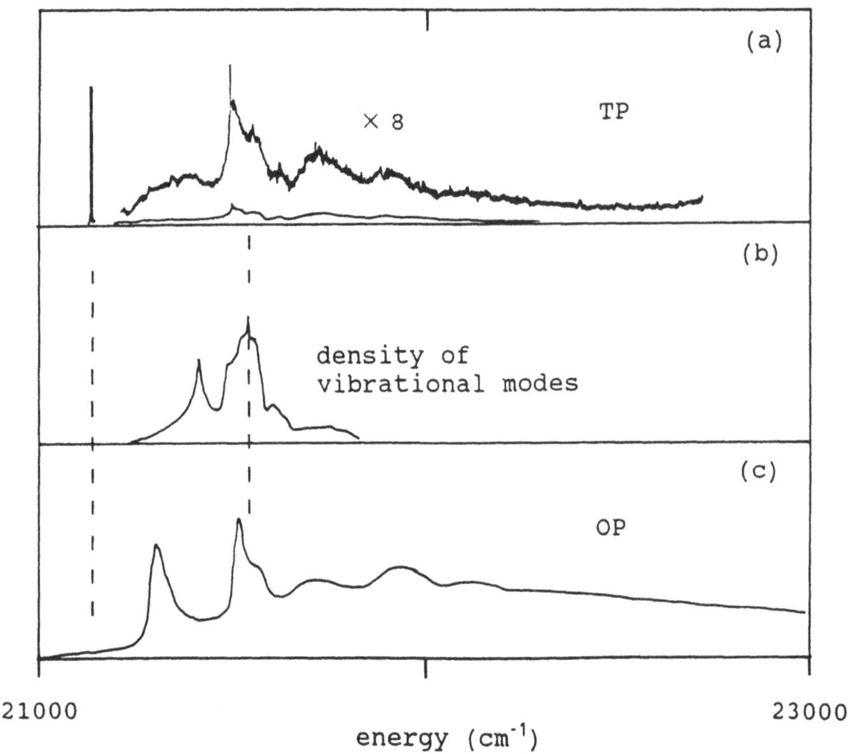

Fig. 12(a) shows the TP $^3A_{2g} \rightarrow {}^1T_{2g}$ absorption spectrum of Ni^{2+} in MgO [51]. Fig. 12(b) shows the density of vibrational modes of MgO [49], and Fig. 12(c) shows part of the OP spectrum of the same transition [48]. (Spectra are redrawn from the references.)

lines. Spin-orbit mixing of $^3T_{1g}$ into $^1T_{2g}$ allows a zero-phonon $^3A_{2g} \rightarrow {}^1T_{2g}$ absorption transition to occur also. The polarizations and relative strengths of the absorption transitions into each spin-orbit split component of the excited states can be calculated theoretically [51], and a reasonable agreement with experiment is found.

Fig. 12(a) shows the $^3A_{2g} \rightarrow {}^1T_{2g}$ TP absorption transitions [51]. There is a single sharp zero-phonon transition, as predicted, and a sideband whose first two peaks show a similarity with the density of vibrational modes of MgO (shown in Fig. 12(b). This is as expected; the even-parity vibrational modes modulate the single pure electronic transition. The degree of modulation is large, and second order overtones can be seen in the sideband. The $^3A_{2g} \rightarrow {}^1T_{2g}$ OP absorption spectrum is shown in Fig. 12(c). There is no zero-phonon line; as predicted, the transition is purely vibronic with a strong peak at 400 cm^{-1} from where the zero-phonon line is expected to be. This reflects the strong odd-parity character of the transverse optical vibrational modes at 400 cm^{-1} . The longitudinal acoustical modes (which peak at 260 cm^{-1}) do not contribute strongly to the OP vibronic process. The strong peak in the OP sideband at around 200 cm^{-1} is attributed to a vibrational mode of the local complex [49,52].

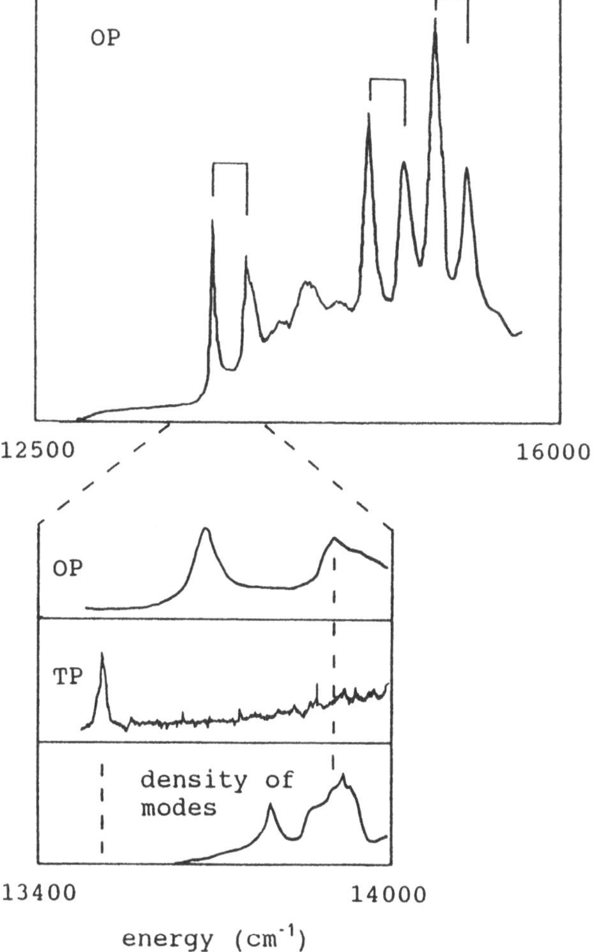

Fig. 13 shows the $^3A_{2g} \rightarrow {}^3T_{1g}^{(a)}$ OP absorption spectrum of Ni^{2+} in MgO [51] as well as the low energy sections of the OP [48] and TP [51] spectra in expanded form. (Spectra are redrawn from the references.)

Fig. 13 shows the $^3A_{2g} \rightarrow {}^3T_{1g}^{(a)}$ OP absorption spectrum [51] as well as the low energy sections of the OP [48] and TP [51] spectra in expanded form. Zero-phonon transitions are not allowed in the OP spectrum, and none are observed; the transition is vibronic. A sharp zero-phonon line is seen in the TP spectrum. Using this as an origin, the first two vibronic peaks in the OP spectrum are seen to occur at approximately 200 cm^{-1} and 400 cm^{-1}, just as was found for the $^3A_{2g} \rightarrow {}^1T_{2g}$ OP spectrum. There are two other pairs of vibronic peaks of similar 200 cm^{-1} separations, these are indicated in the full OP absorption spectrum, and it is believed that these are the analogous 200 cm^{-1} and 400 cm^{-1} vibronic peaks associated with transitions to two other spin-orbit-split sub-levels of $^3T_{1g}^{(a)}$. Thus, although their zero-phonon lines are not allowed in the OP spectrum, their positions can be inferred from the sideband spectrum.

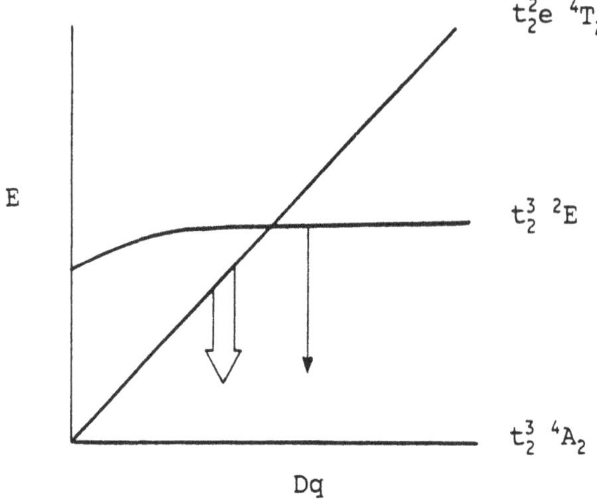

Fig. 14 Variation with increasing octahedral crystal field strength of the low-lying energy levels of Cr^{3+}. At some value of crystal field strength the ^2E and $^4T_2$ levels cross. Above the crossing point, the low temperature luminescence occurs only from the ^2E state. Below the crossing point, a broadband luminescence occurs from the $^4T_2$ state.

A very full analysis of the OP and TP absorption spectra are contained in the paper by Campochiaro et al. [51], and this work, together with the earlier OP study by Bird et al. [48], show a good complementarity in the analyses of the excited states of Ni^{2+} in MgO.

2. Cr^{3+} ions in a crystalline environment

As a general rule, Cr^{3+} in crystalline host materials exhibits strong broad absorption bands and efficient luminescence. It is a much studied ion, and this study has repaid us with a deeper insight into the nature of the radiative process on ions in solids as well as supplying us with a range of useful laser materials.

Cr^{3+} has the $3d^3$ electronic configuration. Dopant Cr^{3+} ions are invariably found in sites of six-fold coordination, and the electrostatic crystal field acting on the ion is predominantly of octahedral character. Often the arrangement of neighbouring ions lacks perfect octahedral symmetry, in which case some additional lower-symmetry crystal field operates, and this causes a splitting of the octahedral levels [53,39]. Fig. 14 shows the low-lying electronic energy levels of Cr^{3+} as a function of octahedral crystal field strength, Dq. [Dq is the parameter which characterizes the strength of the octahedral crystal field.] An important quantity to note here is Δ, the energy difference between the $^4T_2$ and ^2E electronic states : $\Delta = E(^4T_2) - E(^2E)$, as shown in Fig. 15.

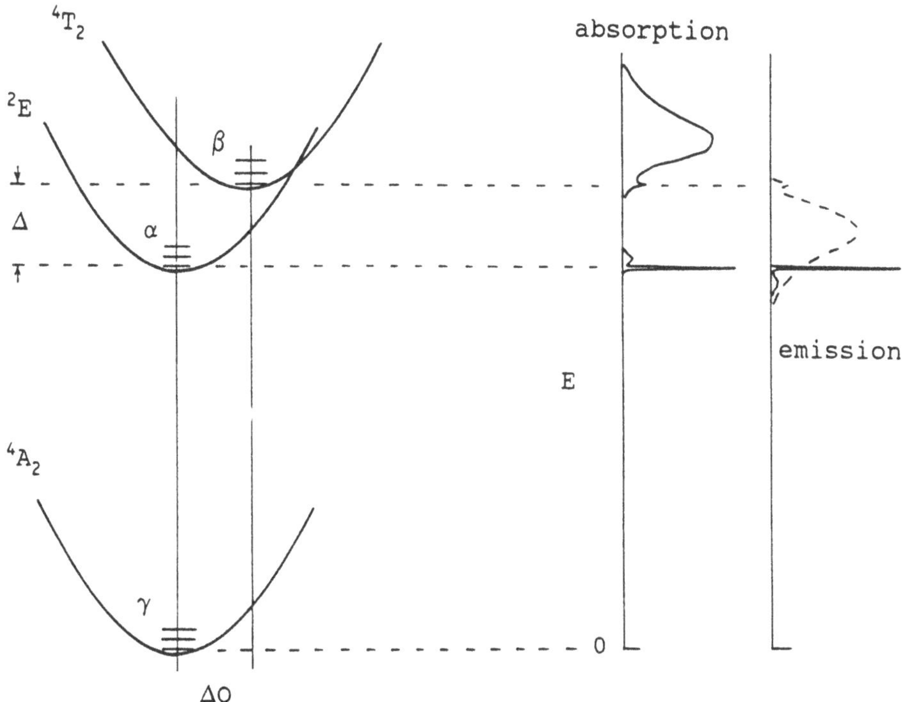

Fig. 15. Configurational coordinate diagram appropriate for the radiative transitions on the Cr^{3+} ion. This is drawn for the case of $^4T_2$ above ^2E. The expected shapes of the $^4A_2 \rightleftarrows {}^4T_2$ and $^4A_2 \rightleftarrows {}^2E$ transitions are sketched. The vibrational levels associated with the ^2E, $^4T_2$, and $^4A_2$ states are labelled α, β, and γ, respectively.

Both the $^4A_2$ ground state and the ^2E excited state are formed from the same t_2^3 strong field orbitals, that is, all three d electrons are in t_2 crystal field orbitals. Consequently, the difference in the ion-lattice coupling in these two electronic states is small. The $^4T_2$ electronic state, on the other hand, is formed from t_2^2e strong field orbitals, thus, the transition from $^4A_2$ to $^4T_2$ involves an electron changing from a t_2 orbital to an e orbital, and as a result, the coupling between the ion and the lattice is different in these two states. We attempt to describe the situation involving the $^4A_2$, ^2E, and $^4T_2$ states using a configurational coordinate diagram. This is shown in Fig. 15. The ionic environment about the Cr^{3+} ion is expected to expand when the ion is raised from the $^4A_2$ ground state to the $^4T_2$ state, and this is represented in the figure by the lateral shift, ΔQ, in the configurational coordinate parabola for this state. The $^4A_2$ and ^2E states, with their similar t_2^3 electronic configurations, have almost the same coupling with the lattice. For simplicity we assume the same vibrational frequency in all three electronic states.

Let us first consider the $^4A_2 \rightleftarrows {}^4T_2$ transition. This is spin-allowed and MD-allowed. Because of the significant shift in configurational coordinate, ΔQ, we expect that this transition will be characterized by a weak MD zero-phonon line and a large and broad multiphonon sideband, the latter being ED in nature, induced by odd-parity vibrations. In cases where the static (or average) crystal field at the site of the Cr^{3+} ion lacks inversion symmetry, then a small amount of odd-parity character is introduced into the even-parity $3d^3$ state, and the zero-phonon lines may have some ED character. The large Stokes shift (2000-3000 cm^{-1} in most crystalline systems) is indicated on the figure. The $^4T_2 \rightleftarrows {}^4A_2$ luminescence transition in Fig. 15 is drawn as a broken curve to draw attention to the fact that luminescence from $^4T_2$ will not occur unless there can be an equilibrium population established in this state by optical pumping. This will occur if $^4T_2$ is below ^2E, but will also occur if $^4T_2$ is above ^2E but with a value of Δ small enough so that a significant Boltzmann equilibrium of population is established in the $^4T_2$ state. When this luminescence transition is observed, the radiative decay time is typically in the range 10 - 100 μs.

Next we consider the $^4A_2 \rightleftarrows {}^2E$ transition. This is spin-forbidden, but this rule is relaxed to some extent by the mixing of $^4T_2$ into ^2E by spin-orbit coupling (H_{SO}). Looking at the configurational coordinate diagram, where the ^2E and $^4A_2$ states have almost identical parabolae, one predicts that this transition should be characterized by a strong zero-phonon line (the R-line) and a weak one-phonon sideband. We must be careful, however, in discussing the shape of this transition, since the transition derives its strength from a mixing of $^4T_2$ into ^2E, and the $^4T_2$ parabola is significantly displaced relative to the $^4A_2$ state. In general, sites of perfect octahedral symmetry are less common than sites in which there is a distorted octahedral field. In such sites, the lower symmetry crystal field causes a splitting into two of the ^2E level, and as a result there are two $^4A_2 \rightleftarrows {}^2E$ zero-phonon lines (R_1 and R_2 lines).

We now look more deeply into the $^2E \rightleftarrows {}^4A_2$ transition. We consider the case where ^2E is below $^4T_2$ with a large value of Δ (> 1000 cm^{-1}). Spin-orbit coupling mixes some $^4T_2$ into ^2E, and it is through this mixing that the transition occurs. The amount of mixing is given by the perturbation formula [54]

$$\frac{\langle {}^4T_2 \mid H_{so} \mid {}^2E \rangle}{\Delta} \tag{7}$$

and the strength of the transition is usually taken to vary as the square of this quantity.

In recent years, attention has been directed at chromium systems in which the room temperature luminescence is dominated by broadband $^4T_2 \rightarrow {}^4A_2$ emission. Let us consider alexandrite ($BeAl_2O_4:Cr^{3+}$) which was the material used for the first tunable solid state laser to operate at room temperature [55]. $^4T_2$ is above ^2E by around 800 cm^{-1}. Under continuous optical pumping at low temperatures only $^2E \rightarrow {}^4A_2$ luminescence occurs, and the spectrum is shown in Fig. 16. It has the sharp zero-phonon R-lines and a one-phonon sideband appropriate to a transition

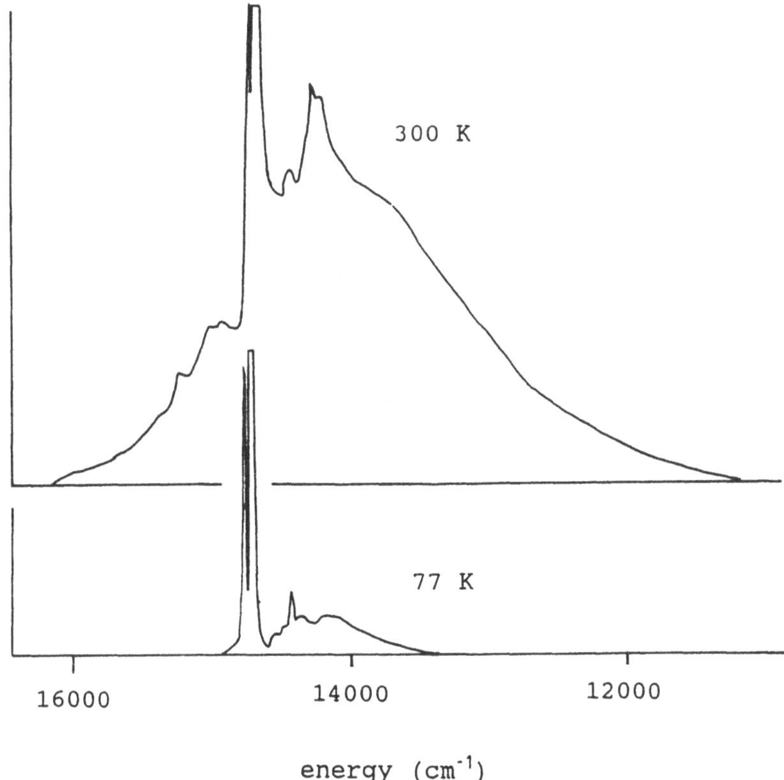

Fig. 16. Luminescence from alexandrite at 77 K and at room temperature.

between two states with similar ion-lattice coupling. The decay time is around 1.5 ms. At room temperature the luminescence spectrum is quite different. This comes about because, through thermalization, about 6% of the excited ions are in the $^4T_2$ state from which broadband luminescence occurs at a much faster rate; the intrinsic luminescence decay time of the $^4T_2$ state is around 10 μs. Thus, the room temperature luminescence shows both sharp $^2E \rightarrow {}^4A_2$ features and broad $^4T_2 \rightarrow {}^4A_2$ features, and the broken curve in Fig. 16 sketches the separation between the two components.

Another important medium for tunable laser action is GSGG:Cr^{3+} [56,57], but here the value of Δ is very small - a value of around 50 cm^{-1} has been quoted for this material [57-59]. It is interesting to enquire into the nature of the excited luminescent state for such a small value of Δ. Now, however, we cannot use the simple perturbation formula (7) for the mixing of the $^4T_2$ into ^2E in the case of small Δ. Further, we cannot ignore the vibrational character of the states which are being mixed together.

Because of the strong coupling between a transition metal ion and the surrounding lattice ions, one should consider the states which are involved in the radiative process as <u>electronic-vibrational states</u>. Let us assume that we can describe these states by crude Born-Oppenheimer product states:

$|\psi,\chi\rangle = |\psi\rangle|\chi\rangle$, where $|\psi\rangle$ refers to the electronic states, ^2E, $^4T_2$, $^4A_2$, and $|\chi\rangle$ refers to the vibrational states $|\alpha\rangle$, $|\beta\rangle$, $|\gamma\rangle$ associated with these electronic states, respectively. These are indicated on Fig. 15. The $|\alpha\rangle$ and $|\gamma\rangle$ sets are very similar, but the $|\beta\rangle$ set is distinct, because of the configurational coordinate shift, ΔQ.

We now consider transitions between the $|^2E,\alpha\rangle$ and $|^4A_2,\gamma\rangle$ electronic-vibrational states. This transition can only occur because of the admixture of $|^4T_2,\beta\rangle$ into $|^2E,\alpha\rangle$; <u>each</u> $|^4T_2,\beta\rangle$ state mixed into $|^2E,\alpha\rangle$ contributes to the dipole moment of the transition [60]. If the separation between the $|^4T_2,\beta\rangle$ states and $|^2E,\alpha\rangle$ is large enough to allow a perturbation approach to be adopted, then the matrix element for the transition between $|^2E,\alpha\rangle$ and $|^4A_2,\gamma\rangle$ is

$$\sum_{\beta} \frac{\langle ^4A_2,\gamma\,|\,D\,|^4T_2,\beta\rangle\langle ^4T_2,\beta\,|\,H_{so}\,|^2E,\alpha\rangle}{\Delta_{\beta\alpha}}$$

$$= \sum_{\beta} \frac{\langle ^4A_2|\,D\,|^4T_2\rangle\langle ^4T_2\,|\,H_{so}\,|^2E\rangle\langle\gamma\,|\,\beta\rangle\langle\beta\,|\,\alpha\rangle}{\Delta_{\beta\alpha}} \qquad (8)$$

where $\Delta_{\beta\alpha} = \Delta + (\beta-\alpha)\hbar\omega$, and Δ is shown on Fig. 15. If Δ is large enough so that we can approximate $\Delta_{\beta\alpha} \approx \Delta$ for the relevant states, then we can use closure, and the matrix element becomes

$$\langle ^4A_2\,|\,D\,|^4T_2\rangle \frac{\langle ^4T_2\,|H_{so}\,|^2E\rangle}{\Delta}\langle\gamma\,|\,\alpha\rangle \qquad (9)$$

If the $|\gamma\rangle$ and $|\alpha\rangle$ sets of states are identical, then $\langle\gamma|\alpha\rangle = \delta_{\gamma\alpha}$, and only a zero-phonon transition can occur. In practice, for a system with a large value of Δ, the spectrum is indeed dominated by a zero-phonon transition. The presence of an accompanying one-phonon sideband indicates that $|\gamma\rangle$ and $|\alpha\rangle$ are not wholly identical. The intensity of the sideband is enhanced by being ED-induced through odd-parity vibrations. Equation (9) is a justification for the use of the simple expression (equation 7) to describe the mixing of $^4T_2$ into ^2E which allows the $^2E \rightleftarrows {}^4A_2$ transition to occur, but we see that equation (7) is really only valid if Δ is very large.

When Δ is small, the values of $\Delta_{\beta\alpha}$ differ significantly from each other, and one cannot use the perturbation approach nor is the closure relationship applicable. There is now a considerable mixing of the $|^4T_2,\beta\rangle$ states into $|^2E,\alpha\rangle$, and vica versa. We write these more mixed states as $\Psi(^2E,\alpha)$ and $\Psi(^4T_2,\beta)$. Evaluation of the strength of the $\Psi(^2E,\alpha) \rightleftarrows |^4A_2,\gamma\rangle$ transition

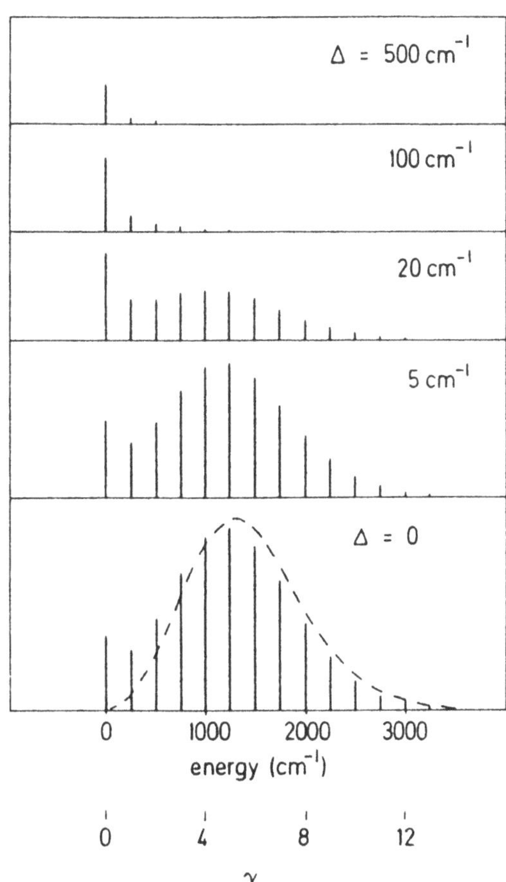

Fig. 17. The vertical lines indicate the strengths of the $\Psi(^2E,0) \rightarrow |^4A_2\,\gamma>$ transitions for various values of γ and Δ. The broken curve gives the envelope of the $\Psi(^4T_2,0) \rightarrow |^4A_2,\gamma>$ transitions for $\Delta = 0$ [61].

involves a complicated summation over the intermediate $|^4T_2,\beta>$ states [60]. Such a calculation has been carried out using approximate methods [61], and the calculated shapes of the $\Psi(^2E,0) \rightarrow {}^4A_2$ transition (i.e. the predicted shape at T = 0 K) for different values of Δ are shown in Fig. 17.

For this calculation, a phonon energy of 250 cm^{-1}, a value of S = 6, and a value of the spin-orbit matrix element $<^4T_2|H_{so}|^2E> = 200\ cm^{-1}$ were assumed. For $\Delta = 500\ cm^{-1}$, the transition is dominated by the zero-phonon line, but as Δ decreases the transition strength as a whole increases, the fractional intensity in the zero-phonon line decreases, and the intensity becomes increasingly concentrated in a broad multiphonon sideband. At very small values of Δ, the overall shape and strength of the transition resembles the broadband $4T_2 \rightarrow {}^4A_2$ transition. This results from the large degree of mixing of the $|^4T_2,\beta>$ and $|^2E,\alpha>$ states. The calculations indicate that for very small values of Δ one should not be able to resolve the luminescence into individual sharp ^2E and broadband $^4T_2$ components, as was done for alexandrite. The observed temperature dependence of the GSGG:Cr^{3+} luminescence spectrum is

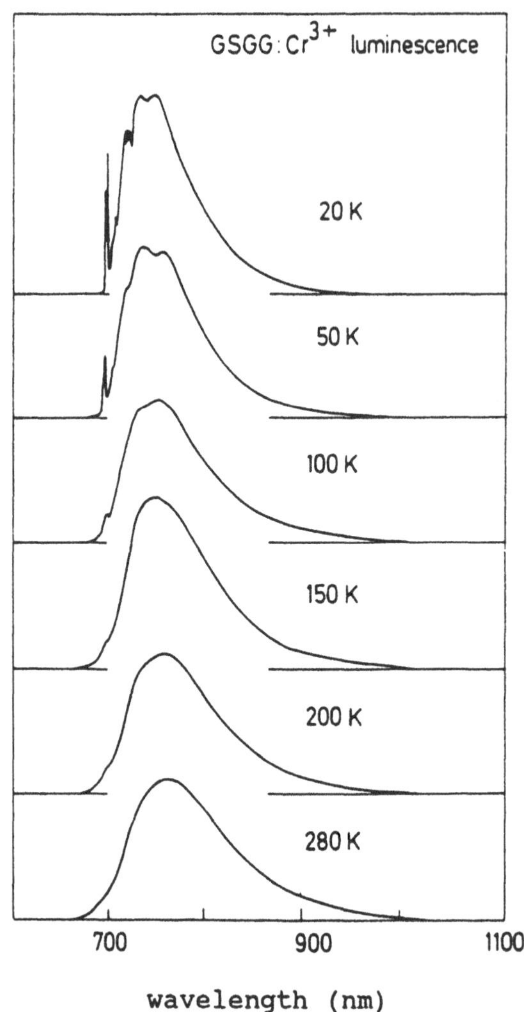

Fig. 18. Variation with temperature of the luminescence from GSGG;Cr^{3+} [62].

shown in Fig. 18. Except for the rapid increase in the broadening of the zero-phonon lines, there is little overall change in the shape, which remains very broadband. The observed spectra seem most consistent with the calculations for a value of $\Delta \approx 0$ at low temperature. The calculations also indicate that there should be only about a factor of two increase in radiative decay rate in going to room temperature [62] rather than the order of magnitude increase in decay rate found for alexandrite [63]. Fig. 19 compares the calculated and observed lifetimes of the GSGG:Cr^{3+} luminescence from 4.2 K to over 600 K. Agreement is good to about 400 K; above that temperature the fall-off in lifetime is likely due to the onset of nonradiative processes out of the $^4T_2$ state at these high temperatures. Calculations of the mixing of $^4T_2$ and ^2E electronic-vibrational states based on a tunnelling process [64] yields a similar explanation for the temperature-dependent lifetime.

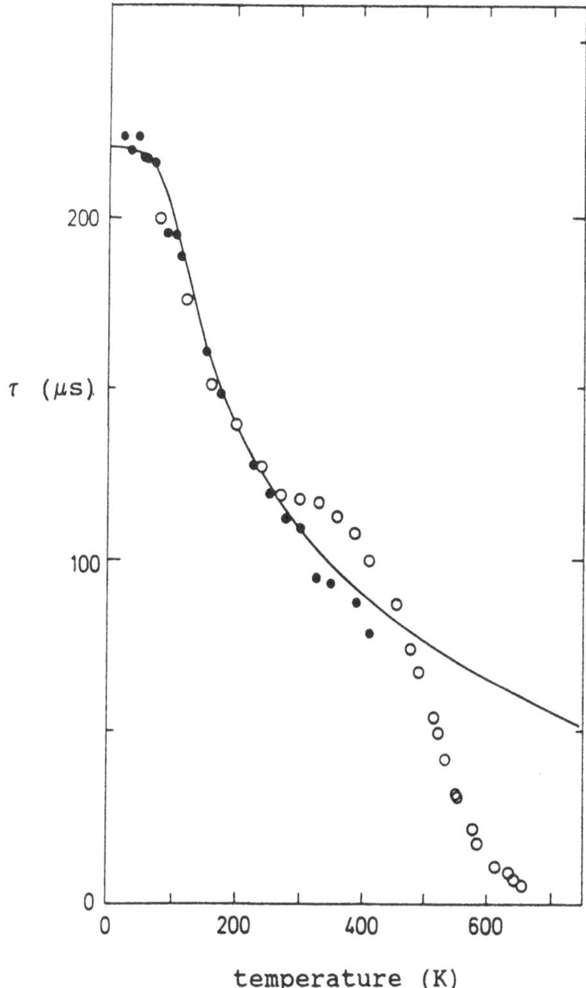

Fig. 19. The closed circles show the decay times measured by Healy et al. [62], the open circles are the measurements of Armagen et al. [65]. The curve is the calculated lifetime [62]. We interpret the fall-off in the measured lifetime above 400 K as due to the onset of nonradiative relaxation processes out of the $^4T_2$ state.

3. Luminescence from Cr^{3+} in glass

There is much interest at the present time in the use of glass as a host material for luminescent ions. Glass is an amorphous material, being a continuous random network of polyhedra (such as SiO_4 tetrahedra in silicate glass, BO_3 triangles in borate glass). In sample glasses these polyhedra are joined at corners to form a random three-dimensional network lacking both symmetry and periodicity. Glasses of spectroscopic interest as host materials contain additional compounds, such as Na_2O, CaO, which disrupt the bonding between the polyhedra, opening up the glass structure and making available a wide variety of sites which can be occupied by luminescent ions.

Such a glass structure is illustrated schematically in Fig. 20. Most glasses in technical use are based on oxide compounds, but in recent years heavy metal fluoride glasses have received a great deal of attention because of their large transmission range extending into the near infrared [67].

When rare earth ions are incorporated in glass, the traditionally-sharp intra-$4f^n$-configurational transitions show strong inhomogeneous broadening. Analysis of the luminescence using fluorescence line narrowing [68,69] demonstrates the very large range of distinct sites available for the rare earth ions in glass.

Transition metal ions, whose outer 3d electrons are more sensitive to variations in the ionic environment than are the rare earth ions, exhibit very broad absorption transitions when in a glass host. Of the transition metal ions in glass, the most studied by far is the Cr^{3+} ion. Whether in crystals or in glasses, Cr^{3+} ions appear to occupy sites of near-octahedral symmetry, and the energy level diagram of Fig. 14 is applicable. In the glass the emission is predominantly from Cr^{3+} ions in low field sites (in which $^4T_2$ is the lowest excited state), but in some glasses a very small fraction of the Cr^{3+} ions occupy high field sites (in which the ^2E level is the lowest excited state). Unfortunately, from the view point of their use as laser media, all Cr-doped glasses so far examined have a low quantum efficiency [70-73]. This is in contrast to the very high quantum efficiency of Cr^{3+} ions in most crystalline materials. Finding such a low quantum efficiency was disappointing, and puzzling.

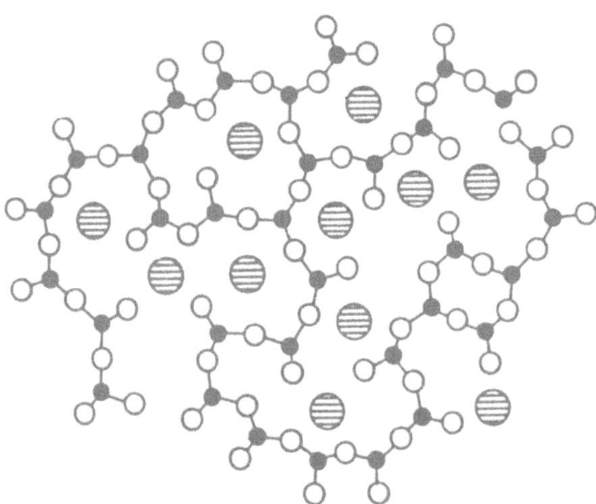

Fig. 20. Two-dimensional representation of a sodium silicate glass. The triangles of small filled and open circles represent the silicate tetrahedra, and the large hatched circles represent Na^+ ions [after Weber, reference 66].

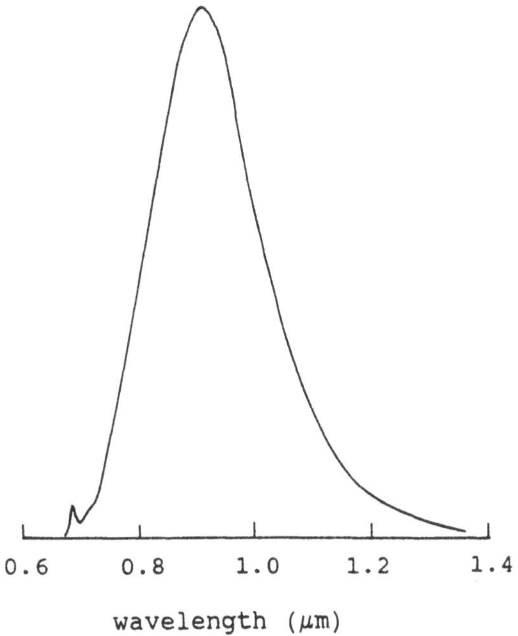

Fig. 21. Luminescence from Cr^{3+} ions in silicate ED-2 glass at 20 K. The small sharp feature at 700 nm is due to $^2E \rightarrow {}^4A_2$ transitions from ions in high field sites.

Fig. 21 shows the luminescence from Cr^{3+} in lithium lime silicate glass (ED-2) at 77 K. The very broad emission is caused by $^4T_2 \rightarrow {}^4A_2$ transitions from ions in low field sites. The weaker sharper component at around 700 nm is caused by $^2E \rightarrow {}^4A_2$ transitions from ions in high field sites. This component has a bandwidth of around 200 cm^{-1}, and it is composed of sharp $^2E \rightarrow {}^4A_2$ luminescence lines from Cr^{3+} ions in different sites in the glass. By applying the technique of fluorescence line narrowing on this inhomogeneously-broadened luminescence component, we can seek information on the range of high field sites occupied by the Cr^{3+} ions. From studies of rare earth ions in glass [68,69] we know that a large range of varieties of sites with different crystal fields are available for the rare earth ions. One might expect a similar situation to apply in the case of Cr^{3+} ions. Since in crystalline materials one finds crystal field splitting of the ^2E state varying in value from 0 to around 300 cm^{-1}, one might expect to find Cr^{3+} ions in glass with ^2E splitting covering this 0 → 300 cm^{-1} range. That is, the energy separation between the R_1 and R_2 lines from Cr^{3+} ions in high-field sites in glass might be expected to have a value anywhere between 0 and 300 cm^{-1}. Fluorescence line narrowing studies of the Cr^{3+} ions in ED-2 glass do not show this effect [74]; instead it is found that all Cr^{3+} ions in high field sites in this glass have a similar R_1-R_2 separation of around 55 cm^{-1}, indicating that in this glass the Cr^{3+} ions in high field sites occupy a single type of site and have a similar symmetry. Recent experiments by Kosch, Seelert, and Strauss [75] on the broadband luminescence from Cr^{3+} ions in ED-2 glass have shown that the broadband emission can be represented as the sum of two Gaussian bands. We take this as an indication that the Cr^{3+} ions in low field sites occupy only two varieties of site, although, as Kosch et al. point out, all the spectroscopic details cannot be entirely fitted to this simple picture - there are complexities due to inhomogeneous broadening, different nonradiative rates, and a dependence of the luminescence shape on excitation wavelength. Nevertheless, it is clear that, rather than occupying a range of different near-octahedral sites, the Cr^{3+} ions in glass only occupy a small number of varieties of site. More recent experiments on Cr^{3+} in a heavy metal fluoride glass (ZBLA) confirm this viewpoint [76].

It is argued [77] that this tendency of the Cr^{3+} ions to occupy a small number of varieties of sites in glass comes about as a consequence of the loose structure of glass and because of the emphasis on local bonding considerations within the glass structure. That glass is a loose structure is evidenced, for example, by the existence of "tunnelling modes" [78,79] which shows that the glass can relax from one quasi-equilibrium state to another. The absence of long range order and the strong coupling within tetrahedra is indicative of the pre-eminence of local bonding in the formation of the glass structure. These properties of the glass structure, it is argued, have consequences for the ionic bonding about the Cr^{3+} ion. Since the Cr^{3+} ion in the $^4A_2$ ground state has a strong preference for six-fold octahedral coordination, it is argued that when a glass containing Cr^{3+} ions is being formed, the Cr^{3+} ion attempts to create a suitable near-octahedral arrangement of local ions about it, and it is capable of doing this in the glass with its emphasis on local bonding considerations. Consequently, in a glass the ions are found to occupy only a small number of varieties of sites. When the ion is raised to the $^4T_2$ excited state, the coupling with the neighbouring ions is changed, the ion does not have the same strong preference for six-fold coordination, and the local complex can adjust to a different configuration of lower energy with much greater facility than in the normal crystalline case. That is, the ΔQ value for the $^4A_2 \rightleftarrows {^4T_2}$ transitions of Cr^{3+} in glass should be larger than that found in crystalline materials.

On theoretical grounds [80] one expects that the larger the value of ΔQ the greater the probability of nonradiative decay. Good experimental evidence for this correlation has been demonstrated by Blasse [81] for many luminescent systems. Hence one can ascribe the low quantum efficiency of Cr^{3+} in glass to this large ΔQ value. Supporting evidence for this viewpoint can be adduced from the $^4A_2 \rightleftarrows {^4T_2}$ Stokes shift. A large ΔQ implies a large Stokes shift, as our earlier discussion showed, so the model being proposed to explain the low quantum efficiency also predicts that the Stokes shift in glass would be noticeably larger than that found in crystals. A survey of a number of Cr^{3+} doped oxide crystals indicates an average Stokes shift of around 3000 cm^{-1}, while for Cr^{3+} in a range of oxide glasses, the average Stokes shift is close to 5000 cm^{-1} [77].

Similar arguments can be made for other transition metal ions in glass, leading to an expectation of an enhanced nonradiative probability and a consequent low quantum efficiency. Experiments on transition metal ions in glass seem to support this conclusion [82].

IV. HIGH CONCENTRATIONS OF OPTICALLY-ACTIVE IONS

Up to now we have considered the optically-active ions as being sufficiently far apart that they do not interact with each other. When, however, the density of optically-active ions is increased, some of the ions may be close enough to interact with each other. As a result, the excitation residing on one ion may be transferred to a nearby unexcited ion, or if the ions are very close together and interact very strongly, they may collectively act as a new electronic species with distinct spectroscopic features. We shall consider some of these phenomena.

IV.A Excitation transfer among optically-active ions

If the density of optically-active ions is sufficiently high, then when an ion is raised to its excited luminescence state it may either (a) release its energy radiatively, or (b) it may act as a sensitizer (or donor) ion and transfer its excitation to a nearly unexcited ion (called the activator or acceptor ion). If the acceptor ion is a distinct luminescent species to the donor ion, then the phenomenon of excitation transfer is easy to detect. Such excitation transfer between various ion types has been studied extensively and has been reviewed, for example, by Di Bartolo [83]. The study of energy transfer in insulators using tunable narrow-band laser spectroscopy has been reviewed by Morgan and Yen [84].

If the transfer occurs between identical ions (resonant transfer) it is more difficult to detect its existence; it does not, for example, show up in the form of additional spectral lines or as a change in the observed macroscopic radiative decay time. One method to detect and parametrize the resonant transfer is the technique of degenerate four wave mixing, also called transient grating spectroscopy [85].

Two laser beams with wavevectors $\mathbf{k}_1$ and $\mathbf{k}_2$, which are derived from the same laser and whose wavelength matches the absorption transition of the ions, enter the same portion of the material where they interfere to produce a periodic spatial variation in the density of excited state ions and a consequent spatial modulation of the refractive index (n) (Fig. 22). If Θ is the angle between $\mathbf{k}_1$ and $\mathbf{k}_2$, then the spatial variation in refractive index creates a refractive index grating with a grating period $d = \lambda/(2\sin\Theta/2)$. We note that by varying the angle Θ the size of the grating period can be changed.

If a third beam, $\mathbf{k}_3$, derived from the same laser and directed along the $-\mathbf{k}_1$ direction (Fig. 22), enters the grating region, it will be diffracted by the grating. By applying the Bragg criterion we find that the resulting diffracted beam, $\mathbf{k}_4$, will travel in the $-\mathbf{k}_2$ direction. The intensity of the diffracted beam $\mathbf{k}_4$ is proportional to the square of the modulation of the refractive index, Δn, that is, the intensity of the diffracted beam varies as the square of the density of excited ions.

When the two laser beams, $\mathbf{k}_1$ and $\mathbf{k}_2$, which create the grating are turned off (while maintaining beam $\mathbf{k}_3$), the grating

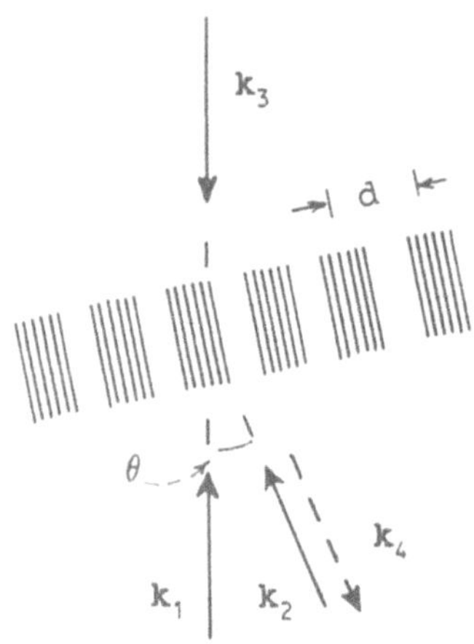

Fig. 22. A diffraction grating of excited ions, represented by the hatched regions, is created by interference between two laser beams $\mathbf{k}_1$ and $\mathbf{k}_2$. A third laser beam, $\mathbf{k}_3$, is diffracted by the grating and scattered into the $\mathbf{k}_4$ direction.

decays with time, and the intensity of the diffracted beam $\mathbf{k}_4$ decreases with time. If the grating fades because of the reduction in the density of excited state ions through their intrinsic decay mechanism, then the strength of the diffracted beam falls exponentially at a rate $2/\tau_F$, where τ_F is the intrinsic decay time of the excited ions. Other mechanisms can also cause a fading of the grating. Principal among these is <u>excitation transfer</u> which can cause a spreading out of the excited ions from the position where they were created, with a resultant washing out of the grating. In general, the expression for the time-dependence of the decay of the scattered beam is quite complicated [86,87]. If the excitation transfer can be described by a diffusion process, then the decay rate of the grating, measured as the decay rate of the scattered beam, is given by $1/\tau$ where

$$1/\tau = 2/\tau_F + 2D(2k\sin\Theta/2)^2 \qquad (10)$$

and where D is the diffusion constant, which characterizes the excitation transfer. k is the magnitude of the wavevector of the laser radiation. Thus, deviations of the observed decay rate from that due to intrinsic decay ($2/\tau_F$) are an indication of the existence of a spatial migration process.

The ability to detect the spatial migration of excitation is dependant upon the diffusion length, $L = (2D\tau_F)^{1/2}$, not being too small compared with the grating spacing. To illustrate this point, we note that we can rewrite equation (10) as

$$\frac{1}{\tau} = \frac{2}{\tau_F}(1 + 2\pi^2(L/d)^2) \qquad (11)$$

As examples of recent analyses of excitation transfer using four wave mixing, we cite the work of the Powell group on garnets [88] and emerald [87].

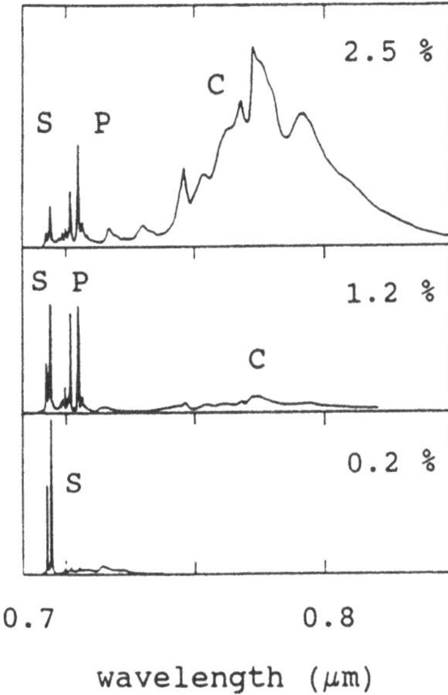

Fig. 23. Luminescence spectra at 77 K of ruby of three different concentrations of chromium.

IV.B <u>Luminescent states of concentrated materials</u>

As the density of optically-active ions increases, the possibility grows that two or more of the ions will be found sufficiently close together to act as a new luminescent species. Fig. 23 compares the luminescence of three ruby samples of different concentrations. In the dilute ruby the luminescence is from single chromium ions (S), whereas in the more concentrated samples one observes, in addition to the single ion luminescence, strong emission from pairs of chromium ions (P) and from clusters of ions (C). The greater intensity of the P and C luminescence arises because of excitation transfer from the single ions to pairs and clusters.

The spectroscopy of exchange-coupled chromium ions in ruby has been the subject of intense study over the past three decades, and a good theoretical analysis is presented by Heber [89]. Heber has recently observed cooperative emission from weakly-coupled pairs in ruby, that is, emission of a single photon from the ion pair when <u>both</u> ions are in their excited ^2E states [90]. A very thorough analysis of the spectroscopic features of chromium pair units in an organic crystalline material is given in the paper of Riesen et al. [91]. In this material the chromium ions do not occur as isolated ions, instead they are found only as a species of exchange-coupled pair, and as a result very clean pair spectra are obtained.

When we come to fully concentrated materials, the nature of the emitting state becomes much more complex. This comes about because neighbouring optically-active ions may be close enough together for very efficient excitation transfer to occur from intrinsic ion to intrinsic ion over microscopically large distances. This process is shown schematically in Fig. 24. This can result in efficient excitation transfer to traps. So efficient can this transfer be that trap densities of a few

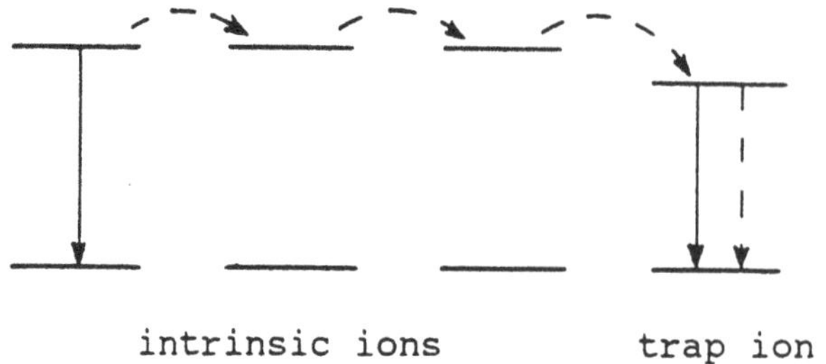

Fig. 24. Rapid migration through the intrinsic ions can result in efficient transfer to traps.

parts per million can draw off all the excitation from theintrinsic ions, resulting in the phenomenon of concentration quenching. The traps themselves may lose the excitation radiatively or nonradiatively.

One can take advantage of this fast excitation transfer in concentrated materials by incorporating a small amount of another appropriate optically-active ion in the concentrated material to act as a luminescent trap. For example, at 100 K MnF_2 doped with trace amounts of Eu^{3+} exhibits strong Eu^{3+} luminescence but no Mn^{2+} luminescence. Although the Mn^{2+} ions are raised to the excited state by optical pumping, all the Mn excitation is transferred to traps, principal of which are the Eu^{3+} ions.

Whether or not excitation transfer rather than intrinsic decay occurs depends on (a) the size of the separation between adjacent intrinsic ions, (b) the nature of the interaction between the adjacent ions, (c) the strength of the radiative transitions on the individual ions, and (d) the amount of lattice relaxation which occurs when the ion is excited. To illustrate this last point, we consider a dipole-dipole interaction between ions A and B in the case (i) where there is little or no lattice relaxation in going to the excited state, and (ii) where there is a considerable lattice relaxation (Fig. 25). The matrix element governing the transition which takes the excitation from the donor ion (D) to the acceptor ion (A) is

$$<g_D\,,\,e_A|\,\frac{D_D\cdot D_A}{4\pi\epsilon_0 R^3}\,|\,e_D\,,\,g_A>$$

$$=\frac{1}{4\pi\epsilon_0 R^3}<g_D|D_D|e_D><e_A|D_A|g_A> \qquad (12)$$

where R is the separation between the ions. Each single ion dipole matrix element in the above equation involves a dipole transition between the ground state and the relaxed excited state. This involves a vibrational overlap integral between the zero-vibrational state associated with the ground electronic state and the zero-vibrational state associated with the

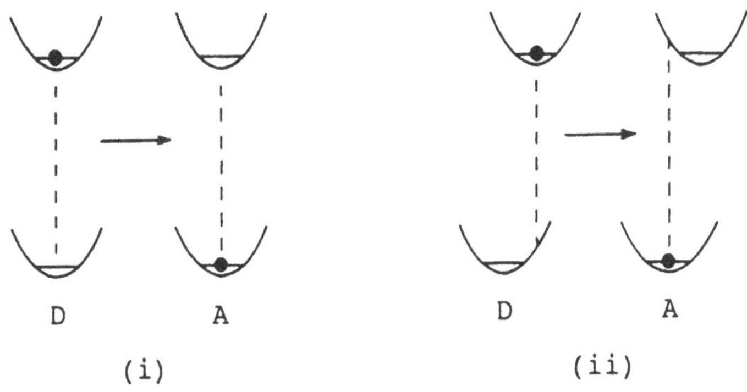

Fig. 25. The resonant excitation transfer from D to A is inhibited if there is considerable lattice relaxation between ground and excited states.

excited electronic state, $\langle 0_g | 0_e \rangle$ In the single configurational coordinate model in the harmonic approximation, this overlap integral has value exp(-S/2),where S is the Huang-Rhys parameter, and the value of this integral can be quite small in the case where there is considerable lattice relaxation in going to the excited state, case (ii) in Fig. 25. Thus, the larger the lattice relaxation in going to the excited state the larger S, and the weaker the transfer process. Further, the larger S the larger the Stokes shift, so we can restate the criterion in another way : if the overlap between absorption and emission is small, then the greater the likelihood that the excitation will stay localized on the individual optically active ions. Unfortunately, the larger S the more efficient is the nonradiative process on the individual ions.

Blasse, in a recent publication [6], cites a number of illustrative examples. Some vanadates and tungstates which possess a very large Stokes shift (~ 15,000 cm^{-1}) show an efficient intrinsic luminescence. There appears to be no excitation transfer among them. In contrast, YVO_4, with its smaller Stokes shift (~ 10,000 cm^{-1}) exhibits luminescence at low temperatures, but excitation migration occurs through the vanadate groups at room temperature. YVO_4:Eu shows strong Eu^{3+} luminescence at room temperature; excitation into the vanadates is followed by excitation migration to the Eu^{3+} ions. The initial classic studies on energy migration to traps in fully concentrated materials were carried out on concentrated manganese systems [92].

V. EFFICIENCY OF LUMINESCENCE FROM EXCITED STATES

For use as a laser medium, it is essential that the quantum efficiency of the material, ϕ_F, which is defined as <u>the probability of emission of a photon per absorbed phonon</u>, be high. In addition to energy transfer and cross-relaxation mechanisms, which can remove the excitation from the ion, a number of intrinsic processes compete with the emission process out of the excited state. The first is nonradiative decay. This is a subject of much study at the present time, and some empirical rules governing the process have been adduced. A second process is excited state absorption (ESA). For laser action to occur, a sizeable density of ions in the luminescent excited state must be created in the material. If these excited ions can absorb light - either from the pump beam or from the stimulated emission - this can be a major drawback to the use of the material as a laser medium. In this section (V) we shall discuss some aspects of quantum efficiency, in particular the new techniques which have been developed for its accurate measurement.

V.A Excited State Absorption in YAG:Ce^{3+}

At low pumping levels the luminescence of YAG:Ce^{3+} is highly efficient, but attempts to achieve laser action using high pump levels failed because of extremely strong excited state absorption (ESA)in the material. In a recent study, this ESA has been analyzed by Hamilton et al. [93].

In the ground state, Ce^{3+} has the outer configuration of a single 4f electron. Two strong 4f → 5d absorption transitions, peaking at 21,750 cm^{-1} and 29,400 cm^{-1}, are found, as illustrated in Fig. 26. ESA out of the lowest 5d level is also shown in the

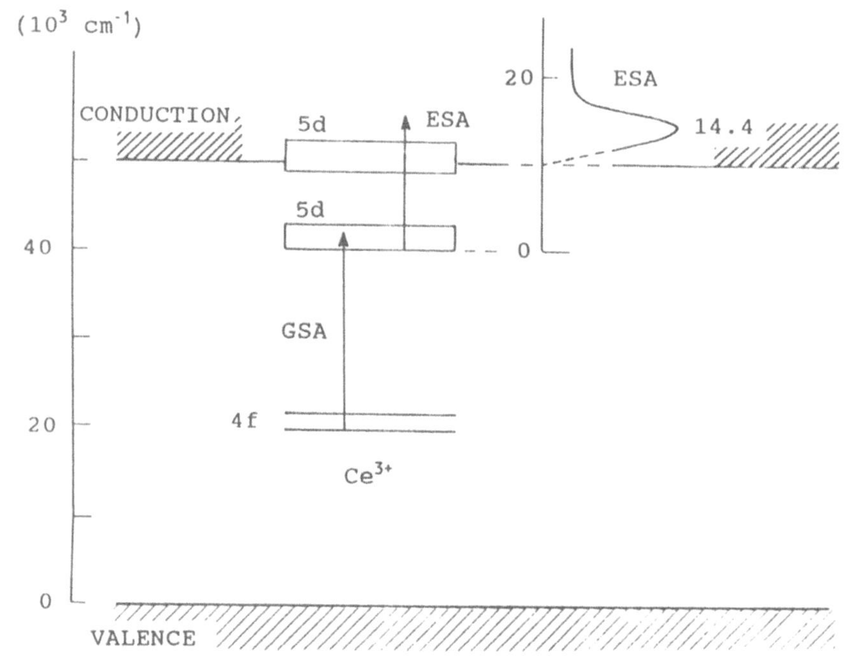

Fig. 26. Energy levels of Ce^{3+} in YAG relative to the conduction and valence bands of YAG.

figure This is an exceedingly strong absorption, much too strong to be assigned to a parity-forbidden transition between crystal field-split states of the 5d electron. Nor does the peak of the ESA correspond to a higher-lying absorption band, as measured from the ground state. Extrapolating the low energy side of the ESA curve to zero (broken line in ESA curve in Fig. 26) shows that the onset of ESA occurs at around 10,000 cm^{-1}. By adding this energy to the energy of the zero-vibrational level of the lowest 5d state (20,440 cm^{-1}), we find that the onset of ESA is about 30,440 cm^{-1} above the ground 4f state. This value is very close to the measured threshold energy for photoconductivity, 30,650 cm^{-1}, [94]. The inference drawn is that the threshold for photoionization of the Ce^{3+} ion is around 30,440 cm^{-1}, and that the strong ESA is due to the photoionization transition from the lowest 5d level. The positions of the Ce^{3+} levels relative to the conduction and valence bands of the YAG host are shown in Fig. 26. The bandgap of YAG is 50,000 cm^{-1}.

V.B Measurement of Quantum Efficiency and ESA

In the past, quantum efficiency has been measured by collecting the emitted photons using an integrating sphere [95]. In recent years new techniques, based on detecting the heat generated as a result of the absorption of light in the solid, have been developed. We describe some of these.

For low incident light levels and for low density of optically active ions (so as to prevent cross-relaxation and energy transfer between ions) the quantum efficiency depends on the competition between radiative and nonradiative decay out of the excited state. In Fig. 27 the various processes are indicated; the transitions for radiative and nonradiative processes are shown by solid and broken arrows, respectively.

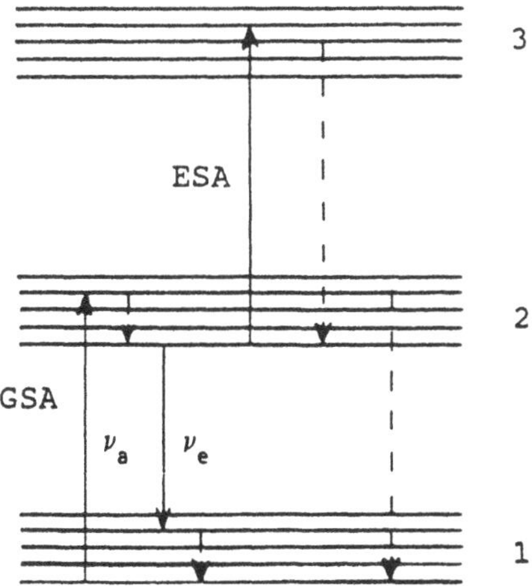

Fig. 27. Transitions between electronic-vibrational states in solids are shown schematically. Radiative transitions are represented by solid arrows, nonradiative transitions by broken arrows.

Absorption of a photon (at frequency ν_a) by the ion in the ground state is represented by the upward solid arrow (GSA). There may be some heat released as the excited ion relaxes to its emitting state; this is represented by the broken downward arrow as the ion decays within the excited state to the relaxed excited state. From the relaxed excited state the ion may decay by emission of a photon of frequency ν_e - and with the possible generation of heat as the ground state relaxes - or it may release all its energy as heat. These two processes are represented by the solid and the broken arrows, respectively, from the excited state to the ground state.

If the flux of absorbing photons, Φ, is high, and if the decay time downwards out of the excited state (τ_2) is long, then a sufficiently large density of ions may be created in excited state 2 by optical pumping so that <u>absorption out of the excited state</u> (ESA) occurs. This is shown by the solid arrow from state 2 to state 3 in Fig. 27. We write σ_2 as the excited state absorption cross section per ion. The probability of ESA increases linearly with Φ, σ_2, and τ_2. ESA is another loss mechanism, and it can be a serious deleterious phenomenon for many otherwise-very-promising laser materials.

A quantity which is closely related to the quantum efficiency is the <u>heat conversion efficiency</u> (μ_H). This is defined as

$$\mu_H = \frac{\text{heat power deposited in the sample } (P_H)}{\text{absorbed optical power } (P_A)}$$

or as

$$\mu_H = \frac{\text{average heat generated per absorbed photon}}{\text{energy of the absorbed photon}} \qquad (13)$$

Let us first consider the case where the photon flux is low so that ESA does not occur. In such a case there is a simple relationship between μ_H and ϕ_F:

$$\mu_H = \frac{\phi_F(\nu_a-\nu_e) + (1-\phi_F)\nu_a}{\nu_a} = 1 - \phi_F\nu_e/\nu_a \qquad (14)$$

Thus, a measurement of $\mu_H = P_H/P_A$, along with the values of ν_a and ν_e, allows the value of ϕ_F to be found.

P_A, the absorbed light power, is easy to measure. Obtaining an accurate measurement of P_H is more difficult. Strauss devised a <u>compensation photocalorimeter technique</u> to obtain an accurate value of P_H [96]. Inside the calorimeter the sample is connected to a thermal bath (whose temperature is lower than that of the sample) through a weak thermal link, and the heat flow from the sample to the bath is replenished by an electronically-regulated heater which maintains the sample at a constant temperature, above that of the bath. When light is absorbed by the sample, heat is generated in the sample, and less electrical power is required to maintain the sample temperature. The difference between the heater power required when the light is <u>off</u> and when it is <u>on</u> provides a precise measurement of P_H, and thus permits an estimate of μ_H, and hence of ϕ_F. Using this technique Seelert and Strauss obtained a value of $\phi_F = 0.73 \pm .03$ for ruby.

A similar very accurate technique has been demonstrated recently by Li, Duncan, and Morrow [93] for the case of ruby. In their method, a laser beam of specific wavelength is first directed onto the ruby sample, whose absorption spectrum is accurately known. Knowing the incident laser power, P_1, and the relevant absorption coefficient, the absorbed optical power, P_A, can be calculated. Because some of the absorbed optical power is converted to heat (P_H), the temperature of the sample rises. To measure P_H accurately Duncan and Morrow carry out a second experiment. They allow the laser beam at optical power P_2 to fall onto a portion of the surface of the sample which is blackened, a known amount of this beam is converted into heat in the sample, and the temperature rises. The values of P_1 and P_2 are adjusted until the temperature rise is the same in both cases. Hence, P_H is equal to the known fraction of second laser beam which is absorbed by the blackened surface. Thus, the measurement of μ_H reduces to the measurement of the <u>relative</u> powers of the laser beam, P_1 and P_2, in the two experiments. As performed by Li, Duncan and Morrow, a measurement of the <u>absolute</u> temperature rise is not necessary, and the technique does not require a knowledge of the thermal properties of the sample. They determined $\phi_F = 0.71 \pm .005$ for ruby pumped in the green and blue absorption bands, in agreement with the value obtained by Seelert and Strauss.

Let us return to the technique of Seelert and Strauss [96]. Their method also allows a measure of σ_2, the excited state absorption cross section at light frequency ν_a, to be measured. If the photon flux is increased, ESA may occur, and the heat conversion efficiency for this ESA process, μ_H(ESA), is equal to unity - all of the ESA laser power is converted to heat (Fig. 27). The overall value of μ_H will thus increase with pump power as ESA becomes larger. By solving the rate equations one finds [96] that for a photon flux density of constant cross section falling on a thin sample

$$\mu_H = \frac{1 + \Phi\sigma_2\tau_2 - \phi_F\nu_e/\nu_a}{1 + \Phi\sigma_2\tau_2} \qquad (15)$$

For a Gaussian beam profile this formula for μ_H must be appropriately modified [96]. Φ can be measured accurately, τ_2 is usually known to high precision, so if ϕ_F has already been measured, then σ_2 can be determined with good accuracy. This is a very accurate technique which, however, only probes ESA at the same frequency as GSA. But the ESA process at this frequency is an important mechanism which can determine the efficiency of conversion of pump lamp power to luminescence efficiency and, ultimately, the possibility of laser action in the material.

V.C Direct measurement of the local distortion about the optically active ion

Throughout these lectures we have used the configurational coordination diagram to describe the shape and width of broadband transitions on optically-active ions in solids. This visualizes a local distortion of the ionic arrangement about the optically-active ion when it is raised to the excited state, and this distortion is represented by the change in the configurational coordinate, ΔQ. The configurational coordinate diagram relates ΔQ to the shape of the transition, and from the analysis of the bandshape a value of ΔQ, the size of the local distortion, is inferred.

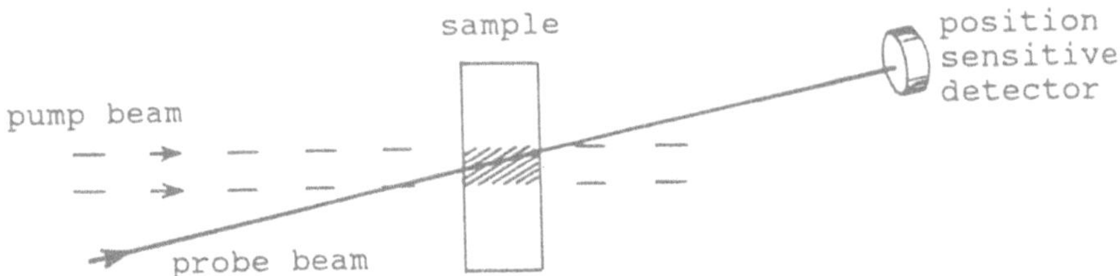

Fig. 28. Representation of the photothermal beam deflection technique used by Strauss to measure the local distortion about the optically-active ions in the sample.

Recently, an ingenious optical technique to measure ΔQ directly has been developed by Eugen Strauss [98,99]. This is the laser photothermal beam deflection technique, which is a variant on the beam deflection spectroscopic technique developed by Boccara and coworkers [100]. This works as follows. A periodically-modulated laser pump beam at the frequency of the absorption transition excites ions in a cylindrical portion of the sample (Fig. 28). The energy absorbed by the excited ions from the pump beam causes a change in the bulk refractive index in the optically-pumped region of the sample (shown hatched in the figure). The induced change in refractive index is measured as a deflection of a probe laser beam (whose frequency need not coincide with the absorption transition) which enters the excited region at a small angle to the pump beam (Fig. 28). Two mechanisms contribute to the change in refractive index. These are (a) a thermal effect due to the release of phonons in the absorption and subsequent re-emission process, and (b) the microscopic lattice change about the excited ions which contribute to a change in the macroscopic expansion of the material. Both effects have distinct time dependences; the thermal effect decays in accordance with a slow diffusive mechanism, while the excited state expansion effect decays with the lifetime of the excited state, generally a faster process. From the variation in the bulk refractive index due to the lattice change about the excited ions, the microscopic local expansion about the excited ion can be determined.

The photothermal technique was applied to, among other systems, the absorption into the $^4T_2$ state of Cr^{3+} in GSGG [101]. From the data it is estimated that each excited ion contributes a total expansion of 0.67 A^3. This must be related to a local volume change. Using an isotropic model and assuming a totally symmetric expansion about the excited ion, one estimates a local expansion of the cage of six oxygen ions of $\Delta Q = 0.12A$. In the framework of the configurational coordinate model, this value corresponds to a Huang-Rhys parameter $S \sim 3$ if the phonon energy of the breathing mode is taken as 350 cm^{-1}.

A similar investigation of the aquo-complex $[Mn(OH_2)_6]^{2+}$ dissolved in water [98] shows that when the Mn^{2+} ion is raised to the $^4T_1$ excited state the Mn—O distance is <u>reduced</u> by 0.14A. The different signs of ΔQ when Cr^{3+} is raised to the $^4T_2$ excited state and when Mn^{2+} is raised to the $^4T_1$ excited state are consistent with the known dependences of the energies of these excited states on octahedral crystal field; for Cr^{3+} the energy of the excited state <u>increases</u> relative to the ground state as the crystal field strength increases, for Mn^{2+} the converse is the case.

Finally, let us consider Strauss's measurement of the lattice relaxation about the excited Cr^{3+} ions in glass [102]. He carried out an analogous measurement of the bulk expansion of the ED-2 silicate glass due to the lattice relaxation about the Cr^{3+} ions which were optically-pumped into the excited state. The bulk effect in the glass was found to be about a factor of five smaller than that measured in GSGG:Cr^{3+}. Strauss attributes this to the ability of the silicate glass to compensate, by local elastic relaxation, for the local lattice distortions about the excited Cr^{3+} ions, and he cites this measurement as evidence to support the model, described earlier [77], to account for the low quantum efficiency of Cr^{3+} ions in glass.

CONCLUSION

In addition to standard spectroscopic methods, new techniques in coherent laser spectroscopy, two-photon spectroscopy, as well as photocalorimetric and photothermal beam deflection techniques, continue to provide us with a deeper understanding of the excited luminescent states of ions in solids.

DEDICATION

During this year (1991), a valued colleague and good friend, Eugen Strauss, whose imaginative development of the photocalorimetric and photothermal beam deflection techniques has been described above, died at a young age. I wish to dedicate this paper to his memory.

REFERENCES

1. G.H. Dieke, Spectra and Energy Levels of Rare Earth Ions in Crystals, Wiley-Interscience, New York (1968).
2. W.T. Carnall, H. Crosswhite, and H.M. Crosswhite, Energy Level Structure and Transition Probabilities of Trivalent Lanthanides in LaF_3, Argonne National Laboratory Report (1977).
3. S. Hüfner, Optical Spectra of Transparent Rare Earth Compounds, Academic Press, New York (1978).
4. G. Blasse and L.H. Brixner, Eur. J. Solid State Inorg. Chem. 26, 367 (1989).
5. G. Blasse, Inorg. Chim. Acta 169, 33 (1990).
6. G. Blasse, Adv. in Inorganic Chem. 35, 319 (1990).
7. J.W.M. Verwey, G.F. Imbusch, and G. Blasse, J. Phys. Chem. Solids 50, 813 (1989).
8. I.N. Chakraborty, J.E. Shelby, and R.A. Condrate, Sr., J. Am. Ceram. Soc. 67, 782 (1984).
9. B. Henderson and G.F. Imbusch, Optical Spectroscopy of Inorganic Solids, Oxford Science Publications (1989).
10. B.R. Judd, Phys. Rev. 127, 750 (1962).
11. G.S. Ofelt, J. Chem Phys. 37, 571 (1962).
12. M.C. Downer, G.W. Burdick, and D.K. Sardar, J. Chem. Phys. 89, 1787 (1988).
13. T.R. Faulkner and F.S. Richardson, Mol. Phys. 35, 1141 (1978).
14. M. Stavola and D.L. Dexter, Phys. Rev. B 20, 1867 (1979).
15. M. Stavola, L. Isganitis, and M.G. Sceats, J. Chem. Phys. 74, 4228 (1981).
16. M. Stavola, J.M. Friedman, R.A. Stepnoski, and M.G. Sceats, Chem. Phys. Lett. 80, 192 (1981).
17. B.R. Judd, Physica Scripta 21, 543 (1980).
18. J. Dexpert-Ghys and F. Auzel, J. Chem. Phys. 80, 4003 (1984).
19. W.T. Carnall, P.R. Fields, and K.J. Rajnak, J. Chem. Phys. 49, 4450 (1968).
20. B.G. Wybourne, Classical Groups for Physicists, Wiley-Interscience (1974).
21. W.F. Krupke, Phys. Rev. B 145, 325 (1966).
22. J. Sytsma, W.van Schaik, and G. Blasse, J. Phys. Chem. 52, 405 (1991).
23. G. Blasse and L.H. Brixner, Inorg. Chim. Acta. 169, 25 (1990).
24. A. Meijerink and G. Blasse, J. Lumin. 47, 1 (1990).
25. F.M. Ryan, W. Lehman, D.W. Feldman, and J. Murphy, J. Electrochem. Soc. 121, 1475 (1974).
26. A. Meijerink and G. Blasse, Phys. Rev. B 40, 7288 (1989).
27. U. Fano and J.W. Cooper, Phys. Rev. A 137, 1364 (1965).
28. M.D. Sturge, H.J. Guggenheim, and M.H.L. Pryce, Phys. Rev. B 2, 2459 (1970).
29. D.S. McClure and C. Pedrini, Phys. Rev. B 32, 8465 (1985).
30. J.D. Axe, Jr., Phys. Rev. A 136, 42 (1964).
31. B.R. Judd and D.R. Pooler, J. Phys. C 15, 591 (1982).
32. M.C. Downer and A. Bivas, Phys. Rev. B 28, 3677 (1983).
33. M.C. Downer, A. Bivas, and N. Bloembergen, Opt. Comm. 41, 335 (1982).
34. M.C. Downer, C.D. Cordero-Montalvo, and H. Crosswhite, Phy. Rev. B 28, 4931 (1983).
35. M.C. Downer in Laser Spectroscopy of Solids II (ed. W.M. Yen), Springer-Verlag, Berlin (1989).

36. J. Sytsma, J.W.M. Vervey, G. Blasse, and G.F. Imbusch, J. Electrochem. Soc. 135, 537C (1988).
37. M. Bouazaoui, B. Jacquier, C. Linares, and W. Strek, J. Phy: Condensed Matter 3, 921 (1991).
38. M. Casalboni, R. Francini, U.M. Grassano, and R. Pizzoferrato, Phys. Rev. B 34, 2936 (1986).
39. S. Sugano, Y. Tanabe, and H. Kamimura, Multiplets of Transition-Metal Ions in Crystals, Academic Press, New York (1970).
40. P.M. Selzer in Laser Spectroscopy of Solids (eds. W.M. Yen and P.M. Selzer), Springer-Verlag, Berlin (1981).
41. S.M. Healy, C.J. Donnelly, T.J. Glynn, G.F. Imbusch, and G.P. Morgan, J.Lumin. 44, 65 (1989).
42. B. Henderson, A. Marshall, M. Yamaga, K.P. O'Donnell, and B. Cockayne, J. Phys. C 21, 6187 (1988).
43. A. Monteil, C. Garapon, and G. Boulon, Proc. Electrochem. Soc. 135, 72 (1988).
A. Monteil, W. Nie, C. Madej, and G. Boulon, Optical and Quantum Electronics 22, S247 (1990).
44. M.D. Sturge, Solid State Physics 20 (eds. F. Seitz, D. Turnbull, and H. Ehrenreich) Academic Press, New York (1967).
45. R. Englman, The Jahn-Teller Effect in Molecules and Crystals, Wiley-Interscience, London (1972).
46. F.S. Ham, in Electron Paramagnetic Resonance (ed. S. Geschwind), Plenum Press, New York (1972).
47. J. Ferguson, H.J. Guggenheim, and D.L. Wood, J. Chem. Phys. 40, 822 (1964).
48. B.D. Bird, G.A. Osborne, and P.J. Stephens, Phys. Rev. B 5, 1800 (1972).
49. M.J.L. Sangster and C.W. McCombie, J. Phys. C 3, 1498 (1970).
50. R. Moncorgé and T. Benyattou, Phys. Rev. B 37, 9186 (1988).
51. C. Campochiaro, D.S. McClure, P. Rabinowitz, and S. Dougal, Phys. Rev. B 43, 14 (1991).
52. N.B. Manson, Phys. Rev. B 4, 2645 (1971).
53. Y. Tanabe and S. Sugano, J. Phy. Soc. Japan, 9, 753 (1954).
54. S. Sugano and Y. Tanabe, J. Phys. Soc. Japan 13, 880 (1958).
55. J.C. Walling, H.P. Jenssen, R.C. Morris, E.W. O'Dell, and O.G. Peterson, Opt. Letters, 4, 182 (1978).
56. B. Struve, G. Huber, V.V. Laptev, I.A. Shcherbakov, and E.V. Zharikov, Appl. Phys. B 30, 117 (1983).
57. B. Struve and G. Huber, Appl. Phys. B 36, 195 (1985).
58. A. Monteil, C. Garapon, and G. Boulon, J. Lumin. 39, 167 (1988).
59. K.P. O'Donnell, A. Marshall, M. Yamaga, B. Henderson, and B. Cockayne, J. Lumin. 42, 365 (1989).
60. C.J. Donnelly, S.M. Healy, T.J. Glynn, G.F. Imbusch, and G.P. Morgan, J. Lumin. 42, 117 (1988).
61. C.J. Donnelly, T.J. Glynn, G.P. Morgan, and G.F. Imbusch, J. Lumin. 48 & 49, 283 (1991).
62. S.M. Healy, C.J. Donnelly, T.J. Glynn, G.F. Imbusch, and G.P. Morgan. J. Lumin., 46, 1 (1989).
63. A. Suchocki and R.C. Powell, J. Chem. Phys. 128, 59 (1988).
64. M. Yamaga, B. Henderson, K.P. O'Donnell, J. Phys: Condensed Matter, to be published.
65. G. Armagen, B. Di Bartolo, and G. Ozen, Proceedings of the First International School on Excited States of Transition Elements (eds. B. Jezowska-Trzebiatowska, J. Legendziewicz, and W. Strek) World Scientific, Singapore, p.31 (1988).

66. M.J. Weber in <u>Laser Spectroscopy of Solids</u> (eds. W.M. Yen and P.M. Selzer), Springer-Verlag, Berlin (1986).
67. A. Bouaggad, G. Fontenau, and J. Lucas, Mat. Res. Bull. <u>22</u>, 685 (1987);
J. Lucas, J. Mater. Sci. <u>24</u>, 1 (1989).
68. S.A. Brawer and M. J. Weber, Appl. Phys. Lett. <u>35</u>, 31 (1979).
69. R.T. Brundage and W.M. Yen, Phys. Rev. B <u>33</u>, 4436 (1986).
70. L.J. Andrews, A. Lempicki, and B.C. McCollum, J. Chem. Phys. <u>74</u>, 5526 (1984).
71. W.R. Seelert, Thesis, University of Oldenburg (1987).
72. R. Reisfeld in <u>Spectroscopy of Solid State Laser-type Materials</u> (ed. B.Di Bartolo) Plenum Press, New York (1987).
73. A. van Die, A.C.H.I. Leenaers, G. Blasse, and W.F. van der Weg, J. Non-Crystalline Solids <u>99</u>, 32 (1988).
74. F.J. Bergin, J.F. Donegan, T.J. Glynn, and G.F. Imbusch, J. Lumin. <u>34</u>, 307 (1986).
75. J. Kosch, W. Seelert, and E. Strauss, J. Lumin. <u>45</u>, 105 (1990).
76. R. Balda, M.A. Illarramendi, J. Fernandez, and J. Lucas, J. Lumin. <u>45</u>, 87 (1990).
77. G.F. Imbusch, T.J. Glynn, and G.P. Morgan, J. Lumin. <u>45</u>, 63 (1990).
78. P.W. Anderson, B.W. Halperin, and C.M. Varma, Phil. Mag. <u>25</u>, 1 (1972).
79. W.A. Phillips, J. Low Temp. Phys. <u>7</u>, 351 (1972).
80. C. Donnelly and G.F. Imbusch in <u>Advances in Nonradiative Processes in Solids</u> (ed. B. Di Bartolo) Plenum Press, New York (1991).
81. G. Blasse, Prog. Solid State Chemistry <u>18</u>, 79 (1988).
82. G.F. Imbusch in <u>Advances in Nonradiative Processes in Solids</u> (ed. B. Di Bartolo) Plenum Press, New York (1991).
83. B. Di Bartolo, <u>Energy Transfer Processes in Condensed Matter</u>, Plenum Press, New York (1984).
84. G.P. Morgan and W.M Yen in <u>Laser Spectroscopy of Solids II</u> (ed. W.M. Yen) Springer-Verlag, Berlin (1990).
85. H.J. Eichler, Optica Acta <u>24</u>, 631 (1977).
86. Y.M. Wong and V.M. Kenkre, Phys. Rev. B <u>22</u>, 3072 (1980);
V.M. Kenkre and D. Schmid, Phys. Rev. B <u>31</u>, 2430 (1985).
87. G.J. Quarles, A. Suchocki, R.C. Powell, and S. Lai, Phy. Rev. B <u>38</u>, 9996 (1988).
88. A. Suchocki and R.C. Powell, Chem. Phys. <u>128</u>, 59 (1988)
89. J. Heber, in <u>Proceedings of the First International School on Excited States of Transition Elements</u> (eds. B. Jezowska-Trzebiatowska, J. Legendziewicz, and W. Strek), World Scientific, Singapore (1989).
90. R. Wannemaker and J. Heber, Z. Phys. B - Condensed Matter <u>65</u>, 491 (1987).
91. H. Riesen and H.N. Güdel, Mol. Phys. <u>60</u>, 1221 (1987).
92. R.L. Greene, D.D. Sell, R.S. Feigelson, G.F. Imbusch, and H.J. Guggenheim, Phys. Rev. <u>171</u>, 600 (1968).
93. D.S. Hamilton, S.K. Gayen, G.J. Pogatshnik, R.D. Ghen, and W.J. Miniscalco, Phys. Rev. B <u>39</u>, 8807 (1989).
94. C. Pedrini, F. Rogemond, and D.S. McClure, J. Appl. Phys. <u>59</u>, 1196 (1986).
95. L.S. Forster and R. Livingstone, J. Chem. Phys. <u>20</u>, 1315 (1952).
96. W. Seelert and E. Strauss, J. Lumin. <u>36</u>, 355 (1987).
97. Y. Li, I. Duncan, and T. Morrow, J. Lumin. (in press).
98. E. Strauss, and S. Walder, Europhys. Lett. <u>6</u>, 713 (1988).
99. E. Strauss, Phy. Rev. B <u>42</u>, 1917 (1990).

100. A.C. Boccara, D. Fournier, W. Jackson, and N.M. Amer, Appl. Opt. 20, 1333 (1981).
101. E. Strauss, Phys. Rev. B 42, 1917 (1990).
102. U. Hömmerich and E. Strauss, J. Lumin. 48 & 49, 574 (1991).

RELAXED EXCITED STATES OF COLOR CENTERS

G. Baldacchini

ENEA, Area INN, Dip. Sviluppo Tecnologie di Punta
Centro Ricerche Energia
C.P. 65, 00044 Frascati (Roma), Italy

ABSTRACT

Color Centers in alkali halides display an optical cycle which has been and it is still today a model case for similar processes in other materials. Moreover the luminescence of some color centers is so efficient that it has been used in laser applications. However the quantum state from which the emission of light is originated, the so called relaxed excited state (RES), is not very well known. Indeed in spite of the wealth of experimental results collected and of the theoretical approaches attempted, an exact description of the RES is still missing.

This paper, confined mainly on F centers which are the simplest point defects in crystals, contains a review of the main experimental evidences which cast some light on the nature of the RES, with special emphasis on the latest magneto-optical experiments. Also a description of the theoretical models has been attempted whenever required by a particular argument.

I. INTRODUCTION

As it is well known from solid state physics, crystalline materials are interesting both in basic and in applied research, mainly because they contain imperfections. Indeed the properties of crystals such as mechanical properties, electrical and thermal conductivity, diffusion, and optical properties are strongly dependent on crystalline irregularities and foreign impurities. For instance impurities in metallic crystals increase their strength, decrease the electrical and thermal conductivity, and influence greatly the diffusion processes. Impurity in insulating crystals give them unusual electrical properties, which are used in semiconductor devices, and last but not least the optical properties are so deeply changed that new absorption and emission effects take place.

The simplest imperfections are the localized ones, i.e., vacancies, foreign atoms in lattice or interstitial positions, and color centers. All these imperfections, called point defects, have strong effects on the properties of the crystal host even when they are present in very small amount. For instance a crystal of KCl, normally transparent when pure, assumes a well defined violet coloration when it contains about 10^{16} F centers/cm^3, which correspond approximately to 1 ppm with respect to the number of potassium or chlorine ions! Because of their intrinsic simplicity

Optical Properties of Excited States in Solids
Edited by B. Di Bartolo, Plenum Press, New York, 1992

the point defects have been systematically studied experimentally and theoretically within the broader field of solid state physics since the beginning of this century. Among them a special attention has been always devoted to color centers in alkali halides [1]. However their systematic study was initiated only around 1920 by R.W. Pohl in Gottingen. At that time there were fondamental and practical reasons for studying color centers in alkali halides, and those reasons, which are given in the following, are still valid today.

The alkali halides crystals, which have a cubic symmetry, are good electrical insulators, are transparent from the ultraviolet to the middle infrared region of the electromagnetic spectrum, and are easily produced pure and with the desired amount of impurities. So they are particularly suitable to study electrical conductivity, diffusion, and optical properties of point defects. Moreover color centers are the simplest among all point defects and as a consequence their properties are more easily understood, at least in principle. Notwithstanding their apparent simplicity they have provided a great deal of basic results, which have shed light on more complex phenomena of point defects in solids.

Still today color centers in alkali halides are actively studied to clarify fundamental processes of radiative and non radiative transitions in solids [2,3], and are used in noteworthy applications [4], like dosimetry, information storage, and tunable near infrared lasers. However a satisfactory description of the emission processes after radiative excitation of color centers or F centers, which can be considered the prototypes of all kind of color centers, is still lacking at moment.

As we will see later in detail the emission is produced by the F center after the electronic excitation has reached the so called relaxed excited state, RES. It is the aim of this paper to describe the known properties of the RES of F centers in alkali halides, and whenever possible to indicate future work to be done in order to improve our knowledge of the RES and the whole optical cycle. In order to accomplish such task it has deemed necessary after a brief introduction describing the more common color centers to give a general account of the F center optical properties in Sec. II. The excited states before and after the relaxation are introduced in Sec. III, while a detailed description of the RES is presented in Sec. IV. In Sec. V a comparison of the experimental results with the outcomes of the vibronic theories is attempted, and eventually a general assessment of the experimental and theoretical knowledge is discussed in Sec.VI.

II. COLOR CENTERS AND OPTICAL CYCLE

II.A. Introduction to color centers

Alkali halide crystals have a cubic symmetry and a face–centered (Li, Na, K, Rb halides and CsF) and simple cubic (Cs halides) space lattice. Figure 1 shows a plane of a face–centered crystal containing several simple color centers. The F center is an electron trapped in an anion vacancy, and, as it is evident by looking at the other electron excess centers in the same figure, it is the simplest center [5]. The F^- is an F center joined by a second electron, the F_2 is formed by two F centers aggregated along the (110) direction, the F_A is an F center associated with a cation impurity, the F_B is an F center associated with two cation impurities, and the F_H is an F center associated with an anion impurity.

These centers and others which will not be described here in detail are not mere scientific curiosities. Indeed most of them have been also used in applied research, for instance as laser active centers [6–9]. The optical properties of color centers are peculiar to each kind of center, but their general behaviors can be fairly well described by an F center–like model.

II.B. Optical cycle of the F center

Figure 2 shows the potential energy of the F center accordingly to a semicontinuum model [5]. The electron of the F center feels a square well potential near the vacancy and a Coulomb field with an effective dielectric constant ε_∞ far away from the vacancy. The transition from the ground state, 1s, to the first excited state, $2p$, corresponds to the main absorption band of the F center, left side of Fig. 2. After the optical excitation, $1s \rightarrow 2p$, the first positive neighboring ions move about 10% outward in order to compensate for the new charge distribution of the electron in the $2p$-like state. The potential energy changes as show in the right side of Fig. 2,

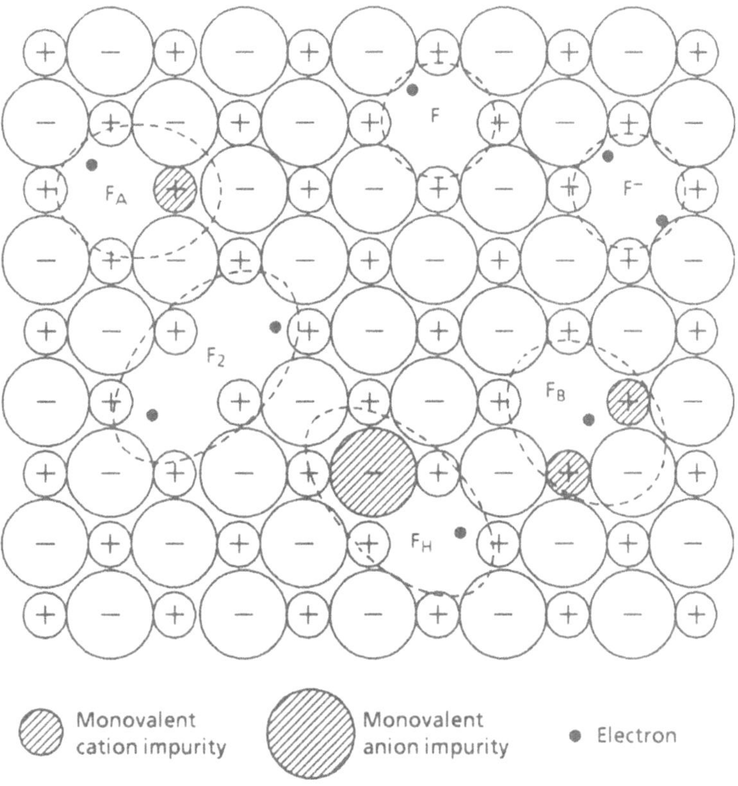

Fig. 1. Structural models of the F center and of simple aggregated centers.

and as a consequence also the energy states undergo a radical variation. The whole process of ion relaxation, as it is commonly known in the literature, takes a very short time, of the order of $10^{-12}s$, to be completed [10,11]. The new $2p$ state is the so called relaxed excited state, and from here the electron decays to the unrelaxed $1s$ state usually with emission of light, which is the well known F center luminescence. At this point of the optical cycle, due to the new charge distribution of the electron, the first positive neighboring ions move back to their original positions with another relaxation process, and the situation depicted on the left side of Fig.1 is restored. It is worthwhile to note at this point that because of the short duration of the two relaxation processes, only the GS and the RES are appreciably populated during the optical cycle.

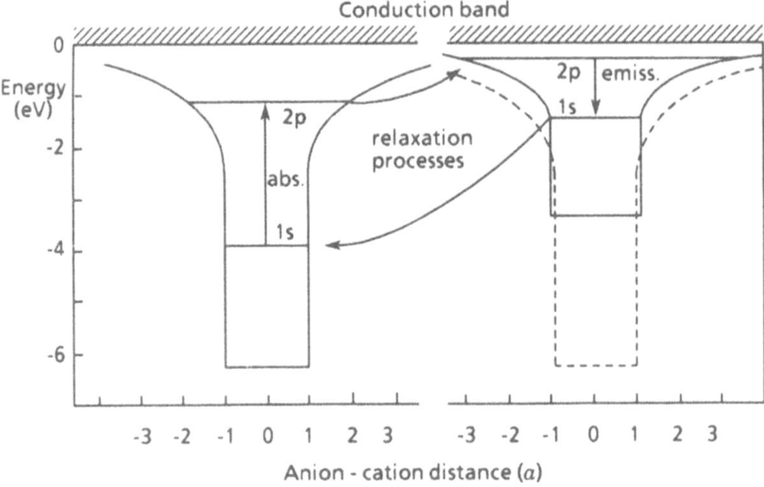

Fig. 2. Semicontinuum model for the F center potential in NaCl, where a is the cation-anion distance. Absorption states (left), emission states (right), and relaxation processes.

The basic idea behind the semicontinuum model works enough well in explaining qualitatively the radiative properties of the optical cycle. Indeed the unrelaxed states of the F center refer to the electron strongly localized around the vacancy, where only a tridimensional box potential is felt. So the separation of the first two quantum states, corresponding to $1s$ and $2p$ levels, is given by:

$$E_{2p} - E_{1s} = E_A = \frac{3\pi^2\hbar^2}{8m\,a^2} = A\frac{1}{a^2} \tag{1}$$

where a is the cation–anion distance. This expression, known as Mollwo–Ivey law, fits the experimental data of the absorption peak energy, E_A, of most of the F centers in alkali halides surprisingly well, with only the exponent of a, n, slightly different from 2, see Fig. 3. The two full curves are the result of a best fit of the experimental data, excluding those of LiI and CsF, to Eq. (1). The parameters of the curves are $A = 17.8$, $n = 1.84$ for the alkali halides with face–centered structure, and $A = 20.6$, $n = 1.80$ for the alkali halides with cesium chloride structure. Analogously the relaxed state corresponds to the electron moving far from the vacancy, i.e. in a Coulomb potential field with effective mass m^* and dielectric constant at high frequency ε_∞. The separation of the first two quantum states can ben derived by the Rydberg law, and in atomic units it is given by:

$$E_{2s} - E_{1s} = E_E = \frac{3e^4m^*}{8\hbar^2\varepsilon_\infty^2} = E\frac{1}{\varepsilon_\infty^2}\ . \tag{2}$$

The experimental data of the emission peak energies, E_E, reported in Fig. 4 show that the functional dependence of Eq. (2) is verified well enough. The constant E derived with a best fit procedure is $E = 5.3$ for the F centers in crystal with NaCl structure, while $E = 7.3$ for the F centers in Cs halides.

In Fig. 2 the electron energy, which increases as the lattice relaxes, is plotted with respect to the conduction band. However the total energy of the system decreases, i.e. the energy of the conduction electron plus vacancy will decrease as

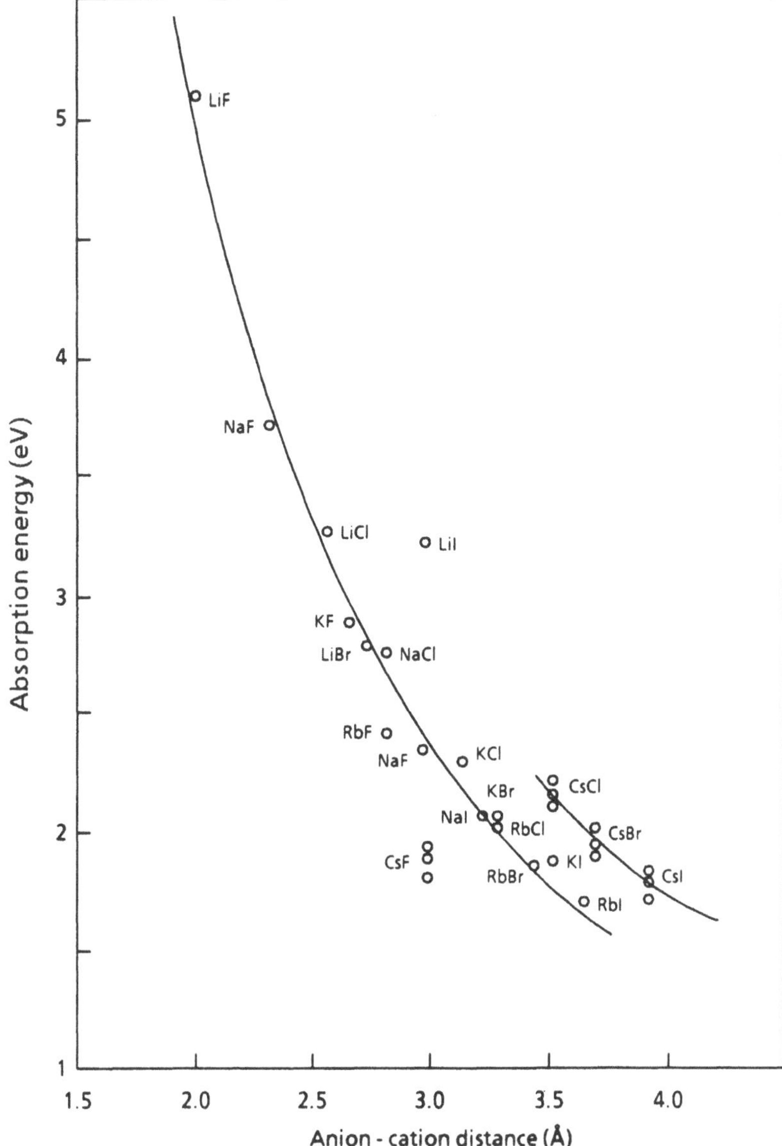

Fig. 3. Peak position at low temperatures of the F center absorptions as a function of nearest-neighbours distance. Continuous lines are best fits of Eq. 1 to the experimental data (see text for details).

the lattice around the vacancy relaxes and phonons are emitted. This aspect of the absorption and emission processes of F centers is better described by the configurational coordinate diagram, which also shows directly the implications of the electron–phonon interaction. The two parabolas of Fig. 5 describe the energy of the F center in the ground, $1s$, and excited, $2p$, state. The normal mode coordinate, Q, represents the ionic displacement as shown previously in Fig.2. It is worthwhile to stress that the amount of horizontal displacement between the two parabolas is a measurement of the electron–lattice coupling, which is quite strong in the case of F centers in alkali halides [12]. Indeed they have very big Stokes shifts $E_A - E_E$, see Figs 3 and 4, which are directly related to the intensity of the coupling. At low temperatures, the electron of the F center occupies the bottom of

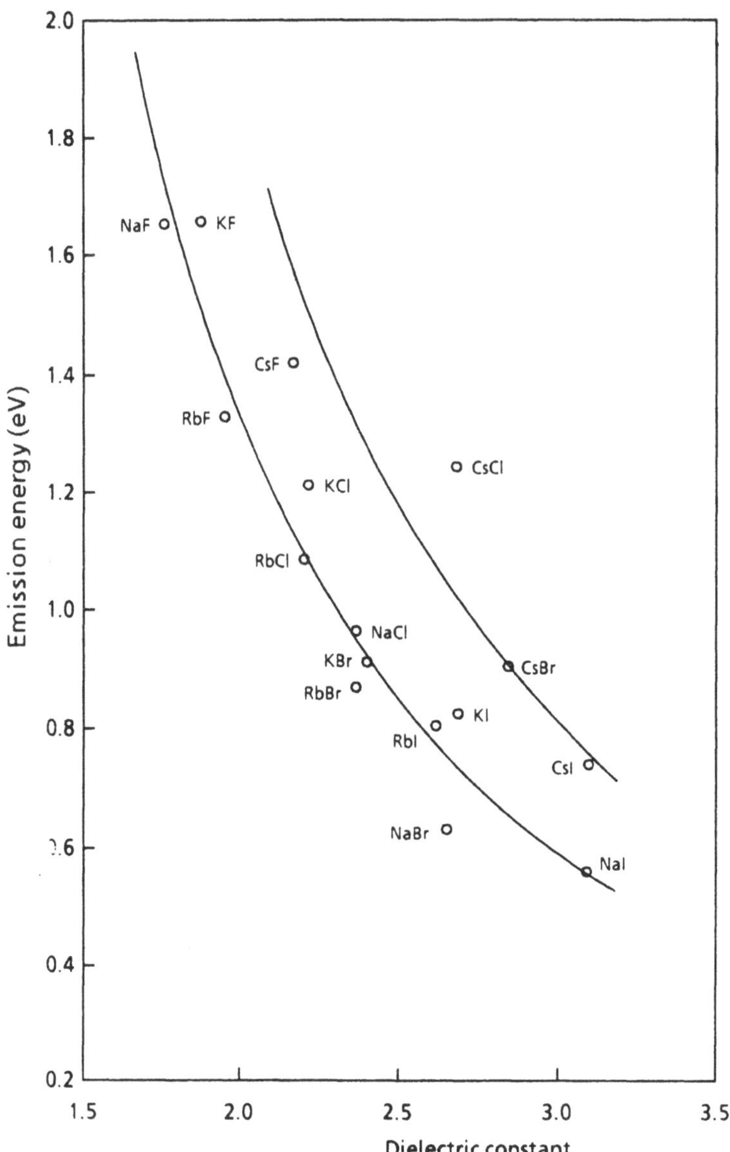

Fig. 4. Peak position at low temperatures of the F center emissions versus the dielectric constant. Continuous lines are best fits of Eq. 2 to the experimental data.

the $1s$ parabola, which is the GS. Optical excitation produces a vertical transition (Frank–Condon principle) to the $2p$ state in a highly excited vibrational level. The vibronic system, electron+phonons, relaxes very quickly, $\sim 10^{-12}s$, to the equilibrium position corresponding to the minimum of the $2p$ parabola, which is the RES. The electron can remain in this state a relatively long time, $\sim 10^{-6}s$, before returning to the unrelaxed GS with a vertical transition accompanied by light emission. In the end, a lattice relaxation completes the optical cycle to the GS. However the emission, which is also called ordinary luminescence, is not the only radiative process which takes place after excitation. Indeed little beyond the obvious Rayleigh line, a structured emission reveals the presence of Raman scattering, which gives detailed information on the lattice modes coupled to the electronic transition [13]. Still beyond that and up to the ordinary luminescence there is a very weak emission tail, known as hot luminiscence, which originates from decaying processes during the relaxation [14,15].

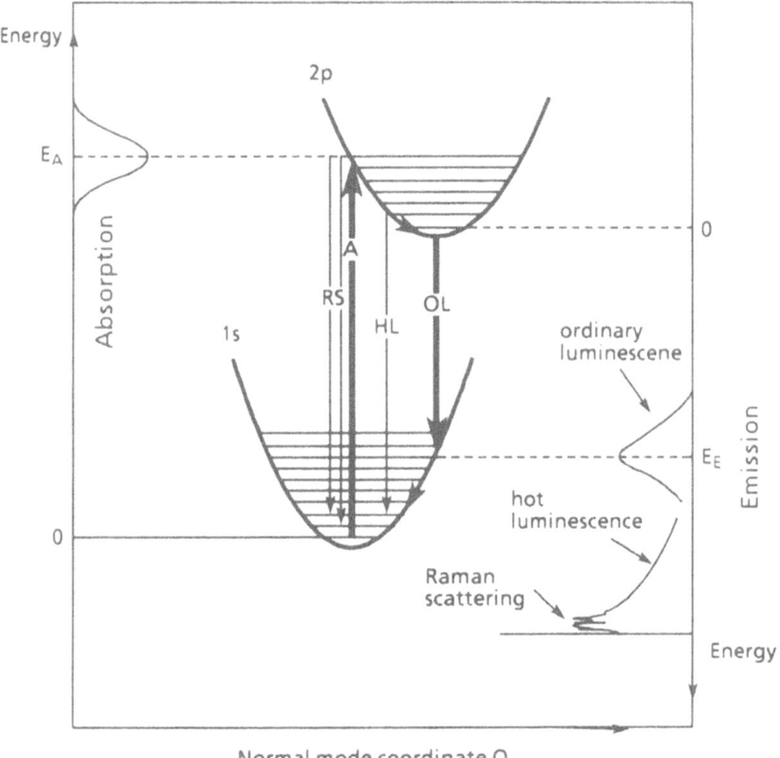

Fig. 5. Configuration coordinate diagram illustrating the absorption and emission phenomena of the F center. The RS and HL is usually few orders of magnitude smaller than the OL.

This complex picture is furtherly complicated by the existence of radiationless processes, which reduce the ordinary luminescence, notably in the case of NaBr and NaI, or quench it completely in other cases [2]. Although the real mechanism of the radiationless transitions is not very well understood at moment, it is clear that the RES is deeply involved in this phenomenon. However by leaving out the uncertain cases where no luminescence was observed up to now, the existence of a relatively long lived RES in the F center optical cycle is a well known experimental fact. But the very nature of this state is still indefinite. More exactly, the whole system of the relaxed excited states is not known experimentally and theoretically as well as the unrelaxed counterpart.

III.A. Unrelaxed Excited States

Figure 6 shows the absorption of a KCl crystal containing *F* centers [16]. In addition to the *F* band there are several other weaker bands on its high energy side, indicated with K, L_1, L_2, and L_3. All these bands are associated with the *F* center, as it has been unambiguously shown experimentally by using the powerful technique of the modulated optical absorption [17], which has been later resumed to identify transitions of other color centers [18]. An intense source of monochromatic *F* light is modulated in amplitude and excites the crystal while a second beam probes the sample at variable wavelengths. The absorption changes measured by the second beam will be in phase with the pump modulation only for absorptions starting from the GS, and so associated with *F* centers. Figure 7 shows the results of the experiment in a colored KCl crystal. The positive peaks of the modulation signal have the same phase as the *F* absorption, and so they are originated from the same pumped state. The *K* band has been associated to transitions from the GS to the whole series of the np ($n = 2, 3, \dots$) states [19], while the final states of the *L* bands are well inside the conduction band, whose minimum is ~ 2.9 eV above the GS in KCl. Magneto–optic experiments have discovered a much more fine structure of the $2p$ state due to the spin–orbit splitting [22], which amounts to 10 meV in KCl and is of a similar order of magnitude in the other alkali halide crystals.

Above the $2p$ state lie other levels like $2s$ or $3d$ that cannot be reached with an allowed optical transition. Such states have been evidentiated by the Stark effect in absorption, which modifies symmetries and selection rules, and makes few transitions partially allowed [20,21]. Figure 8 shows the energy level diagram of the *F* center in KCl as implied by these experiments. On the left side the levels of the hydrogenic model of the *F* center are reported, in the middle the splitting of the cubic field is taken into account, and on the right side the effect of the external electric field has been approximately added.

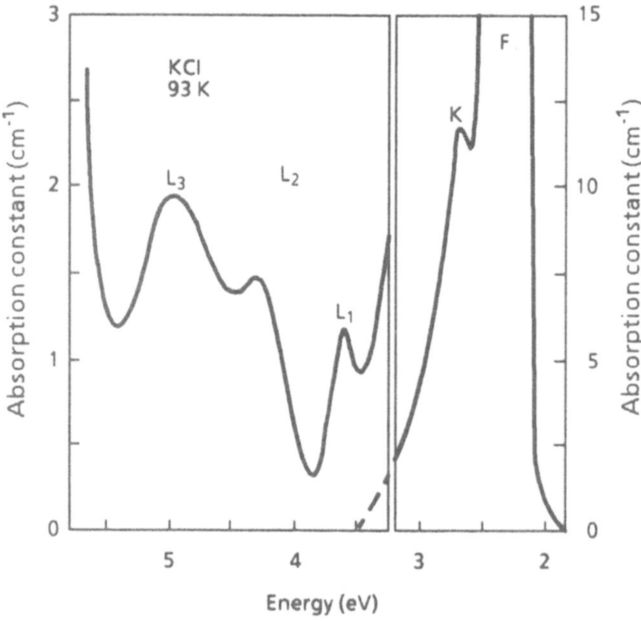

Fig. 6. Optical absorption of the *F*, *K*, and *L* bands in KCl [16].

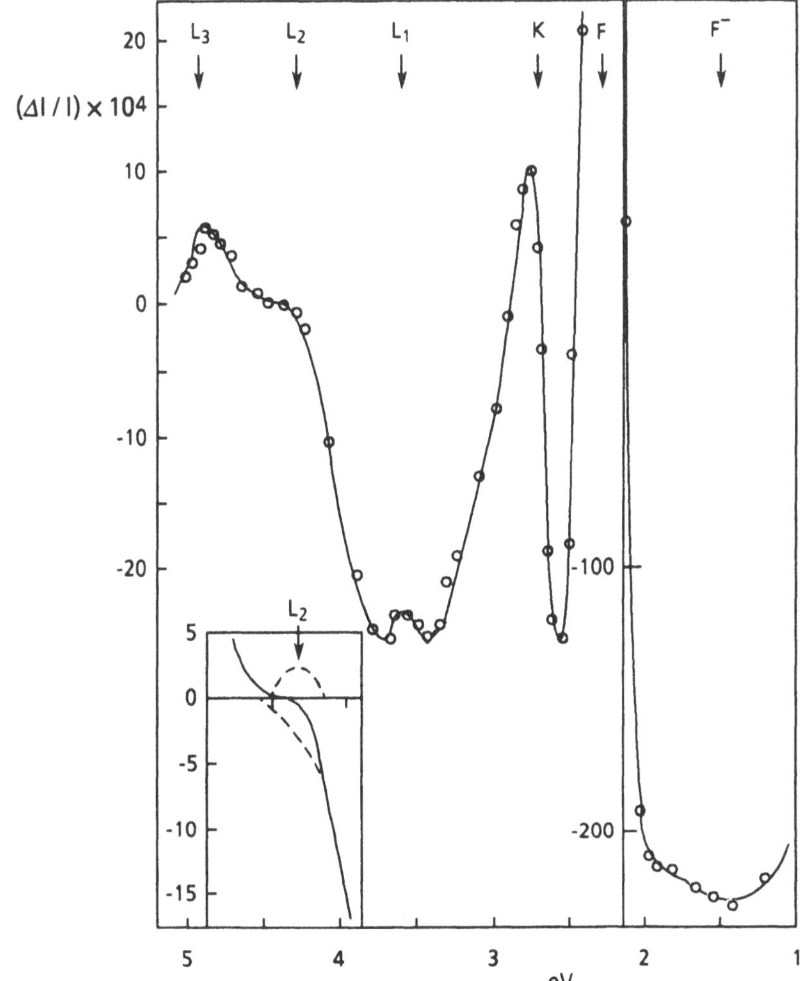

Fig. 7. Modulation index as a function of the photon energy for a crystal of KCl containing $1.4 \cdot 10^{17}$ F centers/cm^3 (chopping frequency 75 Hz, temperature 77 K). In the insert the L_2 band is resolved [17].

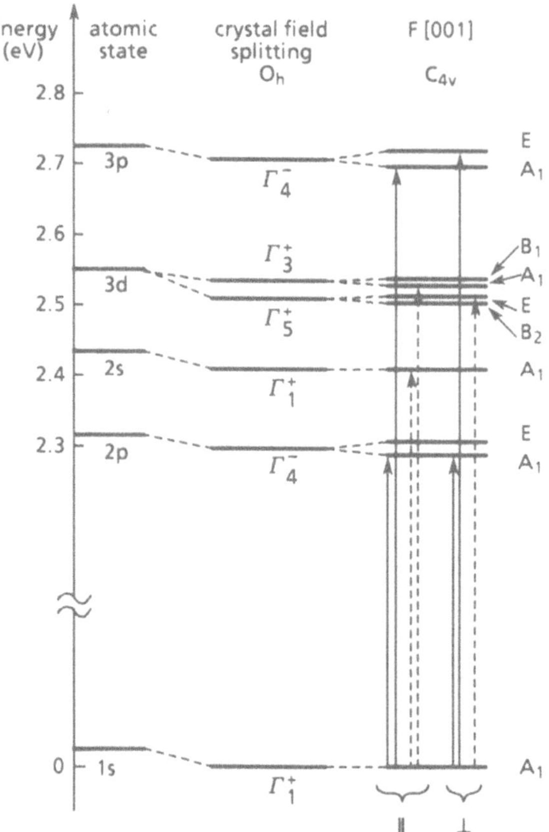

Fig. 8. Energy-level diagram of the F center in KCl in absorption and selection rules for transitions with light polarized parallel or perpendicular to the external electric field applied in the [001] direction. Full lines represent transitions allowed also without field (F and K bands); dashed lines represent field-induced transitions. The position of the atomic states and the amount of the electric field splitting are only approximate [21].

The usual transitions from the GS couples states with odd parity, i.e. $|1s> \rightarrow |2p>$, because they correspond to one phonon absorption. However the technique of two photon absorption could be used in principle to couple states of even parity, i.e. $|1s> \rightarrow |2s>$. However such experiments have not given up to now useful results on the F center system [23], mainly because they have been performed using only a single excitation wavelength. Some experiments, where the one and two photon absorptions in the F band were measured indirectly through the luminescence following the excitation, have given contradictory results as far as the luminescence was concerned [24,25]. So the two photon technique did not add any new information on the level scheme of Fig. 8, while it did show the complexity of the relaxation phenomena after excitation, and the hidden nature of the RES.

In conclusion the unrelaxed excited states of F centers have been fairly well studied experimentally and theoretically up to now, so that a satisfactory description of them exists nowadays.

III.B. Relaxed Excited States

All optical excitations in the F, K, and L bands produce in all F colored alkali halides, with the possible exception of NaBr and NaI [26] which will be dealt with later, only one emission band which is the characteristic F center luminescence. Figure 9 shows the optical cycle of the F center in KCl, where several type of excitations have been taken into account. Although all different excitations get the electrons to the RES, the quantum yield of the subsequent luminescence is not completely indifferent to the various absorption paths. For instance excitation in the high energy side of the F band gives a smaller quantum yield, as measured optically [27] and by photocaloric experiments [28]. Such phenomena are related to the more general field of radiative and non radiative processes, which have been dealt with in details in a previous work [2]. Here the interesting aspect is that the luminescence after whatever excitation is emitted by a well defined state, which we call RES, where the electrons rest for a relatively long time. Figure 10 displays the lifetime of the excited F center in KCl as a function of temperature [29]. Similar results have been obtained for the F center in almost all alkali halide hosts and also for other kind of color centers. As far as the F centers are regarded the lifetime at low temperature, τ_r, ranges from $0,11\ 10^{-6}s$ in NaF [30] to $7\ 10^{-6}s$ in CsI [31]. The values of τ_r were unexpectedly longer than the ordinary atomic lifetimes, of the order of $10^{-8}s$. This discrepancy attracted a lot of interest and moreover it stirred a lively debate on the nature of the RES that it is still going on today.

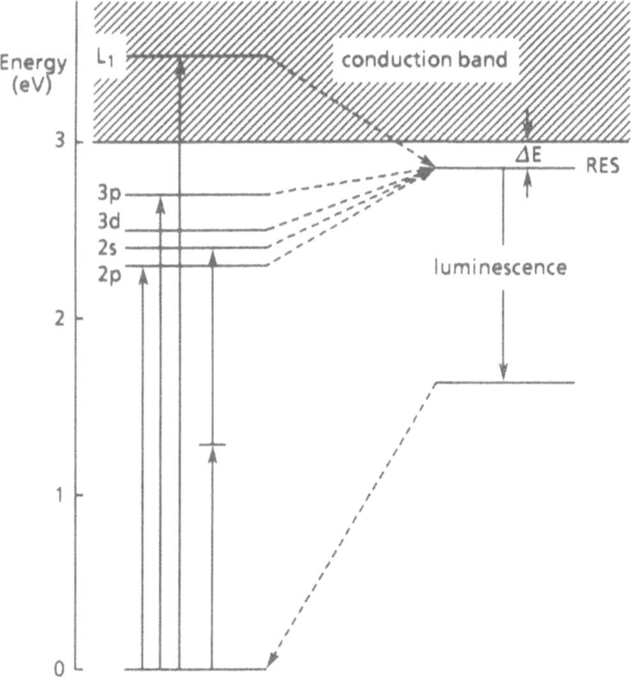

Fig. 9. Absorption-emission scheme of the F center in alkali halides. The energy values hold for KCl and the conduction band has been sketched flat only for convenience.

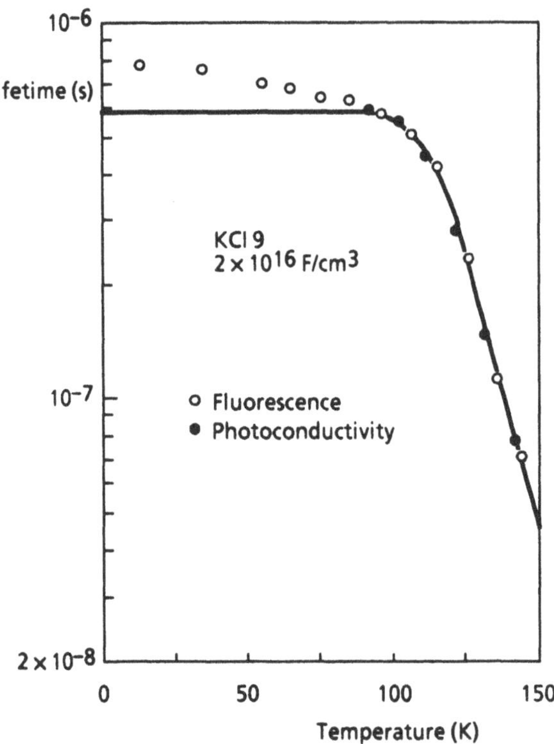

Fig. 10. Lifetime of the excited F center in KCl as a function of temperature. The solid line is a fitting of a simple model which includes the radiative decay and the thermal excitation in the conduction band [29]. See Section VI for details.

In order to explain the long lifetime of the F center, essentially two models for the RES have been taken in consideration with the following features [5]:

1) the large orbit excited state wavefunction
2) the $2s$ state lower in energy then the $2p$ one.

It is well known that the inverse lifetime is given by the radiative transition probability between the initial and final state:

$$\tau_r^{-1} = A| < f|\overline{r}|i > |^2 \tag{3}$$

where A is a constant and $\langle \overline{r} \rangle = \langle f|\overline{r}|i \rangle$ is the dipole matrix element of the transition, given by

$$\langle \overline{r} \rangle = \int_V \psi_f^* r \psi_i \, \mathrm{d}V \tag{4}$$

where $\psi_{i,f}$ is the wavefunction of the initial, final state, and the integral is extended on the volume of definition of $\psi_{i,f}$. By disregarding the case where ψ_i and ψ_f are of the same parity for which $\langle \overline{r} \rangle = 0$, the value of the integral (4) depends strongly on the relative extension of the two wavefunctions. Indeed it is essentially given by their superimposition. This is the idea behind a semicontinuum model developed fully for the F center in NaCl [32]. Figures 11a and b display the calculated electron density relative to the unit cell for the two states $1s$ and $2p$ in absorption and emission, i.e. before and after relaxation respectively. While in the absorption the two states have more or less the same radial extension, after relaxation the $2p$ state, i.e. the RES, is much more diffuse than the $1s$ one. As a

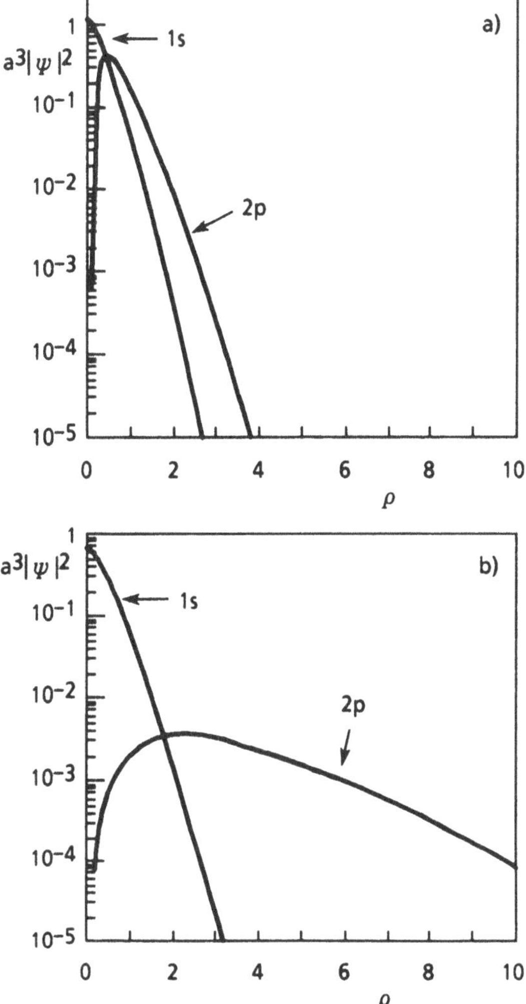

Fig. 11. Calculated probabilities of finding the F center electron in the unit cell (a^3) for the ground state, $|1s\ \rangle$, and the excited state, $|2p\ \rangle$, before (a) and after (b) lattice relaxation in NaCl [32]. Distances r are expressed in units of cation-anion distance a, i.e. $\rho = r/a$.

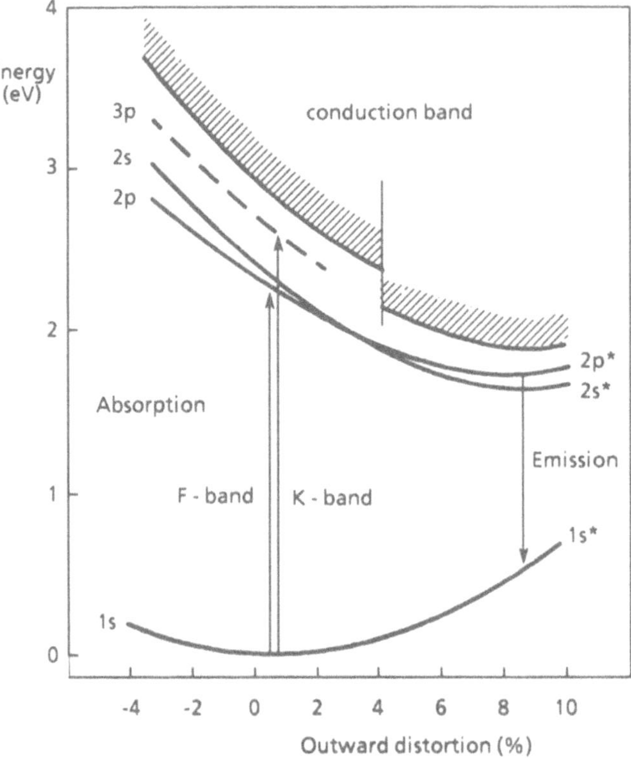

Fig. 12. Energy diagram of the F center in KCl as a function of the positions of the six near neighbouring ions K^+, which shows a possible crossing of $2s$ and $2p$ levels [33].

consequence the dipole matrix element in emission is about an order of magnitude smaller than in absorption, and the calculated lifetime, $\tau_r \leq 134\ 10^{-8}s$, compares well with the experimental value of $100\ 10^{-8}s$ [29].

It has been shown later [33], by using an Hartree–Fock approximation in which the structure of the ions is explicitly taken into account together with the dielectric polarization effects, that a tail develops on the 1s wavefunction in emission. Such variation with respect to the previous calculation tends to increase substantially the value of the dipole matrix element, and so to decrease the calculated lifetime contrary to the experimental observation. However the same approach gives a new and peculiar result shown in Fig. 12, which is practically a configuration coordinate diagram as the one in Fig. 5, with the difference that the present one has been derived by using two different set of parameters, one appropriate for absorption and the other for emission. During the relaxation the $2s$ and $2p$ levels cross each other so that when the equilibrium is reached the $2s$ state lies below the $2p$ one by a few tens of meV. In this case the RES is not anymore a pure $2p$ state but rather a $2s$-like state with some admixture through lattice vibrations with the $2p$ state. Because only the $2p$ state gives a non zero dipole matrix element, see Eq. 4, the small amount of it in the resulting RES can in principle justify the long value of the lifetime. Moreover this calculation [33] shows also that both the $2s$ and $2p$ wavefunctions in emission increase their spatial extension, but not necessarily so much as calculated in the previous model [32].

IV. PROPERTIES OF THE RES

IV.A. Spatial Extension of the Wavefunction

It has been shown in the previous section that the surprisingly long lifetime of the RES can be explained by using two relatively simple models, which can be described in the frame of quantum mechanics. The first one [32] requires a spatially diffuse $2p$ state, Fig. 6b, while in the second one [33] the relaxed $2s$ state lies below the $2p$ state, Fig. 12, both of which can be spatially diffuse in a different way one from the other. So the question of the extension of the RES plays a central role on the nature of this peculiar state.

This information has been obtained through optical-ENDOR measurements which have been successfully performed in KI and KBr [34,35], although it was generally thought that such experiments would have been impossible because of the short lifetime of the RES compared with the inverse nuclear Larmor frequency in the magnetic ratating frame. Figure 13 shows the ENDOR spectrum of the RES of F centers in KI at LHeT, which has been taken from Ref. 34 with some changes in the numbers of equivalent nuclei. The ENDOR frequency is given by:

$$\nu_E = \nu_{HF} + \nu_Z \tag{5}$$

where $\nu_Z = \gamma_n H/2\pi$ is the nuclear Zeeman frequency with γ_n the giromagnetic nuclear ratio and H the external magnetic field, and ν_{HF} is the hyperfine frequency

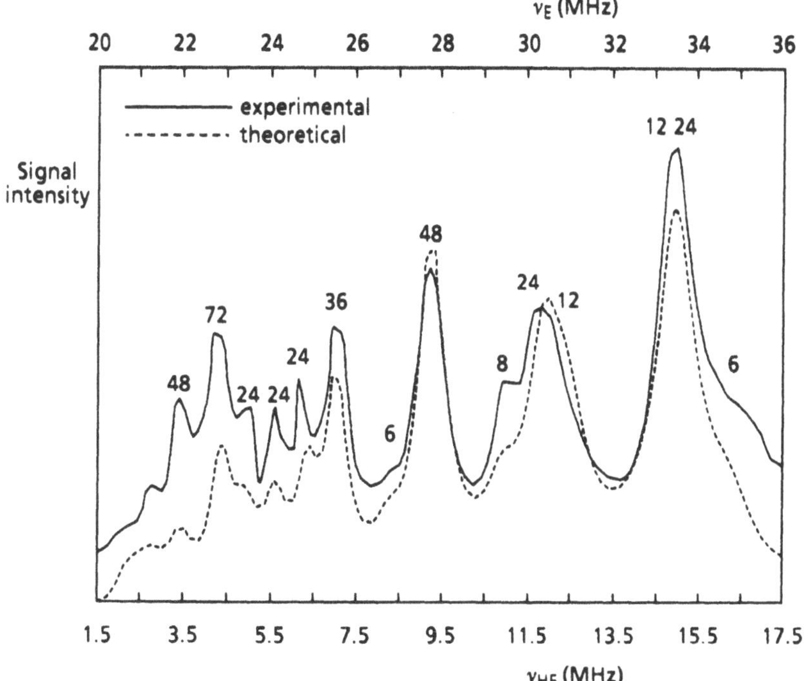

Fig. 13. ENDOR spectrum of the RES of F center in KI at LHeT. The number beside each peak is the number of equivalent nuclei in the shell associated with the peak. The difference between the ENDOR and the hyperfine frequency is the Zeeman nuclear frequency, ν_Z = 18.5 MHz [34].

which, in the case the anisotropic term is negligible, is given by:

$$(\nu_{HF})_i = CA|\psi(\rho_i)|^2 \tag{6}$$

where C is a constant that can be calculated from known parameters, A is the so called amplification factor, and $|\psi(\rho_i)|^2$ is the electron density at the site of the i–th nucleus, where $\rho_i = r_i/a$ [36]. Formula (6) means that the electron wavefunction can be mapped out in the crystal at the position of the ions. In our case the constant C is almost an order of magnitude smaller for posititve ions than for negative ones, so that in practice only the negative ions can be counted for. The ENDOR spectrum in KI is isotropic, and so formula (6) has been used to fit the experimental spectrum of Fig. 13 by using a $2p$ hydrogenic wavefunction without the angular dependence, i.e.

$$|\psi(\rho)|^2 = \left(\frac{\eta^3}{3\pi}\right)(\eta\rho)^2 \exp(-2\eta\rho) \tag{7}$$

the above has been normalized to the unit cell a^3, where a is the lattice parameter. The agreement between the experimental and the theoretical spectrum is quite good for $\eta = 0.515$, which corresponds to a spreading over several lattice distances. The probability density has been reported in Fig. 14 together with the one relative to the GS measured with a pure ENDOR technique [36]. The likeness of the probability densities reported in Fig. 14 with those calculated in Fig. 11b is amazing in view of their completely different derivation. Analogous results have also been obtained in KBr, where $\eta = 0.48$ [35].

ENDOR techniques cannot be applied to F centers in other crystals because of their short lifetime, with the possible exception of CsI where $\tau_r = 7\ 10^{-6}s$. However approximate values of η can be inferred by the width of the electron spin resonance. Indeed the full width at half maximum in our system is given by:

$$\Delta H = D\left[\sum_i (\nu_{HF})_i^2\right]^{1/2} \tag{8}$$

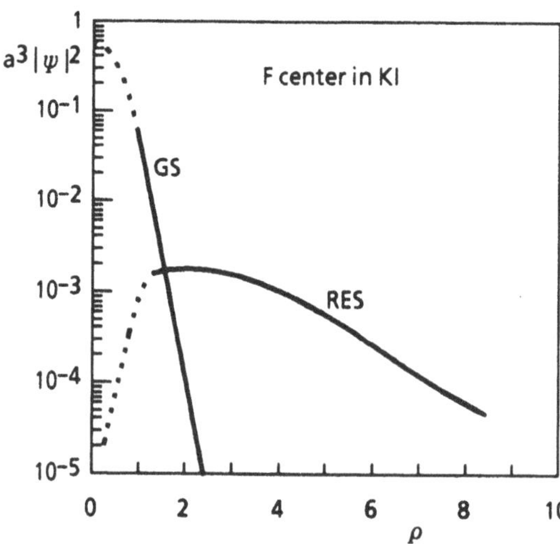

Fig. 14. Probability of finding the electron in the unit cell (a^3) for the ground state, GS, and the relaxed excited state, RES, of F center in KI. The curves are continuous in the region where experimental results from ENDOR measurements have been obtained [36,34].

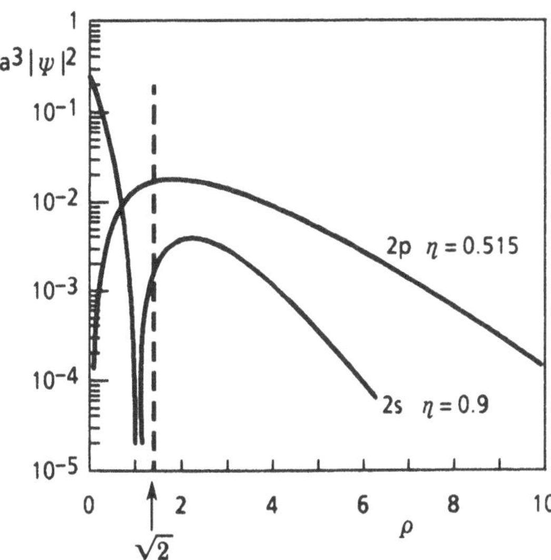

Fig. 15. Radial extension of $2p$ and $2s$ hydrogenic wavefunctions for two different parameters η. ENDOR measurements below $\rho = \sqrt{2}$ do not give any information for the relaxed excited state.

where D is a constant which can be calculated from known parameters [36]. This expression can be fitted to the experimental data by using trial wave functions like in formula (7), obtaining the value of the parameter η. This has been done in a systematic way, and the results have been reported in Table 1 together with those of the GS and with the values of the electron spin resonance g–factor for the RES and GS [36,37]. These results agree fairly well with those obtained for KI and KBr with optical–ENDOR experiments, and indicate again that the wavefunction in the RES is very diffuse, and it is essentially the same in all crystals investigated. On the contrary the wavefunction is very compact in the GS.

All the data presented up to now seem to indicate that the RES wavefunction is a $2p$–like one, but this is not completely true. Indeed the data of Fig. 14 show that no experimental measurements on the electron probability can be made for $\rho < \sqrt{2}$ in the RES, while it has been possible in the GS to make measurements for $\rho = 1$. Figure 15 shows the electron probabilities for a $2p$ and a $2s$ state with different η

Table 1. Wavefunction extension parameter η and electron spin resonance g–factor for the ground and relaxed excited state of F and F_A centers in various crystals [36,37].

Crystal	RES		GS	
	η	g_ρ	η	g
NaCl	0.40	1.966		1.997
KF	0.44	1.994	2.3	1.996
KCl	0.55	1.969	2.45	1.995
KBr	0.34	1.873		1.984
KBr:Li	0.34	1.870		1.982
KBr:Na	0.34	1.868		1.984
KI	0.39	1.686	2.6	1.964
RbCl	0.60	1.930		1.980
RbBr	0.40	1.838		1.967

parameters. Both of them display more or less the same behaviors, such as shape and radial extension for $\rho > \sqrt{2}$, while they are completely different for $\rho < \sqrt{2}$. It is clear that a $2s$ wavefunction properly corrected in the vicinity of the F center, where the potential is different from a Coulomb field, can explain the ENDOR and ESR results very much like the $2p$ wavefunction.

In conclusion these magnetic resonance experiments demonstrate that the RES is a very diffuse state and that it can be a $2s$-like, a $2p$-like, or a mixing of the two. Moreover they also suggest that the real nature of the RES is much more complex than indicated by the two previous models [32,33]. Anyway they support with strong experimental evidences the hydrogenic model for the relaxed excited states, where the electron moves mostly far away the vacancy feeling a Coulumb potential. This is the reason why both the luminescence energy E_E, Eq. (2), and the thermal ionization energy ΔE, Fig. 9, are related to the dielectric constant ε [38] and not to the ions nearest–neighbor distance a.

IV.B. Higher Lying Levels

Up to know the description of the emission after excitation has required only one level, the well known RES, see Fig. 9. However this is an oversimplified picture which was known to be wrong since the first observation of a step rise of the absorption around 0.1 eV measured in presence of excited F centers in KCl [39].

This absorption was interpreted as a transition from the lowest RES to a higher RES situated just below the conduction band, for which $\Delta E = 0.15$ eV. Later on more detailed experiments did show a rather structured absorption spectrum of the RES in several alkali halides [40-42]. Figure 16 shows these kind of spectra in KCl, KBr, and KI [40]. The first narrow absorption has been ascribed to the transition between the $2s$–like lowest RES and a $3p$–like higher state, while the other subsequent peaks correspond to one-, two-, three-phonon (LO) assisted transitions to higher states, which are perturbed and mixed by the vibrational modes of the lattice [43].

Table 2 reports the energy values of the first peak E_0 in various host crystals for F centers [40-42] and F_A centers in KCl [44], together with the values of the thermal ionization energy ΔE_T [45,46].

More recently detailed measurements of luminescence of F centers in NaBr and NaI have shown, besides the strongly quenched emission at 0.63 and 0.56 eV respectively, also a second weak emission at higher energy, more exactly at 1.03 and 0.75 eV respectively [26]. Figure 17 shows the complete emission spectrum in NaI as a function of the temperature, which also implies some correlations among the two components of luminescence. All the observed features and subsequent experiments [47] have indicated that the second peak at $1.65\mu m$ in NaI and $1.2\mu m$ in NaBr occurs from an higher lying level in the relaxed excited configuration.

An ambiguous indication for the existence of other relaxed excited states arises also from electron spin resonance measurements [48,49]. Figure 18 shows the magnetic resonances of the GS, the normal RES 1 and the new RES 2. This new state which has been revealed in NaCl, KCl, KBr and CsBr possesses a g–factor close to that of the free electron and a radiative lifetime an order of magnitude longer with respect to that of RES 1.

At this stage of knowledge it is not yet possible to combine all the previous experimental results in order to sketch a complete picture of the energy levels above the RES. However it is evident that a complex structure above the well known RES exists, and that it may influence more or less deeply the properties of the optical cycle of the F center.

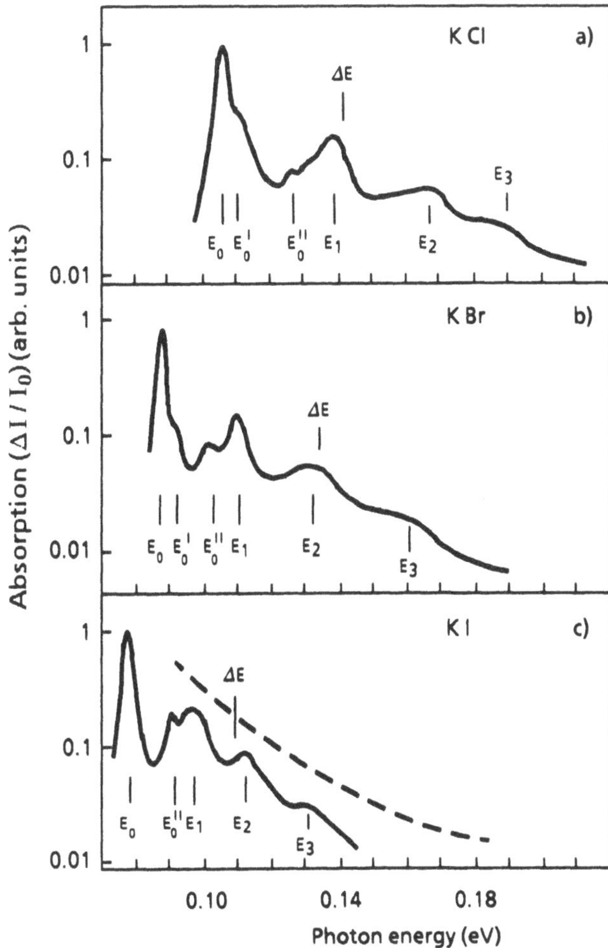

Fig. 16. Transient-absorption spectra of additively colored crystals induced by F-band excitation at 10 K. The induced absorption has been normalized at the maximum value. ΔE represents thermal ionization energy of the excited F center [40]. Previous results in KI [39] are shown in (c) with a dashed curve.

Table 2. Peak energy of zero phonon band E_0 of the transient absorption in the RES measured in F and F_A centers in various crystals. The thermal ionization energies have been taken from Refs. 45 and 46.

Crystal	E_0 (meV)	ΔE_T (meV)
KF	139	138
KCl	106	150
KCl:Na $[F_A(I)]$	104	138
KCl:Li $[F_A(II)]$	no absorption structure	> 270
KBr	87	135
KI	78	110
NaCl	63	72
RbCl	97	119

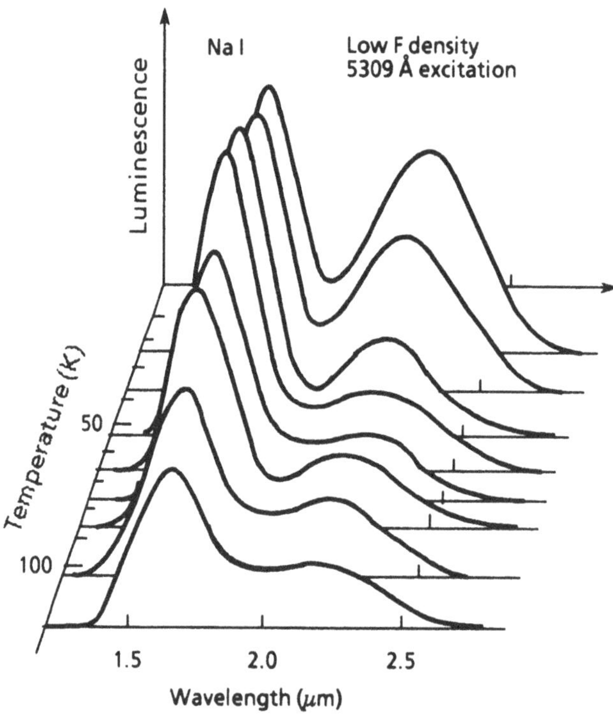

Fig. 17. Temperature dependence of the emission spectra in zone-refined NaI with low F center concentration [26].

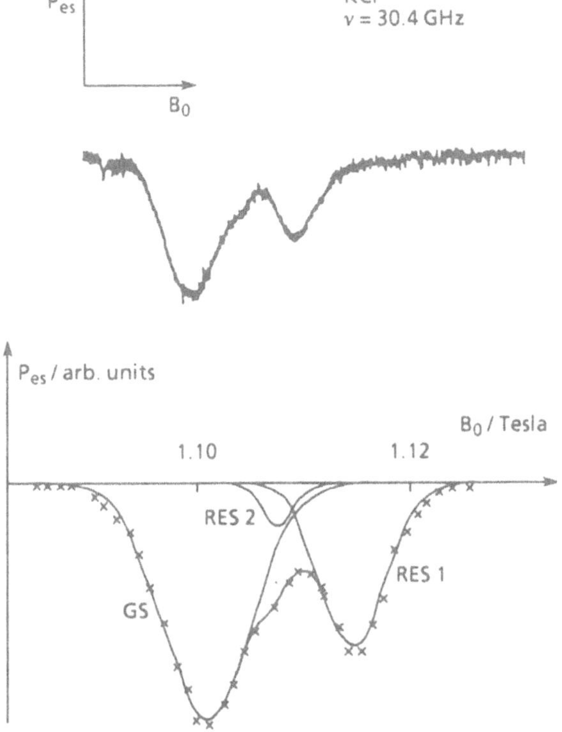

Fig. 18. ESR for ground and relaxed excited states in KCl. The signal at higher field B_0 belongs to the RES, the one at lower field to the ground state. The asymmetry in the ground state signal was a first evidence for the RES 2. The lower part of the figure shows the best fit of the experimental signals by three gaussian curves, the fitted experimental points being marked by x [48].

IV.C. External Field Effects

The hidden structure of the RES has been probed by applying external perturbations, namely electric field, uniaxial stress, and magnetic field. They induce small changes on known properties of F center (electric field and stress) or they produce completely new effects (magnetic field), as it has just been shown in the ESR experiments.

1. Electric Field

The luminescence intensity polarized parallel to the field increases while the perpendicular one decreases, Fig. 19. A red shift of the emission band, Fig. 20, appears together with a small broadening, and eventually the lifetime of the RES decreases Fig. 21. Detailed measurements have been made in the host crystals, NaCl, NaF, KF, KCl, KBr, RbF, RbCl, RbBr, RbI, and CsF [50-54]. All the effects measured show a quadratic dependence from the electric field and are rather small. For

KCl the following values have been found:

$$P_F \text{ (polarization)} = 10 \quad 10^{-6} F^2 \tag{9}$$

$$\Delta E \text{ (shift)} = -0.2 \quad 10^{-6} F^2 \quad \text{eV} \tag{10}$$

$$\frac{\Delta \tau_r}{\tau_r} \text{ (lifetime)} = -5.9 \quad 10^{-6} F^2 \tag{11}$$

where $P_F = [(\Delta I/I)_{\parallel} - (\Delta I/I)_{\perp}]/2$ and the electric field F is expressed in kV/cm. Similar values have been measured for the other crystals except that for CsF, where the induced polarization is more than an order of magnitude smaller than in KCl. This result sets aside CsF from the other alkali halides as far as the nature of the RES is concerned [52]. The previous values refer to low temperature experiments, but a few parameters have also been measured as a function of temperature. For instance the induced polarization is practically independent on the temperature up to $T \simeq 30$ K as it is shown in Fig. 22, while it decreases at higher temperatures.

Stark effects have also been measured in the $F_A(I)$ center in KCl:Na [44], and both polarization and lifetime changes are similar to those of the F center in KCl, showing once more that the $F_A(I)$ center is practically a slight perturbed F center. Recently Stark effects in the $F_A(II)$ center in KCl:Li have also been evidentiated, but not quantified, by measuring the electrical modulation of the output power of a tunable color center laser at $2.7\mu m$ [55].

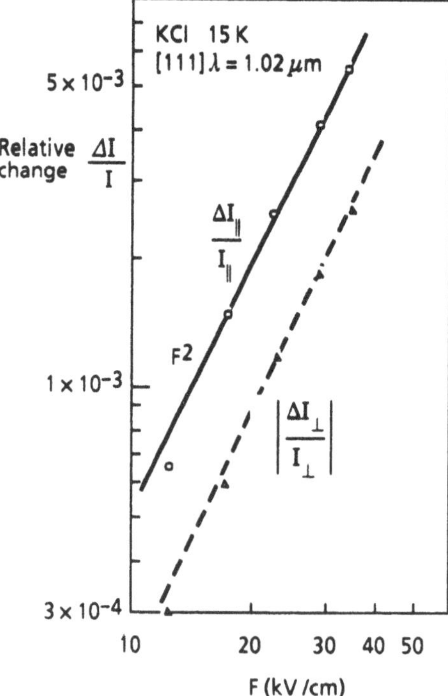

Fig. 19. Dependence on electric field strength F of the polarized luminescence of the F center in KCl [52].

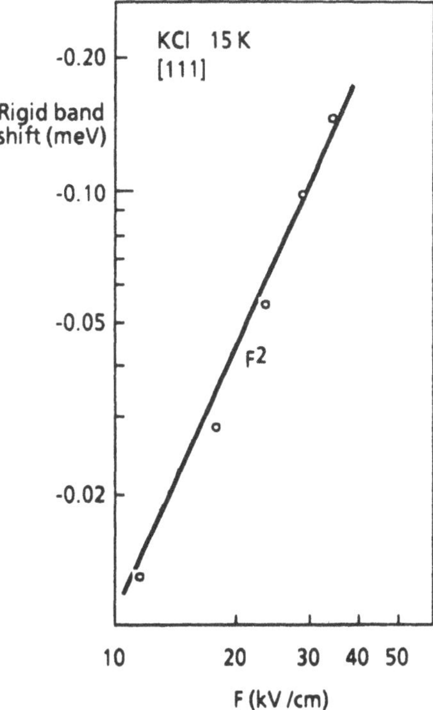

Fig. 20. Rigid shift of the luminescence band of the F center in KCl versus the electric field F [52].

All the Stark effects presented here can be nicely explained in a semiquantitative way by supposing nearly degenerate $2p$- and $2s$-like states mixed by crystal field fluctuations [52], with the $2s$ state lying in the RES below the $2p$ one [33]. The application of an external electric field produces an additional mixing of the states so that there will be an induced polarization, a red shift, and a broadening. Moreover more mixing means more $2p$ character of the resultant RES and so a shorter lifetime. The lifetime can also become shorter by increasing the temperature, because in the process the $2p$ higher state becomes more populated. If α is the mixing coefficient between $2s$ and $2p$ states the two more relevant states can be written as:

$$\begin{aligned} |2s'> &= (1+\alpha^2)^{-1/2}(|2s> +\alpha|2p>) \\ |2p'> &= (1+\alpha^2)^{-1/2}(|2p> -\alpha|2s>) \end{aligned} \quad . \tag{12}$$

Without going in details, for which it is referred to the original papers [51-53] and successive conprehensive review books [56,57], the value of α can be obtained by the following relation

$$\frac{\Delta\tau_r}{\tau_r}\,\frac{1}{P_F} = -\frac{2}{3}\,\frac{1}{(1+\alpha^2)} \tag{13}$$

and the energy separation δE between the $2s'$ and $2p'$ states can be estimated from the temperature dependence of both the polarization and lifetime at low temperatures. For KCl it has been found $\alpha^2 = 0.4$ and $\delta E \simeq 0.017$eV. However these values, which approximately hold also for all the other alkali halides, amidst large experimental uncertainties and simplified assumptions [58], suffer from some inconsistencies. Indeed the big mixing coefficient, which is also too big to explain alone

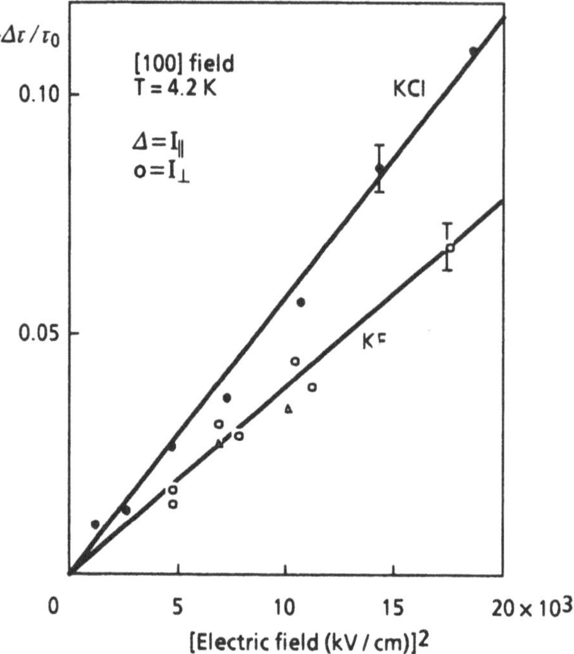

Fig. 21. The electric-field-induced decrease in lifetime for F-center emission in KCl and KF. The zero-field lifetimes were 715 ± 15 ns and 380 ± 20 ns, respectively [51].

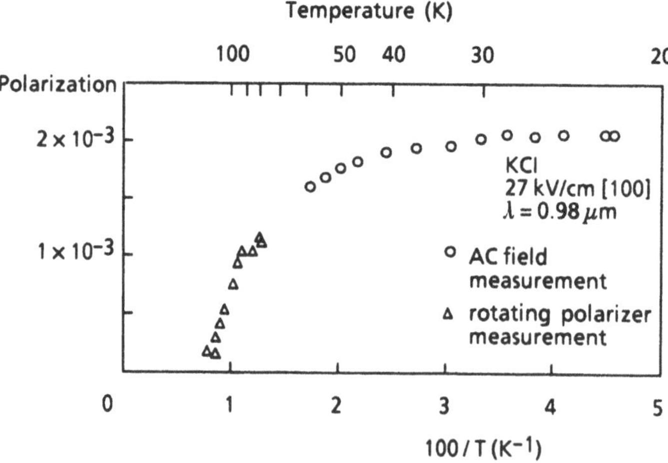

Fig. 22. Temperature dependence of the induced polarization P_F at higher temperature [52].

the long lifetime, and the small spacing energy undermine the principle of the adiabaticity required by the model [53]. However this simple mixed states model gives a reasonable interpretation of the properties of the RES, which would have been hard to be explained within the framework of the large orbit wavefunction alone [32].

2. Uniaxial Stress

Applied uniaxial stress, like the electric field, produces a polarization in the otherwise isotropic luminescence. By using various directions of stress and observations with respect to the crystal axis it is possible to probe lattice deformations of different symmetry, i.e. Γ_3 (tetragonal) and Γ_5(trigonal). The symmetric breathing mode Γ_1 can be activated only by applying an isotropic pressure. Figure 23 shows the induced polarization in the case of Γ_3 symmetry for various crystals [59]. The stress is along the [001] direction and the luminescence observed along [100] direction. The stress induced polarization does not depend on the wavelength of the luminescence, on the F center concentration up to 1.2 10^{17} cm^{-3}, and on the temperature up to 140 K, as it is shown in Fig. 24 for KCl.

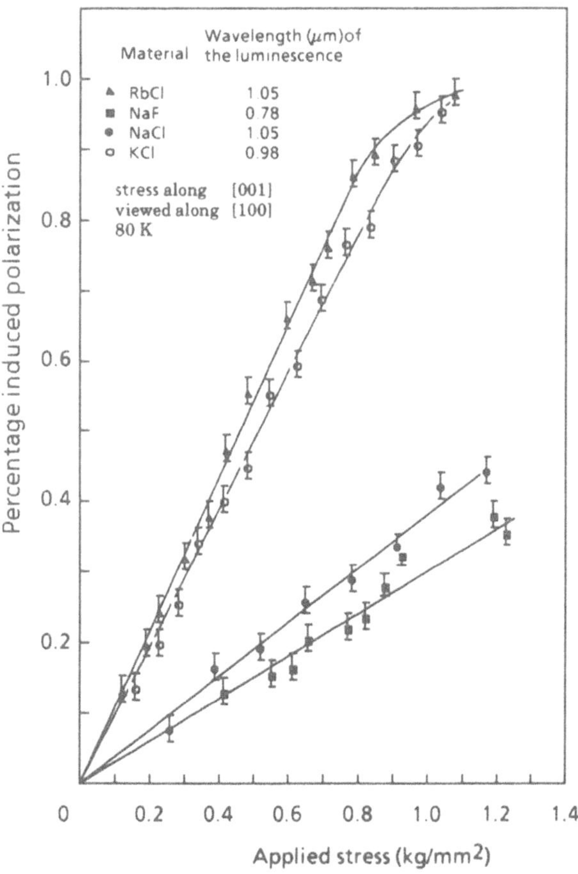

Fig. 23. Percentage induced polarization of F-center luminescence versus applied stress along the [001] direction in kg/mm^2 for NaCl, KCl, RbCl, and NaF. Samples were held at 80 K and measurements were taken at a wavelength near the peak of the luminescence bands [59].

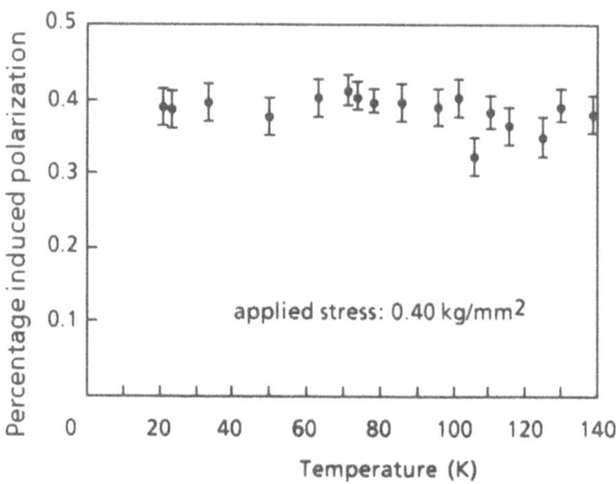

Fig. 24. Percentage induced polarization of KCl F-center luminescence versus temperature for a [001] stress of 0.40 kg/mm² [59].

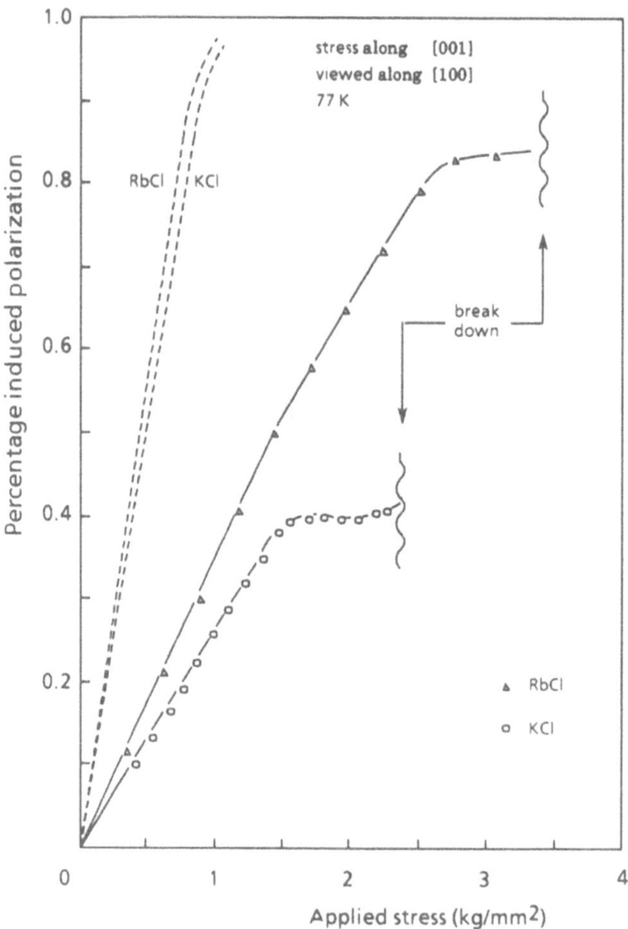

Fig. 25. Percentage induced polarization of F center luminescence, $P_S(\Gamma_3)$, versus stress applied for KCl and RbCl at 77 K [60]. The experimental data from Ref. [59] are also plotted by dotted lines for comparison.

Subsequently these measurements have been repeated in KCl, RbCl and extended to KF, KBr, KI, RbF, RbBr, and CsBr, with results quite different from the previous ones [60]. Figure 25 shows the measurements for RbCl and KCl performed by the two Groups. They differ by a factor 3, much more than the quoted experimental errors, and so only a systematic error can explain such big disagreement. However much more serious, as we will see later in the discussion, is the discrepancy on the temperature dependence. Indeed the more recent measurements show a clear dependence on the temperature practically in all crystals studied. Figure 26 shows the measurements as a function of temperature normalized at $T = 2$ K in KCl. Also here the observed variation from the constant value is bigger than the errors quoted.

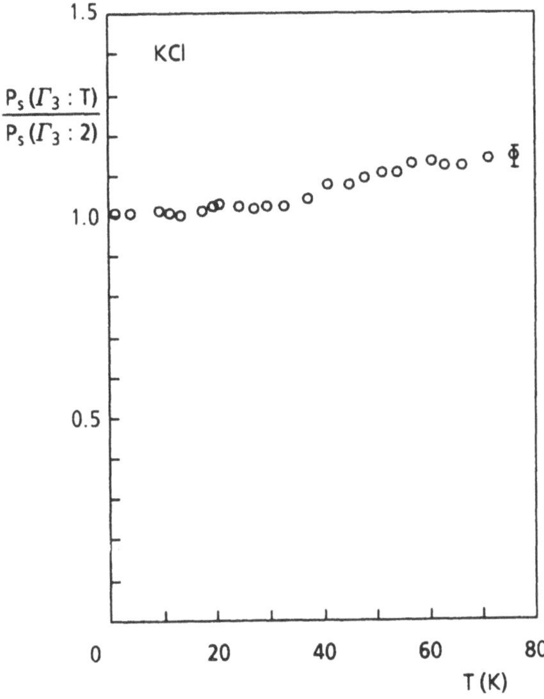

Fig. 26. Temperature variation of stress-induced F center luminescence in KCl. The stress, 1.4 kg/mm^2, is along [001] and polarization is viewed along [100], and so Γ_3 symmetry is probed [60].

The oldest measurements were analyzed in the frame of the very diffuse RES model [59]. In this case the RES should be quite unaffected by the stress and so the induced polarization was ascribed to a mixing in the unrelaxed GS. However a better explanation can be provided by the model of the 2s- and 2p-like states almost degenerate for the RES.

For sake of completeness it is worthwhile to add that also experiments of hydrostatic pressure, which probe the Γ_1 mode, have been performed in KCl, KBr, CsCl, and CsBr [61]. The results are analogous to those obtained in absorption [22] as far as the shift of the emission band is concerned. Moreover a slight quenching of the luminescence and a significant decrease of the lifetime has also been observed. An attempt to explain these behaviors in the frame of the crossover model for the quenching of luminescence [2] has not given satisfactory results [61].

3. Magnetic Field

When a magnetic field is applied to an atomic system the magnetic degeneration of the levels is lifted and the well known Zeeman effect is observed. By keeping in mind that the separation of the levels is $\Delta E = g\beta H \cong 0.12$ meV for a free electron in a field $H = 10$ kG, where β is the Bohr magneton, it is almost impossible to observe such separation directly in the F absorption band, which in KCl is $W_A = 0.163$ eV wide. However by using the Faraday rotation and the magnetic circular dichroic, MCD, techniques, it has been possible to observe these small effects in absorption, and the results obtained have been rewarding [22].

The first successful MCD experiment in emission was obtained in KF [62] and soon after in KCl [63]. Figure 27 shows the results obtained in KF, which consist mainly in a linear dependence of the magnetic circular polarization, defined as the difference between right and left polarized emission normalized to the total luminescence, on the external magnetic field, (a). This effect, which is called diamagnetic because linear in the field, is independent on the temperature, (b), and it amounts to a variation of the emission intensity band, (c), differently to the MCD effect in absorption, where there is a shift of the absorption area. Analogous results were obtained for KCl, but with a different MCD slope [63]. An attempt by the same authors to measure a paramagnetic component, which depends on the polarization of the RES P_ρ, was unsuccessful. They also tried to explain these results in the frame of the model with the $2s$- and $2p$-like states almost degenerate and mixed, eventually finding a strong reduction of the orbital g-factor.

Successively having solved the problem of producing a sizeable value of P_ρ by a special optical pumping [64], accurate measurements of the paramagnetic and also diamagnetic effects were performed in KBr and KI [65]. Figure 28 displays the results obtained for the F center in KI at LHeT. The diamagnetic effect has been measured also with various polarized pumping beams, obtaining the same results for the MCD slope. The paramagnetic effect $S_p = S_p^+ - S_p^-$ is practically independent on the magnetic field up to 30 kG, as it should be because the polarization has been produced by optical pumping out of thermal equilibrium.

The MCD effects can be described by the following simple relation:

$$S_{MCD} = \frac{I^+ - I^-}{I^+ + I^-} = S_d + S_p = C_d H + F_p P_\rho \tag{14}$$

where $I^\pm$ is the intensity of the $\sigma^\pm$ component of the F emission, S_d is the diamagnetic and S_p the paramagnetic signal, and C_d and F_p are the constants which characterize the two effects. Table 3 reports the MCD data in emission which are known up to now. It is not possible here to enter in technical discussions or to describe in detail the lively debate which developed around these experimental values and their interpretation, for which it is referred to the proper pubblications [66-69]. However we will discuss later the meaning and the consequences of the MCD effects on the RES structure.

ESR experiments in the RES have been performed by means of special magneto-optical techniques because a direct magnetic approach would fail for lacking of electron spin population, $10^{12} - 10^{14}$ cm^{-3} at maximum. These experiments have shown that the lowest RES is a Zeeman doublet with $S = 1/2$, it has an isotropic g factor, and the ESR linewidth is due to the hyperfine interaction with the surrounding nuclei [64,70]. By measuring very carefully magnetic resonant and dichroic properties of the RES it has been possible to derive the value of the spin–lattice relaxation time $T_{1\rho}$, which provides the thermalization, given time enough, between the two Zeeman sublevels. Processes which are able to reverse the electron spin are the "direct process" dominant at low temperatures, the "Orbach process" requiring

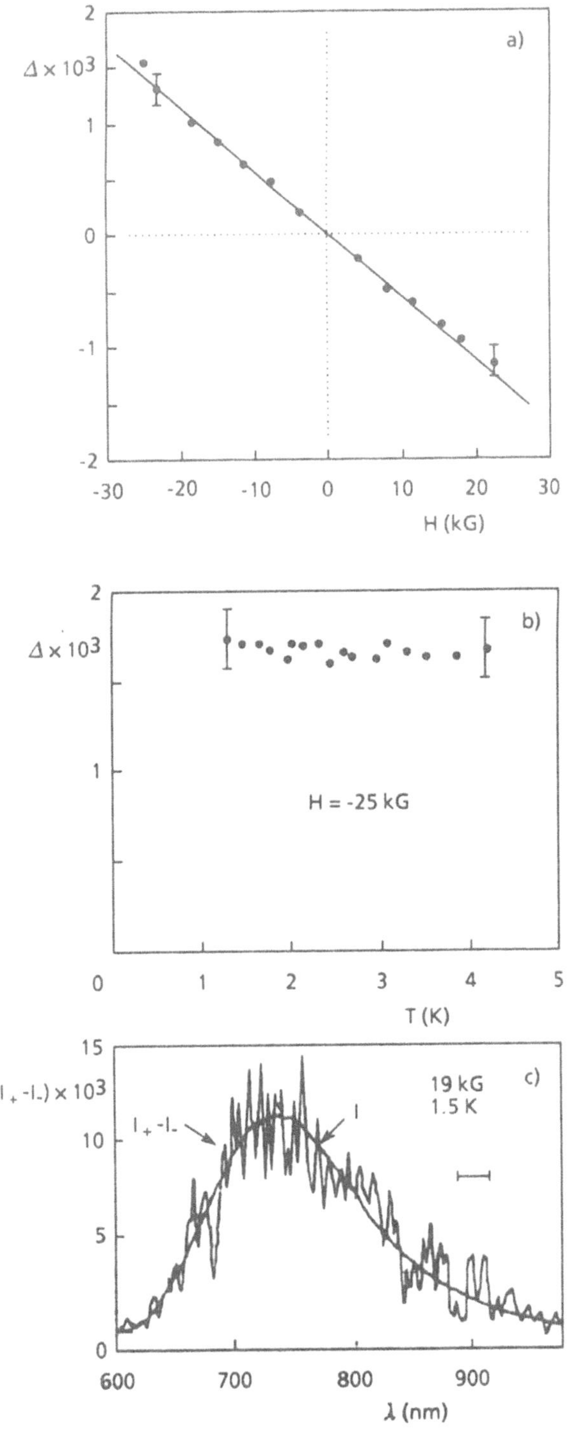

Fig. 27. Magnetic circular polarization, $\Delta = \frac{I_+ - I_-}{I}$, of F center emission in KF as a function of (a) applied magnetic field, (b) temperature, and (c) wavelength [62].

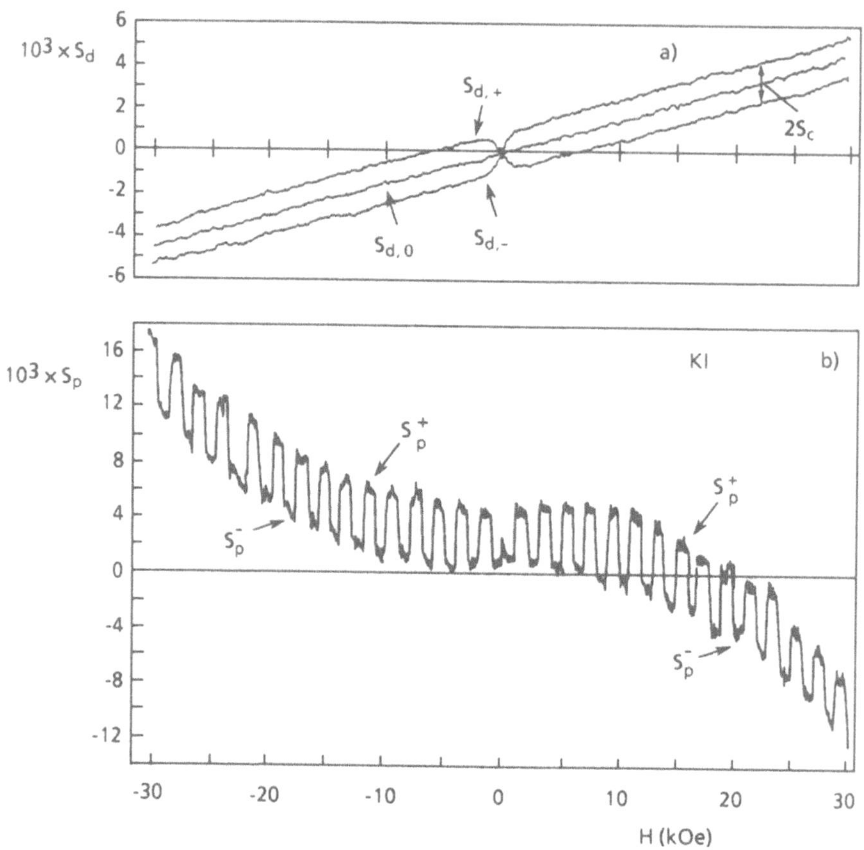

Fig. 28. Magnetic circular polarization of F center emission in KI at T=2 K. (a) Diamagnetic signal for σ^+, π, and σ^- pumping light; (b) Paramagnetic signal for σ^+ and σ^- emission [65].

a third level above the Zeeman doublet by an energy ΔE_0, and the "Raman process" which is a second order process dominant at high temperatures. In general it is possible to write for $T_{1\rho}$ the following expression:

$$\frac{1}{T_{1\rho}} = D \coth\left(\frac{g_\rho \beta H}{2kT}\right) + Oe^{\frac{-\Delta E_0}{kT}} + RT^9 \tag{15}$$

where D, O, and R are constants relative to the above mentioned processes, and k is the Boltzmann constant.

Figure 29 shows the experimental values of $T_{1\rho}$ in KBr and KI measured at relatively high temperature [71]. From the slope of the straight line it is possible to calculate the Orbach excitation energy ΔE_0 which is reported in Table 4 for various crystals. It is whortwhile to note that these values are similar to those calculated from the Stark effect [53,54].

Lately some data have been interpreted as an evidence for the Raman process [61], but the low values of the temperatures involved, 6 ÷ 10 K, cast some doubts on this interpretation, so that a clear evidence of such process is still missing.

At LHeT the spin lattice relaxation in the RES is completely dominated by the direct process, which has been measured in detail in KBr and KI at about 2 K

Table 3. Magnetic circular dichroic parameters in the luminescence of F and F_A centers at LHeT.

Crystal	Magnetic Circular Effects Diamagnetic $\|C_d\|(10^{-8}G^{-1})$	Paramagnetic $\|S_p\|(10^{-3})$	Pumping Wavelength λ (Å)
KF	6^a	–	4360
KCl	9^b	–	5145
	11^d	0.16	5145
	16^e	~ 0	5145
		1.2^e	5750
	13^f	0.1	5145
KCl:Na	12^f	≤ 0.6	5145
KCl:Li	1^f	–	5145
KBr	16.3^c	1.27	6328
	17^d	1.1	6328
	21^e	2.4	6328
KI	18.5^c	2.16	6328
	19^d	2.1	6328
	17^e	1.0	6328
RbCl	22^e	2.3	6328
RbBr	20^e	0.4	6328
NaI (2.25μ)	15^f	≤ 6	5145
NaI (1.65μ)	$\sim 0^f$	0.8	5145

[a] Ref. 62; [b] Ref. 63; [c] Ref. 65; [d] Ref. 66; [e] Ref. 67; [f] Ref. 69.

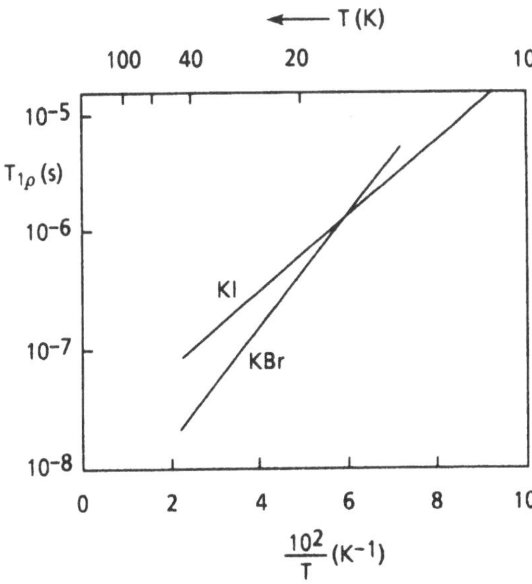

Fig. 29. F center RES spin-lattice relaxation time, $T_{1\rho}$, versus reciprocal temperature in KBr and KI. The straight lines are a clear evidence for an Orbach process [71].

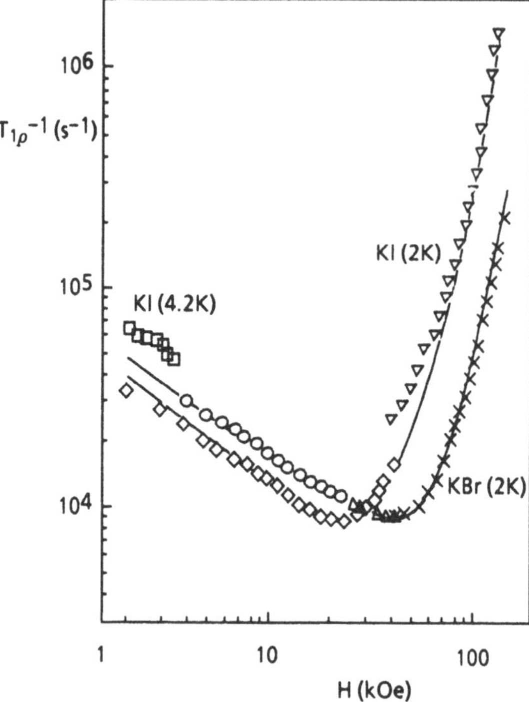

Fig. 30. Inverse of the spin-lattice relaxation time in the RES of F centers in KBr and KI as a function of magnetic field at 2 K. The several symbols represent experimentally obtained points, O △, ◊, □ Ref. [73], x, ▽ Ref. [72]. The continuous line is the best fit of the data with formula (16).

[66,72,73]. All the known data have been reported in Fig. 30, and a best fit of them has been accomplished with the following formula:

$$\frac{1}{T_{1\rho}} = C_\rho H^\alpha + (A_\rho H^3 + B_\rho H^5) \coth\left(\frac{g_\rho \beta H}{2kT}\right) \tag{16}$$

which is similar to the one used for the GS [74], except for the H^α term which takes into account the data at low magnetic fields. This term has some resemblance with some measurements in the GS that are attributed to the exchange interaction between F centers [75,76] or to the presence of impurities, like molecular ions [77]. In the GS the H^3 term is due to the hyperfine interaction of the F center electron with the surrounding nuclei, and the H^5 term reflects the phonon modulation of the spin–orbit coupling through the crystal field. An attempt to calculate

Table 4. Excitation energy of the Orbach process in the RES for various crystals [71].

Crystal	NaCl	KF	KCl	KBr	KI	RbBr	RbI
ΔE_0 (meV)	9.4	7.1	14.2	9.4	6.5	7.0	7.2

Table 5. Fitting parameters to Eq. (16) for $T_{1\rho}$ from Ref. 73. Also the data referring to the spin–lattice relaxation time T_1 in the GS have been reported for comparison.

	KBr	KI
$A(s^{-1}G^{-3})$ $B(s^{-1}G^{-5})$	$1.05\ 10^{-14\ a}$ $1.68\ 10^{-23\ a}$	$9.71\ 10^{-14\ a}$ $10.0\ \ 10^{-23\ a}$
$A_\rho(s^{-1}G^{-3})$ $B_\rho(s^{-1}G^{-5})$	$1.5\ 10^{-11}$ $1.2\ 10^{-11\ b}$ $2.6\ 10^{-21}$ $1.0\ 10^{-21\ b}$	$8.5\ 10^{-11}$ $5.7\ 10^{-11\ b}$ $15.5\ 10^{-21}$ $6.0\ 10^{-21\ b}$
$C_\rho(s^{-1}G^{-\alpha})$ α	$0.50\ 10^7$ -0.613	$1.13\ 10^7$ -0.738

[a] Ref. 74; [b] Ref. 72.

both A_ρ and B_ρ along similar theoretical lines did not give satisfactory results as far as a quantitative comparison with experiments is concerned. Table 5 reports the values of the parameters of Eq. (16) for KBr and KI obtained with the best fit, together with the equivalent values for the GS. There are two or three order of magnitude of difference between the two set of parameters which cannot be accounted for with the actual knowledge on the RES. So a complete explanation of the short value of $T_{1\rho}$ is still missing.

V. DISCUSSION

In the previous Section the main properties of the RES, which have been discovered and studied by using various experimental approaches, have been described in a detailed way. The picture which emerges shows a very peculiar quantum state, or states, possessing several facets which cannot be explained fully by using simple theoretical models like the $2p$ diffuse state [32] or the $2s - 2p$ mixed state [33]. However these two phenomenological models have set the way to more complex theories, which accordingly to their origin have been classified as polaronic and vibronic approaches [56,57]. It is not the aim of this paper to give a detailed description of the successes and shortcomings of each theoretical treatment, and so only a brief discussion on the vibronic theories will be made. Indeed such approach seems to be at moment both rewarding as far as predictions are concerned and relatively simple to be still described with known classical concepts.

The vibronic approaches start with the F center electron interacting with lattice vibrations of different symmetries, i.e. Γ_1^+, Γ_3^+, Γ_5^+, and Γ_4^-. The electronic states taken into consideration are the quasi degenerate $2s$ and $2p$ states in the RES. Usually the $2s$ state is taken to lie below the $2p$ by an energy E_{sp}, but this is not a necessary condition as far as there is some degeneracy between them [78].

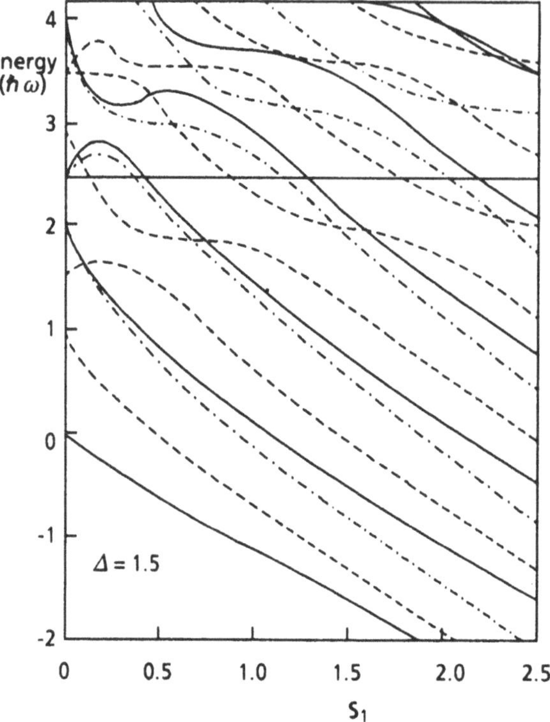

Fig. 31. Vibronic energy scheme as a function of the coupling constant, S_1, with Δ fixed. The various lines correspond to states of different total angular momentum. See text for details [78].

Within this framework the various vibronic approaches differs one from the others because they take into account only one vibrational mode, e.g. the Γ_4^- one [79,80], more than one, e.g. Γ_1^+, Γ_3^+, Γ_4^- [78], and also higher states in the relaxed configuration [81]. Recently the study of the relaxation processes, from Raman scattering, and hot luminescence, and of the ordinary luminescence, did show that the F electron decouples from the ordinary modes in absorption, which are Γ_1^+, Γ_3^+, and Γ_5^+ in the intensity ratio 2:1:1 [82]. At the end of the relaxation only the Γ_1^+ mode is dominant with all the other much more weakier [82], while the $2s$ and $2p$ states are mixed through the Γ_4^- mode vibration [83]. The observation of a zero–phonon line in the transient absorption from the RES [40] is also a proof of the reduced coupling of the F center electron with the lattice, and the subsequent vibronic interpretation [43] further corroborates the Γ_4^- mixing. In conclusion, by excluding also the Γ_1^+ symmetric mode which affects prominently only the red Stark shift [84,85], it is not necessary to take into consideration all the vibrational modes. Indeed the Γ_4^- lattice vibration alone should be enough to describe satisfactorily the experimental data, as it has been assumed previously by using mostly self consistent arguments [86]. It is clear that the inclusion of additional modes of different symmetries can improve the description of the RES properties, but it is not a necessary condition for their explanation.

As a result of the $2s$ and $2p$ states interacting with the Γ_4^- vibrational modes, two series of vibronic states Ψ_I and Ψ_{II} are produced. The first one contains both $2s$ and $2p$ electronic states, while the second one contains only $2p$ electronic states. The lowest energy level of both series belongs to Ψ_I and it is a non degenerate $2s$–like state. The nearest excited level also belongs to Ψ_I series and it is a triplet $2p$–

like state. Figure 31 shows the calculated energy of the lowest vibronic levels of different total angular momentum as a function of $S_1 = E_G/\hbar\omega$ for $\Delta = -E_{sp}/\hbar\omega = 1.5$ [78]. E_G is the vibronic coupling energy, $\hbar\omega$ the energy of the Γ_4^- mode usually the optical longitudinal phonon, and E_{sp} is the energy difference between unperturbed $2s$ and $2p$ electronic states. The values of these various parameters set the energy level scheme of the vibronic model, which for the systems considered here moves between two extreme cases:

1) the strong coupling limit for which

$$S_1 = \frac{E_G}{\hbar\omega} \gg 1 \ , \ |\Delta| = \frac{E_{sp}}{\hbar\omega} \simeq 1 \tag{17}$$

or also $E_G \gg |E_{sp}|$ and $\delta E \ll \hbar\omega$, where δE is the energy separation of the first two levels belonging to the series Ψ_I.

2) the weak coupling limit where

$$S_1 = \frac{E_G}{\hbar\omega} \simeq 1 \ , \ |\Delta| = \frac{E_{sp}}{\hbar\omega} \gg 1 \tag{18}$$

or also $E_G \leq \frac{1}{4}|E_{sp}|$ and $\delta E \leq \hbar\omega$.

In this vibronic picture the RES is the lowest Ψ_I state, which can explain the long lifetime, being a mixture of $2s$ and $2p$ electronic states, and the isotropy of ESR and ENDOR experiments, being a $2s$–like state. Moreover it has also been shown that the shift of the g–factor from that of free electron should be very small, of the order of 10^{-3}, and positive [86]. On the contrary it has been found experimentally relatively big and negative shifts, as shown in Table 1, which so have nothing to do with the vibronic nature of the relaxed states. Indeed they are originated by the same effect as in the GS [36], as it has been suggested by qualitative arguments [72] and lately demonstrated with more rigorous calculations [87]. In this context also the large spatial extension of the RES wavefunction has been confirmed again. In conclusion the measured g shift is due to the spin–orbit coupling in the field of the crystal ion nuclei, because of the p character acquired by the predominanly $2s$–like RES wavefunction after orthogonalization with the crystal ion wavefunctions.

Figure 31 shows such a big number of energy levels above the RES that all the experiments which put in evidence higher lying states, see Section IV B, can be justified in general. In practice every experiment should be studied as a separate case, and this procedure has been performed up to now only for the transient absorption spectrum from the RES in KCl [43]. Figure 32 displays the results of this vibronic calculation, where besides the nearly degenerate $2s$ and $2p$ state also the $3s$, $3p$ and $3d$ electronic states, perturbed and mixed each other through weak vibronic interactions, have been included. The first peak in the figure is due to a zero–phonon transition between the $2s$–like and $3p$–like states, while the subsequent structure E_1, E_2, E_3, formerly E_0'', E_1, in Fig. 16, is originated by one phonon assisted transition to $3s$–, $3p$– and $3d$–like state respectively.

When an external field is applied to the F center the vibronic levels are either mixed according to the symmetries of the interaction and the states involved, either splitted or shifted. An accurate analysis of the previous vibronic approach as far as the lower levels are concerned has shown that the electric field, F, and the uniaxial stress, S, mix and shift only Ψ_I states, while the magnetic field, H, splits the Zeeman levels and mixes Ψ_I and Ψ_{II} states. Hence it is evident that by using the previous external perturbations it is possible to probe levels of different symmetries.

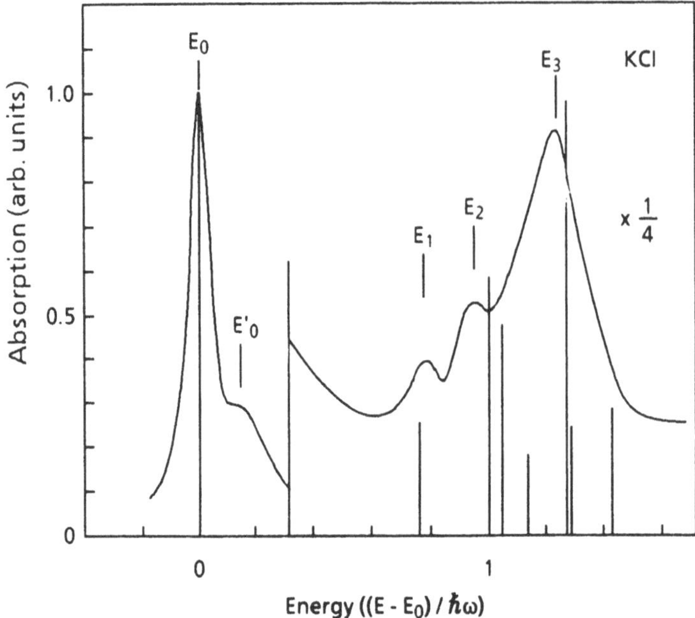

Fig. 32. Transient absorption spectrum for KCl. The line structures are calculated while the solid curve is the observed spectrum. The first peak is due to a zero-phonon assisted transition between 2*s*-like and 3*p*-like states, while the subsequent structure, E_1, E_2 and E_3 is originated by one-phonon assisted transition to 3*s*, 3*p* and 3*d*-like states respectively [43].

In the case of the electric field all Ψ_I states are mixed with the two lowest ones, which differ in energy $E(1) - E(0) = \delta E$, giving the highest contribute. It is found that the induced polarization P_F, (9), the red shift ΔE, (10), and the lifetime variation $\Delta\tau_r/\tau_r$, (11), can be calculated by using consistent vibronic parameters. However the red shift is obtained by taking into account also the Γ_1^+ vibrational mode [84]. In the case where only the two lowest states are taken into account the temperature dependence of the polarization is given by [86,84]

$$\frac{P_F(T)}{P_F(O)} = \frac{1 + 3\exp(-\delta E/kT)}{1 + 3R\exp(-\delta E/kT)} \tag{19}$$

where $R = P_F(O)/P_F(1) = \tau_r(O)/\tau_r(1)$. Figure 33a shows the experimental values of the polarization [50] fitted with Eq. (19). The same expression also holds for the radiative lifetime of the excited F center and in Fig. 33b the experimental data [52] are fitted with Eq. (19), solid line. If more than two vibronic levels are taken into consideration, in this case 30 levels [84], the agreement between experiment and theory improves expecially at higher temperatures, Fig. 33b dashed line. In the calculation also the simmetric Γ_1^+ mode has been taken into account through the coupling parameter S_0, but its influence is insignificant being S_1, the coupling parameter of the Γ_4^- mode, reduced only by 20% for $S_0 = 1.5$.

Anyway the fact that S_1 is very small indicates, see Eqs. (17) and (18), that there is not a strong vibronic coupling between F center and lattice vibration, so that an intermediate or weak coupling limit approach can be used to a good approximation [86]. Indeed in this limit the theory gives

$$\frac{\Delta\tau_r}{\tau_r} \quad \frac{1}{P_F} = -\frac{2}{3} \tag{20}$$

which coincides with Eq. (13) for $\alpha = 0$, namely the mixing of $2p$ state into $2s$ state is practically negligible.

By fitting the temperature dependence of the electric field polarization, Fig. 33a, and of the radiative lifetime, Fig. 33b, one obtains practically the same value for the energy separation, $\delta E \cong 17$ meV. More or less the same value is obtained by the Orbach process of the spin lattice relaxation time in the RES, $\Delta E_0 \simeq 14.2$ meV [70]. So it results $\delta E < \hbar\omega = 26.8$ meV in KCl, which again is an expected result for the weak–coupling limit.

The result of applying uniaxial stresses poses few problems. Indeed the effects measured in two different works [59,60] differ substantially each other when they refer to the same crystal hosts. Not only the measured induced polarizations are different beyond experimental errors, see Fig. 25, but also the induced polarization is independent from temperature for one group [59], see Fig. 24, while it is temperature dependent for the other group [60], see Fig. 26. These experimental discrepancies, expecially the last one, are of paramount importance as far as the vibronic model of the RES is concerned, and they stirred a debate which has yet to be concluded.

The stress induced polarization is a first order effect involving at least at low temperatures, where only the lowest vibronic levels are thermally occupied, exclusively the Ψ_I states. Semiquantitative interpretations of the experimental data at fixed temperatures on the basis of the present vibronic theories are satisfactory [60]. The temperature dependence of P_S can be represented approximately by the

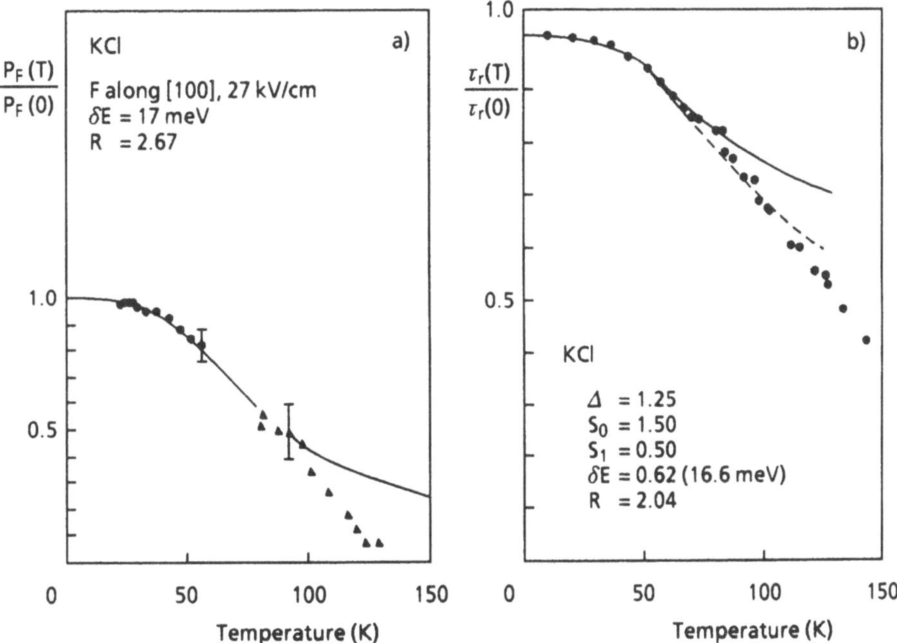

Fig. 33. Temperature dependence of: a) the electric field induced polarization of F center emission in KCl, the solid line represents Eq. (19) with parameters given in figure [51]. b) the radiative lifetime of the excited F center in KCl, the solid line represents Eq. (19) and the broken line is a best fit performed with a more complete vibronic theory, whose parameters are shown in figure [84]. The experimental values (•, ▲) are from Ref. 53.

following formula:

$$\frac{P_S(T)}{P_S(O)} = \frac{1 \;+\; C(T)\exp(\delta E/kT)}{1 \;+\; 3R\exp(\delta E/kT)} \tag{21}$$

which is similar to that used for the Stark effect, Eq. (19), with the same meaning of the various parameters except the new ones $C(T)$ [88]. At low temperatures one has always $C(T) > 3R$, so that there will be always an initial increase of the stress induced polarization as the temperature increases above O K. Figure 34 shows the results of a vibronic calculation including several states in thermal equilibrium [89,88]. The quantity σ is proportional to the stress S applied along [001] and viewed along [100], and so the data refers to Γ_3 symmetry. The curves of Fig. 34 can satisfy the experimental results reported in Fig. 24 only in the case $\Delta > 2.0$, while the results reported in Fig. 26 can be fitted only with $\Delta < 2.0$. This last fitting has been attempted more recently [60] and it has been reported in Fig. 35. The values of Δ and S_1, obtained agree fairly well with those obtained by the temperature dependence of $P_F(T)$ and $\tau_r(T)$, Figs. 33a and b. The inclusion of other vibrational modes, namely Γ_1^+ and Γ_3^+, goes in the direction of enhancing the increasing of $P_S(T)$ with increasing temperature [43], and so this behavior is common to all actual vibronic calculations. This fact seems to assign more credit to the measurements of Ref. 60 than to those of Ref. 59, but at the same time a strong feeling of uncertainty remains on this argument.

From the previous discussion it has been concluded that the vibronic problem of the RES can be treated much more realistically in the weak coupling limit or at maximum in the intermediate coupling. So in order to analyze the MCD data in emission it is possible to use in a first approximation the simple formulation of the weak coupling approach [86]. First of all it should be reminded that in the first order perturbation only Ψ_{II} states are mixed into the lowest Ψ_I state. In this case the magnetic circular dichroic signal is given by:

$$S_{MCD} = \frac{I^+ - I^-}{I^+ + I^-} = -\frac{2(g_L \beta H + \lambda\langle S_z\rangle)}{|E_{sp}| + \hbar\omega} \tag{22}$$

where g_L is the orbital g factor of the electronic p states, λ the spin–orbit splitting, and $2\langle S_z\rangle = P_\rho$. For KCl by using the value $C_d = 11\ 10^{-8} G^{-1}$, which is approxi-

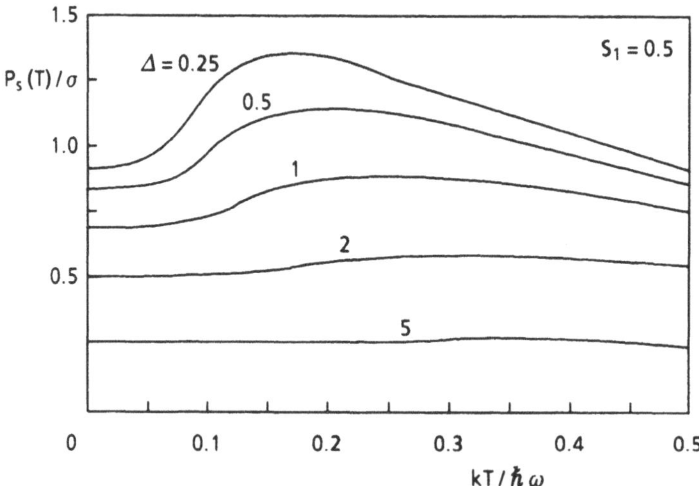

Fig. 34. Stress-induced polarization $P_S(T)$ of luminescence from the RES, for Γ_3 symmetry, as a function of temperature, for different values of $\Delta = |E_{sp}|/\hbar\omega$ and for a fixed value of the vibronic coupling strength $S_1 = E_G/\hbar\omega = 0.5$ [89].

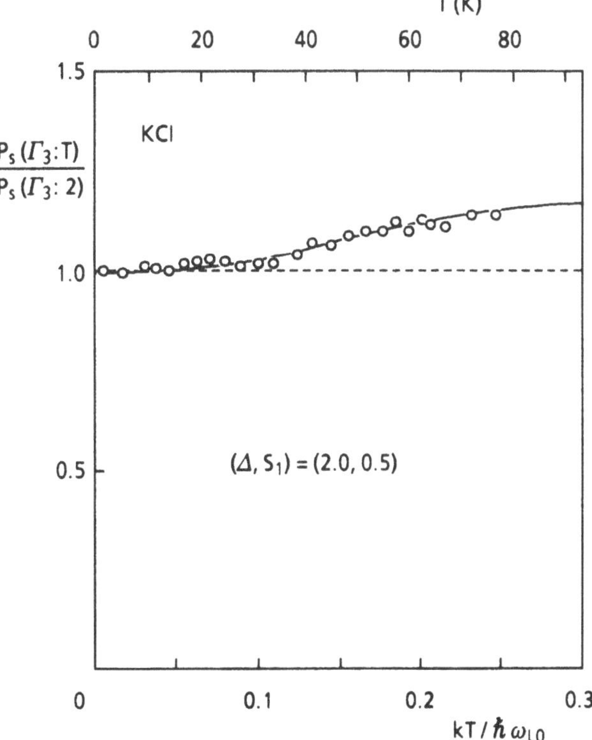

Fig. 35. Temperature dependence of $[P_S(\Gamma_3;T)/P_S(\Gamma_3;T = 2K)]$ versus $(kT/\hbar\omega_{LO})$. White circles are experimental points. Solid line is taken from Ref. 88 in the case when $(\Delta, S_1) = (2.0, 0.5)$ [60].

mately the average value of the three experimental determinations reported in Table 3, and $g_L = 1$, one gets $|E_{sp}| + \hbar\omega = 104$ meV. Being $\hbar\omega = 26.8$ meV, it is $|E_{sp}| = 77$ meV and so $|\Delta| = 2.9$. This value is too big with respect to those found previously, i.e. $\Delta = 1.25$ from the temperature dependence of $P_F(T)$ or $\tau_r(T)$, and $\Delta = 2$ from the tentative temperature dependence of $P_S(T)$ [60].

However the experimental results can also be compared with an exact solution of the vibronic problem performed by taking into consideration the Γ_4^- and also the Γ_1^+ vibration mode [84]. Figure 36 shows the diamagnetic effect in units of $(g\beta H/\hbar\omega)$ as a function of $\Delta = |E_{sp}|/\hbar\omega$ for $S_1 = E_G/\hbar\omega = 0.5$, as determined by the temperature dependence of $\tau_r(T)$, and for two values of the coupling parameter S_0 with the Γ_1^+ mode. In this plot the experimental diamagnetic result for $g_L = 1$ is at the ordinate value 0.47, which corresponds for $S_0 = 0$ to $|\Delta| = 2.75$ very similar to $|\Delta| = 2.9$ determined previously in the limit of weak coupling (this is still a further proof of the validity of this approach!).

So the value of Δ derived from MCD measurements in emission is in any case bigger than those obtained by $P_F(T)$, $\tau_r(T)$ and probably by $P_S(T)$. However by looking carefully to formula (22) and Fig. 36, it is evident that a lower value of g_L can give a lower value of Δ. Indeed in Fig. 36 a second value of the ordinate is reported with a g_L factor perfectly compatible with the actual knowledge of F centers. With the choice $S_0 = 1.5$, $S_1 = 0.5$, and $\Delta = 1.25$ like in Fig. 33b, one gets $g_L \simeq 0.76$ [84]. By using this value in Eq. (22) one gets $|E_{sp}| + \hbar\omega = 79$ meV. With this new value it is possible to calculate the spin–orbit coupling in the RES, which, by taking $|P_\rho| = 2\langle S_z\rangle = 0.05$ [66] and $|S_p| = 0.16\ 10^{-3}$ Table 3, for KCl is

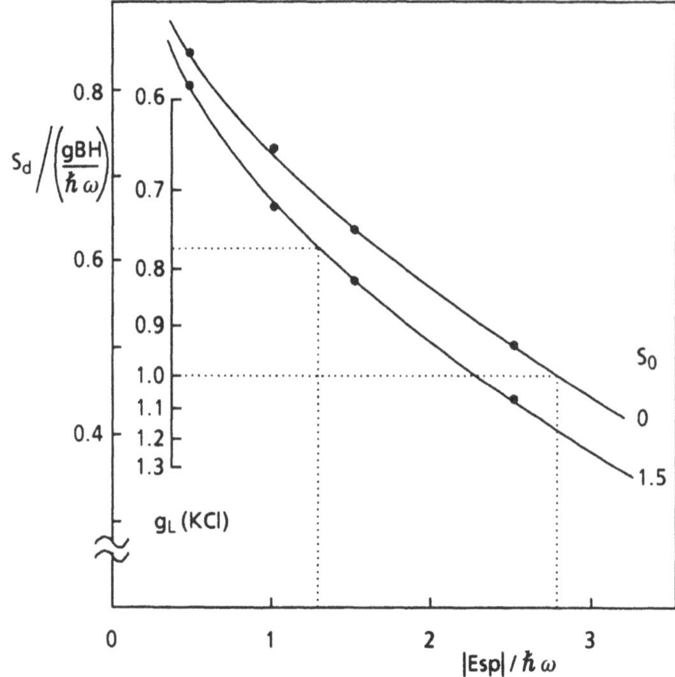

Fig. 36. Diamagnetic effect S_d as a function of $\Delta = |E_{sp}|/\hbar\omega$ for $S_1 = E_G/\hbar\omega = 0.5$ and $S_0 = 0, 15$. This plot has been obtained by the exact vibronic calculations of Ref. 84.

$|\lambda| = 0.25$ meV. This result is much lower than the value of the spin–orbit coupling in absorption, -9 meV [22]. The same striking reduction of $|\lambda|$ hold in all the other measured crystals [66].

Through the MCD experiments in emission also the spin-lattice relaxation time $T_{1\rho}$ has been determined, see Fig. 30. By comparing the value of $T_{1\rho}$ with those of T_1, in the GS, see Table 5, it is at once evident that $T_{1\rho}$ is two or three order of magnitude shorter than T_1. Because of a very similar magnetic field behavior between the two relaxation times it seems logical to assign them to the same basic processes. However an attempt to extend to the RES the theories used for the GS has been unsatisfactory [72]. Indeed for the hyperfine term, H^3, we used the extended wavefunction measured with ENDOR experiment, but the calculated values of A_ρ are more or less similar to those of A for KBr and KI . Most probably a dipole–dipole interaction can play in the RES a role much bigger than in the GS [74]. For the H^5 term both the Kronig–Van Vleck mechanism and the phonon modulation of the g–factor have been taken into account. Calculations show that the first mechanism misses the experimental results by two order of magnitude, while the second one is within a factor two or three in KBr and KI. It is worthwhile here to remind that this second effect [90] is originated by the same mechanism which produces the rather big g–factor shift in the RES [87]. Also it is worthwhile to make a remark about the Kronig–Van Vleck mechanism which describes correctly the magnetic field behavior of $T_{1\rho}$, but it is unable to explain its short values. By using the simplest expression still containing the important parameters it is possible to write:

$$\frac{1}{T_1} \sim \frac{\delta^2}{\Delta^4} \quad H^5 \tag{23}$$

where δ is the spin–orbit coupling and Δ is the energy of the first excited state. Eq. (23) can be applied to both the GS and the RES, and in this last case the results indicate that only small values of the spin–orbit coupling, $\lambda \simeq 0.5$ for KBr and KI, are compatible with the known values of the other parameters [66].

VI. CONCLUDING REMARKS

A comparison between the experimental data of the relaxed excited F center and the vibronic theory mostly in the weak coupling limit has been made in the previous section by considering mainly the KCl crystal . Most of the experiments are accounted for quantitatively by the theory with the exception of a major discrepancy for the uniaxial stress induced effects. In this particular case most of the problems are originated by conflicting experiments.

By taking into account only the earlier results [59] for which the polarization effects are temperature independent, it is found $\Delta = |E_{sp}|/\hbar\omega > 2$ for whatever value of the vibronic coupling $S_1 = E_G/\hbar\omega$. These values agree with those obtained from circular polarization effects, $\Delta \cong 2.9$ with $g_L = 1$, but strongly disagree with those obtained by fitting the temperature dependence of the radiative lifetime and of the polarization induced by an electric field, for which $\Delta = 1.25$ and $S_1 = 0.50$.

On the contrary by taking into account the latest results [60] where the stress polarization effects are temperature dependent, it is found $\Delta \cong 2.0$ and $S_1 = 0.5$. These values agree better with the data of $\tau_r(T)$ and $P_F(T)$, but in order to conciliate them with the circular polarization data it is necessary to have $g_L = 0.76$, and so a consistent quenching of the orbital g-factor, which anyway is still a reasonable result.

However much more critical is the case of KF for which from measurements of $\tau_r(T)$ [50] it has been obtained $\Delta = 0.0$ and $S_1 = 0.7$ [84]. As a consequence a strong temperature dependence of P_S is required by the vibronic theory, as it is shown in Fig. 34 which refers to the case $S_1 = 0.5$ very similar to the previous one. Stress measurements on KF have been done lately [60] and those concerning this particular case are reported in Fig. 37. Practically the measurements of P_S are temperature independent contrary to the theoretical expectations.

In conclusion as far as the stress induced effects are concerned there is not a plausible way to conciliate vibronic theories and experiments at the actual state of knowledge.

Another effect which deserves some comments is the temperature behavior of the RES lifetime expecially below 100 K, see Fig. 10. The total lifetime of the excited F center is usually analyzed with the following relation:

$$\frac{1}{\tau} = \frac{1}{\tau_r} + \frac{1}{\tau_0} \exp[\Delta E_T/kT] \tag{24}$$

which contains the radiative lifetime at low temperature τ_R and a thermally activated excitation process to the conduction band with energy ΔE_T. The full line of Fig. 10 is the best fit of the experimental points with $\Delta E_T = 0.14$ eV, $\tau_R = 0.58$ $10^{-6} s$, and $1/\tau_o = 1.2\ 10^{12}\ s^{-1}$. Eq. (24) cannot account for the small variation of τ below 100 K. However small this amount can be, it has been the subject of an hot dispute among supporters of the large orbit [91,92] and of the $2s - 2p$ degenerate states model for the RES [53,54], which is not yet over today [93].

Without going in too much details eventually the supporters of a $2p$-like diffuse RES claim that the measured lifetime at low temperatures is not an intrinsic

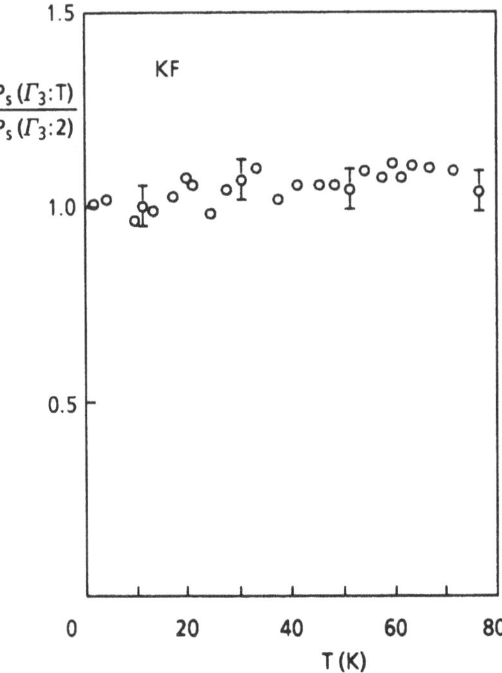

Fig. 37. Temperature variation of stress-induced F Center luminescence in KF. The stress, S=11 kg/mm^2, is directed along [001] and the polarization viewed along [100] in order to probe Γ_3 mode [60].

property [94]. Indeed small amounts of α centers (negative ion vacancies) influence the lifetime through an interaction with F centers, as it is shown in Fig. 38 in the case of KCl [95]. The interaction among F, α, and F^- (formerly F') centers, which are always present in excited F center systems because of the kinetic equation $2F + h\nu \rightarrow F^- + \alpha \rightarrow 2F+$ phonons, is also claimed to be at the origin [93] of a curious nearly oscillatory fine structure of the lifetime as a function of temperature [96]. Figure 39 displays such effect in KCl, which is measurable only when the temperature is kept regulated within $\sim 4\ 10^{-2}$ K. At this point it is worthwhile to emphasize that the small temperature variation of τ_r at low temperatures represents one of the strongest evidence in supporting the $2s$ - $2p$ degenerate states model, and subsequently the vibronic approach. So also small effects on τ_r should not be dismissed without a careful examination, which has not been done yet. Moreover in case the KF crystal is taken in consideration, also here there are contradictory experimental results on the lifetime below 100 K, as it is evidentiated in Fig. 40a and b, where data of different authors are reported.

As it has been said previously the agreement between the vibronic theories, which at moment seem to be more appropriate than others, and expriments is fairly good in KCl, when comparison is made by taking the various properties one by one. The agreement becomes less evident when the various properties are considered all together. This situation becames much worst in the case F centers in other crystal hosts are considered, as it has been already evidentiated in KF. Moreover a careful examination of all the properties shows that the experimental data are far from being complete and reliable.

The intricated situation of the uniaxial stress effects and lifetime measurements has just been examined in part, but, for instance, the situation of the electric field

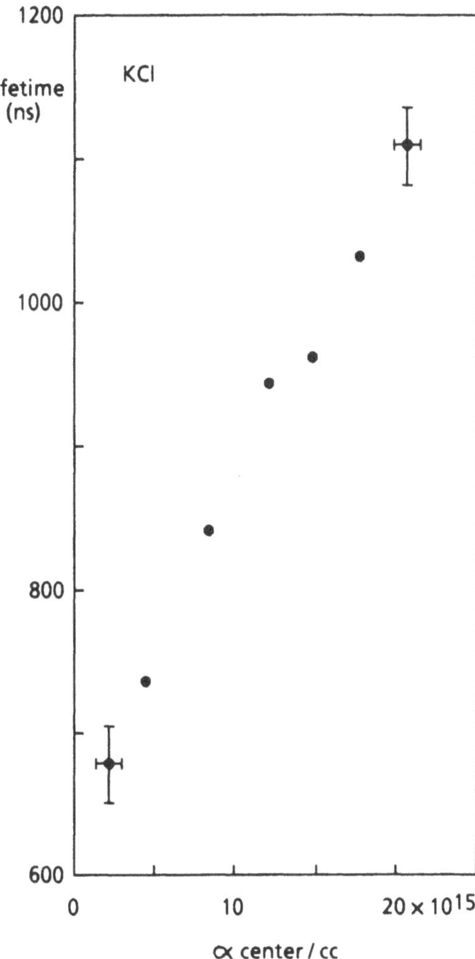

Fig. 38. Luminescence lifetime excited with F light vs the α-center concentration n_α, after different $F \to F'$ conversions in a KCl crystal with $4.97 \times 10^{16} F$ centers/cm^3 initially present. Measurement temperature = 80 K [95].

effects is not much better. Indeed the experimental uncertainties are large and only two authors [52,54] have measured the Stark parameters, and moreover in different crystals, so that it is not possible to make any comparative study [58]. Again Stark effects have been crucial in the choice of the RES model, so that more reliable and extended measurements are needed.

The magnetic circular dichroic effects appear to be more reliable, but however the present measurements cover only a small number of alkali halides. Notable measurements among them are those relative to F_A centers in KCl and F centers in NaI. While the Na ion, $F_A(I)$, does not change appreciably the MCD properties of the F center, a big variation is observed when Li ions, $F_A(II)$, are used as impurities. An analogous situation has been found for the two luminescences in NaI, see Fig. 17. The one at 1.65 μm has the diamagnetic effect completely quenched, see Table 3. These results further indicate that the emitting state of $F_A(II)$ centers and of the RES (2) in NaI are different from the well known RES of F centers, and this still not well known difference influences in a remarkable way the MCD effects

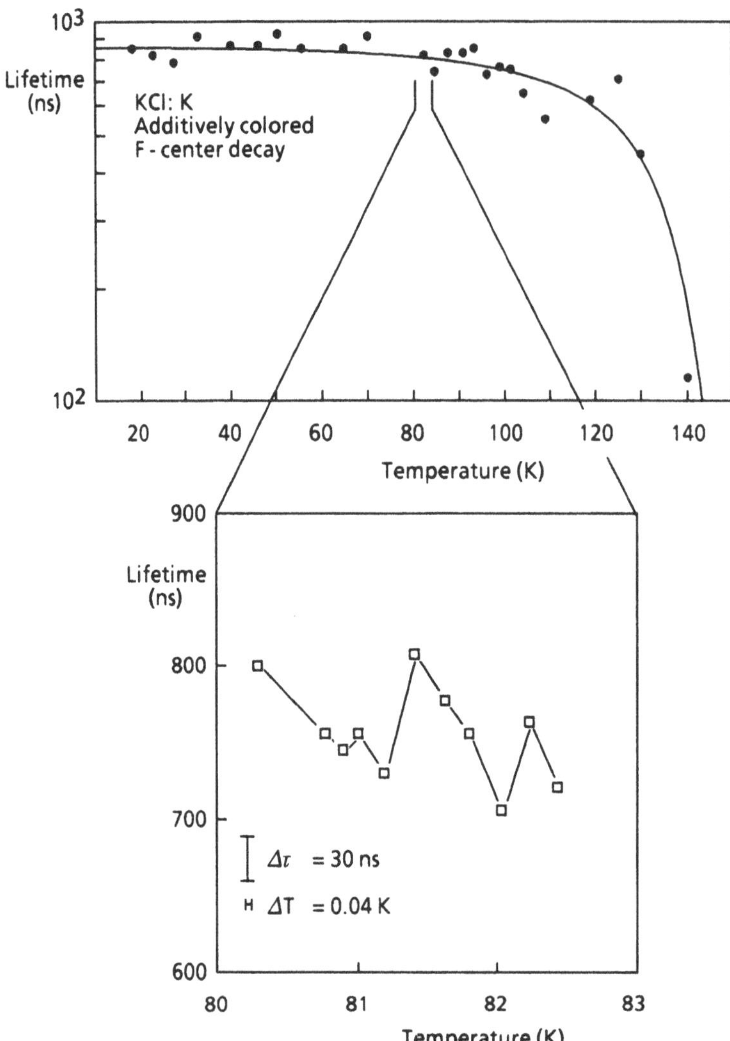

Fig. 39. Lifetime of the excited F center in KCl as a function of temperature, above. The structure of the experimental points is real and has been expanded below, where the error bars on the measurements of time and temperature has also been reported [96].

in emission. Recently the two luminescences in NaI have been assigned to transitions from a true RES and from a bound polaron state respectively [97]. However this assignment seems to be at odd with the values of the emission energy and the MCD signal of the two emissions.

From side effects of MCD in emission the spin-lattice relaxation time in the RES, i.e. the lowest of the vibronic states, has been determined in KBr and KI, see Fig. 30. Up to now it has not been possible to explain satisfactorily the much shorter values of $T_{1\rho}$ with respect to the similar quantity in the GS, T_1, see Table 5. The wavefunction of the RES in deeply involved in determining such shorter values, but the mechanism of interaction with the surrounding ions is still obscure. Most probably an extension of $T_{1\rho}$ measurements in other crystals and possibly in F_A centers would be extremely useful. In general what is missing at moment is a model describing much better the RES wavefunction and its interaction with internal and external field.

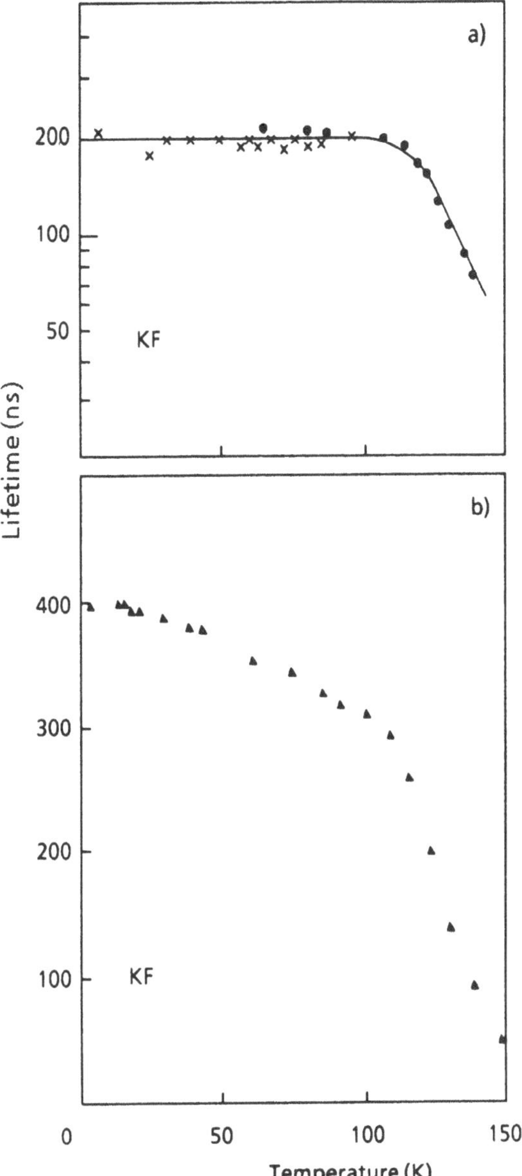

Fig. 40. Lifetime of excited F center in KF as a function of temperature. a) Ref. 30, b) Ref. 53.

Lately the interaction between F centers and molecular ions, which has sizeable effects on the absorption and emission properties, has been studied by using the MCD techniques in absorption [77]. In particular the electron spin memory, ε, during the optical cycle has been determined with and without a nearby molecular ion. A large difference, $\Delta\varepsilon$, has been measured in a few cases, and it seems to be originated by an interaction acting while the electron is around or in the RES. Work is in progress to find out whether this effect can be used as a probe to study in much more detail the nature of the RES.

As far as the experimental data are concerned, it results a need to perform more experiments on new crystals and to repeat few ones on those crystals where errors are too big to be acceptable for a fruitfull comparison with the theory.

To my knowledge there have not been new calculations with old models or new theoretical approaches on this field lately, with the exclusion of a polaronic model [97] and an elegant recursion method [98]. About ten years ago a theoretical paper was closed with the following remarks: "Evidently a vibronic model for the RES that provides a completely consistent quantitative explanation for all data has not been achieved [88]." Today the situation does not look too much different.

In this paper it has been made an attempt to give a concise description of the properties of the RES mainly of the F center, which is the building block of all other more complex color centers. It was not the intention of the author to make a complete review on this matter, so that a few notable contributions have not been discussed in detail or included at all. Instead the main idea was to produce a critical and general study with a few selected arguments treated more deeply, an aim which hopefully has been obtained.

ACKNOWLEDGMENTS Many thanks are due to Dr. M. Cremona and R.M Montereali for discussions and assistance with computer calculations. Also I am indebted to Prof. U.M. Grassano for a number of conversation and a critical reading of the manuscript.

REFERENCES

1. J. Teichmann and K. Szymborski in The History of Solid-State Physics, L. Hoddeson, E. Braun, J. Teichmann, and S. Weart, eds., Oxford University Press, 1990, Chap. 4.
2. G. Baldacchini in Advances in Nonradiative Processes in Solids, B. Di Bartolo ed., Plenum Press, 1991, p. 219.
3. G. Baldacchini in Luminescence: Phenomena, Materials, and Devices, R.P. Rao, ed., Nova Science Publishers Inc., 1991, p.
4. F. Agullo-Lopez, C.R.A. Catlow and P.D. Townsend, Point Defects in Materials, Academic Press, London, 1988.
5. W.B. Fowler in Physics of Color Centers, W.B. Fowler, ed, Academic Press, New York, 1968, Chap. 2.
6. L.F. Mollenauer in Tunable Lasers, L.F. Mollenauer and J.C. White, eds., Spinger-Verlag, Berlin, 1987, Chap. 5.
7. R. German in Handbook of Solid-State Lasers, Peter K. Cheo, ed., Marcel Dekker, Inc., New York, 1989, Chap. 5.
8. G. Baldacchini, Cryst. Lattice Defects Amorph. Mat. 18, 43 (1989).
9. W. Gellermann, J. Phys. Chem. Solids 52, 249 (1991).
10. S. Belke, M. Schubert, and K. Vogler, Opt Commun 38, 369 (1981).
11. J. Wiesenfeld, L.F. Mollenauer, and E.P. Ippen, Phys. Rev. Letters 47, 1668 (1981).
12. W. Hayes and A.M. Stoneham, Defects and Defect Processes in Nonmetallic Solids, J. Wiley & Sons, New York, 1985, Chap. 4.
13. D.S. Pan and F. Luty in Light Scattering in Solids, M. Balkanski, ed. Flammarion, Paris, 1976 p. 539.
14. F. Luty, Semiconductors and Insulators 5, 249 (1983).
15. Y. Mori and H. Ohkura, J. Phys. Chem. Solids 51, 663 (1990).
16. F. Luty, Z. Physik 160, 1 (1960).

17. G. Chiarotti and U.M. Grassano, Phys. Rev. Lett. **16**, 124 (1966) and Nuovo Cimento 46, 78 (1966).

18. L.F. Mollenauer, Phys. Rev. Lett. 43, 1524 (1979).

19. D.Y. Smith and G. Spinolo, Phys. Rev. 140, A2121 (1965).

20. G. Chiarotti, U.M. Grassano, and R. Rosei, Phys. Rev. Lett. 17, 1043 (1966).

21. M. Bonciani, U.M. Grassano, and R. Rosei, Phys. Rev. B8, 5855 (1973).

22. C.H. Henry and C.P. Slichter in Physics of Color Centers, W.B. Fowler, ed, Academic Press, New York, 1968, Chap. 6.

23. U.M. Grassano, private communication.

24. F. De Martini, G. Giuliani, and P. Mataloni, Phys. Rev. Lett. 35, 1464 (1975).

25. K. Vogler, Phys. Stat. Sol.(b) 107, 195 (1981).

26. G. Baldacchini, D.S. Pan, and F. Luty, Phys. Rev. B24, 2174 (1981).

27. S. Wakita, Y. Suzuki, and M. Hirai, J. Phys. Soc. Japan 50, 2781 (1981).

28. Y. Kondo, T. Noto, S. Sato, M. Hirai, and A. Nakamura, J. Luminescence 38, 164 (1987).

29. R.K. Swank and F.C. Brown, Phys. Rev. 130, 34 (1963).

30. L. Bosi, P. Podini, and G. Spinolo, Phys. Rev. 175, 1133 (1968).

31. D. Frohlich and H. Mahr, Phys. Rev. 141, 692 (1966).

32. W.B. Fowler, Phys. Rev. 135, A1725 (1964).

33. R.F. Wood and U. Opik, Phys. Rev. 179, 783 (1969).

34. L.F. Mollenauer and G. Baldacchini, Phys. Rev. Lett. 29, 465 (1972).

35. G. Baldacchini and L.F. Mollenauer, J. de Physique 34, C9-141 (1973).

36. H. Seidel and H.C. Wolf in Physics of Color Centers, W.B. Fowler, ed, Academic Press, New York, 1968, Chap. 8.

37. H.J. Reyher, K. Hahn, Th. Vetter, and A. Winnacker, Z. Physik B33, 357 (1979).

38. G. Spinolo, Phys. Rev. 137, A1495 (1965).

39. K. Park and W.L. Faust, Phys. Rev. Lett. 17, 137 (1966).

40. Y. Kondo and H. Kanzaki, Phys. Rev. Lett. 34, 664 (1975).

41. Y. Kondo, Y. Kayanuma, and H. Kanzaki, Abstracts Conf. on Defects in Insulating Crystals, Gatlinburg, USA, 1977, p. 244.

42. Y. Schneider, Abstracts Conf. on Defects in Insulating Crystals, Gatlinburg, USA, 1977, p. 385.

43. Y. Kayanuma and Y. Kondo, J. Phys. Soc. Japan 45, 528 (1978).

44. H. Ohkura, Y. Mori, Y. Tatsumi, T. Kamimura, K. Imanaka and A. Kobayashi, J. Luminescence 12-13, 435 (1976).

45. Physics of Color Centers, W.B. Fowler, ed, Academic Press, New York, 1968.

46. F. Luty in Physics of Color Centers, W.B. Fowler, ed, Academic Press, New York, 1968, Chap. 3.

47. G. Baldacchini, G.P. Gallerano, U.M. Grassano, and F. Luty, Phys. Rev. B27, 5039 (1983).

48. K. Hahn, H.J. Reyher, Th. Vetter, and A. Winnacker, Phys. Lett. 72A, 363 (1979).

49. N.G. Romanov, Yu.P. Veshchunov, V.A. Vetrov, and P.G. Baranov, Phys. Stat. Sol. b110, 89 (1982).

50. H. Kuhnert, Phys. Stat. Sol. 21, K171 (1967).

51. L.O. Bogan, L.F. Stiles Jr, and D.B. Fitchen, Phys. Rev. Lett. 23, 1495 (1969).

52. L.D. Bogan and D.B. Fitchen, Phys. Rev. B1, 4122 (1970).

53. L.F. Stiles Jr, M.P. Fontana, and D.B. Fitchen, Phys. Rev. B2, 2077 (1970).

54. H. Ohkura, K. Imanaka, O. Kamada, Y. Mori, and T. Iida, J. Phys. Soc. Japan 42, 1942 (1977).

55. P, Minguzzi, F. Pozzi, M. Tonelli, G. Baldacchini, and U.M. Grassano, Opt. Commun. 75, 401 (1990).

56. A.M. Stoneham, Theory of Defects in Solids, Claredom Press, Oxford, 1975, Chap. 15.

57. Y. Farge and M.P. Fontana, Electronic and Vibrational Properties of Point Defects in Ionic Crystals, North-Holland Publ. Company, Amsterdam, 1979, Chap. 4.

58. J.J. Markam, Solid State Commun. 39, 517 (1981).

59. R.E. Hetrick and W.D. Compton, Phys. Rev. 155, 649 (1967).

60. N. Akiyama, K. Asami, M. Ishiguro, and H. Ohkura. J. Phys. Soc. Japan 50, 3427 (1981).

61. K. Asami, K. Iwahana, H. Ohkura, and M. Ishiguro, Radiation Effects 72, 161 (1983).

62. M.P. Fontana and D.B. Fitchen, Phys. Rev. Lett. 23, 1497 (1969).

63. M.P. Fontana, Phys. Rev. B2, 4304 (1970).

64. L.F. Mollenauer and S. Pan, Phys. Rev. B6, 772 (1972).

65. G. Baldacchini and L.F. Mollenauer, Report UCB 34 P20-156, Berkeley, (1972).

66. G. Baldacchini, U.M. Grassano, and A. Tanga, Phys. Rev. B16, 5570 (1977).

67. H. Ohkura, K. Iwahana, M. Tanaka, K. Tara, and K. Imanaka, J. Phys. Soc. Japan 49, 2296 (1980).

68. G. Baldacchini and U.M. Grassano, Phys. Rev. B28, 4855 (1983).

69. G. Baldacchini, U.M. Grassano, A. Scacco, and K. Somaiah, Nuovo Cimento 12D, 117 (1990).

70. Y. Ruedin, P.-A. Schnegg, C. Jaccard, and M.A. Aegerter, Phys. Stat. Sol. b54, 565 (1972).

71. Y. Ruedin, P.-A. Schnegg, C. Jaccard, and M.A. Aegerter, Phys. Stat. Sol. b55, 215 (1973).

72. G. Baldacchini, U.M. Grassano, and A. Tanga, Phys. Rev. B19, 1283 (1979).

73. N. Akiyama and H. Ohkura, Phys. Rev. B40, 3232 (1989).

74. H. Panepucci and L.F. Mollenauer, Phys. Rev. 178, 589 (1969).

75. R.W. Warren, D.W. Feldman, and J.G. Castle Jr., Phys. Rev. 136, A1347 (1964).

76. M.D. Glinchuk, V.G. Grachev, and M.F. Deigen, Sov. Phys. Sol. State 8, 2678 (1967).

77. G. Baldacchini, S. Botti, U.M. Grassano, and F. Luty, Abstracts Int. Conf. on Defects in Insulating Crystals, Parma, 1988, p. 127.

78. Y. Kayanuma and Y. Toyozawa, J. Phys. Soc. Japan 40, 355 (1976).
79. F.S. Ham, Phys. Rev. Lett. 28, 1048 (1972).
80. F.S. Ham, Phys. Rev. B8, 2926 (1973).
81. Y. Kayanuma and Y. Kondo, Solid State Commun. 24, 442 (1977).
82. Y. Mori and H. Ohkura, J. Phys. Chem. Solids. 51, 663 (1990).
83. Y. Kondo, T. Noto, S. Sato, M. Hirai, and A. Nakamura, J. Luminesc. 38, 164 (1987).
84. Y. Kayanuma, J. Phys. Soc. Japan, 40, 363 (1976).
85. K. Ivahana, T. Iida, and H. Ohkura. J. Phys, Soc. Japan 47, 599 (1979).
86. F.S. Ham and U. Grevsmül, Phys. Rev. B8, 2945 (1973).
87. H.J. Reyher and A. Winnacker, Z. Phys. B45, 183 (1982).
88. J. Thomchick and F.S. Ham, Phys. Rev. B22, 6013 (1980).
89. J. Thomchick and F.S. Ham, Sol. State. Commun. 29, 825 (1979).
90. R.A. Carralho, M.C. Terrile, and H. Panepucci, Phys. Rev. B15, 1116 (1977).
91. G. Ascarelli, Phys. Stat. Sol. b63, 349 (1974).
92. L. Bosi, S. Cova, and G. Spinolo, Phys. Stat. Sol. b68, 603 (1975).
93. L. Bosi and M. Nimis, Nuovo Cimento 13D, 1483 (1991).
94. L. Bosi and M. Nimis, Phys. Stat. Sol. b95, 615 (1979).
95. S. Benci and M. Manfredi, Phys. Rev. B7, 1549 (1973).
96. L. Da Silva, G. Aguero, M. Lagos, and E. Caceres, J. Luminesc. 37, 51 (1987).
97. M. Georgiev, Phys. Rev. B30, 7261 (1984).
98. L. Martinelli, G. Pastori Parravicini, and P.L. Soriani, Phys. Rev. B32, 4106 (1985).

PROPERTIES OF HIGHLY POPULATED EXCITED STATES IN SOLIDS : SUPERFLUORESCENCE, HOT LUMINESCENCE, EXCITED STATE ABSORPTION

F. Auzel

C.N.E.T. - Laboratoire de Bagneux
196 Av. H. Ravera 92220 Bagneux - France

ABSTRACT

Besides the well known ways for an excited state to relax down its excitation, I mean fluorescence emissions, non-radiative transitions, energy transfers, there exist a number of other processes which may arise when excited states are highly populated, namely : coherent emission either of the stimulated or of the spontaneous type (superfluorescence), hot luminescence, excited state absorption with or without energy transfers.

The lecture shall present such unusual behaviours starting from the more classical ones which are predominant at lower excitation levels. We shall see how spontaneous emission merges into stimulated one or into coherent spontaneous emission trying to distinguish between amplified spontaneous photons by stimulated emission (ASE), superradiant (SR) and superfluorescent (SF) photons.

Hot luminescence (HL) will be seen as a competition between radiative and non radiative processes before thermal equilibrium. Excited state absorption (ESA) besides its trivial aspect in one center systems, shall be discussed also when high active centers concentration allows fast diffusion between them giving a strongly enhanced ESA.

The implication of such processes for applications shall be discussed on some examples.

I. INTRODUCTION

With the advent of laser excitation bringing its high power density together with its coherence a number of spectroscopic processes otherwise neglected have to be considered. In chapter I we shall deal first with the basic process of stimulated transitions and their intermingling with spontaneous emission right from their common origin : the coupling of the atom state with the quantized electromagnetic field. We shall see how the spontaneous emission appears as a damping term in the Rabi solution of the problem, explaining how some phase memory of the states wave functions may be kept during emission giving rise to coherent spectroscopic properties among them superfluorescence. In fact there are three types of coherent emissions connected with spontaneous emission and which differ from laser emission by the common feature that they do not require a closed optical cavity : Dieke superradiance (SR) [1], superfluorescence (SF) [2] and spontaneous emission amplified by stimulated emission (ASE) [3]. Though the three said effects have sometimes been taken one for the other, conditions to distinguish them have been specified by Bonifacio et al [2]. We shall see how SR comes from an ensemble of coherently prepared two states atoms emitting spontaneously in cooperation within a volume $<\lambda^3$. SF may take place from an ensemble of two states atoms in a volume $>\lambda^3$. Atoms being prepared incoherently in their excited state, emit a cooperative spontaneous emission by a self organization steming from their common emitted

Optical Properties of Excited States in Solids
Edited by B. Di Bartolo, Plenum Press, New York, 1992

field. Though sometimes called superradiance, ASE is just what would happen in a long laser medium without mirror which is very closed to the mirror multifolded laser cavity; propagation then plays an important role which is not necessary for pure SR and SF cases.

Typical of theses coherent spontaneous emissions is the reduced fluorescent lifetime which can become much shorter than the non-radiative decay. Then the radiative process overtake the non-radiative one. This will serve us as a bridge with chapter III where the more general case of hot luminescence shall be dealt with. We will briefly show how hot luminescence, which take place before non-radiative thermalization, can be considered as a special case of "resonant secondary scattering" or resonant Raman effect.

Finally in chapter IV we shall present excited state absorption (ESA) and its enhancement by fast diffusion between ions differencing it from ions cooperative effects.

II. FROM SPONTANEOUS EMISSION TO SUPERFLUORESCENCE

II.A. Induced and spontaneous transitions in the two-level atom [4] [5] [6]

In absence of a strong excitation field, which is the case of classical spectroscopy before the use of lasers, an induced transition between two levels would be described by the well known Golden Fermi Rule based on first order perturbation theory :

$$W_{12}=\frac{2\pi}{\hbar}|<2\,|\,V(t)\,|\,1>|^2\,\delta(E_2-E_1-E) \tag{1}$$

where W_{12} is the probability per unit time for a transition from state 1 to 2 under a time dependant perturbation V(t). However applicability of such relation requires a number of conditions.

In particular the ground state population in level 1 must stay about constant under the interaction, that is strong pumping is avoided; also $W_{12}.t$ should be <<1.

it means : $t \ll \frac{1}{W_{12}}$

a short time interaction adapted to the transition probability.

Since in the following we are interested in large changes in levels population we have to consider the full time dependant Shrödinger equation :

$$i\hbar\frac{\partial}{\partial t}\,|\,\Psi(r,t)>=H\,|\,\Psi(r,t)> \tag{2}$$

where $H = H_0 + V(t)$ is the time dependant Hamiltonien with H_0 the time independant part and V(t) the time dependant one.

Since we are interested mainly by the spontaneous emission at its root we shall quantized the electromagnetic field because we know that the semi-classical description where the atom is quantized and the field is classical, does not provide any spontaneous emission : it is introduced phenomenologically by detailed balance of the population of the two states atom and comparison with Planck law. This procedure introduced by Einstein gave the well known relationship between induced absorption, (or emission) and spontaneous emission probabilities, respectively the B_{12}, B_{21} and A_{21} coefficients [4], but cannot bring the coherent aspect of spontaneous emission.

Solutions to (2) are looked for by projecting the time dependant wavefunctions on the time independant one :

$$\Psi(r,t)=\sum_k C_k(t)\,|\,\phi_k(r)>e^{-i\omega_k t} \tag{3}$$

with

$$\omega_k=E_k/\hbar$$

and (2) becomes :

$$i\hbar\dot{C}_l(t)=\sum_k C_k(t)<\phi_l(r)\,|\,V(t)\,|\,\phi_k(r)>e^{i\omega_{lk}t} \tag{4}$$

with

$$\omega_{lk}=\omega_l-\omega_k$$

For our two states systems, (4) becomes :

$$i\hbar\dot{C}_1(t)=C_2(t)<\phi_1(r)\,|\,V(t)\,|\,\phi_2(r)>e^{i\omega_{12}t} \tag{5}$$

$$i\hbar\dot{C}_2(t)=C_1(t)<\phi_2(r)\,|\,V(t)\,|\,\phi_1(r)>e^{i\omega_{21}t} \tag{6}$$

then $C_1^2(t)$ and $C_2^2(t)$ describe the probabilities for the system to be in state ϕ_1 or ϕ_2 with the normalization :

$$|C_1^2|(t)+|C_2^2|(t)=1 \tag{7}$$

Now the two-levels atom we consider can be conveniently described in term of Pauli's spinors that is the ground and excited states are written as (see for instance [4]):

$$|g>=\begin{pmatrix}0\\1\end{pmatrix} \quad ; \quad |e>=\begin{pmatrix}1\\0\end{pmatrix}$$

with the introduction of "rising" (σ^+) and "lowering" (σ^-) operators :

$$\sigma^+=\begin{pmatrix}0 & 1\\0 & 0\end{pmatrix}=|e><g|$$

$$\sigma^-=\begin{pmatrix}0 & 0\\1 & 0\end{pmatrix}=|g><e|$$

the following relations stand :

$$\sigma^+\sigma^-=\begin{pmatrix}1 & 0\\0 & 0\end{pmatrix}=|e><e|$$

$$\sigma^-\sigma^+=\begin{pmatrix}0 & 0\\0 & 1\end{pmatrix}=|g><g|$$

and

$$\sigma^+\sigma^-+\sigma^-\sigma^+=1$$

Also one can define the following projections :

$$\sigma_x=\sigma^++\sigma^-$$

$$\sigma_y=-i(\sigma^+-\sigma^-)$$

$$\sigma_z=\sigma^+\sigma^--\sigma^-\sigma^+=|e><e|-|g><g|$$

the name of σ^+ and σ^- come from the following relationships :

$$\sigma^+|g>=|e> \quad ; \quad \sigma^-|e>=|g>$$

also : $\sigma^-|g>=0$ and $\sigma^+|e>=0$ since we have only two possible states.

For the steady state Hamiltonian of the atom one has :

$$H_A|e>=\frac{1}{2}\hbar\omega_0|e> \tag{8}$$

$$H_A|g>=\frac{1}{2}\hbar\omega_0|g> \tag{9}$$

with the origin of energy being taken half way between states $|g>$ and $|e>$. By multiplying (8) by $<e|$ and (9) by $<g|$ and substracting one obtains :

$$H_A=\frac{1}{2}\hbar\omega_0\sigma_z \tag{10}$$

Now the field part of the steady state Hamiltonian is written in term of photon creation (a^+) and photon anihilation (a) operators :

$$H_F=\hbar\omega a^+a \tag{11}$$

then the steady state Hamiltonian is :

$$H_0=\frac{1}{2}\hbar\omega_0\sigma_z+\hbar\omega a^+a \tag{12}$$

with the following initial and final states for emission :

$$|1>=|e,n>\rightarrow|2>=|g,n+1> \tag{13}$$

equivalently for absorption it comes :

$$|1>=|g,n>\rightarrow|2>=|e,n-1> \tag{14}$$

when the atom gains energy the number of photon state is reduced by one unit and reciprocally.

The energy difference between the two states of the complete system is :

$$\hbar\omega_{21}=\hbar(\omega_0-\omega) \tag{15}$$

where ω_0 is the atom resonance frequency and ω the electromagnetic field frequency.

The time dependant interaction V(t) promoting the transition is assumed to be only due to the establishment of the coupling of the electric field E(r) with the electric dipole operator (d) of the atom :

$$V(t) = H_{AF} = -d.E(r) \qquad (15bis)$$

using the quantized form of the field [4] :

$$H_{AF} = -i \sum_{k,\lambda} \sqrt{\frac{\hbar\omega_k}{2\varepsilon_0 V}} d.\varepsilon_{k\lambda}(a_{k\lambda}.e^{ikr} - a^+_{k\lambda}.e^{-ikr}) \qquad (16)$$

where k is the wave vector for pulsation ω_k, ε_λ is the polarization vector, V is the volume of the cavity containing the electromagnetic field.

(16) can be written :

$$H_{AF} = -d.(E^+(r) + E^-(r)) \qquad (17)$$

with

$$E^{\pm} = i \sum_{k,\lambda} \sqrt{\frac{\hbar\omega k}{2\varepsilon_0 V}} \varepsilon_{k\lambda} a^+_{k\lambda} e^{\pm ikr} \qquad (18)$$

In general, any operator A can be written with the closure relation :

$$A = \sum_{n,m} | n \rangle\langle n | A |\rangle\langle m |$$

then

$$d = \sum_{\substack{n=e,g \\ m=e,g}} | n \rangle\langle n | d | m \rangle\langle m |$$

or

$$d = | e \rangle\langle e | d | e \rangle\langle e | + | e \rangle\langle e | d | g \rangle\langle g | + | g \rangle\langle g | d | e \rangle\langle e | + | g \rangle\langle g | d | g \rangle\langle g |$$

d being an odd parity operator, transitions are allowed only for $|e\rangle$ and $|g\rangle$ being of opposite parity, then diagonal matrix elements vanish :

$$d = | e \rangle\langle e | d | g \rangle\langle g | + | g \rangle\langle g | d | e \rangle$$

and in term of spinors :

$$d = \sigma^+ \langle e | d | g \rangle + \sigma^- \langle g | d | e \rangle \qquad (19)$$

then from (16) and (19)

$$H_{AF} = -[\sigma^+ \langle e | d | g \rangle + \sigma^- \langle g | d | e \rangle]\,(E^+(r) + E^-(r)) = \hbar(g\sigma^+ - g^*\sigma^-)(a - a^+) \qquad (20)$$

with :

$$\hbar g = -i \sqrt{\frac{\hbar\omega}{2\varepsilon_0 V}} \langle e | d.\varepsilon | g \rangle \qquad (21)$$

and

$$\hbar g^* = i \sqrt{\frac{\hbar\omega}{2\varepsilon_0 V}} \langle g | d.\varepsilon | g \rangle \qquad (22)$$

(21) and (22) are recognized to be the electric dipole matrix elements of the atom and characterize the strength of the possible atomic transitions.

In product (20), terms $-g^*\sigma^- a$ and $-g\sigma^+ a^+$ are assumed to be zero because one cannot "lower" the atom and anihilate a photon nor "rise" the atom and create a photon. Then :

$$H_{AF} = \hbar(ga\sigma^+ + g^* a^+ \sigma^-) \qquad (23)$$

This simplified form of interaction is called the rotating wave approximation [7]. The matrix elements in eq. (5) and (6) are now for the transition between states given in (13) and (14) :

$$\langle \phi_1(r) | V(t) | \phi_2(r) \rangle \equiv \langle 1 | H_{AF} | 2 \rangle$$

$$\langle \phi_2(r) | V(t) | \phi_1(r) \rangle \equiv \langle 2 | H_{AF} | 1 \rangle$$

or for emission :

$$\begin{aligned} \langle 2 | H_{AF} | 1 \rangle &= \hbar \langle g, n+1 | ga\sigma^+ + g^* a^+ \sigma^- | e, n \rangle \\ &= \hbar \langle g, n+1 | g^* a^+ \sigma^- | e, n \rangle \\ &= \hbar g^* \langle g | g \rangle\langle n+1 | a^+ | n \rangle \end{aligned} \qquad (24)$$

since

$$a^+ | n >= \sqrt{n+1} \, | n+1 > \tag{25}$$

then

$$< 2 | H_{AF} | 1 >= \hbar g^* \sqrt{n+1} \tag{26}$$

For absorption it would be :

$$< 1 | H_{AF} | 2 > \quad = \hbar < e, n-1 \, | \, g a \sigma^+ + g^* a^+ \sigma^- \, | \, g, n >$$

$$= \hbar g < e | e >< n-1 | a | n > \tag{27}$$

since

$$| a | n >= \sqrt{n} \, | n-1 > \tag{28}$$

then

$$< 1 | H_{AF} | 2 >= \hbar g \sqrt{n} \tag{29}$$

Comparing eq. (26) and (29) shows that the main difference between emission and absorption is that emission may exist even in absence of external field (n = 0) which is the spontaneous emission we are interested in. The case n≠0 allows the existence of both induced emission and absorption.

1) <u>The Rabi equation in the fully quantized approach.</u> In the following we shall try to derive the coherent part which can exist even in spontaneous emission. We insert matrix elements (26) and (29) into the time dependant Shrödinger equation for our two levels system: eq. (5) and (6).

$$i\hbar \dot{C}_1(t) = C_2(t) \hbar g \sqrt{n} \, e^{i\omega_{12} t} \tag{30}$$

$$i\hbar \dot{C}_2(t) = C_1(t) \hbar g^* \sqrt{n+1} \, e^{i\omega_{12} t} \tag{31}$$

and we shall specify the following initial conditions :

$$C_1(0) = 1 \quad ; \quad C_2(0) = 0 \quad ; \quad C_1(t) \neq 1 \tag{32}$$

that is the system starts from a fully populated state and is allowed to change its state drastically.

Looking for an oscillatory behaviour we set :

$$C_1(t) = e^{i\mu t} \tag{33}$$

which is inserted into (30) :

$$-\mu \hbar e^{i\mu t} = C_2(t) \hbar g \sqrt{n} \, e^{i\omega_{12} t}$$

or

$$C_2(t) = -\frac{\mu}{g\sqrt{n}} e^{i(\mu - \omega_{12})t}$$

deriving and putting into (31) :

$$\frac{\mu(\mu - \omega_{12})}{g\sqrt{n}} e^{i\omega_{12} t} = C_1(t) g^* \sqrt{n+1} \, e^{i\omega_{12} t}$$

giving the equation :

$$\mu^2 - \mu\omega_{12} - | g |^2 \sqrt{n} \sqrt{n+1} = 0 \tag{34}$$

with the solutions :

$$\mu_\pm = \frac{\omega_{12} \pm \sqrt{\omega_{12}^2 + 4 g^2 n^{1/2} (n+1)^{1/2}}}{2} \tag{35}$$

we have the following relations :

$$\mu_+ - \mu_- = \sqrt{\omega_{12}^2 + 4 g^2 n^{1/2} (n+1)^{1/2}} \equiv \Omega \tag{36}$$

Since μ has two solutions $C_1(t)$ can be more generally written :

$$C_1(t) = A e^{i\mu_+ t} + B e^{i\mu_- t} \tag{37}$$

then inserting into (30) :

$$C_2(t) = \frac{-1}{g\sqrt{n}} e^{i\omega_{21} t} \left(\mu_+ A e^{i\mu_+ t} + \mu_- B e^{i\mu_- t} \right) \tag{38}$$

applying the initial conditions (32) :

$$0 = \frac{-1}{g\sqrt{n}}(\mu_+ A + \mu_- B)$$

or $\mu_+ A = -\mu_- B$

and $A = 1 - B$ (39)

then

$$B = \frac{\mu_+}{\mu_+ - \mu_-} = \frac{\mu_+}{\Omega} \quad (40)$$

and

$$A = -\frac{\mu_-}{\Omega} \quad (41)$$

then (39) is written :

$$C_2(t) = \frac{i^2}{g\sqrt{n}} |g|^2 \frac{\sqrt{n}\sqrt{n+1}}{\Omega}\left(\frac{e^{i\mu_+ t} - e^{-i\mu_- t}}{2i}\right) e^{i\omega_{21} t} \quad (42)$$

by definition of (35) and (15) :

$$\mu_+ = \tfrac{-1}{2}(\omega_0 - \omega) + \tfrac{1}{2}\Omega \quad ; \quad \mu_- = \tfrac{-1}{2}(\omega_0 - \omega) - \tfrac{1}{2}\Omega$$

then

$$\omega_0 - \omega + \mu_\mp = \frac{1}{2}(\omega_0 - \omega \mp \Omega) \quad (43)$$

and (42) is written :

$$C_2(t) = \frac{2i\,|g|\sqrt{n+1}}{\Omega} e^{\frac{i}{2}(\omega_0 - \omega)t}\left(\frac{e^{\frac{i\Omega t}{2}} - e^{\frac{-i\Omega t}{2}}}{2i}\right)$$

or

$$C_2(t) = \frac{2i\,|g|}{\Omega}\sqrt{n+1}\, e^{\frac{i}{2}(\omega_0 - \omega)t}\,\mathrm{Sin}\frac{\Omega t}{2} \quad (44)$$

which is the Rabi equation for C_2 (t) in the field quantized case. Ω is called the Rabi frequency or the nutation frequency.

Eq. (36) shows that Ω differs only by the term $\sqrt{n(n+1)}$ instead of n from the one generally considered [4]. In a strong external field n > 1 and (44) is the general result.

2)<u>The synchronous exchange between field and atom.</u> We shall consider first the case of a strong external field in resonance with the atom (n > 1 and $\omega_{12} = 0$).

then :

$$\Omega = \Omega_0 = 2g\sqrt{n} \quad (45)$$

and (44) becomes :

$$C_2(t) = i\,\mathrm{Sin}\frac{\Omega_0 t}{2} \quad (46)$$

the probability of finding the system in state 2 is :

$$|C_2(t)|^2 = \mathrm{Sin}^2\frac{\Omega_0 t}{2} \quad (47)$$

it is also the probability for the transition $1 \to 2$.

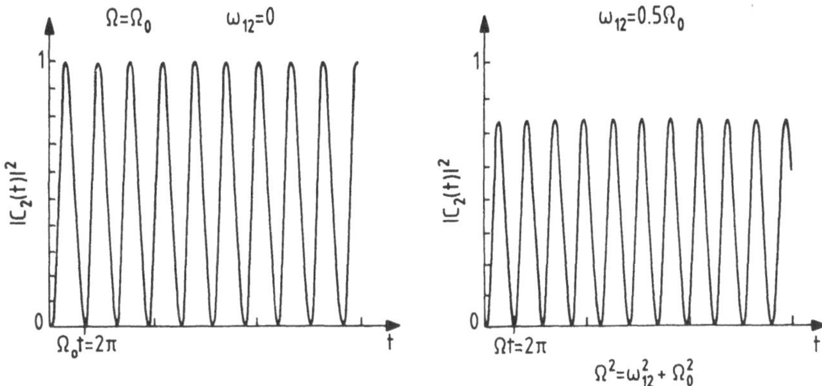

Fig.1. $|C_2|^2$ for a resonant ($\omega_{12} = 0$) and for an off resonant case ($\omega_{12} = 0.5\Omega_0$).

Similarly from (7), one has :

$$|C_1(t)|^2 = 1 - \mathrm{Sin}^2\frac{\Omega_0 t}{2} = \mathrm{Cos}^2\frac{\Omega_0 t}{2} \tag{48}$$

it is the probability for a transition $2 \to 1$. $|C_2(t)|^2$ is shown on Fig 1 for $\omega_{12} = 0$ and for $\omega_{12} = 0.5\Omega_0$.

Let us calculate the dipole moment during the process. It is :
$< \psi(t) | e.d | \psi(t) >$ with from (48), (49) and (3) :

$$\psi(t) = i\,\mathrm{Sin}\frac{\Omega_0 t}{2}\phi_2 + \mathrm{Cos}\frac{\Omega_0 t}{2}\phi_1 \tag{49}$$

showing that the system is in a mixed state except at extrema ($t = 0$; $t = \frac{\pi}{\Omega_0}$; $t = 2\frac{\pi}{\Omega_0}$ ...).

Since the considered interaction H_{AF} was the dipole moment :

$$\begin{aligned} < e.d >= \quad & \mathrm{Sin}^2\tfrac{\Omega_0 t}{2} < 2 | H_{AF} | 2 > + \mathrm{Cos}^2\tfrac{\Omega_0 t}{2} < 1 | H_{AF} | 2 > \\ & + i\,\mathrm{Sin}\tfrac{\Omega_0 t}{2}\cos\tfrac{\Omega_0 t}{2} < 2 | H_{AF} | 1 > e^{i\omega_0 t} \\ & - i\,\mathrm{Sin}\tfrac{\Omega_0 t}{2}\cos\tfrac{\Omega_0 t}{2} < 1 | H_{AF} | 2 > e^{-i\omega_0 t} \end{aligned} \tag{50}$$

then

$< e.d >= \frac{i}{2}\mathrm{Sin}\,\Omega_0 t\, e^{i\omega_0 t} < 1 | H_{AF} | 2 >$ + complex conjugate

and from (26) and (29) :

$$< e.d >= \frac{\hbar g\sqrt{n}}{2}\mathrm{Sin}\,\Omega_0 t\, e^{i\omega_0 t} + C.C \tag{51}$$

assuming g to be real for simplicity :

$$< e.d >= \frac{\hbar\Omega_0}{4}\mathrm{Sin}\,\omega_0 t\;\mathrm{Sin}\,\Omega_0 t \tag{52}$$

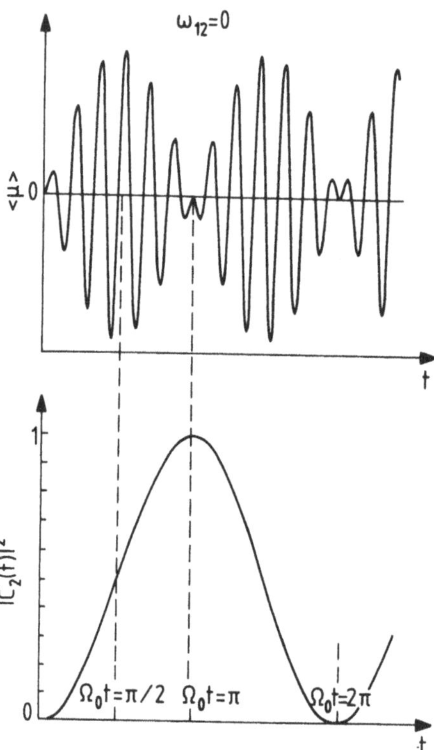

Fig.2. $|C_2(t)|^2$ and < e.d > for the resonant case ($\omega_{12} = 0$).

(52) is plotted on Fig 2 together with $C_2^2(t)$ for the case $\omega_{12} = 0$.

It shows that the dipole moment oscillates at the transition frequency ω_0 but its size increases and decreases at twice the nutation frequency. When the atom is in a well defined state $|e>$ or $|g>$ the dipole does not oscillate, then there is no emission or absorption at corresponding times ($t = 0$; $t = \frac{\pi}{\Omega_0}$; $t = 2\frac{\pi}{\Omega_0}$ ...).

Since up to now, losses or damping have not been introduced, when the external field is suppressed at some intermediate time t, the dipole will continue to oscillate keeping memory of the mixed state in which it was left when the field was suppressed : the field and atom exchanges synchronously and elastically their energy without end at the rate existing at the time of the external field suppression.

This emission is refered to as free induction emission. It is a coherent emission.

3) The pseudo-spin Bloch vector of the two-levels system. [8]. In analogy with the Bloch vector of an electron spin interacting with an oscillating magnetic field superimposed on a permanent one, we have to introduced at this point the "pseudo-spin Bloch vector" because it is the picture the most often referred to in the litterature dealing with superfluorescence.

Let us define a "pseudo-spin Bloch vector" by the following coordinates :

$$S_1 = -\left(C_1^* C_2 e^{i\omega_0 t} + C_1 C_2^* e^{-i\omega_0 t}\right) \tag{53}$$

$$S_2 = i\left(-C_1^* C_2 e^{i\omega_0 t} + C_1 C_2^* e^{-i\omega_0 t}\right) \tag{54}$$

$$S_3 = |C_1|^2 - |C_2|^2 \tag{55}$$

Recalling that : $C_2(t) = i\,\mathrm{Sin}\frac{\Omega_0 t}{2}$

and

$$C_1(t) = \mathrm{Cos}\frac{\Omega_0 t}{2} \tag{56}$$

it comes :

$$S_1 = \mathrm{Sin}\,\Omega_0 t\ \mathrm{Sin}\,\omega_0 t \tag{57}$$

$$S_2 = \mathrm{Sin}\,\Omega_0 t\ \mathrm{Cos}\,\omega_0 t \tag{58}$$

$$S_3 = \mathrm{Cos}\,\Omega_0 t \tag{59}$$

The parametric curve described by eq. (57), (58), (59) is represented on Fig 3.

Starting from the vertical upright position the vector $\vec{S}$ shall go down in the vertical downright position, its extremity having described a spiral on a sphere in the representing space and it shall go up again as time goes on ad infinitum under the external applied field. If field is suppressed the vector describes a circle on the sphere according to the Ω_0 value existing when the field was suppressed.

Comparing (52) and (58) shows that the pseudo-spin Bloch vector can be considered as the unit vector of the dipole electric moment projected on one of the axis of the horizontal plane. This shows that the maximum of radiation is obtained when the Bloch vector is in the horizontal plane which happens for $\Omega_0 t = \frac{\pi}{2}$. If field is suppressed at $t = \frac{\pi}{2\Omega_0}$ there is a maximum of emission. Such excitation is called a "$\frac{\pi}{2}$ pulse".

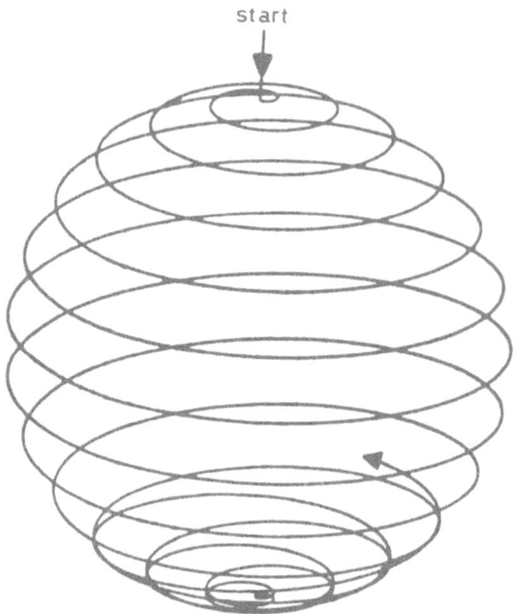

Fig.3. Locus of the pseudo-spin Bloch vector with time.

On the opposite a π pulse alone would leave a zero amplitude dipole. Playing which such pulses in sequence is the root of the photon-echo process which is not however our very subject here.

4) Spontaneous emission and the Weisskopf-Wigner damping. [4]. Spontaneous emission is characterized by its independance from initial photon number that is n = 0. The initial state is $|1> = |e, 0>$ and the final state is $|2> = |g, 1>$. However $|2>$ is not confined to a single mode because the spontaneous photon can be emitted in any direction with any phase and polarization and we have to sum over all theses modes (λ) for photon number $| 1_\lambda >$.

Then eq. (5) and (6) to be considered are now :

$$i\hbar\dot{C}_1(t) = < e,0 \,|\, H_{AF} \,|\, g,1_\lambda > C_2(t) e^{i\omega_{12}t} \tag{60}$$

$$i\hbar\dot{C}_2(t) = \sum_\lambda < g,1 \,|\, H_{AF} \,|\, e,0 > C_1(t) e^{i\omega_{21}t} \tag{61}$$

we solve for C_1 (t); from (61) :

$$C_2(t) = -\frac{i}{\hbar}\sum_\lambda < g,1_\lambda \,|\, H_{AF} \,|\, e,0 > \int_0^t C_1(t') e^{i\omega_{21}t'} dt'$$

which inserted into (60) gives :

$$\dot{C}_1(t) = -\frac{1}{\hbar^2}\sum_\lambda |< g,1_\lambda \,|\, H_{AF} \,|\, e,0 >|^2 \int_0^t C_1(t') e^{i\omega_{12}(t-t')} dt'$$

The integrant is maximum for t = t', then :

$$\dot{C}_1(t) \cong -\frac{1}{\hbar^2}\sum_\lambda |< g,1_\lambda \,|\, H_{AF} \,|\, e,0 >|^2 C_1(t) \int_0^t e^{i\omega_{12}\tau} d\tau \tag{62}$$

with $\tau = t - t'$.

Since we want to know C_1 (t) at any time : $t \to \infty$. The integral can be estimated by the following relation :

$$\lim_{t \to \infty} \int_0^t e^{i\omega_{12}\tau} d\tau = i\frac{P}{\omega_{12}} + \pi\delta\omega_{12}$$

where P/ω_{12} is the function "principal value" [8] :

$$\frac{P}{\omega_{12}} = \frac{1}{\omega_{12}} \qquad for\, \omega_{12} \neq 0$$

$$\frac{P}{\omega_{12}} = 0 \qquad for\, \omega_{12} = 0$$

then we have for (62) :

$$\dot{C}_1(t) + \frac{1}{\hbar^2}\sum_\lambda |< g, 1_\lambda | H_{AF} | e, 0 >|^2 C_1(t)\left[i\frac{P}{\omega_{12}} + \pi\delta\omega_{12}\right] = 0$$

posing :

$$\frac{\hbar\Gamma}{2} = \frac{\pi}{\hbar}\sum_\lambda |< g, 1_\lambda | H_{AF} | e, 0 >|^2 \delta\omega_{12} \tag{63}$$

and

$$\frac{\Delta E}{2} = \frac{1}{\hbar}\sum_\lambda |< g, 1_\lambda | H_{AF} | e, 0 >|^2 \times \begin{cases} \dfrac{1}{\omega_{12}} & for\, \omega_{12} \neq 0 \\ 0 & for\, \omega_{12} = 0 \end{cases} \tag{64}$$

(62) becomes :

$$\dot{C}_1(t) + \frac{\hbar\Gamma + i\Delta E}{2\hbar} C_1(t) = 0 \tag{65}$$

with the initial conditions (32), the solution for (65) is :

$$C_1(t) = e^{\frac{-\Gamma t}{2}} e^{\frac{-i\omega_0 \Delta E t}{2\hbar}} \tag{66}$$

and

$$C_1^2(t) = e^{-\Gamma t} \tag{67}$$

So the probability of being in state |e, 0 > is a decreasing exponential with time; the state has a lifetime :

$\tau_0 = \Gamma^{-1}$ given by (63);

Γ represents a level width called the "homogeneous natural width" because this width is necessitated by the Heisenberg uncertainty relation between time and energy and cannot be reduced.

Eq. (64) gives a shift in the energy separation of the atomic level induced by off resonance photons since it exists only for $\omega_{12} \neq 0$. The experimental shift is however much smaller than the one given by (64) and can be often neglected [4].

(63) can be written :

$$\Gamma = \frac{2\pi}{\hbar^2}\sum_\lambda |< g, 1_\lambda | H_{AF} | e, 0 >|^2 \delta\omega_{12} \tag{68}$$

which is equivalent to the Golden rule for spontaneous emission. Replacing the matrix element by its value (26) :

$$\Gamma = \frac{2\pi}{\hbar^2} \sum_{\lambda} \hbar^2 g^2 (0 + 1_\lambda) \delta\omega_{12} \tag{69}$$

or for one mode :

$$\Gamma_\lambda = 2\pi g^2 \delta\omega_{12} \tag{70}$$

At this point it is interesting to compare this Weisskopf-Wigner damping with what can be obtained from (44).

For spontaneous emission n = 0, then (44) is written :

$$C_2(t) = \frac{2i \mid g \mid}{\Omega} e^{\frac{i}{2}(\omega_0 - \omega)t} \operatorname{Sin} \frac{\Omega t}{2} \tag{71}$$

and the corresponding probability is :

$$\mid C_2(t) \mid^2 = \frac{4g^2}{\Omega^2} \operatorname{Sin}^2 \frac{\Omega t}{2} \tag{72}$$

Let us comment on this apparent oscillatory behaviour : we had found Ω by its definition (36) : $\Omega^2 = \omega_{12}{}^2 + 4g^2 n^{1/2}(n+1)^{1/2}$which reduces to $\Omega^2 = \omega_{12}^2$ for n = 0

Showing that at resonance ($\Omega \to 0$), there is no oscillatory behaviour for $C_2^2(t)$ in spontaneous emission.

This is in contradiction with the result of Allen and Eberly [6] giving :

$\Omega_2 = \omega_{12}^2 + 4g^2(n+1)$, (in fact assuming a spontaneous absorption) (73)

which for spontaneous emission and resonance reduces to $\Omega_0^2 = 4g^2$ the semi-classical result.

Though the contrary is stated in [6], the results of Jaynes and Cummings [9] gives also a non oscillatory behaviour as well as [4] :

$$\Omega^2 = \omega_{12} + 4g^2 n$$

It is sligthly different from (36).

What we can say is that at exact resonance, spontaneous emission by itself keeps no memory at all of the atom wave functions. Outside resonance a slight influence of them are kept through the detunning ω_{12}. On the other hand, (72) with $\Omega \to 0$ reminds us of the Golden Rule formulation :

$$\lim_{\Omega \to O} \mid C_2 \mid^2 (t) = 4g^2 \frac{\pi t}{2} \delta\omega_{12} \tag{74}$$

or

$$W_{spont.} = 2\pi g^2 \delta\omega_{12} \tag{75}$$

the spontaneous emission probability by unit time for one mode , which is just eq. 70.

So, the fully quantized Rabi equation gives the same result for spontaneous emission as the Weisskopf-Wigner approach.

By summing (75) over all modes and polarization, one has from (21)

$$W_{spont.} = \frac{2\pi}{\hbar^2} \sum_{\lambda,\varepsilon} \frac{\hbar\omega}{2\varepsilon_0 V} \mid < e \mid d.\varepsilon \mid g > \mid^2 \delta\omega_{12} \tag{76}$$

$$W_{spont;} = \frac{\pi}{\hbar\varepsilon_0 V} \sum_{\lambda} \omega\delta\omega_{12} [\mid < e \mid d\varepsilon_1 \mid g > \mid^2 + \mid < e \mid d\varepsilon_2 \mid g > \mid^2]$$

with : $\mid < e \mid d \mid g > \mid^2 [\operatorname{Sin}^2\theta(\varepsilon_1^2 \cos^2\Phi + \varepsilon_2^2 \operatorname{Sin}^2\Phi)]$

θ, Φ being the angular coordinate of d along the polarizations :$\varepsilon_1 \operatorname{Sin}\theta \cos\Phi$ and $\varepsilon_2 \operatorname{Sin}\theta \operatorname{Cos}\Phi$

then

$$W_{spont.} = \frac{\pi}{\hbar\varepsilon_0 V} \sum_\lambda \omega |< e \mid d \mid g >|^2 \delta\omega_{12} \operatorname{Sin}^2\theta \tag{77}$$

Summing over modes in all directions :

$$W_{spont.} = \frac{\pi\omega}{\hbar\varepsilon_0 V} |< e \mid d \mid g >|^2 \sum_\lambda \operatorname{Sin}^2\theta\delta\omega_{12} \tag{78}$$

the number of modes in a volume V being :

$$dN = \frac{V}{8\pi^3 c^3}\omega^2 d\omega d\Omega \tag{79}$$

and using the precept [4] : $\delta\omega$ equivalent to spectral mode density :

$$\delta\omega_{12} \rightarrow \frac{dN}{d\omega} = \frac{V\omega^2}{8\pi^3 c^3} d\Omega \tag{80}$$

(78) becomes :

$$W_{spont.} = \frac{\pi\omega}{\hbar\varepsilon_0 V} |<\mid d \mid>|^2 . \frac{V\omega^2}{8\pi^3 c} \int \operatorname{Sin}^2\theta d\Omega \tag{81}$$

$$W_{spont.} = \frac{\omega^3}{\hbar\varepsilon_0 V 8\pi^2 c^3} |<\mid d \mid>|^2 \frac{8\pi}{3}$$

$$W_{spont.} = \frac{\omega^3}{3\pi c^3 \hbar\varepsilon_0} |< d >|^2 = \frac{4\alpha\omega^3}{3c^2} |< r >|^2 \tag{82}$$

where α is the fine structure constant :

$$\alpha = \frac{e^2}{4\pi\varepsilon_0\hbar c} = \frac{1}{137}$$

and d = e.r the dipole moment.
(82) is just the well known Einstein A coefficient for spontaneous emission.

II.B. The merging of coherent emission and spontaneous decay

1) The damping of Rabi oscillation by spontaneous decay. We have just seen that the effect of the coupling of the atom with all the modes of the empty field is to introduce a damping constant.

Neglecting the spontaneous emission shift ΔE, (66) can be written:

$$C_l(t) \cong e^{-\frac{\Gamma_l}{2}t} \tag{83}$$

or :

$$i\hbar\dot{C}_l(t) = -i\hbar\frac{\Gamma_l}{2}C_l(t) \tag{84}$$

Then a general form for the Schrödinger equation (4) with damping can be written :

$$i\hbar\dot{C}_l(t) = \sum_k C_k(t) < \phi_l(r) \mid V(t) \mid \phi_k(r) e^{i\omega_{lk}t} - \frac{i\hbar\Gamma_l}{2}C_l(t) \tag{85}$$

For our two-level system it comes from (30) and (31) :

$$i\hbar\dot{C}_1(t) = C_2(t)\hbar g\sqrt{n}\, e^{i\omega_{12}t} \tag{86}$$

$$i\hbar\dot{C}_2(t) = C_1(t)\hbar g\sqrt{n}\, e^{i\omega_{21}t} - i\hbar\frac{\Gamma_2}{2}C_2(t) \tag{87}$$

where 1 has been removed from (n+1) since as we saw above, Γ_2 replaced it.

Posing

$$C'_2 = C_2 e^{-\Gamma_2 t/2} \tag{88}$$

the system (86), (87) gives a new form for (34) :

$$\mu^2 + \left(\omega_{21} - i\frac{\Gamma_2}{2}\right)\mu - g^2 n - \frac{\Gamma_2}{2} = 0 \tag{89}$$

with solutions :

$$\mu_\pm = -\frac{1}{2}\left(\omega_{21} - i\frac{\Gamma_2}{2}\right) \pm \sqrt{(\omega_{21} - i\Gamma_2)^2 + 4g^2 n} \tag{90}$$

with the Rabi frequency :

$$\Omega = \left(\omega_{21} - i\frac{\Gamma_2}{2}\right)^2 + 4g^2 n \tag{91}$$

and

$$C_2(t) = \frac{i2g\sqrt{n}}{\Omega} e^{\frac{i}{2}(\omega_{21})t} e^{-\frac{\Gamma_2}{2}t} \operatorname{Sin}\frac{\Omega t}{2} \tag{92}$$

and for the probability :

$$|C_2(t)|^2 = \frac{4g^2 n}{\Omega^2} e^{-\Gamma_2 t} \operatorname{Sin}^2 \Omega\frac{t}{2} \tag{93}$$

Showing that spontaneous emission can be viewed as a damping of the Rabi oscillation. Remembering that Rabi oscillation keeps on after the external field has been removed, one

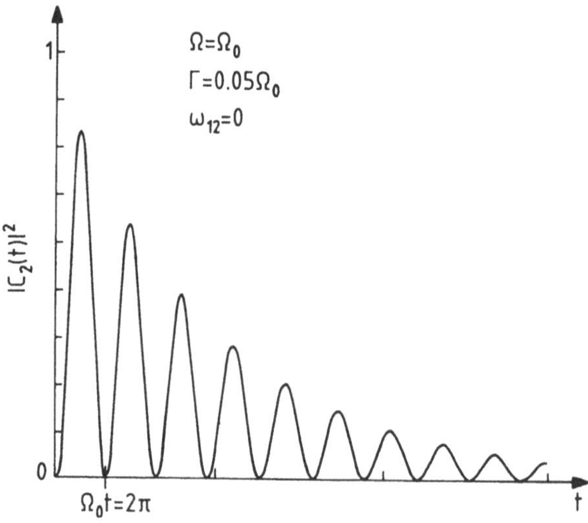

Fig.4. $|C_2(t)|^2$ with spontaneous emission damping.

can see that the basic fact is that the coherence still existing after the removing of the field is the one not yet destroyed by spontaneous emission as shown on Fig 4.

In the Bloch vector representation, it means that instead of circling ad infinitum on a parallel of the sphere of Fig 3, the Bloch vector spirals down with vertical damped oscillation. In case the Bloch vector would fall down without spiraling there would be no coherence left at all. This possibility for a coherence (free induction) is the root of superradiance and superfluorescence.

However, it should be noted that there is no general solution for the problem of the movement of the damped Bloch vector movement [6]. So in general, the Bloch sphere is considered only for time $t << 1/\Gamma$.

2) <u>The link between the various relaxation times.</u> Coming back to the pseudo-spin Bloch vector, its movement of nutation given by (57), (58), (59) in abscence of damping, shows us that it can be described by an equation of the form :

$$\dot{\vec{S}} = \vec{\Omega} \wedge \vec{S} \tag{94}$$

where $\vec{\Omega}$ is the rotation vector axis around which $\vec{S}$ nutates. The coordinates of $\vec{\Omega}$ are $(\Omega_0, 0, \omega_{12})$. Eq. 94 is the "Bloch equation" without damping. Damping has been introduced phenomenologically by Bloch under the following form :

$$\dot{\vec{S}} = (\vec{\Omega} \wedge \vec{S}) - [\vec{S} - \vec{\delta}(1,i)]\overrightarrow{T^{-1}} \quad i = 1,2 \tag{95}$$

With $\vec{T}$ having the coordinates : (T_2, T_2, T_1). T_2 is the so-called "Bloch transverse relaxation time"; it characterizes perturbations on the phase of the dipole not on its energy. T_1 is the "Bloch longitudinal relaxation time"; it characterize perturbations on the energy of the system. The names come from the fact that T_2 apply to the movement perpendicular to $\vec{\Omega}$. Whereas T_1 applies to the movement longitudinal with $\vec{\Omega}$.

In case that the only damping is spontaneous emission, from the definition of $\vec{S}$ in (53) (54) (55) one can see that the damping on S_3 comes from the $|C_1(t)|^2$ and so is just the Γ given by (69), (67). Then T_1 can be identified with τ_0 the spontaneous radiative lifetime.

From the definitions of S_1 and S_2, the link is directly with the $C_2(t)$, not their square, then the corresponding damping is $\Gamma/2$.
one has :

$$\frac{\Gamma}{2} = \frac{1}{T_2} \tag{96}$$

and then :

$$\frac{1}{T_2} = \frac{1}{2T_1} \tag{97}$$

Also because S_1, S_2 are given by products $C_l^*(t).C_k(t)$, the phase of the $C_l(t)$ plays a role and T_2 being linked with such phase is also called the dephasing time.

Up to now we have been dealing with one atom and the field. In the real world we shall have several atoms or ions and there will be ion-ion interaction, ion-phonon interaction, giving contributions to T_2 besides the one from T_1; also there will be ions in differents surroundings with an inhomogeneous spread of energies ω_0 giving also a contribution to the dephasing time. Then generally T_2 will be written :

$$\frac{1}{T_2} = \frac{1}{T'_2} + \frac{1}{T_2^*} \quad \text{and} \quad \frac{1}{T_2} > \frac{1}{T_1} \tag{98}$$

with

$$\frac{1}{T'_2}=\frac{1}{2T_1}+W_{ion-ion}+W_{ion-phonon} \tag{99}$$

affecting all atom homogeneously, and $1/T_2^*$ being the inhomogeneous contribution to the dephasing of the atoms-field interaction. This dephasing comes from the $\omega_{12} \neq 0$ case in Fig 1.

II.C. Field coupled cooperative multiions effects

Because we have looked with some details at the single atom-field interaction, the very root of coherent emissions, we shall in the following consider cooperative multiion effects staying at the "lowest level of subtlety" [6]. The higher level of subtlety may be found in [2], [6], [7], [10] where the construction of the multiion system into Dieke states [1] is analysed, where the statistical dynamics of the whole system is considered through the density matrix dynamics, and where propagation of the interacting field at retarded time is taken into account.

In the following, after following the definition of Bonifacio [2] for superradiance and superfluorescence we shall give the few recent exemples of such effects in solids, the subject of this talk, since most of the first experiments had been performed on gazeous systems.

1) Superradiance. We consider a system of several two-levels atoms occupying a volume the dimension of which is smalled compared to λ^3. In this case the total radiating dipole is given by (52) multiplied by the difference of atoms in the excited state and in the ground state. If the external resonant field is stopped at $t=\frac{\pi}{2\Omega_0}$, we shall create a " macro dipole" given by :

$$<M>=(N_1-N_2)\frac{\hbar\Omega_0}{2}\text{Sin}\,\omega_0 t \tag{100}$$

$(N_1 - N_2)$ is the population inversion per unit volume. The power radiated per unit volume by this dipole is :

$$P=\overline{E(t)\frac{d<M>}{dt}} \tag{101}$$

$$P=\omega_0(N_1-N_2)\frac{\hbar\Omega_0}{2}E_0\overline{\cos^2\omega_0 t} \tag{102}$$

where $E(t)=E_0\,\text{Cos}\,\omega_0 t$ is then the "reaction field", that is the field created by the oscillating dipole itself since the externally applied field has been suppressed. If the small cavity we are considering has a quality factor Q, the field inside , given by the power P is [11] :

$$E_0^2=8\pi QP/\omega \tag{103}$$

combining (102) and (103) gives :

$$P=\pi Q\,\omega(N_1-N_2)^2\frac{\hbar^2\Omega_0^2}{2}$$

or :

$$P=\omega\pi Q(N_1-N_2)^2\hbar^2 g^2 n \tag{104}$$

in term of the dipole matrix element of the atomic transition $\hbar g$.

The square dependance of emitted power on inversion density (N_1 - N_2) is typical of the coherence of the multi-atom emission. It is typical of superradiance as well as we shall see of superfluorescence.

Eq. (104) holds as long as coherence between atoms is not destroyed by spontaneous decay or other dephasing processes.

Experimentally such behaviour was demonstrated in a gaz of OCS at microwave frequencies by Hill et al. [11]. At such low frequency, spontaneous decay is normally very long and the damping shown on Fig 5 for the emission is linked mainly with the dephasing due to collisions between molecules.

The lower part of the figure showing interferences with a 10^{-10} W CW oscillator demonstrates clearly the coherence of the spontaneous decay after the excitation pulse. This decay is also called free induction decay. The synchronism of the many atoms involved is obtained by the coherent pumping pulse, directly exciting the emitting level; in Bonifacio's definition this is superradiance, the ordering being obtained by the coherence of the pump. In Allen and Eberly language [6] this would be "passive superradiance" to distinguish from "active superradiance" where the ordering of the many atoms emission is obtained by their common reaction field. In the example of Fig 5 the observed decay time reflects the pressure dependant T_2 and does not show a time shortened by the superradiance itself.

Such case is called "limited superradiance" by Mac Gillivray and Feld [12] as opposed to "strong superradiance" first reported in 1973 in optically pumped HF gaz [13]. In this case all the energy stored in the sample is emitted cooperatively and so the decay is dramatically accelerated. However, since the macrodipole was not created by the pump but rather by the reaction field it was recognized latter to be a case of superfluorescence [14] as we shall described in next subchapter.

So the first example of real superradiance was in fact the free induction decay and the decay of the photon echo observed in ruby by Kurnit, Abella and Hartmann [15]. In such experiment a $\pi/2$ pulse from a ruby laser was sent onto a ruby crystal, the free induction decay and the echo decay observed were in the range 50 ns to be compared to the usual Cr^{3+} radiative decay time of 4 ms, showing clearly the radiation emission from the macrodipole; see Fig 6.

Returning to eq. (104), we may try to estimate the new decay time resulting from the cooperative emission in strong superradiance. Such decay time is also called "superradiant lifetime" and its inverse "radiative damping constant".

Calculating this lifetime as [5] :

T_{SR} = (stored energy by inversion) / (radiated power), it comes from (104) :

$$\tau_{SR} = \frac{2(N_1 - N_2)\hbar\omega}{\omega\pi Q(N_1 - N_2)^2\hbar^2 g^2 n}$$

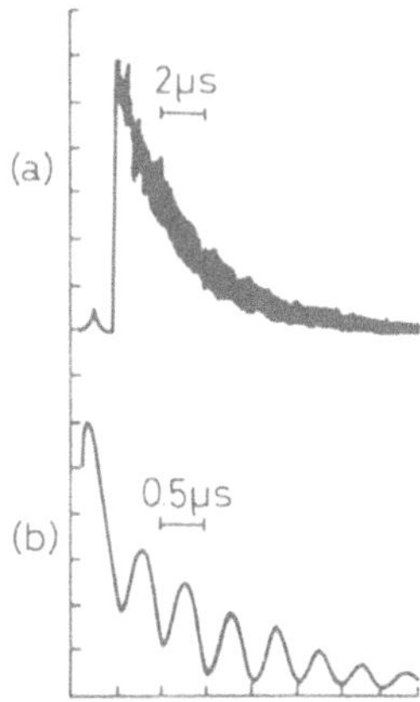

Fig.5. Coherent decay of OCS after coherent excitation (a); interference, beats with a weak CW coherent source give proof of the coherent of emission (b). From [11].

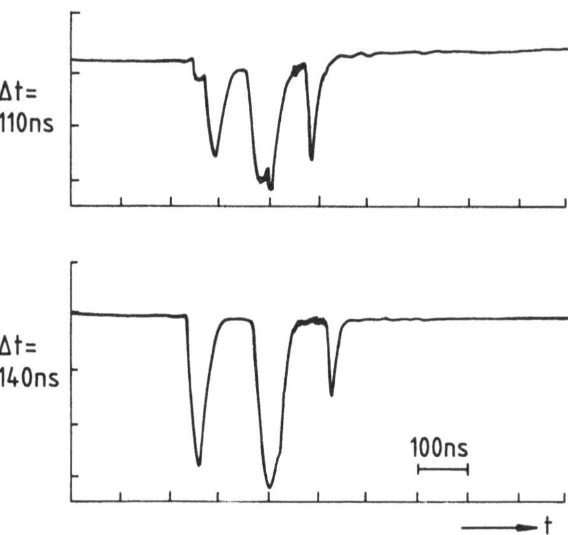

Fig.6. Superradiant emission in the first photon echo emission in a solid : Al_2O_3 : Cr^{3+}. From [15].

$$\tau_{SR} = \frac{2}{\pi Q(N_1 - N_2)\hbar g^2 n} \tag{105}$$

From (75), $2\pi g^2 \delta\omega_{12} = \tau_0^{-1}$ is the spontaneous decay for one mode then :

$$\tau_{SR} = \frac{4\tau_0}{Q(N_1 - N_2)\hbar n}\delta\omega_{12} = \frac{4\tau_0}{Q(N_1 - N_2)n}\delta E_{12} \tag{106}$$

Showing that the spontaneous lifetime τ_0 is reduced in proportion of the initial population inversion everything else being kept constant.

According to geometric conditions for preparation of the excited state of the whole sample, several modes may be allowed to emit then partial summation analoguous to (80), (81) have to be done on (106).

Since the macro-dipole exists right after excitation, the superradiance shortened decay starts without delay and with the $(N_1 - N_2)^2$ behaviour at the beginning. The only condition is that the dephasing rate be slower than the superradiant one so that the macro-dipole keeps on existing during the transition.
That is :

$$\frac{1}{\tau_{SR}} > \frac{1}{T_2} \tag{107}$$

In the optical frequency domaine, superradiance in the Dieke and Bonifacio's sense [1] [2], i.e. with strong radiation damping seems to have been observed in solids up to now only in photon echo experiments as already mentioned. The difficulty is in the preparation of the macro-dipole in a volume smaller than λ^3. However the recent interest in quantum wells allows the construction of microscopic structures with volume $< \lambda^3$ and so we suspect it could be observed in such systems because spontaneous emission is then enhanced [16].

2) Superfluorescence (SF)

α) The basic process. To the difference of the previous case, we consider now the cooperative emission of a large number of initially inverted atoms, without initial macroscopic dipole moment left by the excitation pulse. It means that instead of excitation by a $\frac{\pi}{2}$ pulse creating the Dieke's states, excitation is performed by a π pulse leaving the pseudo spin Bloch vector initially along $\vec{\Omega}$. Then from Fig 3 each atom is in its upper excited state and the whole considered sample is inverted.

Preparation of excitation does not need coherence and in experiments, a phonon emitting step can be used to obtain inversion. Fluctuation in spontaneous emission allows the begining of the effect with the spontaneous lifetime τ_0 of individual atoms initially and with an emitted power proportional to the inverted atoms number N. After a delay t_d the whole system starts radiating coherently with a power proportional to N^2 and the reduced decay time in τ_0/N in analogy with superradiance. Except for its begining, superfluorescence is equivalent to superradiance; it is a self-organized superradiance. Self-organization can be viewed as induced by the field .

This initial interaction between two atoms is depicted on Fig 7.

The building up of the far field of a system of 10^4 individual Bloch vector has been simulated [17]; the result is given on Fig 8 for increasing steps in time.
The weak field emitted by ordinary fluorescence produces a weak polarization of the medium increasing the field which in turn increases polarization and so on up to the superradiant emission. The macroscopic dipole has been spontaneously created by the self-organization of the system alone and not by an external coherent pump.

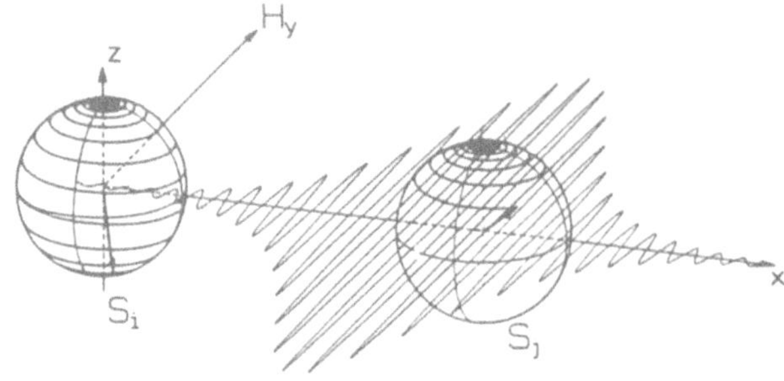

Fig.7. Interaction of two Bloch vectors through their spontaneous electromagnetic field. After [17].

It was recognized also that a macroscopic pencil shaped sample with length >> λ could be used [2] which open the way to experiments in gases at optical frequencies [13] and also in 1982 [18] to the first demonstration of this effect in a solid.

In order to calculate the superfluorescent collective lifetime we just say that the new rate (N/τ_{SF}) is given by the square law behaviour of (104) [7] :

$$N\frac{1}{\tau_{SF}} \equiv N^2\frac{1}{\tau_0}\frac{d\Omega}{\int d\Omega} \tag{108}$$

where $d\Omega$ is the solid angle corresponding to the diffraction lobe of the end surface (S) of the pencil shaped sample into which emission shall go; that is [19] :

$$\frac{d\Omega}{\int d\Omega} = \frac{\lambda_0^2}{S}/\frac{8\pi}{3} \tag{109}$$

the last factor already seen in (81).

Combining (108) and (109) :

$$\tau_{SF} = \frac{8\pi S}{3N\lambda_0^2}\tau_0 \tag{110}$$

This is analoguous to the result (106) found for superradiance in a microcavity.

Posing $\rho = N/V$ for the inverted atoms density, it comes :

$$\tau_{SF} = \frac{8\pi}{3\rho}\frac{\tau_0}{L\lambda_0^2} \tag{111}$$

the most used formula for the fluorescent lifetime [2] [19] [20] in term of the spontaneous atom lifetime (τ_0) for the considered transition.

Eq. (109) assumes that the diffraction angle λ_0^2/S equal the geometric solid angle S / L^2 which corresponds to a Fresnel number :

$$F \equiv S/\lambda_0 L \cong 1 \tag{112}$$

the problem can then be treated as a one dimensional one with two plane waves building up backwards and forwards that is taking into account propagation [19].

The physical process of self organization takes some time corresponding to a delay given by [19] :

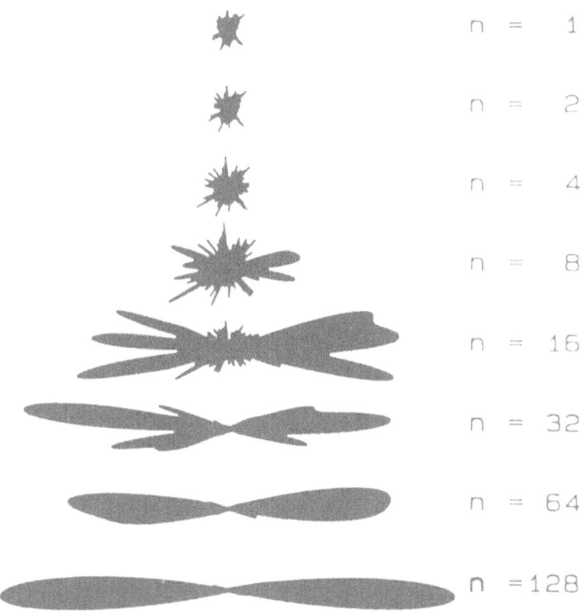

Fig.8. Simulated far field created by an horizontal sheet of 2-level invertaed atoms at various step in time. After [17].

$$t_d = \frac{\tau_{SF}}{16}[\ln 2\pi N]^2 \tag{113}$$

or either by [14] :

$$t_d = \frac{\tau_{SF}}{4}[\ln(2\pi N)^{1/2}] \tag{114}$$

according different calculations.
or either by [12] :

$$t_d = \frac{\tau_{SF}}{4}[\ln(\theta_0)]^2 \tag{115}$$

where θ_0 has the "physical" meaning of the "initial tipping angle" of the Bloch vector in the vertical metastable position as an effect of the vacuum field fluctuation initiating spontaneous decay. It is found to be given by : $\theta_0 \sim = 2N^{-1/2}$ [21]. (113), (114), (115) shows that t_d is linearly linked with τ_{SF} and that delay is reduced when N is increased since, from (110), τ_{SF} decreases more strongly than the logarithmic function. Practically τ_{SF} is one or two order of magnitude smaller than t_d.

Since now we have a large sample, atoms have to be able to communicate by their field more quickly than they can radiate. This condition fixes an upper limit to the increase in the SF rate with the number of inverted atoms N as given by (110) .

When an atom emits with rate $\frac{1}{\tau_c}$, it emits a wave train of length :

$$l_c = C\tau_C \tag{116}$$

When there is coherence in the wave front over the whole end surface of the sample, the wave train shall interact with a maximum of

$$N_c = l_c S \rho \tag{117}$$

atoms. By (110) we have also :

$$\tau_c = \frac{8\pi}{3N_c}\frac{S}{\lambda_0^2}\tau_0 \tag{118}$$

from (116), (117), (118) one gets :

$$\tau_c = \sqrt{\frac{8\pi}{3C}\frac{\tau_0}{\rho\lambda^2}} \tag{119}$$

introducing the lifetime of the photon inside the whole sample or escape time τ_E :

$$\tau_E = L/C \tag{120}$$

(119) is written by (111) :

$$\tau_c = \sqrt{t_{SF}\tau_E} \tag{121}$$

which is the so-called Arecchi-Courtens cooperation time [22]; l_c is the corresponding critical distance.

β) <u>The conditions for SF.</u>

Having built them step by step we have now at hand all the elements involved in the condition of existence of SF :

The conditions for dephasing already stated for SR as (107) may be extended to all dephasing times T_1, T'_2, T^*_2, whatever is the shortest, as :

$$\tau_{SF} < T_{1,} \quad T'_{2,} \quad T_2^* \tag{122}$$

Since the phase has to be kept also during the self-synchronization step, we have :

$$t_d < T_{1,} \quad T'_{2,} \quad T_2^* \tag{123}$$

By (119) we had found the shortest possible τ_{SF} as t_c which by (121) is intermediate between τ_E and τ_{SF}. Summerizing we have :

$$\tau_E < \tau_c < \tau_{SF} < \tau_D < T_{1,} \quad T'_{2,} \quad T_2^* \tag{124}$$

giving the conditions of existence for SF.

Conditions (124) have been stated in simpler form [2] as :

$$L_T \ll L \ll l_c \tag{125}$$

where L_T is a threshold length for SF given by :

$$L_t = \alpha^{-1} \quad \text{and} \quad \alpha = T_2/L\tau_{SF} \qquad (126)(127)$$

(127) is equivalent to $\frac{L\tau_{SF}}{T_2} \ll L$

or $\tau_{SF} \ll T_2$

Also this condition implies

$$\alpha L = \frac{T_2}{\tau_{SF}} \gg 1 \tag{128}$$

which is the "large gain" or "large absorption" condition of Friedberg and Hartmann [23].

3) <u>Amplified spontaneous emission (ASE).</u> ASE is what happened when a long medium with inverted atoms starts amplifying by stimulated emission the spontaneous emission noise of some atoms at the medium extremity. In such a case all atoms of the medium behave individually and there is no self building of a macroscopic dipole. Then intensity keeps being proportional to N or ρ instead of N^2 and ρ^2 for SF. Also there is pratically no delay since there is no need for building up the dipole.

If the number of excited atoms is sufficiently large, ASE emission is still more rapid than single-atom spontaneous emission, because the spontaneous emission is amplified as the radiation propagates. Because of the long shape geometry, radiation is also directional.

The difference with SF stems, in ASE, from the existence of a stronger dephasing which inhibits the self-organization into a macrodipole.

Schuurmans and Polder [24] have given the condition on T_2 at which the transition between these two types of behaviour occurs :

If

$$T_2 \gg \sqrt{\tau_{SF} t_d} \tag{129}$$

SF predominates.

If

$$\tau_{SF} \ll T_2 \ll \sqrt{\tau_{SF} t_d} \tag{130}$$

then ASE predominates.

Also in a very long medium, as in a laser cavity, condition on escape time τ_E and on τ_c are not required. In usual laser amplifiers the condition holds :

$$T_2 \ll \tau_E \tag{131}$$

which is opposite to part of (124).

4) Experimental results for SF and ASE in the solid state

α) Molecular systems

We have seen that superradiance (SR) in solids with radiation damping has been observed in Ruby as the accelerated free induction decay of a photon echo experiment [15]. However it seems that, as well as for the first observation in the gas phase, the strong avalanche emission case is SF and not SR [14].

In solids up to now, , only few cases of SF have been observed. The reason is the difficulty in having long T_2 because in a solid, ions may be subjected to many perturbations : multi-phonon decays, ion-ion energy transfers, besides the inhomogeneous contribution and the strong spontaneous radiative decay at optical frequencies. In gas, it has been assumed that to avoid dipole-dipole interactions, the atom density should be such as [14] :

$$\rho(\lambda_0/2\pi)^3 \ll 1 \tag{132}$$

This condition means that there should be less than one atom per one wavelength volume . In a solid it is never the case. Hopefully we know that dipole-dipole energy transfers [25] in abscence of clusters are important between ions only when ions are at less than about 10 Å in the case of trivalent Rare-earth ions with oscillator of $f \sim 10^{-6}$. For more strongly allowed transition this distance may be enlarged. It means that higher concentrations are possible in solids and consequently smaller dimensions for the sample than for gases.

The first observation of SF in the solid phase was in 1982 by Florian, Schwan and Schmid in KCl:O_2^- [18]. Emission was observed in the visible at 0.63 μm under 30 μJ of 0,265 μm excitation focused on 10 μm diameter (100 MW / cm^2) with sample temperatures between 10 to 30 K. Samples of sizes 5 mm^3 were doped with 0.5 % K O_2. Many characteristics of SF

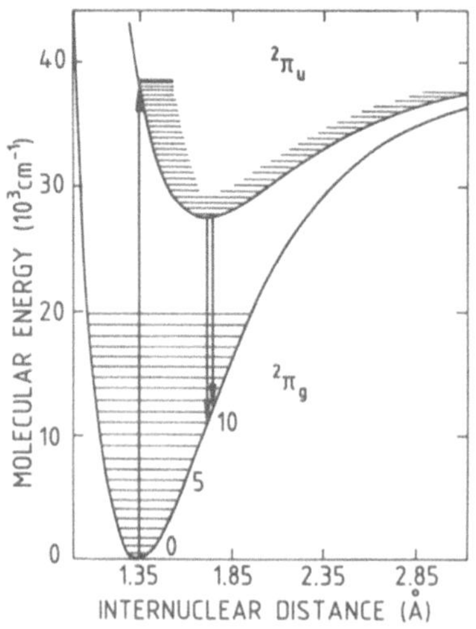

Fig.9. Energy scheme involved in SF for KCl : O_2. After [26].

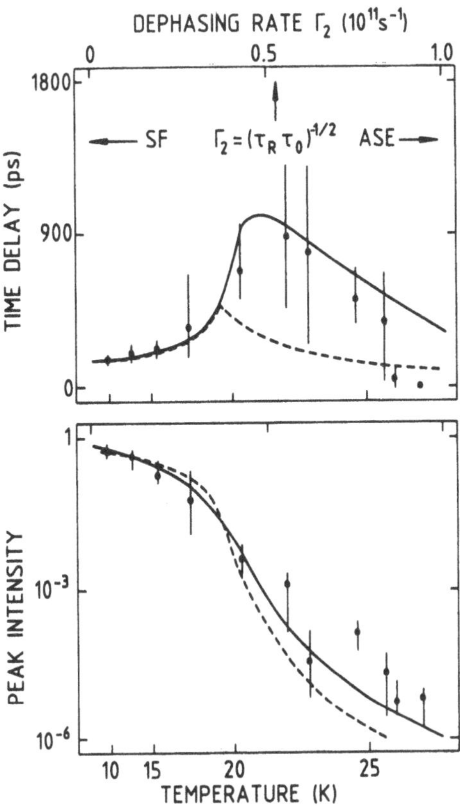

Fig.10. Separation of SF and ASE building up delays by increasing T_2^{-1} with temperature. After [20].

have been observed [26] [27] : line emission narrowing, directivity, threshold, pulse delay behaviour. Also as shown on Fig 9 excitation is through phonon emission steps which preclude the coherent excitation of SR. Same effect has also been observed in K Br : O_2^- and KI : O_2^-.

Emission is distinguished from ASE because it is independant of orientation of the crystal with respect to pumping beam; and even Brewster incidence is used to prevent Fresnel reflexion on the extremities [27]. Also no longitudinal mode stucture as would be expected in a laser is observed; all theses facts point to SF rather than to an ASE process.

Recently in the same KCl : O_2^- system, the smooth evolution from SF to ASE has been investigated by rising slowly the temperature between 10 to 30 K and so increasing the dephasing rate T_2 [20]; the criterion being studied was the Schuurmans and Polder's one as given by (129) and (130). The experimental results are presented on Fig 10, clearly deliminating the two regimes.

The second solid state system having exhibited SF was Pyrene-doped diphenyl crystals [28]. Under nitrogen laser excitation at 337 nm, a crystal of size 1 x 1 x 3 mm exhibited SF at 373.9 nm which seems to be the record for wavelength shortness.

This corresponds to an individual atom radiative lifetime of 100 ns. Other pumping sources with different degrees of coherence have been investigated, showing that results were independant of pumping coherence; this points to an SF process and not to SR as the authors believed. The temperature ranged from 1.5 to 4.2 K, allowing to study the role of non equilibrium phonons and local heating. Estimated τ_{SF} was between 10-100 ps which allowed for a T_2 also in the 100 ps range.

Quite recently [29], the third system into which SF seems to have been observed is the F center / CN^- defect pairs in CsCl. The pecularity of the system is that the effect is seen under CW excitation at threshold level of a few kW / cm^2 at temperature of 15K. Emission is at 5µm and $\tau_{SF} = 2\mu s$ which is rather long. However the individual atom fluorescence is also very long ($\tau_0 = 1ms$) which can explain it. The SF line intensity is about 30 times the normal emission just below threshold. Again the effect is found to be independant of sample orientation which rules out accidental laser activity by unintended specular reflection.

β) Rare earth (R E) doped solids

In 1984, a theoretical investigation by Malikov and Trifonov [29] for YAG : Nd at 100K showed that if superradiance was possible, superfluorescence was impossible to observe in such system because it would be 3 order of magnitude less intense than ordinary fluorescence. Calculation were based on a T_2 = 17 ps and $\alpha L = 5$for an invertion density of 10^{16} cm^{-3}. Superradiance could be obtained by the propagation of a short coherent pulse which could provide pulses with $\tau_{SR} = 3ps$.

The difficulty in observing SF in RE doped systems then seemed to us confirmed by the ASE experiments of Markushev et al [30] on microcrystalline powder of $(Mo\ O_4)_4Na_5Nd$.

They had obtained, under 30 ns pulsed excitation, all the characteristics of ASE (narrowed spectrum, linear dependance of output with ρ) at a threshold of 1.8 MW/cm^2 at 77K. Probably because of the very high concentration of active ions (~5. 10^{21} cm^{-3}), ion-ion interaction had reduced T_2 in a very effective way, so preventing the construction of the macrodipole.

However, we believe to have observed the first superfluorescence effect in a RE doped system [31] under very low CW excitation, much below 1 KW/cm^2.

In our experiment the sample is a single crystal of $LiYF_4$:1% Er^{3+} of thickness 1.5 mm with two unpolished (as cut by a diamond saw) not parallel faces, so that no Fresnel reflexion optical feedback is likely to occur. Excitation was obtained by a CW krypton laser with line at 6470 Å allowing up to 320 mW of incident power on the sample as measured by a pyroelectric radiometer. The pump focalization gave a beam waist diameter of 110 µm. Luminescence observation was in the backward direction with lens system focalizing the sample output face on a monochromator. A room temperature PbS photoresistor used to detect the 2.7 µm emission. Direct recording of the PbS cell output on an oscilloscope allowed to inspect the temporal behaviour of emission.

Below about 80 mW of incident power (842 W/cm^2) with sample at 10K, emission spectra for $^4I_{11/2} \rightarrow {}^4I_{13/2}$, transition was as shown in Fig 11;
above this threshold observed spectra are peaked only at 2.72 µm, showing clearly a spectral narrowing as shown on Fig 11.

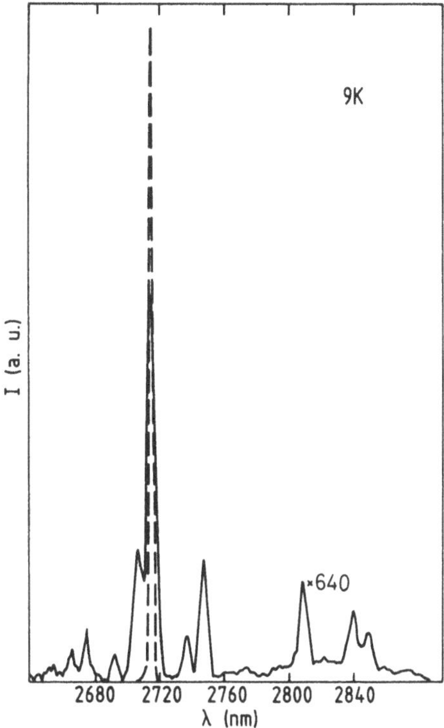

Fig.11. $^4I_{11/2} \rightarrow {}^4I_{13/2}$ emission of $LiYF_4 : Er^{3+}$ at 10K for 70 mW (-) and 144 mW (--) incident power at 0.64 μm. [31].

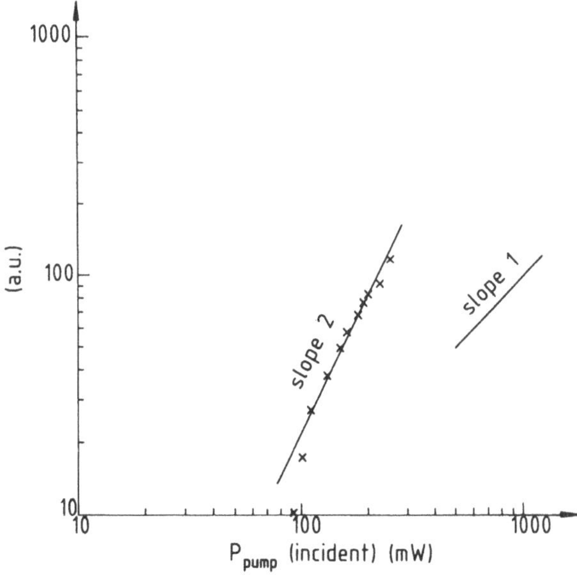

Fig.12. Emission intensity at 2.72 μm vs. pump incident power on $LiYF_4 : Er^{3+}$. [31].

Output intensity vs. incident pump power is shown on Fig 12 on a log-log scale. There is a marked threshold, then emission follows a square law vs. pump power. When temperature is increased up to about 40K, threshold increases up to the 320 mW available for pumping. Because of the mismatch between krypton line at 6470 Å and Er^{3+} $^4I_{15/2} \rightarrow {}^4F_{9/2}$ transition and of the rather low Er^{3+} concentration, the 80 mW threshold corresponds to only 2.8 mW of absorbed power.

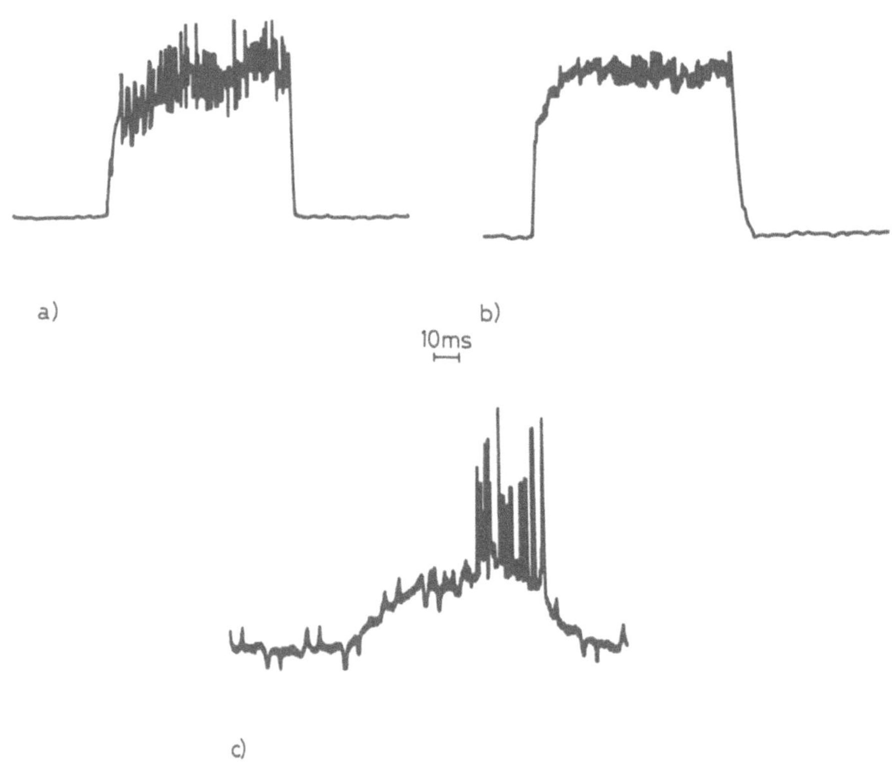

Fig.13. Temporal behaviour of $LiYF_4 : Er^{3+}$ (a) just above threshold; (b) well above; (c) at threshold showing both SF (spikes) and spontaneous emission ($\tau_0 \cong 4ms$).

The temporal behaviours obtained under a quasi-cw excitation square pulse of ~ 50 ms duration are shown in Fig 13 a. just above threshold (107 mW of incident pumping) and Fig 13 b. for emission well above threshold (320 mW incident).

By rising the temperature to 38K it was possible to obtain the behaviour just at threshold (210 mW incident) Fig 13 c.

In [32], the measured τ_{SF} was later found to be detector limited. With an Hg Cd Te (77K) detector with response of 120 ns, τ_{SF} was measured to be 150 ns [32]. A pulsed excitation could provide in some cases a single pulse emission as shown on Fig 14.

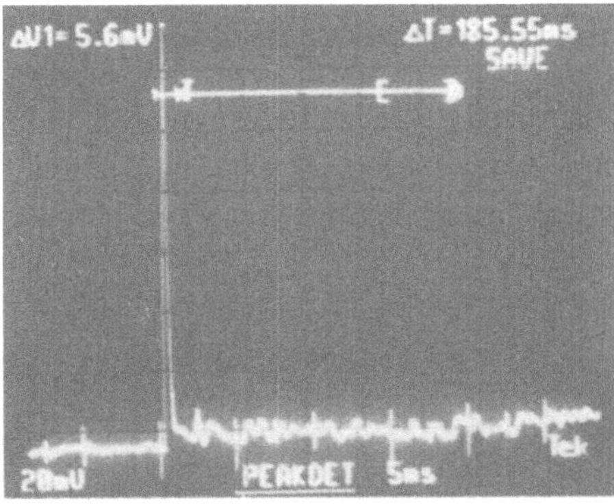

Fig.14. Single SF pulse emission of $LiYF_4 : Er^{3+}$ under pulse excitation. (PbS detector).

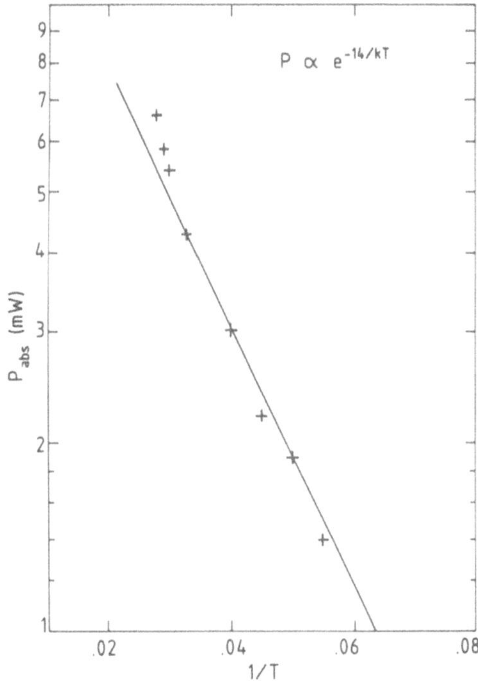

Fig.15. Exponential behaviour for SF threshold with temperature in $LiYF_4 : Er^{3+}$.

In a systematic study of threshold with temperature it was found that this effect could no be observed for T > 60K and that for a given concentration the threshold value behaves exponentially as shown on fig 15 [32] with an activation energy of 14 cm^{-1} corresponding to the first stark splitting of the $^4I_{11/2}$ excited state.

Contrary to the laser action which we could obtained at room temperature with concentration of Er^{3+} = 15% [33], a rather weak Er^{3+} concentration < 5% is necessited.

So we may now list all the arguments in favor of SF in our experiment, which differentiate it from ASE :

- ρ^2 dependance;

- τ_{SF} < 150 ns to be compared to T_2 of about 300 ns to 2 μs [33] in such crystal for RE at low temperature. It has been pointed out that one should consider an effective T_2^* [13] given by :

$$T_2^{*eff} = \alpha L T_2^* \tag{133}$$

in system where $\alpha L \geq 1$. Then condition (128) becomes :

$$\alpha L = \frac{\alpha L T_2^*}{\tau_{SF}}$$

or

$$T_2^* \geq \tau_{SF} \tag{134}$$

which is much less stringent than (128) when dephasing is governed by inhomogeneous broadening as is probably the case.

- the effect has been observed on very small length samples down to 100 μm though the concentration was ≤ 1%. ASE would rather require the opposite.

- the effect needs a weak concentration, pointing also to the necessity to reduced ion-ion dephasing.

- the temperature behaviour points also to the involment of a T_2 variation temperature effect.

Contrary to theses observations, typical ASE is obtained for the same transition of Er^{3+} at 2.7 μm in 3m long fluoride fibers at 300K [34], as shown on Fig 16. The line narrowing is smoothly obtained when pump is increases without any threshold. The narrowing would follow a law of the type :

$$G(\nu) = g(\nu) exp\, g(\nu) L$$

where $g(\nu)$ is the ordinary fluorescence spectrum line shape. In doped fibers this is so much general that the real spectrum shape can be obtained only by looking transversally to the fiber axis to prevent any amplification effect.

5) <u>SF application.</u> Since no mirrors, with of course limited transparency, are needed for SF, it is in principle the most efficient process to produce very narrow pulses of coherent light. In the frequency range where mirrors are difficult to produce it could be the only way to obtain coherent sources : such would be the case for X-rays and γ-rays range. However the drawback is that the spontaneous lifetime of individual atoms becomes very short due to its ν^{-3} and so is probably the most important cause of dephasing.

A more realistic application may be to use it in "triggering spectroscopy" [35] where by injecting a weak resonant field on a prepared SF sample, large change can be induced in the delay and spatial pattern of the emission. By measuring the change in direction of the emission lobes as a function of the injection field frequency the transition of the $7P_{3/2}$ - $7S_{1/2}$ of Cesium has been recorded with only a few photons per pulse; experimentally the spectrum could be recorded down to 8 nW/mm^2.

Along the same lines we are presently working on a solid state amplifier at 2.7 μm for which we expect to have a noise figure below the spontaneous emission noise limit of 3 dB, because noise becomes organized by the signal when it is used to trigger SF.

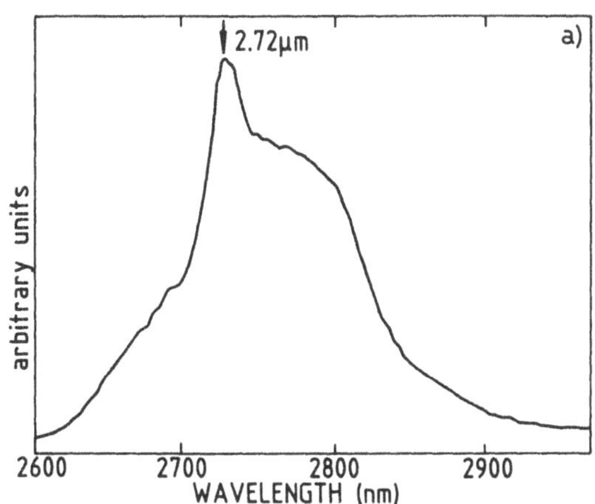

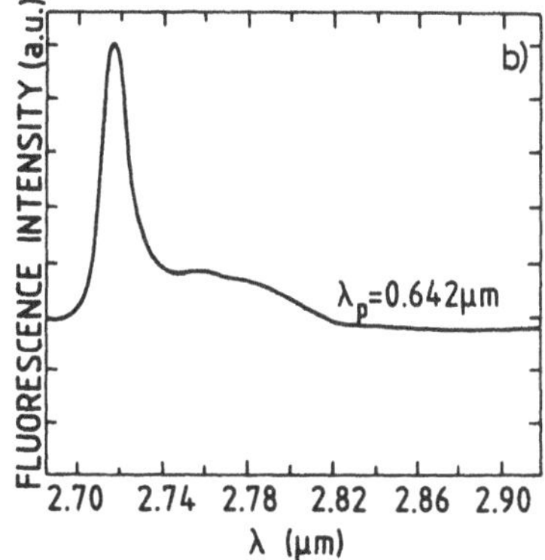

Fig.16. a) normal fluorescence obtained on a small bulk sample of Er^{3+} ZBLAN glass. b) spectrum obtained on same glass draw into a 3m long fiber. [34].

III . HOT LUMINESCENCE (H. L.)

"Hot Luminescence" is a terminology proposed by Rebane [36] to characterize the radiative emission which happens when there is a competition between radiative and non-radiative processes. Since such process has already been studied in this school some years ago [37], we shall present here only some of the general principles.

III.A. A two-levels limiting case

In order to make a connection with the previous chapter, let us first consider the limiting case of the "two-levels atom" again.

As we have seen, under high excitation we may reduce the individual atom radiative lifetime T_1 to the value τ_{SF} by construction of a macrodipole. If this happens, from eq. (98) and (122) we must have :

$$\tau_{SF}^{-1} > W_{ion-phonon} \equiv W_{non\text{-}rad.} \tag{135}$$

We have recently obtain a striking example of such overtaking of radiative decay over the non-radiative one [38] : It is known that anhydrous Lanthanide Chlorides are low threshold laser materials, for instance $CeCL_3 : Nd^{3+}$ [39], and that $NdCl_3$ having low crystal field would be a weak self-quenching type like pentaphosphates [40].

However hydrated Chloride as $NdCl_3 : 6\,H_2O$ is expected to show a strong multivibration quenching by coupling of Nd^{3+} with OH vibration [41]. This is confirmed by the very short fluorescent decay time of ~ 16 ns at 1.05 µm (Fig 17), obtained when a powder of $NdCl_3 : 6\,H_2O$ is irradiated with pulses from frequency doubled YAG : Nd^{3+} laser at levels of a few MW/cm^2 at 0,53 µm.

Of course then, the fluorescent intensity of this material is very weak as it can be expected from a reputedly non-fluorescent material. The lifetime, compared with the anhydrous value (~ 170 µs), is essentially governed by the non-radiative rate (W_{NR}). Then we have : $W_{NR} = 1 / 16 . 10^{-9} = 6.2\ 10^7\ s^{-1}$.

This measured non-radiative rate is consistent with what can be estimated from calculation of a multivibration rate in the "large molecule model" of Haas and Stein [42]. For a small number of simultaneous vibrations [43] and 6 (OH) species coupled to one Nd^{3+} [42].

$$W_{NR} = 6.10^7 e^{-S_0} S_0^{(N-2)}/(N-2)! \tag{136}$$

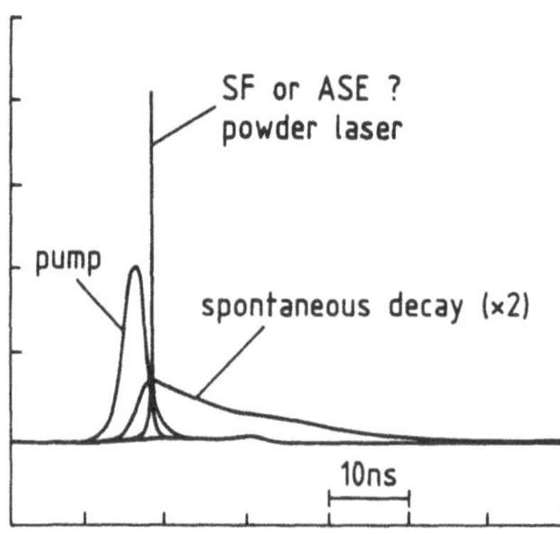

Fig.17. Emission at 1.054 µm of $NdCl_3 : 6H_2O$ powder under strong pulse excitation below and above threshold (300K). [38]

taking the electron-coupling constant as $S_0 \sim 10^{-2}$ [41] and N, the number of simultaneous vibrations spanning the electronic energy gap $^4F_{3/2}$ - $^4I_{15/2}$ of Nd^{3+}, as roughly twice one OH vibrational quantum :

$$W_{NR} = 6.10^7 s^{-1}$$

Though this good fit is certainly fortuitous due to the many approximations behind it, it shows that the non-radiative properties behave as expected. However, when pumping is increased to 1 GW / cm^2 level, a threshold is reached where a strong, narrowed and shortened emission at 1.054 µm is observed, Fig 17.
The observed radiative rate (10^9 s^{-1}) then overtake the nonradiative one by nearly 2 orders of magnitude which in turn allows condition (135) to be more largely fulfilled. So we see that this particular case of hot luminescence is associated with a radiative coherent emission.

III.B. From Resonant Raman scattering (RRS) with ordinary luminescence (O.L), through Hot luminescence (H.L)

In general, the two electronics levels of the atom are associated with vibrations of the solid under the form of vibronic levels as given by the Born-Oppenheimer approximation.

Fig 18 presents in a schematic way what happens when the exciting laser has his frequency progessively matched to the vibronic absorption.

In the case where the excitation frequency is not tuned to the absorption transition, the Raman scattering is given by the well known second order perturbation order formula [44]:

$$S(\omega_1,\omega_2) = 2\pi \sum_n \mid \sum_j H^{+2}_{nj} H^{-1}_{jm} / (E_j - E_m - \omega_1) \mid^2 \delta(E_m + \omega_1 - E_n - \omega_2) \qquad (137)$$

where $H^{\pm 1,2}$ represents the matter-radiation interaction corresponding to the incoming frequency ω_1 and to the emitted one ω_2; j are the intermediate states considered as virtual as long as there is no real population pumped into it by a resonance between ω_1 and a transition m $\rightarrow$ j. If it is the case, the denominator in (137) goes to zero giving very large value for scattering, limited only by a limited lifetime for the intermediate level j.

Such lifetime is due both to spontaneous emission and to coupling with the vibration. The apparent divergence of (137) is removed [44] as already done in chapter II A, by taking the principal value of (137) as in (68). Now, when ω_1 populates levels j, we may ask what kind of ω_2 will be observed. No general answer can be given.

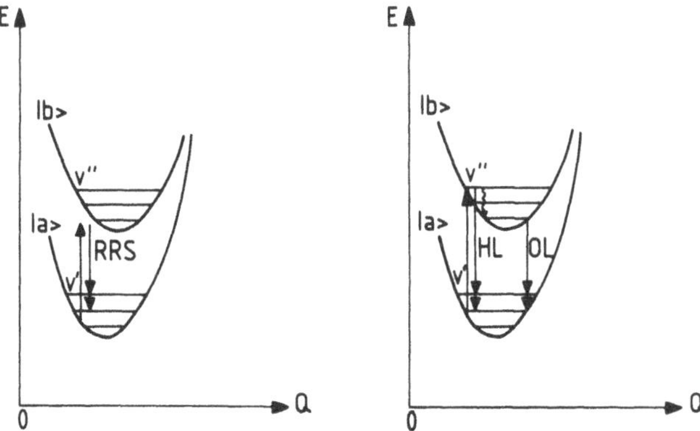

Fig.18. R.R.S., HL, and OL between vibronic states.

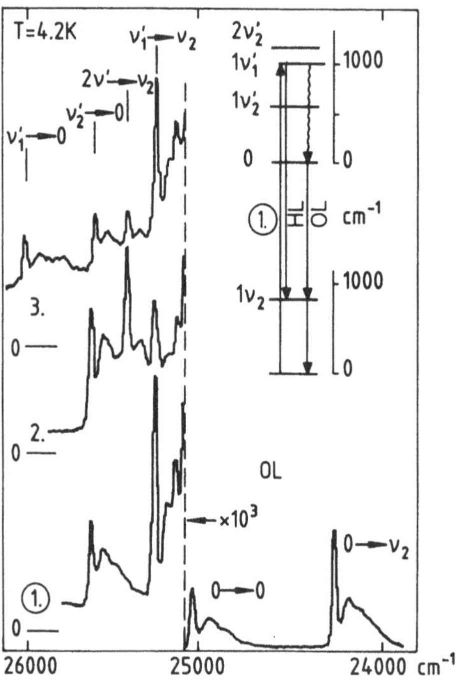

Fig.19. HL emission of KCl-NO_2^- and OL emission at 4.2K; insert shows the vibronic states involved. After [36].

In general, when T_1 for the intermediate state j exceeds T_2, emission at times $t \leq T_2$ describes the part of emission with correlation between primary (ω_1) and secondary photons ω_2, it is RRS; and for time $t >> T_2$, the emission without correlation, that is HL and OL [45].

HL is the radiation after phase relaxation but during energetic relaxation ($t \leq T_1$); OL is the emission after energetic and phase relaxation ($t > T_1, T_2$).

For instance it can be shown that (137) reduces to [44] :

$$S(\omega_1, \omega_2) = \frac{1}{2\gamma} F^a(\omega_1) F^e(\omega_2) \tag{138}$$

that is the product of the absorption spectrum $F^a(\omega_1)$ by the emission spectrum $F^e(\omega_2)$; provided the condition $\gamma \ll W_{ion-phonon}$ holds; γ is the total longitudinal rate $(T_1^{-1} + W_{ion-phonon})$.

An example of HL with OL is presented on Fig 19 for KCl : NO_2^- as seen for the first time [36].

Such HL luminescence being seen at 4.2K, is not a "hot band" excited by thermalization. Note the intensity which is 10^{-3} of OL, and in the order of magnitude of a Raman effect.

IV . EXCITED STATE ABSORPTION (ESA) ENHANCED BY ENERGY TRANSFERS IN RE DOPED SOLIDS

When metastable states (with $W_{spont} < W_{non-rad}$) are highly populated as is usually the case in laser materials for which population inversion is looked for, Excited state Absorption (ESA) may take place. It may happens either at the pumping frequency or at the laser frequency. This is a well recognized drawback which comes as a reduction of the laser cross-section [46]. Such effect being rather trivial shall not be discussed here.

When through high concentration there is a possibility for energy diffusion by energy transfers, ESA can be enhanced with an increase in probability by several order of magnitude. Having already presented the basic processes involved in a previous school [47], we shall described only their gross features.

IV.A. ESA by APTE effect and the role of diffusion

When active ions are situated at sufficiently short distance for interactions between them to take place, two type of up-conversion processes may occur : summation of photon energy through energy transfers (called APTE effect, for Addition de photon par Transferts d'Energie) or cooperative effects either by sensibilization, emission or absorption.

Up to 1966 all energy transfers between rare-earth ions were of the types where the activator ion receiving the energy from a nearly sensitizer (S), was in its ground state. Then we proposed to consider cases where activators (A) were already in an excited state [48] as shown on Fig.20.

The reason for proposing such up-going transfer was to point out that energy transfers then used [49] to improve the laser action of Er^{3+} by pumping Yb^{3+} in a glass matrix could also have the detrimental effect to increase reabsorption [50]. The simple proof of such effect was to look for and up-converted green emission (from $^4S_{3/2}$ of Er^{3+}) while pumping Yb^{3+} ($^2F_{7/2} \rightarrow {}^2F_{5/2}$), which we effectively observed [50] [51]. Of course the situation in Fig.20 could repeat itself several times at the activator.

This means that n-photon up-conversion by energy transfer was possible as demonstrated by the 3-photon up-conversion of 0.97 μm into blue light (0.475 μm) in the Yb^{3+} - Tm^{3+} couple [48]. Independently such IR→blue up-conversion was interpreted by Ovsyankin and Feofilov [52] as a 2-photon effect connected with a cooperative sensitization of Tm^{3+} by two Yb^{3+} ions, because of saturation in an intermediate step reducing intensity from a cubic to a quadratic law.

In order to make the terminology clearer a schematic comparison between APTE effect and other 2-photon up-conversion processes namely : 2-steps ESA, cooperative sensitization, coomperative luminescence is presented on Fig 21, together with typical efficiencies.

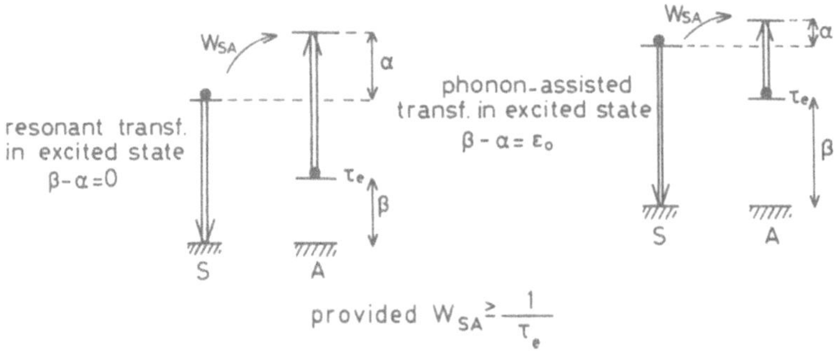

Fig.20. Energy states for ESA by energy transfers.

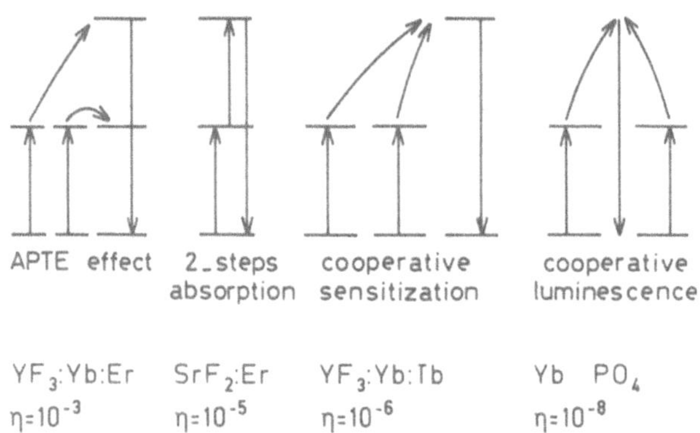

Fig.21. Comparison of level schemes and efficiencies of different up-conversion processes. After [55].

Since we are dealing with non-linear processes, usual efficiency, as defined in percent, has no meaning because it depends linearly on excitation intensity. Value are then normalized for incident flux and given in (cm^2/W) units.

A simple inspection of involved energy schemes shows that they differ at first by the resonances involved for in-and out-going photons : for highest efficiency, photons have to interact with the medium a longer time which is practically obtained by the existence of resonances. As shown, APTE effect is the most efficient because it si the process with approaches more the full resonance case.

However reality is some time not so simple and different up-conversion processes may exist simultaneously or their effects can be tentativelly reinforced reciprocally.

The probability for ESA in a two-step absorption (W_{13}) is just given by the product of the probabilities for each step :

$$W_{13} = W_{12}.W_{23} \tag{139}$$

For ESA by APTE effect we have also the product of two energy transfers probabilities.

Calculating the rate for the same ESA with APTE it comes :

$$N_A W_{13} = (N_A N_A^* W_{SA1}) N_S^* W_{SA2} \tag{140}$$

or

$$W_{13} = N_S^{*2} W_{SA1} W_{SA2} \tag{141}$$

were W_{SA} are the energy transfers probabilities for each step, and N_S^* the concentration of excited sensitizers which are given by :

$$N_S^* = N_S W_{12} \tag{142}$$

Assuming all W_{ij} to have same magnitude and W_{SA} also, we have to compare

$$W_{13} = W_{12}^2 \tag{143}$$

for single ion ESA, and

$$W_{13} = W_{12}^2 N_S^2 W_{SA}^2 \qquad \text{for ESA by APTE.} \tag{144}$$

Clearly the gain by APTE over one ion ESA can come from the product $N_S^2 W_{SA}^2$ which has to be > 1.

This points to and increase in sensitizer concentration (N_S) which leads to "fast diffusion" [46] and allows the use of rate equations in such multiions systems [46].

Many times in litterature up-conversion involving coupled ions are referred to as cooperative effects [53] [54] without demonstration, when in fact as can be guessed from their relative efficiencies, APTE effects are involved [55] [56]. The fact that APTE effect and cooperative ones are often mistaken one for the other is linked to a number of common properties. For instance, for 2-photon up-conversion, both shows an emission lifetime equal to half the absorber lifetime. However as shown in next chapter the difference is more basic, though sometime difficult to establish experimentally except in special cases : when single ion resonances clearly do not exist or when diffusion between ions is prohibited by too small concentration with still an interaction as in clusters.

IV.B. Up-conversion in single ion level description (APTE) and in pair-level one (cooperative effects)

As seen in the introduction up-conversion by energy transfers is just a generalization of Dexter energy transfer [57] to the case of the activator being in a metastable state instead of being in its ground state; this requires that the interaction between S and A (H_{SA}) be smaller than the vibronic interaction of S and A in order that both ions be described by single-ion levels coupled to the lattice. It is generally the case for fully concentrated rare-earth crystals or for clusters, where pair level splitting is of the order of 0.5 cm^{-1} [58]; in host with smaller concentration this interaction can even be weaker, whereas one-phonon or multi-phonon side bands may modulate the levels positions by several 100 cm^{-1}. Further, up-conversion requires that the transfer probability for the second step (W_{SA}) be faster than radiative and non-radiative decay from the metastable level that is $W_{SA} > \tau^{-1}$ with τ the measured lifetime. W_{SA} is obtained from :

$$W_{SA} = \frac{2\pi}{\hbar} |< \psi_S^e \psi_A^0 | H_{SA} | \psi_S^0 \psi_A^e >|^2 \rho(E) \tag{145}$$

where the wave functions are simple products of single ion wave functions; $\rho(E)$ describes, the dissipative density of states due to the coupling with the lattice.

All cooperative processes including up-conversion can be considered as transitions between pair level for both ions as a whole. An electric dipolar transition would be forbidden for such two-center transition and one needs product wave functions corrected to first order to account for interaction for electrons of different centers [59] :

$$\psi_{pair} = \psi^0(S)\psi^0(A) - \sum_{s''\neq 0} \sum_{a''\neq 0} \cdot \frac{< s''a'' | H_{SA} | 00 >}{\delta_{s''} - 0 + \varepsilon_{a''} - 0} \psi_{s''}(S)\psi_{a''}(A) \tag{146}$$

for example for the ground state; s'', a'' are intermediate states for S and A; $\delta_{s''}$, $\varepsilon_{a''}$ are their corresponding energies. Then any one-photon transition in the cooperative description involves already four terms in the matrix element which cannot be reduced to (145).

APTE up-conversion does not correspond to the same order of approximation as cooperative processes which have to be considered practically only when the first type cannot take place.

Such is the case when real levels do not exists to allows energy transfers; it is the case for Yb^{3+} - Tb^{3+} up-conversion [58] or when concentration is to small to allows efficient transfer by energy diffusion between sensitizers. Then cooperative up-conversion is likely within clusters [60]. One may also look for crystal structures where the pairs clustering is built in [61].

A review of the cooperative processes has be given by Ovsyankin [58] in which the conditions for dissipative transfer weakly collectivized states are well stated.

IV.C. Overview of some results in ESA by energy transfers

1) Line-narrowing in n-photon summation as a mean to distinguish between APTE and cooperative processes [58] [62]. Here we discuss of a spectral narrowing effect which is a clear proof of APTE effect for 1.5 μm to visible up-conversion in Er^{3+} ions where both cooperative pair absorption [63] and APTE explanation [64] have been put forward.

Room temperature IR F-center laser excitation between 1.4 and 1.6 μm of 10% Er^{3+} doped vitroceramic and of YF_3 : Er leads to emission bands from near IR to U.V. Such emission may be ascribed to multiphoton ESA excitation respectively of order 1 to 5, either of the APTE or of the cooperative type as depicted respectively with energy levels of single ions (APTE) or with pair levels (Fig 22b and 22a).

Successive absorptions in Fig.22 (a) involve a combination of several J states. APTE effect, because of self matching by multiphonon processes, involves only J = 15/2 and J = 13/2 states.

Excitation spectra on Fig 23 show a striking behaviour : each spectrum presents the same spectral structure with clearly an increasing narrowing with process order. The structure reproduces the Stark structure of the $^4I_{15/2} \rightarrow {}^4I_{13/2}$ first excited terms as can be obtained by a diffuse reflectance spectra. This is a direct proof of the validity of the APTE explanation, since a cooperative effect should show the convolution of all J states involved in the multiple absorption between pair levels [62].

The spectral narrowing can be understood by a rate equation treatment where higher excited population are neglected in front of the lower ones in order to obtained a tractable development (weak excitation assumption).

The emitted power from an n-photon summation is given by :

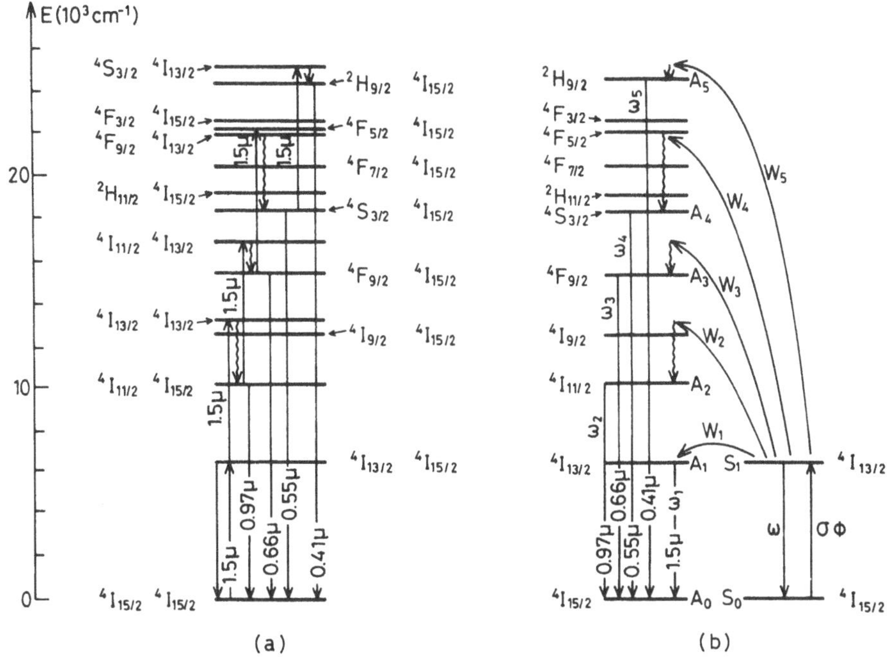

Fig.22. (a) ESA in pair levels (cooperative up-conversion); (b) ESA by APTE effect in Er^{3+}.

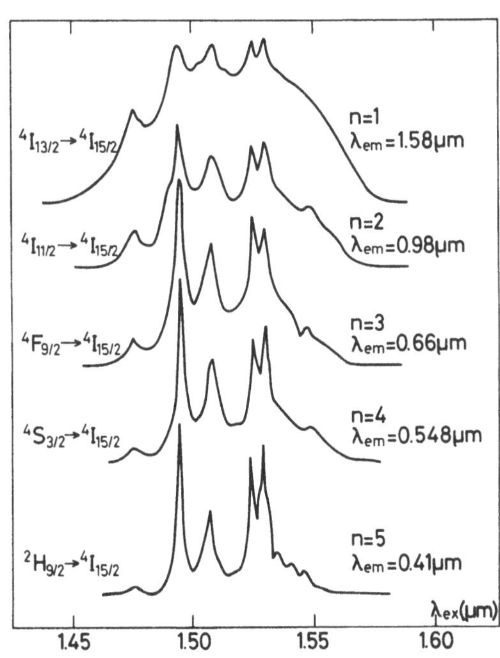

Fig.23. Line narrowing of excitation spectra of n-photon APTE ESA process.

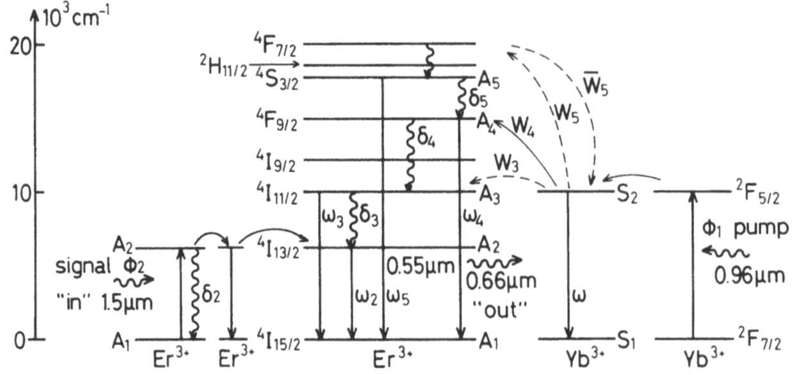

Fig.24. Energy scheme for 1.5 μm detector by APTE ESA pumped at 0.96 μm.

$$P_n(\lambda) = \frac{W_n \cdots W_2}{(\omega_{n-1} \cdots \omega_2)} P_1^n(\lambda) \tag{147}$$

with symbols of Fig 22 and $P_1(\lambda)$ the line shape of $^4I_{15/2} \rightarrow {}^4I_{13/2}$ absorption.

It was found that (147) is effectively verified except for the 2-photon case where the weak excitation assumption behind (147) may not be valid. [62].

2) Use of ESA to detect 1.5 μm radiation. The type of experiment above was reproduced and used as detector screens of 1.5 μm laser emission either using CaF_2 host [65] or YF_3 [53] and by adding Tm^{3+} to reduce the green emission in order to favor the red one. We also obtained a high detectivity 1.5 μm detector (NEP = 10^{-11} W/Hz $^{1/2}$) by adding Yb^{3+} and using an auxiliary steady pumping source at 0.95 μm as shown on Fig 24. [66].

3) Negative roles in applications. Besides the positive role in pumping laser in an anti-stokes way as will be presented in next subchapter, we have seen that APTE effect was found by its negative role in Yb^{3+} - Er^{3+} glass laser at 1.5 μm : being an anti-stokes process it does induce reabsorption from excited states. It has been shown to induce reabsorption in the high concentration Nd^{3+} stoechiometric laser materials [67] : up-conversion feeds non-radiative levels at the expense of the laser efficiency. In Nd P_5O_{14} materials such effect begins to be noticeable when population inversion reaches 10%. We have observed up-conversion saturation in CW tunable laser with $MgF_2 : Ni^{2+}$ [68]; however it seems the ions coupling plays an indirect role, up-conversion being due to reabsorption between single-ion levels. Also we demonstrated it to be a loss in the $LiYF_4 : Er^{3+}$ laser at 2.8 μm [56].

In an other instance where high excitation is used, up-conversion losses and saturation are also found : it is in phosphors for projection cathode ray tubes. Besides ground states depletion saturation, saturation has been recognized to be due to APTE ESA at higher ion concentrations [69] [70] as in Zn_2SiO_4 : Mn, YAG : Tb. Also in electroluminescence phosphors such as ZnS : Mn, APTE has been advocated to explain saturation [71].

4) Positive role in laser anti-Stokes pumping. Because most of the luminescence processes are Stokes type in nature ($\lambda_{exc} < \lambda_{emission}$) lasers are also generally of the Stokes type : $\lambda_{pump} < \lambda_{emiss}$.

Recently, because of the limitation in semiconductor laser diodes to the red or I.R. part of the spectrum, it appeared interesting to look for up-conversion pumped laser : $\lambda_{pump} > \lambda_{emiss}$. This would provide solid state laser operation in the visible using the now well developped IR semi-conductor pumping source; such anti-Stokes laser could be used for high density compact disc memories.

The first results in APTE anti-Stokes laser were obtained for the Yb^{3+} - Er^{3+} and Yb^{3+} - Ho^{3+} systems in BaY_2F_8 crystals [72] a few years after the finding of the APTE effect. Anti-Stokes lasers were then obtained under pulsed flash pumping. More recently CW up-conversion laser in the green has been obtained under IR laser diode pumping [73] of $LiYF_4$: Er^{3+} giving a renewed interest in APTE up-conversion.

Besides this, ESA seems to play a decisive role in the obtention of CW laser action at 2.7 µm in $LiY_{1-x}Er_xF_4$ with large Er^{3+} concentration : due to the long $^4I_{13/2}$ life time, self pulsing only should be obtaine whereas pure CW is in fact observed [56].

Quite recently CW visible up-conversion has been obtained in fibers at room temperature [74]. Though the concentration is small (~ 1000 ppm) the operation may be limited to the existence of clusters of R.E as demonstrated by the existence of energy transfers even at such low concentration [75]. Then in glass fiber, even at low concentration there is always a doubt wether ESA is of the single ion type or of the APTE type. Because of its non-linear nature ESA is found to be very effective in fiber due to the confinement of excitation within the restricted core diameter particularly in single mode fibers. In some case it was even found that ESA can be much larger than ground state absorption which reduces amplification or increases laser threshold in such system.[76].

Due to its efficiency and because of phonon assisted energy transfer, ESA by APTE is very general, though sometimes believed to be of the cooperative type. It is so much likely that considered matrices have smaller phonon energy giving long living metastable levels in rare-earth system. It has then a pervading nature as shown by the many studies that have been devoted to such effect recently in fluoride systems [77] even in glass form [78] [79]. This very general effect appears more efficiently when excitation is concentrated in space, time and spectral domain because of its non-linear nature.

V . CONCLUSION

With the advent of laser excitation sources, the behaviour of ions in solids may show coherent processes which were completely negligeable under incoherent excitation. From this study we understand that because the laser effect itself may be obtained under incoherent pumping, independantly of dephasing consideration, the building up of coherence in the laser material is not based on the coherence of the field-atom interaction but on the macroscopic filtering effect of the multifolded cavity on ASE.

To the opposite, SR and SF are coherent processes right or very soon at the beginning of interaction, memory of the phase being physically linked with the synchronous exchange of energy between field and atom. Though in the laser effect one has to consider spontaneous and induced emissions, theses processes may be considered somewhat separatedly. We have seen how SR and SF are the "inertial" remnants of induced emission before their "dissipation" by spontaneous emission or other dephasing processes. So we would like to stress that, spontaneous and induced emissions cannot be separated as usually accepted in text books, except in the limiting case when dephasing is fast. Effects of this principle are ASE and HL as compared respectively to superfluorescence (SF) and Resonant Raman Scattering (RRS).

Finally, we have shown how ion-ion interactions lead to enhanced ESA by energy transfers (APTE effect). This last process, is to cooperative ion-ion up-conversion, the "damped memory" case for resonant ion-ion interaction where energy could be exchanged synchronously between both ions, in analogy with what we saw for the field-ion interaction.

ACKNOWLEDGEMENT

We would like to thank particularly J. Chavignon for kindly typing the text of this lecture.

REFERENCES

[1] R.H. Dicke, Phys. Rev. 93 99 (1954).
[2] R. Bonifacio and L.A. Lugiato, Phys. Rev. A11 1507 (1975).
[3] T. Waite, J. Appl. Phys. 35 1680 (1964).
[4] M. Weissbluth, Photon - atom Interactionn, Academic Press, London, (1989).
[5] A. Yariv, Quantum Electronics, Wiley, New York, (1967).
[6] L. Allen and J.H. Eberly, Optical resonance and two-levels Atoms, Wiley, New York (1975).
[7] S. Stenholm, Physics Reports 6 1 (1973).
[8] H. Haken, Light, Vo 1 waves, photons, atoms, North-Holland, Amsterdam (1981).

[9] E.T. Jaynes and F.W. Cummings, Proc. IEEE 51, 89 (1963).
[10] M. Gross and S. Haroche, Physics Reports, 93 301 (1982).
[11] R.M. Hill, D.E. Kaplan, G.F. Herrmann, and S.K. Ichiki, Phys. Rev. Lett., 18 105 (1967).
[12] J.C. Mac Gillivray and M.S. Feld, Phys. Rev. 14A 1169 (1976).
[13] N. Skribanowitz, I.P. Herman, J.C. Mac Gillivray and M.S. Feld, Phys. Rev. Lett. 30 309 (1973).
[14] D. Polder, M.F.H. Schuurmans and Q.H.F. Vrehen, Phys. Rev. 19A 1192 (1979).
[15] N.A. Kurnit, I.D. Abella, and S.R. Hartman, Phys. Rev. Lett. 13 567 (1964).
[16] Y. Yamamoto, S. Machida, Y. Horikoshi, K. Igota and G. Björk, Optic Com. 80 337 (1991).
[17] L.O. Schwan, J. of Lum. 48/49 289 (1991).
[18] R. Florian, L.O. Schwan, and D. Schmid, Solid State Commun. 42 55 (1982).
[19] M.F.H. Schuurmans, Optics Comm. 34 185 (1980).
[20] M.S. Malcuit, J.J. Maki, D.J. Simkin, and R.W. Boyd, Phys. Rev. Lett. 59 1189 (1987).
[21] R.F. Malikov, and E.D. Trifonov, Optics Comm. 52 74 (1984).
[22] F.T. Arecchi and E. Courtens, Phys. Rev. 2A 1730 (1970).
[23] R. Friedberg and S.R. Hartmann, Phys. Letters 37A 285 (1971).
[24] M.F.H. Schuurmans and D. Polder, Phys. Letters 72A 306 (1979).
[25] see Ref. [50], p.365.
[26] R. Florian, L.O. Schwan and D. Schmid, Phys. Rev. 29A 2709 (1984).
[27] A. Schiller, L.O. Schwan and D. Schmid, J. of Lum. 38 243 (1987).
[28] S.N. Andrianov, P.V. Zinovev, Y.V. Malyukin, Y.V. Naboikin, V.V. Samartser, N.B. Silaeva, and Y.E. Seibut, Sov. Phys. JETP 64 1180 (1986).
[29] W. Gellermann, Y. Yang, and F. Luty, Optics Comm. 57 196 (1986); Y. Yang and F. Luty, J. of Lum. 40/41 565 (1988).
[30] V.M. Markushev, V.F. Zolin, and C.M. Briskina, Sov. J. Quant. Electron. 16 281 (1986).
[31] F. Auzel, S. Hubert, and D. Meichenin, Europhys. Lett. 7 459 (1988).
[32] S. Hubert, D. Meichenin and F. Auzel, J. of Lum. 45 434 (1990).
[33] G.K. lin, M.F. Joubert, R.L. Cone, and B. Jaquier, J. of Lum. 38 34 (1987).
[34] D. Ronarch, M. Guibert, F. Auzel, D. Meichenin, J.Y. Allain and H. Poignant, Electron. Lett. 27 511 (1991).
[35] N.W. Carlson, D.J. Jackson, A.L. Schawlow, M. Gross and S. Haroche, Optics Comm. 32 350 (1980).
[36] K.K. Rebane and P.M. Saari, J. of Lum. 12/13 23 (1976).
[37] K.K. Rebane, in Luminescence of Inorganic Solids, (B. Di Bartolo, ed.) Plenum, NYC (1978), p. 495.
[38] C. Gouedard, D. Husson, C. Sauteret, F. Auzel, and A. Migus, IAEA Conf. on Drivers for Inertial Confinemant Fusion, Osaka (Japan) April 15-19 1991, paper E04.
[39] S. Singh, R.B. Chesler, W.H. Grokiewicz, J.R. Potopowicz, and L.G. Van Vitert, J. Appl. Phys. 46 436 (1975).
[40] F. Auzel, Mat. Res. Bull. 14 223 (1979).
[41] J. Dexpert - Ghys and F. Auzel, J. Chem. Phys. 80 4003 (1984).
[42] Y. Haas and G. Stein, Chem. Phys. Lett. 11 143 (1974).
[43] F. Auzel, in "Advances in non-radiative processes in solids", (B. Di Bartolo, ed.) Plenum, NYC (1991).
[44] Y. Toyozawa, J. Phys. Soc. Jap. 41 400 (1976).
[45] V.V. Hizhnyakov and I.J. Tehver, J. of Lum. 18/19, 673 (1979).
[46] F. Auzel, in Spectroscopy of Solid State Laser type Materials (B. Di Bartolo, ed.), Plenum, NYC (1987), p.293.
[47] F. Auzel, in Radiationless processes (B. Di Bartolo, ed.), Plenum, NYC (1980), p. 213.
[48] F. Auzel, C.R. Acad. Sci. (Paris) 263 819 (1966).
[49] E. Snitzer and R. Woodcock, Appl. Phys. Lett. 6 45 (1965).
[50] F. Auzel, Ann. Telecom. (Paris) 24 363 (1969).
[51] F. Auzel, C.R. Acad. Sci. (Paris) 262 1016 (1966).
[52] V.V. Ovsyankin and P.P. Feofilov. Sov. Phys. JETP Letters 4 317 (1966).

[53] J.P. Van der Ziel, L.G. Van Vitert, W.H. Grodkiewicz and R.M. Mikulyak, J. Appl. Phys. 60 4262 (1986).
[54] G.J. Kintz, R. Allen and L. Esterowitz, Appl. Phys. Lett. 50 1553 (1987).
[55] F. Auzel, J. of Lum. 31/32, 759 (1984).
[56] F. Auzel, S. Hubert and D. Meichenin, Appl. Phys. Lett. 54 681 (1989).
[57] D.L. Dexter, J. Chem. Phys. 21 836 (1953).
[58] V.V. Ovsyankin, in Spectroscopy of solids containing Rare-earth ions (A.A. Kaplyanskii and R.M. Mac Farlane, eds.) North Holland, Amsterdam (1987) p.405.
[59] M. Stavola and D.L. Dexter, Phys. Rev. B20 1867 (1979).
[60] J.C. Vial, R. Buisson, F. Madeore and M. Poirier, J. Phys. 40 913 (1979).
[61] N.J. Cockroft, G.D. Jones and R.W.G. Syme, J. of Lum. 43 275 (1989).
[62] F. Auzel, in Rare-Earth Spectroscopy, (eds. B. Trzebiatowska, J. Legendziewicz and W. Strek), World Scientific, Singapore (1985), p.502.
[63] M.R. Brown, H. Thomas, J.M. William and R.J. Woodwards, J. Chem. Phys. 51 3321 (1969).
[64] L.F. Johnson, H.J. Guggenheim, J.C. Rich and F.W. Ostermayer, J. Appl. Phys. 43 1125 (1970).
[65] S.A. Pollack, D.B. Chang, I-Fu Shih and R. Tseng, Appl. Optics 26 4400 (1987).
[66] F. Auzel, P.A. Santa-Cruz and G.F. de Sà, Revue Phys. Appl. 20 273 (1985).
[67] M. Blätte, H.G. Damielmeyer and R. Verich, Appl. Phys. (Germ.) 1 275 (1973).
[68] R. Moncorgé, F. Auzel and J.M. Breteau, Phil. Mag. B51 489 (1985).
[69] D.M. de Leeuw and G.W.'t. Hooft, J. of Lum. 28 275 (1983).
[70] W.F. Van der Weg and M.W. Van Tol, Appl. Phys. Lett. 38 705 (1891).
[71] D.H. Smith, J. of Lum. 23 209 (1981).
[72] H.J. Guggenheim and L.F. Johnson, Appl. Phys. Lett. 19 44 (1971).
[73] F. Tong, W.P. Risk, R.M. Mac Farlane and W. Lenth, Electron. Lett. 25 1389 (1989).
[74] J.Y. Allain, M. Monerie, and H. Poignant, Electron. Letters, 27 261 (1990).
[75] D.C. Hanna, R.M. Percival, I.R. Perry, R.G. Smart, A.C. Tropper, Electron. Letters 24 1068 (1988).
[76] M. Monerie, T. Georges, P.L. François, J.Y. Allain, D. Neveux, Electron. Letters 26 320 (1990).
[77] C.Y. Chen, W.A. Sibley, D.C. Yeh and C.A. Hunt, J. of Lum. 43 185 (1989).
[78] D.C. Yeh, W.A. Sibley, M. Suscavage and D.G. Drexhage, J. Appl. Phys. 62 266 (1987).
[79] M.A. Chamarro and R. Cases, J. of Lum. 42 267 (1988).

ADVANCES IN SENSITIZATION OF PHOSPHORS

Bruno Smets

Philips Lighting
5600 JM Eindhoven
The Netherlands

ABSTRACT

Sensitization of the luminescence is mainly applied in fluorescent lamp phosphors, in this way an efficient UV to visible conversion can be realized. The processes, underlying the transfer of the excitation energy from sensitizer to activator, will be discussed in detail. Special attention is paid to the energy migration, as found with concentrated gadolinium phosphors. Thanks to the profound knowledge on the luminescence of rare-earth phosphors, gathered over the last decades, it has become possible to design fluorescent lamp phosphors, which nearly completely fulfil all the requirements as imposed by their application.

I. INTRODUCTION

Luminescent materials are widely applied nowadays. They are mainly used in cathode-ray tubes, in intensifying screens for x-ray diagnostics and in fluorescent lamps. Under electron and x-ray excitation electron - hole pairs are generated. The recombination energy of such an electron - hole pair is eventually transferred to an activator ion present as a dopant in the phosphor, resulting in the emission of visible radiation. For UV photons the energy in most cases is too small to bridge the band gap. Therefore the radiation must directly be absorbed by the activator ion itself or by a sensitizer, which is able to transfer the excitation energy to the activator ion.

The backbone of a fluorescent lamp is the low pressure mercury discharge, emitting mainly 254 nm radiation. This UV radiation is converted into visible radiation by the phosphor layer adherent to the inside of the lamp bulb. A typical example of such a phosphor is $CaWO_4$. Its absorption peaks around 240 nm and its emission around 410 nm (Fig.1). The major drawback of this phosphor is its low quantum efficiency, which only equals 55% at room temperature, the temperature of operation for fluorescent lamps.

Optical Properties of Excited States in Solids
Edited by B. Di Bartolo, Plenum Press, New York, 1992

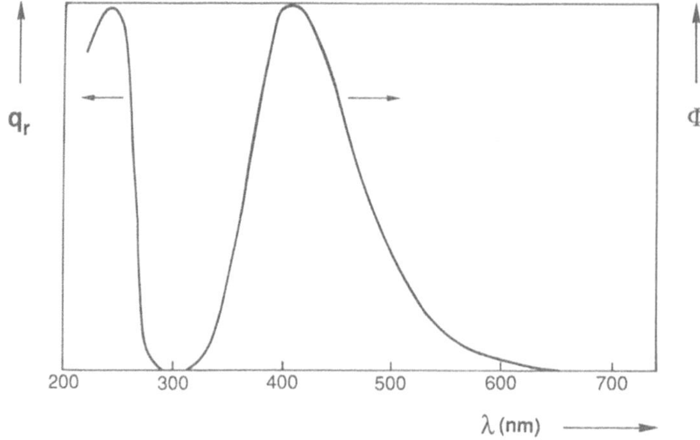

Fig.1
Emission spectrum (Φ equals the radiant power per unit wavelength interval) and ecxitation spectrum (q_r equals the relative quantum output in arbitrairy units) of $CaWO_4$ at room temperature.

The luminescence properties of $CaWO_4$ can easily be understood on the basis of the configurational coordination diagram. Both the absorption and the emission are transitions between the $^1A_1$ ground state and the $^3T_1$ excited state of the WO_4^{2-} group, showing a nearly tetrahedral configuration [1]. In Fig.2 the potential energy of the $^1A_1$ and $^3T_1$ state is plotted as a function of the tungsten oxygen distance r. Note that the equilibrium distance of the excited state is shifted towards much larger distances. Because the Born - Oppenheimer approximation is valid for optical transitions, the latter transition can be represented by vertical lines in the

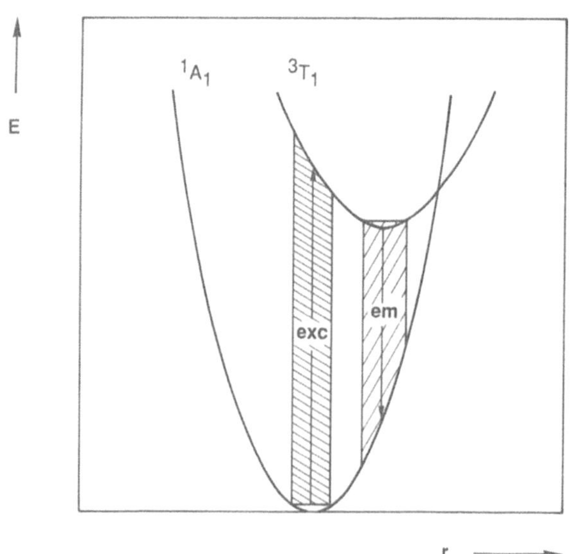

Fig.2
Schematic configurational coordination diagram of $CaWO_4$.

configurational coordination diagram. At zero degrees Kelvin only the lowest vibrational level of the ground state will be populated, resulting in small differences in the tungsten oxygen distance between the different tungstate groups at the moment the excitation energy is absorbed. In view of the offset between the $^1A_1$ and $^3T_1$ state the differences in energy absorbed will be substantial and band absorption will be observed. After excitation there will be a strong relaxation towards the zero-phonon level of the excited state, followed by emission. Also in this case a band emission is found. Due to this relaxation the energy difference between excitation and emission, the Stokes shift, will be substantial.

During the excitation process vibrational levels of the excited state are populated with at an energy, which comes close to the energy of the crossing point between the two parabolas, inducing non-radiative transitions from the $^3T_1$ state to the $^1A_1$ state. Consequently the efficiency of $CaWO_4$ will be low, even at low temperatures. At higher temperatures the situation becomes even worse, due to the thermal activation high lying vibrational levels will indeed become populated. This decrease in probability of radiative processes with increasing temperature is often called thermal quenching.

In the case of Eu^{3+} phosphors a completely different situation is found. For trivalent europium the equilibrium distance of the ^7F ground state and the $^5D_0$ excited state is quite the same (Fig.3). This results in line emission (Fig.4) and absorption, showing no Stokes shift, a very high efficiency and a high quenching temperature. In the case of a two-level system ultraviolet to visible conversion and a high efficiency are often incompatible. An efficient UV to visible conversion can, however, be

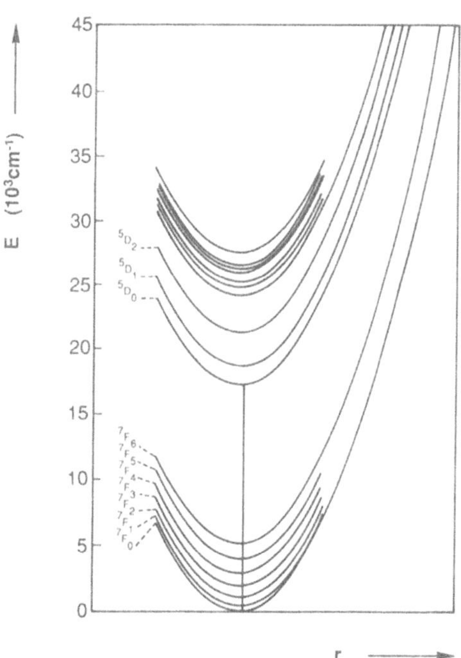

Fig.3
Configurational coordination diagram of the $4f^6$ states of Eu^{3+}.

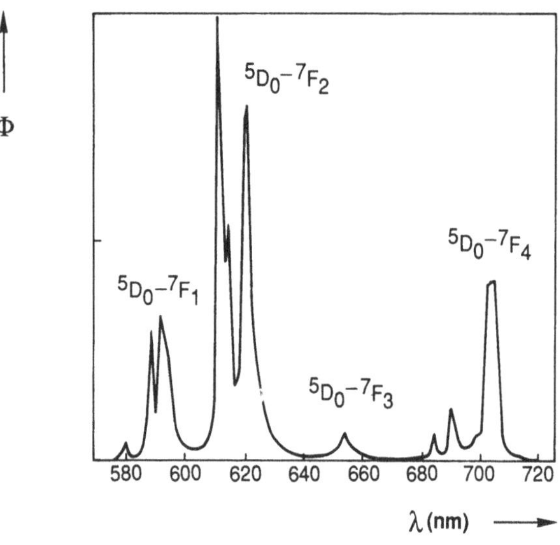

Fig.4
Emission spectrum of $Ba_2Gd_{1.8}Eu_{.2}Si_4O_{13}$

obtained when other energy levels are involved, either belonging to the same luminescent center or to an other luminescent center, i.e.: the sensitizer ion. In efficient lamp phosphors sensitization of the luminescent level therefore will be a prerequisite.

In the first part of my contribution we will consider in more detail the luminescence of isolated ions and the energy transfer between these centers. In recent years energy migration has gained a lot of scientific as well as technological interest. The energy migration over Ln^{3+} sublattices will be discussed in the second part of this series of lectures. The final part will be devoted to the application of phosphors in fluorescent lamps.

II. EFFICIENCY OF LUMINESCENT CENTERS

II.A. Dynamic Jahn-Teller Effect

It was outlined in the introduction that a large Stokes shift is not compatible with a high efficiency at room temperature, at least if only two levels are involved in the luminescence process. There is however one exception to this rule, when the dynamic Jahn-Teller effect is acting on the excited state a much more complex behaviour is found. This effect is well known for s^2 ions in alkali halides [2], where the luminescent center is octahedrally coordinated with six halogen ions. Due to coupling of the excited state with vibrational modes the symmetry of the AX_6 octahedron is lowered, resulting in the crystal field stabilization of the excited state. Recently it was found that the dynamic Jahn-Teller effect also governs the luminescence of $Ba_6Y_2Al_4O_{15}:Sn^{2+}$ [3] and of $YPO_4:Sb^{3+}$ [4].

In Fig.5 the excitation spectrum of $Ba_6Y_2Al_4O_{15}:Sn^{2+}$ is shown. The

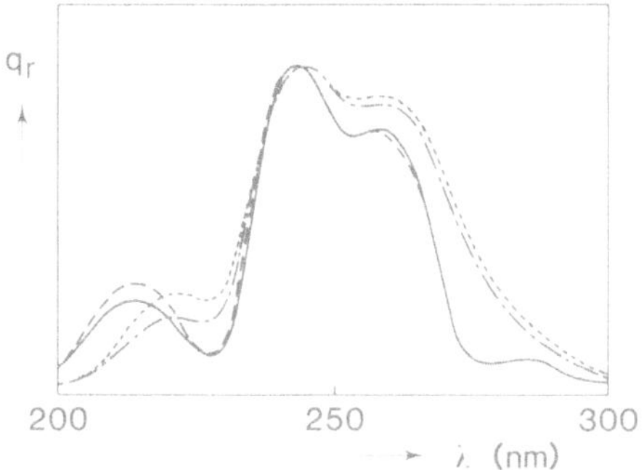

Fig.5

Excitation spectrum of the 460 nm emission of $Ba_6Y_2Al_4O_{15}$: $Sn^{2+}_{0.006}$; ——— 77k and -.-.-.- at room temperature, and of $Ba_6Y_2Al_4O_{15}$: $Sn^{2+}_{0.012}$ — — — at 77k and - - - - - at RT.

transition located around 250nm corresponds with the transition from the $^1S_0$ ground state to the $^3P_1$ first excited state of the Sn^{2+} ion. This band shows a pronounced doublet structure, its shape being highly temperature dependent. The energy splitting ΔE(T) originates from the dynamic Jahn-Teller effect, and can be expressed as [2]:

$$\Delta E(T)=\Delta E(0)\left\{\coth\frac{h\nu_{eff}}{2kT}\right\}^{1/2} \qquad (1)$$

where ΔE(0) is the Jahn - Teller splitting at zero Kelvin and $h\nu_{eff}$ the average phonon energy of the vibrations involved. Such a behaviour is indeed found for $YPO_4:Sb^{3+}$, as is shown in Fig.6. In contrast to the alkali

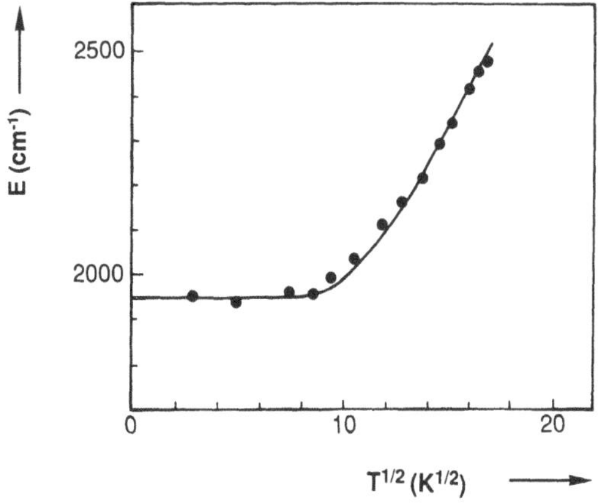

Fig.6

Jahn-Teller splitting of the lowest excitation band of $YPO_4:Sb^{3+}$ [4]. The drawn line is the best fit to equation (1).

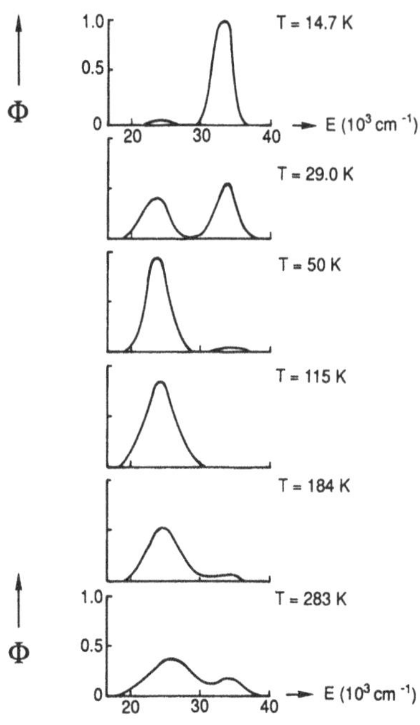

Fig.7
Emission spectrum of $YPO_4:Sb^{3+}$ at different temperatures [4]. The excitation wavelength equals 245 nm.

halides the Sb^{3+} ion in this case does not enter a site with O_h symmetry but one with D_{2d} symmetry.

On excitation of $YPO_4:Sb^{3+}$ in the $^3P_1$ antimony level two emission bands are found. Their intensity ratio turns out to be highly temperature

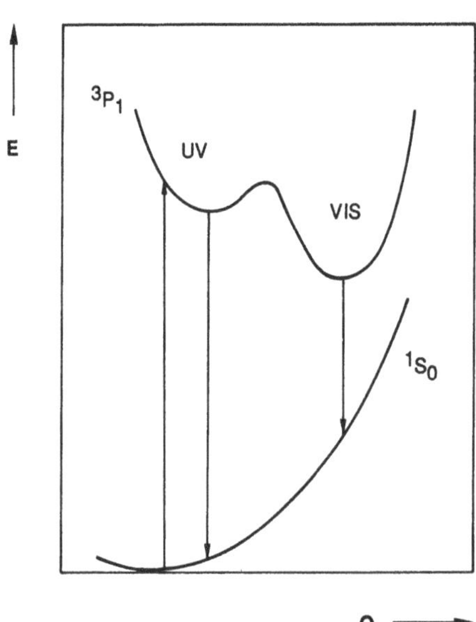

Fig.8
Schematic configurational coordination model for $YPO_4:Sb^{3+}$.

dependent (Fig.7). These data can be explained on the basis of the presence of two minima in the adiabatic potential energy surface of the relaxed excited state. These minima arise from the Jahn-Teller interaction of the ^3P state with vibrational modes. Whether one or two minima will occur in the adiabatic potential energy surface will largely be determined by the ratio between the spin-orbit coupling and the electron-lattice interaction. Therefore in the case of Pb^{2+} and Bi^{3+}, showing more pronounced spin-orbit coupling, the Jahn-Teller effect has not been observed until now in emission [5].

In Fig.8 a schematic configurational coordination diagram is drawn for YPO_4:Sb^{3+}, showing two distinct minima. On excitation in the ^3P level, the system will rapidly relax to the highest of the two minima at low temperature, resulting in UV emission. At higher temperatures the energy barrier, separating both minima, can be surmounted and the second visible emission band will also be observed. Its intensity will increase with increasing temperature, until the first band has completely vanished. At still higher temperatures both minima will become thermally populated and the UV emission band reappears.

For the isostructural compounds $LuPO_4$ and $ScPO_4$ the dynamic Jahn-Teller effect will also act on the ^3P excited state. In table 1 the quenching temperature T_q and the Stokes shift are given for the three $LnPO_4$ compounds with the zircon structure. T_q is defined as the temperature at which the total quantum efficiency of the Sb^{3+} emission starts to decrease. In contrast to normal two-level systems the quenching temperature increases with increasing Stokes shift in the case of s^2 ions. Therefore a high efficiency and a large Stokes shift are compatible for luminescent centers showing Jahn-Teller interactions.

Table 1
Luminescence properties of zircon-type $LnPO_4$:Sb^{3+}

	T_q (K)	Stokes shift (cm^{-1}) of visible emission
YPO_4	270	15400
$LuPO_4$	240	14600
$ScPO_4$	130	12400

II.B. Multi-Level System

Normally a large difference in equilibrium distance between excited and ground state will induce radiationless transitions and consequently a low yield. When more levels are involved a different situation is found, this is the case for Ln^{3+} ions. The energy levels belonging to the well shielded $4f^n$ configuration have the same equilibrium distance. In the case of $4f^{n-1}5d^1$ and charge-transfer states a different equilibrium distance will be found, because outer 5d electrons or electrons from the surrounding anions are involved.

In Fig.9 the energy level diagram is shown for Eu^{3+} in Y_2O_3. The

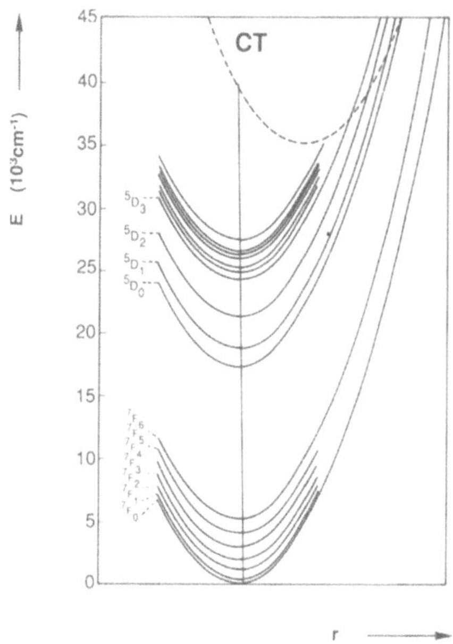

Fig.9
Configurational coordination diagram for $Y_2O_3:Eu^{3+}$.

charge-transfer band is an allowed transition located in the ultraviolet. In view of the offset between the charge-transfer level and the $4f^6$ levels non-radiative transitions from the charge-transfer level to the $^5D_3$, $^5D_2$, $^5D_1$, and $^5D_0$ state will be highly probable. This is indeed confirmed by the emission spectrum of $Y_2O_3:Eu^{3+}$ (Fig.10). If the charge-transfer band is

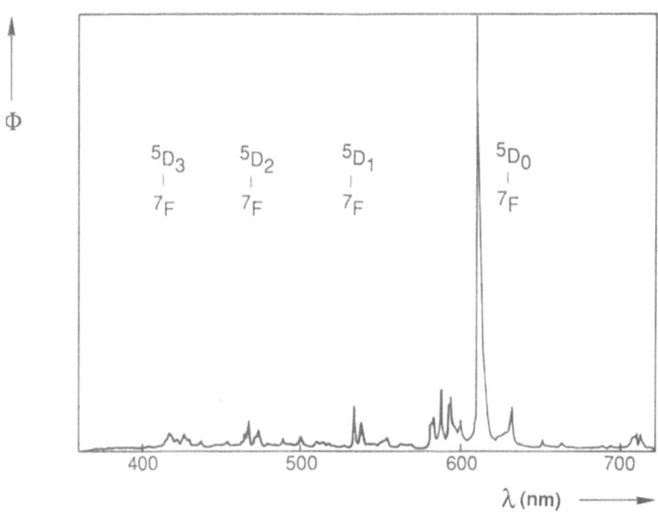

Fig.10
Emission spectrum for $Y_2O_3:Eu^{3+}_{0.002}$ on excitation in the charge-transfer band.

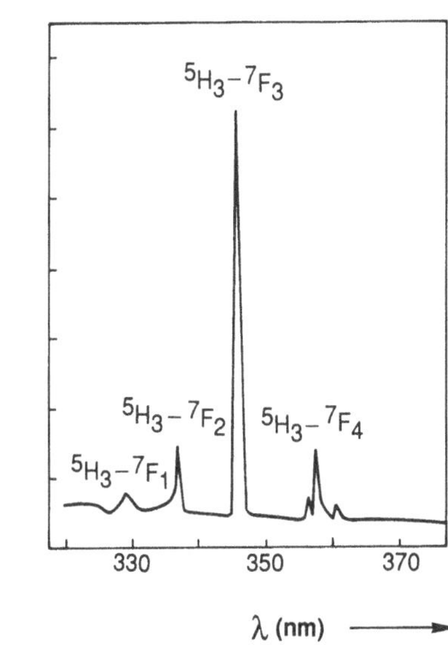

Fig.11

Eu^{3+} UV emission on excitation at 250 nm observed for $NaGdF_4{:}Ce_{0.01},Eu_{0.01}$ by Kiliaan et al. [6].

located at even higher energies, as is the case in $NaGdF_4{:}Eu^{3+}$, also higher $4f^6$ levels can be populated. For this compound even emission from the $^5H_3$ level can be observed (Fig.11). The transitions within the $4f^6$ states will of course show a high efficiency in view of their small Stokes shift.

For Tb^{3+} phosphors a quite similar behaviour is found. Absorption will take place in the $4f^75d^1$ band, this also is an allowed transition. Both the terbium $^5D_3$ and $^5D_4$ level will be feeded by the $4f^75d^1$ level, resulting in blue and green luminescence (Fig.12).

II.C. Multi-Phonon Transitions

The two terbium spectra in Fig.12 show a completely different ratio between the blue $^5D_3$ and the green $^5D_4$ emission. The reason for this difference is that in the case of the borate phosphor, although the minima in the parabola are located at the same value of r, non-radiative transitions will occur. The energy difference between both terbium levels is indeed taken up by lattice vibrations [7]. Due to this multi-phonon transition the $^5D_4$ level is populated at the expense of the $^5D_3$ level.

The probability W_{nr} of this type of non-radiative transition can be described by an energy gap law [8]:

$$W_{nr} = \beta \exp\left[-\alpha(\Delta E - 2h\nu_{max})\right] \qquad (2)$$

where ΔE equals the energy gap to be bridged and $h\nu_{max}$ the energy of the promoting vibrational mode. This equation only holds when at least three phonons are involved in the process. In table 2 typical values are given for α, β and $h\nu_{max}$.

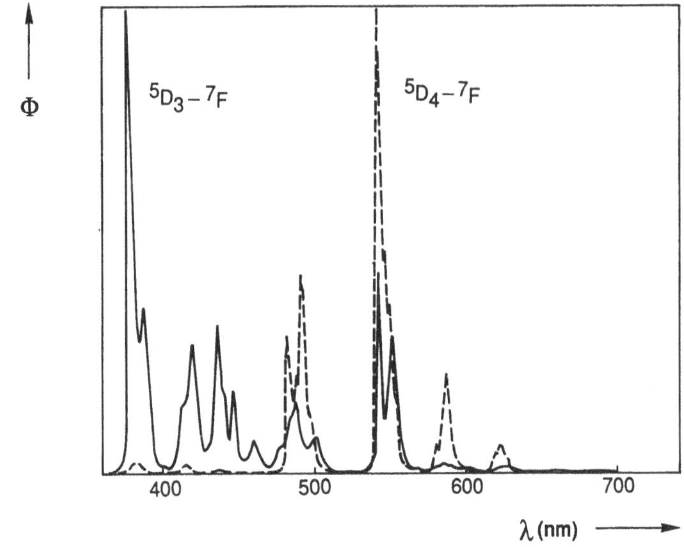

Fig.12

Emission spectrum of ——— $(Ba,Ca)_2SiO_4:Tb_{0.01}$ and of — — — $(Gd,La)B_3O_6:Bi_{0.03},Tb_{0.01}$ under 254 nm excitation.

The energy difference between the $^5D_3$ and $^5D_4$ level equals about 5600 cm^{-1}. When the appropriate values are inserted in equation 2, it follows that W_{nr} equals 1800 s^{-1} in the borates and 10 s^{-1} in the silicates. These data should be compared with the probability for radiative decay, which equals about 1250 s^{-1} for the $^5D_3$ level [12]. In the case of the borate phosphor the probability for radiative and non-radiative decay is roughly the same, in the case of the silicate phosphor however the probability for radiative decay of the $^5D_3$ level is much higher. Therefore in the latter compound emission from the $^5D_3$ level will be observed.

The non-radiative rate decreases exponentially with the energy difference to be bridged. For ΔE values up to about 5000 cm^{-1} the rate

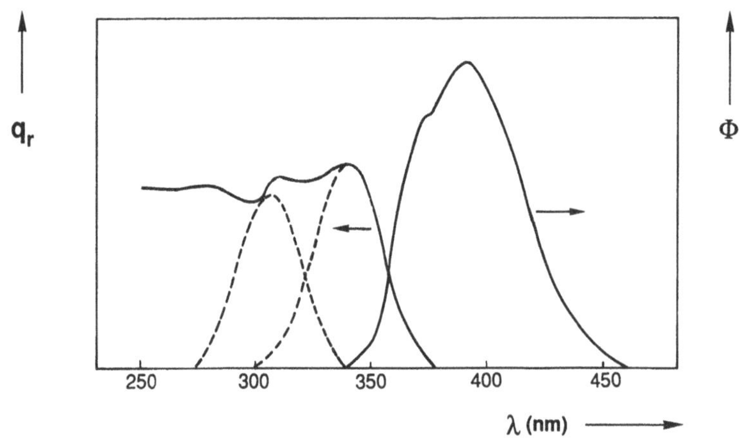

Fig.13

Cerium emission and excitation spectrum for $KCaLa_{.4}Ce_{.6}(PO_4)_2$ [15].

Table 2
Energy gap law parameters

matrix	β (s^{-1})	α (cm)	$h\nu_{max}$ (cm^{-1})
borates [9]	7.8 10^7	3.8 10^{-3}	1400
silicates [10]	7.8 10^7	4.7 10^{-3}	1100
Y_2O_3 [11]	4.3 10^6	3.8 10^{-3}	550

might equal the one of the competiting radiative transitions. For larger gaps as found in $Gd^{3+}(^6P-^8S)$, $Tb^{3+}(^5D_4-^7F)$ and $Eu^{3+}(^5D_0-^7F)$ the radiative processes will largely dominate.

III ENERGY TRANSFER

III.A. Resonant Transfer

When two luminescent centers are close together, the excitation energy may be transferred from one center, the sensitizer ion S, to the second center, the activator ion A. This process has been considered in detail by Dexter [13], following the classical work of Forster [14]. A stringent condition for transfer is that both S and A are in resonance, i.e.: the energy difference between their ground and excited state should be equal. Secondly S and A should show some type of interaction with one another in order that energy transfer can take place. The probability for transfer is given as:

$$P_{SA} = \frac{1}{h} \left| < S, A * \mid H_{SA} \mid S*, A> \right|^2 \int f_S(E)\ f_A(E)\ dE \qquad (3)$$

The integral stands for the spectral overlap between the emission of the sensitizer and the absorption of the activator, H_{SA} is the interaction Hamiltonian, $|S>$ or $|A>$ and $|S*>$ or $|A*>$ are the electronic ground state and excited state of sensitizer and activator respectively.

Transfer can either occur by multipolar or by exchange interaction. The latter type of interaction requires wave function overlap and therefore is confined to short distances. The exact distance dependence is largely determined by the type of interaction. In the case of exchange interaction the distance dependence is given as:

$$P^{ex}_{SA} = C\ e^{-2R/L}, \qquad (4)$$

where L equals the effective Bohr radius of the sensitizer. For multipolar interactions the following relation holds:

$$P^{mm}_{SA} = \frac{1}{\tau_s} \left(\frac{R_c}{R}\right)^n, \qquad (5)$$

where n=6 for dipole-dipole interaction, n=8 for dipole-quadrupole interactions and n=10 for quadrupole-quadrupole interactions. τ_s equals the radiative lifetime of the sensitizer excited state and R_c is the critical distance, defined as the distance for which the transfer probability equals the probability for radiative decay of S.

1. Multipole Interaction. The cerium-terbium transfer in $KCaLa(PO4)_2$ is a typical example of dipole-dipole interaction [15]. The Ce^{3+} emission in the ultraviolet (Fig.13) overlaps very well with a number of Tb^{3+} absorption lines (Fig.14). the transfer probability between Ce^{3+} and Tb^{3+} can be expressed as:

$$P_{SA}^{dd} = \frac{3\hbar^4 c^4}{4\pi K^2} \frac{Q_a}{R^6} \frac{1}{\tau_s} \int \frac{f_S(E)\ f_A(E)}{E^4} dE \qquad (6)$$

Q_a equals the absorption cross section of the activator, c the velocity of light and K the dielectric constant of the host lattice. Because the critical distance R_c is defined as the distance for which P_{SA} = $1/\tau_S$, equation 6 can be rewritten as:

$$R_c^6 = \frac{3\hbar^4 c^4}{4\pi K^2} Q_a \int \frac{f_S(E)\ f_A(E)}{E^4} dE \qquad (7)$$

The overlap integral can be deduced from the spectral data. Because transistion within the $4f^8$ configuration of Tb^{3+} are largely forbidden Q_A will be low, and R_c only equals 5 Å.

On the basis of the crystallographic data and the value of R_c derived above it is expected that the transfer efficiency will equal 0.16 for $KCaLa_{.9}Ce_{.05}Tb_{.05}(PO_4)_2$, in excellent agreement with the experimental data. When the terbium content is increased, the probability increases that a

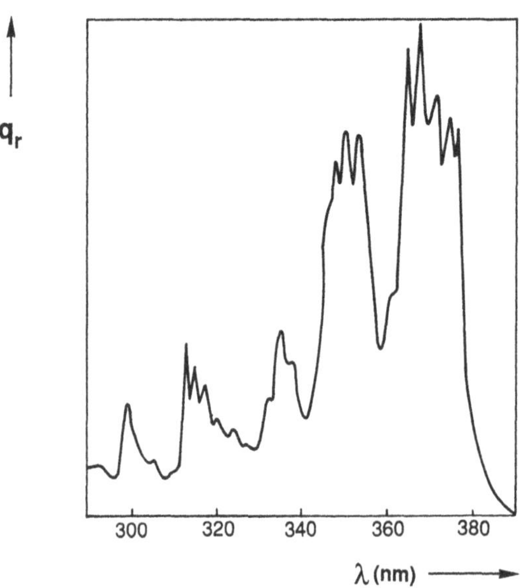

Fig.14
$^5D_4$ - $^7F_5$ terbium excitation spectrum of $KCaLa_{.95}Tb_{0.05}(PO_4)_2$ [15].

terbium ion is present within a distance of 5 Å from a cerium ion, resulting in more efficient energy transfer.

Oddly enough a higher transfer efficiency is also found when the cerium content is increased, the latter equals about 0.25 for the composition $KCaLa_{.35}Ce_{.6}Tb_{.05}(PO_4)_2$. As can be seen from Fig.13 for this cerium content their exists a spectral overlap between the cerium emission and absorption (or excitation) band, the latter being an allowed transition. Therefore the excitation energy is also able to migrate through the cerium sublattice, resulting in a more efficient overall transfer from cerium to terbium.

2. Exchange Interaction. Often it is fairly difficult to determine the exact nature of the interaction responsible for energy transfer between S and A. Only on the basis of very detailed calculations it is possible to reach a decisive conclusion. The work of Soules et al. [16] on the energy transfer between Sb^{3+} sensitizer ions and Mn^{2+} acceptor ions in $Ca_5(PO_4)_3F$ is a good example of such an investigation.

After insertion of the appropriate data in equation 6, the following expression was derived for the transfer probability on the basis of dipole-dipole interaction:

$$P_{SA}^{dd} = \frac{1}{\tau_s} \left(\frac{4.02}{R} \right)^6 , \qquad (8)$$

yielding a critical distance of 4 Å. In a similar way an expression was deduced for dipole-quadrupole interaction:

$$P_{SA}^{dq} = \frac{1}{\tau_s} \left(\frac{4.20}{R} \right)^8 , \qquad (9)$$

in this case R_c equals 4.2 Å. The authors also derived an expression for exchange interaction:

$$P_{SA}^{ex} = K.R^{16} e^{-2R/L} \sin^2\theta \cos^2\varphi , \qquad (10)$$

where K= 48.7(Å)$^{-16}$ μs^{-1} and L= 0.55 Å. As can be seen from the occurrence of polar coordinates in equation 10 the exchange interaction is highly anisotropic.

The dependence of the Sb^{3+} emission on the Mn^{2+} concentration as given in Fig.15 can be explained on the basis of the three types of interaction considered. However, in case of dipole-dipole interaction a critical distance of 9.75 Å has to be invoked, and in the case of dipole-quadrupole interaction R_c would equal 10.4 Å. Because fitting the concentration dependence results in much higher R_c values than derived theoretically, it is clear that only exchange interaction can be responsible for the transfer process. This is also confirmed by the decay behaviour of the Sb^{3+} luminescence.

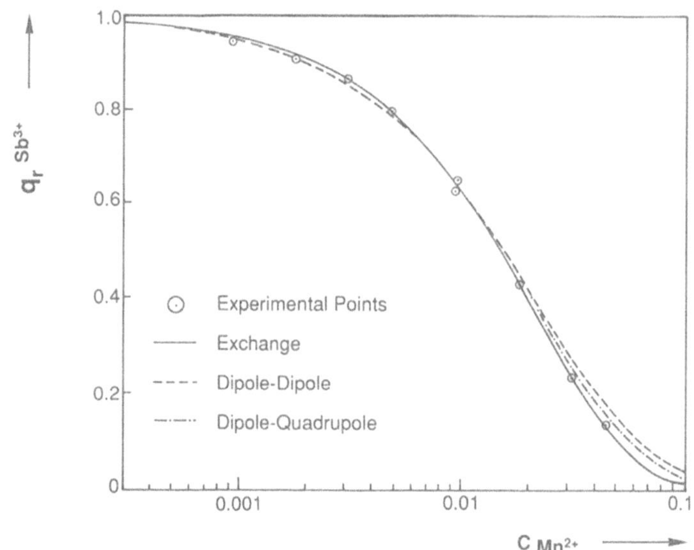

Fig.15
Relative Sb^{3+} quantum yield as a function of the Mn^{2+} concentration in $Ca_5(PO_4)_3F$.

3. <u>Cross-Relaxation</u>. In all systems discussed until now the resonance condition is fulfilled, because the energy difference between ground state and excited state is the same for sensitizer and activator. It is, however, not necessary to transfer all of the excitation energy. Typical examples of this cross-relaxation process are shown in Fig.16.

Due to the complexity of most of the rare-earth energy level diagrams it may occur that the energy difference between two sets of levels is almost identical. This holds for example for the energy difference between the highest Stark levels of the europium $^5D_0$-$^7F_1$ and terbium $^5D_4$-$^7F_4$

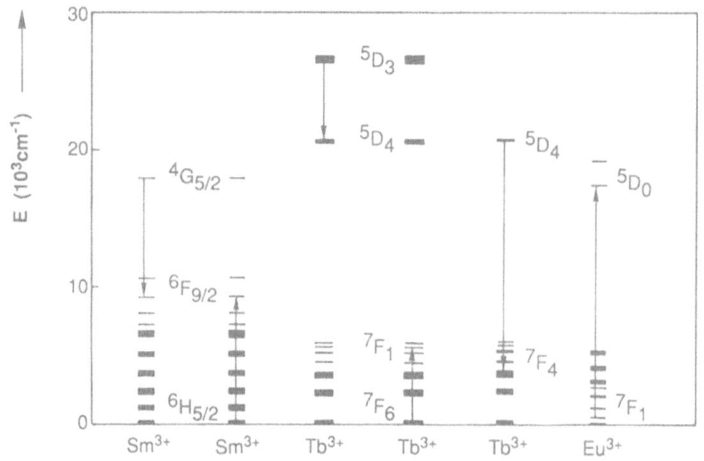

Fig.16
Cross-relaxation between similar and dissimilar ions.

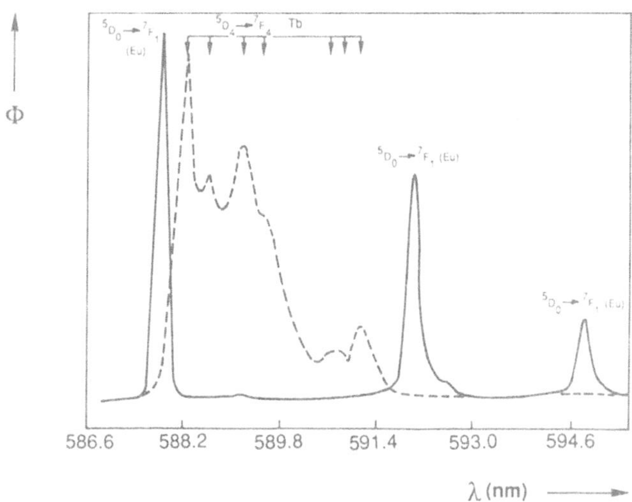

Fig.17
High resolution emission spectrum of $Tb_{0.66}Eu_{0.33}P_5O_{14}$ (λ_{ex}=485.3 nm, T=41 K) [17].

transitions in $(Eu,Tb)P_5O_{14}$ (Fig.17), which show a small but noticeable overlap. Therefore at higher temperatures, when the $^7F_1$ level of Eu^{3+} becomes thermally populated, energy transfer from Tb^{3+} to Eu^{3+} becomes feasible.

In the case of Sm^{3+} the energy of the $^4G_{5/2}$-$^6F_{9/2}$ transition matches the one of the $^6H_{5/2}$-$^6F_{9/2}$ transition. At higher samarium concentrations the $^6F_{9/2}$ level will become populated at the expense of the $^4G_{5/2}$ level. The first one is a non-emitting level, in view of the small separation between

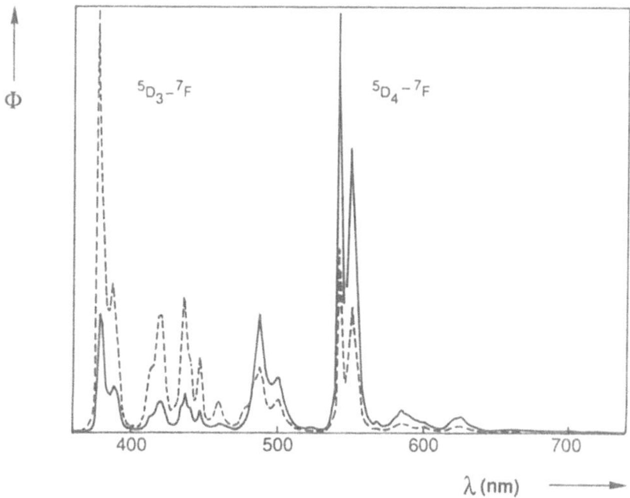

Fig.18
Emission spectrum of $(Ba,Ca)_2SiO_4$:$Tb_{0.01}$,$Li_{0.01}$ - - - - - and $(Ba,Ca)_2SiO_4$:$Tb_{.2}$,$Li_{.2}$ ———— (254 nm excitation).

this level and the lower-lying levels, the second one is an emitting level. Therefore at higher Sm^{3+} concentrations the luminescence will be quenched by cross-relaxation.

For Tb^{3+} cross-relaxation will also occur, the transitions involved are $^5D_3$-$^5D_4$ and $^7F_6$-$^7F_1$. In most systems both the $^5D_3$ and $^5D_4$ state are emitting levels. As a consequence of the occurrence of cross-relaxation the terbium emission will change from blue to green with increasing terbium concentration, this is exemplified for $(Ba,Ca)_2SiO_4$:Li^+,Tb^{3+} in Fig.18.

III.B. Non-Resonant Transfer

1. Multi-Phonon Assisted Processes. As was discussed in section II.C the energy gap between the different $4f^n$ levels of a rare-earth ion can be bridged by the emission of a number of phonons. In a similar way energy can be transferred from a sensitizer level to a lower-lying level of the activation. The transfer probability W_t shows an exponential energy gap dependence:

$$W_t = W_t\ (0)\ e^{-\beta\Delta E} \qquad (11)$$

This equation only holds when the gap is larger than the energy of the phonons involved.

Yamada et al [18] have studied in detail the multiphonon assisted transfer processes in Y_2O_3 doped with high concentrations in rare-earth

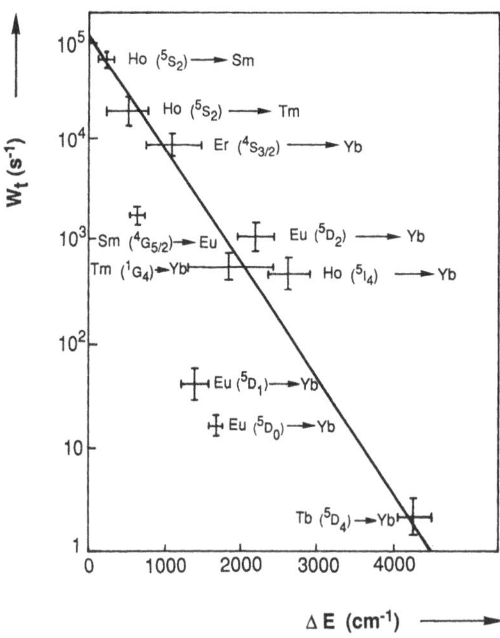

Fig.19
Dependence of the transfer probability between S and A on their energy mismatch in Y_2O_3 [18].

ions. Fig.19 clearly shows that equation 11 is indeed obeyed. In view of its exponential dependence the transfer probability drastically levels off with increasing energy mismatch.

2. <u>One- and Two-Phonon Assisted Processes</u>. Due to local distortions of the luminescent site the energy levels will be slightly shifted from one activator ion to the other. One and two phonon-assisted processes will be responsible for the energy transfer under these non-resonant conditions. This type of process has been extensively studied by Holstein et al. [19]. For very small mismatches in energy two phonon processes are more likely than one phonon processes.

In the case of $EuMgB_5O_{10}$ [20] the energy mismatch will be very small, because only one crystallographic rare-earth site is present. The temperature dependence of the transfer probability as derived by Buijs and Blasse is shown in Fig.20. At low temperatures the data can be fitted to

$$B^3=4.2\ 10^{-4}\ T^3 \qquad (12)$$

where $B^3=9/4\ \pi^2 P_{ss}.C_A$ and C_A stands for the killer concentration. According to Holstein such a behaviour will be found for a two-site non-resonant two-phonon assisted process. In this process one phonon is absorbed at one site and the other is emitted at the second site, in this way the energy mismatch is compensated for (Fig.21). At higher temperatures the temperature behaviour can best be described by the following relation:

$$B^3=2.9\ 10^4\ e^{-(229cm^{-1}/kT)} \qquad (13)$$

This type of dependence is predicted for a one-site resonant two-phonon process. The activation energy of 229 cm^{-1} corresponds very well with the energy of the lowest F_1 Stark level, which equals 243 cm^{-1}. In this process a phonon is absorbed, populating the $^7F_1$ level, followed by the emission of a phonon of somewhat different energy by the same europium

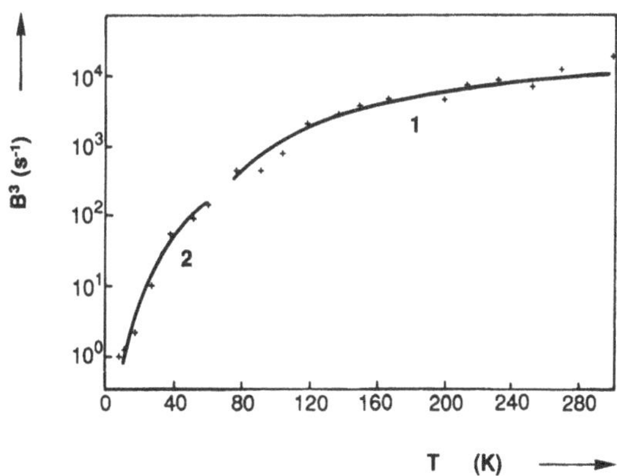

Fig.20
Temperature dependence of the transfer probability between Eu^{3+} ions in $EuMgB_5O_{10}$ [20].

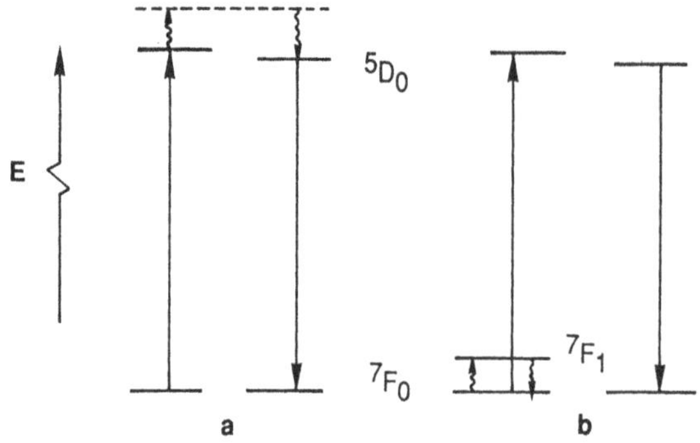

Fig.21
Schematic diagram of the phonon-assisted processes responsible for the energy transfer in $EuMgB_5O_{10}$
(a) two-site non-resonant two-phonon process
(b) two-site resonant two-phonon process.

ion. In this way the energy difference between the $^5D_0$ levels is bridged, enabling the energy transfer from one europium to the other. In cubic Gd_2O_3:Eu^{3+} two distinct crystallographic sites are present: one with C_2 and one with S_6 symmetry. Energy transfer is observed from europium ions in the S_6 site to the europium ions in the C_2 site [21]. The temperature dependence of this process is depicted in Fig.22. This temperature dependence can best be described as:

$$P = 5.10^4 + 20.10^4\ e^{-(139cm^{-1}/kT)} \tag{14}$$

The exponential term can be explained on the basis of a resonant one site two-phonon assisted transfer, involving the lowest $^7F_1$ Stark level of Eu^{3+}.

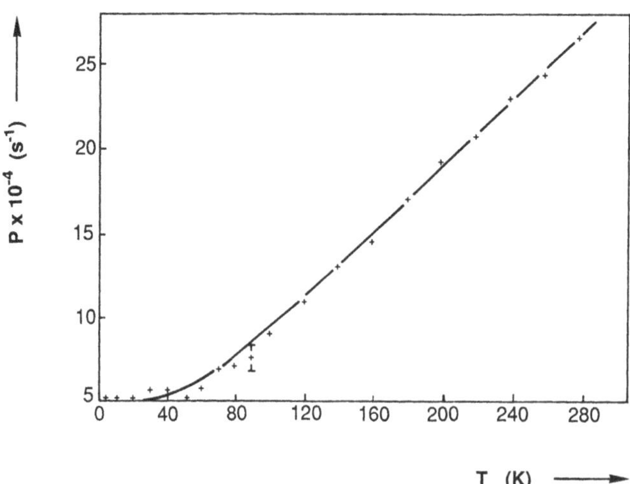

Fig.22
Temperature dependence of the $Eu(S_6)$ - $Eu(C_2)$ transfer rate in Gd_2O_3:$Eu_{0.1}$ [21].

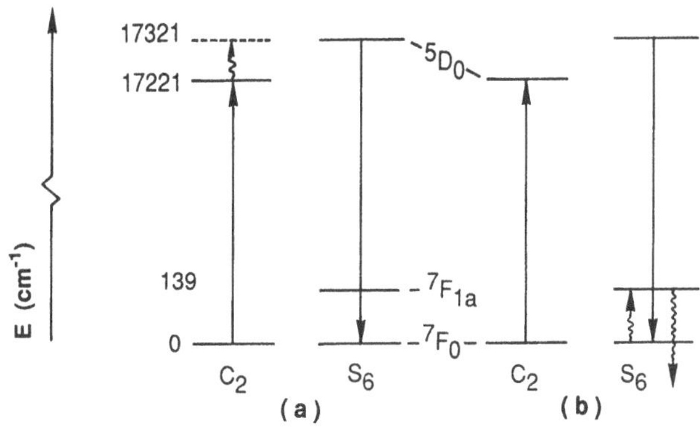

Fig.23

Schematic diagram of the phonon-assisted processes responsible for the Eu(S_6) to Eu (C_2) energy transfer in cubic Gd_2O_3.

(a) one phonon process

(b) two-site resonant two-phonon process.

At lower temperatures the transfer rate is much higher than expected on the basis of this process. Because in this case the energy difference between donor and acceptor $^5D_0$ levels is fairly high, i.e.: 100cm^{-1}, one phonon assisted processes will have a rather high probability (Fig.23).

IV. ENERGY MIGRATION

Energy migration is highly probable under resonance and near-resonance conditions. Therefore, in concentrated Ln^{3+} compounds the excitation energy will be able to jump from one Ln^{3+} ion to another. This process can be repeated several times, resulting in energy migration through the rare-earth sublattice.

Gadolinium compounds are typical examples of compounds showing such an energy migration. As a consequence of this migration the excitation energy may reach an activator ion. Because Gd^{3+} only shows transitions in the ultraviolet, originating from the 4f configuration, its absorption strength will be very low. In order to absorb the UV efficiently a sensitizer ion, showing allowed UV transitions, should be incorporated in the phosphor. The overall transfer process can be represented as follows:

$$S \rightarrow (Gd)_n \rightarrow A\cdot$$

IV.A. Gd^{3+} Mediated Transfer

Let us consider for example $LiY_{(1-x)}Gd_xF_4:Ce_{0.01},Tb_{0.01}$. [21]. From Fig.24 it follows that for x=0 almost only Ce^{3+} emission will be found. Due to the low Ce and Tb concentration, the separation between these ions will be large, hampering direct energy transfer. When the gadolinium content is increased the Gd^{3+} emission increases at the expense of the Ce emission. At even higher gadolinium contents the gadolinium emission levels off again, until almost solely Tb^{3+} emission is found.

It is clear from Fig.24 that the transfer to Tb^{3+} is almost complete

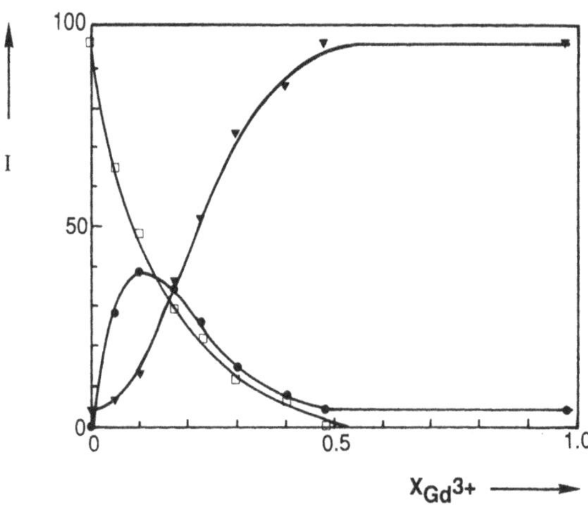

Fig.24
The relative emission intensity for Ce^{3+} (□), Gd^{3+} (•) and Tb^{3+} (▾) as a function of the Gd^{3+} concentration x_{Gd} for $LiY_{0.98-x}Gd_xCe_{.01}Tb_{.01}F_4$ at 100 K. Excitation occurs in the Ce^{3+} fd band.

for x=0.5, this concentration is designated as the critical concentration x_{cr} for energy migration through the lattice. The energy is indeed able to migrate through the complete lattice, when each Gd ion has two gadolinium neighbours. If N equals the number of sites to which energy can be transferred, it follows that:

$$x_{cr} = 2/N \qquad (15)$$

In $LiGdF_4$ each gadolinium ion has four nearest neighbours at 3.8 Å and four next nearest neighbours at 5.2 Å. Because x_{cr} = 0.5 energy is transferred to the four nearest neighbours and not to the next nearest neighbours. Therefore the critical distance for Gd-Gd transfer will be larger than 3.8 and smaller than 5.2 Å. The transfer probability was estimated to equal 10^7 s^{-1} [22].

On the basis of this transfer probability it is possible to derive whether we are dealing with exchange or multipole interaction. Using the distance dependencies given in section III.A. and an effective Bohr radius of 0.29 Å for Gd^{3+} [23], the transfer rate to the next nearest neighbours can be estimated (table 3). These data should be compared with the probability for radiative decay, which equals 2.10^2 s^{-1} for Gd^{3+}. If the transfer is governed by multipole-multipole interaction, the probability for energy transfer both to the nearest neighbours and next-nearest neighbours would be much higher than the probability for radiative decay, resulting in a much lower x_{cr}. Therefore, we can reach the conclusion that the energy transfer is $LiGdF_4$ is governed by exchange.

In a similar way the critical distance for Gd-Gd transfer was derived for a number of compounds [24]. There appears to be a striking difference in R_c between the fluorides and oxides, R_c is about 5 Å in the former case and around 6 Å in the latter case. The critical distance of 5 Å for $LiGdF_4$

Table 3
Estimated distance dependence of the energy transfer probability in $LiGdF_4$:Ce,Tb; d-d (dipole-dipole) d-q (dipole-quadrupole), q-q (quadrupole-quadrupole); ex (exchange) [22].

	interaction			
R(Å)	d-d	d-q	q-q	ex.
3.8	10^7	10^7	10^7	10^7
5.2	2.10^6	8.10^5	4.10^5	10^3

could well be explained on the basis of exchange interaction. It seems that in the case of oxides additional interactions play a role. For oxides the Gd $4f^65d^1$ level is located at lower energies than for the fluorides

Table 4
Estimate of R_c for different Gd^{3+} compounds [22].

	R_c (Å)	
	minimum value	maximum value
GdF_3	4.4	5.5
$LiGdF_4$	3.8	5.2
$NaGdF_4$	3.8	6.0
Gd_2O_3	5.4	6.1
$GdAl_3B_4O_{12}$	5.9	7.3

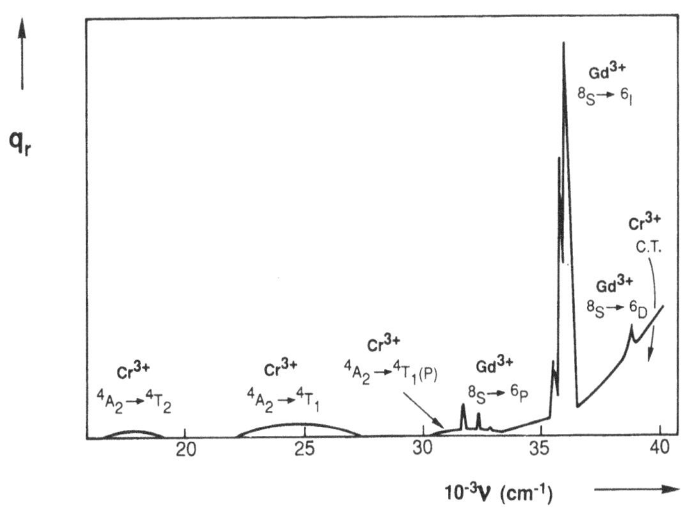

Fig.25
Excitation spectrum of Cr^{3+} at $13570cm^{-1}$ in $GdAlO_3$:$0.01Cr^{3+}$ at 298 K [25].

[24]. This will favour the admixture of 5d wavefunctions into the 4f wavefunctions, resulting in an enhancement of the forced electrical dipole character of the transitions. Due to these dipole interactions the transfer probability in oxides will increase, explaining the larger critical distance with respect to the fluorides.

In the next sections we will consider in more detail the sensitization of the Gd lattice, the energy migration and the trapping of the excitation energy. We will start at the end of the chain and work through the overall process from trapping to sensitization. Indeed, in order to get insight in one step of the chain, the consecutive steps should to be known in detail.

IV.B. Trapping Efficiency

The trapping efficiency of Cr^{3+} and several rare-earths dopants in $GdAlO_3$ has extensively been studied by de Vries et al. [25]. After excitation in the Gd^{3+} ^6I level, the following sequence in trapping efficiency was found:

$$Er^{3+} < Tm^{3+} < Dy^{3+} \leq Sm^{3+} << Eu^{3+} \approx Tb^{3+} < Cr^{3+}$$

The trapping efficiency is much higher in the case of the last three compounds than in the case of the other four compounds.

For Cr^{3+} there is a substantial overlap between the $^4A_2$ - $^4T_1$ transition and the gadolinium ^8S - ^6P transition, its charge transfer band overlaps with the ^8S - ^6I Gd^{3+} transition (Fig.25). The europium charge-transfer band as well as the terbium $4f^75d^1$ band both overlap with the ^8S - ^6I Gd^{3+} transition. Because in the case of Cr^{3+},Eu^{3+} and Tb^{3+} we are dealing with allowed transitions, energy transfer by multipole-multipole interaction is favoured, the transition probaility equals about 10^6 s^{-1}.

The other rare-earth ions do not show an allowed broad band absorption in the wavelength range 250-320 nm. Obviously there does not exist a relation between the intensity of their forbidden transitions around 310 nm and the trapping rate derived (table 5). Consequently the trapping rate

Table 5
Oscillator strength of activator ion absorption lines around 32 100 cm^{-1} [26]

Ion	Oscillator strength
Er^{3+}	$9\ 10^{-8}$
Tm^{3+}	$2\ 10^{-6}$
Dy^{3+}	$2\ 10^{-6}$
Sm^{3+}	$5\ 10^{-7}$

can not be governed by multipole-multipole interactions. The results can, however, be explained by assuming an exchange interaction due to the admixture of X^{4+}-Gd^{2+} or X^{2+}-Gd^{4+} levels. The energy of these charge-transfer states can be estimated, using the values of the standard reduction potential [27] (table 6). The more negative the potential the higher the energy of the charge-transfer state and the lower the transfer probability.

Table 6
Estimated energy of charge-transfer states.

Ion	Redox couple	standard potential (eV)
Er^{3+}	Er^{4+} - Gd^{2+}	- 10
Tm^{3+}	Tm^{2+} - Gd^{4+}	- 10
Dy^{3+}	Dy^{4+} - Gd^{2+}	- 8.9
Sm^{3+}	Sm^{4+} - Gd^{2+}	- 9.1

IV.C. Dimensionality of the Energy Migration

Time resolved spectroscopy is an ideal tool in order to study the energy migration through the rare-earth lattice. On excitation in the rare-earth ion S, the energy might be transferred to another S ion or to the trap A. We will only consider the two limiting situations: fast diffusion, i.e. $P_{SS} \gg P_{SA}$, and diffusion limited energy migration, i.e. $P_{SS} \ll P_{SA}$.

In the case of fast diffusion the decay rate of S is determined by the concentration C_A of A, as well as by the transfer probability P_{SA} [28]:

$$I = I_o \exp(-t/\tau_s) \exp(-C_a P_{SA} t) \qquad (16)$$

Where τ_s equals the radiative lifetime of S. Under these conditions always an exponential decay will be found. The decay in a concentrated system will however be much faster than the decay of an isolated ion:

$$I = I_o \exp(-t/\tau_s) \qquad (17)$$

When the energy migration is diffusion limited, the time dependence will be determined by the dimensionality of the diffusion process. In the case of diffusion in three dimensions the following exponential dependence will hold for $t \to \infty$ [28]:

$$I = I_o \exp(-t/\tau_s) \exp(-11.4 C_A C^{1/4} D^{3/4} t) \qquad (18)$$

Where C is a parameter describing the SA interaction and D the diffusion constant for the migration of the excitation energy.

For diffusion in two dimensions the time dependence will never become exponential and equals in the limiting case of $t \to \infty$ [29]:

$$I = I_o / 4\pi C_A a^{-2} Dt \exp(-t/\tau_s) \qquad (19)$$

Here a represents the trapping radius of the activator ion A. Assuming a high trapping rate, the equation for one-dimensional diffusion becomes [30]:

$$I = I_0 \exp(-t/\tau_S) \exp[-3(\pi^2 C_A P_{SA} t/4)]^{1/3} \qquad (20)$$

Also in this case no exponential time dependence will be found.

Energy migration in concentrated Eu systems has been studied in detail [21,31,32]. The energy migration over the C_2 sites in Eu_2O_3 shows a typical three-dimensional behaviour [21]. As can be seen from Fig.26 for $t \to \infty$ a logarithmic behaviour is indeed found. Each $Eu^{3+}(C_2)$ ion is surrounded by six C_2 nearest neighbours at a distance of 3.6 Å, the transfer probability equals $3.10^5\ s^{-1}$ at 90 K. If we compare this value with the radiative decay probability of the $^5D_0$ level, which equals $10^3\ s^{-1}$, we come to the conclusion that in the absence of traps on the average 300 steps may be involved in the energy migration process.

$NaEuTiO_4$ is built up of rare-earth double layers separated by a distance of about 10 Å [31]. In this double layer each Eu^{3+} ion has four nearest neighbours at a distance of 3.7 Å. The time dependence of the luminescence does not show an exponential behaviour for long decay times, as expected for two-dimentional diffusion (see Fig.27). In this case P_{ss} equals $5.10^7\ s^{-1}$, therefore, up to 5.10^4 hops may be involved in the energy migration.

Also in the case if $EuMgB_5O_{10}$ a non-exponential decay is found. This compound contains linear europium chains [32]; the interchain Eu separation equals 4.0 Å, the intrachain separation 6.0 Å. In Fig.28 a fit to equation (19) is shown, proving that transfer is only observed along the rare-earth chains. In view of the fact that transfer probability equals $10^7\ s^{-1}$, on the average 10^4 steps can be involved in the energy migration at low trap concentrations.

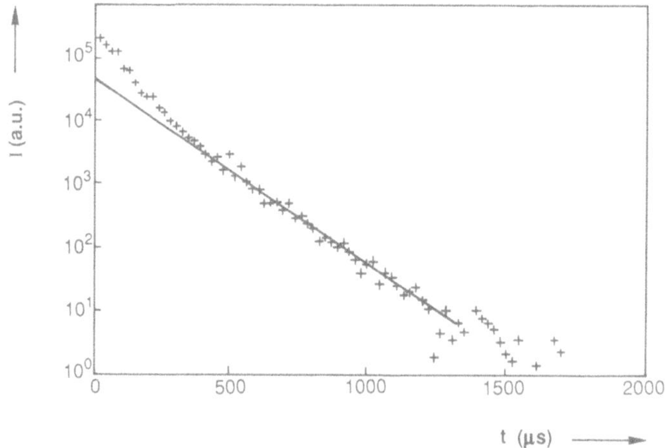

Fig.26

Decay curve of $^5D_0 \to {}^7F_2$ $Eu^{3+}(C_2)$ emission in Eu_2O_3 at 90 K. Solid line is the best fit to equation 16 [21]. I equals the intensity in arbitrary units.

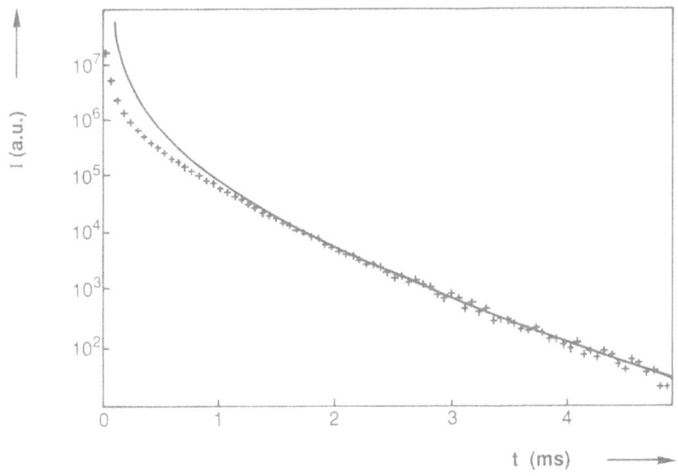

Fig.27
Decay curve of the intrinsic Eu^{3+} emission at 27.5 K in $NaEuTiO_4$ [31]. Solid line is the best fit to equation 17.

In all examples mentioned we are dealing with diffusion limited migration, i.e.: $P_{ss} << P_{sa}$ for trap concentrations in the order of 10^{-2} - 10^{-3}. This situation is generally found in concentrated Ln^{3+} phosphors. In the case of europium phosphors the critical distance for Eu-Eu transfer seems to be in the order of 4 Å, a value which is much smaller than the one derived for Gd-Gd transfer in oxides. In view of the much shorter critical distance, one- and two-dimensional diffusion can much more easily be observed in the concentrated europium phosphors than in concentrated gadolinium phosphors.

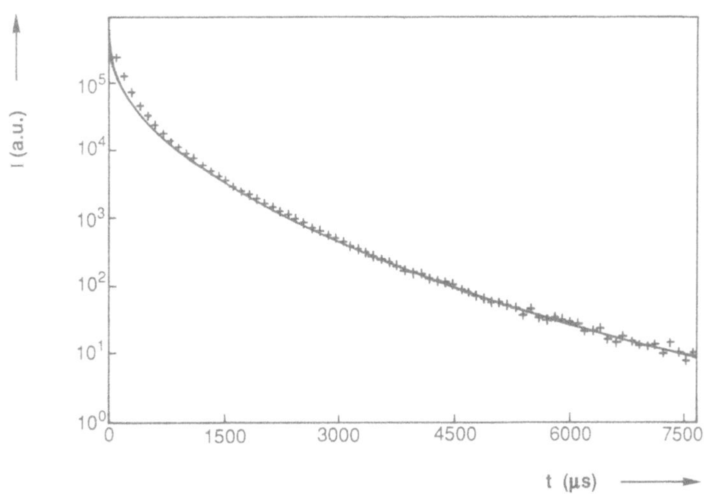

Fig.28
Decay curve of $^5D_0$ - $^7F_2$ Eu^{3+} emission in $Eu_{0.999}Nd_{0.001}MgB_5O_{10}$ at 112 K [32]. Solid line is the best fit to equation 18.

IV.D. Sensitization of the Gd^{3+} Sublatice

Gadolinium does only show two transitions with an energy below 40.000 cm^{-1}, i.e. the ^8S - ^6I transition around 270 nm and the ^8S - ^6P transition around 310 nm. The oscillator strength of the first transition is about one order of magnitude higher than the one of the second transition. An efficient sensitizer, therefore, should be able to absorb the 254 nm radiation and to convert this radiation into an emission, which preferentially overlaps with the ^8S - ^6I transition. Good candidates for gadolinium sensitization consequently are Ce^{3+}, Pr^{3+}, Bi^{3+} and Pb^{2+}. The advantages and disadvantages of applying these sensitizers will be discussed in more detail in the next subsections.

1. Sensitization with Ce^{3+}. The luminescence of $Li(Gd,Y)F_4:Ce^{3+},Tb^{3+}$ has already been discussed to some extent in one of the preceeding sections (IV.A). In Fig.29 the relative intensity of the Ce^{3+}, Gd^{3+} and Tb^{3+} emission is plotted as a function of the gadolinium content. These data relate to room temperature and deviate substantially from the data obtained at 100 K (Fig.24).

In this compound each rare-earth ion is surrounded by four nearest neighbours at 3.8 Å and four next-nearest neighbours at 5.2 Å. On the basis of dipole-dipole interaction (equation 7) the critical distance for Ce^{3+}-Tb^{3+} transfer was estimated to equal 4.3 Å [27]. Therefore, the transfer to the nearest neighbours will almost be complete and the interaction with the next-nearest neighbour will hardly contribute to the

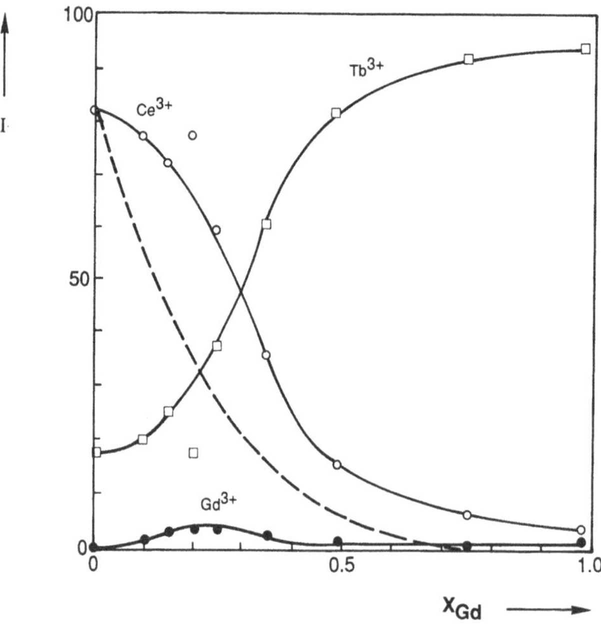

Fig.29
The relative emission intensity for Ce^{3+} (o), Gd^{3+} (•) and Tb^{3+} (□) in $Li\,Y_{.98-x}Gd_xCe_{.01}Tb_{.01}F_4$ at RT. The excitation occurs in the Ce^{3+} $5d^1$ band [23]. The dashed line corresponds to the relation $I = I_0(1-x)^4$.

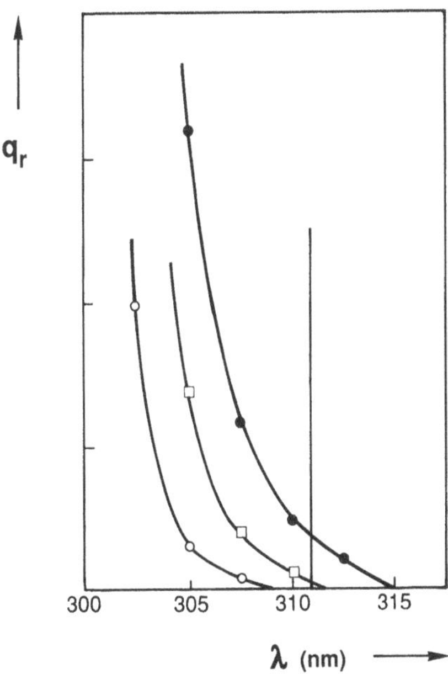

Fig.30
Ce^{3+} excitation spectrum in $LiYF_4:Ce^{3+}$ as a function of the temperature at (O) 150 K, (□) 200 K and (•) 275 K. The vertical line indicates the position of the Gd^{3+} emission line at 311 nm.

transfer process. The intensity of the Ce^{3+} emission is consequently expected to depend in the following way on the gadolinium content x:

$$I_{Ce} = I^{o}_{Ce}(1 - x)^{4}. \qquad (21)$$

Where (1 - x) gives the probability that a nearest neighbour site is occupied by a Gd^{3+} ion. It is clear from Fig.29 that this relation definitely does not hold for $Li(Gd,Y)F_4:Ce,Tb$, at least around room temperature. The cerium to gadolinium transfer seems to be far less pronounced than expected on the basis of equation 20.

In Fig.30 it is shown that, due to broadening of the Ce^{3+} absorption band with temperature, the Ce^{3+} absorption slightly overlaps the Gd^{3+} emission for temperatures above 200 K. Although the spectral overlap is rather small, R_c will equal 11.5 Å at room temperature [22] due to the high Ce^{3+} absorption strength. This will result in a substantial back-transfer from gadolinium to cerium.

In order to be an effective sensitizer for the gadolinium sublatice the Ce^{3+} luminescence should fulfil the following conditions :

1) The cerium emission should show a substantial overlap with the Gd^{3+} ^8S - ^6I transition;
2) The lowest cerium absorption band on the contrary should not have any overlap with the Gd^{3+} ^8S - ^6P transition;
3) The cerium absorption bands, however, must match the emission characteristics of the excitation source and peak around 254 nm.

2. Sensitization with Pr^{3+}. Praseodymium is known to show line emission in the visible part of the spectrum. Under specific conditions it may also show broad band emission in the ultraviolet, originating from the 4f5d - $4f^2$ transition. This will only occur when the offset between the parabolas for both levels is small, i.e.: the Stokes shift should be smaller than 3000 cm^{-1} [33]. The luminescence properties of the 4f5d level of Pr^{3+} can be predicted on the basis of the ones of the Ce^{3+} 5d level in the same host lattice (table 7). The energy difference between the lowest absorption band of Pr^{3+} and Ce^{3+} seems indeed to be around 12500 cm^{-1} and the Stokes shift is obviously smaller in the case of Pr^{3+} than in the case of Ce^{3+}.

Table 7
Luminescence properties of Ce^{3+} and Pr^{3+} in different host lattice [33].

	Ce^{3+}		Pr^{3+}	
host	lowest absorption band (cm^{-1})	Stokes shift (cm^{-1})	lowest absorption band (cm^{-1})	Stokes shift (cm^{-1})
$Y_3Al_5O_{12}$	22000	3800	34500	3000
$Gd_{9.33}(SiO_4)_6O_2$	31000	9000	43000	7000
BaY_2F_8	34500	3000	47000	2000

On the basis of these data Srivastava et al. [34] came to the conclusion that YBO_3 would be an excellent host lattice for 4f5d Pr^{3+} emission stimulable by a mercury discharge. Indeed the lowest Ce^{3+} absorption band in YBO_3 is located at 27500cm^{-1} and the Stokes shift of the 5d emission equals only 2000cm^{-1} [35]. The excitation and emission spectrum of $YBO_3:Pr^{3+}$ are shown in Fig.31. As was expected on the basis of the Ce^{3+} data the lowest Pr^{3+} excitation band is located at 250 nm (40000 cm^{-1}) and the Stokes shift equals about 2000 cm^{-1}.

The emission spectrum consists of different well separated bands, originating from the transition from the lowest 4f5d level to the different levels of the $4f^2$ configuration. From Fig.31 it is also obvious that the praseodymium ion in YBO_3 fulfils all conditions to be an effective sensitizer of the gadolinium sublattice. The excitation spectrum overlaps with the 254 nm Hg radiation, but does not show any overlap with the ^8S - ^6P transition at 313 nm. The emission spectrum on the contrary shows a substantial overlap with both the 313 nm ^8S - ^6I and the 270 nm ^8S - ^6P gadolinium lines.

On the basis the data presented one would expect that $GdBO_3$ co-doped with Pr^{3+} and Eu^{3+} would be a highly efficient phosphor. In reality the efficiency is fairly low and equals only 15% [36]. In Fig.32 the decay of the Eu^{3+} $^5D_0$ - $^7F_2$ emission is shown on excitation in the $^5D_2$ level of $YBO_3:Eu^{3+}$ and $YBO_3:Pr^{3+},Eu^{3+}$. In the case of $YBO_3:Eu^{3+}$ the decay time of 4.15 ms corresponds with the radiative lifetime of the Eu^{3+} ion. In the

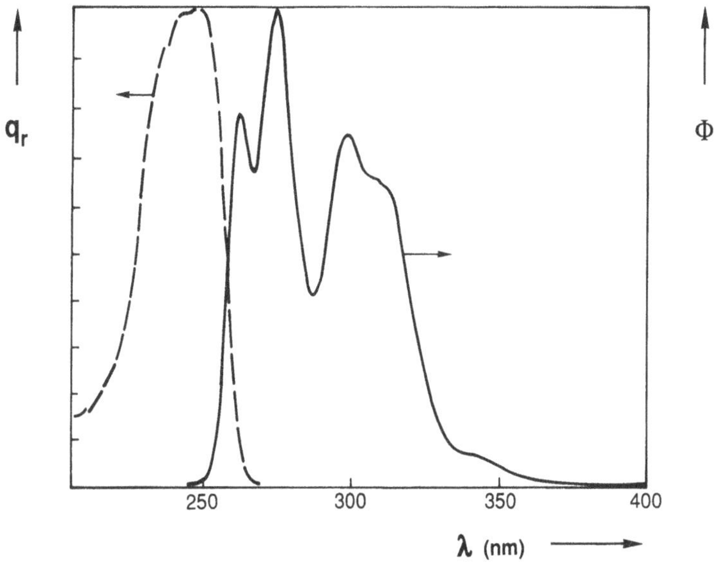

Fig.31
Room temperature excitation and emission spectra of YBO_3:0.001Pr^{3+} [36].

case of the Pr^{3+} co-doped sample a non-exponential decay is found, pointing towards energy transfer from Eu^{3+} to Pr^{3+}. This is indeed confirmed by the observation of very weak emission lines at 16530 and 16750 cm^{-1}, originating from the $^1D_2$-$^3H_4$ Pr^{3+} transition. The $^1D_2$ level of praseodymium is nearly resonant with the europium $^5D_0$ level, making transfer by phonon-assisted processes highly probable. A similar situation

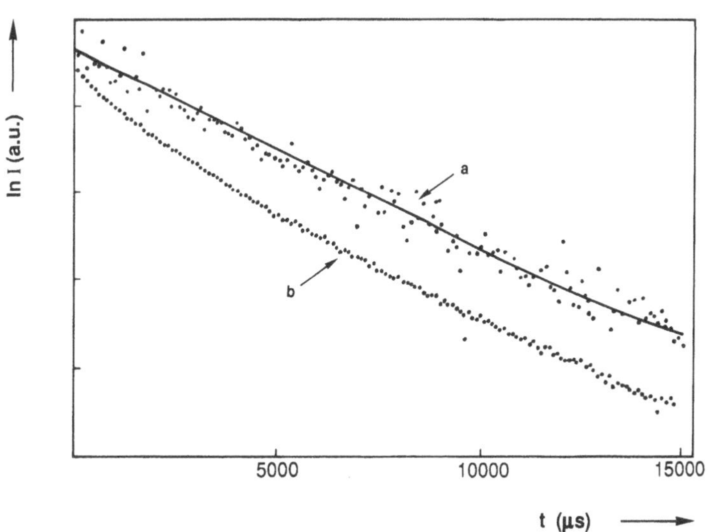

Fig.32
Decay of the Eu^{3+} $^5D_0$-$^7F_2$ emission in (a) $Y_{0.99}Eu_{0.01}BO_3$ and (b) $Y_{0.98}Pr_{0.01}Eu_{0.01}BO_3$ on excitation in the Eu^{3+} $^5D_2$ level [36].

will also occur, when the sample is co-doped with terbium. In this case the Tb^{3+} $^5D_4$ level is nearly resonant with the Pr^{3+} $^3P_0$ level.

It will be clear from this discussion that Pr^{3+} can be a highly efficient sensitizer for gadolinium emission. When gadolinium is however applied as an intermediate in order to transfer excitation energy to other Ln^{3+} activator ions, praseodymium will in most cases also act as a luminescence killer. Praseodymium consequently can never be applied in efficient phosphor, emitting visible radiation.

3. Sensitization with Bi^{3+}. The sensitization of gadolinium phosphors with Bi^{3+} has been studied by Kiliaan et al. [37]. $Gd_2O_2SO_4$:Bi^{3+},Tb^{3+} proved to be a fairly efficient phosphor, the efficiency equals 40%. Its excitation spectrum is shown in Fig.33. It consists of a broad band due to the Bi^{3+} $^1S_0$-$^3P_1$ transition located at 265 nm and of some Gd^{3+} lines around 275, 280, 307 and 310 nm. In the visible part of the spectrum apart from the Tb^{3+} transitions a broad band is found around 520 nm.

In the case of $Y_2O_2SO_4$:Bi^{3+},Tb^{3+} this band is even more prominently present (Fig.34). Other features in the emission spectrum are the Bi^{3+} ^1S_o - $^3P_1$ emission at 320 nm and the Tb^{3+} transitions superimposed on the band structure. The weak intensity of the terbium lines confirms that the Bi^{3+} to Tb^{3+} emission is indeed mediated by Gd^{3+}. The observation of two bismuth emission bands is rather surprising in view of the fact that only one crystallographic cation site is available to accommodate the Bi^{3+} ion. A similar behaviour has also been reported for LaOCl:Bi^{3+} [38]. It is well known that s^2 ions tend to form pairs or ion clusters [39]. Therefore, the UV emission of LaOCl:Bi^{3+} was ascribed to Bi^{3+} ions on a regular site and the less efficient visible band was attributed to the emission originating from bismuth clusters. Probably the same situation holds for $Ln_2O_2SO_4$.

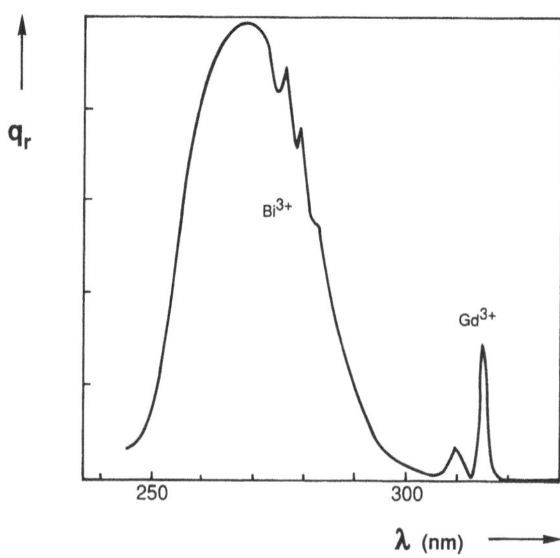

Fig.33
Room temperature excitation spectrum of Tb^{3+} $^5D_4$ emission in $Gd_2O_2SO_4$:$Bi^{3+}_{0.01}$,$Tb_{0.01}^{3+}$ [37].

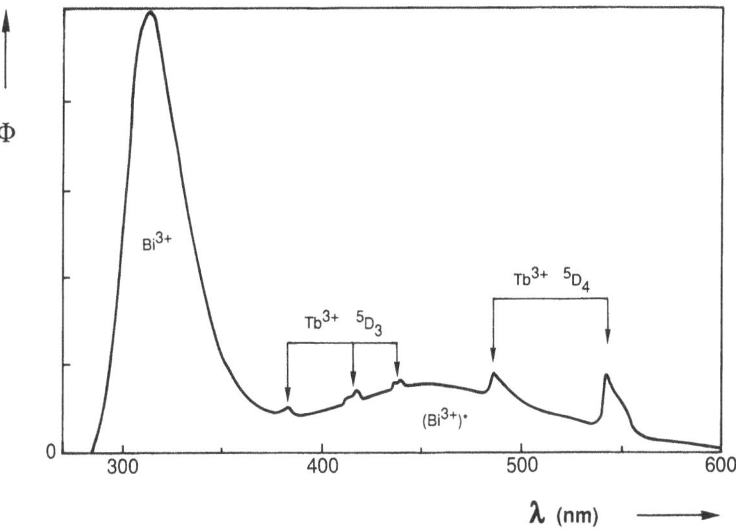

Fig.34
Emission spectrum upon excitation into Bi^{3+} ions (λ_{ex} = 265 nm) at room temperature for $(Y_{0.98},Bi_{0.01},Tb_{0.01})_2O_2SO_4$ [37].

The efficiency of $Gd_2O_2SO_4$:Bi^{3+},Tb^{3+} equals about 40%, because only part of the excitation energy is indeed fed into the gadolinium sublattice. A substantial fraction of the Hg radiation is absorbed by the bismuth clusters. It will be clear from their emission spectrum that the Bi^{3+} clusters are not able to transfer excitation energy to the gadolinium ions and most of the energy will be lost in non-radiative processes.

4. Sensitization with Pb^{2+}. In the case of the green emitting phosphor $Sr_2Gd_8(SiO_4)_6O_2$:Pb^{2+},Tb^{3+} an efficiency of around 75% is found on excitation in the Pb^{2+} $^1S_0$ - $^3P_1$ band [40]. In the apatite structure of this compound two distinct crystallopgraphic cation sites are available, i.e.: the 4f and 6h site. The Sr^{2+} ions enter two of the 4f sites, the gadolinium ions enter the two 4f sites as well as the 6h sites.

From the luminescence data of $Sr_2Y_8(SO_4)_6O_2$:Pb^{2+} is clear that Pb^{2+} enters both sites (Fig.35). Indeed two emission peaks, with distinct excitation spectra, are found. In the 6h site one of the coordinating O^{2-} ions does not belong to the silicate groups and its Ln^{3+} - O^{2-} distance is much smaller than for the other oxygen ions, i.e.: 2.3 Å instead of 2.6 Å. For s^2 ions in a asymmetrically coordinated site always large Stokes shifts are found [38]. Therefore the long-wavelength UV emission is ascribed to Pb^{2+} ions in the 6h site and the short wavelength UV emission to Pb^{2+} in the 4f site. At room temperature the long-wavelength emission will partly be quenched, explaining why the efficiency of $Sr_2Gd_8(SiO_4)_6O_2$:Pb^{2+},Tb^{3+} is only about 75%.

V LAMP PHOSPHORS

In the previous sections we have focussed our attention on the

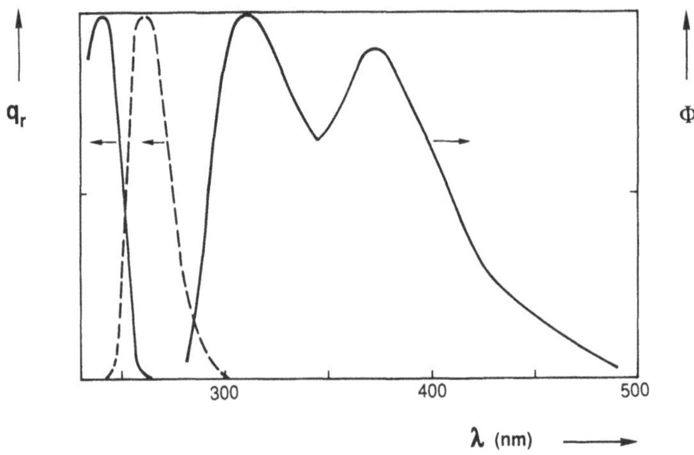

Fig.35
Emission spectrum of $Sr_{1.9}Pb_{0.1}Y_8(SiO_4)_6O_2$ and excitation spectrum of the 310 nm emission (——) and 380 nm emission (-----) at liquid helium temperature [40].

luminescence properties of UV excitable phosphors. The last part of my contribution will be devoted to the application of these phosphors in lamps. On the basis of the profound knowledge gathered during the previous decades it has become possible to design phosphors, which fulfil nearly all the requirements as imposed by their application.

The major fields of lamp phosphor application are: general lighting and UV irradiation. In general lighting application one aims at an optimum combination of lamp efficiency, colour rendition and cost price. UV emitting lamps are often used to stimulate specifically a given (bio)chemical reaction, typical examples are the sun-tanning lamps and the lamps for medical application.

V.A. General Lighting

Fluorescent lighting already started before the second world war. A blend of phosphors was applied to the inner surface of the lamp bulb, this blend was able to convert the ultraviolet emission generated by the mercury discharge into white radiation. The mercury discharge emission mainly consists of 254 nm (85%) and 185 nm (12%) radiation, the remaining 3% is emitted in the long wavelength UV and the visible part of the spectrum. The original blend was quickly replaced by a single phosphor, i.e. $Ca_5(PO_4)(Cl,F):Sb^{3+},Mn^{2+}$. These halophosphate lamps dominated the fluorescent lighting market for many decades and remain an important product even today.

The advent of rare-earth based phosphors in 1975, however, has drastically revolutionized the field of fluorescent lighting. Higher efficiencies and a much better colour rendition are obtained with the so-called tricolour lamp, based on a blend of three rare-earth phosphors: one emitting in the blue, one in the green and one in the red part of the spectrum. An almost perfect colour rendition can be obtained with the Special de Luxe lamps, also based on rare-earth phosphors. Another advantage of these new phosphors is that they can be applied in lamps with a smaller bulb diameter, this ultimately resulted in the introduction of compact fluorescent lamps.

1. Halophosphate Lamps. Jenkings et al. [41] discovered that when $Ca_5(PO_4)_3(Cl,F)$ is doped with Sb^{3+} an efficient blue phosphor is obtained. As was discussed before co-doping with Mn^{2+} will result in sensitization of the orange manganese emission (III.A.2). By combining blue-green with orange radiation white light is obtained.

In colorimetry it is normal practice to represent the different colours in a chromaticity diagram (Fig.36). Each colour is represented by its colour coordinates. Not all the combinations of x and y can be obtained, only the combinations falling within the colour chart are allowed. The border of this colour chart is formed by the colour points of monochromatic radiation. With the aid of such a diagram it becomes fairly easy to predict the colour point of a blend. When two colours are mixed, the colour point of the blend will always lie on the line connecting the colour points of its constituents. In a similar way for a three component blend the colour point of the mixture will fall within the triangle suspended by the points of the single components.

By definition the emission of a light source is considered to be white if its colour coordinates correspond with the one of a black body radiator. The colour appearance of such a radiatior will largely depend on its temperature and will change from red, over orange, to white and even to bluish white with increasing temperature. The corresponding colour points are located on the black-body locus (BBL). It is general practice

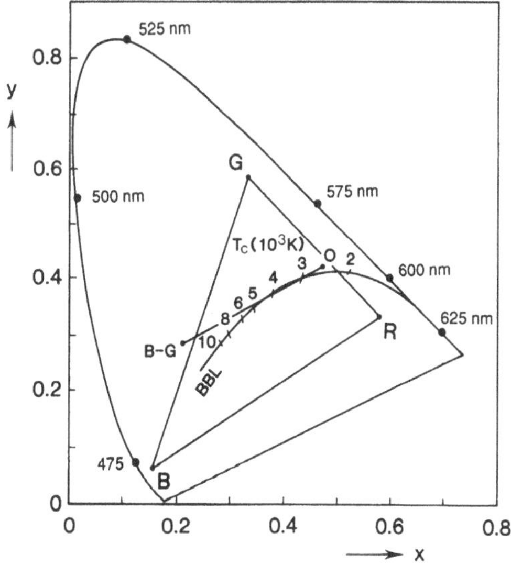

Fig.36
CIE chromaticity diagram with the black-body locus (BBL). White light can be generated by blending blue-green (B-G) with orange (O) light or by mixing blue (B), green (G) and red (R).

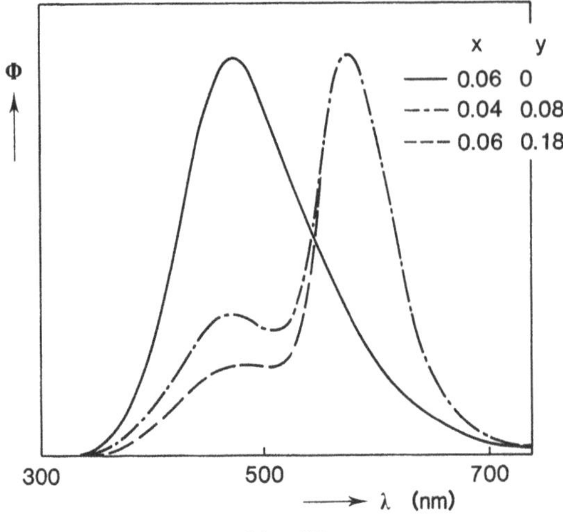

Fig.37
Emission spectra of $Ca_5(PO_4)_3(Cl, F):Sb^{3+}{}_x$, $Mn^{2+}{}_y$ on excitation in the Sb^{3+} band at room temperature.

to characterize a lamp by its correlated colour temperature, defined as the temperature of the black-body radiator with the same colour point.

In Fig.36 the colour points of the Sb^{3+} (B-G) and manganese emission (O) in calcium halophosphate are given. By carefully adapting the ratio between both emissions, lamps with a correlated colour temperature ranging from 2700 to 6500 K can be made. The ratio between both emissions can easily be adjusted by changing the dopant level of both activator ions (Fig.37).

The application of calcium halophosphate in a fluorescent lamp results in an efficacy of 80 lm/w and a colour rendition of 50 to 60. The colour rendition is largely determined by the spectral energy distribution of the lamp. It is established by comparing the colour coordinates of a set of test colours under illumination with the lamp to be tested and with the corresponding black-body radiator. The colour rendering index (CRI) will equal 100 when all the colour point are the same in both cases. The lower the CRI of a light source, the less natural the appearance of a coloured body will be to the human eye. In this respect the halophosphate lamp shows a rather moderate performance.

2. Tricolour Lamp. As is shown in Fig.36, it is also possible to create white light by blending blue with green and red radiation. On the basis of computer calculations Koedam and Opstelten [42] came to the conclusion that a fluorescent lamp, combining a high efficiency with a good colour rendition, could be realized by mixing three phosphors emitting in a narrow wavelength interval centred around 450, 550 and 610 nm respectively. Narrow band emission is typically found with rare-earth doped phosphors. Indeed the trivalent rare-earths, such as Tm^{3+}, Sm^{3+}, Dy^{3+}, Eu^{3+} and Tb^{3+}, all show line emission, whereas phosphors doped with divalent lanthanides, i.e. Sm^{2+}, Yb^{2+} or Eu^{2+}, will show band emission. The half width of this emission is, however, appreciably smaller

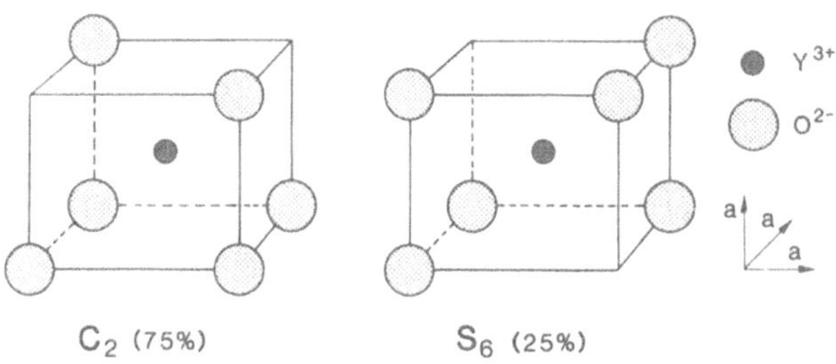

Fig.38
The two rare-earth sites in Y_2O_3:Eu^{3+}

than for Sb^{3+} and Mn^{2+} in calcium halophosphate. Only three years later such a lamp was realized by Verstegen et al. [43] and the tricolour lamp became commercially available within one year thereafter. The phosphor blend consisted of the blue emitting phosphor $BaMgAl_{10}O_{17}$:Eu^{2+}, the green emitting phosphor $CeMgAl_{11}O_{19}$:Tb^{3+} and the red emitting phosphor Y_2O_3:Eu^{3+}.

Y_2O_3 has the same cubic crystal structure as Eu_2O_3. In this structure two distinct crystallographic Y sites are present : 75% of the sites show a C_2 symmetry and 25% an S_6 symmetry (Fig.38). For trivalent europium ions in a C_2 symmetry the forced electrical dipole transitions, $^5D_0$ - $^7F_2$ at 610 nm and $^5D_0$ - $^7F_4$ at 700 nm, are favoured, in a S_6 symmetry only the magnetic dipole transition, $^5D_0$ - $^7F_1$ at 590 nm, is allowed.

From the emission spectrum it follows that almost only the 610 nm transition is observed (Fig.39). The nearly complete absence of magnetic dipole transitions is caused by the efficient transfer between Eu^{3+} in the S_6 site to Eu^{3+} in the C_2 site, this transfer was discussed before in the

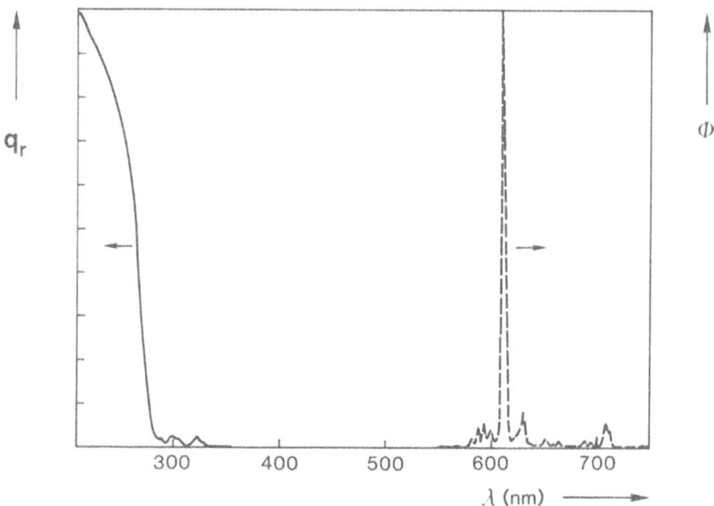

Fig.39
Emission and excitation spectrum of Y_2O_3:3%Eu^{3+} at room temperature.

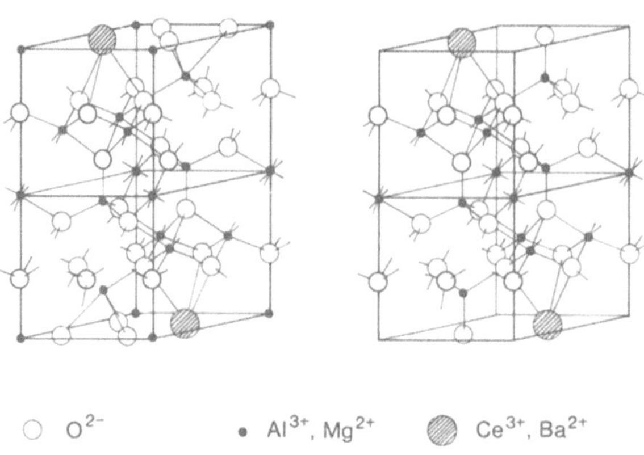

Fig.40
Half the unit cell of $CeMgAl_{11}O_{19}$ (a) and $BaMgAl_{10}O_{17}$ (b).

case of cubic Gd_2O_3 : Eu^{3+} (section III.B.2). In the case of europium entering the C_2 site of Y_2O_3 the $^5D_0$ - $^7F_2$ transition is largely favoured over the $^5D_0$ - $^7F_4$ transition. The intensity of the $^5D_0$ - ^7F transitions is known to be highly sensitive towards small changes in the environment of the luminescent ion [44]. From the excitation spectrum it follows that the europium charge-transfer band, located at high energies overlaps rather well with the 254 nm Hg radiation. The 185 nm Hg radiation will be absorbed by the host lattice. Under both modes of excitation the quantum efficiency equals about 95%.

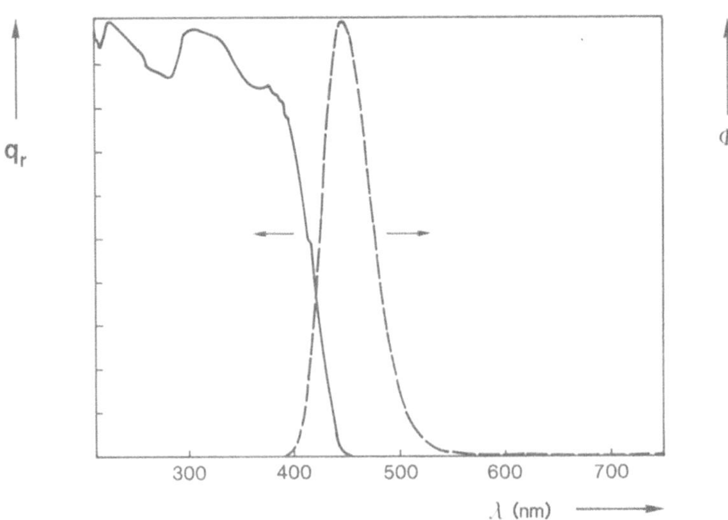

Fig.41
Room temperature excitation and emission spectrum of $BaMgAl_{10}O_{17}$:$0.1Eu^{2+}$.

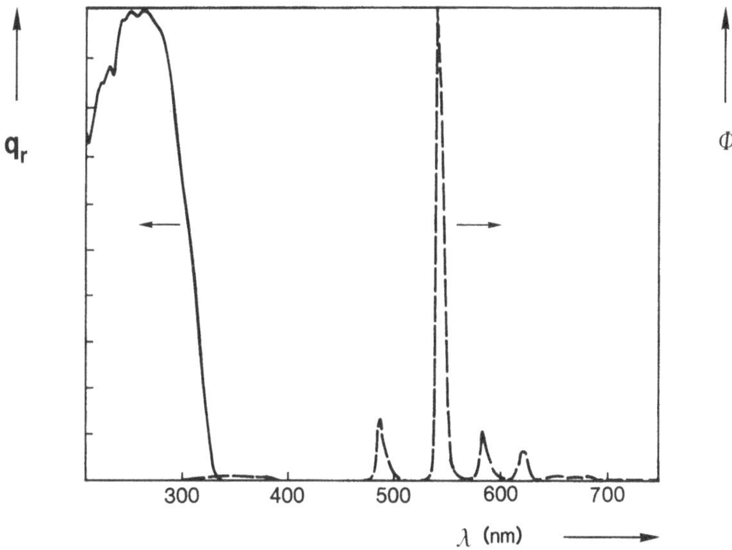

Fig.42
Room temperature excitation and emission spectrum of $CeMgAl_{11}O_{19}$:0.33Tb^{3+}.

$BaMgAl_{10}O_{17}$ and $CeMgAl_{11}O_{19}$ are closely related compounds, both are built out of large spinel blocks, separated by an intermediate layer (Fig.40). In the barium compound this layer contains one Ba^{2+} ion and one O^{2-} ion. Because Ce^{3+} is a much smaller ion, the intermediate layer in the case of the cerium compound also contains one Al^{3+} and three O^{2-} ions in addition to the Ln^{3+} ion.

The excitation spectrum of $BaMgAl_{10}O_{17}$:Eu^{2+} is dominated by the crystal field terms of the $4f^65d^1$ level (Fig.41). Since we are dealing with only

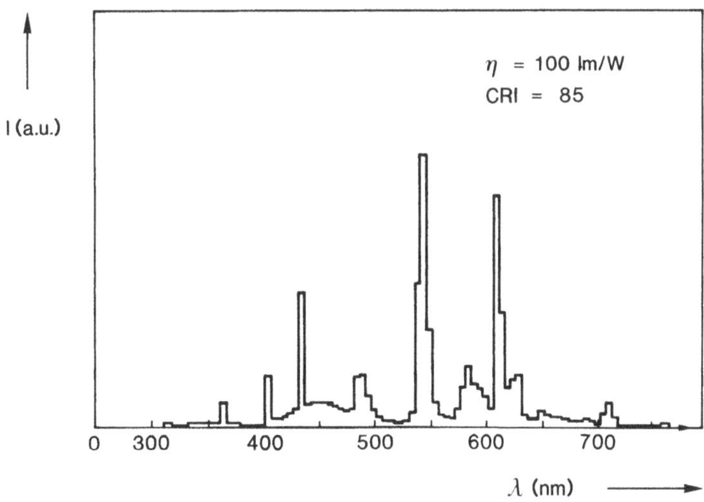

Fig.43
Emission spectrum of a tricolour lamp with correlated colour temperature of 4000 K.

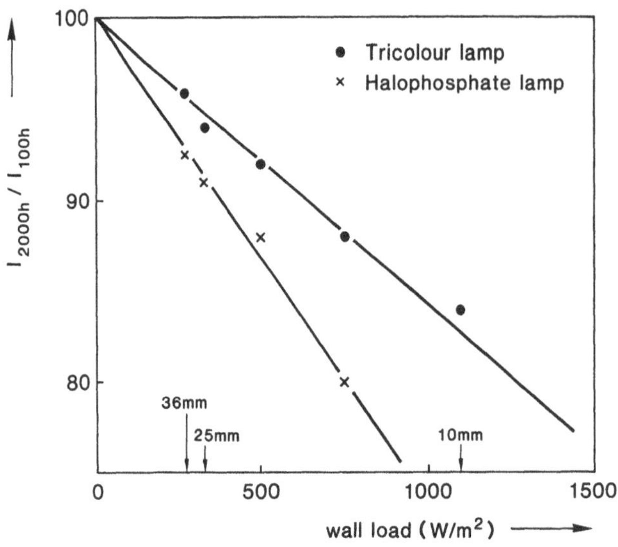

Fig.44
Decrease in output after 2000 hrs. of burning, normalized to the output after 100 hrs. of burning, as a function of the wall load for (x) a halophosphate lamp and (•) a tricolour lamp.

one crystallographic site, in which Eu^{2+} is able to enter, a narrow band emission is found originating from the same $4f^65d^1$ level. The Stokes shift for Eu^{2+} in this compound is rather small, therefore an efficiency of 90% can be obtained.

The luminescence spectra of $CeMgAl_{11}O_{19}:Tb^{3+}$ are shown in Fig.42. The absorption of the 254 nm radiation occurs in the Ce^{3+} $5d^1$ band, the 185 nm radiation will be absorbed by the Tb^{3+} $4f^75d^1$ level. The emission spectrum consists mainly of Tb^{3+} $^5D_4$ emission, but also some Ce^{3+} emission is observed. Because the probability for radiative transitions is around 10^7 s^{-1} in Ce^{3+}, even at high terbium concentrations the radiative process will be able to compete with the Ce^{3+} - Tb^{3+} transfer over a distance of 5.6 Å [45]. In view of the large Stokes shift of the Ce^{3+} emission no energy transfer will occur in $CeMgAl_{11}O_{19}$. Almost no emission from the $^5D_3$ Tb^{3+} level is observed, at these high Tb^{3+} contents. The $^5D_4$ level becomes indeed populated at the expense of the $^5D_3$ level by multipolar cross-relaxation.

The emission spectrum of a tricolour lamp is shown in Fig.43. Apart from the Eu^{2+}, Tb^{3+} and Eu^{3+} emission the mercury lines at 405, 436 and 545 nm also contribute to the spectral energy distribution in the visible spectrum part of the spectrum. The cerium emission of $CeMgAl_{10}O_{19}:Tb^{3+}$ as well as the 311 and 365 nm Hg radiation can also be distinguished in the UV. This lamp combines an efficacy of 100 lm/w with a colour rendition index of 85, a much higher value than found for halophosphate lamps. In most applications a CRI of 85 guarantees a quite natural illumination.

Unexpectedly the tricolour lamp shows another advantage over the halophosphate lamp, notably a much better maintenance during lamp life is found. The output of a fluorescent lamp shows a complex exponential decay

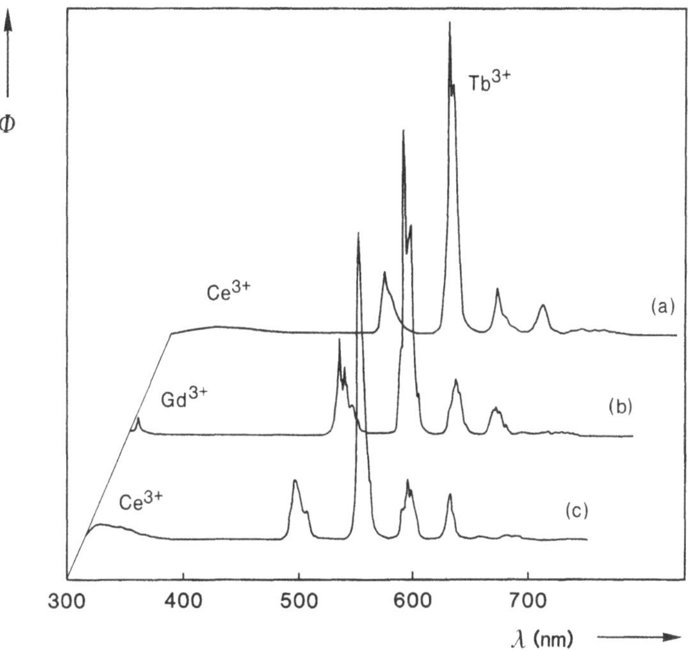

Fig.45
Emission spectrum of $Ce_{.67}Tb_{.33}MgAl_{11}O_{19}$ (a), or $Ce_{.3}Gd_{.5}Tb_{.2}MgB_5O_{10}$ (b) and $Ce_{.45}La_{.4}Tb_{.15}PO_4$ (c) at room temperature.

as a function of the burning time. Therefore in Fig.44 the output decreaseafter 2000 hrs. of burning is plotted as a function of the wall load. The slope of this curve is much steeper in the case of a halophosphate than in the case of a tricolour lamp. The wall load of a fluorescent lamp is almost solely determined by the tube diameter. From the data presented it will be clear that it is possible to realize narrow diameter tricolour lamps (26 mm) with the same economical life as halophosphate lamps with a standard diameter (36 mm). It even proved to be feasible to reduce the bulb diameter to 10 mm. With this small diameter the discharge tube can be folded up, in this way the compact fluorescent lamp is born.

Although the tricolour lamp turned out to be a commercial success, even this magnificent lamp concept has its limitations. First of all the luminous flux proved to be somewhat lower than anticipated. A colour rendition of 85 is sufficient in most applications; under some conditions a higher CRI is however required, i.e.: flower displays and museum illumination. Due to the use of large amounts of lanthanides, the cost price of a tricolour layer is about a factor of 20-30 higher than of a halophosphate layer. Most efforts in phosphor research during the last decade therefore have concentrated on these three items.

3. Second Generation Tricolour Lamps. Because the human eye is most sensitive to green radiation, the use of an improved green phosphor will result in a higher lamp efficiency. Almost simultaneously two new green

tricolour phosphors where developed. $(La,Ce)PO_4:Tb^{3+}$ is based on the monoclinic modification of lanthanum orthophosphate, better know as the mineral monazite. The second phosphor $(Ce,Gd)MgB_5O_{10}:Tb^{3+}$ is derived from a new compound, which was first described by Saubat et al. [46]. $EuMgB_5O_{10}$ (section IV.C) belongs to the same class of compounds, built up out of linear Ln chains.

The emission spectrum of the green phosphors is shown in Fig.45. Apart from the fine structure, the visible spectrum is quite the same for the three compounds. This fine structure is of no importance whatsoever, as far as the illumination properties of the tricolour lamp is concerned. In the UV spectrum of $(La,Ce)PO_4:Tb^{3+}$ a cerium emission band is found, which is located at a much shorter wavelength than the cerium band in the aluminate. In the case of $(Ce,Gd)MgB_5O_{10}$ region Gd^{3+} emission is found instead of Ce^{3+} emission in the UV.

In both the phosphate and borate phosphor energy migration will occur. Due to the small Stokes shift energy migration in $(La,Ce)PO_4:Tb^{3+}$ will occur over the Ce^{3+} ions [47], a similar situation as found in $KCaLn(PO_4)_2$ (section III.A.1). In the case of the pentaborate energy migration will occur over the gadolinium sublatice [48]. In contrast to $EuMgB_5O_{10}$ (section IV.C), where the energy migration was purely one-dimensional in the case of $GdMgB_5O_{10}$ intrachain as well as interchain energy migration is found.

More details on the efficiency of the green phosphors are given in table 8. Although both in the phosphate and the borate a lower terbium content is applied than in the aluminate, a higher visible efficiency is found. Thanks to the occurrence of energy migration in these compounds more excitation energy will end up at a Tb^{3+} site. In the case of the gadolinium phosphor the branching between visible and UV emission apparently is more favourable than in the case of the phosphate. The high

Table 8
The quantum efficiency of green emitting tricolour phosphors.

	quantum efficiency		
compositions	U.V.	Tb^{3+}	overall
$Ce_{.67}Tb_{.33}MgAl_{11}O_{19}$	5	85	90
$Ce_{.45}La_{.4}Tb_{.15}PO_4$	7	88	95
$Ce_{.3}Gd_{.5}Tb_{.2}MgB_5O_{10}$	2	90	92

degree of energy migration in $GdMgB_5O_{10}$ facilitates the transfer of the excitation energy to Tb^{3+}, but also to the ubiquitously present traps. Indeed, the overall quantum efficiency turns out to be lower for the pentaborate than for the orthophosphate.

4. Special de Luxe Lamp. We have already pointed out that in some applications a higher colour rendering index than 85 is required. By shifting the emission wavelength of the blue phosphor to longer

wavelengths a higher colour rendition can indeed be obtained [49]. In Fig.46 the lamp properties are plotted as a function of the emission wavelength of the blue phosphor for a blend of this phosphor with $Y_2O_3:Eu^{3+}$ and with $(Ce,Gd)MgB_5O_{10}:Tb^{3+}$. The highest output is indeed obtained for an emission wavelength of 450 nm as found in $BaMgAl_{10}O_{17}:Eu^{2+}$. When the emission wavelength is shifted towards 490 nm the CRI can be increased to a value of 92. This increase in CRI will, however, occur at the expense of the lamp output. Efficient blue-green emission around 490 nm is found in the phosphor $Sr_4Al_{14}O_{25}:Eu^{2+}$ [50].

This new aluminate consists of layers containing AlO_6 octahedra separated by twinned layers of AlO_4 tetrahedra (Fig.47). Both layers are interconnected by additional AlO_4 tetrahedra [51]. It can concluded from Fig.47 that in this structure two distinct crystallographic sites are present. In the emission spectrum of $Sr_4Al_{14}O_{25}:Eu^{2+}$ two emission bands are observed : a small one around 420 nm and a large one at 490 nm (Fig.48). This observation is consistent with the presence of two distinct crystallographic Sr^{2+} sites. The number of both sites is however the same and consequently the intensity of both emissions is expected to be identical. The explanation for the low intensity of the 420 nm band is found in the excitation spectra. The excitation band of the 490 nm emission indeed completely overlaps the 420 nm emission band. Because the $4f^65d^1$-$4f^7$ transitions in Eu^{2+} are allowed the critical distance for the transfer between these europium ions will be very high, i.e. 20-30 Å. Such efficient energy transfer between inequivalent Eu^{2+} site has also been observed in other Eu^{2+} phosphors [52]. Due to this efficient transfer even at a fairly low europium content almost only 490 nm emission will be found.

In order to increase the colour rendition even further, one has to step down from the narrow-band concept. When broad band green and red phosphors are used, more or less continuous spectra can be generated,

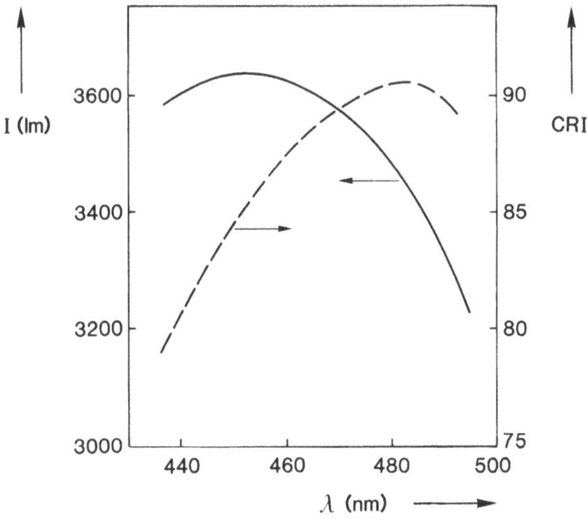

Fig.46
Lumen output and CRI of a TLD 36W T8 lamp as a function of the emission wavelength of the blue component.

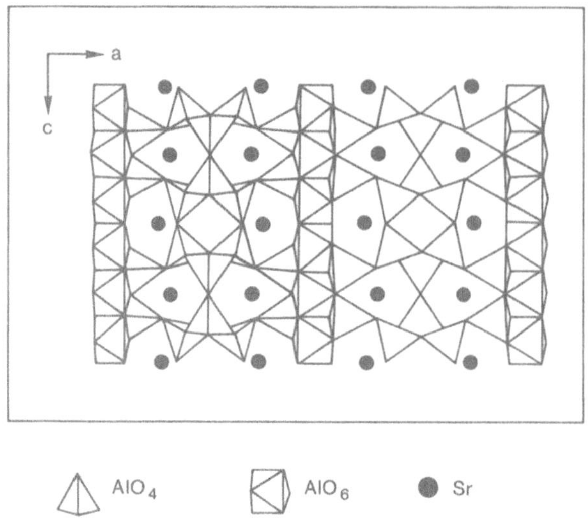

Fig.47
Projection of the crystal structure of $Sr_4Al_{14}O_{25}$.

resembling the visible emission of a black-body radiator. Red band emission is found in $GdMgB_5O_{10}$ doped with Mn^{2+} (Fig.49). The small Mn^{2+} ions are accommodated into the octahedral Mg^{2+} sites. For such a coordination the Mn^{2+} $^4T_1$-$^6A_1$ emission is mostly located in the red part of the spectrum [53]. It is of course also possible to dope the pentaborate simultaneously with Tb^{3+} and Mn^{2+}. Although for the compositions shown in Fig.49 the manganese concentration is smaller than the terbium concentration, the

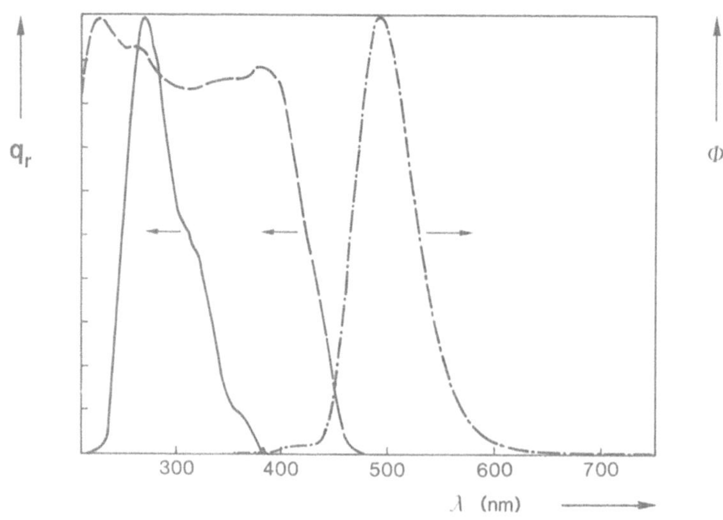

Fig.48
Room temperature excitation and emission spectrum of $Sr_4Al_{14}O_{25}$:0.1Eu^{2+}; ——— excitation spectrum of 420 nm emission; - - - - excitation spectrum of 490 nm emission and - . - . - emission spectrum under 254 nm excitation.

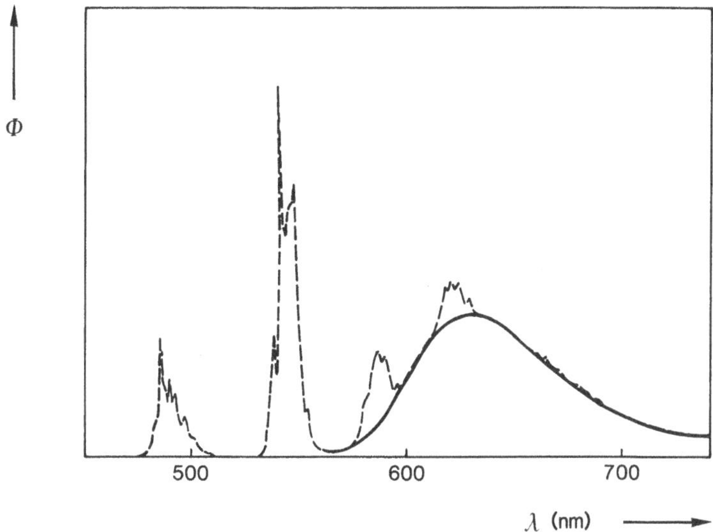

Fig.49
Emission spectrum of : ——— $Ce_{.2}Gd_{.8}Mg_{.9}Mn_{.1}B_5O_{10}$ and of - - - $Ce_{.3}Gd_{.5}Tb_{.2}Mg_{.9}Mn_{.1}B_5O_{10}$ at room temperature.

integrated Mn^{2+} emission turns out to be higher than the integrated terbium emission. This is consistent with the fact that transition metals are more efficient traps for the excitation energy, migrating through the Gd^{3+} lattice, than terbium ions are (section IV.B).

Until now we do not have at our disposal a suitable broad band phosphor, emitting in the green-yellow part of the spectrum. In order to cover this part of the spectrum the Tb^{3+} emission of the pentaborate is combined with the manganese emission of the halophosphate. By blending three phosphors, i.e. $Ca_5(PO_4)_3(Cl,F):Sb^{3+},Mn^{2+}$, $(Ce,Gd)MgB_5O_{10}:Tb^{3+},Mn^{2+}$ and $Sr_4Al_{10}O_{25}:Eu^{2+}$ a more or less continuous spectrum is obtained, extending

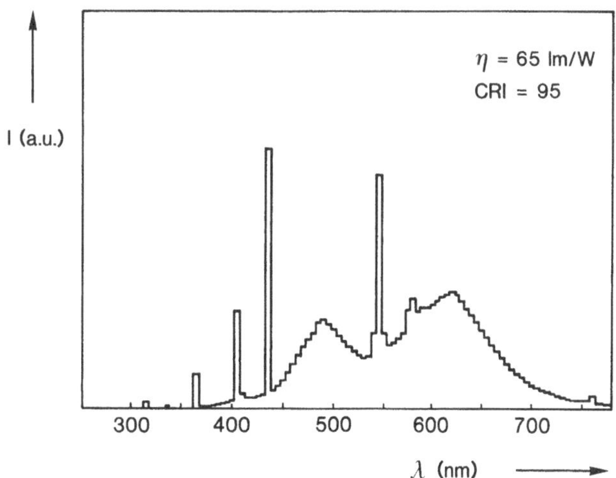

Fig.50
Emission spectrum of a Special de Luxe lamp with a correlated colour temperature of 4000 K.

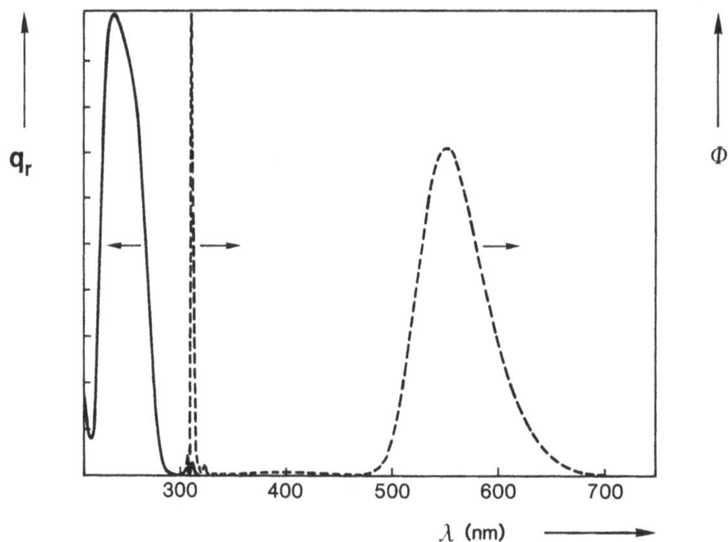

Fig.51
Room temperature excitation and emission spectrum of
$Sr_3Gd_2Si_6O_{18}:Pb^{2+}{}_{.06},Mn^{2+}{}_{.05}$.

from the blue to the deep red (Fig.50). The Special de Luxe lampcontaining this particular blend of phosphors yields a CRI of 95 and an efficacy of 65 lm/w.

5. Cost-Price Reduction. Rare-earth oxides, except for lanthanum and cerium, are expensive materials, because in their preparation a painstaking separation by solvent - solvent extraction is required. In order to reduce the cost price of the phosphor layer the content in rare-earths, therefore should be kept to a minimum. As we can deduce from table 8 the Tb^{3+} content of the second generation of green phosphors is substantially reduced with respect to $CeMgAl_{11}O_{19}:Tb^{3+}$.

One could go even one step further by eliminating terbium completely. The terbium emission consists of a main line at 545 nm and three major side bands at 490, 580 and 620 nm. This emission can also be replaced by a narrow band emission. Such an emission located at 550 nm is indeed found in $Sr_3Gd_2Si_6O_{18}$ doped with Pb^{2+} and Mn^{2+}. Also in this phosphor the energy transfer occurs via the Gd^{3+} sublattice, as is apparent from its excitation and emission spectrum (Fig.51). Excitation takes place in the $^1S_0$ - $^3P_1$ absorption band of the Pb^{2+} ion. Around 400 nm a shallow band is observed in emission, which is probably caused by the presence of Pb^{2+} clusters. The overall efficiency of this phosphor is about 85%, a sufficienct high value to replace the Tb^{3+} phosphors presently in use. This phosphor suffers however from radiation damage; under irradiation with 185 nm mercury defects are generated, lowering the output of the phosphor by about 30% [54].

As far as the red phosphor is concerned both the host lattice Y_2O_3 and the europium dopant largely contribute to the costs price. In the case of

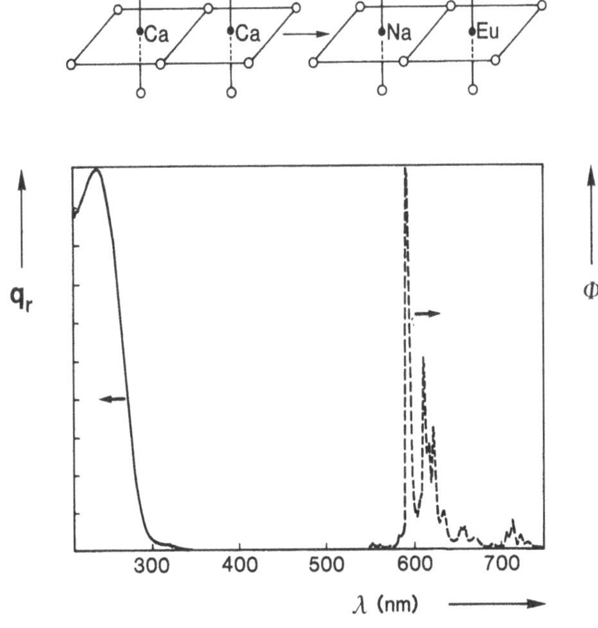

Fig.52
Luminescence spectrum of CaO:$Na^{+}_{.03}$,$Eu^{3+}_{.03}$ along with a tentative description of the europium ion coordination.

the red phosphor the use of a line emission is a prerequisite, the use of a band will always result in a drastic drop in lamp output. Line emission in this spectral region is solely found with Sm^{3+} or Eu^{3+}. Only by using a cheap host lattice, the cost price of the red phosphor can be reduced substantially.

Calcium ions nearly have the same ionic radius as trivalent Eu^{3+} ions, i.e. 1.00 and 0.95 Å respectively [55]. Therefore, Eu^{3+} can easily be incorporated in a calcium compound such as CaO. In order to compensate for the charge imbalance, an equimolar amount of sodium, with an ionic radius of 1.02 Å, can also be incorporated [55]. A typical emission spectrum of CaO:Eu^{3+},Na^{+} is shown in Fig.52; the $^5D_0$ -$^7F_1$ magnetic dipole transition apparently dominates this spectrum, indicating that the Eu^{3+} site shows inversion symmetry. The inversion symmetry of the calcium site in the rock salt structure of CaO is preserved, when the calcium ions are replaced by ions with nearly the same ionic radius. If CaO is solely doped with Eu^{3+}, a completely different emission spectrum emerges (Fig.53). In this case cation vacancies will compensate for the charge imbalance. At higher concentrations, the vacancies will tend to cluster with the europium ions, lowering the symmetry of the Eu^{3+} site. This distortion from the inversion symmetry results in a drastic increase in the intensity of the $^5D_0$ - $^7F_2$ hypersensitive transition.

In view of the position of its emission maximum at 590 nm, CaO:Na^{+},Eu^{3+} is not a good candidate for application in a tricolour lamp. In CaO:Eu^{3+}, however, the emission is located at the desired wavelength.

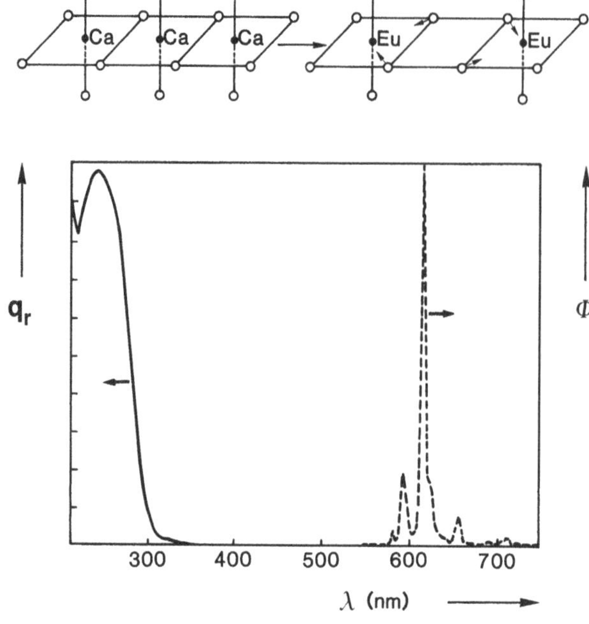

Fig.53
Luminescence spectrum of $CaO:Eu^{3+}_{.03}$ along with a tentative description of the europium ion coordination.

It should be noted that in this compound the intensity of the $^5D_0$ -$^7F_4$ transition is even lower than in the case of $Y_2O_3:Eu^{3+}$. Although the europium charge transfer band is located at nearly the same energy as in $Y_2O_3:Eu^{3+}$, the efficiency of the $CaO:Eu^{3+}$ only equals about 50%. Probably in the latter system the relaxation of the excited state is more pronounced. The larger shift of the minimum of the charge transfer state, with respect to the minimum in the $4f^6$ states, will induce non-radiative losses in $CaO:Eu^{3+}$.

All attempts to come to a drastic reduction in the costprice of the tricolour phosphor have been unsuccessful until now. Whether or not it is possible to obtain the outstanding luminescence properties of the present tricolour phosphors in much cheaper systems, is a point which only the future is able to prove.

V.B. SPECIAL APPLICATIONS

Most sensitizer ions, we have been dealing with, show efficient UV luminescence. This type of emission can be used to stimulate a number of chemical reactions. In industrial applications it is often possible to tune the stimulation spectra to the 254 or 365 nm radiation emitted by high pressure mercury lamps. This lamp type has a much higher intensity than the low pressure mercury lamps. For biochemical reactions, such as the ones induced in phototherapy, the emission of the lamp should be tuned

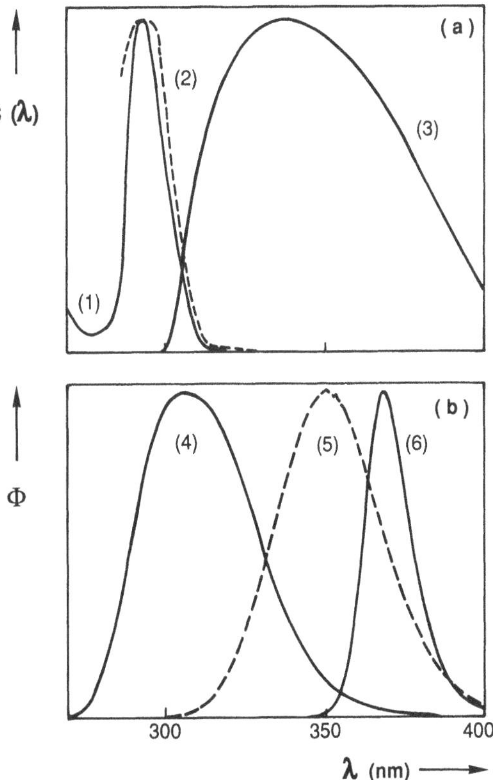

Fig.54

a. Relative spectral sensitivity S(λ) of the human skin for : (1) erythema [56] (2) delayed pigmentation [57] (3) direct pigmentation [58].

b. Emission spectrum of $Sr(Al_{11.9}Mg_{.1})O_{19}:Ce^{3+}{}_{.1}$ (4), $BaSi_2O_5:Pb^{2+}{}_{.03}$ (5) and $SrB_4O_7:Eu^{2+}{}_{.02}$ (6) under 254 nm excitation at room temperature.

exactly to the stimulation spectrum of the human skin. We will deal with two examples : the sun-tanning lamps and the psoriasis lamp.

1. Sun-Tanning Lamps. When the human skin is exposed to UV radiation with a wavelength larger than 280 nm three short-term processes can be induced : erythema or sunburn, skin darkening or direct pigmentation and melanin production or indirect pigmentation. Direct and indirect pigmentation are completely distinct processes. In the direct pigmentation process the melanin grains of the basal cells tend to concentrate on the cell surface. The delayed pigmentation process induces the generation of melanin, resulting in a more permanent effect.

The stimulation spectra for these three processes are given in Fig.54. Direct pigmentation is found in the long wavelength UV. The sensitivity curves for erythema and delayed pigmentation nearly coincide in the short-wavelength UV, although in the 310 to 330 nm region the sensitivity for delayed pigmentation turns out to be much higher than the sensitivity for erythema.

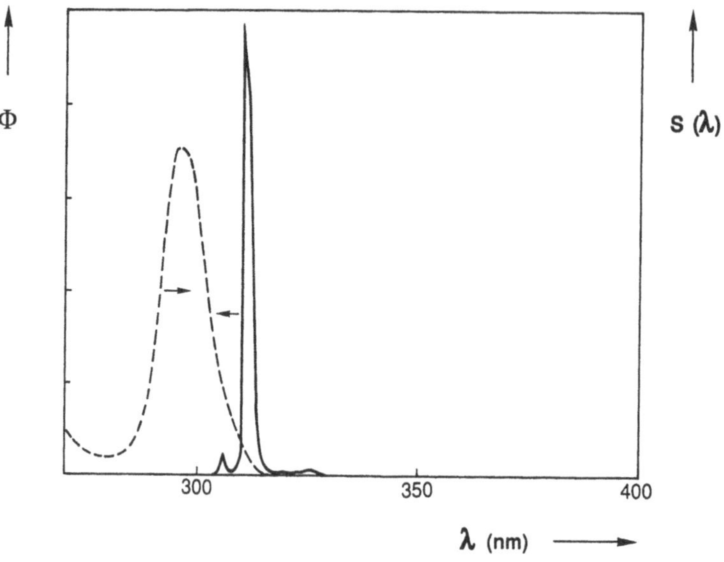

Fig.55
Room temperature emission spectrum of $(Gd_{.5},La_{.5})B_3O_6:Bi^{3+}_{.03}$ under 254 nm excitation along with the erythema sensitivity curve S(λ).

From these data it will be clear that the emission of $Sr(Al_{11.9}Mg_{.1})O_{19}:Ce^{3+}_{.1}$ will induce severe sun burn. No erythemal effect is expected from $SrB_4O_7:Eu^{2+}_{.02}$; this phosphor will induce direct pigmentation, but no delayed pigmentation whatsoever. For a long lasting tan the emission of the phosphor should also cover the 310 to 330 nm spectral region; such an emission is found for $BaSi_2O_5:Pb^{2+}_{.03}$. The radiation emitted by this phosphor hardly shows any erythemal effect but stimulates the direct as well as delayed pigmentation.

2. Psoriasis. Psoriasis is a common skin disease marked by red scaly patches. This disease can not be cured, but can be controlled using UV treatment. The gadolinium emission as found in $(Gd,La)B_3O_9:Bi^{3+}$ and $GdBO_3:Pr^{3+}$ was discovered to be highly effective [59]; short wavelength UV is indeed more effective than long wavelength UV in this respect. For wavelengths lower than 310 nm erythema starts to develop. Therefore, the gadolinium line emission is at the right position to control effectively the spread of psoriasis, without inflicting sun burn (Fig.55).

REFERENCES

1. G. Blasse, Structure and Bonding 42, 1 (1980)
2. A. Ranfagni, D. Mugai, M. Bacci, G. Viliani and M. Fontana, Adv. in Phys. 32, 823 (1983)
3. B. Smets, J. Verlijsdonk and J. Rutten, Mat. Res. Bull. 24, 431 (1989)
4. E. Oomen, W. Smit and G. Blasse, Phys. Rev. B. 37, 18 (1988)

5. G. Blasse, Prog. Solid St. Chem. 18, 79 (1988)
6. H. Kiliaan, J. Kotte and G. Blasse, Chem. Phys. Lett. 133, 425 (1986)
7. C. Struck and W. Fonger, J. Chem. Phys. 64, 1784 (1976)
8. J. van Dijk and M. Schuurmans, J. Chem. Phys. 78, 5317 (1983)
9. R. Reisfeld, in "Radiationless Processes" (Eds. B. di Bartolo and V. Goldberg), Plenum Press, New York (1980) p. 489
10. L. Riseberg and H. Moos, Phys. Rev. 174, 429 (1968)
11. C. Lange and M. Weber, Phys. Rev. B 16, 3259 (1977)
12. W. van de Weg and M. van Tol, Appl. Phys. Lett. 38, 705 (1981)
13. D. Dexter, J. Chem. Phys. 21, 863 (1953)
14. T. Förster, Ann. Phys. 2, 55 (1948)
15. C. Parent, P. Bochu, G. le Flem, P. Hagenmuller, J.C. Bourcet, and F. Gaume-Mahn, J. Phys. Chem. Solids 45, 39 (1984)
16. Th. Soules, R. Bateman, R. Haves and E. Kreidler, Phys. Rev. B 7, 1657 (1973)
17. I. Laulicht and S. Meirman, J. Lumin 34, 287 (1986)
18. N. Yamada, S. Shionaya and T. Kushida, J. Phys. Soc. Jap. 32, 1577 (1972)
19. T. Hoslstein, S. Lyo and R. Orbach, "Laser Spectroscopy of Solids" (Eds. W. Yen and P. Selzer), Springer Verlag, Berlin (1981), p. 38
20. M. Buijs, and G. Blasse, J. Lumin 34, 263 (1986)
21. M. Buijs, A. Meijerink and G. Blasse, J. Lumin 37, 9 (1987)
22. H. Kiliaan, A. Meijerink and G. Blasse. J. Lumin. 35, 155 (1986)
23. L. Elias, W. Heaps and W. Yen, Phys. Rev. B, 8, 4989 (1973)
24. A. de Vries, H. Kiliaan and G. Blasse, J. Solid State Chem. 65, 190 (1986)
25. A. de Vries, W. Smeets and G. Blasse, Mat. Chem. Phys., 18, 81 (1987)
26. W. Carnall, P. Fields and K. Rajnak, J. Chem. Phys., 49, 4412 (1968)
27. W. Carnall, Handbook on the Physics and Chemistry of Rare-Earth, (eds. K. Gschneider and L. Eyring), North Holland, Amsterdam, (1979). Vol. 3, p. 203
28. D. Huber, Laser Spectroscopy of Solids, (eds. W. Yen and P. Selzer), Springer Verlag, Berlin, (1981), chapter 3
29. D. Huber, Phys. Rev. 20, 2307 (1979)
30. B. Movaghar, G. Sauer and D. Wurtz, J. Stat. Phys. 27, 473 (1982)
31. P. Berdowski and G. Blasse, J. Lumin. 29, 243 (1984)
32. M. Buijs and G. Blasse, J. Lumin., 34, 263 (1983)
33. A. de Vries and G. Blasse, Mat. Res. Bull., 21, 68, (1986)
34. A. Srivastava, M Sobierja, S. Ruan and E. Banks, Mat. Res. Bull., 21, 1455 (1986)
35. G. Blasse and A. Bril, J. Chem. Phys., 47, 5139 (1967)
36. A. de Vries, G. Blasse and R. Pet, Mat. Res. Bull., 22, 1146 (1987)
37. H. Kiliaan, F. van Herwijen and G. Blasse, Mat. Chem. Phys., 18, 351 (1987)
38. A. Wolfert and G. Blasse, Mat. Res. Bull., 19, 67 (1984)
39. T. Tsuboi, Phys. Rev. B29, 1022 (1984)
40. H. Kiliaan and G. Blasse J. Electrochem. Soc. 136, 562 (1989)

41. H. Jenkings, A. McKeag and P. Ranby, J. Electrochem. Soc., 96, 1 (1949)
42. M. Koedam and J. Opstelten Light. Res. Technol., 3, 205 (1971)
43. J. Verstegen, D. Radielovic and L. Vrenken, J. Electrochem. Soc., 121, 1627 (1974)
44. R. Peacock, Struct. Bonding, 22, 83 (1975)
45. J. Verstegen, J. Sommerdijk and J. Verriet, J. Lumin., 6, 1425 (1973)
46. B. Saubat, M. Vlasse and C. Fouassier, J. Solid State Chem., 34, 271 (1980)
47. J.-C. Bourcet and F. Fong, J. Chem. Phys., 60, 34 (1974)
48. M. Saakes, M Leskelä and G. Blasse, Mater. Res. Bull. 19, 83 (1984)
49. B. Smets, J. Rutten, G. Hoeks and J. Verlijsdonk, J. Electrochem. Soc., 136, 2119 (1989)
50. J. v. Kemenade and G. Hoeks, Electrochem. Soc. Spring Mtg., extendend abstract no. 607, San Fransico, 1983.
51. T. Nadezhina, E. Pobedimskaya and N. Belov, Kristallografiya, 21, 826 (1976)
52. G. Blasse, J. Solid State Chem., 62, 207 (1986)
53. M. Leskelä, M. Saakes and G. Blasse, Mater. Res. Bull., 19, 151 (1984)
54. H. Verhaar and W. van Kemenade, Submitted for publication in Mater. Chem. Phys.
55. R. Shannon and C. Prewitt, Acta Crystallogr. B, 25, 925 (1969)
56. A. Mc. Kinlay and B. Diffy, "Human Exposure to Ultraviolet radiation - Risks and Regulations" Excerpta Medica, Oxford (1987)
57. J. Parrish, K. Jaenicke and R. Anderson, Photochem. Photobiol. 36, 187-189 (1982)
58. W. Henscke and R. Schulze, Strahlentherapie 64, 14 (1939)
59. H. van Weelden, H. Baart de la Faille, E. Young and J. van der Leun, Brit. J. Dermatol., 119, 11 (1988)

LASER SPECTROSCOPY INSIDE INHOMOGENEOUSLY BROADENED LINES

Roger M. Macfarlane

IBM Research Division
Almaden Research Center
650 Harry Road
San Jose, California 95120-6099

ABSTRACT

Spectral lines in solids are often inhomogeneously broadened at low temperatures and this obscures important information such as the homogeneous linewidth, hyperfine structure, ion-pair interactions and tunneling splittings. We describe the use of frequency domain (e.g., spectral holeburning) and time domain (e.g., photon-echo, optical free induction decay (FID) and quantum beat) techniques to eliminate the effects of inhomogeneous broadening. In this way spectral resolution can be increased by three to eight orders of magnitude. These techniques are illustrated by numerous examples taken from the $f^n \rightarrow f^n$ spectroscopy of trivalent rare-earth ions doped into insulating solids. For metastable levels at low temperatures homogeneous linewidths of $\sim 1 - 10$ kHz can be measured. Mechanisms responsible for spectral holeburning are classified according to the level which stores population in the bleaching process. We discuss the almost universal role of fluctuations of host-lattice nuclear spins in determining the homogeneous linewidths of metastable optical levels at low temperatures ($T < 4K$), and point out that the dynamics of these nuclear spins are strongly modified by the presence of the rare-earth ions. The result is a spread of nuclear spin fluctuation times ($10^{-5} - 10^{-2}$ sec.) which causes spectral diffusion resulting in non-exponential photon-echo decays and time dependent holewidths. Another consequence of spectral diffusion is that the measurement of the "homogeneous linewidth" by photon echo or FID, and holeburning techniques can give different values in apparent contradiction. The difference, however, is due to the different time scale of the two measurements with respect to the nuclear spin-flip times. The use of sharp probes such as spectral holes to sensitively measure small splittings and the effect of external perturbations is also illustrated.

Optical Properties of Excited States in Solids
Edited by B. Di Bartolo, Plenum Press, New York, 1992

I. INTRODUCTION

Optical spectroscopy is fundamentally important to our understanding of excited states in solids. The advent of lasers, with their properties of coherence and monochromaticity, dramatically changed the face of optical spectroscopy and the kind of information that can be obtained from it [1-6]. For the case of solids, the f-f transitions of rare-earth ions in crystals have so far provided one of the best demonstrations of the richness available from modern laser spectroscopy [5], and these will be discussed here. The spectroscopy of rare-earths has a long and distinguished history and has received much attention in the last two or three decades because of the importance of rare-earth doped materials as solid state lasers. The techniques and the general principles we describe can, however, be applied to a much wider range of materials.

To provide an overview of the f-electron energy levels of rare-earths, the famous diagram first constructed by Dieke and Crosswhite [7] is reproduced in Fig. 1. In a crystalline environment, the J-multiplets of the free-ion are split into several components, typically $\sim$100 cm^{-1} apart. Since these splittings are of the same order as phonon frequencies, the upper components of a manifold are often broadened by up to several cm^{-1} due to spontaneous phonon emission even at liquid helium temperatures. On the other hand, optical transitions between the ground state and the lowest level of metastable multiplets can be very sharp at low temperatures, and these are of particular interest for high resolution laser spectroscopy.

The crystal-field theory describing the energies of these f-electron levels has been extensively developed and is treated in a number of books and reviews [8-11]. The energy levels are expressed in terms of a limited set of free-ion and phenomenological crystal field parameters, and this has led to a very successful systematic understanding of the spectroscopy of rare-earth ions in a variety of insulating and semiconducting crystals. An approach to understanding the origins of the crystal field parameters has been elaborated in the superposition model of Newman and co-workers [12]. In this model phenomenological crystal-field parameters are separated into geometrical parts determined by the coordination geometry, which can be calculated, and a more intrinsic part which remains phenomenological.

In rare-earth ions the 4f electrons are shielded from the crystalline environment by the outer 5s and 5p electrons resulting in crystal field splittings which are less than those due to the spin-orbit interaction, which separates states of different J, or the Coulomb interaction which separates terms of different S and L. The crystal field mixes states of different J; the good quantum numbers are now the irreducible representations of the point symmetry group of the rare-earth site. In the actinide series, the 5f electrons are less shielded and the effects of the crystal field are stronger. Optical transitions within the f-electron manifolds have magnetic dipole character when the rare-earth site is centrosymmetric and electric dipole when the site is non-centrosymmetric. In the latter

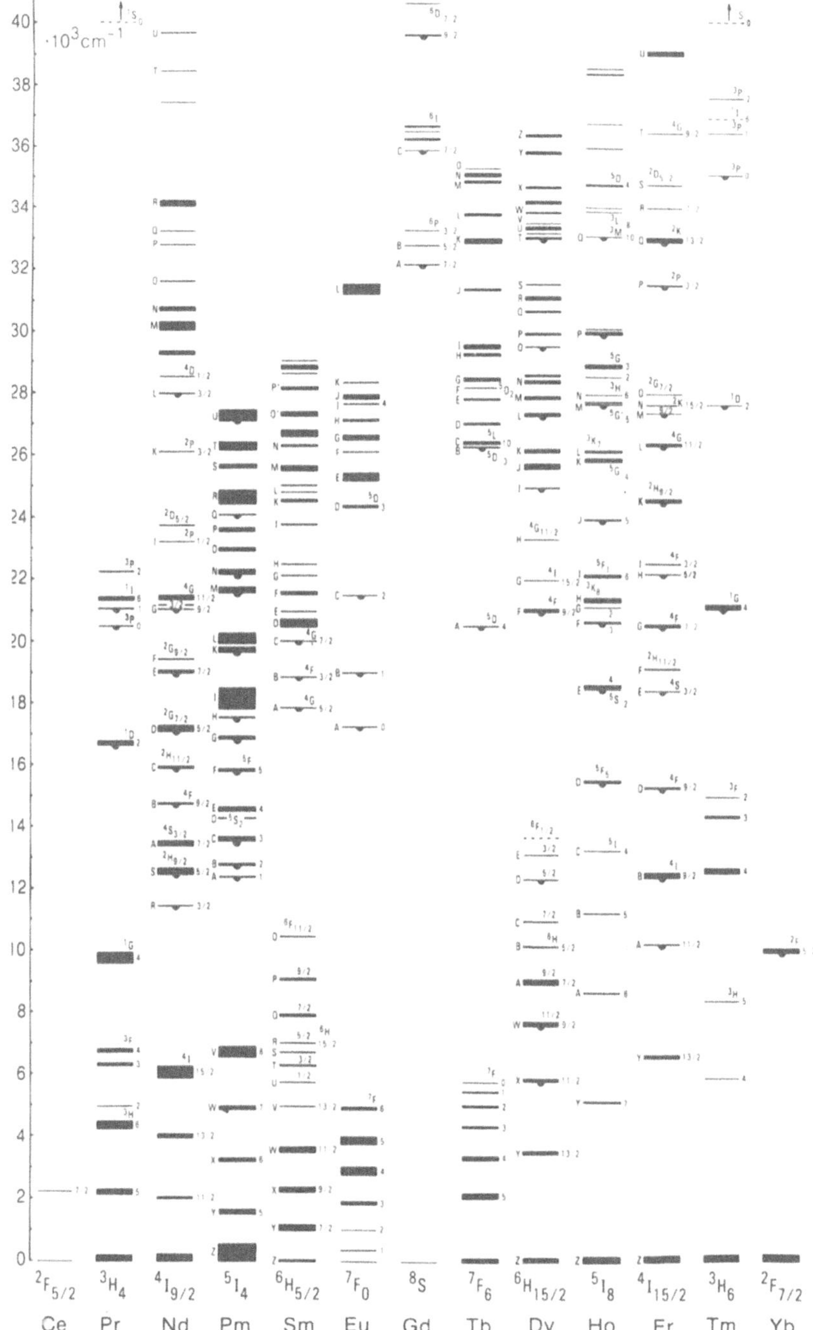

Fig. 1. Energy levels of the triply ionized rare-earths in $LaCl_3$ (after [7])

case the crystal field mixes in odd parity states from other configurations, a process analyzed in some detail by Judd [13] and Ofelt [14], whose formalism has been widely applied since that time [15].

Here we describe the application of laser spectroscopy to ionic solids, with particular reference to techniqes for achieving resolution many orders of magnitude greater than the limits imposed by inhomogeneous broadening. This increased resolution has been used to measure, for example, homogeneous line broadening, hyperfine interactions, spectral diffusion and the effects of small external perturbations. The techniques used involve both frequency and time domain measurements. Numerous examples will be given, together with references to original work for further details and discussion.

II. HOMOGENEOUS BROADENING

Line broadening limits the resolution available in spectroscopic measurements but it also contains a considerable amount of information about the interaction of optical centers with their environment. In a simple view this can be divided into homogeneous broadening, an intrinsic broadening due to dynamical effects which is experienced equally by all the optical centers in a solid, or inhomogeneous broadening which is static, and arises from the spread of resonance frequencies resulting from the distribution of micro-environments such as defects, disorder or lattice strains. This separation is not always possible as it depends on the time-scale of the measurement process, but it remains a useful simplified view [16]. In an equivalent time domain description, the homogeneous linewidth Γ is related to a phase coherence time T_2 and the population decay time T_1 by

$$\Gamma = (2\pi T_1)^{-1} + (\pi T_2)^{-1} \tag{1}$$

The concept of T_1 and T_2 has its origins in the nuclear magnetic resonance (NMR) of 2-level nuclear spin systems [17]. In many respects these are analogous to 2-level optical systems and the coherent dynamics can be described by the equation of motion of the density matrix ρ for the energies of the 2-level system (levels a,b) and the interaction of the radiation field with the material system:

$$i\hbar\dot{\rho} = [\mathscr{H}, \rho] + \text{relaxation terms} \tag{2}$$

where $\mathscr{H}$ is the Hamiltonian of the system including the unperturbed 2-level system and the light-matter interaction. Equation (2) leads to the optical Bloch equations (3), for the Bloch vector $\mathbf{R} = (u, v, w)$ whose motion in a three-space $(\vec{1}, \vec{2}, \vec{3})$ describes the dynamics of the coherent light-matter interaction. The motion is simplified by choosing a coordinate system rotating about the $\vec{3}$ axis at the optical frequency.

$$\dot{u} + \Delta v + \frac{u}{T_2} = 0,$$
$$\dot{v} - \Delta u - \chi w + \frac{v}{T_2} = 0, \qquad (3)$$
$$\dot{w} + \chi v + \frac{(w - w_{eq})}{T_1} = 0,$$

where

$$u = \rho_{ab} + \rho_{ba},$$
$$v = i(\rho_{ba} - \rho_{ab}), \qquad (4)$$
$$w = \rho_{bb} - \rho_{aa}$$

The damping terms in (3) are introduced phenomenologically. The relaxation time T_2 corresponds to motion of the Bloch vector **R** in the 1-2 plane and T_1 to a reduction in the length of this vector. For the lowest level of an SLJ manifold, T_1 is normally much greater than T_2 and Γ is limited by pure dephasing processes. Neglecting relaxation, (3) is equivalent to

$$\frac{d\mathbf{R}}{dt} = \mathbf{\Omega} x \mathbf{R} \qquad (5)$$

where the driving vector $\mathbf{\Omega}$ = $[-\chi, 0, \Delta]$ is expressed in terms of the optical Rabi frequency χ = $(\mu.\mathbf{E})/\hbar$ (μ_{ab} is the transition dipole moment, **E** the electric field of the light), and Δ the frequency difference between that of the driving field and the 2-level resonance ω_{ab}.

These equations are the cornerstone of coherent spectroscopy and their derivation is described in many works, see for example references [1-3]. Equation (5) is the basis of the vector model in which the driving field exerts a torque on the Bloch vector. This allows a geometrical interpretation of the equations of motion [18] as summarized in Fig. 2. The description of coherent optical phenomena in terms of the 2-level Bloch equations breaks down in two important regimes. The first is when three or more levels are present, as a result of hyperfine interactions for example, and the radiation field couples more than two levels. The second is when the radiation field is strong and it partially decouples the electronic system from the relaxation bath [19].

At room temperature most of the spectral lines in solids are homogeneously broadened by the absorption, emission or scattering of phonons. When the electron-phonon coupling is strong it is common to observe broad bands corresponding to electron-plus-phonon transitions which are also homogeneously broadened, appearing as a convolution of one or more phonon density of states distributions. Such features are commonplace in the spectra of transition metal ions and color centers for example.

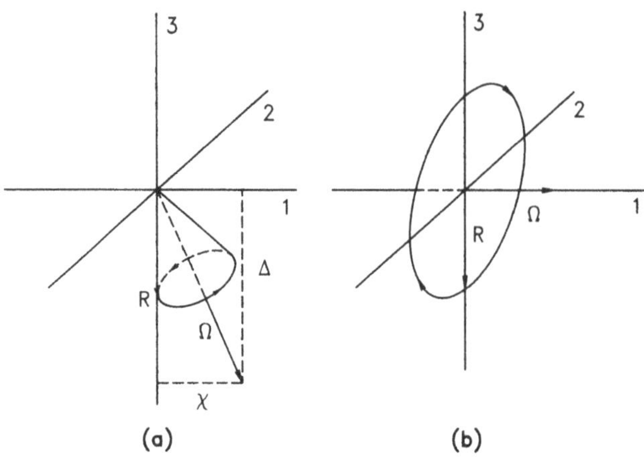

Fig. 2. Vector model of a 2-level transition. In (a) the driving vector Ω causes the Bloch vector R to precess in a cone about Ω. (b) At exact resonance the cone becomes a circle in the 2-3 plane.

The subject of phonon-induced broadening of the zero-phonon lines of defect centers in this context was first elaborated by McCumber and Sturge [20] and McCumber [21], and later generalized by Osad'ko [22] and Skinner and co-workers [23,24]. A very useful, didactic treatment has been given by DiBartolo [25]. Phonon broadening has two main origins, direct processes where a single phonon is emitted or absorbed, and phonon scattering processes which may be quasi-elastic or inelastic. Higher order multiphonon processes may be important in determining relaxation rates from metastable levels, or cross-relaxation rates, but they are rarely important in limiting the linewidth. The contribution from direct processes connecting a pair of electronic levels separated by a frequency ω has the form:

$$\Gamma(T)_{dir} = C_1[1 + n(\omega,T)] \tag{6}$$

where $n = 1/[\exp(\hbar\omega/kT - 1)]$ is the occupation number of phonons of frequency ω at temperature T. If C_1 is known from the lifetime of the upper level determined by spontaneous phonon emission, $C_1 n(\omega)$ can be obtained and this gives the temperature dependence of the width of the lower level.

With the assumption of a Debye phonon spectrum and a small inelasticity in the scattering process, the phonon scattering contribution can be expressed as [20]:

$$\Gamma_{scat} = C_2\left(\frac{T}{T_D}\right)^7 \int_0^{T_D/T} \frac{x^6 e^x dx}{(e^x - 1)^2} \tag{7}$$

where C_2 is the electron-phonon coupling strength and T_D the Debye temperature. The value of the integral in (7) has been tabulated by Di Bartolo [25]. This contribution is proportional to T^7 at low temperature and T^2 at high temperature. In a real system it is often difficult to extract the different contributions to the overall phonon-induced broadening because there are numerous direct and scattering channels and the phonon density of states can deviate significantly from a Debye spectrum [23]. At low temperatures (below about 5–10K), homogeneous broadening due to the fluctuating local fields of electron spins or nuclear spins on surrounding ions can compete with or even dominate phonon contributions. The nuclear spin contribution in particular is almost always present and provides the limiting linewidth below 2K. This was clearly demonstrated by experiments in which the nuclear spins were decoupled from the electrons by coherent averaging [26,27]. To observe these non-phonon line broadening mechanisms, techniques must be used which eliminate inhomogeneous broadening. Some of these are discussed below. In the case of disordered solids such as glasses there are additional low frequency excitations described as tunneling states or two level systems, which dominate the low temperature thermal broadening and give rise to a temperature dependence with an exponent between linear and quadratic [28].

III. INHOMOGENEOUS BROADENING

As we have seen, at room temperature, homogeneous broadening is normally the dominant source of linewidth. At low temperatures however several situations obtain. When the separation between electronic energy levels is of the same order as the phonon frequencies, spontaneous emission from upper levels can occur, producing a broadening due to T_1 processes, which for rare-earth levels varies from ~MHz to ~cm^{-1} depending on the energy separations and the corresponding density of states at this separation. The levels of most interest to us here are relatively metastable, with lifetimes of μsec to msec; for example isolated electronic levels or the lowest level of a manifold well separated from others. In this case the lines exhibit inhomogeneous broadening, i.e., the spectral lines are an envelope over much narrower homogeneous linewidths. Traditionally, a distinction has been made between the static distributions of local environments due to strains, defects and magnetic spin configurations etc. which give rise to inhomogeneous broadening, and fluctuating components which cause homogeneous broadening. The transition from inhomogeneous broadening of a metastable level to homogeneous broadening of higher lying levels by spontaneous phonon emission at low temperatures (2K), is illustrated by measurements of the total linewidth of six transitions from the $^5I_8$ ground state of Ho^{3+} in LaF_3 to components of the $^5F_5$ term (Fig. 3). These measurements were made in a fluorescence excitation spectrum by scanning a dye laser with a width of ~1 MHz. The transition to the lowest component at 6411.48Å is inhomogeneously broadened by ~1 GHz due to crystal strains and shows partially resolved hyperfine splittings of the $^5I_8$ ground state. The lifetime of this level is 550μsec which contributes only 290 Hz to the total linewidth. The transitions to higher lying levels become increasingly broader due to spontaneous phonon emission. There is a dramatic increase for the level at 51 cm^{-1} where the lowest optical phonon mode

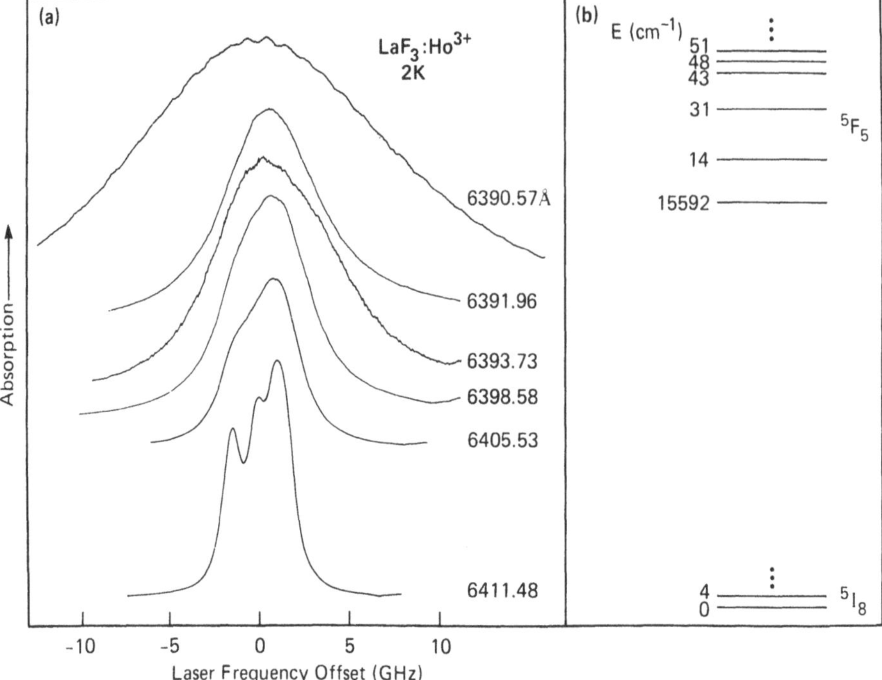

Fig. 3. (a) Excitation spectrum of absorption to several excited components of the $^5F_5$ manifold of $LaF_3:Ho^{3+}$ at 2K showing the transition from inhomogeneous broadening of the lowest component to homogeneous broadening of the upper levels. (b) The relevant energy levels.

contributes a peak to the density of states. The linewidth of this transition at 6390.57A is 15 GHz which corresponds to a lifetime for spontaneous phonon emission of 10 psec.

It is normally assumed that inhomogeneously broadened lines have a smooth, near Gaussian shape. However, the nature of the inhomogeneity also contains discrete components due to defects, ion-pairs etc. which give additional structure, especially in the wings of the line [29]. More generally the line profile shows statistical fine structure due to fluctuations in the number of atoms absorbing in a given homogeneous packet [30]. When a very small spatial volume is excited, the dopant concentration is low and the homogenous linewidth narrow, the discreteness of the inhomogeneous broadening in the wings of the line reaches its ultimate limit of individual absorptions due to single molecules [31,32]. This structure contains a great deal of detailed information on the nature of real solids but much more work has to be done before it is completely understood.

IV. HYPERFINE INTERACTIONS

IV.A. The Hyperfine Hamiltonian

Hyperfine interactions play a very important role in the high resolution spectroscopy of rare-earth ions because they give rise to resolvable structure and are often the source of line broadening at low temperatures. For the most part they are adequately handled by effective, or spin, Hamiltonians starting from a zeroth order basis of the crystal field Hamiltonian. In most of the experiments to be reviewed below, the results are well accounted for by the Hamiltonian,

$$\mathcal{H} = [\mathcal{H}_{FI} + \mathcal{H}_{CF}] + [\mathcal{H}_{HF} + \mathcal{H}_{Q} + \mathcal{H}_{Z} + \mathcal{H}_{z}]. \tag{8}$$

The first two terms are the free ion Hamiltonian including spin-orbit coupling, and the crystal field Hamiltonian. The remaining terms have much smaller contributions of the order of optical inhomogeneous broadening and are, in order of appearance, the hyperfine coupling between the 4f electrons and the rare earth nucleus, the nuclear electric quadrupole interaction, the electronic Zeeman interaction, and the nuclear Zeeman interaction.

In the calculation of crystal field energy levels and electronic g-values, the first two terms of (8) are often diagonalized within the $4f^n$ configuration, thus taking into account intermediate spin-orbit coupling, and crystal field J-mixing. A simpler treatment of the zero-order Hamiltonian is, however, almost always used for a discussion of hyperfine interactions, i.e., the use of L-S coupling and a treatment of the crystal field interaction diagonal in J. This is generally a good approximation since the separation between $^{2S+1}L_J$ manifolds is larger than crystal field splittings.

Ions with an odd number of f-electrons, i.e., Ce^{3+}, Nd^{3+}, Sm^{3+}, Gd^{3+}, Dy^{3+}, Er^{3+} and Yb^{3+} have electronic Kramers' doublet levels with magnetic moments of the order of a Bohr magneton. For even numbers of f-electrons, i.e., Pr^{3+}, Pm^{3+}, Eu^{3+}, Tb^{3+}, Ho^{3+} and Tm^{3+} the levels are electronic singlets where the orbital angular momentum is quenched by the crystal field except in sites of axial or higher symmetry where some levels are non-Kramers' doublets with nonzero angular momentum along the symmetry axis.

The crystal-field wavefunctions, i.e., the eigenstates of the first two terms in (8), determine the hyperfine structure and magnetic response due to the last four terms. The results are most simply expressed in terms of an effective Hamiltonian for the electronic level of interest. The explicit forms of these four terms beginning with the magnetic hyperfine interaction are:

$$\mathcal{H}_{HF} = 2\mu_B \hbar \gamma_N < r^{-3} > \vec{N} \cdot \vec{I} \tag{9}$$

where μ_B is the Bohr magneton, γ_N is the nuclear gyromagnetic ratio, $2\mu_B < r^{-3} > \vec{N}$ represents the field at the nucleus due to the f electrons and $< r^{-3} >$ is the expectation value of the inverse cube electron-nuclear distance. Within one LSJ state,

$$\mathscr{H}_{HF} = \Lambda_J \vec{I} \cdot \vec{J} \tag{10}$$

where

$$\Lambda_J = 2\mu_B \gamma_N \hbar < r^{-3} > < J\|N\|J > . \tag{11}$$

The quadrupole Hamiltonian has the form:

$$\mathscr{H}_Q = P[(I_{z'}^2 - I(I+1)/3) + (\eta/3)(I_{x'}^2 - I_{y'}^2)], \tag{12}$$

where P is the quadrupole coupling constant, η the electric field gradient (EFG) asymmetry parameter and x'y'z' are the principal axes of the EFG tensor. $\mathscr{H}_Q$ has contributions from the lattice, from the f electrons and from coupling to closed shells whose charge distribution is distorted by the electric field of the lattice. This latter, the so-called Sternheimer anti-shielding effect [33], often dominates.

The interaction of the electrons with an external magnetic field is given by:

$$\mathscr{H}_Z = \mu_B \vec{H} \cdot (\vec{L} + 2\vec{S}) \tag{13}$$

where $\vec{H}$ is the external magnetic field, $\vec{L}$ and $\vec{S}$ are the electronic orbital and spin angular momenta, and μ_B is the Bohr magneton. Within a single LSJ state, this can be written as

$$\mathscr{H}_Z = g_J \mu_B \vec{H} \cdot \vec{J} \tag{14}$$

where g_J is the Lande g-value, $g_J = (3/2) - [L(L+1) - S(S+1)]/[2J(J+1)]$. The nuclear Zeeman interaction is

$$\mathscr{H}_z = -\hbar \gamma_N \vec{H} \cdot \vec{I} \tag{15}$$

with γ_N and I being the nuclear gyromagnetic ratio and nuclear spin, respectively.

IV.B. Electronic Singlets

For non-Kramers' ions in sites of low symmetry (e.g., the C_2 site in LaF_3 or the C_s site in $YAlO_3$) the electronic degeneracy of a given J-state is completely lifted, and the electronic angular momentum is quenched to first order. Hyperfine and electronic magnetic effects appear in second order of perturbation theory, and the effective Hamiltonian including nuclear Zeeman and nuclear quadrupole interactions, is [34]:

$$\mathscr{H}_{eff} = -g_J^2\mu_B^2\vec{H}\bullet\tilde{\Lambda}\bullet\vec{H} - (2\Lambda_J g_J\mu_B\vec{H}\bullet\tilde{\Lambda}\bullet\vec{I} - \mathscr{H}_Z) - (\Lambda_J^2\vec{I}\bullet\tilde{\Lambda}\bullet\vec{I} + \mathscr{H}_Q)$$
$$\equiv -g_J^2\mu_B^2\vec{H}\bullet\tilde{\Lambda}\bullet\vec{H} + \mathscr{H}_Z' + \mathscr{H}_Q' \tag{16}$$

where the tensor $\tilde{\Lambda}$ is given by

$$\Lambda_{\alpha\beta} = \sum_{n=1}^{2J+1} \frac{<0|J_\alpha|n><n|J_\beta|0>}{\Delta E_{n,0}} \tag{17}$$

with $|0>$ being the level to which $\mathscr{H}_{eff}$ applies, $|n>$ the other crystal field levels of the LSJ term involved, and $\Delta E_{n,0}$ is the energy difference $E_n - E_0$. The first term in (16) is the quadratic electronic Zeeman shift, $\mathscr{H}_Z'$ is the enhanced nuclear Zeeman Hamiltonian [35], i.e.,

$$\mathscr{H}_Z' = -\hbar[\gamma_x H_x I_x + \gamma_y H_y I_y + \gamma_z H_z I_z] \tag{18}$$

where x, y and z are the principal axes of the Λ-tensor and the effective enhanced nuclear gyromagnetic ratios are given by

$$\gamma_\alpha = \gamma_N + \frac{2g_J\mu_B\Lambda_J\Lambda_{\alpha\alpha}}{\hbar}. \tag{19}$$

The term $\Lambda_J^2\vec{I}\bullet\tilde{\Lambda}\bullet\vec{I}$ in (16) is the second order magnetic hyperfine interaction, or pseudo-quadrupole interaction [36], which can be written in the same form as $\mathscr{H}_Q$, i.e.,

$$\mathscr{H}_{pq} = D_{pq}[I_z^2 - I(I+1)/3] + E_{pq}[I_x^2 - I_y^2]. \tag{20}$$

In terms of the Λ tensor, the zero field parameters D_{pq} and E_{pq} can be written:

$$D_{pq} = \Lambda_J^2 \bullet \left[\frac{1}{2}(\Lambda_{xx} + \Lambda_{yy}) - \Lambda_{zz}\right] \tag{21}$$

$$E_{pq} = \Lambda_J^2 \bullet \frac{1}{2}[\Lambda_{yy} - \Lambda_{xx}]. \tag{22}$$

An accurate calculation of the enhanced nuclear Zeeman and pseudo-quadrupole contributions requires accurate crystal field wavefunctions. These are rarely available for ions in low symmetry sites but in some cases it is possible to obtain good agreement with measured values. When combining the pseudo-quadrupole and pure quadrupole terms into one effective quadrupole Hamiltonian, it must be kept in mind that the principal axes of the Λ-tensor (x,y,z), and of $\mathscr{H}_Q(x', y', z')$ are in general different, although they coincide in cases of axial or higher symmetry. When the sum of these two terms is diagonalized, a third set of principal axes, x″, y″ and z″ will result for which:

$$\mathscr{H}'_Q = D[I^2_{z''} - I(I+1)/3] + E[I^2_{x''} - I^2_{y''}]. \tag{23}$$

D and E in this equation correspond to the combined quadrupole and second order hyperfine coupling constants in the double primed axis system. For certain rare earth ions, the enhanced nuclear Zeeman and pseudo-quadrupole coupling are much larger than the nuclear Zeeman and electric quadrupole interactions. In this case the axes x, y, z and x″, y″, z″ may nearly coincide.

IV.C. Non-Kramers' Doublets

In axial or higher symmetry, not all of the levels have their electronic angular momentum along the symmetry axis quenched by the crystal field. As a consequence of time reversal symmetry, the angular momentum perpendicular to the symmetry axis is zero to first order. The resulting effective Hamiltonian is dominated by the parallel electronic Zeeman effect and a hyperfine term containing only S_zI_z. The second order terms discussed in the preceding section are present as well and can be treated in the form of enhanced nuclear Zeeman and pseudo-quadrupole spin Hamiltonians. Thus, a suitable effective Hamiltonian for a non-Kramers' doublet is

$$\mathscr{H} = g_z\mu_B H_z S_z + \Lambda S_z I_z + \mathscr{H}'_Q + \mathscr{H}'_z \tag{24}$$

where S_z is an effective spin quantum number which can take the values $\pm 1/2$. The quantities g_z and Λ are given by $g_z = 2g_J < + |J_z| + >$ and $\Lambda = 2\Lambda_J < + |J_z| + >$ with $| + >$ and $| - >$ being the two components of the doublet. $\mathscr{H}'_z$ and $\mathscr{H}'_Q$ are defined as in (16) and (23), respectively.

IV.D. Kramers' Doublets

For ions with an odd number of electrons, Kramers' theorem dictates that all levels have electronic degeneracy which can only be lifted by time-odd interactions such as the Zeeman interaction with an external magnetic field. This electronic degeneracy and associated magnetic moment leads to large first order Zeeman and hyperfine interactions. Thus, an appropriate spin Hamiltonian is similar to (24) except that the g-values and hyperfine interactions are anisotropic and nonzero in all three principal directions:

$$\mathscr{H} = \mu_B \vec{H} \cdot \tilde{g} \cdot \vec{S} + \vec{I} \cdot \tilde{\Lambda} \cdot \vec{S} + \mathscr{H}'_Q + \mathscr{H}'_z. \tag{25}$$

Since Kramers' degeneracy arises from time reversal symmetry, random strains cannot lift this degeneracy but they may contribute to inhomogeneous broadening of transitions between different Kramers' doublets.

V. SPECTRAL HOLEBURNING

V.A. Introduction

Spectral holes can be bleached in inhomogeneously broadened line profiles by irradiation with a narrow band laser source which selectively removes population from the ground state for those ions resonant with the laser. This is shown schematically in Fig. 4. The absorption which is lost can, in principle, appear at other frequencies as 'antiholes'. However this is normally not observed since the transition corresponding to the antihole is no longer selectively narrowed since the inhomogeneous broadening acts differently on different transitions. One notable exception to this is when the population reservoir is a hyperfine component since lattice strains and other crystal field effects shift the hyperfine components in a similar way [38]. Spectral holes provide a very useful sharp feature for a wide variety of high resolution spectroscopic measurements. In addition there are many mechanisms for spectral holeburning which are interesting in themselves involving selective photophysical and photochemical processes. The relaxation of the holes can provide convenient or even unique ways to measure population relaxation from otherwise inaccessible levels.

V.B. Mechanisms for Holeburning

Figure 5 summarizes the most important mechanisms for rare-earth ions. The majority are applicable to other systems such as transition metal ions, color centers, defects in semiconductors and organic molecules. The general principle is based on the existence of a population reservoir above the ground state and isolated from it by a

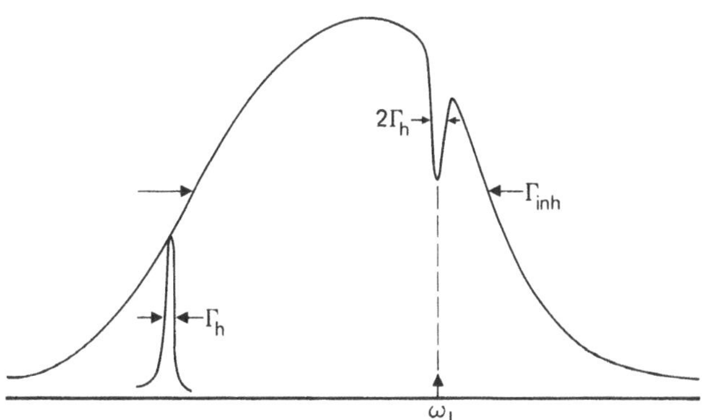

Fig. 4. Inhomogeneous line as the envelope of much narrower homogeneous components. Selective irradiation with a narrow band laser can lead to holeburning with a hole-width equal to twice the homogeneous linewidth.

relaxation rate which is longer than the excited state lifetime. The separation in energy of this reservoir from the ground state has to be larger than the laser bandwidth or the homogeneous linewidth of the optical transition so that the laser does not pump ions from the reservoir. Conversely as lasers become more frequency stable new population reservoirs become available. It is convenient to describe the mechanisms of holeburning in terms of these reservoir levels.

1. Population Saturation. The simplest conceptually, and in principle the most widely applicable mechanism, involves a relatively long-lived excited electronic state. We first consider two-level saturation (Fig. 5a) where population is transferred from the ground state to the excited state coupled by the incident laser field. The frequency dependence of the population in the vicinity of the steady-state saturation hole is given by:

$$I(\omega) = \frac{1 + \omega^2 T_2^2}{\omega^2 T_2^2 + (1 + \chi^2 T_1 T_2)} \tag{26}$$

where the width of the hole is:

$$\Gamma = (\pi T_2)^{-1}[1 + \chi^2 T_1 T_2] \tag{27}$$

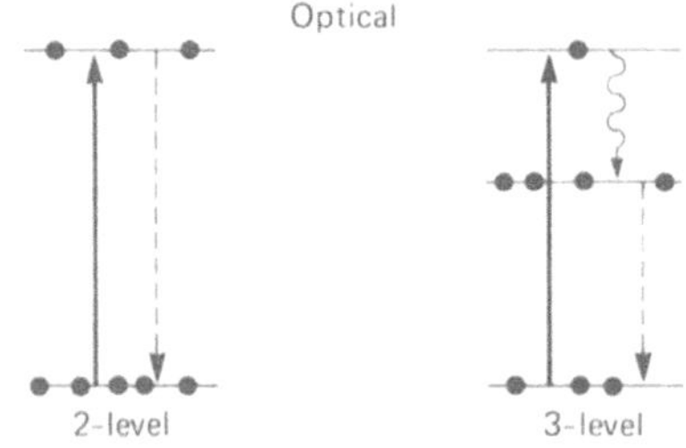

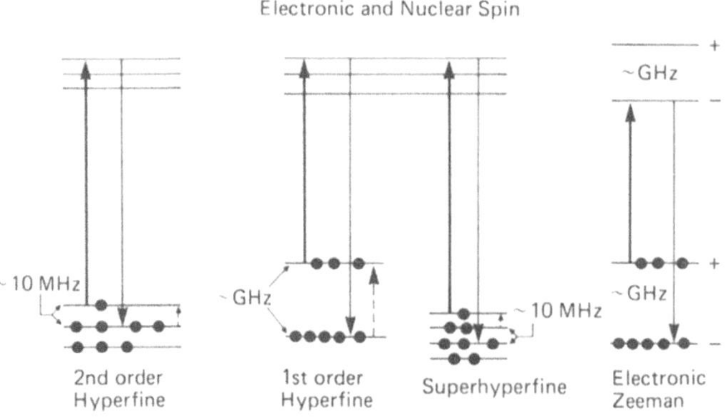

Fig. 5. Schematic illustration of mechanisms for spectral holeburning showing many of the population reservoirs which lead to the bleaching process.

The term in square brackets represents the power broadening which can be a pronounced effect when T_2 is long. This mechanism was first invoked for solids by Szabo [37] who measured holeburning in the $^4A_2 \rightarrow ^2E(\bar{E})$ level of Cr^{3+} in ruby. Holes of this kind should be observed in transmission since the emission from the upper level occurs on the same timescale as the hole lifetime and hence is too slow as a probe of the hole in a fluorescence excitation spectrum. A very wide range of systems is expected to exhibit two-level saturation and the introduction of high powered ring dye lasers and Ti:sapphire lasers will extend this even further. The holes can be probed using a frequency ramp which is fast compared to the excited state lifetime as described above. An alternative is to use continuous saturation by a carrier with the addition of frequency-scannable sidebands, or two lasers--one to saturate and the other to probe the resulting hole.

We have been assuming that the ions do not interact so that the group of ions for which the ground state has been depleted (i.e. a hole burned), do not transfer this selectively excited population to a neighbor. At a higher concentration of dopant ions this transfer may occur and the hole will broaden since the resonance frequency of the ion receiving the excitation is shifted due to inhomogeneous broadening. This spectral diffusion leads to a reduction in hole depth or a broadening of the hole, depending on the magnitude of the frequency shift.

More generally, the population reservoir can be a second excited state not directly coupled by the laser field (Fig. 5b). In rare-earth systems this is provided by lower lying metastable optical levels such as $^5I_7$ in Ho^{3+}, $^4I_{13/2}$ in Er^{3+} and $^3F_4$ in Tm^{3+}. These typically have storage times of tens of milliseconds. Again, as in two level saturation, if energy transfer occurs while ions are in the metastable storage level, the hole will broaden since the depleted population no longer resides on the group of ions initially selected by the laser. An example of this three-level holeburning mechanism is provided by $LaF_3:Ho^{3+}$ [16], where the $^5I_7$ level, which has a lifetime of approximately 20 msec, acts as the storage level. Holes burned in the $^5I_8 \rightarrow ^5F_5$ transition are shown in Fig. 6. In this case side-holes are observed due to hyperfine splittings in the $^5F_5$ level, but there are no antiholes because the nonequilibrium population is not stored in the hyperfine components.

2. Hyperfine Holeburning. Population storage in ground state hyperfine levels is also a rather general mechanism for rare-earth ions since the hyperfine splittings are usually greater than the optical homogeneous linewidth at low temperatures. In addition, the nuclear spin-lattice relaxation times of seconds to minutes provide a very convenient experimental timescale for the observation of holes. Two situations obtain. For singlet electronic states the quadrupole and second order hyperfine, or pseudoquadrupole splittings (see (23)) are typically 5 – 100 MHz which is smaller than the inhomogeneous broadening. A laser at a particular frequency is resonant with all of the possible optical

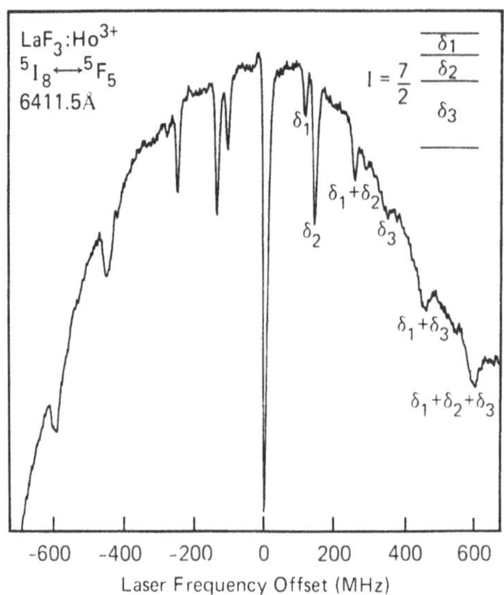

Fig. 6. Holeburning in LaF_3:Ho^{3+} using the long lived $^5I_7$ level as a reservoir. The side holes are due to excited state hyperfine splittings.

transitions between hyperfine components in the ground and excited states, but for different subsets of ions. Optical pumping of the hyperfine levels transfers population among the ground state components, emptying the levels resonant with the laser (Fig. 5c). In contrast to the case for different electronic transitions, it is expected (and found) that the inhomogeneous broadening is strongly correlated for optical transitions between different hyperfine levels of a given pair of electronic states. Thus, in addition to a hole at the laser frequency, side-holes are expected at the excited state splitting frequencies. Similarly, the ground state levels which receive greater than equilibrium populations show narrow enhanced absorptions (anti-holes) [38]. The resulting spectrum of holes and antiholes is a superposition of those for each of the subsets of ions. The frequencies of the sideholes give excited state hyperfine splittings while those of the antiholes give the ground state splittings and all sums and differences of ground and excited state splittings. Figure 7 gives details of the resulting spectrum for the case of a nuclear spin of 5/2 applicable to trivalent praseodymium or europium, A typical example is provided by the much studied Pr^{3+} ion in LaF_3 whose energy levels are given in Fig. 8, and Fig. 9 shows a hole/antihole spectrum for this material at 2K. The side holes at 3.7 MHz and 4.7 MHz were not resolved due to the laser frequency jitter in these measurements.

The second situation involves doubly degenerate electronic states which exhibit a first order hyperfine splitting, typically of several GHz (Fig. 5d). In this case it is possible for the hyperfine splitting to be resolved outside of the inhomogeneous broadening, enabling selective excitation of individual hyperfine transitions to be carried out. This is shown for CaF_2:Pr^{3+} in an axially symmetric site in Fig. 10. Here there is a first order

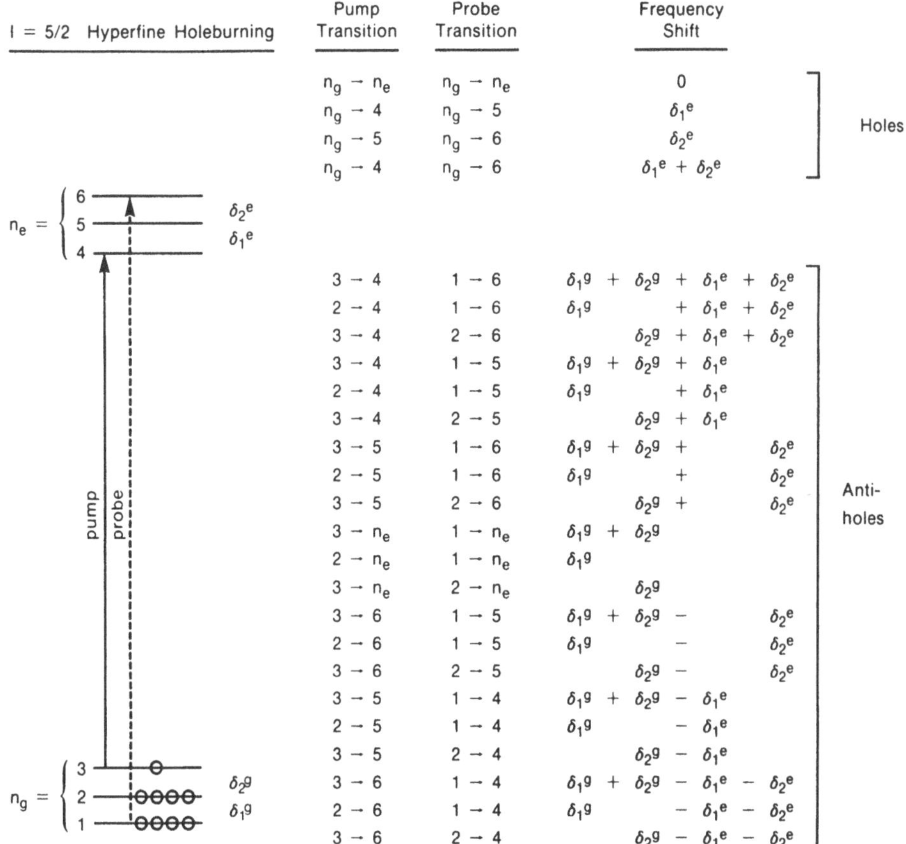

Pump Transition	Probe Transition	Frequency Shift	
$n_g \rightarrow n_e$	$n_g \rightarrow n_e$	0	Holes
$n_g \rightarrow 4$	$n_g \rightarrow 5$	δ_1^e	
$n_g \rightarrow 5$	$n_g \rightarrow 6$	δ_2^e	
$n_g \rightarrow 4$	$n_g \rightarrow 6$	$\delta_1^e + \delta_2^e$	
$3 \rightarrow 4$	$1 \rightarrow 6$	$\delta_1^g + \delta_2^g + \delta_1^e + \delta_2^e$	Anti-holes
$2 \rightarrow 4$	$1 \rightarrow 6$	$\delta_1^g + \delta_1^e + \delta_2^e$	
$3 \rightarrow 4$	$2 \rightarrow 6$	$\delta_2^g + \delta_1^e + \delta_2^e$	
$3 \rightarrow 4$	$1 \rightarrow 5$	$\delta_1^g + \delta_2^g + \delta_1^e$	
$2 \rightarrow 4$	$1 \rightarrow 5$	$\delta_1^g + \delta_1^e$	
$3 \rightarrow 4$	$2 \rightarrow 5$	$\delta_2^g + \delta_1^e$	
$3 \rightarrow 5$	$1 \rightarrow 6$	$\delta_1^g + \delta_2^g + \delta_2^e$	
$2 \rightarrow 5$	$1 \rightarrow 6$	$\delta_1^g + \delta_2^e$	
$3 \rightarrow 5$	$2 \rightarrow 6$	$\delta_2^g + \delta_2^e$	
$3 \rightarrow n_e$	$1 \rightarrow n_e$	$\delta_1^g + \delta_2^g$	
$2 \rightarrow n_e$	$1 \rightarrow n_e$	δ_1^g	
$3 \rightarrow n_e$	$2 \rightarrow n_e$	δ_2^g	
$3 \rightarrow 6$	$1 \rightarrow 5$	$\delta_1^g + \delta_2^g - \delta_2^e$	
$2 \rightarrow 6$	$1 \rightarrow 5$	$\delta_1^g - \delta_2^e$	
$3 \rightarrow 6$	$2 \rightarrow 5$	$\delta_2^g - \delta_2^e$	
$3 \rightarrow 5$	$1 \rightarrow 4$	$\delta_1^g + \delta_2^g - \delta_1^e$	
$2 \rightarrow 5$	$1 \rightarrow 4$	$\delta_1^g - \delta_1^e$	
$3 \rightarrow 5$	$2 \rightarrow 4$	$\delta_2^g - \delta_1^e$	
$3 \rightarrow 6$	$1 \rightarrow 4$	$\delta_1^g + \delta_2^g - \delta_1^e - \delta_2^e$	
$2 \rightarrow 6$	$1 \rightarrow 4$	$\delta_1^g - \delta_1^e - \delta_2^e$	
$3 \rightarrow 6$	$2 \rightarrow 4$	$\delta_2^g - \delta_1^e - \delta_2^e$	

Fig. 7. Frequencies of holes and anti-holes expected for holeburning using hyperfine reservoirs for a nuclear spin of 5/2.

hyperfine splitting in the electronic doublet ground state described by (24) and an unresolved, second order splitting in the singlet excited state described by (23). Holeburning in one hyperfine line can produce holes or antiholes on the other lines depending on the relaxation rates between the components (Fig. 10c). The hole burned in line 5, for example, produces a hole in line 3 which has the same projection of the nuclear spin but different electronic components. Since this is a non-Kramers' doublet, strain fields can couple the two electronic components leading to a rapid relaxation. Relaxation from line 5 to line 6 on the other hand is nuclear-spin forbidden and hence slower. In this case an antihole is seen on line 6.

3. Superhyperfine Holeburning. For ions with an electronic magnetic moment the superhyperfine interaction between the optical center and neighboring nuclear spins can produce splittings of ~10 MHz and the resulting levels can act as population reservoirs for holeburning [39] (Fig. 5c). The optical pumping mechanism is similar to the hyperfine case above, but the complexity of the superhyperfine level structure, compounded by the laser frequency jitter, makes a resolution of individual antiholes difficult. However, the redistribution of population produced by optical pumping can be

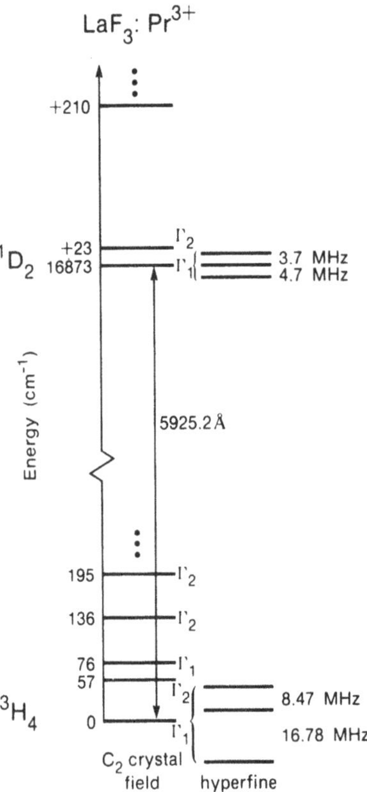

Fig. 8. Energy levels of the ground and $^1D_2$ manifolds of LaF_3:Pr^{3+}

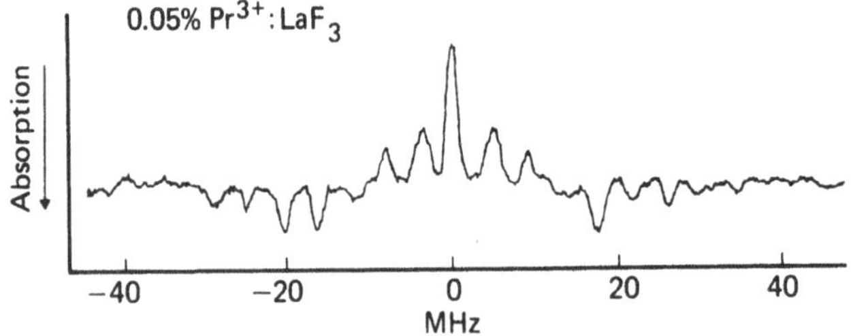

Fig. 9. Pattern of holes and antiholes observed in the $^3H_4(1) \leftrightarrow {}^1D_2(1)$ transition of LaF_3:Pr^{3+} at 2K.

reversed by applied rf fields, and this forms the basis for the observation of optically detected nuclear resonance of specific nuclear spins in the vicinity of an optical center. This was first demonstrated in CaF_2:Pr^{3+} where ^{19}F resonances were observed as changes in the fluorescence of Pr^{3+} produced by modification of the superhyperfine pumping cycle [40]. In this case the optically active center itself need not have a nuclear moment.

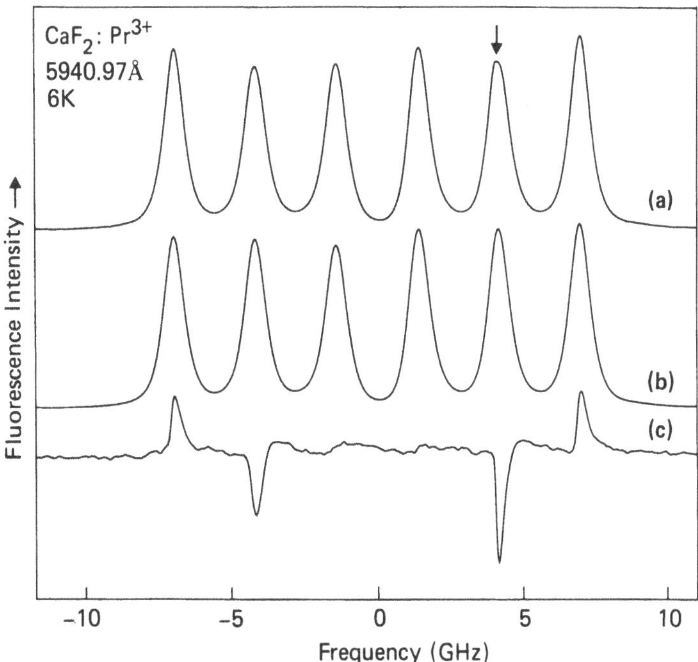

Fig. 10. $^3H_4(1) \leftarrow\rightarrow {}^1D_2(1)$ absorption of $CaF_2:Pr^{3+}$ showing resolved first order hyperfine structure. (b) before holeburning, (a) laser irradiation in line 5 leads to bleaching which shows more clearly in the difference spectrum (c), where there are holes in lines 5 and 2 and antiholes in lines 1 and 6.

4. Zeeman Sub-level Holeburning. The relaxation time between the two components of a Kramers' doublet in an external magnetic field can be rather long (msecs-secs) since direct phonon relaxation is forbidden between the time reversed states of the doublet. This makes the Zeeman components practical population storage reservoirs (Fig. 5f). The relaxation between these components varies as the fourth power of the field for small fields due to the admixture of other Kramers' doublets by the external magnetic field [41]. Holeburning can be done at very low fields since it is only necessary to split the components by several homogeneous linewidths. It is therefore expected that the hole lifetimes observed would be much longer than those seen in electron-spin resonance. In practice however, other processes such as spectral diffusion to paramagnetic impurities can be important at low fields. The process of spectral diffusion itself i.e. mutual spin-flips between the spins of the active paramagnetic center will lead to hole filling since the depleted population giving rise to holeburning will no longer reside on the initially selected ions.

A demonstration of this storage mechanism was made in $LaF_3:Nd^{3+}$ [42]. Holes were burned in the $^4I_{9/2}(1) \rightarrow {}^4G_{5/2}(1)$ absorption at 2K in a sample doped with 0.005% Nd. The characteristic feature of this holeburning mechanism is the appearance of antiholes at the ground-state splitting frequency as shown in Fig. 11. In addition a hole on either side of the central hole is observed at the excited state Zeeman splitting. Thus

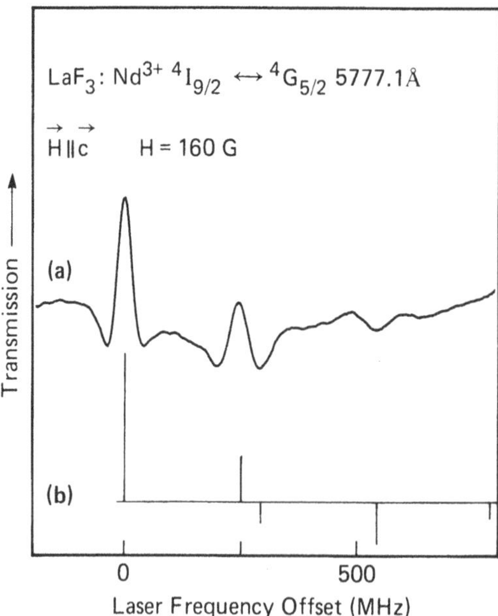

Fig. 11. (a) Holeburning in LaF_3:Nd^{3+} using Zeeman sublevels of the ground state as a population reservoir. (b) shows the pattern of holes and antiholes expected. Some additional antiholes are seen due to optical pumping of Nd-F superhyperfine levels.

this technique can be used to identify and make a separate measurement of excited and ground state Zeeman effects and, from the hole lifetime, the spin-lattice relaxation or the concentration dependent spin-spin relaxation times can be determined. Additional features are seen in the spectrum of Fig. 11. which are ascribed to optical pumping of the superhyperfine levels arising from Nd-F coupling (see subsection 3 above).

5. Persistent Spectral Holeburning. This can be loosely defined as holeburning having a lifetime of a least several hours. More characteristic are the mechanisms, which include photoionization [43,44], light induced proton tunneling [45 – 48], photochemistry involving bond breaking [49] and photophysical processes in which the environment, often in a glass or polymer, rearranges slightly in the presence of light [50,51]. Since the holes are often very sharp (~ tens of MHz) the frequency shift produced by the photoreaction can be small so that subtle, often unexpected, changes in the environment of a photoactive species can lead to holeburning. This mechanism is not illustrated in Fig. 5 but should rather be thought of as requiring a barrier between two long lived groundstates which is surmounted with the aid of light.

Historically this kind of holeburning was first observed in organic molecules either due to local deformations in a glass [50] or due to proton tautomerism in porphyrin-like molecules [45,46]. The latter in particular, form a prominent class of systems amenable to detailed study and which provide a link to biologically important molecules and systems [52]. Figure 12 shows a simple example of free-base porphin (H_2P) in n-hexane

[53], where the host matrix creates a spectral distinction between the two tautomeric forms. This produces two absorption lines that can be transformed into one another by light, leading to holeburning at low temperatures where the lines are inhomogeneously broadened. Studies of the temperature dependence of the holewidths of H_2P in n-octane [54] showed that thermal broadening involves the excitation of low frequency librational motion of the molecule in the matrix.

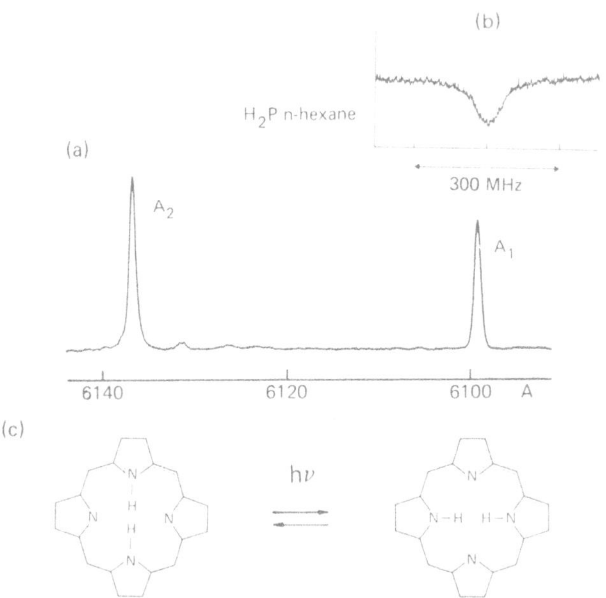

Fig. 12. Optical spectrum of the $S_0 \rightarrow S_1$ transition of free-base porphin in n-hexane showing the two tautomeric forms which can be interconverted by resonant excitation. (b) Conversion with a narrow-band laser leads to persistent holeburning.

The observation of persistent holeburning in inorganic systems began with the bleaching of color centers [43] in alkali halides and rare-earth absorption in glasses [55]. Later the photoionization of divalent rare-earth ions led to the demonstration of an important new mechanism of holeburning [56] *viz* photon-gated holeburning in which the selective bleaching occurs only in the presence of a second, or gating, light source. This process is shown schematically in Fig. 13. The significance of photon-gated holeburning, in addition to opening up a new class of materials for study, was that the holes are very resistant to being bleached in the process of reading. Persistent holeburning has been proposed as a mechanism for frequency domain optical storage [57] and for this application the stability of the holes is very important [58]. Organic molecules also exhibit numerous examples of photon-gated holeburning [58,59].

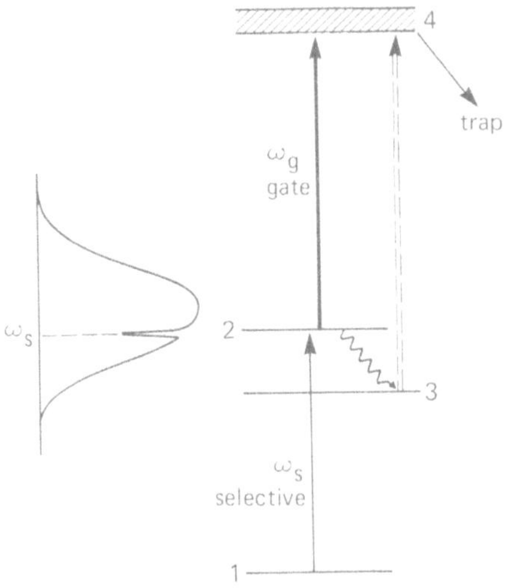

Fig. 13. Schematic diagram of photon-gated holeburning in which selective bleaching at ω_s occurs only in the presence of gating light at ω_g.

V.C. Measurement Techniques for Spectral Holes

The techniques used for measuring the holes depend to a large extent on the hole lifetimes. For long lived holes, a single narrow-band cw laser can bleach the hole and then, after attenuation to avoid further bleaching, can be scanned to probe either the reduced absorption or reduced fluorescence in an excitation spectrum. Commercially available cw dye lasers with a bandwidth of ~1 MHz can typically scan 100 MHz/msec so this works well for hole lifetimes $\gtrsim$100 msec. If attenuation of the beam during the probe cycle is not essential, holes living 1 – 10 ms can be scanned by applying an external voltage ramp to the galvanometer plate which changes the effective cavity length of the laser and hence its frequency. Faster frequency scans can be accomplished by applying a voltage ramp to an intracavity phase modulator; for example one made of AD*P can scan ~1 MHz/V and this can easily be done for a ±100 MHz scan on the timescale of microseconds. A better solution is to use a pair of extra-cavity acousto-optic modulators driven by a chirped rf source which can ramp the frequency of the Bragg diffracted beam about the original burning frequency. This has the advantage of not introducing losses into the laser cavity itself, and not interfering with the frequency locking servo loop of the laser which often has a bandwidth comparable with the frequency scan times. Again scan widths of ±100 MHz can readily be obtained. Another class of hole-scanning techniques involves scanning the resonance frequency of the optical transition with an electric or magnetic field applied to the sample. An application of these fast hole scans is to the measurement of time dependent holebroadening which often occurs due to spectral diffusion. In this case a hole burned at short times is probed at varying delays and the shape, width and area of the hole recorded (see section VIII).

VI. COHERENT TRANSIENT TECHNIQUES

Time domain measurements fall naturally into two categories; coherent transient effects where the time evolution of the optical polarization of the sample gives information about optical dephasing, and incoherent fluorescence decay and transient absorption measurements of population dynamics. We concentrate here on the former. Many of the coherent transient effects have been introduced to the optical field from NMR where they are of fundamental importance for determining dynamical and structural properties of materials.

In general, the aim of a coherent transient experiment is to measure the optical dephasing time T_2. The dephasing exhibited by the sample is not always characterized by a single parameter T_2 due to a non-exponential decay of the coherence. This is discussed further under photon echoes. The concept of a dephasing time is, however, very useful and we continue to use it here.

VI.A. Optical Free Induction Decay

This is perhaps the simplest coherent transient. As normally practised, a cw laser at frequency ω_1 irradiates a sample for a time long compared to T_2 and thereby induces a coherent polarization. This partially saturates a narrow frequency packet in the inhomogeneous line and when the laser is switched off, the sample continues to emit coherently (the free induction decay by analogy with NMR) with its characteristic dephasing time T_2 which is related to the width of the saturation hole by (1). In terms of the vector model the preparation pulse tips the Bloch vector and gives a component in the $1-2$ plane which corresponds to a coherent sample polarization. The polarization so produced is proportional to the cube of the driving field χ, and has the form

$$P_3(t) \propto \chi^3 \exp\{(-t/T_2)[1 + (1 + \chi^2 T_1 T_2)^{1/2}]\} \tag{28}$$

The term proportional to χ^2 is the power broadening term, emphasizing that this third order FID should be measured in the low power regime. In addition it has been found that at high intensities the solutions based on the Bloch equations break down and the power broadening can saturate at a value below that given by (28) [19]. If the preparation pulse is short, it may prepare a packet larger than the homogeneous linewidth. The resulting coherence decays as the different oscillators in the inhomogeneous ensemble get out of step and the coherent emission has a shorter decay given by the inverse of the frequency width prepared. In NMR the preparation pulse is usually made short enough to prepare the entire inhomogeneous line and the decay of the FID then gives the inhomogeneous linewidth. This so called first order FID has been observed optically but has not found widespread use.

A convenient way to detect the optical free decay is to switch the frequency of the laser from ω_1 to ω_2 at the end of the preparation time [60,61], and use the laser as a local oscillator for the detection at a beat frequency $\omega_1 - \omega_2$. The frequency switching of the laser has been achieved with an intra-cavity phase modulator [61] or an extra-cavity acousto-optic modulator [62]. This involves a relatively simple experimental set-up with a cw laser transmitted through a sample and detected with a p-i-n diode for example. An alternative approach is to switch the frequency of the sample to detune the group of atoms or ions in the frequency packet from the incident laser. A Stark field [63] is very convenient for this purpose and for systems with a large electronic magnetic moment, Zeeman switching of an external magnetic field has been used [64]. In practice a limitation in the applicability of the free-induction decay is the frequency jitter of the laser which contributes to the width of the saturation hole and shortens the decay. For commercial cw dye lasers this contribution is typically several hundred kHz, but stability of better than 1kHz can be obtained and in expert hands less than 1Hz has been achieved [65]. Note that it is the short term frequency jitter which is important since the experiment is completed in a time of the order of T_2 which is rarely longer than 10μsec. Dye lasers suffer from high frequency noise because of the free surface of the liquid jet, whereas solid-state lasers such as Ti:sapphire are quieter and better suited to the measurement of long coherence times.

VI.B. Delayed Optical Free Induction Decay

The presence of a long-lived coherence decay in the nonlinear FID experiment, is intimately related to the presence of the narrow spectral hole that is burned into the inhomogeneous absorption line during the excitation pulse. Muramoto et al. [66] carried out a Stark switching experiment in ruby in which the laser was turned off after the excitation pulse, and after a delay time τ such that $T_2 << \tau << T_1$ was pulsed back on followed by a Stark switch after an additional time much less than T_2. During the delay time any coherence produced by the excitation pulse decayed, but the saturation hole remained. The short excitation pulse was found to excite an FID beat signal which exhibited a decay corresponding to the width of the hole, whereas if the initial long excitation pulse was omitted, only a very weak, short-lived FID could be observed. Furthermore, if the second, short pulse was frequency shifted, the beat frequency corresponded to the position of the hole, not that of the short pulse.

As noted in the discussion of the FID technique, any spectral diffusion occurring on the timescale of the experiment can affect the width of the saturation hole and thus the decay time. In particular, in the pulse sequence described above, any spectral diffusion occurring during the delay period will cause the hole to broaden, and the broadening will be read out by the short reading pulse as a shortened FID. The full pulse sequence for this experiment is shown in Fig. 14. By performing this delayed FID experiment as a

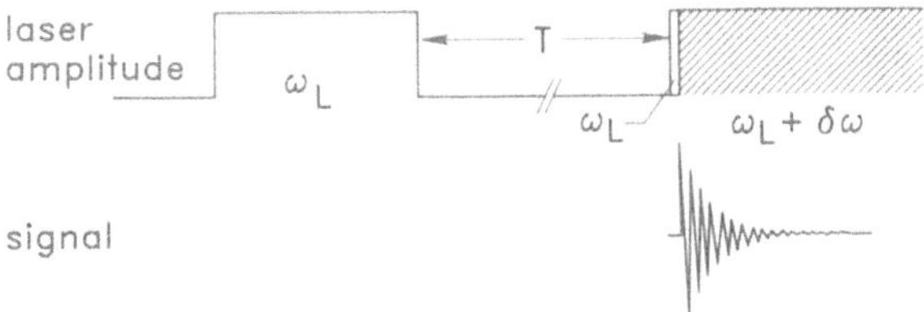

Fig. 14. Pulse sequence for observation of a delayed optical FID. The laser is frequency switched after the delay T to measure the width of the hole burned at the beginning of the cycle.

function of delay time, Shelby and Macfarlane [67] measured the timescale of spectral diffusion in $CaF_2:Pr^{3+}$ due to spin flips on fluorine nuclei surrounding the Pr^{3+} ion. Holebroadening began at ~10 μsecs, the flip rate of distant nuclei and continued to several msecs for near neighbor nuclei.

1. The Case of $LaF_3:Ho^{3+}$. A clear example of the use of the delayed FID to study spectral diffusion is provided by $LaF_3:Ho^{3+}$ [68]. The Ho^{3+} site has C_2 symmetry, so that all the levels are nondegenerate. The first excited electronic state is only $4cm^{-1}$ above the ground state and this results in a large enhanced magnetic moment in the singlet ground state of ~1 MHz/G due to second order hyperfine interaction. The transition from the ground state to the lowest component of $^5F_5$ shows a large difference in linewidth measured by optical FID and holeburning. At 1.4K the FID gives a width of 540kHz, detected by frequency switching the laser (Fig. 15). This is attributed to dephasing by the ^{19}F nuclear spins in the bulk i.e. outside the frozen core. Measurements using a delayed FID are shown in the right side of Fig. 15. For delays up to 600 μs after a 20 μs burning pulse, the holewidth increases from 1.08 MHz to 5.2 MHz. Using frequency domain techniques to extend these measurements to longer time it was found that this spectral diffusion continues for at least 15 ms at which time the holewidth is 9 MHz. From the enhanced magnetic moment of Ho^{3+}. in this system we expect a nearest neighbor holmium-fluorine coupling of 1-2 MHz. The hole broadening is therefore ascribed to frozen core nuclear spin flips of the fluorine ions

VI.C. Photon Echoes

Photon echoes, the optical analogue of spin echoes long known in NMR [69], were first observed by Kurnit et al. [70] in ruby using a Q-switched ruby laser. Since that time photon echoes have found wide application in the study of coherence decay of optical transitions in gas phase atoms and molecules, and in solids [71], and are still the subject of active experimental and theoretical investigations.

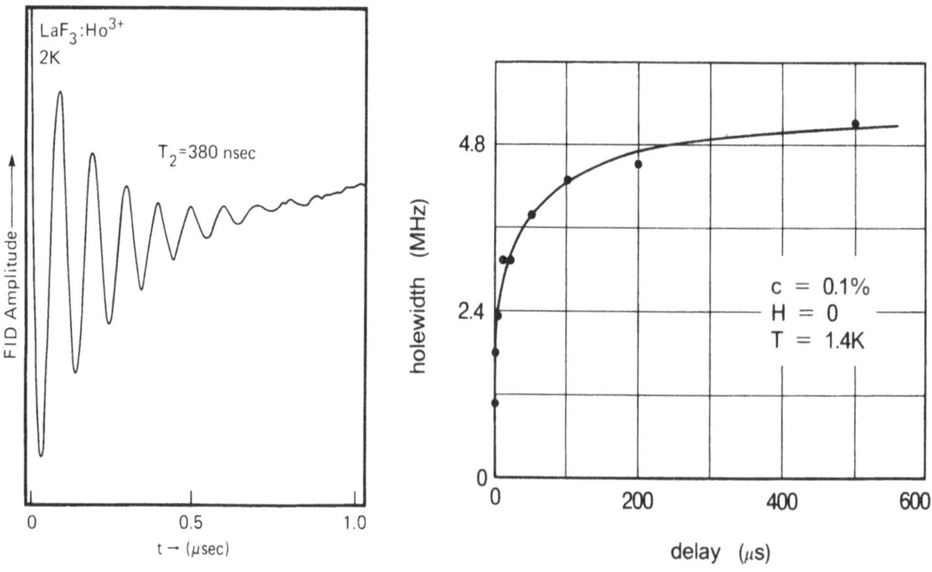

Fig. 15. Left: Optical FID on the $^5I_8(1) \leftrightarrow {}^5F_5(1)$ transition in $LaF_3:Ho^{3+}$.
Right: Delayed FID showing line broadening due to spectral diffusion.

The photon echo technique is quite different from that described above for the FID in that the sample is excited by short pulses whose spectral widths are large compared to the homogeneous width of the optical transition. No significant spectral selection by the laser occurs. Rather, the effect of inhomogeneous broadening is removed by the pulse sequence itself. The two-pulse echo experiment is carried out by exciting the sample with two laser pulses, separated by a delay τ (Fig. 16b). The first pulse creates a coherent superposition of the ground and excited state wave functions. This coherence is detectable because it leads to a macroscopic oscillating dipole which is capable of emitting coherent radiation and it rapidly decays on a timescale given by the inverse of the broad portion of the inhomogeneous line that was excited. The second pulse has the effect of exchanging to some degree the amplitudes of the ground and excited states in the coherent superposition, i.e the sign of this accumulated phase is reversed for each given ion. Following an additional evolution period of time τ, the net phase shift is cancelled for every ion, leading to a rephasing of the coherence and emission of a burst of coherent radiation (Figs. 16a, 16c). This burst of light is the photon echo. The intensity or amplitude of the echo reflects the decay of coherence due to homogeneous relaxation processes during the time 2τ, and a plot of echo amplitude versus 2τ, yields the dephasing time, T_2. Ideally, one wants to excite the sample to a coherent superposition state with equal amplitudes for being in the ground and excited states, i.e., the first pulse should have pulse area, $\Theta = \int \mu_{12} E(t)dt = \pi/2$ a so-called $\pi/2$ pulse which rotates the Bloch vector of Fig. 2 up into the $1-2$ plane. The second pulse should be a $\Theta = \pi$ pulse, to perfectly interchange the amplitudes and phase factors for the ground and excited states. This reverses the signs of the Bloch vectors which are arrayed in the 1-2 plane. However, echoes can be observed with non-optimal pulse areas. An important case is that of small area pulses for which the echo amplitude scales as $\Theta_1\Theta_2^2$ where $\Theta_{1(2)}$ refers to the area of the first (second) pulse. In many rare earth systems at low temperatures the dephasing times are $1-10\,\mu$ sec and this is the time scale of the pulse sequence required.

The actual experimental methods for observing echoes have taken several forms. Pulses can be generated by gating a cw laser [72] by frequency switching the laser [61] or the sample [63] into resonance with a portion of the inhomogeneous line as described under optical FID, or a combination of these [73]. The use of frequency switching allows heterodyne detection of the echo amplitude rather than the intensity [61,73], so that the signal decays as $\exp(-2\tau/T_2)$. Alternatively, one or more pulsed lasers can be used. The delay can be generated by optical delay lines which are particularly convenient on the picosecond timescale [74] and can also be used with nanosecond pulses but here the adjustable delays obtained by two separate lasers are more versatile [75]. In these cases the echo is detected directly with a suitably gated detector, or by angular resolution of the echo pulses using non-colinear excitation pulses. The intensity of the echo then decays as $\exp(-4\tau/T_2)$.

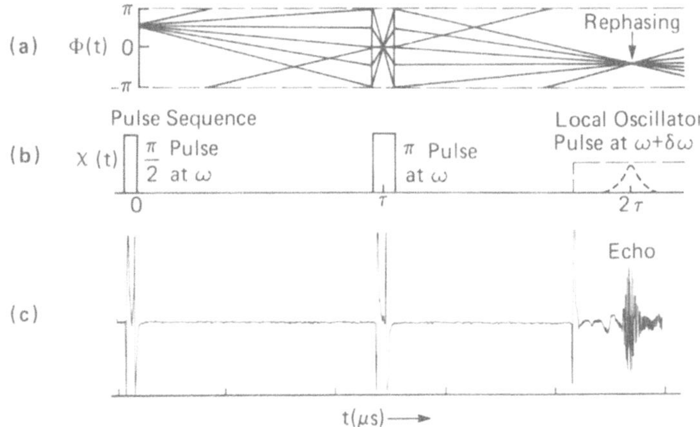

Fig. 16. Pulse sequence for a two pulse echo. (a) shows the time evolution of the phase of the coherent superposition produced by the pulses in (b). In (c) an echo is heterodyne detected by turning on a frequency shifted laser as a local oscillator.

Because short pulses are used for excitation, and the elimination of inhomogeneous broadening does not depend on selection of a single homogeneously broadened packet, the photon echo experiment is relatively immune from laser jitter problems. In fact, even with pulses from multi-mode pulsed lasers, echoes can be observed and used for accurate T_2 measurements. In the case of pulses gated from a cw laser, the effect of laser frequency jitter on the decay time is essentially removed if the Fourier transform width of the pulse is much greater than the jitter width. However, if heterodyne detection is used, the phase diffusion of the laser causes the phase of the heterodyne beat within the echo envelope to be randomized, making signal averaging techniques more difficult to apply.

A significant difference between the use of pulses of length ~ 1 μsec gated from a cw laser, as opposed to nsec pulses from a Nd:YAG or nitrogen pumped dye laser, arises when hyperfine splittings or other sub-level structure is present. If, in such a case, the splittings are less than the Fourier width of the excitation pulses, a quantum-beat-like interference can occur between coherences excited on two transitions which share a common level [76]. The result is an echo decay which exhibits beats at frequencies determined by the ground and excited state hyperfine splittings and their sums and differences. The Fourier transform of the decay curve yields the hyperfine splittings [77].

A new method of measuring photon echoes was recently demonstrated [78]; the so called cw photon echo. Here a multimode cw laser with wavevector $\mathbf{k}_1$ and a relatively short coherence time τ_c, effectively supplies a series of excitation pulses of width τ_c and continuously variable delays. A non-copropagating pulsed laser of wavevector $\mathbf{k}_2$ produces a series of echoes in the direction $2\mathbf{k}_2 - \mathbf{k}_1$. This scheme is equivalent to performing a series of ordinary photon echo experiments with varying delays in real time. Again, because the excitation pulses have the small width τ_c, strong hyperfine modulation is observed on the echo decay.

The above discussion has assumed a simple model in which dephasing can be accurately represented by a single exponential decay with time constant T_2. If the dephasing is due to interaction with a reservoir with a correlation time on the order of the echo decay time, this approximation may fail, and very nonexponential decays result. Such a situation indeed occurs in the context of optical dephasing of rare earth ions due to the perturbation of the bulk nuclear spin dynamics by the strong magnetic moment of the rare earth. The resulting spin diffusion barrier leads to the presence of a frozen core of nuclear spins surrounding the rare earth and to slow spectral diffusion resulting in a nonexponential decay. These effects are well known in electron spin echo measurements [79,80] and have recently been observed in photon echoes [81].

To observe the effects of spectral diffusion due to the frozen core of nuclear spins it is necessary to eliminate electron spin-spin coupling by working at a high dilution of active ions and a magnetic field strong enough to significantly depopulate the electron spin Zeeman levels. Ganem et al. [81] showed for ruby and $YLiF_4:Er^{3+}$ that at a low concentration of Cr^{3+} ions and high magnetic fields (>20 kG) the photon echo decay becomes independent of field and concentration. In this limit, superhyperfine interactions with the ^{27}Al or ^{19}F nuclei limit the decay rate. There is a transition from an exponential photon echo decay which is field and temperature dependent to a nonexponential decay which is independent of temperature and field (see Fig. 17).

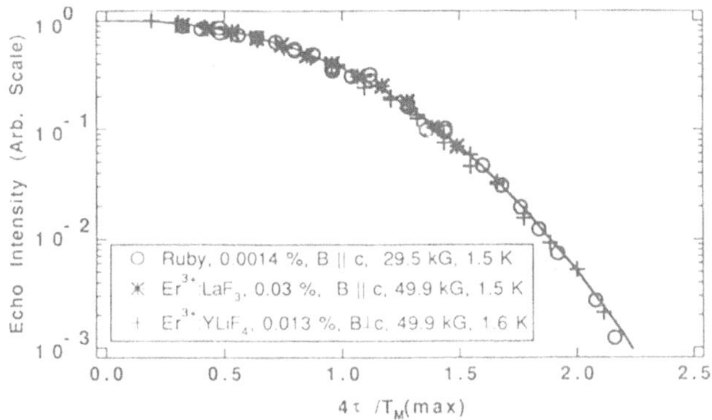

Fig. 17. Photon echo decays in ruby as a function of external magnetic field and concentration. As the Cr^{3+} concentration is reduced dephasing is dominated by frozen core nuclear spin dynamics and the echo decay becomes nonexponential.

The nonexponential decays can be described in a form proposed by Mims [79] for ESR:

$$I = I_0 e^{-(4\tau/T_M)^x} \tag{29}$$

The shape of the echo decay envelope is characterized by two parameters, the phase memory time T_M and an exponent x which describes the echo decay shape. When $x = 1$ the decay is exponential and $T_M = T_2$. Measurements of this kind were also made on $LaF_3:Er^{3+}$. In each sample the echo decay became independent of field for H/T greater than some critical value which depends on concentration. When this occurs, x = 2.4 and

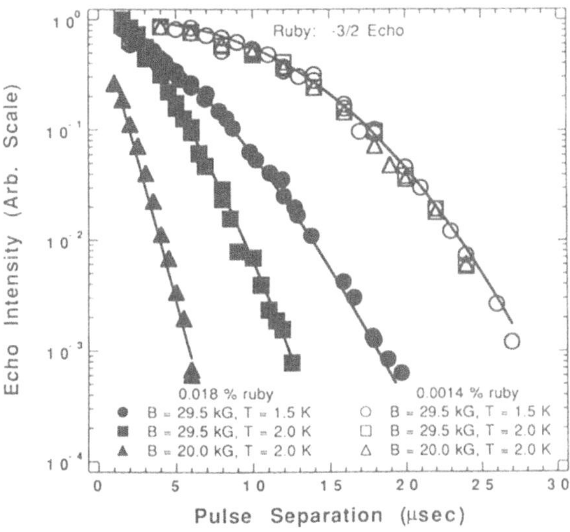

Fig. 18. Nonexponential photon echo decay in three materials with the time axis normalized to T_M (see text). The shape of the decay is identical.

a maximum value of the phase memory time, T_M(max) is achieved. The universal character of the nonexponetial decay in these three systems is shown in Fig. 18 where the echo decay is plotted as a function of the dimensionless parameter $4\tau/T_M$(max). The shape of the echo decay appears to be determined by the way in which the interactions fall off with distance in the frozen core. Modelling of the spin dynamics in the core to describe this phenomenon, is in progress.

VI. Stimulated Photon Echoes

The stimulated, or three pulse photon echo [69] can be thought of as a ($\pi/2$, π) photon echo pulse sequence in which the π pulse is broken into two $\pi/2$ pulses by the insertion of a long delay, T in its center (Fig. 19). The resulting pulse sequence is two $\pi/2$ pulses separated by a time τ followed by a third $\pi/2$ pulse after an additional delay T. The first pulse creates a coherent superposition, and after an evolution time τ, the phases of these superposition states accumulate according to the offset from the laser frequency within the inhomogeneous line. The second pulse puts ions with in-phase coherence into the excited state, and those that have accumulated a π phase shift back into the ground state, thus storing the accumulated phase information in the form of a population difference which varies periodically as a function of $\Delta\omega\tau$ across the inhomogeneous line. This frequency grating is subject to decay due to population decay (T_1 for a two-level system) *and* due to spectral diffusion which tends to smear out the grating. This effect is analogous to the delayed FID experiment in which the hole is filled in by spectral diffusion. Again by analogy with the delayed FID, the amplitude of the grating can be read out after a delay T by applying a short pulse. The peaks in the population distribution will all have accumulated a multiple of 2π of phase after an additional delay time τ following the readout pulse. At that time a refocusing of the coherence takes place, and an echo is emitted. The amplitude of the echo gives information about dephasing during the two τ delays and spectral diffusion or population decay during the T delay. Stimulated echoes have been reported in $LaF_3:Pr^{3+}$ [82,83] and have recently been used to measure spectral diffusion in $YAlO_3:Pr^{3+}$ [84].

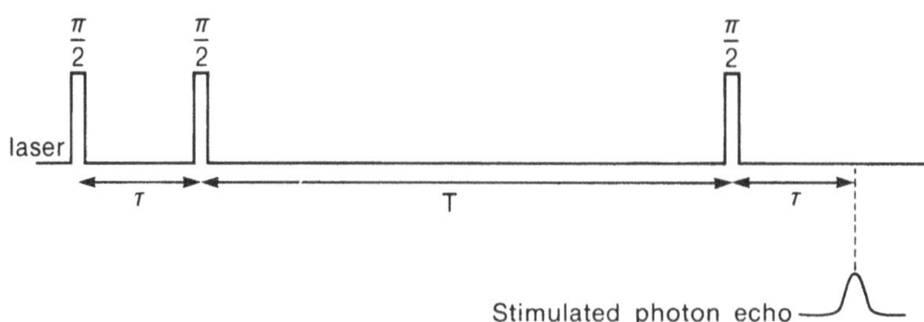

Fig. 19. Pulse sequence for a stimulated photon echo.

VII. OTHER TECHNIQUES

VII.A. Accumulated Photon Echoes

If an excited state has a long lifetime or relaxes to a long lived intermediate state, the resulting population storage time, makes it possible for multiple repetitions of the first two pulses of a stimulated echo sequence to accumulate a much larger population frequency grating than that from an individual pair of pulses [85,86]. If such pulse pairs with delay time τ and repetition period T are applied, each of the two pulses can act as a readout pulse to stimulate an echo from the grating which is built up. The amplitude of the echo still reflects the coherence lost during the two evolution periods of time τ, and thus can be used to measure T_2.

The main utility of this type of echo arises from the ability to build up large population differences via the accumulation effect and thus to observe relatively intense echo signals, even though only small area pulses might be available. This has been found particularly useful when picosecond pulses are used. If slow spectral diffusion is present on the timescale of the storage level relaxation time, however, it can cause the accumulated frequency grating to be washed out, and lead to a value of T_2 from the accumulated echo decay that is shorter than that which would be measured with a two pulse echo or optical FID. In this sense this experiment is equivalent to the determination of T_2 from a holeburning experiment which uses the same storage level as a population reservoir.

VII.B. Photon Echo Nuclear Double Resonance (PENDOR)

This technique involves the detection of a reduction in photon echo amplitude when an rf field is applied on resonance to a sub-level (e.g., hyperfine [87,88], or superhyperfine [89]) transition. There are two somewhat different experimental realizations of this idea.

In the first, [90,91] an rf pulse is applied during the time between the second and third pulses of a *three pulse* stimulated echo sequence. Resonances are observed as reductions of the echo signal as the frequency of the rf is swept. There are two possible mechanisms for signal generation. First, in the case of resonances of host nuclei, i.e., superhyperfine resonances, the rf-induced spin flips of these nuclei tend to randomize the local field strengths. This will smear out the population frequency grating, just as spectral diffusion would do, and thus lead to a loss of echo amplitude. Second, the rf can remove population from the sublevels which form the echo, to ones which are not in resonance with the laser.

The second type of PENDOR experiment consists of the application of an rf pulse during the entire *two pulse* photon echo sequence [88]. The predominant mechanism for generation of the double resonance signal is the generation of sidebands on the sublevels at the rf Rabi frequency resulting in a modulation of the echo decay curve. As the rf frequency is swept, the echo amplitude exhibits resonances at the sublevel frequencies. The results obtained for the $^3H_4 \leftrightarrow {}^1D_2$ transition of Pr^{3+} in $YAlO_3$ is shown in Fig. 20 [88]. Both ground and excited state hyperfine splittings were obtained.

The main advantage of PENDOR techniques over holeburning ODMR is that they are generally applicable to both excited and ground state resonances irrespective of branching considerations. Like ODMR, they measure sub-level spectra with a resolution determined by the inhomogeneous width of hyperfine transitions, not by optical T_2 or laser frequency jitter. However, they do require a reasonably long optical T_2 in order that echoes can be observed, and also that the rf transitions can be expected to substantially affect local fields on the scale of the homogeneous linewidth on a timescale reasonable for rf transition rates.

VII.C. Quantum-beat Free Induction Decay

This experiment [92] obtains sub-level spectra by monitoring coherent Raman scattering from coherence between sublevels. Rather than using rf to induce the coherence, a short optical pulse is applied in resonance with the optical transition. If the pulse is short enough that its Fourier width is greater than the sub-level splittings, coherence is excited between sublevels when both sublevels have allowed optical

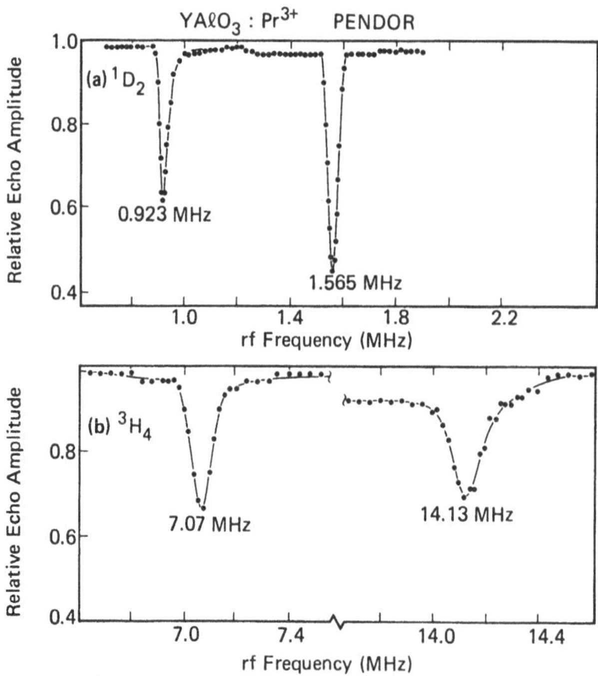

Fig. 20. PENDOR spectrum for $YAlO_3{:}Pr^{3+}$ showing the resonances at ground and excited state hyperfine splittings.

transitions to a common level. The coherence is excited via a two photon Raman excitation process and does not require a population difference between the sublevels. It can yield high resolution spectra, even in the presence of fast optical dephasing or laser frequency jitter. The time evolution of the sub-level coherence can be monitored by coherent Raman scattering of a weak probe beam (Fig. 21). The Raman scattering is detected by observing the beats between the Raman scattered light and the laser beam with a photo-diode.

With sufficiently short pulse excitation, all ground and excited state coherences can be excited and the Fourier transform of the quantum beat FID contains all of the sublevel resonances. The results of a measurement on the $^3H_4 \longleftrightarrow {}^1D_2$ transition of Pr^{3+} in YAG is shown in Fig. 22 [92]. A comparison with the energy level diagram of YAG:Pr^{3+} in Fig. 23 shows that the excitation pulse width of 0.2μsec was not sufficiently short to excite ground state sublevel coherence and only those of the excited state were measured in this experiment.

VIII. TIME RESOLVED HOLEBURNING

The results of a measurement of the homogeneous width of an optical transition by spectral holeburning and by coherent transient techniques often appeared to be in conflict because they yielded different values for this width. As the quality of the data improved it became clear that this difference was not due to systematic experimental errors, but that the two classes of technique often provide different information about the

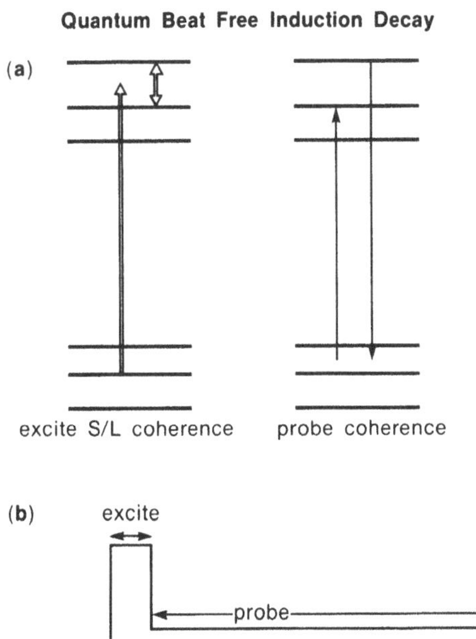

Fig. 21. (a) excitation and probing steps for the observation of quantum-beat free induction decay. (b) The optical pulse sequence.

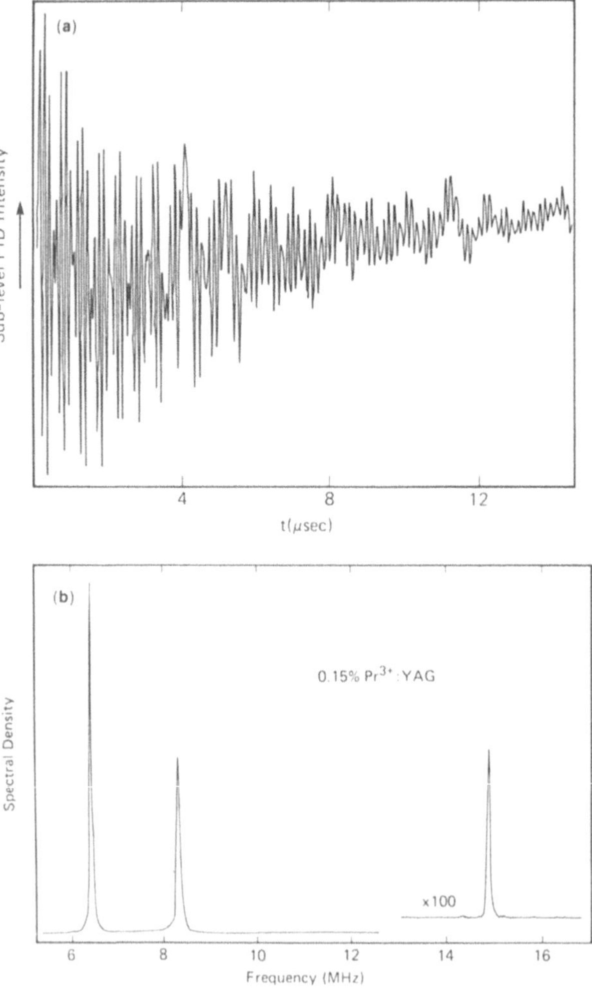

Fig. 22. Quantum-beat FID on the $^3H_4(1)\leftarrow\rightarrow{}^1D_2(1)$ transition of YAG:Pr^{3+} (a) the beat signal and (b) the Fourier transform showing the two excited state pseudo-quadrupole splittings and their sum.

dynamical environment of the optical center. This difference comes about from the different timescales of observation in the two cases. As we have seen, spectral holeburning often involves population storage in long lived reservoirs with lifetimes of μsecs to hours. The holeburning process itself then involves optical pumping over many cycles so that while, in principle, it is possible to burn holes quickly, in practice long holeburning times are common. The holes thus provide a frequency memory which records the dynamics of the environment on this timescale. On the other hand, coherent transient measurements are made on the timescale of the dephasing itself and slow changes in the local fields do not affect the coherence loss, being viewed rather as part of the inhomogeneous broadening. This situation is well illustrated by rare-earth impurity systems. The source of optical dephasing at low temperatures is often the fluctuating local fields due to mutual flips of the nuclear spins of the host. The rate of these spin

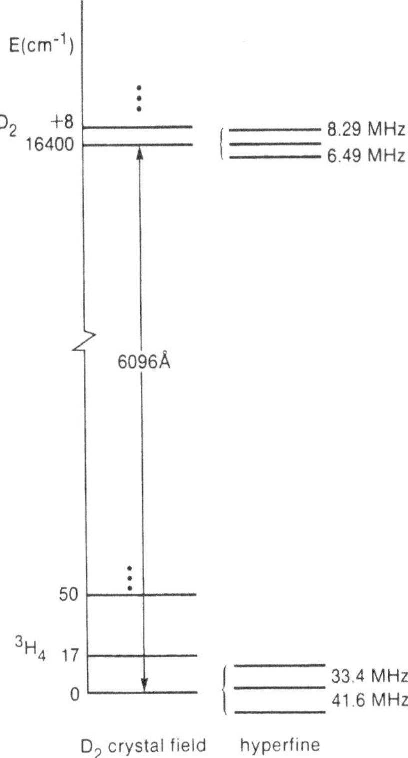

Fig. 23. Energy levels of YAG:Pr^{3+}

flips is strongly affected by the presence of the rare-earth ion, since this usually has a magnetic moment which is much larger than that of a nuclear spin. The result is that nuclei in the vicinity of the impurity ion are detuned from the bulk and flip much more slowly. For a fast probe such as a photon echo these "frozen" spins do not contribute to the dephasing, whereas for holeburning, the distribution of slow spin flips will often cause spectral diffusion [67,68]. Amorphous materials and polymers also exhibit a wide range of timescales associated with the motion of molecules or groups of molecules at low temperatures. Fayer and co-workers have demonstrated the importance of spectral diffusion in holeburning experiments in these materials and compared these measurements with the results of psec photon echoes and a theoretical analysis of the problem [93 – 95].

The frequency domain measurement of time resolved holeburning can be made by applying a frequency ramp at varying delays after a burning pulse. This ramp may be of the laser frequency, for example using an intra-cavity phase modulator or extra-cavity acousto-optic modulator, or of the sample frequency using an applied electric or magnetic field ramp. Scans of several hundred MHz in tens of microseconds are typical. One point to keep in mind is that the laser frequency jitter itself is a function of the timescale, so ideally this should be characterized separately. For a commercial dye laser we found from measurements of jitter limited optical FID's and the transmission through a high finesse cavity, that the jitter was ~100 kHz at $1\mu s$, ~1 MHz at $15\mu s$ and ~2.5 MHz at 1 min. In many cases better frequency stability is necessary. An example of the application of these techniques was the measurement of spectral diffusion due to fluorine nuclear spin flips in $YLiF_4:Er^{3+}$ [68]. On the time scale between that of the photon echo decay (~10 μsecs) and that of the hole lifetime (~ 10ms) the hole broadened by about an order of magnitude. The frequency domain measurement of hole broadening is complementary to the time domain measurements discussed in section VI.B.

IX. SPECTROSCOPY IN EXTERNAL FIELDS

Spectral holeburning provides a sharp optical probe giving orders of magnitude improvement in resolution, and hence sensitivity, in the measurement of frequency shifts induced by external fields. A unique characteristic of persistent holeburning is the long term frequency memory provided by the hole. This has been exploited to measure the shifts induced by slowly varying perturbations such as large magnetic fields [96], external stress [97] or temperature [54]. The ability to measure thermally induced shifts with high precision is particularly useful since this can be done at low temperatures where conventional spectroscopy fails and yet where a theoretical understanding of the shifts is easier. Some examples of the use of holeburning and coherent transients to measure small frequency shifts and splittings will be given below.

IX.A. Nonlinear Zeeman Effect

The linear Zeeman effect is only observed for degenerate states, while the nonlinear Zeeman effect i.e. the off diagonal magnetic coupling between different electronic states, occurs quite generally for singlets or doublets of the appropriate symmetry. Since the magnetic coupling is almost always small compared to the level spacings, the resulting shifts are very small and rarely observable using conventional spectroscopy. Their measurement is however important for level assignments and for an understanding of enhanced nuclear moments. (see (19)). A good example is provided by $BaClF:Sm^{2+}$ where the Sm^{2+} ions exhibit persistent spectral holeburning by selective photoionization of Sm^{2+} in a photon-gated process. A schematic energy level diagram as a function of magnetic field is shown in Fig. 24. Many levels exhibit small quadratic Zeeman shifts which are too small to be measured by conventional spectroscopy at reasonably accessible magnetic fields. A hole is burned in zero field and provides a sharp, long lived frequency marker which shifts with the absorption line in the external field H_0. A typical set of data is shown in Fig. 25 where the quadratic nature of the $^7F_0 \rightarrow ^5D_0$ frequency shift is clearly seen. For $H_0 \| c$ the nonlinear coefficient is ± 0.89 Hz/G^2 and for $H_0 \perp c$ it is +0.80 Hz/G^2. Thus it would require fields of almost 200 kG to shift the line by its inhomogeneous width. Almost all of the shift comes from the interaction between the $^7F_0$ and $^7F_1$ levels. The coefficients are very sensitive to the nature of the electronic wavefunctions and it was found that nonlinear Zeeman coefficients calculated using free-ion wavefunctions gave agreement with the experimental values within 5% [96].

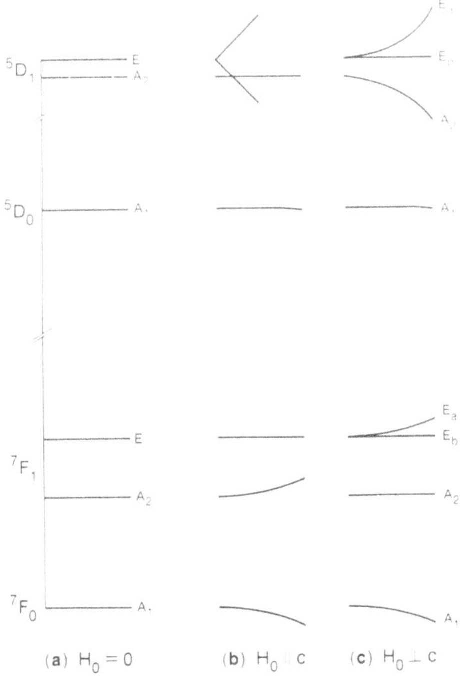

Fig. 24. Energy level diagram for BaClF:Sm^{2+} as a function of magnetic field.

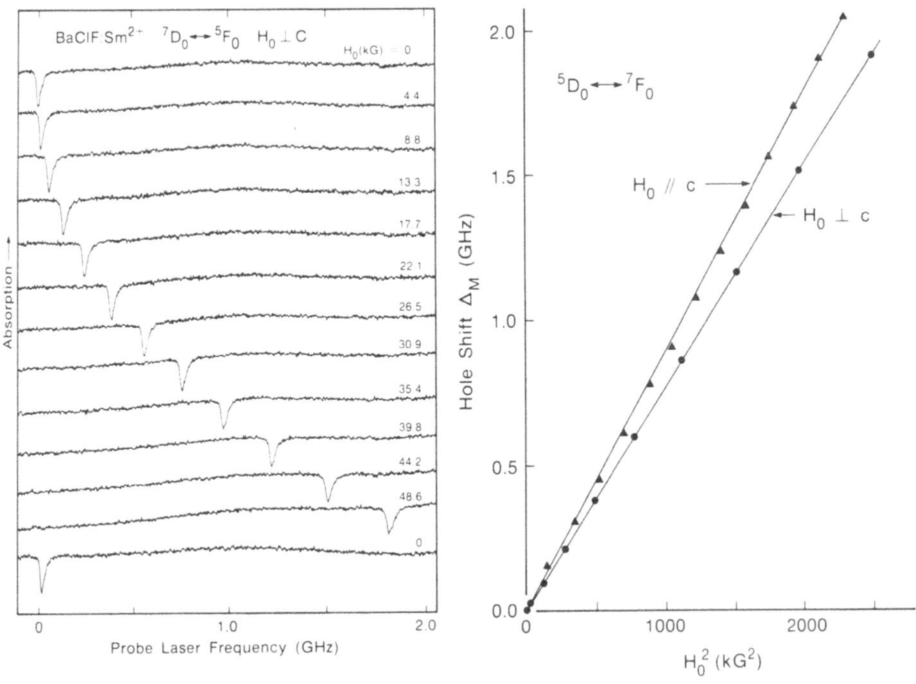

Fig. 25. Quadratic Zeeman effect for the $^7F_0 \leftrightarrow {}^5D_0$ transition of BaClF:Sm^{2+} measured from the shift of persistent spectral holes burned using a photon gated process. Left, the hole shift and right a plot showing the quadratic nature of the shift.

In contrast to the Zeeman effect, an electric field can produce a linear shift of a non-degenerate level as well as a linear splitting of degenerate levels. The magnitude of this Stark effect is difficult to predict theoretically since the response of the electrons to an applied electric field is very complicated and local field corrections are problematical. Nevertheless the Stark effect remains very powerful in determining the symmetry of defect centers, and hence probable physical models for them. It also offers a convenient means of rapidly tuning or switching energy levels which is often used in the observation of coherent transients, and in experiments on energy transfer and diffusion.

The use of holeburning to measure the linear Stark effect is most beautifully shown by color centers in cubic crystals such as NaF [98]. Here the defect, or vacancy, lowers the symmetry of a given center but the cubic symmetry of the crystal is retained overall because of the presence of equivalent centers with different orientations. When an external electric field is applied, the individual levels shift or split and, in addition, the orientational degeneracy of the centers is lifted because the electric field breaks the cubic symmetry. Figure 26 shows this effect for a color center in NaF which has a prominent zero phonon line at 6070Å with an inhomogeneously broadened linewidth of about 2 cm^{-1}. The holes are broadened somewhat by the fact that they were burned deeply, but there is still very good resolution for modest applied fields, providing a level of detail which is not possible with conventional spectroscopy. Using the results of similar measurements in other geometries it was determined that this center has C_s symmetry, with a permanent dipole moment along [mml] with $l/m = 1.44$. Further details are given in [98].

A very different kind of Stark effect measurement can be made using coherent transients, for example an optical FID in which the resonant frequency of the ions is switched by a Stark field. As an example we consider the $^3H_4 \rightarrow {}^1D_2$ transition of $LaF_3{:}Pr^{3+}$ [99]. In this crystal there are three pairs of sites (labelled $\pm\alpha$, $\pm\beta$, $\pm\gamma$) related by inversion. Each site has a local C_2 axis along a crystal a-axis. The optical transition and the Stark effect are non-zero only for electric fields along the local C_2 axis. Figure 27 shows the results of experiments with the dc electric field along the α direction and the laser field either perpendicular to α (when only the β and γ ions are probed) or parallel to α. In the first case, the beat frequency in the heterodyne detected FID results from the coherent emission of the β and γ ions which had the same Stark shift, and in the second case there are two different Stark shifts due to α ions and $(\beta + \gamma)$ ions and hence two different beat frequencies. In an experiment such as this, the Stark shift of different sets of ions can be simultaneously measured by Fourier transformation of the FID oscillations.

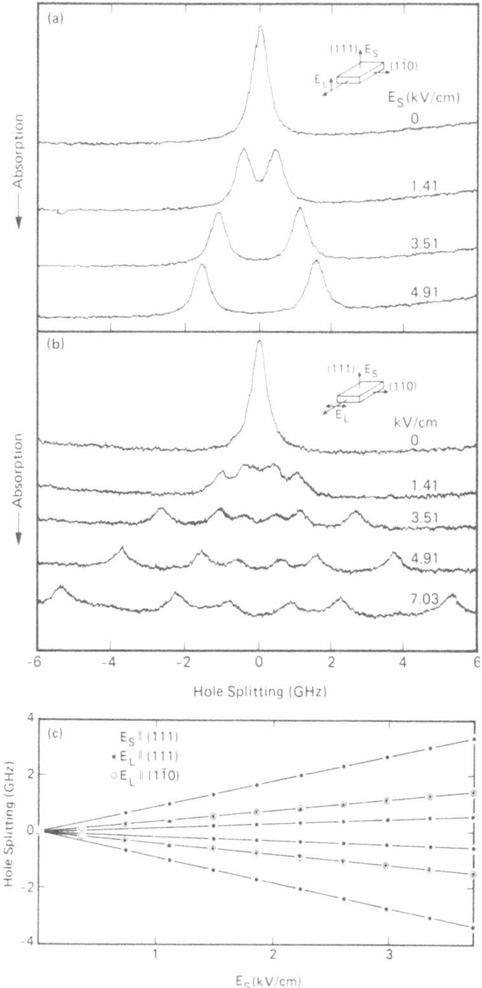

Fig. 26. Linear Stark effect measurement on the 6070Å zero-phonon line of a color center in NaF, using persistent holeburning.

IX.C. Nuclear Zeeman Effect

Since the nuclear magnetic moment is more than three orders of magnitude smaller than the electronic moment, optical measurement of the nuclear Zeeman effect in solids requires techniques to eliminate inhomogeneous broadening. An example of the use of spectral holeburning is again provided by $LaF_3:Pr^{3+}$ as shown in Fig. 28 [100]. Holeburning occurs due to optical pumping of the ground state hyperfine levels giving both side-holes and anti-holes. In the presence of a magnetic field, side-holes due to the nuclear Zeeman effect in the excited $^1D_2$ electronic state are seen to split out and are clearly resolved. At high fields, the pseudoquadrupole coupling becomes a small perturbation on the splittings produced by the enhanced nuclear Zeeman effect and the

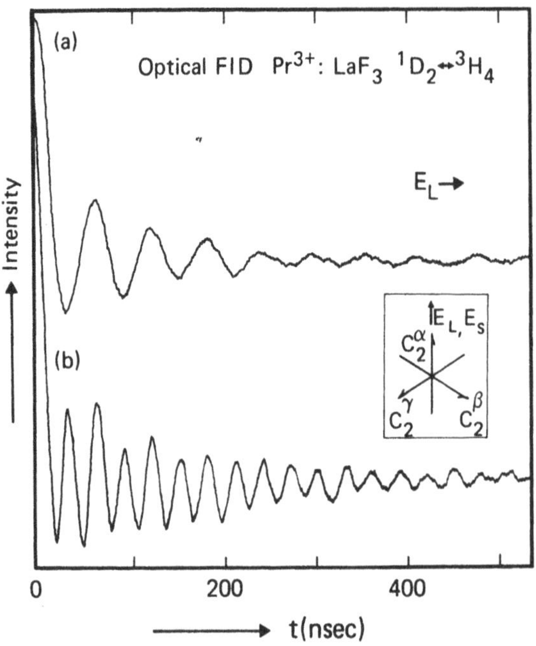

Fig. 27. Optical free induction decay in LaF_3:Pr^{3+} using Stark switching of the energy levels. In (a) the level shift is the same for all ions and in (b) there are two beat frequencies arising from the different Stark shifts of inequivalent sets of ions.

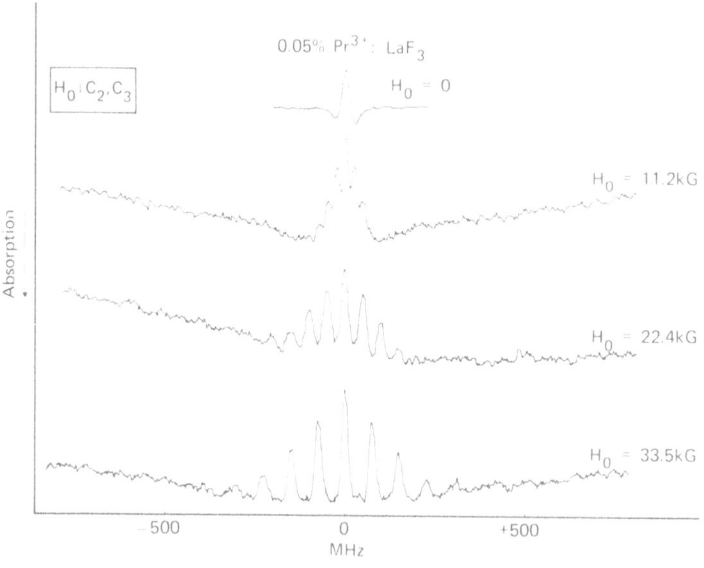

Fig. 28. Holeburning measurement of the nuclear Zeeman effect in LaF_3:Pr^{3+}. The side holes come from splittings in the excited electronic state.

latter become almost equally spaced. The nuclear moment in the excited state is enhanced by only about a factor of two: $\gamma_x/2\pi = 2.2$ kHz/G, $\gamma_y/2\pi = 1.9$ kHz/G, $\gamma_z/2\pi = 3.6$ kHz/G, compared to the bare moment of 1.2 kHz/G, whereas in the ground state it is enhanced by about a factor of ten. This reflects the different magnitudes of the Λ tensor in (17). The ground state splittings would appear as anti-holes, but due to the branching ratios in a field and the complex structure, such antihole patterns are not seen. In this case another technique such as optically detected nuclear resonance or quantum beat free-induction decay should be used.

X. CONCLUSION

The application of the many techniques of high resolution laser spectroscopy makes it possible to eliminate the effects of inhomogeneous broadening which often dominates spectral linewidths in solids at low temperatures. This typically improves resolution by three to four orders of magnitude over conventional spectroscopy, but it can be greater than six orders as for example in Y_2O_3:Eu^{3+} where a homogeneous linewidth of less than 1 kHz was measured by a 2-pulse photon echo [101]. With this increase in resolution, it is possible to measure the homogeneous linewidth, hyperfine structure, spectral diffusion and the effect of small external perturbations with great precision. To obtain the homogeneous linewidth, techniques which measure on the timescale of the optical dephasing itself e.g. photon-echo or free induction decay, are strongly preferred because they are less affected by spectral diffusion due to a time varying local environment which can produce extra broadening in a holeburning experiment. The spectral diffusion itself, on the other hand, can often best be measured by time resolved holeburning although it also shows up in nonexponential photon-echo decays. Of the coherent transient techniques the photon echo has the advantage that it is relatively insensitive to laser frequency jitter and power broadening effects which can complicate FID measurements. For rare-earth ions, the mechanisms responsible for homogeneous line broadening are summarized in Fig. 29. At room temperature absorption, emission and scattering of phonons are the dominant broadening processes. Upon cooling, inhomogeneous broadening limits the total linewidth around 100K or so. Below this, where laser techniques become necessary, phonon broadening continues to dominate down to a few Kelvin. There it is replaced, first by electron spin fluctuations if the ion density is sufficiently high, then by the almost universal nuclear spin fluctuation mechanism. Of central importance is the role of the optical center in strongly modifying the dynamics of the nuclear spins which surround it. It does this through the large local magnetic field accompanying the electronic moment or the enhanced nuclear moment which is characteristic of rare-earth and transition metal ions. This produces a "frozen core" of slowly fluctuating spins which makes them less effective in producing homogeneous broadening and very effective in producing spectral diffusion. This often gives rise to a large difference in the linewidth measured by fast coherent transients and the slower holeburning techniques. Thus some caution is needed in interpreting data on homogeneous linewidths. As laser and detector technology advances, both in spectral

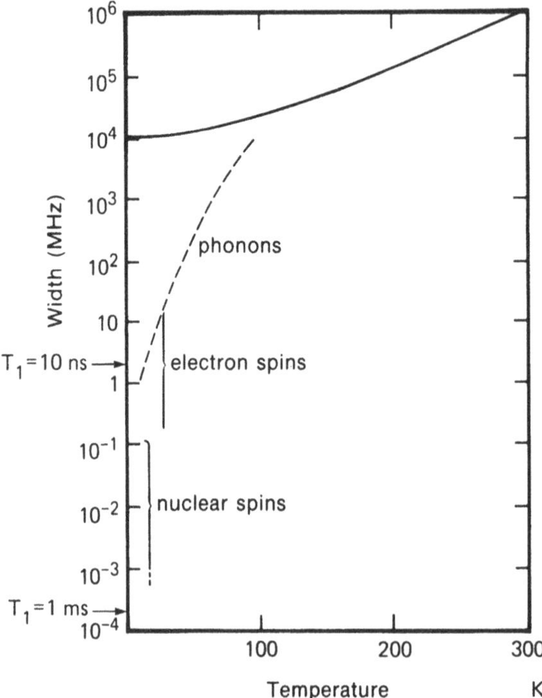

Fig. 29. Schematic illustration of the mechanisms for homogeneous line broadening in trivalent rare-earth systems. The solid curve shows the total linewidth including the inhomogeneous component which is assumed independent of temperature.

coverage and time resolution, the techniques described here will be applicable to a wider and wider variety of materials. The dynamics of excited states; the decay of both populations and coherence, remains of fundamental importance for our understanding of the nature of solids and for their technological applications.

REFERENCES

1. L. Allen and J. H. Eberley, "Optical Resonance and Two-Level Atoms" (Wiley, New York), 1975.
2. R. L. Shoemaker, *Ann. Rev. Phys. Chem.* **30**, 239 (1979).
3. M. D. Levenson and S. Kano, "Introduction to Nonlinear Laser Spectroscopy," Academic Press, New York, 1988.
4. V. M. Agranovitch and R. M. Hochstrasser (eds.), "Spectroscopy and Excitation Dynamics of Condensed Molecular Systems," North Holland, Amsterdam, 1983.
5. R. M. Macfarlane and R. M. Shelby in "Spectroscopy of Solids Containing Rare Earth Ions," eds., A. A. Kaplyanskii and R. M. Macfarlane (North Holland, Amsterdam), p. 51, 1987.
6. W. M. Yen and M. D. Levenson (eds.), "Lasers, Spectroscopy and New Ideas: A Tribute to Arthur Schawlow," Springer-Verlag, Berlin, 1987.
7. G. H. Dieke and H. M. Crosswhite, *Appl. Opt.* **2**, 681 (1963).

8. B. G. Wybourne, "Spectroscopic Properties of Rare Earths", Wiley Interscience, New York, 1965.

9. G. H. Dieke, "Spectra and Energy Levels of Rare-Earth Ions in Crystals", Wiley Interscience, New York, 1968.

10. C. A. Morrison and R. P. Leavitt in Handbook of the Physics and Chemistry of Rare Earths, K. A. Geschneidner, Jr. and L. Eyring, eds., Amsterdam, Vol. 5, p. 463 (1982).

11. W. T. Carnall, G. L. Goodman, K. Rajnak, and R. S. Rana, *J. Chem. Phys.* **90**, 3343 (1989).

12. D. J. Newman, *Adv. Phys.* **20**, 197 (1971); D. J. Newman and B. Ng, *Repts. Progr. Phys.* **52**, 699 (1989).

13. B. R. Judd, *Phys. Rev.* **127**, 750 (1962).

14. G. S. Ofelt, *J. Chem. Phys.* **37**, 511 (1962).

15. R. D. Peacock, *Structure and Bonding* **22**, 83 (1975).

16. R. M. Macfarlane in "Lasers Spectroscopy and New Ideas: A Tribute to Arthur L. Schawlow" Springer-Verlag, Berlin, 1981.

17. F. Bloch, *Phys. Rev.* **70**, 460 (1946).

18. R. P. Feynman, F. L. Vernon and R. W. Hellwarth, *J. Appl. Phys.* **28**, 49 (1957).

19. R. G. DeVoe and R. G. Brewer, *Phys. Rev. Lett.* **50**, 1269 (1946); P. R. Berman and R. G. Brewer, *Phys. Rev. A* **32**, 2784 (1985).

20. D. E. McCumber and M. D. Sturge, *J. Appl. Phys.* **34**, 1682 (1963).

21. D. E. McCumber, *J. Math. Phys.* **5**, 221 (1964); *Phys. Rev.* **133**, A163 (1964).

22. I. S. Osad'ko in "Spectroscopy and Excitation Dynamics of Condensed Molecular Systems," eds., V. M. Agranovich and R. M. Hochstrasser (North Holland, Amsterdam), 1983.

23. D. Hsu and J. L. Skinner, *J. Chem. Phys.* **81**, 1604 (1984); *ibid* **81**, 5471 (1984); *J. Chem. Phys.* **83**, 2097 (1985); *ibid* **83**, 2107 (1985); *ibid* **87**, 54 (1987).

24. J. L. Skinner, *Ann. Rev. Phys. Chem.* **39**, 463 (1988).

25. B. DiBartolo, "Optical Interactions in Solids" (John Wiley, New York), 1968.

26. S. C. Rand, A. Wokaun, R. G. DeVoe and R. G. Brewer, *Phys. Rev. Lett.* **43**, 1868 (1979).

27. R. M. Macfarlane, C. S. Yannoni and R. M. Shelby, *Opt. Commun.* **32**, 101 (1980).

28. R. M. Macfarlane and R. M. Shelby, *J. Lumin.* **36**, 179 (1987).

29. J. C. Vial and R. Buisson, *J. Phys. Lett.* **43**, L399 (1982).

30. W. E. Moerner and T. P. Carter, *Phys. Rev. Lett.* **59**, 2705 (1987).

31. W. E. Moerner and L. Kador, *Phys. Rev. Lett.* **62**, 2535 (1989).

32. M. Orrit and J. Bonard, *Phys. Rev. Lett.* **65**, 2716 (1990).

33. R. M. Sternheimer, *Phys. Rev.* **146**, 140 (1966).

34. M. A. Teplov, *Sov. Phys. JETP* **26**, 872 (1968).

35. B. Bleaney, *Physica* **69**, 317 (1973).

36. J. M. Baker and B. Bleaney, *Proc. Roy. Soc. Lond. Ser A* **245**, 156 (1958).
37. A. Szabo, *Phys. Rev. Lett.* **25**, 924 (1970).
38. L. E. Erickson, *Phys. Rev.* **B16**, 4731 (1977).
39. R. M. Macfarlane, R. M. Shelby and D. P. Burum, *Opt. Lett.* **6**, 593 (1981).
40. D. P. Burum, R. M. Shelby and R. M. Macfarlane, *Phys. Rev.* **B25**, 3009 (1982).
41. J. H. van Vleck, *J. Chem. Phys.* **7**, 72 (1939); *Phys. Rev.* **57**, 426 (1940).
42. R. M. Macfarlane and J. C. Vial, *Phys. Rev.* **B36**, 3511 (1987).
43. R. M. Macfarlane and R. M. Shelby, *Phys. Rev. Lett.* **42**, 788 (1979).
44. R. M. Macfarlane and R. M. Shelby, *Opt. Lett.* **9**, 533 (1984).
45. A. A. Gorokhovskii, R. K. Kaarli and L. A. Rebane, *JETP Lett.* **20**, 216 (1974).
46. S. Völker and J. H. van der Waals, *Mol. Phys.* **32**, 1703 (1976).
47. R. M. Macfarlane, R. J. Reeves and G. D. Jones, *Opt. Lett.* **12**, 660 (1987).
48. C. von Borczyskowski, A. Oppenlander, H. P. Trommsdorff and J.-C. Vial, *Phys. Rev. Lett.* **65**, 3277 (1990).
49. H. de Vries and D. A. Wiersma, *Phys. Rev. Lett.* **36**, 91 (1976).
50. B. M. Kharlamov, R. I. Personov and L. A. Bykovskaya, *Opt. Commun.* **12**, 191 (1974).
51. G. J. Small in "Spectroscopy and Excitation Dynamics of Condensed Molecular Systems", eds. V. M. Agranovitch and R. M. Hochstrasser, North Holland, Amsterdam, (1983) p. 515.
52. V. G. Maslov, A. S. Chunaev, and V. V. Tugarinov, *Mol. Biol.* **15**, 788 (1981); J. K. Gillie, B. L. Fearey, J. M. Hayes and G. J. Small, *Chem. Phys. Lett.* **134**, 316 (1987).
53. S. Völker and R. M. Macfarlane, *IBM J. Res. Dev.* **23**, 547 (1979).
54. S. Völker, R. M. Macfarlane and J. H. van der Waals, *Chem. Phys. Lett.* **58**, 8 (1978).
55. R. M. Macfarlane and R. M. Shelby, *Opt. Commun.* **45**, 46 (1983).
56. A. Winnacker, R. M. Shelby and R. M. Macfarlane, *Opt. Lett.* **10**, 350 (1985).
57. A. Szabo, U.S. Patent 3,896,420 (1975); G. Castro, D. Haarer, R. M. Macfarlane and H. P. Trommsdorff, U.S. Patent 4,101,976 (1978).
58. W. E. Moerner, W. Lenth and G. C. Bjorklund in "Persistent Spectral Holeburning: Science and Applications" ed. W. E. Moerner, Springer-Verlag, Berlin (1988) p. 251.
59. H. W. H. Lee, M. Gehrtz, E. Marinero and W. E. Moerner, *Chem. Phys. Lett.* **118**, 611 (1985).
60. J. L.Hall in Atomic Physics 3 eds., S. J. Smith and G. K. Walters (Plenum, New York), p. 615 (1973).
61. R. G. Brewer and A. Z. Genack, *Phys. Rev. Lett.* **36**, 959 (1976).
62. R. G. DeVoe, A. Szabo, S. C. Rand and R. G. Brewer, *Phys. Rev. Lett.* **42**, 1560 (1979).

63. R. G. Brewer and R. L. Shoemaker, *Phys. Rev.* **A6**, 2001 (1972).
64. D. M. Boye, R. Wannemacher, Y. P. Wang, W. Grill, J. E. Rives and R. S. Meltzer, *J. Lumin.* **45**, 431 (1990).
65. J. L. Hall and T. W. Hansch, *Opt. Lett.* **9**, 502 (1978); M. Zhu and J. L. Hall, Conference on Quantun Electronics and Laser Science QELS'91 Technical Digest p.232, Optical Society of America (1991).
66. T. Muramoto, S. Nakanishi and T. Hashi, *Opt. Commun.* **24**, 316 (1978).
67. R. M. Shelby and R. M. Macfarlane, *J. Lumin.* **31/32**, 839 (1984).
68. R. Wannemacher, R. S. Meltzer and R. M. Macfarlane, *J. Lumin.* **45**, 307 (1990).
69. E. L. Hahn, *Phys. Rev.* **80**, 580 (1950).
70. N. A. Kurnit, I. D. Abella and S. R. Hartmann, *Phys. Rev. Lett.* **13**, 567 (1964).
71. W. H. Hesselink and D. A. Wiersma, in "Spectroscopy and Excitation Dynamics of Condensed Molecular Systems", eds. V. M. Agranovitch and R. M. Hochstrasser, North Holland, Amsterdam, (1983) p. 249.
72. P. F. Liao, R. Leigh, P. Hu and S. R. Hartmann, *Phys. Lett. A* **41**, 285 (1972).
73. R. M. Macfarlane, R. M. Shelby and R. L. Shoemaker, *Phys. Rev. Lett.* **43**, 1726 (1979).
74. W. H. Hesselink and D. A. Wiersma, *Chem. Phys. Lett.* **54**, 227 (1978).
75. J. B. Morsink, T. J. Aartsma and D. A. Wiersma, *Chem. Phys. Lett.* **49**, 34 (1977).
76. D. Grischkowsky and S. R. Hartmann, *Phys. Rev. B* **2**, 60 (1970).
77. Y. C. Chen, K. P. Chiang and S. R. Hartmann, *Phys. Rev. B* **21**, 40 (1980).
78. M. Mitsunaga, *Phys. Rev. Lett.* **63**, 754 (1989).
79. W. B. Mims, *Phys. Rev.* **168**, 370 (1968).
80. P. Hu and S. R. Hartmann, *Phys. Rev. B* **9**, 1 (1974).
81. J. Ganem, Y. P. Wang, D. Boye, R. S. Meltzer, W. M. Yen and R. M. Macfarlane, *Phys. Rev. Lett.* **66**, 695 (1991).
82. Y. C. Chen, K. P. Chiang and S. R. Hartmann, *Opt. Commun.* **29**, 181 (1979).
83. J. B. Morsink and D. A. Wiersma, *Chem. Phys. Lett.* **65**, 105 (1979).
84. Y. S. Bai and R. Kachru, preprint.
85. W. H. Hesselink and D. A. Wiersma, *Phys. Rev. Lett.* **43**, 1991 (1979).
86. W. H. Hesselink and D. A. Wiersma, *J. Chem. Phys.* **75**, 4192 (1981).
87. K. Chiang, E. A. Whittaker and S. R. Hartmann, *Phys. Rev. B* **23**, 6142 (1981).
88. R. M. Shelby, R. M. Macfarlane and R. L. Shoemaker, *Phys. Rev. B* **25**, 6578 (1982).
89. D. P. Burum, R. M. Macfarlane and R. M. Shelby, *Phys. Lett. A* **90**, 483 (1982).
90. P. Hu, R. Leigh and S. R. Hartmann, *Phys. Lett. A* **40**, 164 (1972).
91. P. F. Liao, P. Hu, R. Leigh and S. R. Hartmann, *Phys. Rev. A* **9**, 332 (1982).
92. R. M. Shelby, A. C. Tropper, R. T. Harley and R. M. Macfarlane, *Opt. Lett.* **8**, 304 (1983).

93. C. A. Walsh, M. Berg, L. R. Narasimhan and M. D. Fayer, *J. Chem. Phys.* **86**, 77 (1987).

94. M. Berg, C. A. Walsh, L. R. Narasimhan, K. A. Littau and M. D. Fayer, *J. Chem. Phys.* **88**, 1564 (1988).

95. Y. S. Bai and M. D. Fayer, *Phys. Rev.* **B39**, 1066 (1989).

96. R. M. Macfarlane, R. M. Shelby and A. Winnacker, *Phys. Rev.* **B33**, 4207 (1986).

97. R. J. Reeves and R. M. Macfarlane, *Phys. Rev.* **B39**, 5771 (1989).

98. R. T. Harley and R. M. Macfarlane, *J. Phys. C* **16**, 1507 (1983).

99. R. M. Shelby and R. M. Macfarlane, *Opt. Commun.* **27**, 399 (1978).

100. R. M. Macfarlane and R. M. Shelby, *Opt. Lett.* **6**, 96 (1981).

101. R. M. Macfarlane and R. M. Shelby, *Opt. Commun* **31**, 169 (1981).

EXCITED-STATE DYNAMICS AND ENERGY TRANSFER IN DOPED-SUBSTITUTED GARNETS

A. Brenier, C. Madej, C. Pédrini and G. Boulon

Laboratoire de Physico-Chimie des Matériaux Luminescents
Université Claude Bernard Lyon I
Unité de Recherche Associée CNRS N° 442
Bât. 205 - 69622 Villeurbanne Cedex - France

ABSTRACT

Progress in the study of solid-state laser materials stimulate spectroscopic research especially with crystals of garnet structure. The most often used dopants are transition metal ions like Cr^{3+} for tunable systems and rare-earth ions like Nd^{3+}, Tm^{3+}, Ho^{3+}, Er^{3+} for line spectra lasers. In some hosts such as $Gd_3Ga_5O_{12}$(GGG) and $Gd_3Sc_2Ga_3O_{12}$(GSGG) Cr^{3+} ions are also excellent sensitizers for rare-earth activators. So, a good deal of research has been directed towards the study of material characteristics essential to lasing which can be altered by changing the material's composition. Our approach has been to substitute Ca^{2+} and Zr^{4+} ions for some of the Gd^{3+} and Ga^{3+} ions in GGG to give a mixed garnet structure designated (Ca,Zr)-substituted garnet. Such substitution results in a larger lattice constant and then in a weaker crystal field strength allowing more efficient energy transfer between Cr^{3+} and rare-earth ions.

The main goal of the lecture is to illustrate both the optical properties of excited-states in solids by giving a few examples of the excited-state dynamics of Cr^{3+}, Tm^{3+} and Ho^{3+} ions in (Ca,Zr)-substituted garnets and, also, the energy transfer mechanisms between these ions.

I. INTRODUCTION

Among all types of laser materials, garnet crystals have attractive properties and a good deal of research has been directed towards the study of material characteristics essential to lasing (1-2-3). These characteristics can be altered by changing the material's composition: $Y_3Al_5O_{12}$ (YAG), $Gd_3Ga_5O_{12}$ (GGG), $Gd_3Sc_2Ga_3O_{12}$ (GSGG), La_2Lu_3 Ga_3O_{12} (LLGG) are very well known. Our own attempt at the latter approach is substituting Ca^{2+}, Mg^{2+} and Zr^{4+} ions for some of the Ga^{3+} and Ga^{3+} ions in GGG to give a mixed garnet structure designated (Ca,Zr)-substituted GGG [4-5].

Chromium ion plays an important role in laser-type solid-state materials. It can be used directly as active center in tunable laser materials in the near infrared spectral region as well as sensitizor in Rare-Earth doped host materials. The evolution of the optical properties of Cr^{3+} ion can be easily observed depending on the energy gap ΔE between $^4T_2$ and ^2E excited electronic levels. Usually if $\Delta E >> o$ we say that Cr^{3+} ion is inserted with high crystal field like in Cr^{3+}-doped YAG where $\Delta E = 1043$ cm^{-1} [6]. At the opposite side of the scale if $\Delta E << o$ like in Cr^{3+}- doped LLGG for which $\Delta E = -1000$ cm^{-1}we say that Cr^{3+} ion is inserted with low crystal field. Between these two cases ΔE is close to zero for many crystals and we speak of an intermediate crystal field. It is the case of Cr^{3+}- doped GSGG with $\Delta E \simeq 50$ cm^{-1}[7] which gives very interesting tunability capability and also strong sensitization effect on Nd^{3+} ion so that the conversion efficiency is twice of that of YAG-Nd^{3+}. However GSGG suffers of Scandium presence and it should be better to find new materials with a much lower cost such as (Ca,Zr)-substituted GGG. On the other hand (Ca,Zr)- sustituted GGG containing different concentrations of optically inert ions allows us to play with ΔE: from the samples we have used ΔE evaluated to 100 cm^{-1} whereas $\Delta E = 335$ cm^{-1} in Cr^{3+}-doped GGG. The first studies were devoted only to Cr^{3+} [5] because like in the mentioned garnets it is a potential tunable laser ion and, moreover, an efficient sensitization both of Nd^{3+} [4] and Tm^{3+} [8] ions have been obtained as a result of larger lattice constant and then a weaker crystal field strength allowing good overlapping between the emission spectrum of Cr^{3+} ion and the absorption spectra of Nd^{3+} or Tm^{3+} ions.

The main goal of this paper is to illustrate both the optical properties of excited-states in solids by giving a few examples of the excited-states dynamics of Cr^{3+}, Tm^{3+} and Ho^{3+} ions in (Ca,Zr)-substituted garnets and, also, the energy transfer mechanisms between these ions. Another important feature is emphasized, corresponding to the systematic presence of multisites or nonequivalent centers playing a role in the optical properties.

II. EFFECTS OF DISORDER ON EXCITED-STATE DYNAMICS PROPERTIES OF Cr^{3+} IONS

A special attention has been paid in our Laser Program on the effects of disorder on spectroscopic properties. How optical properties are they affected by random or non-random distribution of ions ? Can we separate the different contributions of the causes of inhomogeneity to experimental data?

Some contributions answering to these questions have been given in our previous works by selecting $Y_3Al_5O_{12}$ (YAG) host which may be doped by several ions such as Cr^{3+} [6-9], Nd^{3+} [10-11], Ce^{3+} [12], Tm^{3+} [13] and many others by conserving excellent optical properties and so allowing reliable conclusions. The most attractive ideas concern the presence of multisites not only in YAG but also in GSGG [7-14], GGG [7], and more especially non-random distributions of activator centers imbedding the networks [15-16-17-18].

This part has been chosen to develop the unusual disordered characteristics of Cr^{3+}-doped (Ca,Zr)-substituted GGG single crystals within garnets family we are growing in our Laboratory by the Czochralsky method [4-5].

Let us recall that the chemical formula of garnet tructure is $A_3B_2C_3O_{12}$ where A, B and C are dodecaedral, octaedral and tetraedral sites respectively. The presence of three kinds of sites is already a great advantage for research allowing multiple substitution. Cr^{3+} ions go to B sites but it has been shown there is an inversion ratio between ions in GGG, a little amount of Gd^{3+} ions occupying also some B sites in place of Ga^{3+} ions [19]. Because ionic radius of Gd^{3+} is much larger than Ga^{3+} ionic radius, there is creation of defects at a random distance of a given Cr^{3+} site. The situation is worse in (Ca,Zr)-substituted GGG due to a great deal of Ca^{2+}-Zr^{4+} defects introduced in larger quantity (0.45 mol/formula) in order to have a congruent composition [20] and, then, the disorder is expected to be much higher. By consequence, Cr^{3+} multisites will result from these randomly defects, distributed in the host, each site being characterized by a particular value of ΔE, like in Cr^{3+}- doped glasses but with a lower distribution of sites. So that (Ca,Zr)-substituted GGG single crystal has a fundamental interest for spectroscopy by its intermediate disordered structure between pure crystal and real glass.

The purpose of this work is to study how the Ca^{2+}- Zr^{4+} ion pairs may modify both the main spectroscopic properties, the dependence of the degree of crystal field strength reflected by the average value of ΔE and also the energy transfer mechanisms from Cr^{3+} multisites revealing more strongly their existence. So that we will study their dynamics by mean of a simple model based on a local dilatation of the lattice around each defect. It should be reminded that the special structure of ^2E and $^4T_2$ excited states, with low and strong electron-phonon coupling then no Stokes shift for $^2E \leftrightarrows {}^4A_2$ and a Stokes shift for $^4T_2 \leftrightarrows {}^4A_2$, lead to an exponential decay profile per site. Decay time τ is so given by the following relation deduced from the 3-levels scheme $^4A_2$, ^2E and $^4T_2$:

$$\frac{1}{\tau} = \frac{1}{\tau(^2E)} \, \frac{n(^2E)}{n(^2E)+n(^4T_2)} + \frac{1}{\tau(^4T_2)} \cdot \frac{n(^4T_2)}{n(^2E)+n(^4T_2)}$$

$n(^2E)$ and $n(^4T_2)$: population of ^2E and $^4T_2$ levels
By taking into account of degenerencies and Bolzmann's law at thermal equilibrium we can write a

$n(^2E) \simeq 4$ and $n(^4T_2) \simeq 12 \exp(-\Delta E/kT)$

(k: Boltzmann constant and T: temperature)

The presence of multisites gives more complexity to the study of decay profiles which may be associated to more or less crystal field strengths. One additional difficulty is due to the occurence of energy transfer mechanisms between strong crystal field sites and low crystal field sites, already observed in glasses and glass-ceramics [21-22] contributing to another cause of non-exponentiality of decays strongly enhanced by the large difference of the intrinsic lifetimes of each ^2E and $^4T_2$ levels belonging to the two types of site. Moreover depending of the nature of the interaction between the ions (dipole-dipole, dipole-quadripole, quadripole-quadripole) the decay's curvature may change notably. Another cause of variation of the profiles has been advanced from non-random distribution of Cr^{3+} ions [15-16-17-18]. So we must investigate different materials in order to try to separate if possible the different contributions of the causes of inhomogeneity to experimental data like decays. We report only results given by Cr^{3+}-doped substituted GGG.

II.A. Effect of Ca-Zr ion pairs on main spectroscopic properties of Cr^{3+} doped substituted-GGG

The main parameters have been reported in Fig. 1 by using configurational coordinate curves for ${}^4A_2$, ${}^4T_2$ and ${}^2E$ levels of Cr^{3+} ion. They have been evaluated from :

- Emission Peak P_{em} and absorption peak P_{abs} of ${}^4T_2 \rightarrow {}^4A_2$ transition determinated from emission and absorption spectrum
- Zero phonon energy E_o of ${}^4T_2$ level calculated by assuming an equal distance from P_{em} and P_{abs}.
- Energy gap ΔE between ${}^4T_2$ and ${}^2E$ levels, the position of this last one being the one measured at liquid helium temperature.
- Energy $\hbar\omega$ of one phonon of the vibration associated to ${}^4T_2$ and ${}^4A_2$ levels and the stoke shift S_o (Huang-Rhys parameter) of the twoparabolas associated with ${}^4T_2$ and ${}^4A_2$ levels. These two parameters are determinated from a fit of the ${}^4A_2 \rightarrow {}^4T_2$ absorption band with STRUCK and FONGER model [19].
- Radiatif lifetime τ_R of the ${}^4T_2$ / ${}^2E \rightarrow {}^4A_2$ transition.

The fluorescence of this transition, selected at 750 nm, is almost exponential at room temperature.

All these parameters are listed in Table 1. The rôle of Ca-Zr ions is clear: the crystal field which acts on Cr^{3+} ions is decreased, ΔE is decreased and radiatif lifetime τ_R also because the thermal equilibrium ${}^2E$ / ${}^4T_2$ is moved towards the more radiatif ${}^4T_2$ level. This fact is reflected in the stimulated emission cross-sections which are given in Fig. 2 for the two materials pure GGG and (Ca,Zr)-substituted GGG.

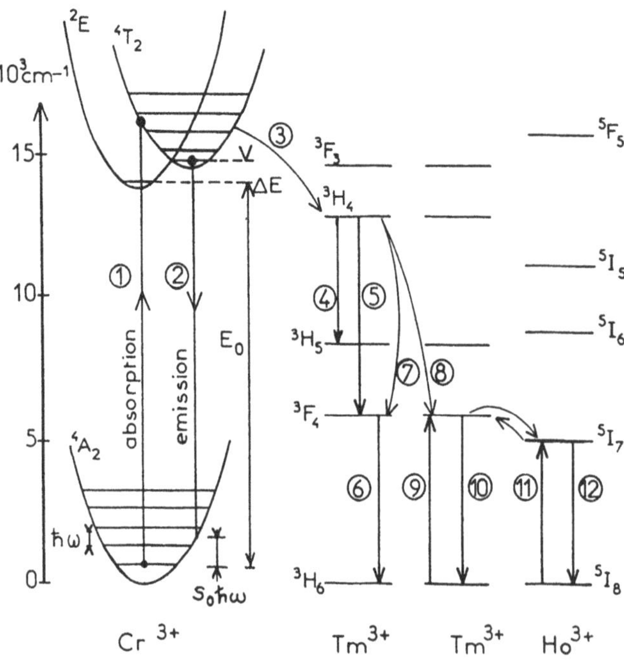

Figure 1 - Energy level schemes of Cr^{3+}, Tm^{3+} and Ho^{3+} ions and main energy transfer or relaxation processes. The curved thin lines are representative of non radiative energy transfer possibilities and the vertical full lines symbolise absorption and emission transitions.

With garnets, many possibilities occur for laser emissions

Cr^{3+} alone : tunable emission in the near infra red by $^4T_2 \rightarrow {}^4A_2$ vibronic transition (1,2)

Cr^{3+}-Tm^{3+} : $^3H_4 \rightarrow {}^3H_5$ at 2.3 µm (1,3,4)

$^3H_4 \rightarrow {}^3F_4$ at 1.5 µm (1,3,5)

$^3F_4 \rightarrow {}^3H_6$ at 1.8 µm (1,3,7,6) and 1,3,8,10)

Cr^{3+}-Tm^{3+}-Ho^{3+} :

$^5I_7 \rightarrow {}^5I_8$ at 2.1 µm (1,3,7,6,9,10,11,12)

and (1,3,8,9,10,11,12)

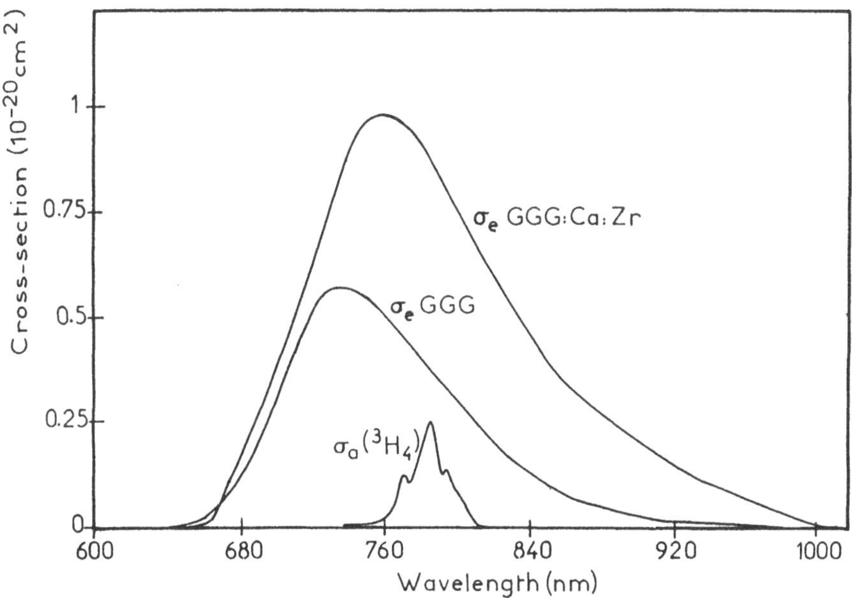

Figure 2 - Stimulated emission cross-sections of Cr^{3+}-doped GGG and Ca,Zr)-substituted GGG noted σ_e and absorption cross-section of $^3H_6 \rightarrow {}^3H_4$ transition in Tm^{3+}-doped (Ca,Zr)-substituted GGG at room temperature.

II.B. Analysis of the fluorescence decays in relation with multisites

The fluorescence decays become strongly non exponential at low temperature as can be seen in Fig. 3. Time resolved spectroscopy reveals in Fig. 4 that the fast decays are originating from the sites emitting mainly in the broad band and that the longer ones are coming from the sites emitting mainly in the $^2E \rightarrow {}^4A_2$ line associated with its vibronic sideband in good agreement with the small and high ΔE energies respectively of these two limit cases.

A simple model has been given to connect the Cr^{3+} multisites and the dynamics of fluorescence [24]. The inversion ratio of garnets between some Gd^{3+} ions from dodecaedral sites to octaedral sites in place of Ga^{3+} ions and the introduction of Ca-Zr ion pairs inside the GGG host create defects which are located at random distance to a Cr^{3+} ion position. Of course the inversion ratio has a negligeable influence in (Ca,Zr)-substituted GGG with respect of Ca-Zr ion pairs as shown by the resolved emission spectrum of Cr^{3+}-doped GGG [7] and by the unresolved broad emission spectrum of

Table 1 - Spectroscopic parameters of Cr^{3+}-doped (Ca,Zr) substituted GGG and in Ca-Zr GGG at room temperature

Parameters	
$P_{abs.}$ (nm)	639
$P_{em.}$ (nm)	759.5
Eo (cm^{-1})	14417
ΔE (cm^{-1})	10
$h\omega$ (cm^{-1})	650
So	2.4
δ_R (μs)	94

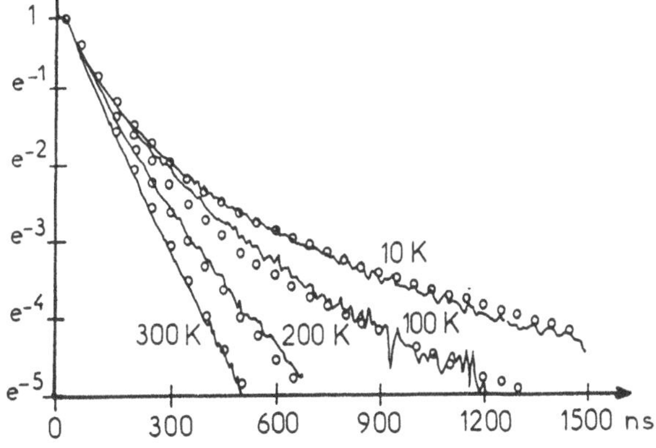

Figure 3 - Fluorescence decay of Cr^{3+} ion in (Ca,Zr)-substituted GGG under 532 nm laser excitation at different temperatures. The full lines represent experimental data and the circles the calculated values from paragraph II.B.

Cr^{3+}-doped (Ca,Zr)-substituted GGG [5]. We have assumed that it exits a volume V around each Cr^{3+} ion, the fluorescence of which is affected only if one or n defects are inside V. V is then the only free parameter of the model concerning the multisites to be determined by fitting the fluorescence decays of Fig. 3. The number of defects inside V may be connected with the local expansion of the lattice constant a(x).

The lattice constant a(x) can be calculated by using an empirical formula deduced for garnets due to B. Stroka [25] :

$$a(x) = 12.3844 + 0.150\ x \qquad (1)$$

we find x =0.46 for (Ca,Zr)-substituted GGG and

$$a(x) = 12.4535\ \text{Å}$$

x can be deduced from n following the relation :

$$8x \cdot \frac{V}{a^3(x)} = n \quad \text{(there are 8 formulas/unit cell)} \qquad (2)$$

On the other hand, the probability to find n randomly distributed defects inside V is:

$$P(n) = \frac{\bar{n}^n}{n!} e^{-\bar{n}},$$

$\bar{n}$ being the average value of n such that : (3)

$$a(\bar{n}) = \bar{a} = 12.4535\ \text{Å}$$

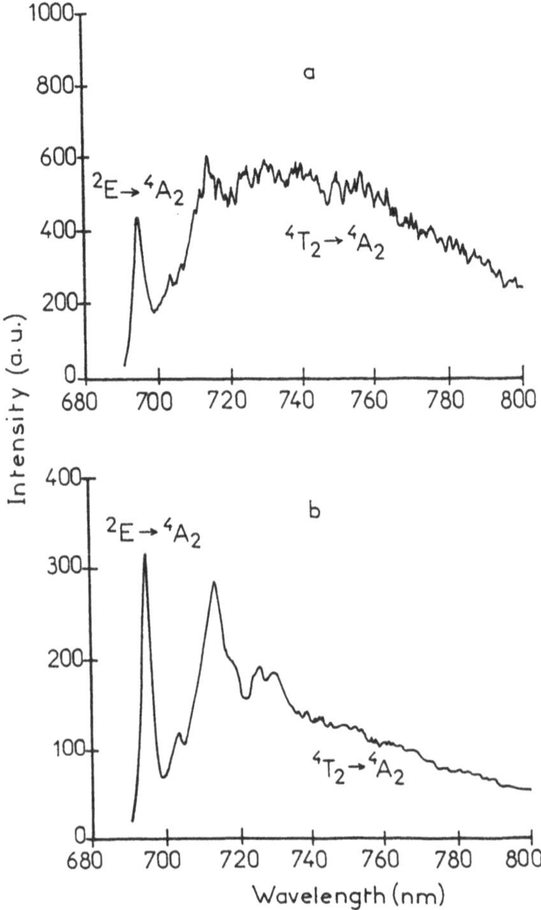

Figure 4 - Time resolved emission spectra of Cr^{3+}-doped (Ca-Zr)-substituted GGG at 10 K under 532 nm laser excitation
a - delay t=O, gate width 3μs
b - delay t=300μs, gate width 6μs

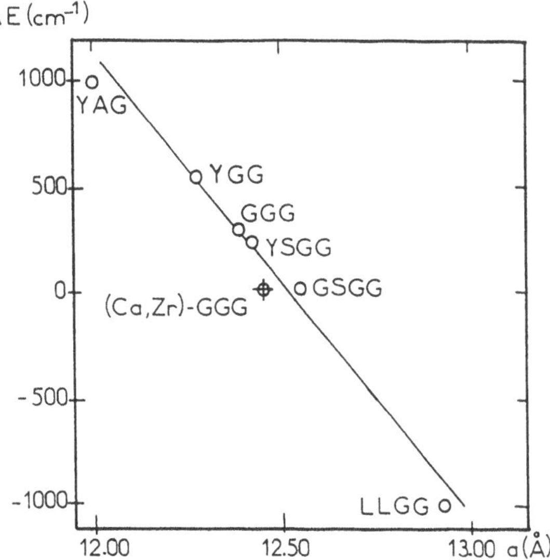

Figure 5 - Dependence of ΔE energy gap between $^4T_2$ and ^2E excited states of Cr^{3+} ion and a lattice constant size for garnets

From the study of several Cr^{3+}-doped garnets, [26] it has been shown that a linear decreasing occurs between ΔE and a, as seen in Fig. 5. The main consequence is the lattice constant depends both on the number n of defects and on temperature :

$$\Delta E(n,T) = \overline{\Delta E} + \frac{d\Delta E}{da} \; [a(n,T)-\bar{a}] \qquad (4)$$

$$\Delta E = 10 \text{ cm}^{-1} \qquad \bar{a} = 12.4535 \text{ A}$$

The thermal expansion of garnets a(n,T) has been studied on a relatively small range of temperature around room temperature [27]. We have assumed that :

$$a(n,T) = a(n,300) + \frac{da}{dT} \; (300-T) \qquad (5)$$

if $10 < T < 300$ K.

Table 2-a summarizes data we have obtained concerning Cr^{3+} multisites by fitting the fluorescence decays of Fig. 3 from the total fluorescence of $^2E+^4T_2 \rightarrow {}^4A_2$ transition from 680 to 850nm under $^4T_1$ level excitation.

Table 2a. Cr^{3+} multisite characteristics at 2 different temperatures in (Ca,Zr)-substituted GGG.

			10 K			300 K	
n	P(x)	a(n)	ΔE(n)	$\tau_{2E}(n)$	a(n)	ΔE(n)	$\tau_{2E}(n)$
		Å	cm^{-1}	μs	Å	cm^{-1}	μs
0	0.000	12.3677	499	2094	12.3844	404	1454
1	0.04	12.3819	418	1540	12.3986	323	1013
2	0.10	12.3963	336	1079	12.4130	241	667
3	0.16	12.4107	254	714	12.4274	159	413
4	0.18	12.4252	171	445	12.4419	76	248
5	0.17	12.4399	88	267	12.4565	-7	170
6	0.14	12.4546	4	167	12.4713	-91	273
7	0.09	12.4694	-81	256	12.4861	-176	459
8	0.05	12.4844	-166	432	12.5011	-261	743
9	0.03	12.4994	-252	707	12.5161	-347	1135
10	0.01				12.5313	-433	1639

$$I(t) = \sum_{n \geq o} P_{(n)} \frac{e^{-t/\tau_{(n)}}}{\tau(n)} \qquad (6)$$

$\tau(n)$ corresponds to the radiative lifetime we took from spin-orbit coupling dependence of multisites by taking into account of Boltzmann's law [28-29].

The best fits in Fig. 3 give the following parameters [24] :

$$V = 1.3\ a^{-3} \qquad \tau(^4T_2) = 82\ \mu s$$

$$\frac{d\Delta E}{da} = 5700\ cm^{-1}\ \mathring{A}^{-1} \qquad \frac{da}{dT} = -5.75\ 10^{-5}\ \mathring{A}\ K^{-1}$$

To summarize, we have shown that this simple model describes quite well the increasing of disorder by introducing (Ca,Zr) ion pairs in GGG. There are a lot a sites with a smooth variation of probability versus n which has been reported in Table 2-a. Moreover, the validity of the model has been probed by treating GGG where only two main sites appear [24] as can be seen in Table 2-b.

Table 2b. Cr^{3+} multisite characteristics at 2 different temperatures in GGG.

n	P(n)	10 K			300 K		
		a(n)	ΔE(n)	$\tau_{^2E}(n)$	a(n)	ΔE(n)	$\tau_{^2E}(n)$
		Å	cm^{-1}	ms	Å	cm^{-1}	ms
0	0.63	12.3594	447	1.48	12.3761	372	1.08
1	0.29	12.3774	367	1.06	12.3940	291	0.74
2	0.07	12.3955	285	0.72	12.4122	210	0.48
3	0.01	12.4137	202	0.46	12.4304	128	0.29

III. ENERGY TRANSFER IN (Ca,Zr)-SUBSTITUTED GGG DOPED WITH Cr^{3+} AND Tm^{3+} IONS [8]

The Cr^{3+} ions have been excited at 640 nm in the $^4A_2 \rightarrow {}^4T_2$ absorption band. The chromium fluorescence $^4T_2 \rightarrow {}^4A_2$ decays exponentially in the absence of Tm^{3+} ions, but non-exponentially when Tm^{3+} ions are present. The non-exponential character of the decays in Fig. 6 increases with the Tm^{3+} concentration and the decays strongly shorten. This is of course related to the $Cr^{3+} \rightarrow Tm^{3+}$ energy transfers via $^4T_2$ (Cr^{3+} ion) and $^3H_4(Tm^{3+})$ excited states represented in Fig. 1 and clearly seen in Fig. 2 by the overlapping of stimulated emission cross section of $^4T_2 \rightarrow {}^4A_2$ and absorption cross section $^3H_6 \rightarrow {}^3H_4$ of Tm^{3+}.

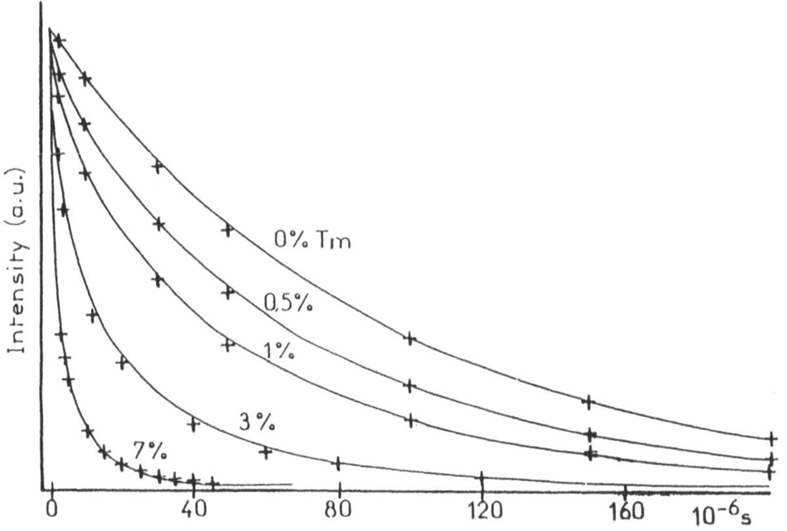

Figure 6 - $^4T_2 \rightarrow {}^4A_2$ fluorescence decays of Cr^{3+} at room temperature for various Tm^{3+} concentrations. Solid lines represent the best fits to experimental data (+).

In total order to describe the transfers, we have supposed that each $^4T_2$ Cr^{3+} excited ion was surrounded by a statistically uniform distribution of Tm^{3+} ions in their $^3H_6$ ground state, and that a dipole-dipole type interaction was responsible of the $Cr^{3+} \rightarrow Tm^{3+}$ energy transfer. This case was first studied by FÖRSTER [30] and is usually treated using the well-known INOKUTI and HIRAYAMA expression [31]:

$$N_1(t) = N_1(0) \exp \left[- \frac{t}{\tau_1} - \frac{4\Pi^{3/2}}{3} C R'^3_o \times \left(\frac{t}{\tau_1}\right)^{1/2} \right] \quad (7)$$

where N is the $^4T_2$ Cr^{3+} population, C is the Tm^{3+} acceptor concentration given in table 3, τ_1 represents the lifetime of the $^4T_2$ Cr^{3+} donor and R'_o is the radius of the sphere of influence around the Cr^{3+} donor.

The best fit to the five experimental decays in Fig. 6 has been obtained with the values $\tau_1 = 90 \pm 1\mu s$ and $R'_o = 9.4 \pm 0.1Å$. The quantum yield η_T of the $Cr^{3+} \rightarrow Tm^{3+}$ energy transfer is given in Table 4. η_T is close to 1 for C = 7%.

Let us examine now the excited state dynamics of the $^3H_4$ state of the Tm^{3+} acceptor. This dynamics is complicated by the $Tm^{3+} \rightarrow Tm^{3+}$ cross-relaxation process in which Tm^{3+} ion acts as a donor. This means that each Tm^{3+} ion belongs with a $W(\phi)d\phi$ probability to a class of ions characterized by a deexcitation rate ϕ due to the cross-relaxation. The normalized function $W(\phi)$ ($\int_0^\infty W(\phi)d\phi = 1$) can be calculated by the MARKOV's procedure [8]. In the case of dipole-dipole interaction, this function is given by:

$$W(\phi) = \frac{b}{2\sqrt{\pi}} \phi^{-3/2} \exp\left(- \frac{b^2}{4\phi}\right) \quad (8)$$

where b is given by (13).

In order to simplify the model, we have neglected the energy diffusion among the Tm^{3+} ions but taken into account the cross-relaxation. In this case, the variation rate of the $N_{2\phi}(t)$ population of the class of Tm^{3+} ions characterized by the cross-relaxation rate ϕ is:

$$\dot{N}_{2\phi}(t) = - N_{2\phi} \left(\frac{1}{\tau_2} + \phi\right) - \left(\frac{N_1}{\tau_1} + \dot{N}_1\right) W(\phi) \quad (9)$$

where N_1 is given by (7).

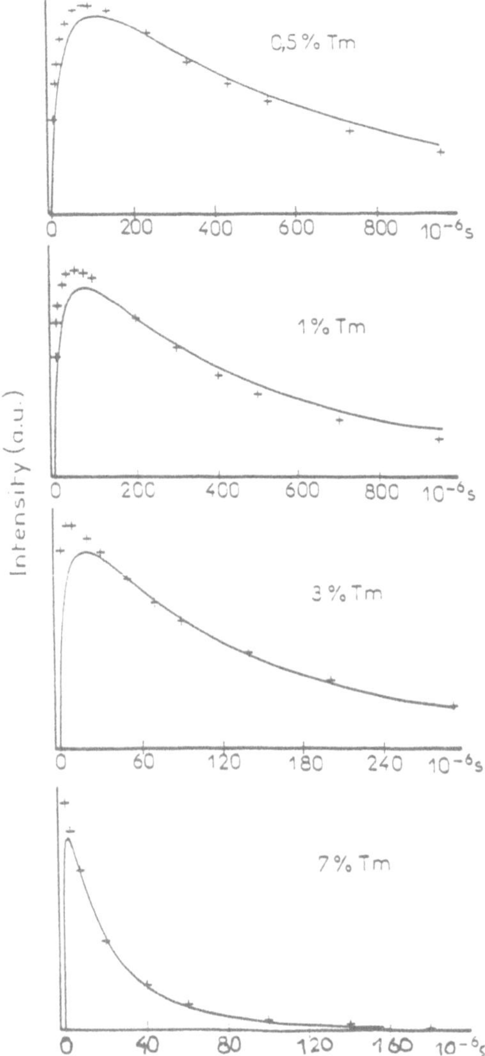

Figure 7 - Time-dependence of the $^3H_4 \rightarrow {}^3H_6$ fluorescence at room temperature unde the $^4T_2$ Cr^{3+} excitation for various Tm^{3+} concentrations.

The solution of (9) is then :

$$N_{2\phi}(t) = N_1(0)W(\phi)\exp\left[-t\left(\frac{1}{\tau_2} + \phi\right)\right] - W(\phi)N_1(t)$$

$$+\left(\frac{1}{\tau_2} + \phi - \frac{1}{\tau_1}\right) \times \int_0^t w(\phi)N_1(t')\exp\left[-(t-t')\left(\frac{1}{\tau_2} + \phi\right)\right]dt' \qquad (10)$$

The integration of $N_{2\phi}(t)$ over ϕ gives the total population $N_2(t)$ of the $^3H_4$ level. Taking into account of Appendix I in reference [8] we obtain :

$$N_2(t) = N_1(0)\exp\left(-\frac{t}{\tau_2} - bt^{\,1/2}\right) - N_1(t) +$$

$$\int_0^t N_1(t')\exp\left[-\frac{(t-t')}{\tau_2} - b(t-t')^{1/2}\right]$$

$$\left[\left(\frac{1}{\tau_2} - \frac{1}{\tau_1}\right) + \frac{1}{2}b(t-t')^{-1/2}\right]dt' \qquad (11)$$

This equation (11) is used to fit the experimental data illustrated in Fig. 7. One can see that the dynamics is correctely described only for weak Tm^{3+} concentrations where the $Tm^{3+} \rightarrow Tm^{3+}$ energy diffusion is weak. It is evident that for Tm^{3+} concentrations greater than 1%, the diffusion must be involved but then the solution of the rate equation is not easy.

IV. CROSS-RELAXATION MECHANISM BETWEEN Tm^{3+} IONS [8]

The time-dependence of the $^3H_4$ population when this level is directly excited has been studied by considering the fluorescence of the $^3H_4 \rightarrow {}^3F_4$ transition (Fig. 8) which can be well separated by interference filters from the excitation wavelength. We do not observe risetimes and the decays are non-exponential and shorten strongly when the Tm^{3+} concentration increases. These results are interpreted by the $^3H_4$, $^3H_6 \rightarrow {}^3F_4$, $^3F_4$ cross-relaxation mechanism represented in Fig. 1 and which is usually involved and has been proved to occur in many Tm^{3+}-doped crystals [32-33-34-35-36]. In the present case, this transfer is phonon-assisted since the gaps $^3H_4 - {}^3F_4 = 6600\ cm^{-1}$ and $^3F_4 - {}^3H_6 = 6000\ cm^{-1}$ are slightly different.

Because of the non-exponential character of the decays whatever the Tm^{3+} concentration considered (0.5; 1; 3; 7%), we are not dealing of a very

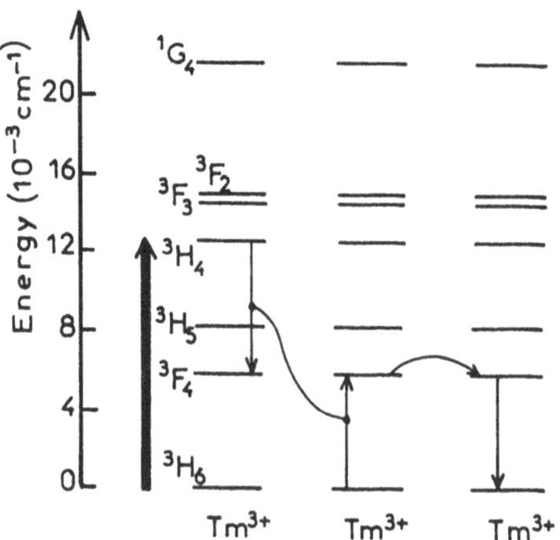

Figure 8 - Energy level schemes of Tm^{3+} ions and main energy transfer and relaxation processes.

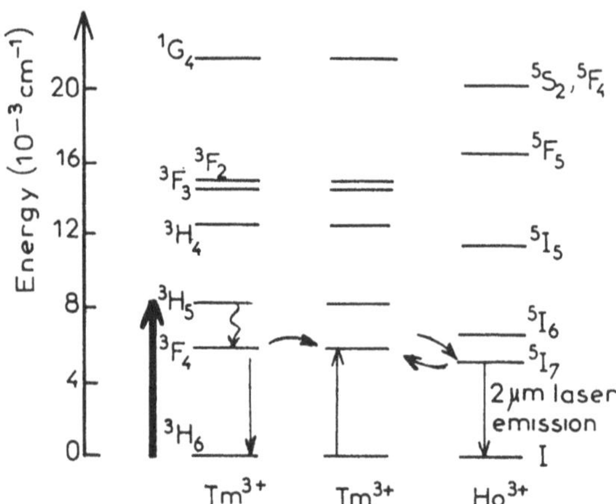

Figure 9 - Energy level schemes of Tm^{3+} and Ho^{3+} ions and main energy level transfer and relaxation processes.

fast energy diffusion among Tm^{3+} ions. Therefore all Tm^{3+} sites are not equivalent and the GRANT's procedure [37] leading to the description of energy transfers by rate equations cannot be applied. Considering that each Tm^{3+} ion in the $^3H_4$ excited state is surrounded by a uniform statistical distribution of Tm^{3+} ions in the ground state $^3H_6$ with which it is in dipole-dipole interaction, and supposing a diffusion limited transfer regime, we have used the YOKOTA and TANIMOTO expression [38] to describe our experimental data:

$$N_2(t) =$$

$$N_2(0)\exp\left[-\frac{t}{\tau_2} - b\, t^{1/2}\left(\frac{1 + 10.87x + 15.50\, x^2}{1 + 8.743\, x}\right)^{3/4}\right] \qquad (12)$$

with $$b = \frac{4}{3}\,\pi^{3/2}\, C\, \frac{R_o^3}{\tau'^{1/2}_2} \qquad (13)$$

where τ_2 is the lifetime of the $^3H_4$ level (in the absence of transfer), C is the concentration of Tm^{3+} in the ground state, R_o is the radius of the "sphere of influence" defined as the distance at which the energy transfer rate W_{DA} (R) is equal to the decay rate τ_2 of the donor:

$$W_{DA}(R) = \frac{C^{(6)}_{DA}}{R^6_{DA}} = \frac{1}{\tau_2}\left(\frac{R_o}{R_{DA}}\right)^6 \qquad (14)$$

R_o can be calculated, if the dipole-dipole term $C^{(6)}$ is known, from

$$R_o^6 = C^{(6)}_{DA} \times \tau_2 \qquad (15)$$

$$x = D\,(C^{(6)}_{DA})^{-1/3}\, t^{2/3} = DR_o^{-2}\,\tau_2^{1/3}\, t^{2/3} \qquad (16)$$

where D is the diffusion coefficient.

The long time behaviour of (12)is referred to as diffusion-limited relaxation, and is characterized by exponential decay at long times. The fluorescence decay function becomes

$$N_2\,(t) = N_2(0)\,\exp\left[-\frac{1}{\tau_2} - \frac{1}{\tau_D}\right] \qquad (17)$$

$$\text{with } \frac{1}{\tau_D} \simeq 4\,\pi\, C\, (C_{DA}^{(6)})^{1/4}\, D^{3/4} \qquad (9)$$

The diffusion coefficient is related to the rate of energy transfer from D ion to nearest neighbor D ion [39]:

$$D = W_{DD}\, R_{min}^{2} \qquad (18)$$

Then

$$D \propto \frac{C_{DD}}{R^6} \times R^2 = \frac{C_{DD}}{R^4} \propto C^{4/3} \qquad (19)$$

since $C \propto R^{-3}$.

Finally the diffusion coefficient D is found to be proportional to $C^{4/3}$ and we can write:

$$D = k\, C^{4/3} \qquad (20)$$

The best fit to experimental data (Fig. 10-a) has been obtained with the three free parameters of the model:

$\tau_2 = (1.065 \pm 0.01) \times 10^{-3}$s, $R_o = 10.5 \pm 0.1$Å and
$k = (1.55 \pm 0.01) \times 10^{-39}\ cm^6\ s^{-1}$.

The quantum efficiency η_{CR} of the cross-relaxation process is given by:

$$\eta_{CR} = 1 - \frac{1}{N_2(o)\tau_2} \int_o^{\infty} N_2(t)dt \qquad (21)$$

The values of η_{CR} calculated from (21) are gathered in Table 3. One can see that the cross-relaxation becomes the dominant process when the Tm^{3+} concentration exceeds 1% and completly governs the excited state dynamics at higher concentrations above 7%.

Table 3 - Quantum efficiencies of the $Tm^{3+} \rightarrow Tm^{3+}$ cross-relaxation η_{CR} and $Cr^{3+} \rightarrow Tm^{3+}$ energy transfer η_T processes (± 1%).

Tm^{3+} concentration				
x%	0.5	1	3	7
η_{CR}(%)	36	59	89	98
η_T(%)	27	45	77	94

The excitation of Tm^{3+} ions in the $^3H_4$ excited state also leads to the fluorescence emission of the $^3F_4$ level corresponding to the $^3F_4 \rightarrow {}^3H_6$ transition in Fig. 8. The fluorescence decay, illustrated in Fig. 8b, is preceded by a rise the time-constant of which strongly shortens when the Tm^{3+} concentration increases. The time-dependence of the $^3F_4$ population $N_3(t)$ may be written as:

$$\dot{N}_3(t) = - \frac{N_3}{\tau_3} + AN_2 -(2 \frac{N_2}{\tau_2} + \dot{N}_2) \qquad (22)$$

with the initial condition $N_3(o) = 0$, where $N_2(t)$ is given by (12)and τ_3 is experimentally known (11.25 ms). The first term represents the relaxation of the $^3F_4$ level and the second one describes the feeding of $^3F_4$ by the $3H_4$ deexcitation directly or via the $^3H_5$ level (A stands for the $^3H_4$ deexcitation probability). The third term is a way to represent the contribution of the cross-relaxation process to the time-dependence of N_3. In the identity equation

$$\dot{N}_2 \equiv - \frac{N_2}{\tau_2} + (\frac{N_2}{\tau_2} + \dot{N}_2) \qquad (23)$$

the second term actually describes the cross-relaxation. The factor 2 in (22) takes into account the fact that one $^3H_4$ Tm^{3+} excited ion leads to two $^3F_4$ Tm^{3+} excited ions. Obviously, the second and third terms both contribute to the rise of the fluorescence. We have solved the differential equation (22) by neglecting the second term. For high Tm^{3+} concentrations

(3 and 7%), it is clear from Table 3 that the cross-relaxation process is predominant and this approximation is justified. For lower Tm^{3+} concentrations, it should be noted that the fluorescence decays of the $^3H_4$ level appear to be more exponential. This means that N_2 and $\dot{N}_2$ are nearly proportional and that the second and third terms of (22) are also themselves proportional.

The solution of (22), in which A = 0 is the following:

$$N_3(t) = 2N_2(0)\exp(-t/\tau_3) - 2N_2(t) + 2\left(\frac{1}{\tau_3} - \frac{1}{\tau_2}\right) \int_0^t N_2(t')\exp\left(\frac{t'-t}{\tau_3}\right)dt' \quad (24)$$

$N_2(t)$ is of course calculated from (12) and the integral is numerically solved. One can see in Fig. 10-b that the experimental data are rather well described by (24). It can be noted that no new fitting parameters had to be used for this description.

V. EXCITED STATE DYNAMICS AND ENERGY TRANSFERS BETWEEN Tm^{3+} AND Ho^{3+} IONS [40]

Let us consider now the energy transfer occurring from the $^3F_4$ lowest excited state of Tm^{3+} to the $^5I_7$ lowest excited state of Ho^{3+} as represented in Fig. 9.This transfer is very efficient and can be studied after either $^3H_5$ or $^3F_4$ excitation of Tm^{3+} ions, since the first excitation is followed by a very fast (<1μs) $^3H_5 \rightarrow {}^3F_4$ relaxation and leads to an exponential $^3F_4 \rightarrow {}^3H_6$ emission (≃ 1.8μm) with no risetime in the Cr,Tm-codoped crystals [1]. The introduction of Ho^{3+} ions is responsible of a non-exponential and faster decay of this fluorescence and simultaneously gives rise to a strong 2μm $^5I_7 \rightarrow {}^5I_8$ emission transition of the Ho^{3+} ions. In order to describe the $^3F_4(Tm^{3+}) \rightarrow {}^5I_7(Ho^{3+})$ transfer, it is very convenient to follow the CHANDRASEKHAR's procedure [41] described in appendix when applied to the case of energy transfers.

Applications of the Markov's method to this problem leads, in the case of dipole-dipole interaction, to a simple expression of the stationary distribution $\omega(\phi)$ of the transfer rate of a given donor ion D to have the prescribed value ϕ:

$$\omega(\phi) = (b/2\pi^{1/2})\ \phi^{-3/2} \exp(-b^2/4\ \phi) \quad (25)$$

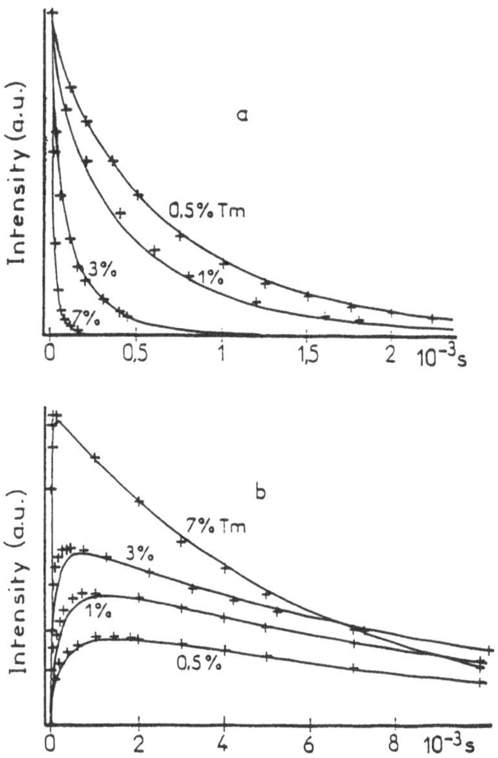

Figure 10 – Time-dependence of (a) : the $^3H_4 \rightarrow {^3F_4}$ fluorescence and of (b) : the $^3F_4 \rightarrow {^3H_6}$ fluorescence in (Ca-Zr)-substituted $Gd_3Ga_5O_{12}$ at room temperature for various Tm^{3+} concentrations. Solid lines represent the best fits to experimental data (+).

with $\quad b = \frac{4}{3} \pi^{3/2} C_A \sqrt{P}$ (26)

and $\quad \phi = \sum^{N_A} T_{DA_i}(r_i)$ (27)

F_A is the concentration of acceptor ions A_i, $P \equiv C_{DA}$ is the microscopic constant of the D → A transfer, N_A is the total number of acceptor ions, r_i is the distance between D and A_i ions and T_{DA_i} represents the transfer rate on each acceptor A_i.

To describe the transfers, it is therefore rightful to use the rate equations for a class ϕ of ions and their complete resolution is possible knowing $\omega(\phi)$. In simple cases, the resolution leads to the well-known Inokuti and Hirayama's expression [31]. However, the method is of most interest in more complicated cases where the Inokuti and Hirayama formula is not further valuable.

Coming back to the particular case of the $^3F_4(Tm^{3+}) \rightarrow {}^5I_7(Ho^{3+})$ energy transfer, the population $N_\phi(t)$ of Tm^{3+} ions belonging to the class ϕ and in the $^3F_4$ excited state decays according to the rate equation:

$$\dot{N}_\phi(t) = - N_\phi(t) \left(\frac{1}{\tau} + \phi\right) \tag{28}$$

where τ represents the lifetime of $^3F_4$ without any Ho^{3+} ions.

The solution of (28) is obvious:

$$N_\phi(t) = N_\phi^o \exp\left[- t \left(\frac{1}{\tau} + \phi\right)\right] \tag{29}$$

with $\quad N_\phi^o = N^o \omega(\phi)$ at $t = O$ (30)

N^o is the overall initial population of the $^3F_4$ level.

The total population N(t) of the $^3F_4$ level is then:

$$N(t) = \int_o^\infty N_\phi(t)\, d\phi \tag{31}$$

and taking into account (12), we find

$$N(t) = N^o \exp\left[-\frac{t}{\tau} - b\sqrt{t}\right] \tag{32}$$

with $$b = \frac{4}{3}\pi^{3/2} C_{Ho} \sqrt{P''} \tag{33}$$

C_{Ho} is the concentration of Ho^{3+} in the ground state and P" stands for microscopic constant of the $^3F_4 \rightarrow {}^5I_7$ transfer.

As previously predicted, in this simple case, equation (32) is simply the Hinokuti and Hirayama formula. It fits quite well the beginning of the experimental $^3F_4 \rightarrow {}^3H_6$ fluorescence (first hundred of microseconds in Fig. 11 curves a). However the rest of the decay is not correctly described. In fact, in $YLiF_4$ and in indium-based fluoride glasses, we had clearly identified an efficient $^5I_7$ (Ho^{3+}) $\rightarrow$ $^3F_4$ (Tm^{3+}) back-transfer [35-42]. The final part of the decay seen in Fig. 11 for both $^3F_4$ and $^5I_7$ levels are similar indicating that such a mechanism actually occurs in the samples presently studied and it has to be included into expression (32). The back-transfer is proportional to the number $\omega(\phi)$ of Tm^{3+} ions of the class ϕ and to the total number N'(t) of Ho^{3+} ions in the $^5I_7$ excited state, so that equation (28) is replaced by

$$\dot{N}_\phi(t) = -N_\phi(t)\left(\frac{1}{\tau} + \phi\right) + k\,\omega(\phi)\,N'(t) \tag{34}$$

with k is the ϕ independant back-transfer constant. N'(t) is given by a rate equation analog to (34) and it turns out that two coupled equations have to be solved. In order to avoid the difficult resolution of such a system, it is better to proceed according to the following way. A preliminary fit of N'(t) to the experimental data is first obtained with the expression

$$N'(t) \simeq A_1\,[\exp(-t/\tau') - \exp(-t/\tau_1)] + A_2\,[\exp(-t/\tau') - \exp(-t/\tau_2)] \tag{35}$$

τ' (9.4 ms) is the lifetime of $^5I_7$ measured after a direct excitation of Ho^{3+}; A_i and τ_i are adjustable parameters without real physical meaning. The best fits lead to curves b in Fig. 11. The solution of equation (34) is then

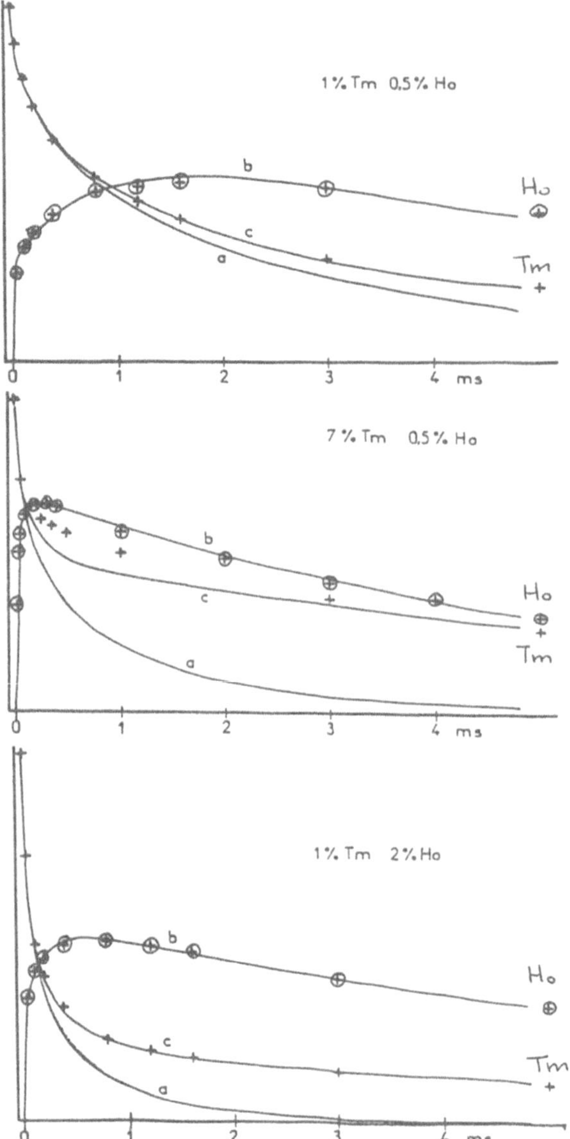

Figure 11 - Experimental decays of :

+ : ${}^{3}F_{4} \rightarrow {}^{3}H_{6}$ fluorescence of Tm^{3+}

O : ${}^{5}I_{7} \rightarrow {}^{5}I_{8}$ fluorescence of Ho^{3+}

The solid lines are the best fits to experimental data obtained with models described in the text.

$$N_\phi(t) = N^o\, \omega(\phi) \exp(-t/\tau - t\,\phi) + k\,\omega(\phi) \left[\frac{A_1 + A_2}{1/\tau + \phi - 1/\tau'} \left(\exp(-t/\tau') - \exp(-t/\tau - t\phi)\right) - \sum_{i=1}^{2} \frac{A_i}{1/\tau + \phi - 1/\tau_i} \left(\exp(-t/\tau') - \exp(-t/\tau_i - t\phi)\right) \right] \quad (36)$$

The total population N(t) of the $^3F_4$ level is given after integration over ϕ:

$$N(t) = N^o \exp(-t/\tau - b\sqrt{t}\,) + I \quad (37)$$

where the first term is the Hinokuti and Hirayama's expression and I is the result of a numerical integration of the second term of (36) over ϕ for each value of t.

Expression (37) leads to curves c of Fig. 11. The best fits were obtained for values of P" indicated in Table 4 (P" is the microscopic constant of the $^3F_4$ (Tm) $\rightarrow$ $^5I_7$ (Ho) transfer). Note that P" is the only free physical parameter of the model because the intrinsic lifetime τ of $^3F_4$ was separately measured in samples free of Ho^{3+} ions [36]. It should be remarked that in samples containing 1 at % Tm, the best fits were obtained with comparable values of P" and well describe the experimental results. On the other hand, in the case of the Tm heavily doped sample (7 at %), the best fit is less satisfactory and was obtained with a much higher value of P". This is likely due to the fact that our model does not take into account the energy diffusion among the Tm^{3+} ions which is certainly effective in the case of high concentrated compounds.

Table 4 gives also the quantum yield η_T of the $^3F_4 \leftrightarrows {}^5I_7$ transfer when the $^5I_7$ population is maximum. It was deduced experimentally (and not from the theoretical model) for the two samples containing the same amount of 0.5 at % Tm and different Tm concentrations (1 and 7 at %). It is clear that the back-transfer induces a strong decrease ($\simeq$ 30%) of the quantum yield in going from 1 to 7 at % Tm.

Table 4 - $^3F_4(Tm^{3+}) \rightarrow {}^5I_7(Ho^{3+})$ transfer parameters

crystal	P'' (10^{-39} cm^6 s^{-1})	τ (10^{-3} s)	η_T (%)
1 at % Tm; 0.5 at % Ho	3.0	11.25	52
7 at % Tm; 0.5 at % Ho	14.3	6.0	35.5
1 at % Tm; 2 at % Ho	2.1	11.25	/

VI. QUANTUM YIELD OF $^4T_2$ $(Cr^{3+}) \rightarrow {}^5I_7$ (Ho^{3+}) TRANSFER [40]

The schematic representation of the absorbed energy conversion in Fig. 1 for (Ca-Zr) - substituted $Gd_3Ga_5O_{12}$ crystals shows that the $^4T_2(Cr^{3+}) \rightarrow {}^5I_7$ (Ho^{3+}) transfer takes place following three stages: $^4T_2(Cr^{3+}) \rightarrow {}^3H_4$ (Tm^{3+}) energy transfer, $^3H_4$ $(Tm^{3+}) \rightarrow {}^3F_4$ (Tm^{3+}) cross- relaxation exchange of excitation between Tm^{3+} ions and $^3F_4$ $(Tm^{3+}) \rightarrow {}^5I_7$ (Ho^{3+}) transfer which can be preceeded by energy migration between Tm^{3+} ions. The quantum yields of the steps were calculated for samples of constant Cr^{3+} and Ho^{3+} concentrations (0.5 at %) and containing different amounts of Tm^{3+} ions. Related curves are displayed in Fig. 12. The efficiency of the first process $^4T_2 \rightarrow {}^3H_4$ was reported in a previous paper [8]. A good description of the $^3H_4 \rightarrow {}^3F_4$ de-excitation of the Tm^{3+} ions must include the cross-relaxation type energy transfer among the Tm^{3+} ions, the Tm^{3+} intra-center de-excitation and the $^3H_4$ $(Tm^{3+}) \rightarrow Ho^{3+}$ direct energy transfer [35]. In order to take into account of all these phenomenons, the quantum yield of the second $^3H_4 \rightarrow {}^3F_4$ mechanism was evaluated by measuring the relative $^3F_4$ emission intensities obtained after excitations of $^3H_4$ and $^3F_4$, of samples placed inside an integrating sphere. Note that the value of the quantum rate at high Tm concentration is nearly 2 since one Tm^{3+} ion in its $^3H_4$ excited state may lead to two Tm^{3+} ions in the $^3F_4$ state. The efficiency of the third $^3F_4 \rightarrow {}^5I_7$ step was determined in the previous section. Since we have only two experimental points, we have supposed a linear dependence of the quantum yield with Tm concentration. The product of the efficiencies of the three stages gives finally the quantum yield of the $^4T_2(Cr^{3+}) \rightarrow {}^5I_7$ (Ho^{3+}) energy transfer. The related curve in Fig. 12 exhibits a maximum for 5 at % Tm concentration. This result is important since it shows that due to the $^5I_7$ $(Ho^{3+}) \rightarrow {}^3F_4$ (Tm^{3+}) back transfer, a further increase of Tm^{3+} concentration is not efficient.

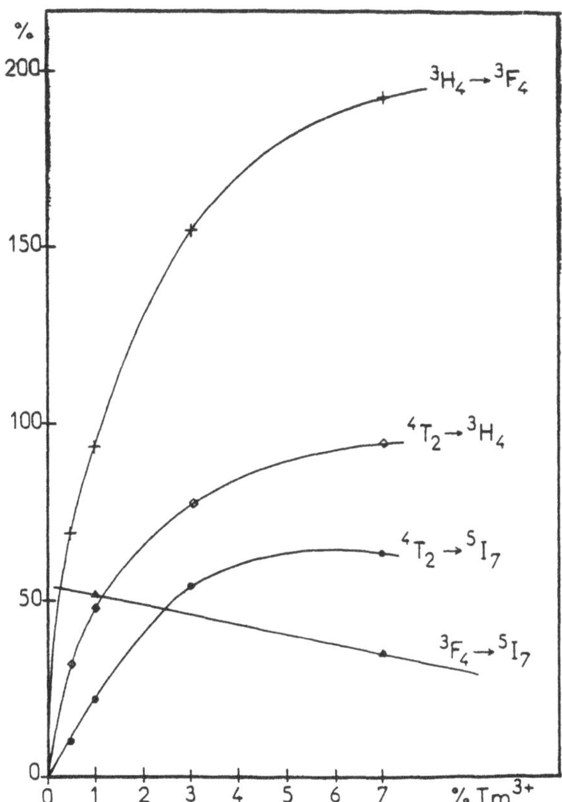

Figure 12 – Quantum yields of various transfers in samples containing O.5 at % Cr^{3+} and Ho^{3+} and different amounts of Tm^{3+}. The concentrations are nominal.

VII. CONCLUSION

Excited-state dynamics, energy transfers and spectroscopic parameters of Cr^{3+}, Tm^{3+}, Ho^{3+} ions in (Ca,Zr)-substituted GGG single crystals have been studied by using laser Spectroscopy Techniques such as Site Selective Spectroscopy, Time-Resolved Spectroscopy and analysis of the fluorescence decays. As examples, $Cr^{3+} \rightarrow Tm^{3+}$ energy transfers, $Tm^{3+} \rightarrow Tm^{3+}$ cross-relaxation processes and $Cr^{3+} \rightarrow Tm^{3+} \rightarrow Ho^{3+}$ energy transfers were analyzed in details and their efficiencies calculated as a function of the dopant concentrations. Nonequivalent centers are clearly observed with Cr^{3+} activators and they play important role for optical properties.

Cr^{3+}-doped (Ca,Zr)-substituted GGG is especially interesting due to high crystalline disorder allowing both ^2E and $^4T_2$ emitter excited states giving possible theoretical studies on the fluorescence decay curvatures. A reasonable agreement between experiment and theory was achieved with a simple model based on thermal expansion of the lattice, presence of defects connected with Ca^{2+} and Zr^{4+} pair ions, Boltzmann's law and spin-orbit coupling but with a final description including only one parameters the volume V around a given Cr^{3+} ion inside of which a defect affects fluorescence properties. This material has been chosen to show Cr^{3+} multisites influence on the decays without appreciable energy transfer between Cr^{3+} ions.

Our Laser Program includes new spectroscopic investigations on transition metal ions and rare-earth ions in GGG, substituted-GGG and YAG. We think so to bring a contribution to knowledge especially of solid-state laser materials by growing crystals and analyzing optical properties.

ACKNOWLEDGMENTS

We are grateful to Misses M. Blanchard, H. Defour, D. Léon and Mrs A. Lagriffoul, J.Y. Rivoire, B. Varrel and R. Perrin for efficient technical assistance. This work is supported in part by The Defense Ministry DGA/DRET under Grant 88/133 and by Région Rhône-Alpes.

REFERENCES

1. B. Struve et al, Appl. Phys. B 28 (1982) 235
2. B. Struve and G. Huber, J. Appl. Phys. 57 (1985) 45

3. E.V. Zharikov et al, Sov. J. Quantum Electron. QE 12 (1982) 338 and 1124
4. G. Boulon, C. Garapon and A. Monteil, in Advances in laser Sciences II edited by M. Lapp, W.C. Swatley and G.A. Kenney-Wallace (AIP-New-York 1987) p. 87
5. A. Monteil, C. Garapon and G. Boulon, J. of Luminescence 39 (1988) 167
6. W. Nie, A. Monteil and G. Boulon, Optical and Quantum Electronics 22 (1990) S227
7. A. Monteil, W. Nie, C. Madej and G. Boulon, Optical and Quantum Electronics 22 (1990) S247
8. A. Brenier, C. Madej, C. Pédrini and G. Boulon, J. of Physics C : Condens. Matter 3 (1991) 203-214
9. W. Nie, G. Boulon and A. Monteil, Chem. Phys. Letters 164 (1989) 106
10. W. Nie, G. Boulon and A. Monteil, J. Phys. (France) 50 (1989) 3309
11. J. Mares, W. Nie and G. Boulon, J. Phys. (France) 51 (1990) 1655
12. J. Mares, B. Jacquier, C. Pédrini and G. Boulon, Revue Phys. Appl. 22 (1987) 145
13. W. Nie, Y. Kalisky, C. Pédrini, A. Monteil and G. Boulon, Optical and Quantum Electronics 22 (1990) S 123
14. O.J. Donnelly, T.J. Glynn, G.P. Morgan and G.F. Imbusch, J. of Luminescence 48-49 (1991) 283
15. S.R. Rotman, F.X. Hartmann, Chem. Phys. Letters 152 (1988) 311
16. S.R. Rotman, Appl. Phys. B 49 (1989) 59
17. S.R. Rotman, Appl. Phys. Letters 54 (1989) 2053
18. J. Mares, Z. Khas, W. Nie and G. Boulon, J. Phys. (France) June 1991
19. M. Allibert et al, J. of Crystal. Growth 23 (1974) 289
20. D. Mateika et al, J. of crystal growth 30 (1975) 311
21. A. Vand Die, G. Blasse and W. Van der Weg, J. Phys. C 18 (1985) 3379
22. G. Boulon, Materials Chemistry and Physics 16 (1987) 301
23. W.H. Fonger, C.W. Struck, J. Luminescence 10 (1975) 1-30
24. A. Brenier, G. Boulon, C. Pédrini, C. Madej, J. of Applied Physics to be published (1992)
25. G. Strocka, P. Holst, W. Tolks Dorf, Philips J. Res. 33 (1978) 186
26. E.V. Zharikov, Proceedings of the ISSL (1986)
27. S. Geller, G.P. Espinosa, L.D. Fullmer, P.B. Crandall, Mat. Res. Bull. vol 7 (1972) 1219

28. B. Struve, G. Huber, Appl. Phys. B36 (1985) 195

29. M. Yamaga, B. Henderson, K. P. O'Donnel, J. Phys. Condens. Matter I (1989) 9175

30. Th. Förster, Z. Naturforsch. 4a, (1949) 321

31. M. Inokuti and F. Hirayama, J. Chem. Phys. 43 (1965) 1978

32. E.W. Duczynski, G. Huber, V.G. Ostroumov and L.A. Shcherbakov Appl. Phys. Lett. 48 (1986) 1562

33. T. Becker, R. Clausen, G. Huber, E.W. Duczynski and P. Mitzecherlich, Tunable Solid State Lasers, Technical Digest, (Optical Society of America, Washington, DC 1989) pp. 183-185

34. B.M. Antipenko, J. Tech. Phys. 54, (1984) 385

35. A. Brenier, J. Rubin, R. Moncorgé and C. Pédrini, J. Phys. France 50 (1989) 1463

36. G. Armagan, B. Di Bartolo and A.M. Buoncristiani, J. of Luminescence 44 (1989) 129 and (1989) 141

37. J.C.W. Grant, Phys. Rev. B4 (1971) 648

38. M. Yokota and O. Tanimoto, J. Phys. Soc. Japan 22 (1967) 779

39. N. Bloembergen, Physica (Utr.) 15 (1949) 386

40. A. Brenier, C. Madej, C. Pédrini and G. Boulon, J. of Physics C Condensed Matter to be published (1991)

41. S. Chandrasekhar, Rev. Mod. Phys. 15 (1943) 1

42. A. Brenier, C. Pédrini, B. Moine, J.L. Adam and C. Pledel, Phys. Rev. B41 (199O) 5364

APPENDIX

The total transfer rate ϕ of a donor ion is the sum of the transfer rates T_{DA_i} on each acceptor ion A_i of the sample:

$$- \phi = - \sum^{N_A} T_{DA_i} (r_i) \qquad \text{(A1)}$$

Where N_A is the total number of acceptor ions and r_i is the distance between D and A_i ions. Because r_i is a random number, ϕ is also a random number that extends between zero and infinity. The question here is to calculate the probability W (ϕ) dϕ for the transfer rate of a given ion D to have the preassigned value ϕ, since the probability $\tau(r)$ to find an acceptor ion A at the distance r of D is known.

Following CHANDRASEKHAR we shall use Markov's method. It consists of calculating firstly the Fourier transform A_{N_A} (ρ) of W(ϕ) with the formula:

$$A_{N_A} (\rho) = \left[\int_0^R \exp (i\rho T_{DA} (r)) \tau (r) dr \right]^{N_A} \qquad \text{(A2)}$$

where R is the radius of the sample supposed to be a sphere.

Assuming an uniform random distribution for the acceptor ions A_i around the donor ion D: $\tau(r) = \frac{3}{R^3}$, we shall let R and N_A tend to infinity simultaneously in such a way that the concentration C_A of acceptor ions A_i is constant:

$$C_A = \frac{(4/3)\pi R^3}{N_A}$$. In the case (A10) becomes:

$$A(\rho) = \exp (- 4\pi C_A \int_0^\infty (1 - e^{i\rho T_{DA}(r)}) r^2 dr) \qquad \text{(A3)}$$

We now consider the case of a multipolar interaction between D and A: $T_{DA}(r) = \frac{C_{DA}}{r^s}$ (s = 6, 8, 10 for dipole-dipole, dipole-quadrupole, quadrupole-quadrupole interaction).

By using residu Cauchy's theorem the result will be:

$$A(\rho) = e^{-b(-i\rho)^{3/5}} \tag{A4}$$

with $b = \frac{4\pi}{3} C_A C_{DA}^{3/5} \Gamma(1 - \frac{3}{5})$

and where Γ is the Euler function defined by

$\Gamma(t) = \int_0^\infty e^{-x} x^{1-t} dx.$

The probability $w(\phi)$ is then given by the Fourier transform of $A(\rho)$

$$W(\phi) = \frac{1}{2\pi} \int_{-\infty}^{\infty} e^{-i\rho\phi - b(-i\rho)^{3/5}} d\rho \tag{A5}$$

In the case of interest here s = 6 and if we change the variable of integration $x = \rho^{1/2}$ and by using Cauchy's theorem we find:

$$W(\phi) = \frac{b}{2\ \pi} \phi^{-3/2} e^{-b^{2/4}\phi} \text{ if } \phi \geq 0 \text{ and} \tag{A6}$$

$W = 0$ if $\phi < 0$.

The time evolution of the donor ion population $N_D(t)$ is the superposition of the exponential evolutions of each class of donor ions characterized by its transfer rate ϕ:

$$N_D(t) = N_D^\circ \int_0^\infty W(\phi) e^{-t(\frac{1}{\tau_D} + \phi)} d\phi \tag{A7}$$

Expression (A5) leads to the following properties of $W(\phi)$:

$$\int_0^\infty w(\phi) d\phi = 1 \tag{A8}$$

$$\int_0^\infty w(\phi)\, e^{-t\phi}\, d\phi = e^{-bt^{1/2}} \tag{A9}$$

$$\int_0^\infty w(\phi)\, e^{-t\phi}\, d\phi = \frac{1}{2}\, b\, t^{-1/2}\, e^{-bt^{1/2}} \tag{A10}$$

By taking account of (A9), expression (A7) leads to the well known Hinokuti and Hirayama formula (7).

STUDIES OF THE CHARGE TRANSFER STATES OF CERTAIN RARE-EARTH ACTIVATORS IN YTTRIUM AND LANTHANUM OXYSULFIDES

C.W. Struck

GTE Laboratories
40 Sylvan Road
Waltham, MA 02254

W.H. Fonger

174 Guyot Avenue
Princeton, NJ 08540

ABSTRACT

The lowest charge transfer state(CTS)of Eu^{3+} and, to a lesser extent, of Yb^{3+} and Tm^{3+}, in Y and La oxysulfides, can be placed in energy by fitting their absorption bands, their emission bands (if present), their feeding behavior into lower rare-earth free-ion-like states, and their role in the sequential quenchings of the activator line emissions with increasing temperature. The CTS is found to be a source, at relatively low temperatures, of free holes which can be trapped for months and then released thermally or with IR irradiation to give the line emissions of the activator. A study of the fraction of CTS excitations which do create free holes as a function of excitation temperature allows determining the bonding energy of the hole to the Eu^{2+} left behind. It is found to be about 1000 cm^{-1}. The binding energy of Eu^{3+} for a free electron can also be determined. It is found to be near 8000 cm^{-1}. The current knowledge of the CTS is examined and many unanswered questions are raised. Some approaches to exploring these questions are suggested.

I. INTRODUCTION

States of a system in zero order are built of molecular orbitals pictured as linear combinations of atomic orbitals, i.e., are configurations. The occupied atomic orbitals yield effective charges on the constituent atoms, e.g., $M^{+}\_X^{-}$. Even when the true ground state is a linear combination of configurations, the effective charges on the constituent atoms is a useful construct. In many cases, excited states involve effective charges on the atoms similar to those

Optical Properties of Excited States in Solids
Edited by B. Di Bartolo, Plenum Press, New York, 1992

for the ground state, e.g., $Ce(^2F_{7/2})^{3+}-F^-_3$ and $Ce(^2F_{5/2})^{3+}-F^-_3$ or $Tb(f^8:{}^5D_4)^{3+}-F^-_3$ and $Tb(f^7:{}^8S_{7/2}),5d:{}^9D_4)^{3+}-F^-_3$. An excited state which derives from significantly different effective charges on its constituent atoms than does the ground state can be labeled a "charge-transfer" state.

With this definition, charge transfer states include free excitons, self-trapped excitons, impurity-trapped excitons, a cation species with an electron transferred from the cation to a neighboring host cation (e.g., $Tb^{3+}-La^{3+} \rightarrow Tb^{4+}-La^{2+}$), and a cation species with an electron transferred from the lattice anion to the cation (e.g., $Eu^{3+}-S^{2-} \rightarrow Eu^{2+}-S^{1-}$).

Here, we describe the last of the above types. Indeed, we restrict ourselves to trivalent rare-earth ions in La and Y oxysulfides, because we have studied only this series in some detail.

Charge-transfer states of rare-earths in oxysulfides manifest themselves as broad-band absorptions, in favorable cases as broad-band emissions, as the state determining the sequence of quenching temperatures of the rare-earth line emissions, as the state determining the pattern of direct feeding into the many trivalent rare-earth free-ion-like states, and finally as the state inducing long term storage of the excitation energy. We take up each of these in turn.

The work reported here is explained fully in a newly published book by the authors[1], which also lists all pertinent references.

II. CURRENT KNOWLEDGE

IIA. Broad-Band Absorptions

A broad-band absorption of Eu^{3+} in Y_2O_2S is shown in Figure 1. This figure shows the excitation spectra of an emission from $^5D_0$ and an emission from $^5D_3$. The spectra are uncorrected, and the peak of the broad band is at higher energy than shown here.

The single-configurational-coordinate model of figure 2 is a useful tool for understanding this broad band. The broad band absorption is from the initial $^7F_0$ state into the offset CTS at low temperatures, but at higher temperatures it includes lower-energy absorptions from thermally populated ^7F_j states. This diagram is useful also because it allows picturing the important readjustments in vibrational energy which occur after photon absorption.

The model appropriate for quantitatively fitting the band shape and position is the W_p function expression developed by Struck and Fonger. This is the sum of the Franck-Condon factors, i.e., the squared overlap integrals between the vibrational wavefunctions, thermally weighted for the Boltzmann population of the initial vibrational state.

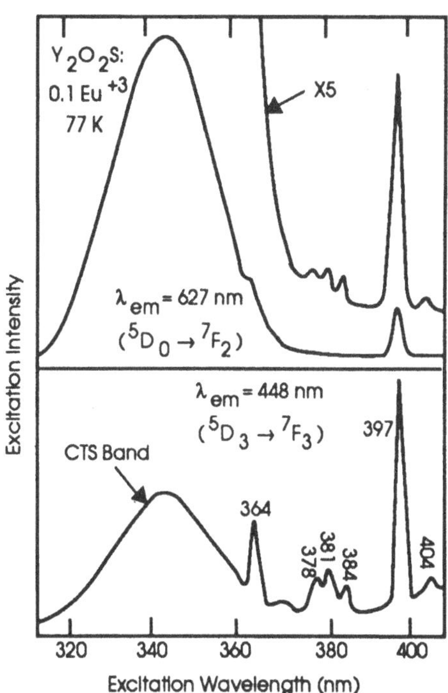

Figure 1. Excitation spectra of Eu^{3+} in Y_2O_2S:0.1%Eu at 77K. The spectra are for a $^5D_0$ and a $^5D_3$. The spectra are uncorrected for the spectral distribution of the excitation source which peaked near 80 nm.

$$W_p = \sum_{n=\max(o,-p)}^{\infty} (1-r)r^n \langle v_{n+p}|u_n\rangle^2 \tag{1}$$

Here, r=exp(-ħw/kT) and v_m and u_n are the vibrational wavefunctions for the excited (offset) and the ground electronic states, respectively. This expression can be evaluated directly by summing Eq.(1), using Manneback recursion formulas to obtain the overlap integrals. Alternatively, one can derive and use its infinite series form,

$$W_p(S,\langle m\rangle) = \exp(-S\langle 2m+1\rangle) \sum_{j=\max(o,-p)}^{\infty} \frac{(S\langle m\rangle)^j)\ (S\langle 1+m\rangle)^{p+j})}{j!(p+j)!}. \tag{2}$$

A third method is through its recursion formula,

$$S\langle m\rangle W_{p+1} + pW_p - S\langle 1+m\rangle W_{p-1} = 0. \tag{3}$$

For equal force constants, W_p has been shown identical to the expression derived by Huang and Rhys and by Pekar for a single-frequency multiple-coordinate (Einstein) model of a crystal. The key is that every transition weight summed in Equation (1) pertains to the same photon energy, namely that displaced to higher energies from the zero-phonon line by p excited-state phonon energy units ħw. This is true for both the SCC and the Einstein models.

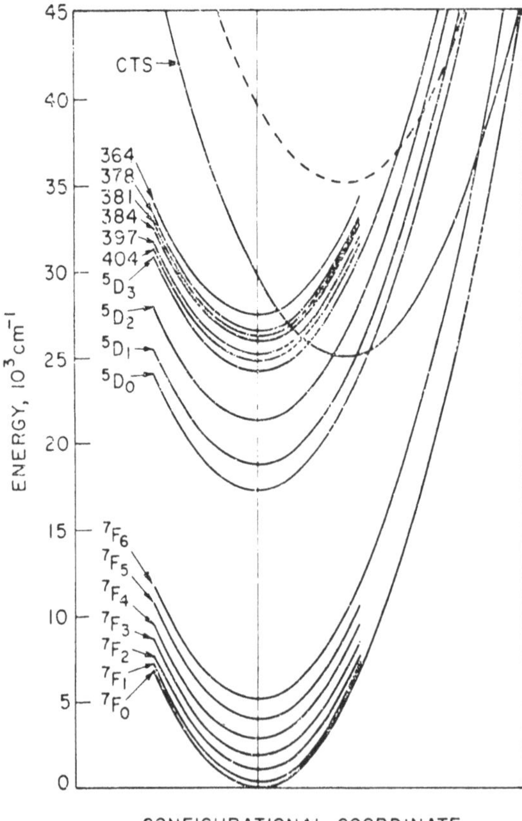

Figure 2. Single configurational coordinate model of the $Y_2O_2S:Eu^{3+}$ system. The 4f states above $^5D_3$ are indexed by their abosrption wavelengths in nm from $^7D_0$.

The proof uses the reproductive property of the W_p functions,

$$W_p(S_1+S_2,\langle m\rangle)=\sum_{p_1=-\infty}^{\infty} W_{p_1}(S_1,\langle m\rangle)W_{p-p_1}(S_2,\langle m\rangle). \qquad (4)$$

which can be proved with the help of the recursion formula for W_p,Eq.(3). It is easy to generalize Eq.(4) and link the SCC model to the Einstein model. There is thus an exact equivalence in the predictions of the band shapes for these two models.

Equation (1) for unequal-force-constants does not give a one-dimensional discrete set of allowed transition energies. Struck and Fonger give a numerical rather than an analytic handling of the shape function for unequal force constants in one coordinate. The crucial approximation is to transfer transition weight at any demanded energy to transition weight at nearby members of a set of equally spaced energies, in particular, those for the final-state phonon energies at 0 K. Thus one obtains the unequal-force-constants overlap integrals from their

Manneback recursion formulas, and then combines them after having weighted them for the thermal population of their initial vibrational level and after having transferred the resulting transition weight to the adopted set of energies. The result is a distribution that goes smoothly to the correct equal-force-constants distribution as the parabolas approach equal phonon energies, and that also goes smoothly to the correct and known distribution at zero temperature.

The unequal-force-constants case have two natural parameters, given by Manneback, Θ and a_{uv}. These are related to the vibrational energies, force constants, and displacements, via

$$A_{nm}=\int_{-\infty}^{\infty} u_n(z_u)v_m(z_m)dx=\langle u_n|v_m\rangle.$$

$$z_u=x|x_u \qquad z_v=(x-a)/x_v$$

$$x_u^2=\hbar|(Mk_u)^{1/2} \qquad x_v^2=\hbar|(Mk_v)^{1/2}$$

$$W_u=(k_u|M) \qquad W_v=(k_v|M)$$

$$E_u=[n+(1/2)]\hbar W_u \qquad E_v=[m+(1/2)]\hbar W_v$$

$$(1/x_{uv}^2)=(1/x_u^2)+(1/x_v^2)$$

$$a_{uv}=a/x_{uv}$$

$$\tan\Theta = (\frac{k_v}{k_u})1/4=(\frac{W_v}{W_u})1/2=(\frac{1/x_v}{1/x_u})=\frac{x_u}{x_v} \qquad (5)$$

One can also combine two or more unequal-force-constants expressions in different frequencies into a one-dimensional shape function with some chosen discrete ladder of allowed transition energies.

With these model expressions, the placement of a pair of parabolas in a single configurational coordinate space affords an understanding of the position and the shape of the broad band absorptions. This diagram can represent either the SCC or the multiple coordinate model, and it can have either equal or unequal force constants. The diagram in Figure 2 places the Eu^{3+} CTS so as to fit appropriate W_p functions to the absorption band shapes and (with the correct zero-phonon-line position) the band positions as a function of temperature in Y_2O_2S. The $La_2O_2S:Eu^{3+}$ CTS is about 2000 cm^{-1} lower in energy.
The fits obtained for both hosts are shown in Figure 3. The points are the experimental data. The full and the dotted lines are calculated as sums of V_{pv} functions, which are the unequal-force-constants analogues of W_p functions describing the absorption process in our notation.

The sums are over the initial ^7F_j index, where the $^7F_j \rightarrow$ CTS absorption band shapes are weighted by the Boltzmann population of the initial state.

The general broadening with increasing temperature, the gradual shift to lower energies with increasing temperature, and the 2000 cm^{-1} difference in placement of the CTS, with La_2O_2S lower than Y_2O_2S, are evident in this figure.

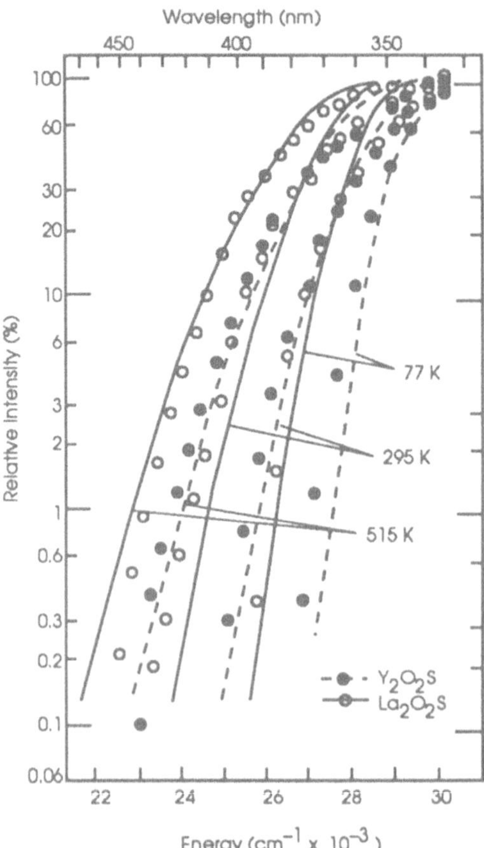

Figure 3. Fit of the absorption bands of $Y_2O_2S:Eu^{3+}$ and $La_2O_2S:Eu^{3+}$. The fit uses V_{PV} functions, the unequal-force-constants analogue of the W_P functions of Eq.(1). The points are experimental excitation spectral data, the lines are combinations of the $^7F_j \rightarrow$ CTS absorption band shapes weighted by the Boltzmann population of the initial state.

II.B. Broad-Band Emissions

In one system, namely oxysulfide:Yb^{3+}, absorption into the CTS leads to emission from the CTS. The emission is easily identified because it has in it the structure of the Yb^{3+} free-ion-like energy levels, i.e., it consists of two broad bands separated by the 10000 cm^{-1} energy difference between the excited $^2F_{7/2}$ state and the ground $^2F_{5/2}$ state of the Yb^{3+} ion. This pattern is shown in Figure 4.

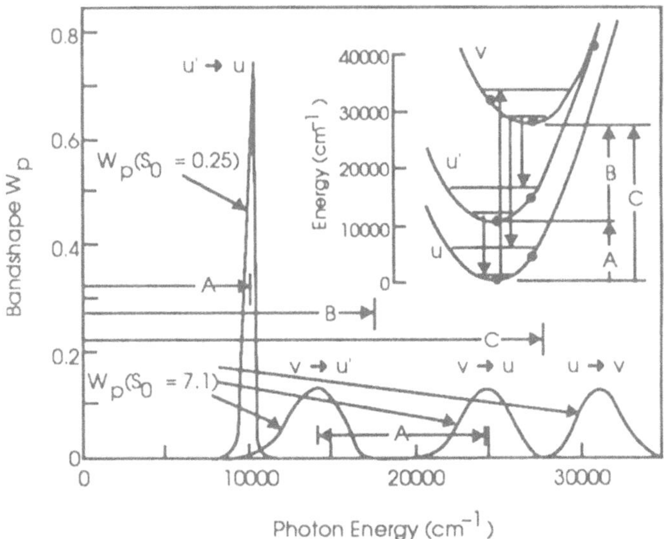

Figure 4. The Oxysulfide:Yb^{3+} system. In the energy level diagram, u, u′, and v correspond to the $^2F_{5/2}$, $^2F_{7/2}$, and CTS states, respectively, and A,B,C correspond to the zero-point energies for $^2F_{5/2} \rightarrow {}^2F_{7/2}$, $^2F_{7/2} \rightarrow$ CTS, and $^2F_{5/2} \rightarrow$ CTS absorptions, respectively. The absorption band near 31000 cm^{-1}, the two broad band emissions near 25000 and 15000 cm^{-1}, and the narrow line emission near 1000 cm^{-1} are shown. The accuracy of the fit obtained is not shown.

Moreover, at low Yb concentrations, it is found in addition that the number of photons emitted in the lower-energy broad band is exactly the number emitted in the narrow line $^2F_{7/2} \rightarrow {}^2F_{5/2}$ emission. One is therefore seeing the competition between one-step, radiative, depopulation of the CTS to $^2F_{5/2}$ and two-step, both radiative, depopulation of the CTS to $^2F_{7/2}$ initially and from there to $^2F_{5/2}$.

The same W_p functions used for absorption are used to understand the energy and the bandshape of these emissions. However, in Equation (1) the sum is over the vibrational quantum numbers of the excited state rather than the ground state, and the transition weight W_p is associated with the energy displaced to lower energies by p ground-state phonon energies. The placement of the CTS in Figure 4 is such as to provide an understanding of both the absorption and the emission band shapes and positions.

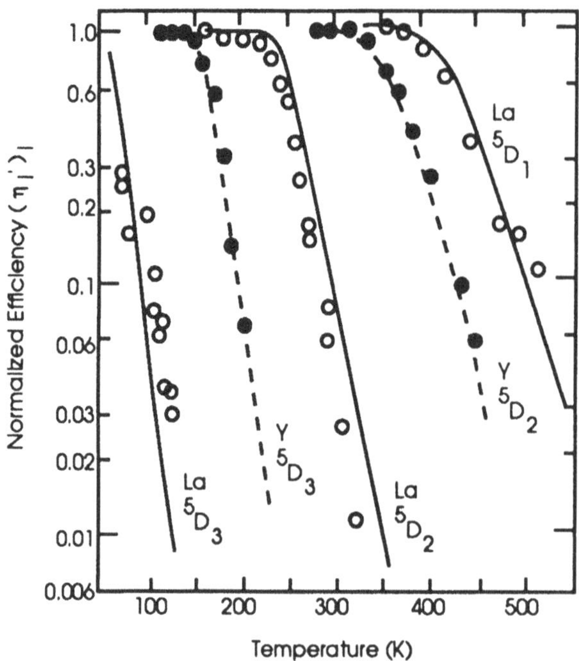

Figure 5. Fit using W_p-type functions of the thermal quenchings in Oxysulfide:Eu^{3+} systems. The points are the experimentally determined quenching efficiencies, i.e., the ratio of the line emission intensities observed to what they would be if we discount the temperature dependence of the $^5D_j \rightarrow {}^5D_{j-1}$ nonradiative rate. The lines are the fits to these quenching efficiencies using the same parabola placements and shapes which gave the fit to the absorption spectra shown in Figure 3. The functions used are U_{PU} functions, our unequal-force-constants analogues for W_p functions for describing emission spectra. These emission-spectra functions, used at zero photon energy, have been shown adequate for understanding the temperature dependence of nonradiative rates.

II.C Sequential Quenching of Rare-Earth Line Emissions

Figure 5 shows the thermal quenchings of the $^5D_j \rightarrow {}^7F$ line emissions of Eu^{3+} in yttrium and lanthanum oxysulfides. This pattern, of the higher states quenching at lower temperatures, is understandable quantitatively by postulating that the upper level is depopulated thermally into the offset CTS and that the CTS in turn feeds the lower ^5D_j levels through crossovers nearer the CTS minimum.

The W_p functions with negative p's have been used also to assess the thermal quenching rates. The p index is selected by the proper energy-balance equation with zero photon energy. The curves in Figure 5

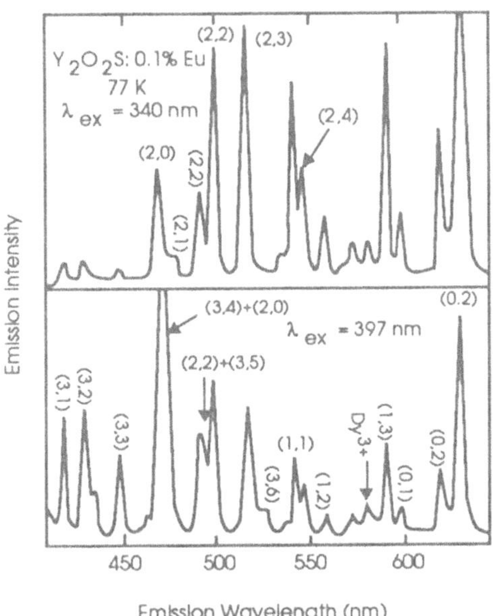

Figure 6. Emission spectra for Eu^{3+} in Y_2O_2S exciting at 397 nm (i.e., effectively into $^5D_3$) and into the CTS. The (a,b) notation is for $^5D_a \rightarrow {}^7F_b$ transitions. The $^5D_3$ emissions are prominent for 397 nm excitation but are not for the higher-energy 340 nm excitation.

are such fittings with the same placement of the CTS which simultaneously affords an understanding of the absorption band energy and shape.

II.D. Level Skipping and Feeding the ^5D_J States

In Eu^{3+} in oxysulfides, populating the CTS leads to skipping certain ^5D_J states and directly populating lower ^5D_J states which have crossovers near the bottom of the CTS parabola.

This tendency is already evident in Figure 1, where the intensity of the 397 excitation peak to the CTS peak is small for $^5D_0$ emissions but large for $^5D_3$ emissions. Such energy-level skipping is also demonstrated in Figure 6. This figure shows prominent $^5D_3$ emissions when this state is directly populated (397 → $^5D_3$ transitions are extremely fast), but distinctly less prominent $^5D_3$ emissions for excitation into the higher energy CTS.

One can determine the line spectrum obtained after directly populating each of the ^5D_j states and then analyze the line spectrum obtained for excitation into the CTS as the sum of the feeding fraction into each ^5D_j state times the pattern emitted from that state. These feeding fractions are in direct proportion to the CTS → ^5D_j feeding rate. Thus the W_p functions allow a quantitative assessment of these feeding fractions.

Table 1. Observed and Calculated Feeding Fractions in LaOCl:Eu^{3+}, La_2O_2S:Eu^{3+}, and Y_2O_2S:Eu^{3+}

host	T,K	θ	$\hbar\omega_v$ cm^{-1}	$\hbar\omega_u$ cm^{-1}	α_3	α_2	α_1	α_0
LaOCl	77				.01	.04	.45	.50
	77	45°	169.7	169.7	.00	.00	.47	.52
La_2O_2S:Eu^{3+}	77				.02	.30	.65	.04
	77	44°	295	275	00	.31	.63	.06
	295							
	295	44°	295	275		.32	.57	.11
Y_2O_2S:Eu^{3+}	77				.20	.65	.15	.02
	77	44°	295	275	.13	.83	.04	.00
	295					.59	.37	.03
	295	44°	295	275		.85	.14	.01

Table 1 gives the feeding fractions seen for Y_2O_2S:Eu^{3+}, La_2O_2S:Eu^{3+}, and LaOCl:Eu^{3+}. It also gives the feeding fractions calculated using those parameters, whether equal or unequal force constants, which give adequate understandings of the broad band absorptions into the CTS and the thermal quenchings of the Eu^{3+} line emissions. The fit is deemed adequate, but not perfect.

The CTS feeds the ^5D_j states in about 10^{-13} seconds, and thermal equilibrium among the vibrational levels of the CTS seems dominantly established within this time. There is evidence in these feeding fractions, however, of some non-thermal feeding events, a few percent of those described by the thermal equilibrium, into ^5D_j states with higher crossovers. These are the analogue of "hot luminescence" and might well be the object of further sub-picosecond spectroscopic investigations.

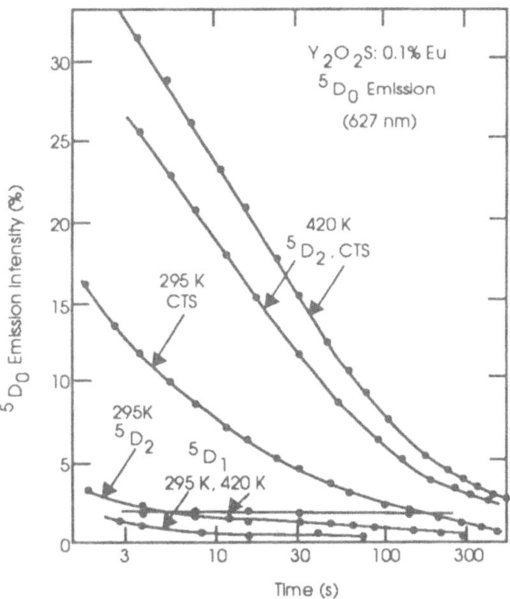

Figure 7. Phosphorescence traces for CTS excitation of Eu^{3+} in Y_2O_2S:0.1%Eu. The emission monitored is the strongest line corresponding to $^5D_0 \rightarrow {}^7F_2$ transitions. The temperatures are noted in the figure.

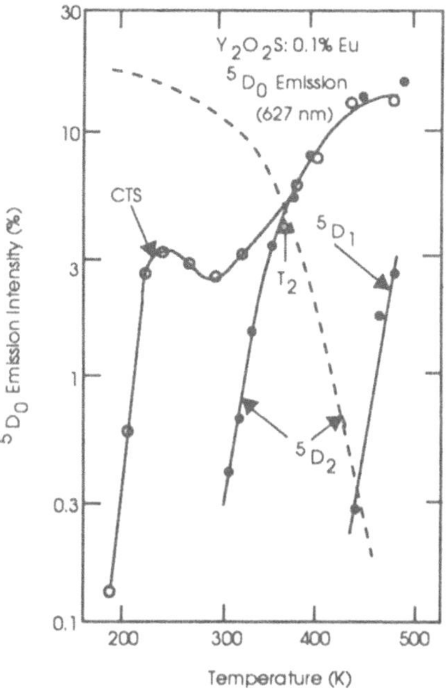

Figure 8. 15-sec. phosphorescence intensity for several excitation energies of Eu^{3+} in Y_2O_2S:0.1%Eu as a function of excitation temperature. Also shown as a dotted line is the quenching of $^5D_2$ emissions with temperature, as studied in Figure 5.

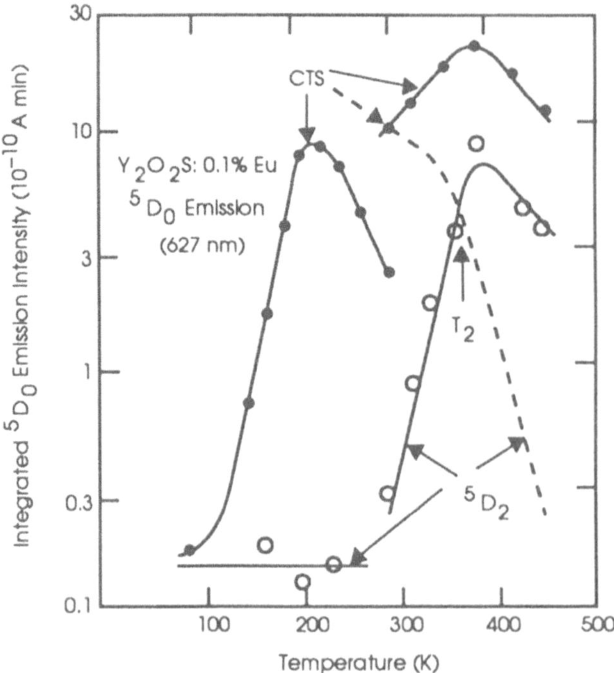

Figure 9. Thermal-glow intensities for several excitations of Eu^{3+} in Y_2O_2S:0.1%Eu as a function of excitation temperature. The phosphor had been previously exhausted of storage, either by IR irradiation or by heating to 500K, then cooled to the abscissa temperature, exposed to a fixed dose of excitation at the wavelength indicated for each curve, then heated at a fixed rate. The emission from $^5D_0 \rightarrow {}^7F_2$ tansitions at 627 mn was observed and the integrated intensity under the glow peak measured. This is plotted in the figure.

II. E. The Breakup of the CTS into a free hole and a trapped electron

Every Eu^{3+} transition discussed so far has a radiative lifetime no longer than a few milliseconds. Nevertheless phosphorescence is sometimes seen lasting many minutes. Figure 7 shows this phosphorescence. It is there even at room temperature.

Figure 8 shows the strength of this phosphorescence 15 seconds after a long fixed excitation at several different excitation energies.

Moreover, energy can be stored for months and then released, either by IR or thermal stimulation, as $^5D \rightarrow {}^7F$ line emissions. Figure 9 shows the magnitude of the thermal glow peak obtained by a fixed excitation dose into a previously exhausted phosphor and its dependence on the temperature of the excitation.

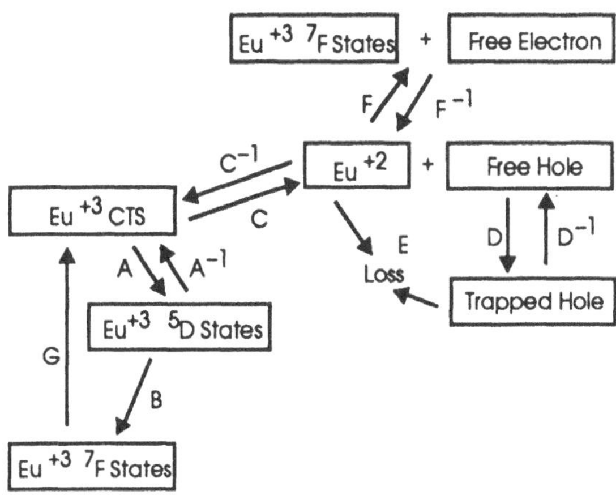

Figure 10. Schematic model of the breakup of the CTS into Eu^{2+} and a free hole with subsequent trapping of this hole. The new processes are C (CTS breakup), C^{-1} (CTS re-formation), D (trapping of the free hole), D^{-1} (detrapping of this hole), E (energy loss), F (dissociation of the Eu^{2+} into a free electron and a ground-state $^7F_0$ Eu^{3+} ion), and F^{-1} (trapping of an electron by a Eu^{3+} ion). The D, E, and F processes and their inverses will not be discussed in this paper.

In both Figure 8 and 9 one sees that excitation even into $^5D_2$ leads to both phosphorescence and storage. The crucial clue is that excitation into $^5D_2$ does so only at temperatures for which $^5D_2$ is thermally quenching into the CTS.

Thus we need an enhancement of the Figure 2 model. The CTS can be thought of as an electron and a hole bound to an Eu^{3+} ion and to a nearby S^{2-}, respectively. Thus we are led to Figure 10. We postulate that the hole as S^{1-} and the electron as Eu^{2+} are held together with only a modest binding energy. At some temperature, the hole can wander to further-off S sites, leaving a trapped electron as Eu^{2+}. The hole may get trapped elsewhere.

This is the state responsible for the glow peaks. A month later, one can free the trapped hole, either thermally or with IR, and the free hole can find a Eu^{2+} and recombine with it to form the CTS, which then feeds the ^5D states and creates the $^5D \rightarrow {}^7F$ line emissions.

The phosphorescence is due to the same breakup of the CTS into Eu^{2+} and a free hole, the transient trapping of this hole in some shallow traps, the thermal emptying of these shallow traps, and the subsequent recombining of free hole and Eu^{2+} to form the CTS.

Table 2. Rates of Some of the Processes Diagrammed in Figure 10

process	Rate
$^7F \rightarrow$ CTS excitation	G
CTS $\rightarrow$ ^5D feeding	$A = ap_0$
$^5D \rightarrow$ CTS quenching	$A_{-1} = a^{-1}n_D$
$^5D \rightarrow$ ^7F emission	$B = bn_D$
CTS dissociation	$C = \sigma_{p0}v_T\bar{p}_0p_0$
free hole capture by Eu^{2+}	$C^{-1} = \sigma_{p0}v_Tn_0p$

Table 3. Notation Used in Table 2

quantity	Symbol
Rate constant for CTS $\rightarrow$ ^5D feeding	a
Rate constant for $^5D \rightarrow$ CTS quenching	a^{-1}
Rate constant for $^5D \rightarrow$ ^7F emission	b
Concentration of Eu in the CTS	p_0
Concentration of Eu^{2+}	n_0
Concentration of free holes	p
$N_v exp(-E_{p0}/kT)$	$\bar{p}_0$
Hole binding energy to Eu^{2+} to form CTS	E_{p0}
Hole capture crosssection for Eu^{2+}	σ_{p0}
Thermal velocity of the free carriers	v_T

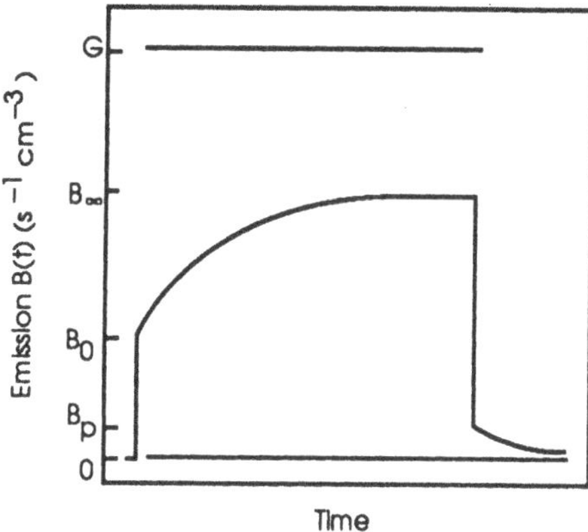

Figure 11. Schematic of the transient behavior of a oxysulfide:Eu^{3+} phosphor previously exhausted of stored energy. There is an instantaneous rise to a level B_0 (instantaneous is fast on a millisecond time scale), a gradual slow increase to a steady state level, an instantaneous fall to a level lower than B_0, and then a slow phosphoresence. The steady state level reached depends on the excitation intensity and tends towards the excitation level G at high excitation intensity. We discuss only B_0/G in this paper.

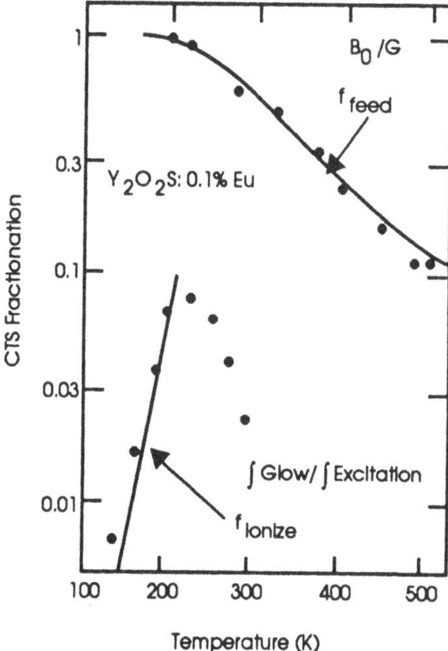

Figure 12. B_0/G (upper curve) and glow (lower curve) vs. excitation temperature. The points are experimental values. The curves are fits of Eq. (5) (upper curve) and of $1-f_{feed}$ (lower curve) for the parameter values in Table 3.

Table 4. Parameters describing CTS dissociation in oxysulfides

host	$\sigma_{p0}v_T N_V/a$	$E_{p0}(\mathrm{cm}^{-1})$	$\sigma_{p0}v_T N_V/a$	$E_{p0}(\mathrm{cm}^{-1})$
Y_2O_2S	230	1200	45	900
La_2O_2S	900	1700	75	1300

One can measure the binding energy of the hole to the Eu^{2+}. One finds that, for a previously exhausted phosphor, the response to a long pulse of CTS excitation is as shown in Figure 11. There is an immediate buildup to an emission level B_0, followed by a very slow increase to some saturation level B_∞. B_∞ for low Eu concentrations is the excitation rate G.

The $B_0|G$ ratio is found to be temperature dependent. It is measurable by displaying this transient for a phosphor cooled to a given temperature after having been exhausted of storage. Figure 12 shows this $B_0|G$ ratio as a function of the excitation temperature for Y_2O_2S:0.1%Eu. Also shown in this figure is the area under the thermal glow curve as a function of the excitation energy. This area is corrected for the phosphorescence, which also results from the breakup of the CTS. The curve through the higher data points in this figure is our fit to f_{feed}, the fraction of CTS excitations which directly feed the ^5D states. The curve through the lower data points is our fit to $f_{ionize}=1-f_{feed}$.

These fits are derived using the rate expressions listed in Table 2, where the notation is given in Table 3.

At low concentrations and at low temperatures, when no ^5D state is being quenched to the CTS, i.e., $a^{-1}=0$,

$$f_{feed}=\frac{B}{B+C}=\frac{B}{G+C^{-1}}. \tag{6}$$

The second form follows from equating the rate of populating and depopulating the CTS. These rates are very nearly equal for any time in the Figure 11 response curve, because the $[Eu^{2+}]$ is always small. At the onset of excitation, C^{-1} is zero and thus

$$f_{feed}=\frac{B}{G}=\frac{B}{C+A}=[1+(\sigma_{p0}v_T N_V|a)\exp(-E_{p0}|kT)]^{-1}. \tag{7}$$

The solid curves in Figure 12 are Eq. (7) for f_{feed} and its associated f_{ionize}. The pre-exponential factor in this equation is taken as 230 and the dissociation energy of the CTS, E_{p0} taken as 1200 cm^{-1}. A more precise model would recognize that the net A rate for the CTS → ^5D transition falls as each particular ^5D_J state quenches to the CTS. If one approximates this A behavior as a step function downwards from a to 3a/4 to a/2 to a/4 as $^5D_3$, $^5D_2$, and $^5D_1$ is quenched, respectively, one gets an equally good fit with the parameters shown in the rightmost two columns of Table 4 for $Y_2O_2S:Eu^{3+}$ and $La_2O_2S:Eu^{3+}$. One can also model the effect of the quenching of $^5D_0$ to the CTS, seen above 450 K for $La_2O_2S:Eu^{3+}$.

There is, by the way, evidence that at high Eu concentrations, the electron trapped as Eu^{2+} can hop from Eu to Eu and recombine nonradiatively with the trapped hole. The hopping is a type of impurity-band conductivity. The loss process introduces an excitation-intensity-dependent luminescence efficiency, which will not be explored further here.

Having now a measure of the dissociation energy of the CTS, we are in the position of placing this state into the SCC diagram of Figure 13, namely, the curve labeled "Eu^{2+}+h(∞)". We place it above the CTS by the dissociation energy. We place it also with less offset than the CTS, because we believe the equilibrium distance for Eu^{2+}-S^{2-} will be shorter than that for Eu^{2+}-S^{1-}=CTS.

We also place the host band-gap at its known position, the curve labeled "$^7F_0$+e(∞)+h(∞)". This is shown with zero offset because influences of happenings at ∞ on a local Eu^{3+} should be negligible.

At once we have a new understanding, namely that the attractive energy of Eu^{3+} for an electron is about 8000 cm^{-1}. This new understanding comes from the somewhat unconventional picture of the energy levels of Figure 12.

The usual single-electron-orbital-energy picture is, in our opinion, somewhat awkward in providing this understanding. This usual band-gap picture would have the top of the valence band of the host at a reference energy of zero, and the bottom of the conduction band at 35000 cm^{-1} for La_2O_2S, 37000cm^{-1} for Y_2O_2S. A Eu^{2+} impurity conduction band must be shown about 8000 cm^{-1} below the conduction band, or at 27000 or 29000 cm^{-1}, respectively, for La_2O_2S and Y_2O_2S. About 1000 cm^{-1} lower, or at 26000 or 28000 cm^{-1} respectively, is the CTS. However, the CTS must simultaneously be shown as an empty acceptor level 1000 cm^{-1} above the valence band, since it can place a hole into the valence band with only a 1000 cm^{-1} thermal barrier. This is the first problem, but there is another. The CTS exists only after absorbing a 30000 cm^{-1} photon and after a 4500 cm^{-1} readjustment downwards in the lattice energy. the $^7F_0$ ground state must therefore be 25500 cm^{-1} below the valence band. There are indeed many ^5D and ^7F states, each with an incompletely filled 4f shell, at energies below the reference zero at the top of the valence band and this ground-state energy, at-25500 cm^{-1}. What then is the meaning of the Fermi level?

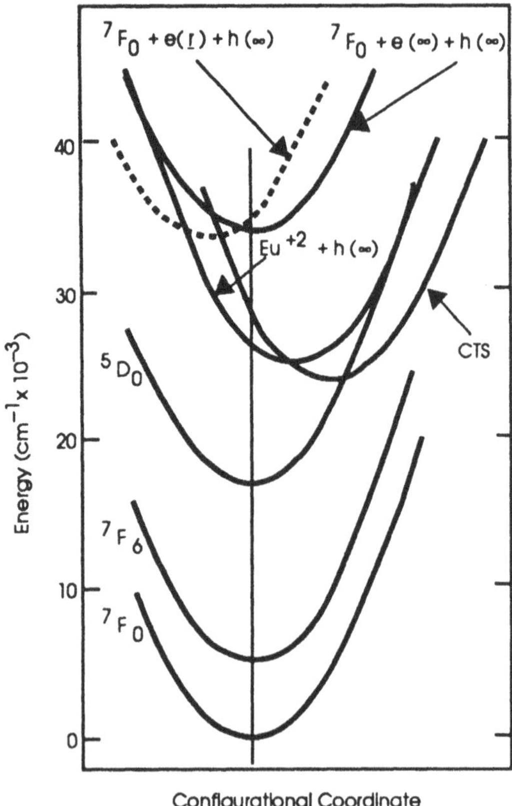

Figure 13. Configurational coordinate model of oxysulfide:Eu^{3+} phosphors which includes the breakup of the CTS and the feeding mechanism for host-band-gap excitation.

III. FUTURE WORK

What has been presented here of our knowledge about CTS's is nevertheless very rudimentary. In general, we know only the position of the lowest CTS. We have therefore a spectroscopic understanding of these states similar to what our understanding of the H atom would be if we knew only the 1s → 2s excitation energy.

There is one experiment in which a two-photon excitation into the Tm^{3+} CTS in La_2O_2S was found which placed the CTS higher than the host band-gap. There is some question whether this is the lowest CTS. Whether it is or not, the appropriate experiment for uncovering higher-than-band-gap energy levels surely is excited-state absorption using a strong laser to populate a higher energy M^{3+} free-ion-like state and another laser to probe for both narrow-line (free-ion-like) and broad-band (CTS) absorption/excitation bands peculiar to the M impurity in the host and at higher-than-band-gap energy.

Even when such experiments have afforded the understanding of the CTS energy levels, many questions will come to mind. What can be measured about the thermalization rate and about the faster-than-thermalization feeding processes occurring? How long do CTS's with higher-than-band-gap energies live against creating a host-lattice hole, or electron, or pair? How long does a high-energy free-ion-like state live against creating a hole-electron pair?

There is the question of understanding the dissociation energy of the CTS. Why does La_2O_2S:Eu have a smaller CTS dissociation energy than Y_2O_2S:Eu? Though one can understand the order-of-magnitude found for these energies, how can one understand the actual magnitudes of the two energies and especially their difference? One can see similar storage effects in these hosts with Yb^{3+} activator. What is the Yb^{3+} CTS dissociation energy?

There is the similar question of understanding the blinding energy of an electron to Eu^{3+}.

In Tb^{3+}, the CTS would have an electron switched from the Tb^{3+} inner 4f shell of the ion to the conduction band of the host. What is the expected f number of this transition?

There is the question of clarifying the theoretical description of luminescence centers. The usual cluster description in the Hartree-Fock or X-a approach is related to the band-gap picture sketched out in this paper. How can such a model be forced to the correct energy placement of the CTS, to the correct low thermal barrier for releasing free holes, to the correct binding energy of a free electron to the Eu^{3+}? It seems that one must find directly the set of allowed total cohesive energies of the system, rather than get excitation energies from promoting an electron into the lowest unoccupied one-electron-orbital calculated for the ground state.

There is the question of the temperature dependence and the concentration dependence of the conductivity seen in the Eu^{2+} impurity band after the CTS dissociation. What governs this conductivity? Is the electron migration accompanied by a phonon cloud?

Finally, there is the need to identify the trapping centers for holes and the nonradiative-recombination centers. Are they confined to surfaces, or to grain-boundaries, or to dislocations, or are they impurity related?

For all these studies, better samples are needed, including single crystals, with controlled concentrations of known and characterizable impurities and defects.

IV. CONCLUSIONS

Experimental determinations of the energy of the lowest CTS in oxysulfides with several rare-earth activators are afforded by fitting the shape and position of the broad-band absorption band due to the activator, using the W_p or related functions. In a few favorable cases the CTS is depopulated radiatively; the shape and the position of the emission band from the CTS can then provide additional information on its placement, again through W_p-like functions. In other cases, the feeding of the several possible lower rare-earth free-ion like states can be experimentally determined and this pattern can be fit using the same W_p-like functions. Likewise, the emptying of these rare-earth free-ion-like states thermally can be shown to take place through these CTS and the sequential quenchings of the narrow-line emissions with increasing temperature also afford a placement of the CTS.

The CTS is found to dissociate measurably at relatively low temperatures. For Eu^{3+}, this dissociation is pictured as freeing the hole from the CTS understood as a trapped excitonic level, leaving Eu^{2+} and allowing the hole to be trapped elsewhere. A study of the temperature dependence of the fraction of CTS excitations which produce such free holes shows that the binding energy of this hole to the Eu^{2+} is about 1000 cm^{-1}.

Placing the determined states into a configurational-coordinate diagram gives then that the binding energy of an electron by Eu^{3+} to form Eu^{2+} is of the order of 8000 cm^{-1}.

Many interesting questions remain.. Excited-state absorption spectroscopy is most likely needed for elucidating the electronic structure of the CTS. The understanding of the two measurable bonding energies, of an electron to Eu^{3+} to form Eu^{2+} and of a hole to Eu^{2+} to form the CTS, would be a real challenge for our current understanding of deep impurity levels in semiconductors. Transient studies, probably in the sub-picosecond regime, would be necessary for the lifetime of the higher CTS's against host hole-electron pair generation.

REFERENCE

1. Struck, C.W. and Fonger, W.H., Understanding Luminescence Spectra and Efficiencies Using W_p and Related Functions, Springer-Verlag, Heidelberg, 1991.

PHOTOCHEMISTRY, CHARGE TRANSFER STATES AND LASER APPLICATIONS OF SMALL MOLECULES IN RARE GAS CRYSTALS

N. Schwentner and M. Chergui

Institut für Experimentalphysik, Freie Universität Berlin
Arnimallee 14, D-1000 Berlin 33, F.R.G.

ABSTRACT

A microscopic theoretical and experimental description of an elementary solid state photoreaction, i.e. the dissociation of a small molecule in a rare lattice is presented. The ingredients and results of molecular dynamics calculations for the dissociation of F_2, Cl_2, HI and H_2O molecules are discussed and the concepts of cage effect, impulsive and delayed exit are illustrated. Examples for the determination of the structure and dynamics of educts and products in the matrix from ESR, infrared and neutron experiments are combined with experimental investigations of the barrier heights of the cage and the dependence of the dissociation efficiency on excess energy, deuteration, guest-matrix pair potentials and temperature. A long range migration of F fragments is demonstrated and applied to produce XeF molecules in rare gas crystals. The concept of solid state excimer lasers and the achieved large gain coefficients and stored energy densities in the UV spectral range are discussed.

I. INTRODUCTION

Most chemical processes in Nature and industry occur in the condensed phase, especially in liquids. Over the past two to three decades, the understanding of chemical dynamics in the gas phase has strongly sharpened, especially that of photochemical reactions, and is gaining refinement with the advent of new techniques such as "real-time" spectroscopy. With some delay, the study of chemical dynamics in liquid solutions has followed thanks to the advent of picosecond spectroscopy, new theories and large-scale computer simulations /1,2/. Although a huge amount of data has been collected on chemical dynamics in liquids and that this field is rapidly expanding /3/, attempts to reveal the details of chemical reactions at the molecular level are hampered by the bulk properties of the liquids, the fluctuations of the environment at the reaction site and by back-reactions which ensue a chemical event. The drawback of time-evolving structures is avoided by studying chemical dynamics in dense media with constant equilibrium positions. Over the past 5 years or so, several groups have been exploring two new avenues, i.e. clusters and inert gas crystals. Both aim at preparing the system with a minimal density of states in a better defined geometry.

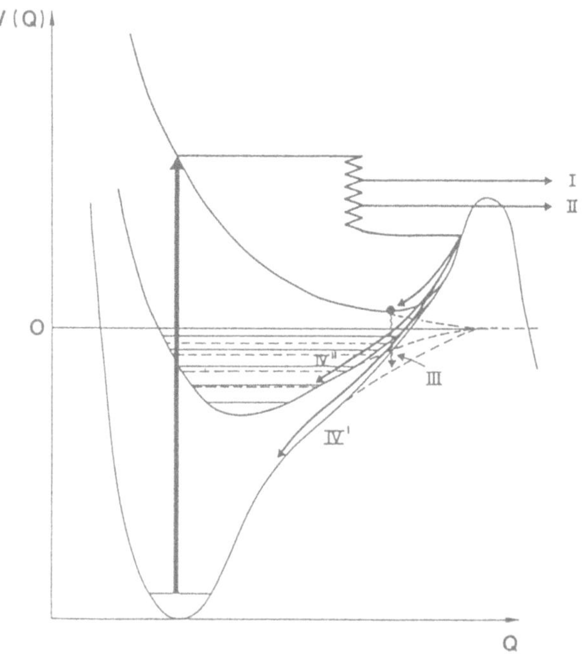

Fig. 1 Intramolecular potentials (solid lines) with matrix perturbation potentials (dashed lines) along the reaction coordinate Q. Excitation leads to: I) cage exit after minimal scattering within cage, II) tunneling through the barrier, III) stabilisation of fragments at large internuclear distances followed by fluorescence, IV') cage recombination and energy relaxation in ground state, IV") same as VI' but with energy relaxation in bound excited state.

To date, a clearcut description of the simplest of all chemical events: the dissociation of a molecule in a solvent cage, is still missing. The so—called "cage effect", was initially proposed by Franck and Rabinowitsch /4/ in order to explain the reduced photochemical yield of free radicals in solutions as compared to the gas phase. The presence of solvent atoms or molecules is to create a barrier in the assymptotic limit of intramolecular potentials of the guest molecule /5/ as shown in fig. 1. The height and width of the barrier, along the reaction coordinate Q, depend on the guest–host intermolecular potential, the fragment–host molecule (atom) potential and on the host–host potential. Excitation of a molecule on the free molecule purely repulsive potential surface leads to different scenarios which are depicted in fig.1: cage exit by overcoming the barrier (I), cage exit by tunneling through the barrier (II), stabilisation of fragments at an equilibrium position defined by the repulsive surface and the barrier (III) and cage (geminate) recombination (IV). Process I may occur impulsively (direct cage exit) or after scattering at the cage walls (indirect or delayed cage exit). Process III may lead to light emission to high lying vibrational levels of the ground state while in process IV, recombination may lead to population of high vibrational levels of the ground state (IV'), or to population of excited states which correlate with the same assymptotic limit (IV"), or to both. In this case chemiluminescence is to be expected from the near UV to the mid–IR range.

Obviously, a solvent (liquid or solid) induced cage barrier is not a stable entity and evolves in space and time according the nature of the system and to physical parameters such as density, temperature, etc.. A steady state experiment probes a barrier which is averaged over time and space and to date this is the only type of experiments which are performed in the solid phase.

There is a long history of photochemical reactions studied in rare gas matrices, which dates back to the birth of the matrix-isolation method /6-8/. The description of dissociation dynamics and transport of fragments in inert gas crystals is of interest not only with respect to liquid state chemistry but has recently gained momentum in the search for new solid state chemical lasers /9/ and acts as model for understanding the generation of color centres in Alkali Halide crystals. Finally, whereas we have reached a high level of understanding of the vibrational, electronic and intermolecular non- radiative relaxation of molecules embedded in inert gas crystals /10,11/, the one non-radiative channel that has so far escaped quantitative description is the dissociative channel.

In this contribution, we will review the state-of-art of studies on photodissociation dynamics of molecules embedded in rare gas crystals. Permanent dissociation of a molecule in condensed matter implies cage exit of the fragment and their stabilisation away from the reaction site, thus related phenomena such as photoinduced mobility and charge transfer reactions will be invoked.

II. STRUCTURE AND DYNAMICS OF EDUCTS AND PRODUCTS IN THE CRYSTALS

The best way to study the cage effect on molecular dissociation is to start off with a model system, i.e. a molecule dissociating along a repulsive surface (fig.1) in a chemically inert medium having a simple structure. The description of the cage effect therefore rests on a number of experimental and theoretical prerequisites which we discuss thereafter.

II.A Potential Surfaces of Parent Molecule

The requirement of a purely repulsive potential surface is met in a large number of small molecules (diatomic and triatomic) and aims at reducing the number of degrees of freedom and reaction cordinates. The nature of the electronic states (Rydberg or valence) and that of the potential surfaces (dissociative, predissociative, etc...) is known from gas phase studies, molecular theory and ab-initio calculations /12/. The way the excited states are affected by the environment is investigated by absorption or emission spectroscopy. From a theoretical point of view, the description of the molecule-matrix interaction, and in general of the many-body interactions, rests on the additivity of pair potentials. Indeed, the weak van der Waals forces binding the crystal atoms together, can be regarded as a perturbation to the stronger intramolecular potentials. The availability of the guest molecule - host atom, fragment-host atom and host atom-host atom intermolecular pair potentials is therefore a fundamental ingredient.

II.B Properties of Rare Gas Matrix

As crystal medium, we use rare gas crystals which are particularly attractive for studying photoreactions, since /13/:

- They are chemically inert
- They have the simple face-centered-cubic (f.c.c) lattice structure
- Their physical properties are well known
- They are transparent in the VUV up to a range lying between 8 eV (Xe) and 17 eV (Ne). This is an important property considering the fact that most simple molecules have their dissociative continua lying above ~5 eV
- In such low temperature solids, rotation as well as large amplitude motions are often hindered and the density of states is reduced compared to the liquid or gas phase

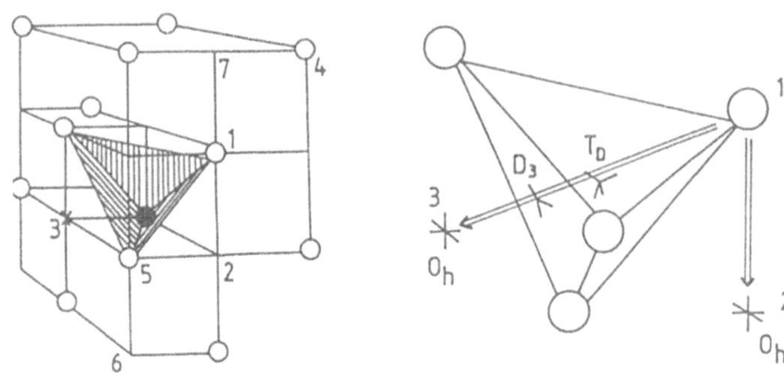

Fig.2. Part of the fcc lattice of a rare gas crystal. Sites of O_h, D_3 and T_d symmetry are indicated in the enlarged tetrahedron on the right hand side.
For the numbering of sites see text.

- It is in principle possible to establish a link between reaction site geometry and reactivity
- Reactions with long time constants (e.g. tunnel effects) can be observed
- Reactivity of "forbidden" states (e.g. triplet states) can be enhanced in heavy matrices /14/.

II.C Site Geometry of Educt (Parent Molecule) and Product (Fragment)

The parent molecule can occupy different sites in the fcc lattice of the rare gas crystal depending on size and the pair potential. Several important site geometries are indicated in the scheme of the fcc lattice of fig.2. 1 represents a substitutional site, 3 an interstitial site with octahedral (O_h) symmetry and in addition sites with the lower tetragonal (T_d) and trigonal symmetries (D_3) are shown. Insertion of a guest molecule causes in general a local rearrangement of the environment as compared to the undistorted lattice and can lead even to the coexistence of different crystal phases especially of the fcc with the hcp (hexagonal cubic) phase /15/. The site geometry can be experimentally determined by ESR /16/, neutron diffraction, x-ray diffraction and EXAFS studies. The dynamics of the molecule-crystal lattice interaction at the trapping site (local modes, etc...) /17/ can be studied by IR, Raman /18/ or electronic spectroscopy as well as neutron scattering /15/. An example for the demonstration of the nearly free rotation of an H_2O molecule in an Ar lattice by infrared spectroscopy is given in Fig.3a,b together with neutron diffraction patterns (Fig.3 c,d) which illustrate the preparation dependent deformation of the unit cell of the Kr lattice around the H_2O guest molecule. Obviously the trapping site geometry can be influenced by preparation conditions like condensation temperature and speed, by concentration, further guest molecules etc. The site distribution of fragments after dissociation is also an important piece of information. The dissociation event is usually accompanied by a large release of heat in the form of translational energy and local deformations or rearrangements are thus likely to happen. The"permanent dissociation" implies stabilisation of fragments outside the cage. Identification of fragments is done in photolysis experiments by detecting them using IR, Raman or electronic spectroscopies, which also help in identifying their site distributions although ESR spectroscopy is more powerful in this respect /16/.

III. SPECTROSCOPY OF DISSOCIATION BARRIERS

III.A Sample Preparation

In general a mixture of the matrix gas and the parent molecule is

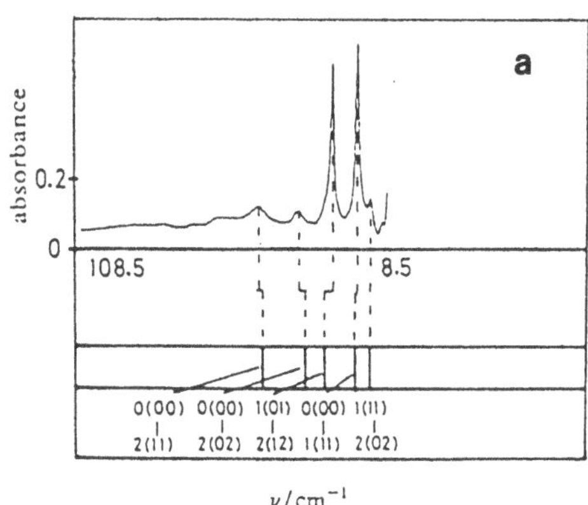

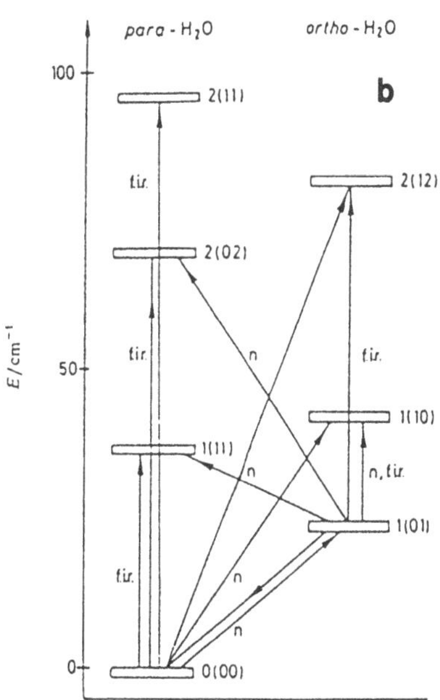

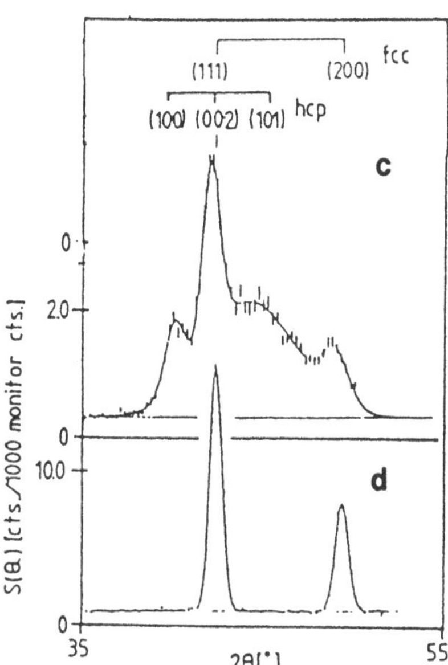

Fig. 3.

a) Far infrared absorption spectrum of D_2O in Ar with D_2O rotational transitions

b) Rotational level diagram with sets of the three quantum numbers $J(K_a, K_c)$ of an antisymmetric top. J: total angular momentum and difference of K_a and K_c is even (odd) for symmetric (antisymmetric) rotational states.

c) Neutron diffraction pattern of a not annealed H_2O in Kr sample showing a superposition of fcc and hcp contibutions.

d) as c) annealed at 100K for 1h (After ref.15)

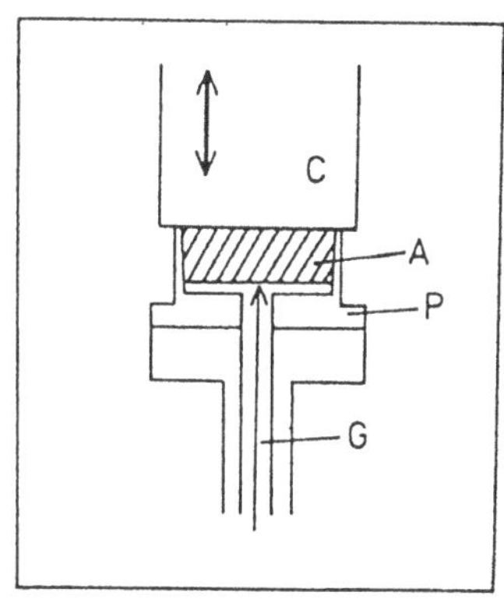

Fig.4.

Crystal growth device with movable cooled copperblock (C) growing crystal (A), plexiglass box (P) and gas inlet tube (G).

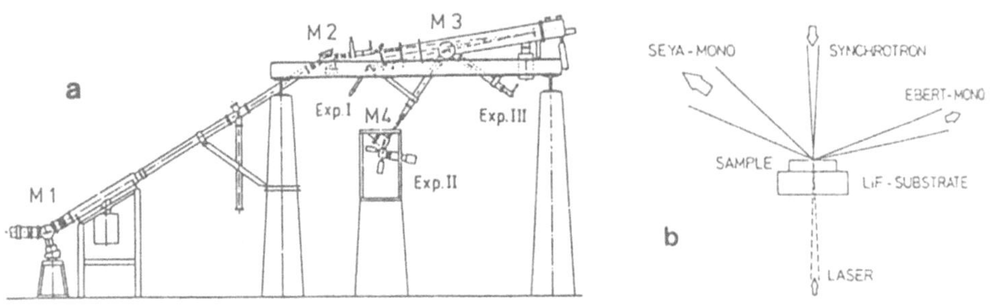

Fig.5. a) Beam line for Synchrotron radiation (range 40–400 nm) at BESSY with mirrors M1 to M4 and matrix experiment II at exit slit of monochromator. b) The counterpropagating Synchrotron radiation and laser beams are focussed in the sample and fluorescence is dispersed by a VUV–(Seya) and UV–visible–monochromator (Ebert).

condensed on a transparent window (quartz, MgF_2, LiF) at thicknesses of some microns which can be measured by interference fringes. Cooling is provided either by a liquid He cryostat for temperatures down to 4K or if 12K are sufficient a closed cycle refrigerator can be used /7,8,19/. In a similar way it is possible to grow polycrystals of about 1 cm^3 volume in cuvettes of quartz and for very slow growing speed also single crystals can be obtained. The crystals are especially important for the laser applications (section VII). For spectroscopy in the VUV spectral range /20/ a free standing crystal can be produced by removing the growing box (Fig.4). In all cases the samples have to be kept in excellent vacuum to prevent condensation of residual gas.

III.B Dissociation and Fragment Detection

Mapping of the cage effect goes by determining: 1) energy thresholds to permanent dissociation, 2) the energy dependence of permanent dissociation efficiency and, 3) the dependence of these observables with respect to temperature, crystalline environment, isotopic substitution of the guest molecules, sample preparation conditions, etc...

The first and second point can best be met by the use of tunable light sources such as dye lasers or synchrotron radiation. Whereas dye lasers are limited in tunability (visible – uv), their high fluences are convenient for the detection of low intensity signals, in particular to study processes with low rate constants (e.g. tunnelling). However, most simple molecules have their first dissociation continua lying at $\lambda \leq 300$ nm and synchrotron radiation is the adequate tool in this case /21/. It has however a lower fluence than lasers and is therefore less sensitive to weak signals. Finally, it is possible to combine synchrotron radiation for say dissociation, with a laser for the detection of fragments or vice–versa (Fig.5) /21/.

The experimental procedure goes by detecting for a given irradiation dose at a given photon energy, the amount of fragments permanently produced in the sample. The different methods are illustrated for the dissociation of H_2O to H and OH /21–24/ on the repulsive A $^1B_1$ surface shown in Fig.6a. Irridiation within the X $^1A_1$→A $^1B_1$ absorption band (140–200 nm, Fig.6b) above a barrier discussed in section VI leads to destruction of H_2O molecules, their concentration is reduced and the sample becomes more transparent (top of 6b). Simultaneously new absorption bands appear which result from the produced OH and H fragments. The changes in the transmission spectra are in general small and the fragments are detected with better sensitivity by their fluorescence bands shown for H and OH in Fig.7a and 7b respectively. A plot of the intensity in the fluorescence band versus the exciting photon energy is called

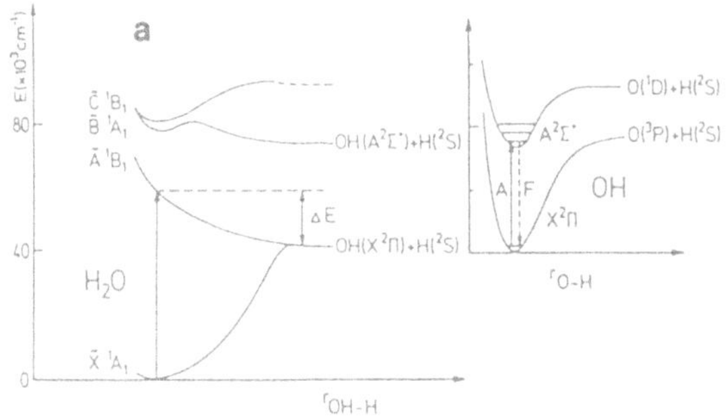

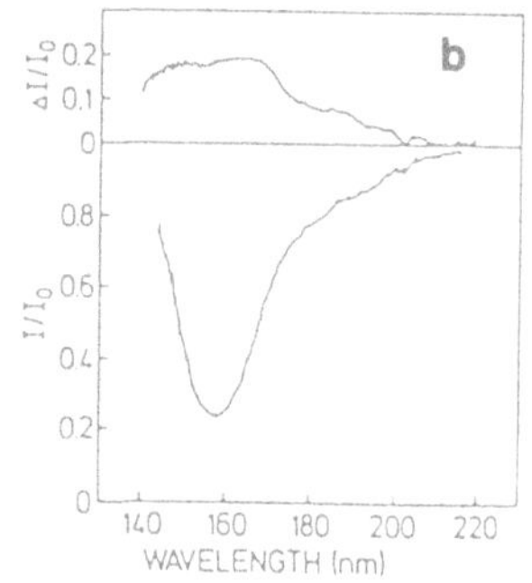

Fig.6. a) Potential surfaces of H_2O along the OH–H coordinate with the excess kinetic energy ΔE on the repulsive A $^1B_1$ surface. In addition the X–A transition of OH is shown which is used in detection by absorption A and fluorescence F.
b) X→A $^1B_1$ absorption band of H_2O in Ar (bottom I/I_0) and its bleaching shown by the increasing transmission $\Delta I/I_0$ due to dissociation by irradiation (top) (Ref.21).

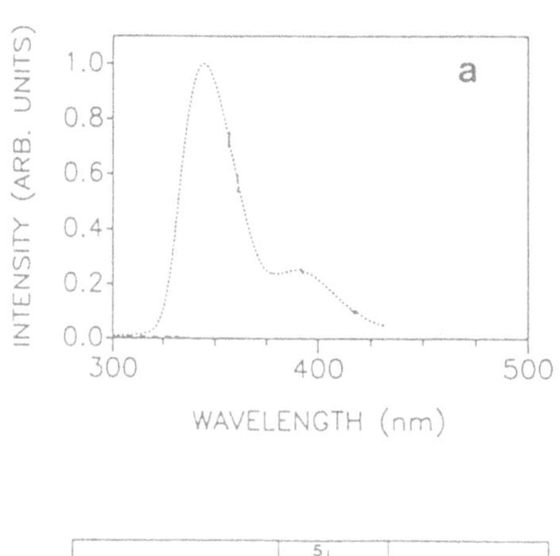

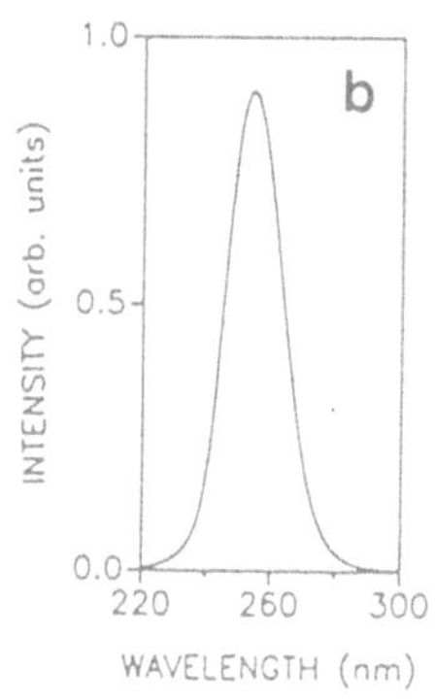

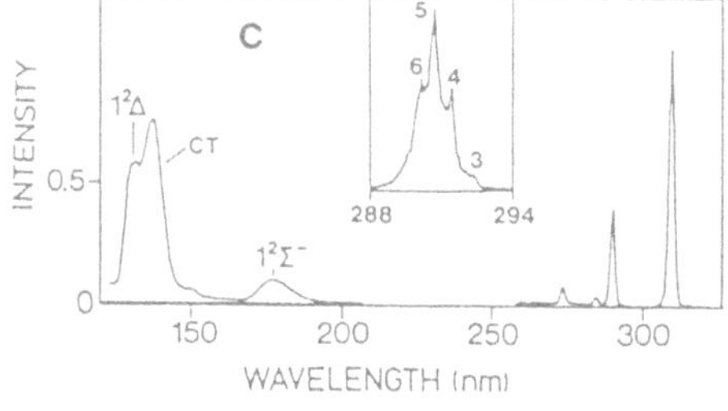

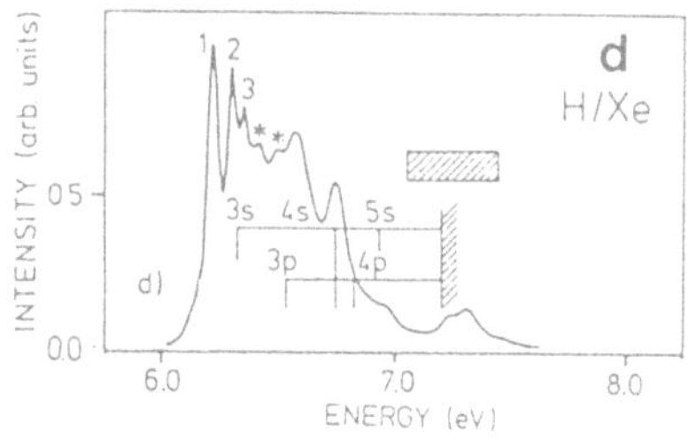

Fig.7. Fluorescence bands from fragments of H_2O dissociation with the A $^2\Sigma^+ \rightarrow X^2\Pi$ emission band of OH in Ar (7a) and the emission of H in Xe (7b) which is attributed to an Xe_2H excimer. Excitation spectra (intensity in a fluorescence band versus excitation energy) for the OH emission band (7c) with A $^2\Sigma^+$, $1^2\Sigma^-$, $1^2\Delta$ intramolecular states and a matrix induced charge transfer state (C.T.) and for Xe_2H excimer band (7d) with delocalized charge transfer Rydberg states (3s to 5s, 3p and 4p) (Ref. 37).

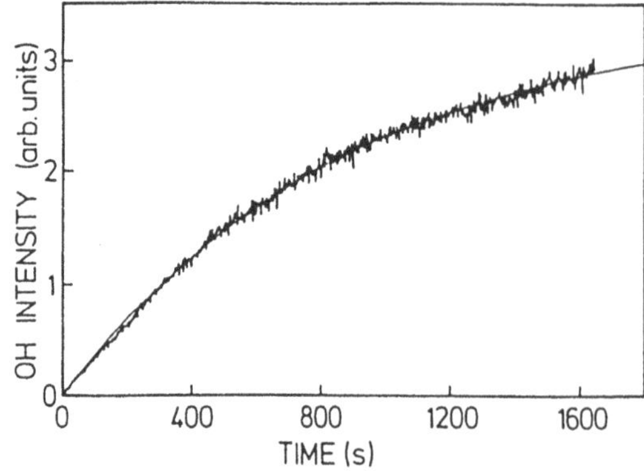

Fig.8. Growth of the OH content in an H_2O doped Ar sample. Continuous dissociation with Synchrotron radiation at λ=155 nm and simultaneous monitoring of the OH fluorescence by excitation of the OH (X–A) transition with the frequency doubled dye laser (Ref.21).

excitation spectrum and represents essentially the absorption spectrum of the fragment (Fig.7c,d) provided the quantum efficiency is independent of excitation energy. An optimum wavelength for fragment detection is chosen from the excitation spectrum and the growth of the fragment content in the sample for successive dissociation cycles is monitored by the stepwise increase in the fragment fluorescence. If two light sources are available than the growth of the fragment content in time (Fig.8) during dissociation with synchrotron ratiation (155 nm) can be recorded simultaneously by permanent excitation of the generated fragments with a dye laser (311 nm).

These experiments give information about the system before and once dissociation has been completed. With this in hand, one tries to reconstitute the history of the system between those two extreme situations, i.e. to describe the "transition state" representing the cage exit mechanism. Is the cage exit mechanism direct (ballistic) or indirect (delayed)? Is it mediated by pure scattering or by more specific processes such as "harpoon" reactions with cage atoms or molecules? Is there a possibility of back scattering after cage exit or diffusive recombination? etc... All those questions can indirectly be answered and the answers are cross–checked by Molecular Dynamics simulations, which when available, provide a "real–time" picture of the cage exit mechanism, pending the advent of "real–time" spectroscopy applied to solid–state reactivity studies.

IV. DISSOCIATION IN MOLECULAR DYNAMICS CALCULATIONS

Molecular dynamics (MD) calculations are a wide spread technique to solve many particle problems and are applied to free molecules, liquids and solids /25/. Up to now the dissociation of Cl_2 /26,27,28/, HI /29,30/, H_2O /31/ and F_2 /32,33/ in rare gas matrices has been treated by MD calculations. The parent molecule is placed in the center of a cluster consisting of typical N=90, 256, 1098, .. rare gas atoms. In principle the classical trajectory of each of the particles due to the interaction with all the other particles is calculated numerically by a stepwise integration of the classical Newton equations. For N particles a set of 3N Hamilton differential equations for the conjugate generalized coordinate q_k and momenta p_k has to be solved /25/

$$\dot{q}_k = -\partial H/\partial p_k \qquad k=1,\dots 3N \qquad (1)$$

$$\dot{p}_k = -\partial H/\partial q_k \qquad (2)$$

with the Hamiltonian

$$H = \Sigma \frac{p_k^2}{2m_k} + V(q) \tag{3}$$

The potential V(q) is calculated from a sum of pair potentials over all particles. Convenient and fast algorithms have been developped for the time consuming step by step numerical integration of the equations of motions /25/.

A specific problem arises from the fact that in principle dissociation in an infinitely large crystal should be simulated. In practice clusters of finite size have been treated and it has been checked by enlarging the cluster that the results do not depend on cluster size. In another approach periodic boundary conditions have been applied. In this case elongations of particles at the border of the central cluster can be focussed back into the cluster due to the periodicity. Thus the dissipation of the excess energy to the bulk may not be treated correctly. Further improvements can be achieved by introducing artifical dissipative terms in the Hamiltonian or by more complicated shell structures in which the interactions of the central part with the remote parts are reduced.

More fundamental restrictions are imposed by the classical treatment. A typical dissociation cycle consists of excitation from the ground state to the repulsive state (Fig.1) and either dissociation across the barrier or curve crossing back to one of the bound states during the energy dissipating rattling in the cage. Neither the initial nor the final curve crossing is inherently included in a classical description and both have to be added artificially in a way which is consistent with the correct quantum mechanical treatment. Furthermore zero point motion, dispersion of wave pakets and tunneling are missing in a classical description. The classical approximation is most crude for light particles. Therefore a combination of a classical and quantum mechanical treatment is being developped consisting in a classical MD description of the motion of heavy particles especially from the cluster and a solution of the Schrödinger equation for a light particle (H,D) in the cage. The method has been already applied to simulate the tunneling contribution to the dissociation of HI in Xe /30/.

For a typical MD calculation it is necessary to average the initial conditions correctly. Therefore the parent molecule is placed in its electronic ground state into the cluster and the system is allowed to evolve in time at the chosen temperature. The lattice rearrangement is taken into account in this way and the calculation delivers in addition the dynamics of the parent molecule in the ground state. A Cl_2 molecule in a Xe crystal has a preferential orientation and at low temperature the molecule cannot rotate freely but shows an oscillatory librational motion /27/ with an amplitude of only 1 to 2 degrees (fig.9a). At higher temperatures the amplitudes increase. Around 95K the molecule can point into every direction and starts to rotate (Fig.9b). After a statistically long duration of the time evolution on the ground state the parent molecule is switched instantaneously to the dissociating potential surface conserving the positions and momenta of all particles. This corresponds to a vertical transition from the ground state to the excited state. The choice of the initial position and momentum on the repulsive surface has a strong influence on the following trajectories. Therefore only a sufficiently large ensemble of different trajectories which represent the correct statistical distribution of initial conditions delivers a reliable description of the dissociation process. The system evolves now on the excited surface and typical trajectories will be presented in the following sections. The calculation is terminated either by a successful escape (taking care of reentries)

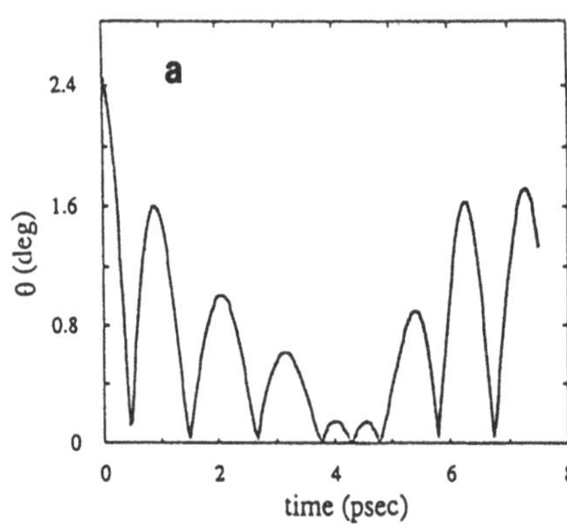

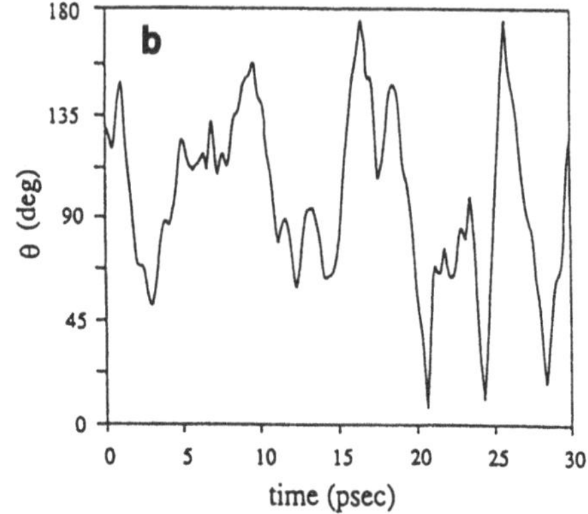

Fig.9. Variation in time of the Cl_2 orientation in Xe (angle Θ) from molecular dynamics calculations (ref.27) for 2K (a) and 110K (b).

due to a low kinetic energy close to the thermal energy. The MD calculations predict also the lattice rearrangement around an escaped fragment and the dynamics and the thermalization in the new site.

The ratio of the excess kinetic energy E (Fig.1) and the energy loss ΔE of the fragment per scattering event determines the mean number of attempts available for the fragment to cross the barrier. The intuitive notion that this ratio should have a strong influence on the dissociation efficiency and kinetics is confirmed by the MD calculations. It plays such a central role that it has been used to classify different types of dissociation. The maximum fractional energy which is transfered in a head–on two particle elastic scattering event is:

$$\Delta E/E = 4\, m\, m'/(m + m')^2 \tag{4}$$

for particles with masses m and m'. ΔE is large for particles with similar masses. In this case the fragment has only few attempts and the MD calculations predict the so called "<u>impulsive</u>" or "<u>ballastic exit</u>" which will be discussed in section V for Cl_2 in rare gas matrices. ΔE is small for scattering of a light particle at a heavy particle, the excess energy E is dissipated only by many scattering events and the so–called "<u>delayed exit</u>" will be discussed for the escape of H and D in a rare gas lattice in section VI. Finally the barriers for special pair potentials of fragment and host are low in certain "channeling" directions of the crystal lattice and the resulting "long range transport" is shown for F_2 dissociation in section VII.

V. IMPULSIVE EXIT: Cl_2 DISSOCIATION

The MD calculations predict an impulsive exit for Ne, Ar /26,28/ and Xe matrices /27/, due to the efficient energy exchange. A typical trajectory is shown in fig.10. After one or two collisions bringing the Cl atom out of the original cage, thermalisation occurs and the Cl fragment is stabilised in the product site. Cage exit occurs in less than 1 psec but only when the Cl_2 molecule is oriented in the direction of the "transition state" at the moment of photodissociation. This is evident from fig.11, because the temperature at which dissociation becomes effective coincides with the onset of free rotation of Cl_2 in Xe matrices near 95K (Fig.9). For T>95K, the quantum efficiency grows, then goes through zero at ~125K and rises again as T approaches melting. The increase above 95K is attributed to larger vibrational amplitudes of matrix atoms so that the potential barrier between them is decreased and the probability for cage exit grows. For T≃125K, it seems that energy transfer

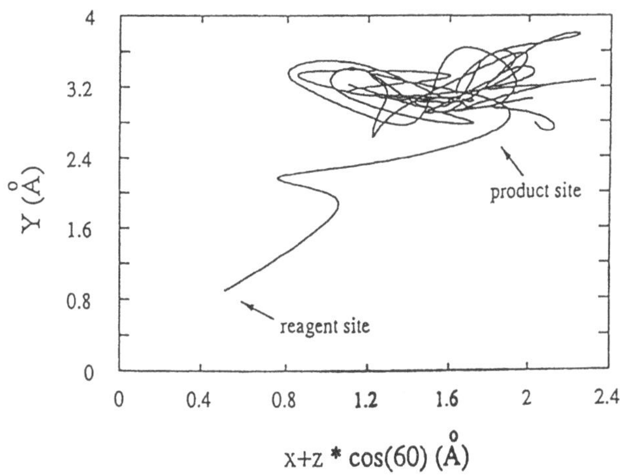

Fig. 10. Projection in a plane (y and x+1/2 z axis) of a calculated Cl trajectory for Cl_2 dissociation in Xe matrix at 110K and 8ps time duration (Ref. 27).

from the strongly vibrating Xe atoms to the Cl fragments occurs, deviating the latter from their ballistic trajectories and leading eventually to recombination. The melting point of solid Xe is approached for T>125K and the vibrations of Xe atoms are very large, dissociating trajectories are no longer limited to small solid angles, which could explain the rise in quantum efficiency (fig.11).

Experimentally the cage effect has been demonstrated by the very low quantum yield of less than 10^{-6} for the permanent dissociation of Cl_2 in Ar even for an excess energy of more than 2 eV above the gas phase dissociation limit /34/. Recombination leads to efficient population of the bound $A'^3\Pi$ (2u) state and subsequent radiative decay, a situation analogous to route IV" in Fig.1. The caging for Cl_2 and similar results for ICl /35/ have been confirmed for Br_2 and I_2 /36/ and a model has been proposed based on the cage size in order to explain the different post–recombination populations of the $B^3\Pi$ (0^+_u), $A^3\Pi$ (1u) and $A'^3\Pi$ (3u) states upon excitation in the dissociative continuum of the B state.

A spectroscopic determination of the barrier heights in the case of Cl_2 dissociation has been carried out only recently because the VUV radiation from a Synchrotron has to be exploited (Fig.5). The experimental results for dissociation in a Xe lattice are more complicated than described in the original MD calculation /27/ because Cl atoms are produced in two types of surroundings which can be distinguished by their stability with respect to annealing /37/. Therefore the results for an Ar matrix will be presented /37/.

According to its size a Cl_2 molecule replaces two Ar atoms substitutionally and occupies for example the sites 1 and 4 in Fig.2. Dissociation leaves one Cl atom well accomodated in site 4, the other is transfered to the tight interstitial site 3 and Cl atoms in two very different surroundings could be expected. This is not the case in a well ordered sample according to the experiments because the stress induced by the interstitial Cl atoms initiates a rearrangement in which the Ar atom at site 5 is shifted to the now empty site 1 and the interstitial Cl atom takes its place. In this way two Cl atoms are sitting in substitutional sites and are separated by one Ar atom. As a matter of fact the absorption spectra of Cl atoms from photolysing Cl_2 in well

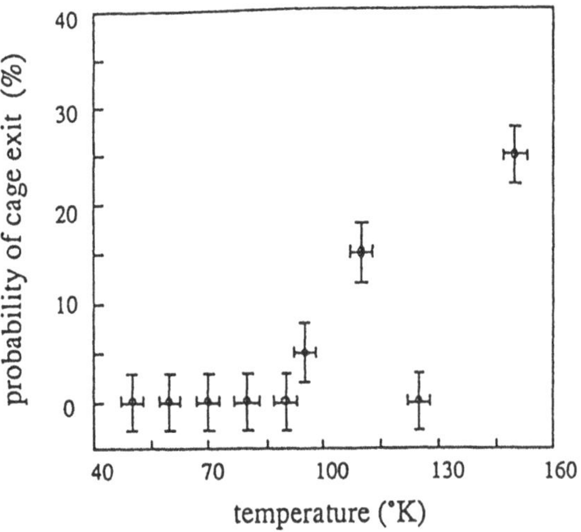

Fig. 11.
Calculated temperature dependence of cage exit probability for Cl_2 dissociation in Xe (Ref.27)

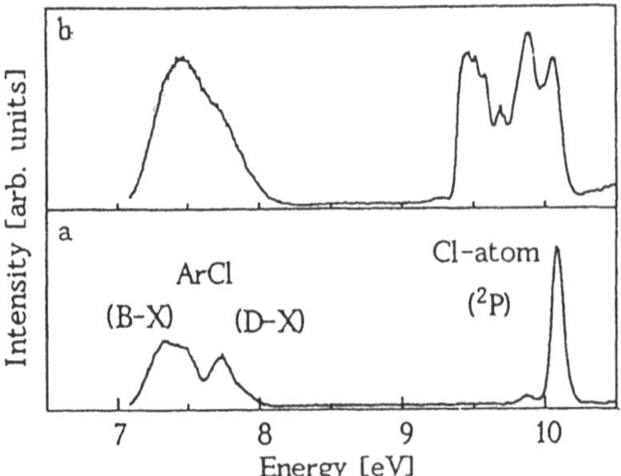

Fig. 12.
Intensity of the Ar_2Cl emission band (260 nm) versus excitation energy for a crystalline sample (a) and a non-crystalling sample (b) form ref. 37.

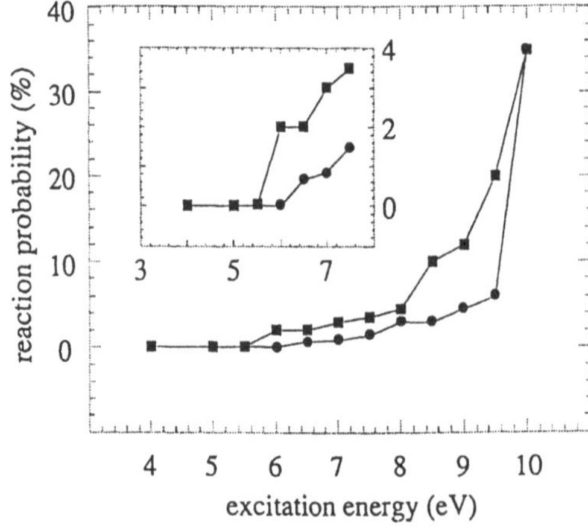

Fig. 13. Comparison of experimental (dots) and calculated (squares) dissociation probability versus photon energy for Cl_2 in Ar and 5K. Insert: Onset with enlarged scale (Ref. 28, 37).

annealed Ar samples are dominated by a single line which is attributed to the substitutional Cl atoms (Fig. 12.a). This picture is confirmed by photolysing low temperature condensates (5 K) in which defects and also single substitutional Cl_2 sites are stabilized. In this case additional lines due to these different surroundings of the generated Cl atoms are observed (fig.12.b). The metastable sites disappear by annealing and again the most favourable geometry is recovered /37/.

The dissociation efficiency versus photon energy (Fig.13) shows a prominent threshold around 9.3 eV which corresponds to a matrix induced barrier of about 6.8 eV /37/ based on the Cl_2 binding energy of 2.5 eV. This tremendous barrier is in accord with the earlier experiments /34/ and it underlines the strength of the steric hindrance imposed by the matrix. Ejection of a Cl atom on the path 1 to 3 (fig.2) from a Cl_2 molecule which is oriented along the sites 1 and 4 leads to crossing of the D_3 barrier. Summation of Cl–Ar pair potentials yields an estimate of the barrier of 7.7 eV /37/ in reasonable agreement with the experimentally determined barrier. A closer inspection of Fig.13 indicates a second but much less efficient threshold at photon energies of 6.2 eV corresponding to a barrier of 3.7 eV. The first MD calculations predicted the steric hindrance /26,27/ but the threshold was not included due to a too limited range of excess energies. Very recent and more advanced MD calculations /28/ confirm that a relaxed divacancy site is the lowest energy configuration for Cl_2 in Ar and that permanent dissociation results in two Cl atoms at substitutional sites separated by one Ar atom. The recent MD calculations /28/ predict the high and low energetic thresholds for dissociation of Cl_2 molecules and reproduced also the low quantum efficiency between 6 eV and 8 eV photon energy (Fig.13). According to the MD simulations, a large excitation energy of at least 6 eV is necessary to displace an Ar atom in the event of an impulsive exit of Cl. But recombination is still very likely below 8 eV excitation energy, yielding the low quantum efficiency. Stabilisation of Cl fragments becomes efficient only above 8 eV, due to insertion of the displaced Ar atom after its transport through the heated region around the cage on a psec time scale.

A rather different and interesting process which leads to dissociation of Cl_2, HCl and other halogens in Xe matrices is a two–photon excitation combined with an harpooning reaction /38–40/. A two–photon induced charge transfer from the rare gas matrix to a halogen molecule X_2^- potential surface is repulsive in the X–X coordinate, a halogen atom X is ejected from the complex and a Rg^+X^- species is formed /38/ according to

$$Rg + X_2 + 2h\nu \rightarrow Rg^+X_2^- \rightarrow Rg^+X^- + X \qquad (5)$$

The observed threshold for generation of Cl atoms of 3.5 eV is situated 0.8 eV below the HCl dissociation limit of 4.4 eV. The growth curves for Cl atoms are of second order in laser fluence and only a two–photon transition either to the repulsive $^1\Pi_u$ state of HCl or to the charge transfer state $Xe^+(HCl)^-$ initiates the dissociation. The cross section for the $^1\Pi_u \leftarrow X$ transition in gas phase experiments is smaller than 10^{-50} $cm^4.s$ whereas a lower bound of 10^{-43} $cm^4.s$ is obtained for the dissociation in Xe matrix from the growth curves assuming a quantum efficiency of unity. Thus the charge transfer state is established as precursor for dissociation and the high quantum efficiency of the process is illustrated /38/.

The concept of a "negative cage effect" has been introduced to explain the high quantum efficiency of the harpooning mechanism despite the low kinetic energy of the Cl atom of about 1 eV. The Cl_2^- ion is created on the repulsive part of its internal potential while the potential along the molecule cage coordinate becomes attractive due to the $Xe^+Cl_2^-$ interaction. The rest of the cage atoms is polarized by the dipole and is attracted to the ionic moiety.

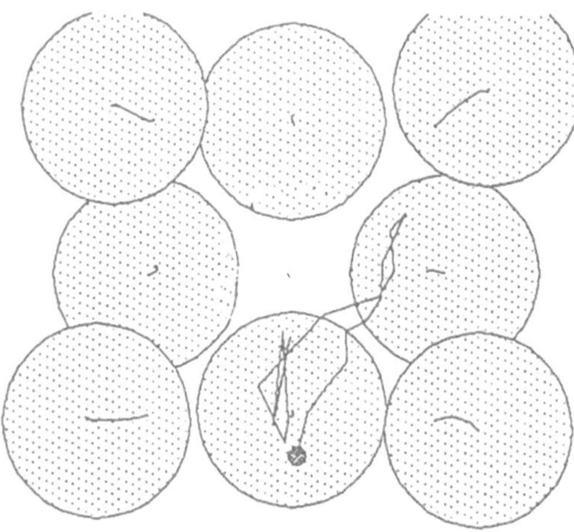

Fig.14.
Calculated time development of H trajectory (wiggly line) and Ar cage atom positions (centers of balls) for photodissociation of H_2O in Ar at 5K and 2eV excess kinetic energy, stopped after 200 femtoseconds (Ref.31).

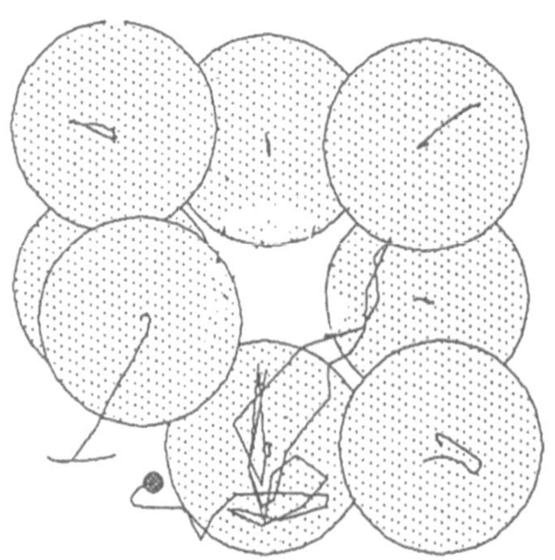

Fig. 15.
Continuation of Fig. 14 and stopped after 500 femtoseconds (Ref. 31).

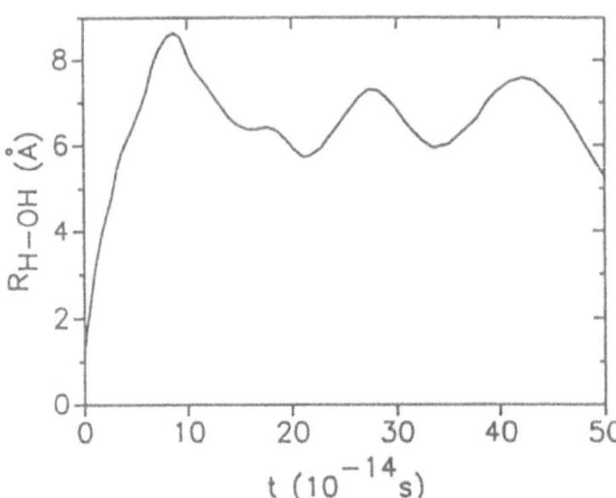

Fig.16. H–OH distance versus time for H_2O dissociation in Ar matrices. Similar conditions as in figs. 14 and 15 (Ref. 31).

A stretch of the repulsive Cl^-–Cl coordinate in concert with a collapse of the cage around the Xe^+Cl^- exciplex completes the dissociation and the $Xe_2^+Cl^-$ configuration will be formed /38,39/.

VI. DELAYED EXIT: H–ABSTRACTION

MD calculations for the abstraction of H atoms have been carried out for HI in Xe /29,30/, and for H_2O in Ar /31/ matrices. The conclusions are in marked contrast to the impulsive exit discussed in section V as has to be expected from the large mismatch in the masses of the H fragment and the matrix atoms which results in a small energy loss per collision (see equ.4 in section IV). Excess kinetic energies of 1.5 eV to 3 eV were used for the simulations which were run for 4 to 10 psec. After this time, if the H atom was found to be located outside the original site, permanent dissociation was assumed. The energy loss per collision is small due to the low H atom–to–matrix atom mass ratio (<<1) but cage exit occurs contrary to the Cl_2 case even after many collisions when the H fragment has lost most of its kinetic energy. As can be seen in figs.14 and 15, there is considerable rattling in the original cage and the matrix atoms are displaced as a result of collisions. Cage exit occurs then and is followed by stabilisation in a nearby interstitial site. This process occurs in a few hundred fsec for H_2O in Ar matrices (fig.16) /31/. The rapid loss of kinetic energy of the H fragment versus time is illustrated in fig.17 for the case of HI in Xe matrices. The kinetic energy reduces within 0.3 psec to ~10–20% of its original value of 1.5 eV. A temperature dependent dissociation quantum efficiency was reported for HI in Xe matrices being 0 at 0K, 0.2 at 17K and 0.1 at 35K /29/. This temperature effect was attributed to a lowering of the effective barrier by increasing the amplitudes of Xe vibrations. As temperature is further increased, energy transfer from the Xe vibrations to the moving H atom becomes more important, raising the probability of recombination. The HI/Xe photodissociation simulation was further improved /30/ by consideration of quantum effects treating the H atom by time–dependent wavepackets and the Xe atoms classically. The main conclusions are that: a) there is a non–zero cage exit probability ($\leq 10^{-4}$) at T=0K which is due to tunneling, b) The loss of kinetic energy of the H fragment is not as rapid as in the classical calculation.

The most detailed experimental studies for H abstraction have been reported for the dissociation of H_2O on the purely repulsive A $^1B_1$ state (Fig.6) which leads to OH and H fragments. This process is very well characterized in the gas phase both experimentally /41/ and theoretically /41/ and it is known that the H atoms take over more than 90% of the excess energy as kinetic energy in the relevant energy range. It is therefore particularly suited for a case study of the cage effect on the H atom abstraction which was carried out in Ar, Kr and Xe /21–24/ using tunable synchrotron radiation. The OH and H fragments have been recorded by absorption (Fig.6) and fluorescence spectroscopy (Fig. 7,8). The relative quantum efficiency vs. dissociating energy is obtained by dividing the fragments content by the separately measured transmission of the samples. Fig.18 shows the dissociation efficiency of H_2O (D_2O) in the threshold region in Ar, Kr and Xe, respectively, as a function of dissociation energy and for various temperatures. The dissociation threshold lies at higher energy than the threshold of absorption and the efficiency increases with excess kinetic energy above threshold. The experimental points in fig.18 were fitted with the analytical expression:

$$q(\hbar\omega) = A(\hbar\omega - E_{th})^n \qquad (6)$$

where the threshold energy E_{th}, the exponent n and the prefactor A, (which we call the "characteristic efficiency") were treated as free parameters. The

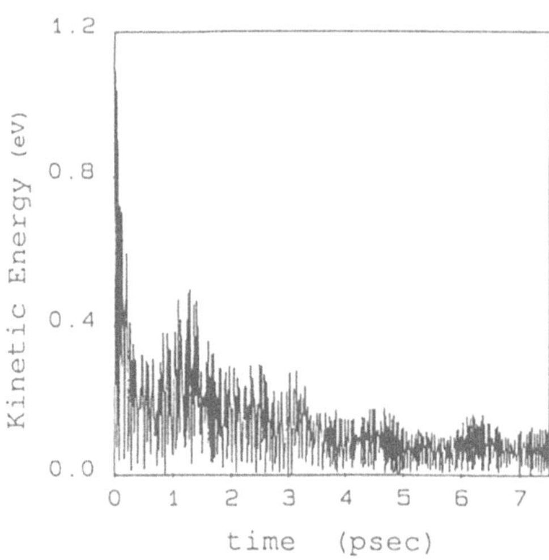

Fig. 17. Kinetic energy of H fragment vs. time in the photodissociation of HI in Xe matrices at 5 K and an excess kinetic energy of 1.5 eV (After ref.29).

results of the fit are shown as curves in fig.18 and the fit parameters are given in table 1. For all matrices, n lies in the vicinity of ~2. The strongest variations are due to A and E_{th} in agreement with the trend apparent in the curves (fig.18), i.e.:

a) E_{th} corresponds to barrier heights of E_b=1.3–1.8 eV, relative to the gas phase adiabatic dissociation energy of 5.118 eV.
b) The threshold energies (E_{th}) shift slightly red from Ar to Kr, then abruptly to Xe.
c) There is also a small red shift of E_{th} from D_2O to H_2O and with increasing temperatures.
d) The characteristic efficiency A is larger for H_2O as compared to D_2O.
e) It increases from Ar to Xe.
f) It increases strongly with temperature for Ar, milder for Kr and in Xe, no temperature effect is observed.

According to mid–IR /42/, far–IR and neutron scattering studies /43/, H_2O occupies a substitutional site of the f.c.c. rare gas lattice and is a free rotor. From ESR experiments, it is known that H sits on an octaheadral (O_h) interstitial site in rare gas crystals /44/. In fig. 2, H_2O sits at site 1, the H fragment has to end up in site 3 in order to ensure permanent dissociation /21/ and the minimal energy pathway is that going through the symmetry site D_3 (fig.2). Thus the barrier energies E_b have been estimated by summing the three H–Rg pair potentials at the center of the tetrahedron basis (D_3 site). The agreement is satisfying in view of the uncertainty in the pair potentials. Concerning point c), the weak red shifts are due to the higher zero point energy of H_2O compared to D_2O and to the dilatation of the lattice with temperature (neglecting all dynamical effects) /22/. Eq.6 was rationalised, in ref. 22, in terms of the integral over the exit probabilities of the H atom for successive scattering events in the original cage, so long its energy lies above E_b. The characteristic efficiency A in eq. 6 is then proportional to the inverse fractional energy loss $E/\Delta E$ per collision (equ.4).In table 1, it can be seen that the observed trends on A are very well reproduced by those of $E/\Delta E$.

Table 1.

Parameters derived from fits of the spectra in Fig.18 with equ.6. The barrier height E_b follows form E_{th}–5.118 eV. (Ref. 22–24). The $E/\Delta E$ values (equ. 4) are normalized to 1 for Ar matrix.

	T(K)	n	A	E_{th}(eV)	E_b(eV)	$E/\Delta E$
H_2O/Xe	5–40	2.5	3.7	6.4	1.28	3.3
H_2O/Kr	5	1.76	1.4	6.81	1.71	2
	40	1.81	2	6.78	1.68	
H_2O/Ar	5	2.1	1	6.9	1.8	1
	30	2.1	4.1	6.8	1.7	

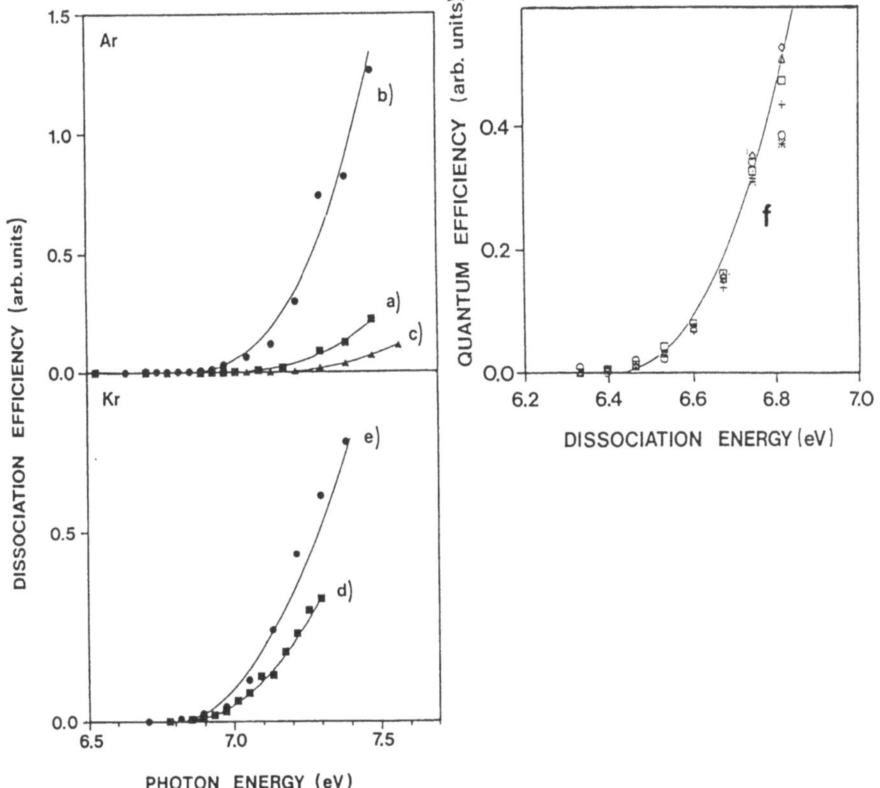

Fig. 18. Points: Relative quantum efficiencies for dissociation of 0.2% H_2O (D_2O) in Ar, Kr and Xe matrices. a) H_2O/Ar at 5K, b) H_2O/Ar at 30K, c) D_2O/Ar at 5K, d) H_2O Kr at 5K, e) H_2O/Kr at 40K, f) H_2O/Xe in range 5 to 40K. Solid lines represent fits with equ. 6 (see table 1) (after ref. 22,24).

Some notable contradictions show up with the MD simulations which predict delayed cage exit for the H atom:

- The barrier energy calculation are based on the assumption of a direct cage exit and reproduce well the expected experimental energies.
- For the predicted delayed cage exit no well defined barrier energy is expected and the calculations did not derive thresholds. Furthermore, an anticipated threshold for delayed exit should depend on energy dissipation per scattering event. Although $E/\Delta E$ changes drastically from Ar to Xe (tab.1), this trend is not supported by the experimental values.
- Scattering with OH, which would weaken the matrix effect, can be ruled out as a major contribution on the basis of the D_2O results /22/.

Well above the threshold, the permanent dissociation efficiency reaches a plateau with an absolute quantum efficiency of ~20% – 30% in good agreement with the MD simulations /29,31/.

VII. LONG RANGE TRANSPORT: F_2 DISSOCIATION

The fragments stabilized in the previous MD calculations in the vicinity of the original site. An interesting long range transport has been found for simulations of F_2 dissociation in Ar matrices /32/. The calculations were in part motivated by the high dissociation quantum efficiency of F_2 in Ar crystals and the long range migration of F atoms observed in experiments on XeF laser emission in Xe:F_2:Ar crystals /9,45/ (section VIII). The simulations show that:

a) The dissociation quantum efficiency has a threshold at ~0.5 eV excess kinetic energy, is energy dependent and reaches 100% for high kinetic energies.

b) There is an inverse temperature effect with the dissociation efficiency being somewhat lower at 12K compared to 4K contrary to the case of Cl_2. The interpretation is that the F_2 molecule occupies a loose–fit substitutional site at 12K and excutes rotation–translation coupled motion. At 4K, F_2 freezes in a librational well pointing at one of the cones of reaction.

c) Cage exit is impulsive. The threshold energy of ~0.5 eV coincides with the barrier of passage of F atoms through the D_3 site of the argon lattice (corresponding to a window of the cage).

d) In a number of trajectories, the F atom undergoes long range migration (up to ≃ 30Å) after cage exit. Such a trajectory is shown in fig.19, where the F atom accomplishes wigly motion through the lattice without any sidestep. The main loss of energy occurs during the collisions that direct the hot atom into the channel. Once in the channel, the atom follows a curvilinear path very efficiently. This motion is not due to thermal diffusion but is explained by the short range of the repulsive F–Ar potential and a stabilisation of the trajectory via the attractive part of the F–Ar potential.

Experiments on F_2 dissociation /46–49/ were carried out on F_2 doped free standing Ar and Kr crystals at 12K. F_2 was dissociated along the purely repulsive $^1\Pi_u$ potential curve with a first laser and the F fragments were monitored via the argon fluoride emissions upon excitation over the ArF (B←X) transition at 193 nm by a second laser. The main observations are:

a) there exists a threshold for permanent dissociation at an excess kinetic energy of ~1.1 eV with a quantum efficiency of ~5%. This onset

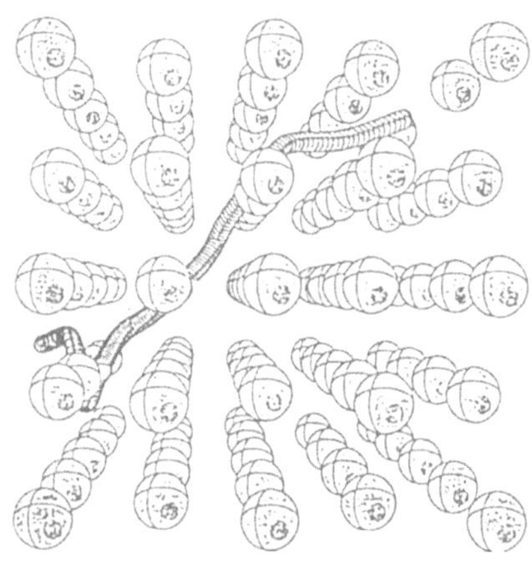

Fig. 19. Calculated long range trajectory of an F atom in an Ar lattice from F_2 dissociation (Ref.32).

corresponds to the threshold of F_2 absorption. The dissociation quantum efficiency increases with excess kinetic energy reaching 100% for energies $\geq$2.5 eV.

b) Upon initial impulse, F atoms undergo long range migration due to photodissociation of F_2 or relaxation of argon fluorides. This migration (over 10–15 lattice sites) was verified in doubly doped F_2:Xe(Kr):Ar /48/ or F_2:Xe:Kr crystals /49/. Shuttling of the F atom from Ar to Kr(Xe) or Kr to Xe centers was demonstrated and the dependence of migration upon initial impulse was also studied and shows a threshold of 1.9 eV in the case of Kr /48/.

c) The inverse temperature effect predicted by the MD simulations was confirmed for the case of F_2 photodissociation in Kr crystals /47/ where at low excess kinetic energies, the dissociation efficiency is 2 to 3 times larger at 4.5K as compared to 12K.

These observations clearly agree with the trends predicted by MD simulations and in a detailed investigation it has been shown that two orientations of the escaping F atoms relative to the cage have to be taken into account /33/. the long range transport allows for the formation of reaction products across several lattice constants and leads us to the photochemical preparation of radicals such as XeF or Xe_2H for solid state laser applications.

VIII. LASER APPLICATIONS

Many gas–phase lasers are operated with rare gases as a buffer. The rare gases serve for an efficient energy deposition and an optimization of the energy funnelling processes. In rare–gas excimer lasers and in rare–gas–halogen excimer lasers they are even part of the lasing compounds. In principle, these gas–phase lasers could be converted to solid–state lasers by using rare–gas crystals /9/. This concept has been applied at present to lasers which operate in the infrared (IR) /50/, the visible, ultraviolet (UV) /45,51–55/ and vacuum–ultraviolet (VUV) /56,57/ spectral range and even in the windowless region below 100 nm /56/.

Several advantages can be anticipated for the solid–state lasers. The density exceeds that of the gas phase by several orders of magnitude and, therefore, much higher densities of the excited species are attainable, resulting, at least in principle, in orders–of–magnitude–larger gain coefficients. Furthermore, the excited centres are spatially fixed in the rare–gas lattice. Quenching processes by mutual annihilation of excited species cause severe losses in the gas phase due to efficient scattering of the mobile particles, especially at high excitation densities. These processes can be strongly suppressed in the solid phase by the spatial localization.

In addition, it is possible to stabilize metastable compounds in the lattice. Better efficiences can be expected, for example, for rare–gas–halogen excimers because the required halogen atoms can be permanently bound to a rare–gas atom in the lattice. The halogen atom concentrations are depleted in the gas phase by recombination and must be regenerated by dissociation in each excitation event. There is still a lack of intense and tunable UV and VUV laser sources. Dye lasers are limited to wavelengths larger than 320 nm and the $CeLaF_3$ laser at 286 nm is the shortest solid state laser using conventional crystals /58/. Rare gas crystals are transparent far into the VUV spectral range and can serve as hosts for molecules with laser transitions up to the vacuum ultraviolet. Excited state densities of the order of $N=10^{18}$ cm^{-3} can be prepared in the crystals with high quantum efficiencies. At present crystals with a length of 1 cm and good optical quality are grown routinely. The amplification of a beam with intensity I_0 to an intensity I by a laser amplifier with a cross section σ_s for stimulated emission and a population inversion ΔN is given by

$$I/I_0=\exp(\sigma_s \cdot \Delta N \cdot l) \tag{7}$$

for an amplifying length l /59/. For excimers the excited state density N corresponds to the inversion ΔN due to the fast relaxation on the repulsive ground state potential. Strong amplification occurs already in a single pass for electronic transitions with σ_s of 10^{-16} to 10^{-18} cm^2 and the assumed density $N=10^{18}$ cm^{-3} and length l=1 cm. Therefore dopands with electronic transitions of this type bare the potential for efficient lasing with small signal gain coefficients $\alpha=N\cdot\sigma$ of 1–100 cm^{-1} provided there are no competing loss processes.

VIII.A Spectroscopy and Preparation of XeF in Ar and Kr Crystals

Rare gas halide excimers provide in general transitions with high cross sections σ_s for stimulated emission and a large spectral range can be covered by different rare gas – halogen combinations /60/. Lasing has been achieved on several transitions of XeF in Ar and Kr crystals /45,51–55/ and it will be used as an example. The ground state $X^2\Sigma^+$ of XeF is weakly bound /61/ and the repulsive $A^2\Pi$ state shows the same dissociation limit (Fig.20). The lowest excited states are charge transfer states F^-Xe^+ and are denoted by C, B and D (Fig.20). The emission bands of XeF molecules isolated in solid Ar correspond to the transitions from C→A (536 nm, fwhm=60 nm), B→X (411 nm, fwhm=2) D→A (350 nm) and D→X (286 nm, fwhm=6nm) which are shifted to the red compared to the gas phase due to the interaction of the excited charge transfer states with the matrix (Fig.21). The linewidth and the lifetimes are little affected by the matrix. B→X and D→X are bound–bound transitions and represent envelopes of several transitions to vibrational levels of the ground state broadened by the interaction with the matrix. C–A and D–A are bound–free transitions and the large width results from the steep slope in the repulsive ground state A. Only the C–A, B–X and D–X transitions have large cross sections for stimulated emission σ_s (Tab.2).

The XeF centers in Ar and Kr crystals are prepared by dissociation of F_2 in crystals which have an original composition of typically F_2: Xe:Ar=1:1: 3000. For dissociation gas phase excimer lasers with wavelengths of 248 and 308 nm are used. These wavelengths lie within a broad absorption band of the F_2

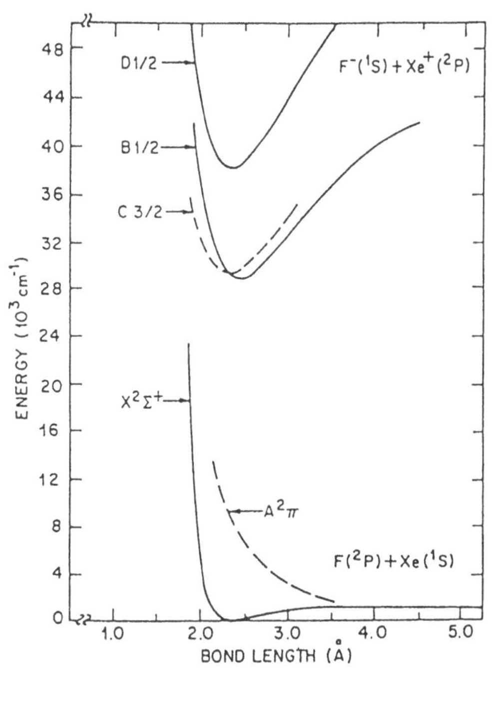

Fig. 20.
Potential energy diagram of F_2 (Ref. 61)

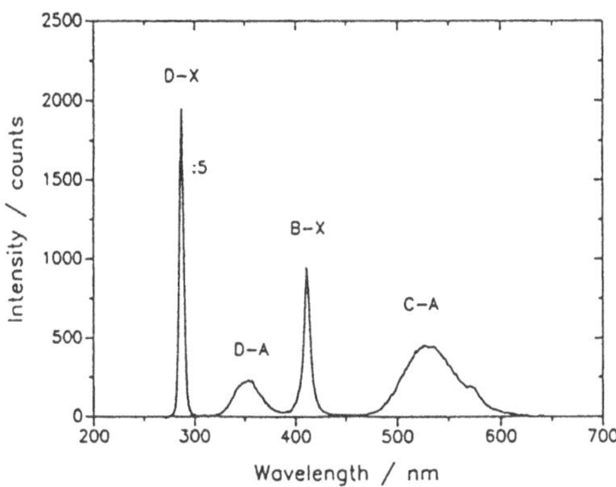

Fig. 21.
Emission spectrum of XeF in an Ar crystal for an original composition of $Ar/Xe/F_2$=2000/1/1. The D–X intensity has been reduced by a factor of 5 (Ref.53).

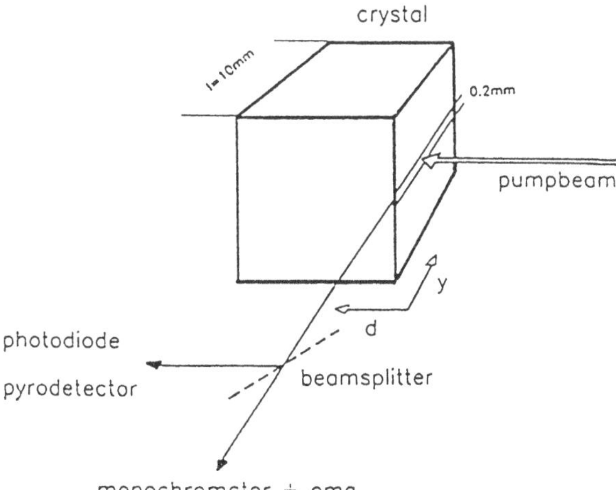

Fig. 22. Scheme for transversal pumping of XeF/Ar solid state laser with detection by monochromator with optical multichannel analyser or photodiode or pyrodetector (Ref. 54).

molecule due to a transition to a repulsive state. The corresponding absorption cross-section is very small. Therefore, the crystal is initially nearly transparent. The excess energy in the repulsive state relative to the adiabatic dissociation limit amounts to about 2.4 eV. In section VII it has been discussed that this excess energy leads to a dissociation efficiency near unity and to hot F atoms which can migrate long distances of the order of 10 lattice constants. Xe atoms on the way of these hot F atoms act as deep traps due to the deeper XeF potential depth and stable XeF molecules are formed. An initial concentration of 10^{19} cm^{-3} of F_2 and Xe particles leads to a concentration of about $2x10^{17}$ cm^{-3} of XeF molecules formed directly by "hot" F atoms and the reminder trapped F atoms can be thermally activated by a heat cycle leading to a final amount of $2x10^{18}$ cm^{-3} of XeF molecules. Prolonged irradiation at 308 nm results also in a destruction of XeF due to the strong C–A emission which leads to a dissociation of XeF on the repulsive A state. Thus the XeF concentration depends on a dynamic equilibrium of several aggregation and dissociation processes.

From the XeF concentration of $N=2x10^{-18}$ cm^3, the known branching ratio γ for the populations of the C,B and D state and from σ_s a maximal gain coefficient

$$\alpha_m = N \cdot \gamma \cdot \sigma \tag{8}$$

has been derived (Tab.2) These calculated gain coefficients of the order of 10 to 100 cm^{-1} exceed the gas phase values by two to three orders of magnitude and it is interesting to see if they are confirmed by gain measurements.

VIII.B Gain Measurements for XeF

The crystals are pumped with gas phase KrF (248 nm) or XeCl (308nm) excimer laser radiation in an area of 0.2x10 mm^2 at the front surface of the crystal (Fig.22) by means of crossed cylindrical lenses and the light penetrates about 1 to 2 mm into the crystal. Spontaneous emission will be amplified preferentially along the pumped line if there is sufficient gain. Strong amplification was demonstrated for C–A, B–X and D–X in Ar /45,51–54/ crystals and for D–X in Kr crystals /55/ from the observation of (i) a marked decrease of the divergence of the emitted light resulting in a well defined beam along the amplying direction (ii) an exponential growth up to 4 orders of magnitude in amplification direction with deposited energy, and (iii) a strong spectral narrowing of the emission line in amplifying direction.

The spectral narrowing is directly related to the gain coefficient α_l and corresponds in the limit of small signal gain and a Gaussian line shape to /59/

$$\alpha_l \cdot l = \ln 2 \, \{1-\exp[-\ln(\Gamma/\Gamma')^2]\}^{-1} \tag{9}$$

with the linewidth Γ for spontaneous emission, the reduced linewidth Γ' and the amplification length l=10 mm. For large $\alpha \cdot l$ it can be approximated by

$$\alpha_l \cdot l = (\Gamma/\Gamma')^2 \tag{10}$$

As an example line narrowing by more than a factor of 10 is illustrated for the C–A transition in Fig.23 and in this way α_l values of up to 300 cm^{-1}, 150 cm^{-1}, and 400 cm^{-1} have been determined for the D, B, and C state respectively. These α_l values seem to be too high compared to the α_m values. In addition the amplification would be more than e^{100} and saturation would be reached in a single path. Equ.10 is valid only for amplification by a single path and for no feedback by cavity effects, otherwise the gain coefficient will be overestimated. To test the fullfillment of this condition the pumped length l in the crystals

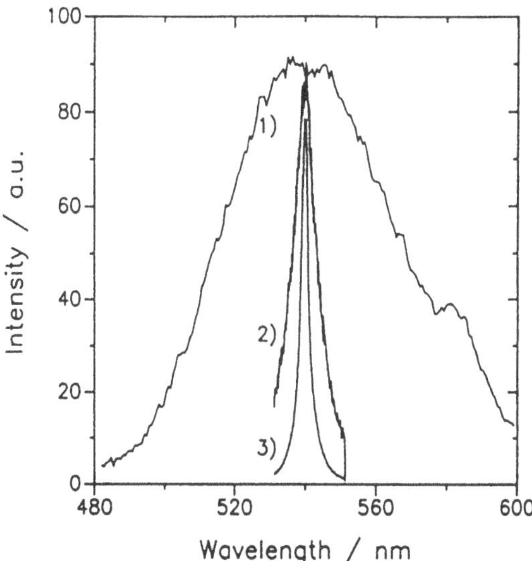

Fig. 23. Line narrowing due to amplification in the C–A transition of XeF in Ar for pump energies at 308 nm of 2 mJ (1), 35 mJ (2) and 58 mJ (3) and a pumped length of 10 mm (Ref. 54).

Table 2.

Cross section σ_s for stimulated emission, quantum efficiency q, calculated small signal gain coefficients α_m (equ.8) and from amplification measurements α_a (equ.11) (Ref.54).

	$\sigma_s(cm^2)$	q(%)	$\alpha_m(cm^{-1})$	$\alpha_a(cm^{-1})$
D–X	$1.2\cdot 10^{-16}$	7.2	140	8.6
B–X	$1\cdot 10^{15}$	0.19	60	8
C–A	$1\cdot 10^{17}$	8.1	20	8

has been varied by means of an aperture. The growth in intensity along the amplifying direction and the increase in line narrowing with pumped length l have been recorded for all three transitions. The intensity grows exponentially in the small signal gain limit. Thus the gain coefficient α_a follows from the increase in intensity from I_1 to I_2 for a length increment Δl from

$$\alpha_a = \Delta l^{-1} \ln(I_2/I_1) \tag{11}$$

The intensities in Fig.24 increase linearly on a logarithmic scale with the pumped length as expected from equ.11 up to about l=4 mm. This indicates that we are in the small signal gain limit up to this pumped length with a constant amplification coefficient α_a of about 6.5 cm^{-1} for the chosen pump energy of 9 mJ. The gain coefficient increases with pump energy. Similar results are obtained for the D–X and B–X transitions and are collected in Tab.2.

A jump in the intensities by about two orders of magnitude is observed by increasing the pumped length to 5 mm (Fig.24). The system has crossed the threshold for laser oscillations and saturates already for pumped length of 7 to 8 mm. The l dependence of the intensity in Fig.24 can be modelled by a summation of many passes of the light with a gain coefficient of 6 cm^{-1}, a reflectivity of about 4% for the quartz windows of the cell, scattering losses

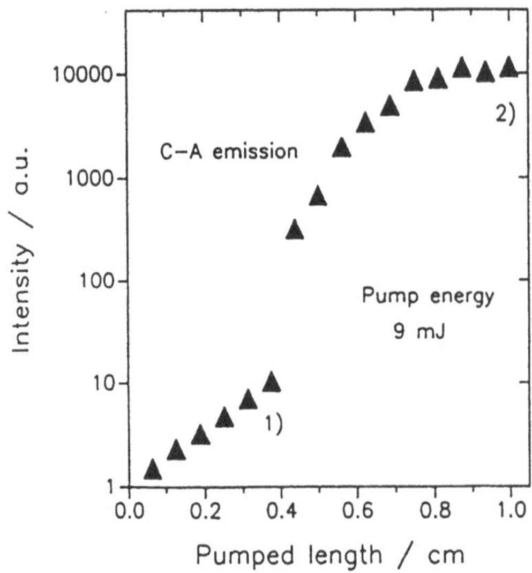

Fig. 24. Intensity of C–A transition as in Fig.23 versus pumped length (Ref.54).

of 0.5 cm^{-1} and a saturation of the amplification. The gain just compensates at l=5 mm the large losses by the 4% cavity reflectivity. The quantum efficiencies for the conversion from pump energy to the XeF laser energy follow from absolute energy measurements with the pyrodetector. The quite high values of 7 to 8% for the D–X and C–A transitions (Tab.2) could be improved with better adapted cavity reflectivities probably by more than a factor of two which has been proved recently.

The high gain values and the stored energy densities of 100 J cm^{-3} are impressive. On the other hand strong losses due to transient absorption have been reported for several other excimer systems in rare gas crystals and each system has to be characterized individually. But the broad range of applicability of the concept is clearly demonstrated by the successful operation of a CO/N_2 laser in the infrared at 5μ /50/ and by frequency tripling in Ar and Kr crystals /56/ resulting in coherent radiation down to 96 nm.

This work has been supported by the Deutsche Forschungsgemeinschaft via the Sonderforschungsbereich 337 and by the Bundesministerium für Forschung und Technologie via contract 05413 AXI7 TP5.

References

1. J. Schroeder and J. Troe, Ann. Rev. Phys. Chem. 38, 163 (1987)
2. J.T. Hynes, Ann. Rev. Phys. Chem. 36, 573 (1985)
3. G.R. Flemming and P.G. Wolynes, Physics Today, May 1990, p. 36
4. J. Franck and E. Rabinowitsch, Trans. Faraday Soc. 30, 120 (1934)
5. B. Dellinger and M. Kasha, Chem. Phys. Letters 36, 410 (1975) and 38, 9 (1976)
6. I. Norman and G. Porter, Nature 174, 508 (1954)
   E. Whittle, D.A. Daws and G.C. Pimentel, J. Chem. Phys. 22, 1943 (1954)
   G.C. Pimentel, Spectrochim. Acta 12, 94 (1958)
7. M. Moskovits and G.A. Ozin, eds, Cryochemisty (Wiley, New York 1976)
8. M.E. Jacox in Chemistry and Physics of matrix–isolated species, eds: L. Andrews and M. Moskovits (North–Holland, 1989), p. 139
9. N. Schwentner, J. Mol. structure 222, 151 (1990)

10. H. Dubost and F. Legay in Inert gases, ed.: M.L. Klein (Springer-Verlag, Berlin 1984), p. 145
11. N. Schwentner, E.-E. Koch and J. Jortner in Energy transfer in condensed matter, ed.: B.Di Bartolo (Plenum Press, New York 1984) p. 417
12. G. Herzberg in "Molecular Spectra and Molecular Structure" (Robert E. Krieger Pub. Comp. Malabar 1989)
13. see also Rare gas solids vols. I&II, eds: M.L. Klein and J.A. Venables (Academic, London 1976)
14. S. Collins, J. Phys. Chem. 94, 5240 (1990)
15. E. Knözinger, P. Hoffmann, M. Huth, H. Kollhoff, V. Langel, O. Schrems and V. Schuller Mikrochimica Acta III, 123 (1987) and references therein
E. Knözinger, V. Schuller and V. Langel, Faraday-Discuss. Chem. Soc. 86, 1 (1988)
V. Langel, V. Schuller, E. Knözinger, H.V. Fleger and H.J. Lauter, J. Chem. Phys. 89, 1741 (1988)
16. H. Coufal, E. Lüscher and H. Micklitz: Electron spin and Molecular Gramma Resonance Studies of Rare gas Matrix-Isolated Atoms and Ions in Springer Tracts on Modern Physics, vol. 103 (Springer, Berlin 1984)
17. S.S. Cohen and M.L. Klein in Inert gases, ed.: M.L. Klein (Springer-Verlag Berlin 1984) p.87
18. H.J. Jodl in "Chemistry and Physics of matrixisolated species" eds. L. Andrews, M. Moskovits (North Holland, 1989) p. 343
19. L. Andrews ibid ref. 18, p.3
20. V. Rudnick, R. Haensel, H. Nahme and N. Schwentner, Phys. Stat. Sol., Ser. A, 87, 319 (1985)
21. R. Schriever, M. Chergui, H. Kunz, V. Stepanenko and N. Schwentner, J. Chem. Phys. 91, 4128 (1989)
22. R. Schriever, M. Chergui. Ö. Ünal, N. Schwentner, U. Stepanenko, J. Chem. Phys. 93, 3245 (1990)
23. R. Schriever, M. Chergui and N. Schwentner, J. Chem. Phys. 93, 9206 (1990)
24. R. Schriever, M. Chergui and N. Schwentner, J. Phys. Chem. (August 1991)
25. M.P. Allen and D.J. Tildesley "Computer Simulations of liquids", (Clarendon Press, Oxford 1987)
26. R. Alimi, A. Brokman and R.B. Gerber in Stochasticity and Intramolecular Redistribution of Energy, eds: R. Lefebvre and S. Mukamel (D. Reidel Pub. 1987) p. 233
27. R. Alimi, A. Brokman and R.B. Gerber, J. Chem. Phys. 91, 1611 (1989)
28. R. Alimi, B. Gerber, H. Kunz, J. McCaffrey and N. Schwentner, to be published
29. R. Alimi, R.B. Gerber and V.A. Apkarian, J. Chem. Phys. 89, 174 (1988)
30. R.B. Gerber and R. Alimi, Chem. Phys. Lett. 173, 393 (1990)
31. I. Gersonde and H. Gabriel, private communication
32. R. Alimi, R.B. Gerber and V.A. Apkarian, J. Chem. Phys 92, 3551 (1990)
33. R. Alimi, R.B. Gerber and V.A. Apkarian, Phys. Rev. Letters 66, 1295 (1991)
34. V.E. Bondybey and C. Fletcher, J. Chem. Phys. 64, 3615 (1976)
35. V.E. Bondybey and L.E. Brus, ibid 62, 620 (1975)
ibid 64, 3724 (1976)
36. P. Becken, M. Mandich and G. Flynn, J. Chem. Phys. 76, 5995 (1982)
M. Mandich, P. Becken and G. Flynn, J. Chem. Phys. 77, 702 (1982)
P. Becken, E.A. Hanson and G.V. Flynn, J. Chem. Phys. 78, 5892 (1983)
37. H. Kunz, J.G. McCaffrey, M. Chergui, R. Schriever, Ö. Ünal and N. Schwentner, J. Lumin. 48/49, 621 (1991)
H. Kunz, J.G. McCaffrey, R. Schriever and N. Schwentner, J. Chem. Phys. 94, 1039 (1991)
H. Kunz, J.G. McCaffrey, M. Chergui, R. Schriever, Ö. Ünal, V. Stepanenko and N. Schwentner, J. Chem. Phys. (August 1991)
H. Kunz, J.G. McCaffrey and N. Schwentner, to be published

38. M. Fajardo and V.A. Apkarian, J. Chem. Phys. 85, 5660 (1986), ibid. 89, 4102 (1988)
39. M. Fajardo, R. Withnall, J. Feld, F. Okada, W. Lawrence, L. Wiechman, Laser Chem. 9, 1 (1988)
40. A. Katz, V.A. Apkarian, J. Chem. Phys. (in press 1990)
41. P. Andresen and R. Schinke in Molecular Photodissociation dynamics eds: M.N.R. Ashfold and J.E. Baggott (Roy. Soc. Chem. London 1987), p. 61
V. Engel, R. Schinke and V. Staemmler, J. Chem. Phy. 88, 129 (1988)
42. R.L. Redington and D.E. Milligan, J. Chem. Phys. 39, 1276 (1963)
43. W. Langel, E. Knözinger and H. Kollhoff, Chem. Phys. Letters 124, 44 (1986)
44. J.R. Morton, K.F. Preston, S.J. Strach, F.J. Adrian and A.N. Jette, J. Chem. Phys. 70, 2889 (1979)
45. N. Schwentner and V.A. Apkarian, Chem. Phys. Letters 154, 413 (1989)
46. K. Kunttu, J. Feld, R. Alimi, A. Becker and V.A. Apkarian, J. Chem. Phys, 92, 4856 (1990)
47. K. Kunttu and V.A. Apkarian, Chem. Phys. Letters 171, 423 (1990)
48. J. Feld, K. Kunttu and V.A. Apkarian, J. Chem. Phys. 93, 1009 (1990)
49. K. Kunttu, E. Sekreta and V.A. Apkarian, ibid (submitted)
50. H. Dubost, R. Charneau, M. Chergui and N. Schwentner, J. Luminesc. 48/49, 853 (1991)
51. A.I. Katz, J. Feld, and V.A. Apkarian, Optics Lett. 14, 441, (1989)
52. V.A. Apkarian, Proceedings of the International Conference on Lasers, p. 121, 1989
53. G. Zerza, R. Kometer, G. Sliwinski, and N. Schwentner, J. Luminescence 48/49, 616 (1991)
54. G. Zerza, F. Knopp, R. Kometer, G. Sliwinski and N. Schwentner, SPIE Proceedings, "Symposion on High Power Lasers, Jan. 91, Los Angeles 1410–1421
55. H. Kunttu, W.G. Lawrence and V.A. Apkarian, J. Chem. Phys. 94, 1692 (1991)
56. G. Schilling, W.E. Ernst and N. Schwentner, Opt. Commun. 70, 428 (1989)
57. H. Nahme and N. Schwentner, Appl. Phys. B51, 177 (1989)
58. D.J. Ehrlich, P.F. Moulton and R.M. Osgood, Opt. Letters 5, 339 (1980)
59. A. Yariv, "Quantum Electronics", chapt. 12, 13 (Wiley, New York, sec. ed. 1975)
60. Ch.K. Rhodes, "Excimer Lasers", p. 88 ff, (Springer Verlag Berlin 1984 sec. ed.)
61. J. Goodman and L.E. Brus, J. Chem. Phys. 65, 3808 (1976)

THE STUDY OF PARAMAGENTIC EXCITED STATES BY ELECTRON PARAMAGNETIC RESONANCE

J.H. van der Waals

Huygens Laboratory
Leiden University
Leiden, The Netherlands

ABSTRACT

After a short introduction in which the classic optical pumping experiment on mercury vapour by Kastler and collaborators is reviewed, the discussion centers on excited states with S ≥ 1 of molecules and polyatomic ions in solids. The first successful experiment of this kind was a detection of the metastable triplet state of the naphthalene molecule with a 2.45 s lifetime by Hutchinson and Mangum. In the experiment cw EPR was used, but for short-lived excited states this method fails because of an insufficient steady-state population in the excited state. Two alternatives are examined: optical detection and, in particular for nonradiative excited states, electron-spin-echo detection.

Although the Kastler experiment has served as a source of inspiration for the optical detection of EPR transitions in excited states in solids, the low symmetry of a polyatomic system (as opposed to a free atom) has fundamental consequences. In EPR experiments with optical detection on spin triplet and quartet states in a solid one utilizes the great disparity in the radiative decay rates from the individual spin components, a disparity that arises from the anisotropy of spin-orbit coupling in situations of low symmetry. Because of the existence of a zero-field splitting in multiplets with S ≥ 1, the experiment also can be carried out in the absence of an external magnetic field ("zero-field ODMR"). In a spin-echo experiment one samples a transient magnetization in the excited state - usually created by laser-flash excitation - on a μs time scale. Both types of experiment cannot only be used to identify the excited state via a determination of

Optical Properties of Excited States in Solids
Edited by B. Di Bartolo, Plenum Press, New York, 1992

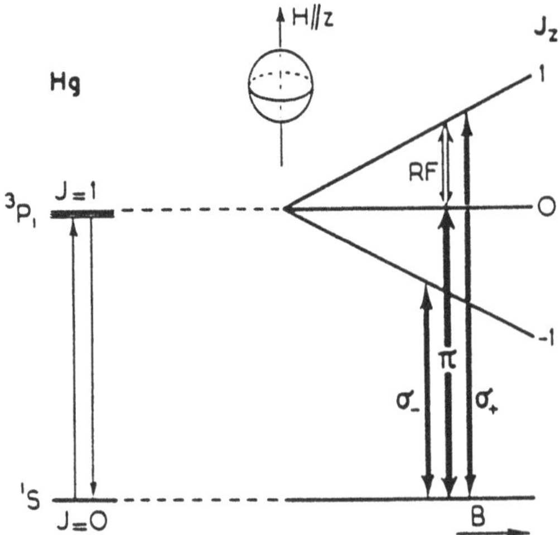

Figure 1. Schematic representation of the first optical pumping experiment on mercury vapour proposed by Kastler and Brossel, and performed by Brossel and Bitter [1]. For the free atom the three sublevels of the $^3P_1$ excited state are degenerate, but in a magnetic field they split as indicated on the right. R.F. transitions between the Zeeman components are detected optically via changes in the polarization of the fluorescence.

the energy parameters in the spin hamiltonian, but they also are well-suited to study the dynamics of the generation and decay of the excited state.

I. INTRODUCTION

I.A. Kastler, Brossel, and Bitter's Optical Pumping Experiment on Mercury

The investigation of excited states by magnetic resonance has its roots in the famous optical pumping experiment on mercury proposed by Brossel and Kastler [1].

The gist of this experiment is sketched in Fig.1. When dilute mercury vapour is excited by a (low-pressure) mercury lamp, atoms are excited from the 6^1S_0 ground state to the 7^3P_1 excited state. Subsequently, after a mean lifetimes of 0.1 μs, the excited atoms decay back to the ground state by the emission of light. The excited state of the atom with J=1 is triply degenerate, but by the application of a magnetic field, say $\vec{B}\|\vec{z}$, it splits into three Zeeman components with magnetic quantum number m = 1,0, - 1, respectively. Because the z-component of the total angular momentum of the atom plus radiation field is conserved, the radiation from the m = ± 1 levels will be circularly polarized (σ ± components) and that from the m = 0 level linearly polarized parallel to $\vec{z}$ (π component). In a field of the order of 100 G used in Brossel and Bitter's experiments [1] the Zeeman splitting between two adjacent components is a few hundred MHz.

Now suppose that the sample is selectively "pumped" into the m = 0 state by z-polarized light propagating along the $\vec{x}$ direction, while an observer equipped with a detector sensitive to the σ ± components looks at the emission emitted along $\vec{z}$. Because in the absence of collisions m is conserved in the excited state, the observer then will detect a zero signal. But, if with help of an oscillator connected to a coil wrapped around the sample cell, he now switches on an r.f. field resonant with the $|\Delta m|$ = 1 Zeeman transition, a signal will appear.

The magnetic resonance experiment of Fig. 1 proved the source of inspiration for a remarkable variety of experiments on the interaction between atoms and radiation fields; of these the Ph.D. thesis of Cohen-Tannoudji represents an early example [2]. As a means for identifying paramagnetic excited species by magnetic resonance, the chief characteristics of a Kastler-type experiment on an atomic vapour may be summarized as follows:

(1) The radiative decay channels of the three Zeeman components have equal probabilities, but mutually orthogonal polarizations -- a state of affairs imposed by the spherical symmetry of the atom.

(2) For a given experimental arrangement, the intensity of the detected signal is independent of the lifetime of the excited state and primarily determined by the efficiency of the optical pumping process. Because of the selective nature of this process, the resonance can be arranged to occur between two levels with large population differences ("spin polarization"), far in excess of that prevailing at thermal equilibrium.

(3) The r.f. resonance is not detected as such but at a much higher frequency as an optical signal against a background that may be reduced to almost zero by a suitable experimental arrangement (good polarizers, low vapour pressure, and coated cell walls to suppress m-changing collisions). This method of detection is an example of what is often referred to as "trigger detection".

It took a decade before magnetic resonance experiments on paramagnetic excited states in solids were crowned with success. In 1958 Hutchison and Mangum succeeded in the detection of the phosphorescent triplet state of the naphthalene molecule in a durene crystal by electron paramagnetic resonance (EPR) [3]. In the next year Geschwind, Collins, and Schawlow reported a successful EPR experiment with optical detection on the $\bar{E}(^2E)$ excited state of Cr^{3+} in ruby [4]. The two experiments show essential differences and each served as a prototype for the further investigation of excited states in solids by magnetic resonance. Whereas the first experiment is on an $S = 1$ state which, as we shall see, looses its degeneracy in the solid, the second is on a Kramers-degenerate ^2E state. When optical detection is used, the two types of systems need different approaches. In this review I will concentrate on metastable spin triplets. The methods developed for triplet states, however, have also proved their worth in some studies on quartets and, in principle, they should be applicable to any multiplet with $S \geq 1$.

For the study by means of EPR of excited states in which the spin degeneracy is preserved in the solid, such as $S = \frac{1}{2}$ excited states (e.g. Cr^{3+}), or Γ_8 multiplets of rare earth ions in cubic crystals, the reader is referred to the chapter by Geschwind in the book "Electron Paramagnetic Resonance" [5].

II. METASTABLE TRIPLET STATES IN SOLIDS AND THEIR DETECTION BY CONVENTIONAL EPR

II.A. The Optical Pumping Cycle in Polyatomic Molecules and Hutchison and Mangum's Experiment

Many organic molecules, in particular those with series of double bonds, exhibit two luminescent decay channels when excited by U.V. light. These channels differ in their spectra as well as in the rates of decay. The origin of this phenomonon remained controversial until Jablonski's work

[6,7] provided the first clue. He proposed a scheme, the modernized version of which I would like to call the optical pumping cycle of a phosphorescent organic molecule. Fig. 2 represents this cycle for the specific case of quinoxaline (1,4 diazanaphthalene) as an example. On excitation of the (diamagnetic) molecule in one of its singlet-singlet absorption bands part of the energy is degraded by what is now called "intersystem crossing" into a metastable state T.

Whereas in a dilute solid solution at low temperature the lifetime of the first excited singlet state S_1 is of the order of 10^{-7} - 10^{-9} s, the metastable state has a lifetime between, say 10s (benzene) to 10^{-4}s (p-benzoquinone). The emission from S_1 now is generally referred to as "fluorescence" and the afterglow from the metastable state as "phosphorescence". In addition to these radiative decay processes there may be radiationless decay from S_1 and (or) T. For some time the physical nature of Jablonski's metastable state remained elusive, but then both G.N. Lewis and his school [8,9,10] and Terenin [11] suggested that it has to be identified with the lowest spin triplet state (S = 1) of the molecule. In particular, the systematic investigation of a large number of luminescent compounds by Lewis and Kasha [9,10] lent strong support to this idea. The final proof was provided by Hutchison and Mangum's successful EPR experiment on phosphorescent naphthalene in a durene host crystal, which proved its S = 1 character [3,12]. Since then EPR has become a common technique for the study of photoexcited triplet states in crystals and glasses [13,14,15,16].

In Fig. 3 I have sketched a Zeeman diagram for phosphorescent naphthalene as derived from the EPR experiments of Hutchison and Mangum. It is the analogue of that for the $^3P_1$ state of mercury of Fig. 1, but differs from the earlier figure in two important, interrelated aspects. First, in the absence of a field, there is a "zero-field splitting" (ZFS) of the triplet sublevels which amounts to about 0.1 cm^{-1} for naphthalene. This splitting arises because in a polyatomic molecule composed of relatively light atoms, like naphthalene (point group D_{2h}) or quinoxaline (C_{2v}), as opposed to an atom, the orbital angular momentum is quenched, $\langle\vec{L}\rangle = 0$, while the spin angular momentum is still approximately conserved. However, owing to the low symmetry, different orientations of the spin angular momentum $\vec{S}$ are physically distinct and thus have different energies [17]. Second, as a consequence of this zero-field splitting, the Zeeman diagram becomes anisotropic, and for a fixed microwave quantum the fields-for-resonance vary with the orientation of the crystal in the applied field. Moreover, because of the ZFS (or, in other terms, as a consequence of the break-down of the cylindrical symmetry of the Zeeman problem for the Hg atom of Fig. 1), the component of the spin angular momentum along the magnetic field is no longer

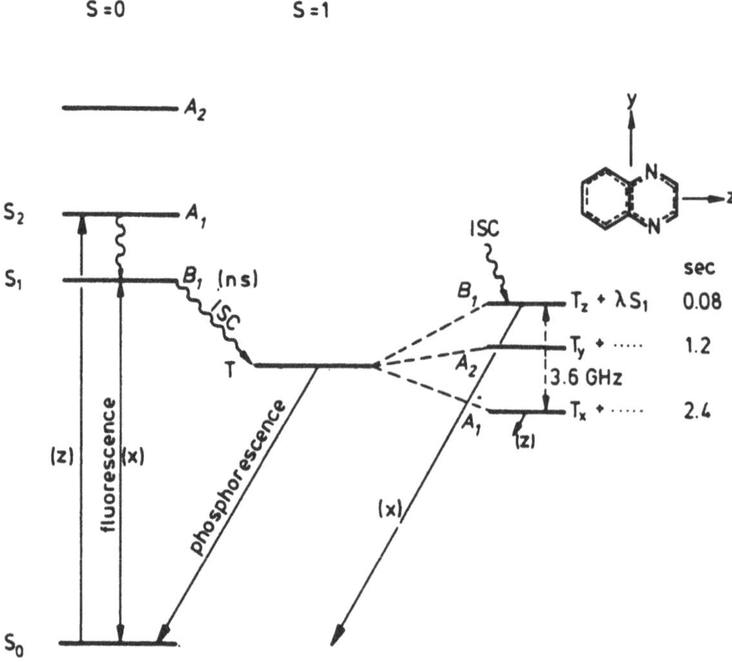

Figure 2. The optical pumping cycle of a phosphorescent organic molecule. The singlet states are on the left, the metastable triplet state is at the centre. Because of the relatively low symmetry of a polyatomic system the spin degeneracy is lifted and a zero-field splitting results, indicated at the right on an enlarged scale. The data refer to 1,4 diazanaphthalene (quinoxaline) as an example. The total orbital ⊗ spin symmetries of the states are labelled in italics according to the irreducible representations of the point group C_{2v} and the polarizations of the 0-0 transition refer to the axis system indicated [21].

strictly conserved and the magnetic quantum number m is not a "good" one unless the Zeeman energy is much larger than the ZFS. As a result of this, one can observe not only the $\Delta m = \pm 1$ transitions, but also the $\Delta m = 2$ transition. (The latter proves to have a much smaller anisotropy than the former and was the first to be observed in a randomly oriented sample) [13,14].

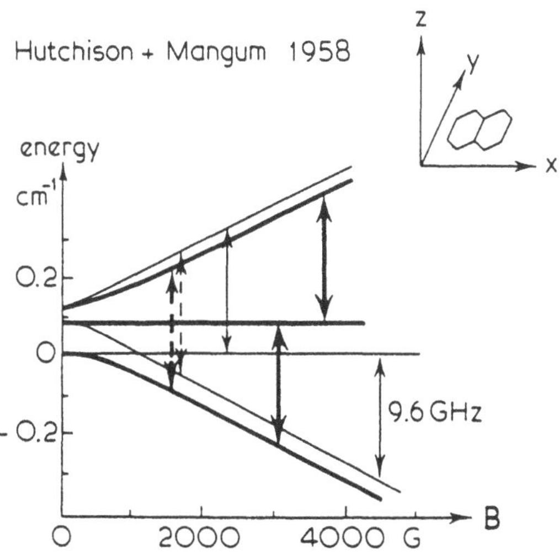

Figure 3. The first EPR experiment on a photoexcited triplet state: naphthalene as a guest in a durene single crystal [3,12]. Zeeman diagrams are shown for two orientations of the naphthalene molecule in a magnetic field ($\vec{B} \parallel \vec{y}$ heavy lines, $\vec{B} \parallel \vec{z}$ thin lines). The solid arrows represent the $\Delta m = \pm 1$ transitions first observed by Hutchison and Mangum, the broken arrows the $|\Delta m|=2$ transition [13]. Note the zero-field splitting and the resulting magnetic anisotropy.

II.B. Spin Hamiltonian and Zero-field Splitting

In an EPR experiment like that of Hutchison and Mangum the ZFS may be determined by studying the fields for resonance as a function of the orientation of the crystal in the field. In practice this is done by fitting a "spin-Hamiltonian" of the form

$$\mathcal{H} = \vec{S}\cdot\vec{\vec{D}}\cdot\vec{S} + \beta\vec{B}\cdot\vec{\vec{g}}\cdot\vec{S} \qquad (1)$$

to the experimental data. The first term represents the zero-field (or "fine-structure") splitting and the second term the magnetic energy in the applied field $\vec{B}$; β is the Bohr magneton and $\vec{\vec{D}}$ and $\vec{\vec{g}}$ are second rank tensors to be determined from experiment.

The use of a spin Hamiltonian of the form (1) for the analysis of magnetic resonance data goes back to ideas first introduced by Pryce [18,19] and discussed in detail by Abragam and Bleaney [20]. Instead of expressing the spin-dependent interactions by a representation of the appropriate Hamiltonian operator in a Hilbert space spanned by a large number of spin-orbital states, one restricts the function space to the three components of the triplet studied and treats the tensors $\vec{\vec{D}}$ and $\vec{\vec{g}}$ as empirical parameters. This approach is valid provided that the energy span of the triplet is much smaller than the energies separating it from other electronic states -- i.e. any orbital degeneracy must have been removed by a crystal field or Jahn-Teller effect. For systems of a relatively high symmetry, for instance those considered in figures 2 and 3, the tensors $\vec{\vec{D}}$ and $\vec{\vec{g}}$ are diagonal in a common principal axis system $\vec{x},\vec{y},\vec{z}$ fixed by molecular symmetry. The principal values of $\vec{\vec{D}}$, to be denoted as X,Y,Z, are equal to the energies of the three triplet components in zero field.

If we think of our triplet as consisting of three levels, uncoupled from all other states, the ZFS tensor $\vec{\vec{D}}$ arises from the dipolar interaction of the two parallel spins in the triplet state while the Zeeman tensor $\vec{\vec{g}}$ is equal to $g_e\vec{\vec{1}}$, where $g_e = 2.0023$ is the g-value for a free electron. The interaction of the triplet levels with other spin-orbital states through spin-orbit coupling (SOC), however, gives additional contributions to both tensors which may be estimated by perturbation theory [18,20]. For most organic molecules, in which SOC is weak, the assumption of an "uncoupled" triplet is close to reality and the higher-order terms may be neglected. The spin Hamiltonian is then conventionally written in the form

$$\mathcal{H} = D(S_z^2 - \tfrac{1}{3}\vec{S}\cdot\vec{S}) + E(S_x^2 - S_y^2) + g_e\beta\vec{B}\cdot\vec{S}. \qquad (2)$$

The parameters D and E are related to the diagonal elements of $\vec{\vec{D}}$ according to

$$D = -\frac{2}{3}Z, \quad E = \frac{1}{2}(Y - X). \tag{3}$$

For a proper understanding of the situation in zero magnetic field group theory is a useful tool [21]. For an atom in a state with no orbital angular momentum and S = 1, say a ^3S state of helium, the three (degenerate) spin states may be represented by the eigenfunctions T_1, T_0, T_{-1} of the operator S_z, with eigenvalues 1,0 and -1, respectively. This particular choice is favoured because, if a magnetic field $\vec{B}$ is switched on and $\vec{z}$ chosen along the direction of $\vec{B}$, then these functions T_1, T_0, T_{-1} remain the proper eigenfunctions for the atom in the field. This conclusion follows from the Hamiltonian of the Zeeman problem, here equal to the last term of (2). But, even without specifying the exact form of the Hamiltonian, one is forced to this conclusion by the observation that each of the three functions belongs to a different irreducible representation of the two-dimensional rotation group governing the symmetry of an atom in a magnetic field.

If we now turn to a molecule like quinoxaline or naphthalene in zero field, where the cylindrical symmetry is lost and the appropriate point group is C_{2v} or D_{2h}, the functions T_1, T_0, T_{-1} no longer are the eigenfunctions in zero field. One can show however that in this situation the linear combinations

	C_{2v}	D_{2h}
$T_x = 2^{-\frac{1}{2}} (T_{-1} - T_1)$	B_2	B_{3g}
$T_y = 2^{-\frac{1}{2}} (T_{-1} + T_1)$	B_1	B_{2g}
$T_z = T_0$	A_2	B_{1g}

(4)

belong to different irreducible representations of the point group and, therefore, must diagonalize the zero-field Hamiltonian $\mathcal{H}(B = 0) = \vec{S}\cdot\vec{\vec{D}}\cdot\vec{S}$. (The proof is simple: the three eigenfunctions of a spherical system in a state with S = 1 -- just like the components of angular momentum -- span the representation $D^{(1)}$ of the rotation group in three dimensions. If the symmetry is lowered, the representation $D^{(1)}$ is resolved into $A_2 + B_1 + B_2$ for C_{2v}, or into $B_{1g} + B_{2g} + B_{3g}$ for D_{2h} [21]). Actually, by analogy with an atomic state with one unit of orbital angular momentum (L=1) the result (4) is not surprising. Because, for say, a carbon atom in an environment of orthorhombic symmetry the appropriate 2p atomic orbitals (a.o.'s), likewise, are written as

$$p_x = 2^{-\frac{1}{2}}(p_{-1} - p_1),\ p_y = 2^{-\frac{1}{2}}\, i(p_{-1} + p_1),\ p_z = p_0. \tag{5}$$

In the molecules here taken as examples the axis system $\vec{x},\vec{y},\vec{z}$ is predetermined by symmetry. For molecules of lower symmetry the eigenfunctions in zero field, T_x, T_y, T_z, can still be related to those of the operator S_z as in (4), but the directions $\vec{x},\vec{y},\vec{z}$ that diagonalize the tensor $\vec{\vec{D}}$ now are unknown a priori and have to be determined from experiment.

The three zero-field eigenfunctions of our triplet (S = 1) transform under rotations like p-type a.o.'s (ℓ=1), i.e. like the components of angular momentum. Hence, we have by analogy to $\langle p_u|\ell_v|p_w\rangle = i\varepsilon_{uvw}$ for the matrix elements of the spin operators

$$\langle T_u|S_v|T_w\rangle = i\varepsilon_{uvw},\quad u,v,w, \in (x,y,z). \tag{6}$$

and in particular

$$\langle T_u|\vec{S}|T_u\rangle \equiv 0. \tag{7}$$

Because the expectation value (7) vanishes, one is wont to say that "the spin angular momentum is quenched". But, since $S_u T_u = 0$, it is more accurate to say that in the state T_u the angular momentum S = 1 is known to lie in the plane u = 0 [21]. Further, because the electron-spin magnetic moment $\vec{\mu} = \beta g\vec{S}$, eq. (6) implies that magnetic dipole transitions may be induced in all three pairs of zero-field levels [21,22]. With the application of a magnetic field the three zero-field transitions go over continuously into the $\Delta m = \pm 1$ and $|\Delta m|=2$ transitions of the Zeeman diagram of Fig. 3.

Before leaving the subject of the spin states in zero field, I have sketched in Fig. 4 the relationship which exists between the symmetry of a triplet species and its zero-field-splitting pattern. The pictures are dictated by group theory and should be self-explanatory. For comparison analogous patterns for a spin quartet ($S = \frac{3}{2}$) have been drawn in the bottom row; because of the odd number of electron spins involved, Kramers degeneracy persists, irrespective of the spatial symmetry of the species. The sketches convey an important message: even

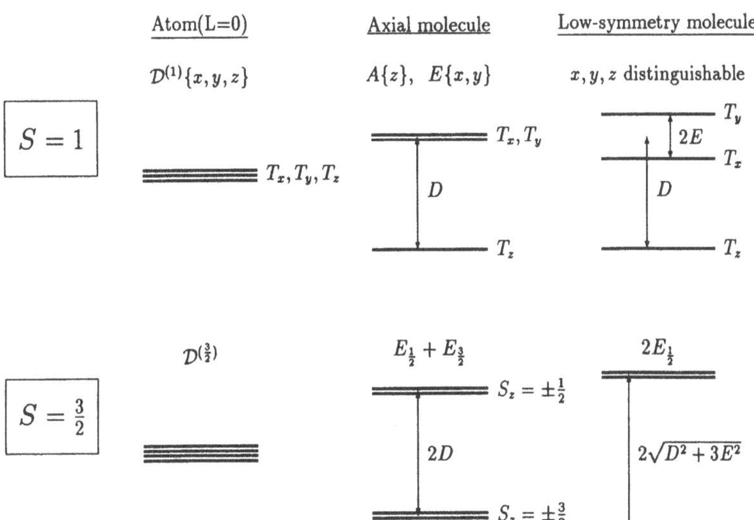

Figure 4. The relation between molecular symmetry and the zero-field splitting patterns of spin triplet (S = 1) and quartet ($S = \frac{3}{2}$) states. The first row of each entry shows the irreducible representations of the relevant point group to which the three components of the angular momentum operator belong. The splittings are expressed in the zero-field parameters D and E of the spin Hamiltonian (2). The patterns as sketched are for positive D and E, if D and/or E are negative the order of the corresponding levels is reversed.

a qualitative resonance experiment may suffice to draw a definite conclusion about the symmetry of an excited triplet species. Thus, EPR experiments have revealed that, for instance, the phosphorescent triplet state of triphenylene remains very nearly threefold axially symmetric [14], whereas benzene on excitation is markedly distorted by a pseudo Jahn-Teller effect [23].

II.C Limitations of Conventional CW EPR for the Detection of Excited States - Alternatives

The detection by EPR of metastable triplet states of naphthalene and a few related molecules in crystals [3,12] and glasses [13,14] in the late 1950's led to a flurry of similar experiments on a wide variety of substances. All these experiments were carried out in more-or-less conventional cw EPR spectrometers designed for the study of ground-state paramagnetic species and provided with facilities for exciting the sample in the cavity with intense uv light. It was soon discovered, however, that with such an instrument the successful experiments, with some rare exceptions, are restricted to those on aromatic hydrocarbons. The reason for this is simple: most aromatic hydrocarbons, as compared to other excited systems, have a particularly long phosphorescence lifetime (generally of the order of seconds), and because of this, a relatively high steady-state concentration of excited molecules can be generated in the cavity.

From their signal intensity Hutchison and Mangum [12] estimated that with a 1000 W AH-6 exciting source they had a steady-state of $N_s \approx 1.10^{14}$ triplet spins in their cavity; with a signal-to-noise ratio of 20 this corresponds to a detection limit of 5.10^{12} spins. From a consideration of detailed balance, on the other hand, it follows that

$$N_s = P_\tau \qquad (8)$$

where P is the number of molecules pumped into the triplet state and τ the triplet lifetime (see also below under eq. (9)). For continuous excitation $P = N_{abs}\phi_{isc}$ in which N_{abs} is the number of photons absorbed by the sample per second and $\phi_{isc} \approx 0.4$ and $\tau = 2.55$ s, from which one concludes that to attain the above value of N_s about 0.1 mW of light with a wavelength below 300 nm must have been abosrbed by the sample. From this example we conclude that a conventional cw-EPR experiment with microwave detection offers little hope for the study of excited states with ms lifetimes and/or low intersystem crossing yields. In the subsequent sections of this chapter two alternatives to conventional cw EPR spectroscopy will be considered that, each in its own domain, have proved particularly successful: optical detection and pulsed, "electron-spin-echo" (ESE) experiments. Moreover, I will show that, because of the existence of a zero-field splitting in molecular systems with $S > \frac{1}{2}$, many of these experiments may be carried out with advantage in the absence of an applied field.

III. OPTICAL DETECTION OF EPR IN EXCITED STATES OF POLYATOMIC SYSTEMS IN SOLIDS

III.A. Experiments in a Magnetic Field

Because of the limitations of cw EPR spectroscopy for the study of short-lived excited states, the obvious question arose: can one, perhaps, detect the microwave transitions between the spin levels of a phosphorescent triplet through changes in the phosphorescence? Such optical detection promised to offer two advantages: 1) enhancement of the sensitivity because of the far greater energy of the phosphorescnece quantum as compared with the microwave quantum; 2) a signal that no longer depends on the number N_S and thus decreases with τ, but that, instead, should be proportional to the light flux emitted by the sample, $N_{abs}\phi_{ph}$, where ϕ_{ph} is the phosphorescence quantum yield. However, there are some subtle differences between the atomic and molecular situations and it took until 1967 when Sharnoff [24] and two other groups [25,26] realized EPR with optical detection for a phosphorescent organic molecule. Consider how such an experiment might be realized for quinoxaline of Fig. 5. The $T \rightarrow S_0$ phosphorescence transition, although spin-forbidden, acquires a small probability because of spin-orbit coupling (SOC) of the triplet state to singlet states. As noted, the three zero-field components belong to different irreducible representations of the point group C_{2v}, as indicated in Fig. 2. (These spin $\otimes$ orbital symmetries are obtained as the direct products of the $^3B_2$ orbital symmetry times the relevant spin symmetry). Each of the three zero-field triplet components may acquire some singlet character by spin-orbit coupling, but SOC can couple each of them *only to singlet states of the same symmetry*. In quinoxaline with its highly non-spherical electronic structure, however, SOC is very anisotropic and in the axis system of Fig. 5 the only effective SOC is that of T_x with S_1. Consequently, more than 95 per cent of the phosphorescence here

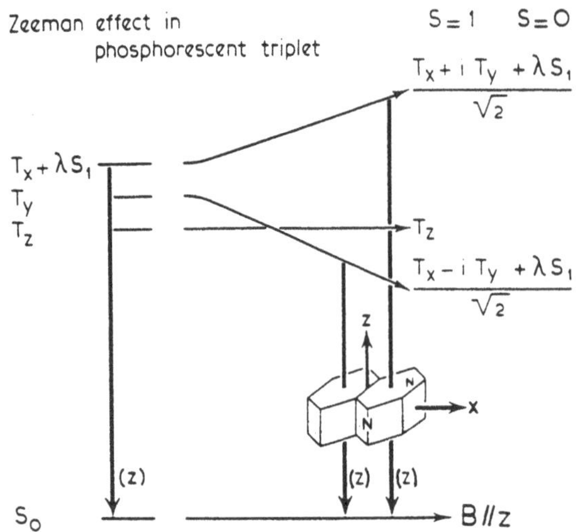

Figure 5. Principle of OD-EPR experiment on phosphorescent quinoxaline. The phosphorescence owes its intensity to SOC of the triplet sublevels with excited singlet states; in this process the coupling of T_x to S_1 proves dominant and hence the phosphorescence is z-polarized. By application of a magnetic field (here $\parallel \vec{z}$) the radiative character is distributed over several Zeeman levels with preservation of its z-polarization. EPR transitions may now be detected through a change in phosphorescence intensity caused by hitting one of the Zeeman transitions with resonant microwaves. (The axes here have been labelled by analogy to those of naphthalene in Fig. 3, i.e. different from the convention used in Fig. 2).

originates from T_x [27], and in accordance with the symmetry of this sublevel it must be z-polarized. We further note that the transition from the Y_y level of A_2 symmetry is electric-dipole forbidden, whereas the very weak transition from T_z must be x-polarized. The great disparity in the radiative decay rates is reflected in the lifetimes of the individual levels quoted at the right in Fig. 2, even though the real situation is more complicated than sketched because of vibronic coupling and additional radiationless decay.

Now the great discovery of the early 1970's was that the situation of widely varying radiative decay rates within a triplet is by no means an exception but of general validity for the large number of phosphorescent molecules investigated. Initially these were almost exclusively organic with conjugated double bonds [28], but recently the same has been found for a variety of polyatomic inorganic ions such as NO_2^- [29], $Cr_2O_7^{2-}$ [30], VO_4^{3-} [31,32], and MoO_4^{2-} [31,32]. The inequality of the radiative rates is also understood theoretically on the basis of McClure's pioneering paper on SOC in aromatic molecules [33] and subsequent work in this field (see [34] for the specific case of quinoxaline and further references on the subject). Hence, on inducing a microwave transition between two Zeeman levels $|i\rangle$ and $|j\rangle$ the optically detected EPR signal, instead of being determined by N_s eq. (8) will be given by

$$S_{i,j}^{\vec{E}} \propto P \frac{|k_i^{T,\vec{E}} - k_j^{T,\vec{E}}|}{\sum k_i}. \tag{9}$$

Here $k_i^{T,\vec{E}}$ is the radiative decay rates of level $|i\rangle$ for light polarized parallel to $\vec{E}$, k_i in its total decay rate, and the sum is taken over the three Zeeman components of the triplet (or the 2S + 1 components of a higher spin multiplet). For the time being we think of experiments carried out under circumstances where thermal equilibrium between the spin levels exists. Though for most systems, as we shall see laer, one can take advantage of the fact that by lowering the temperature a regime can be attained in which the levels are isolated from one another and high degrees of spin alignment can be realized.

We see from the expression (9) that successful EPR experiments with optical detection may be carried out on photo-excited triplet states of polyatomic molecules, provided one chooses an orientation of $\vec{B}$ for which the sublevels have different probabilities for radiative decay. Consider the system of Fig. 5, for instance, when continuously illuminated in a relatively strong field and at a temperature where thermal equilibrium between the three levels exists. Instantaneous "tickling" with microwaves that are resonant with the $m_s = 0 \leftrightarrow m_s = 1$ transition will then result in an increase of the phosphorescence intensity, while resonance with the $m_s = 0 \leftrightarrow m_s = -1$ transition will produce a decrease.

In Fig. 5 I have also tried to visualize the fundamental difference between the Kastler-type experiment on an atomic triplet state and the present type of experiment on a polyatomic system. For the latter the emission from the triplet which is detected, with very few exceptions, is "stolen" through SOC from a single, linearly polarized electric-dipole transition in the singlet system. When a strong Zeeman field is applied, the resultant electron spin will become quantized relative to the field and precess about the field -- just as in an atom. (See the triplet part of the eigenfunctions at the right in Fig. 5.) But, nevertheless, the emission from the molecule will stay linearly polarized. The situation, therefore, is identical to that for an observer in a Kastler-type experiment who looks along the magnetic field at the $\sigma_{\pm}$ components through a linear polarizer. In the molecular case, this polarizer is intrinsic in the molecule. Thus, whereas an EPR transition in an excited atom has to be detected as a change in polarization of the emission, an analogous transition in a molecular system manifests itself through a change in intensity of the emission. For this reason, and also to avoid complications due to birefringence, OD-EPR experiments on polyatomic ions and molecules in solids are generally carried out by looking at the total emission intensity, without polarizers. Accordingly in equations like (9) the reference to a specific decay mode $\vec{E}$ is omitted and k_i^r now refers to the radiative decay rate of level $| \, i >$ summed over different polarizations. Of course, in subtle experiments where one wants to study the decay modes of the individual spin sublevels of a triplet, the use of polarizers may still be quite helpful [32].

An example of an optically-detected EPR spectrum of the luminescent MoO_4^{2-} ions in a $CdMoO_4$ crystal recorded by Van Tol is shown in Fig. 6. The transitions induced by the 75 GHz microwaves are indicated in the Zeeman diagram at the bottom; because of a Jahn-Teller distortion there are four magnetically inequivalent excited triplet species.

In conventional EPR where microwave power is detected one is used to work at a fixed frequency in a high-Q cavity and to tune the system into resonance by varying the strength of an applied field. With optical detection things are different. There is no need for a cavity tuned to a specific microwave frequency and in several ways the applied field may be considered an unnecessary evil: it scrambles the zero-field states with their characteristic decay modes and introduces spatial anisotropy.

Because of this scrambling, one might expect that for the study of metastable triplet states by EPR distinct advantages could be gained by carrying out the experiments in zero-field with optical detection. As we shall see in the next section, this intuitive reasoning has proved

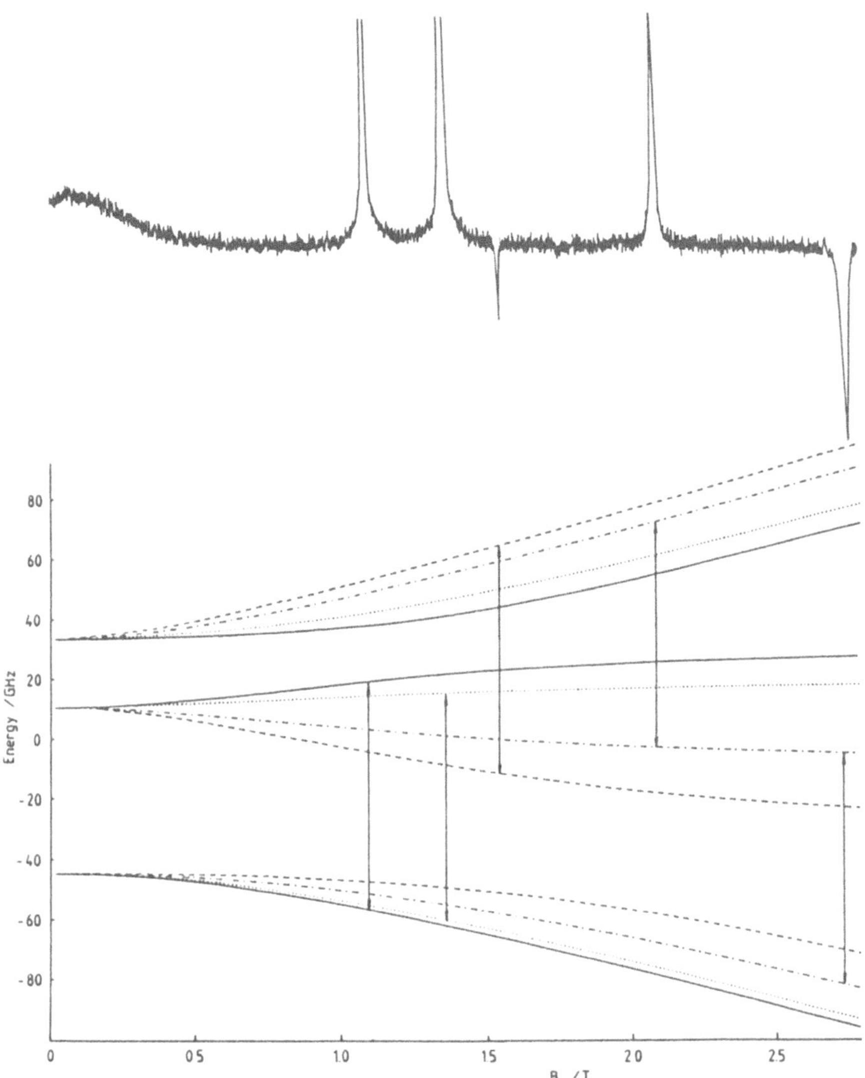

Figure 6. OD-EPR spectrum of $CdMoO_4$ at 75 GHz and 1.2 K recorded by Van Tol [32]. As illustrated by the four sets of energy levels in the Zeeman diagram for the chosen direction of the magnetic field with respect to the crystal, the observed resonances can be attributed to four equivalent triplet states which differ only in the orientation in the field $\vec{B}$. The "positive" resonances correspond to an increase in the detected emission, while the "negative" signals correspond to a decrease in the emission upon the application of microwaves.

correct, and experiments in zero field are now widely used. If one wishes to obtain quantitative, spatial information on the structure of an excited state, however, the experiment has to be done in an applied magnetic field. The field then serves as a probe for determining the directions of the principal axes of the various tensors in the spin Hamiltonian.

An example is provided by the recent EPR and ENDOR experiments in a magnetic field -- all with optical detection -- on the metastable triplet state of the VO_4^{3-} ion in YPO_4 by Van Tol [32]. The VO_4^{3-} ions in this system suffer a Jahn-Teller distortion on excitation, much like the MoO_4^{2-} ions in $CdMoO_4$ of Fig. 6. From Van Tol's determination of the principal values and principal axes of the three relevant spin-dependent tensors in the problem ($\vec{\vec{D}}$, $\vec{\vec{g}}$ in Eq. (1), and the hyperfine tensor for the coupling of the electron spin to the ^{51}V nucleus) a remarkably detailed insight into the structure of the excited state was obtained. It appears that the Jahn-Teller distortion is related to a localization of the excitation with one unpaired spin residing on the central metal and the other on a single oxygen ligand [32].

A further example is reproduced in Fig. 7 showing an OD-EPR signal of the luminescent triplet state of MoO_4^{2-} ions in a $CaMoO_4$ crystal recorded by Barendswaard in our laboratory. The remarkable structure results from two effects: a slight variation of the zero-field splitting parameter D with isotopic mass, and in addition a magnetic hyperfine coupling for the two odd isotopes. The top curve represents the experiment and the bottom curve a theoretical fit by Van Tol [31,32].

Thus far I have discussed triplet (S = 1) excited states because for these a vast amount of experimental material is available. This does not imply, however, that optical detection of EPR is restricted to triplet states only. The technique can be used for the study by EPR of any excited state which decays to a state of different multiplicity under the emission of light. Successful experiments have, for instance, been carried out on the luminescent ^4E_u state ($S = \frac{3}{2}$) of copper porphine [35]. For spin-allowed transitions, for instance in $S = \frac{1}{2}$ systems, optical detection cannot be used because the radiative transition probabilities for the transitions from the different sublevels then tend to be equal.

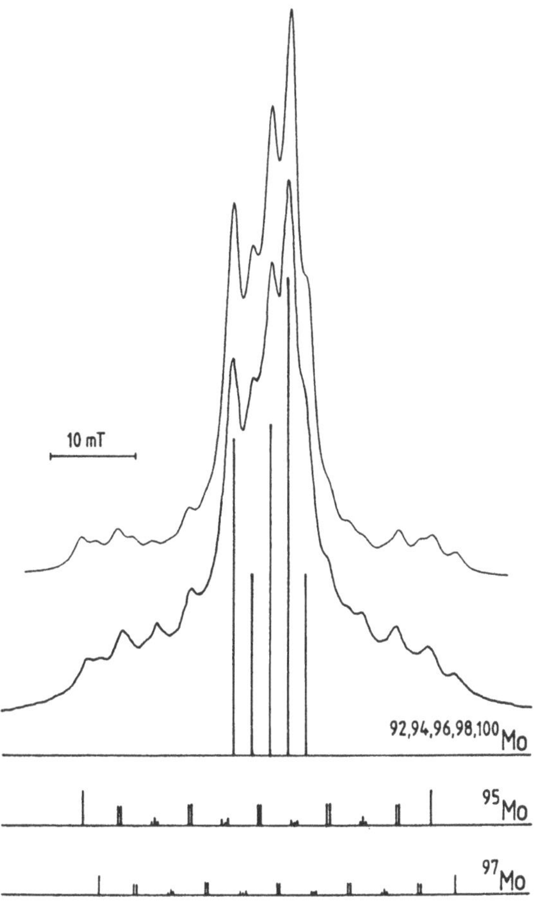

Figure 7. OD-EPR transition in the metastable triplet state of an $MoO_4^=$ ion in $CaMoO_4$ recorded by Barendswaard with $\vec{B}$ in the xz plane 50° from $\vec{z}$ towards $-\vec{x}$, B = 1.9 T, ν = 75 GHz. The spectrum at the top is a simulation of the experimental line profile shown beneath it. The splitting arises from two effects: a slight variation of the zero-field splitting with nuclear mass, and hyperfine coupling of the electron spin with the magnetic moments of the odd isotopes (for all details see [32]).

III.B. Experiments on Triplet States in Zero Field

In 1967 we realized an EPR experiment on a phosphorescent triplet state in zero-field using microwaves of variable frequency and optical detection. The first experiments concerned the tetramethylpyrazine and quinoxaline molecules in a durene host crystal [36], and soon the method was widely applied. I will not reproduce the older results here as excellent reviews of what has been achieved with this technique (now often referred to as zero-field optically detected magnetic resonance, ZF-ODMR) have been given by Kinoshita, et.al. [28] and Chan [37].

In principle the technique is simple. In the 1-12 GHz region -- in which the zero-field transitions of most organic triplet states are located -- microwaves may be applied to the sample by placing it in a small helical coil connected to a variable frequency oscillator [36,37]. At higher frequencies, important for the study of inorganic ions and defects in semiconductors, a Fabry-Perot cavity has been used with advantage [32,38]. In both instances the open structure facilitates optical excitation and detection.

The AF-ODMR technique allows one to determine the zero-field splitting of a luminescent triplet state with great precision, even in a random sample. But I think its two virtues of greatest practical use are: (1) the possibility to perform transient experiments by which the kinetics of populating via intersystem crossing and decay (radiative and radiationless) of the triplet state may be unravelled [27,39,40]; (2) its combination with optical spectroscopy as a tool for understanding phosphorescence excitation and emission spectra [41]. In these two areas EPR in a magnetic field is inferior because of the "scrambling" by the field of the populating and decay channels, mostly with unknown relative phases [42].

It is only fair also to weigh the disadvantages. If one wishes to study magnetic hyperfine structure or do ENDOR experiments, EPR in an external magnetic field is nearly always superior. By the application of a field to an oriented system one not only restores the magnetic hyperfine splitting of the triplet sublevels (which vanishes in zero field, except as a higher-order effect), but as we have seen, also obtains precise information on the spatial characteristics of various interactions.

On the use of zero-field OD-EPR for the determination of the kinetics of populating and decay of metastable triplet states extensive reviews are available [28,37]. Therefore, I shall merely sketch the general principles and consider one specific example of the many technical variations that have been used. In Fig. 8 the optical pumping cycle involving a metastable triplet state has been redrawn. As a five-level system: the ground state S_0, the first excited singlet state S_1, and the three sublevels T_u of the metastable triplet state ($u = x,y,z$). In contrast to the triplet levels, the state S_1 is very short-lived (less than $10^{-\tau}$s) and on continuous uv excitation it will carry a negligible fraction of the total population; the same holds a fortiori for other, more highly excited (singlet and triplet) states that have been omitted from the figure. Thus, for the experiments presently to be discussed with a time resolution of the order of 10^{-5}s, our system behaves like a pseudo-four-level system, and the optical pumping process may be described in terms of a set of linear differential equations in which the following rate parameters appear ($u,v = x,y,z$):

1. The populating rate P_u of each individual component T_u on excitation "via" S_1,
2. The rates k_u for the decay $T_u \to S_0$,
3. The rates W_{uv} of spin reorientation via the process $T_u \to T_v$ (not drawn).

Whereas, in the absence of quantum yield experiments, the populating rates can only be determined up to a common unknown constant that depends on the condition of excitation, the k_u and W_{uv} represent absolute rates; the problem in its most general form thus contains eight unknown parameters.

Each of the decay rates k_u is the sum of a radiative and a radiationless part $k_u = k_u^r + k_u^d$, and the phosphorescnece intensity I for light of a given polarization $\vec{E}$ is given by

$$I^{\vec{E}}(t) = c^{\vec{E}} \sum_u k_u^{\tau,\vec{E}} N_u(t), \tag{10}$$

where the $c^{\vec{E}}$ are instrumental constants and the N_u denotes the population

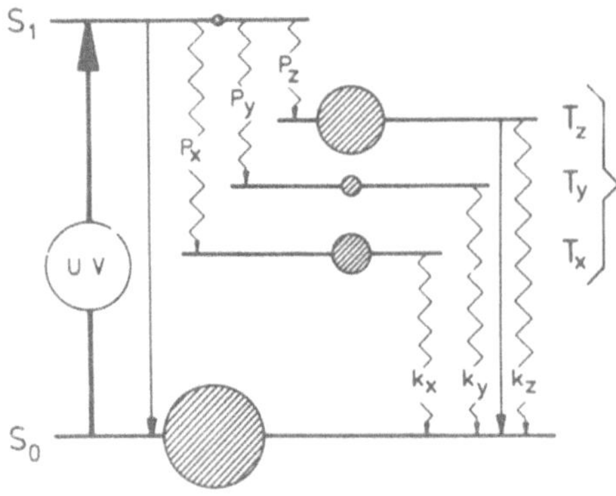

Figure 8. Schematic drawing of the optical pumping cycle of a metastable triplet state. P_x, P_y, P_z are the populating rates of the three sublevels and k_x, k_y, k_z their decay rates. Further it is assumed that the radiative decay is almost exclusively from the top component T_z, as for qunoxaline in Fig.2. The picture represents the situation at low temperature where the spin sublevels are "isolated" from each other because spin-lattice relaxation is slow compared to the decay, $W_{uv} << k_u$ (u,v ∊ x,y,z). Because of this isolation optical pumping will lead to spin alignment, with relative populations that depend on the condition of the experiment and are symbolized by the discs. For qunoxaline in durene flash excitation will yield relative populations equal to $P_x = 0.015$, $P_y = 0.02$, $P_z = 1$, whereas the steady-state populations $N_u^0 = P_u/k_u$ under continuous excitation are in the ratio $N_x^0 = 0.68$, $N_y^0 = 0.42$, $N_z^0 = 1$. Microwave contact between two of the levels will cause a shift in the populations which manifests itself through the intensity of the radiative decay.

of T_u at a particular time t. In the great majority of experiments to date it has been standard (though not always sound!) practice to observe an average over different polarizations and to drop the superscripts $\vec{E}$ in eq. (10). The success of the dynamic experiments derives from the circumstance that the k_u^r tend to be greatly different for the three sublevels and, consequently, when by the application of resonant microwaves, molecules are transferred from one triplet sublevel to another, appreciable changes of the phosphorescence intensity may occur. For quinoxaline, for instance, the spin-orbit coupling between the top component of the lowest $\pi\pi^*$ triplet state and the $n\pi^*$ state S_1 is by far the most important and, as a result, this component carries almost the entire activity for decay to the (vibrationless) ground state.

The great variety of experiments designed to unravel the dynamics of the optical pumping cycle all have in common that one analyzes the change in phosphorescence intensity that occurs in response to a "tickling" of the system of excited molecules by resonant microwaves. But, although the mathematics of the problem is well understood, the accuracy with which the rate parameters may be determined for a given system from its response to a number of imposed perturbations (switching the exciting source on or off, and microwave contacts) will depend greatly on the characteristics of the system. In particular one should realize that the response of the system is monitored via a single property only: the phosphorescence intensity. The design of the experiment, therefore, must be tailored to the properties of the system under examination and no universal "best" method can be prescribed. Further, it will be an exception if one succeeds in determining all eight independent rate parameters with reasonable accuracy without making some a priori assumptions.

In the microwave-induced delayed phosphorescence (MIDP) method [27,39], the first of its kind, one takes advantage of the circumstance that for many systems, by lowering the temperature to below a few Kelvin, one can reach a situation where relaxation practically has ceased. That is, one has "isolation" of the three spin levels whenever the condition $W_{uv} \ll k_x, k_y, k_z$ (for all u,v = x,y,z) is met. In this manner the size of the problem is reduced and it then proves feasible, in general, to measure the absolute decay rates k_u, the relative radiative decay rates k_u^r, and the relative populating rates P_u with adequate precision.

Let me illustrate the above general ideas with a recent experiment by Coremans [43] on the luminescence of a $Ba_2TiOSi_2O_7$ crystal. In this minderal, "fresnoite", the strong blue luminescence originates from a metastable triplet state centered on titanium bound in a square pyramidal conformation to five oxygens, four of which are shared with the pyrosilicate groups. The zero-field splitting of this triplet is given by $|D| \approx 70$ GHz, $|E| = 3.02$ GHz, and the MIDP experiments to be discussed were on the $|2E| = 6.04$ GHz transition between the T_x and T_y spin sublevels.

With a 10 ns laser-pulse at 283 nm the triplet sublevels are populated. After a certain delay, t_1, a 100 μs resonant microwave pulse (i.e. short compared to the millisecond lifetimes) is given with the sample placed in a helix, and the resulting change of the phosphorescence is recorded in a signal averager. After a number of such experiments the microwave power is turned off and the same number of shots of the decaying phosphorescence unperturbed by a microwave pulse is subtracted. One thus obtains an MIDP signal on a straight baseline as shown in the insert of Fig. 9. The height of the leading edge of the MIDP signal as a function of the delay t_1 is given by

$$h(t_1) = f[N_x(t_1) - N_y(t_1)]c[k_y^\tau - k_x^\tau]. \tag{11}$$

The first bracket in this equation represents the population difference between the two levels at the moment the microwave pulse is given and f is the fraction of this difference transferred by the pulse; the second bracket represents the difference in radiative decay rates and the simple constant of proportionality c holds under the assumption that polarization effects may be neglected. In order to reduce the error due to fluctuations in the laser power, the height of the MIDP signal actually measured in the experiment was scaled to the intensity of the laser shots by taking $h(t_1) = \Delta U(t_1)/I(0)$.

Because the laser flash is very short relative to the decay, eq. (11) may alternatively be written as

$$h(t_1) = c'\left[P_x e^{-k_x t_1} - P_y e^{-k_y t_1}\right], \tag{12}$$

where the proportionality factor $c' = fc[k_y^\tau - k_x^\tau]$. The decay rates of the triplet sublevels involved in the transition can be determined by fitting eq. (12) to a set of experimental data. The pre-exponential factors of the fit yield the relative initial populations of the two sublevels, which are directly proportional to the populating rates P_x and P_y. In Fig. 9 a fit of the height of a set of MIDP signals as a function of t_1 is shown. From such fits Coremans obtained at 1.2 K: $k_x = 310 \pm 30$ s^{-1}, $k_y = 25 \pm 10 s^{-1}$, $P_x/P_y = 0.75$, and $k_x^\tau/k_y^\tau \approx 7.6$. The ratio of the radiative rates given follows from the analysis of the shape of the

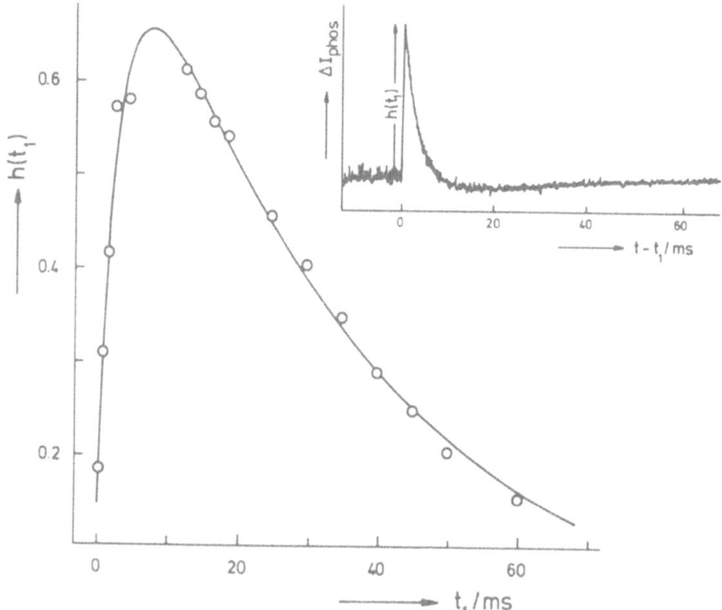

Figure 9. Microwave-Induced-Delayed-Phosphorescence (MIDP) experiments on the luminescent spin triplet in $Ba_2TiOSi_2O_7$ recorded by Coremans at 1.2 K [43]. The curve represents the height $h(t_1)$ of the MIDP signals at 6.1 GHz as a function of the delay t_1. An individual MIDP signal taken with t_1 = 13 ms is shown in the insert.

individual MIDP transients shown in the insert of Fig. 9. The change in phosphorescence intensity caused by the microwave pulse with time is given by an equation similar to (11)

$$\Delta I(t - t_1) = h(t_1) \left[k_y^{\tau} e^{-k_y(t-t_1)} - k_x^{\tau} e^{-k_x(t-t_1)} \right], \quad (t \geq t_1). \qquad (13)$$

A fit of change of ΔI with time thus yields the relative radiative rate of the two levels connected by the microwave transition.

A frequently used alternative to MIDP is the microwave-saturation-recovery technique [40]. In a sample exposed to continuous uv excitation, a steady-state distribution over the sublevels is established, which is then suddenly perturbed by a resonant microwave pulse. Instead of following the decay of the luminescence after this perturbation as in MIDP, one here studies the return to the steady state, which when isolation prevails is again described by a bi-exponential.

Before leaving the subject of EPR in zero field, I may mention that luminescent triplets in the absence of an external magnetic field have proven an elegant playground for a variety of experiments in which the coherent excitation of spin transitions in a triplet is followed optically. A review of earlier work on organic molecules is given in [44]; more recently extensive work has been done by Glasbeek et.al. on spin triplet and quartet defects in CaO [45] and on the triplet ground state of the N-V centre in Diamond [46].

IV. ELECTRON-SPIN-ECHO EXPERIMENTS FOR THE IDENTIFICATION OF NON-RADIATIVE EXCITED STATES

IV.A. Electron-Spin-Echo Detected EPR Spectra

As we saw, optical detection opened the way for the study of short-lived, luminescent excited states. But how to tackle nonradiative systems? Amongst organic molecules, in particular, there were many intriguing systems in which one suspected the presence of low-lying metastable triplet states which had thus far escaped identification by either optical spectroscopy or EPR because of their nonluminescent nature attributed to dominant radiationless decay. In the past decade a break-through has occurred in this field, made possible by the availability of pulsed lasers for the (near-) instantaneous generation of large numbers of excited species and the development of electron-spin-echo (ESE) equipment into practical instruments for the detection of paramagnetic species on a microsecond time scale. Unfortunately, the intricacies of the coherent phonomena that underlie the spin-echo techniques fall outside the scope of this introduction. I shall refrain from giving a watered-down explanation and just state the sequence of events in the experiment. A good introduction can be found in the report of a recent workshop on pulsed EPR [47]; those who are interested in the fundamentals should consult the contribution by Mims -- who for many years was a lonely pioneer in this field -- to Geschwind's Electron Paramagnetic Resonance [5].

In Fig. 10b I have sketched one of the two most common pulse sequences used in laser-flash-ESE experiments on metastable triplet states. With the laser flash one first generates a substantial number of excited species in, say, 10 ns. Important is further that, owing to the sublevel selectivity of the intersystem-crossing process, one obtains the ensemble of spins in a high degree of polarization. Hence, if the EPR experiment can be carried out in a time short relative to the time T_1 governing the thermalization of the spin distribution over the sublevels, the sensitivity of the experiment may be greatly enhanced relative to that attainable in a cw experiment. This, now, is realized in the ESE detection of EPR. With two microwave pulses of the proper lengths and resonant with one of the Zeeman transitions, an "echo" is generated. If we skip certain complications that may arise from hyperfine interaction, the intensity of this echo is a measure for the EPR signal that would be observed at the given microwave frequency and magnetic field strength in a conventional EPR experiment if this were feasible.

An ESE-detected EPR spectrum is generated by recording the echo height whilst the magnetic field strength is slowly swept through resonance, keeping the time intervals t_d and τ constant. An example is shown in Fig. 10a for CrO_4^{2-} ions incorporated in a $CaSO_4$ (anhydrite) host [48]. These signals settled the controversial existence of the metastable triplet states of the CrO_4^{2-} ion. Just as in a conventional EPR experiment, the fine-structure and Zeeman tensors, $\overset{\rightarrow}{\vec{D}}$ and $\overset{\rightarrow}{\vec{g}}$, in the spin Hamiltonian (1) may be determined from a study of the resonance fields as a function of the orientation of the crystal in the field. In this particular instance, the realization of such an orientational study met with a severe obstacle; for most orientations the echoes disappeared for the time interval τ = 600 ns used in the experiment of Fig. 10. The reason lies in the intrinsic limitations on the length of the interval τ for the ESE experiment to be successful. On the one hand, τ should be shorter than, or at most comparable with, the dephasing time (or "spin memory time") T_2 of the transition studied. On the other hand, the echo cannot be detected before the microwave field set up in the cavity by the two microwave pulses has decayed sufficiently. This instrument-limited "dead time" is of the order of 30 Q/ν, where Q is the quality factor of the cavity and ν the microwave frequency; in the experiments on CrO_4^{2-} in $CaSO_4$, for which Q = 75, ν = 9 GHz, whence 30 Q/ν = 250 ns, distorted echoes could still be discerned on a background resulting from cavity-ringing for τ = 60 ns. In addition, the delay t_d cannot be taken

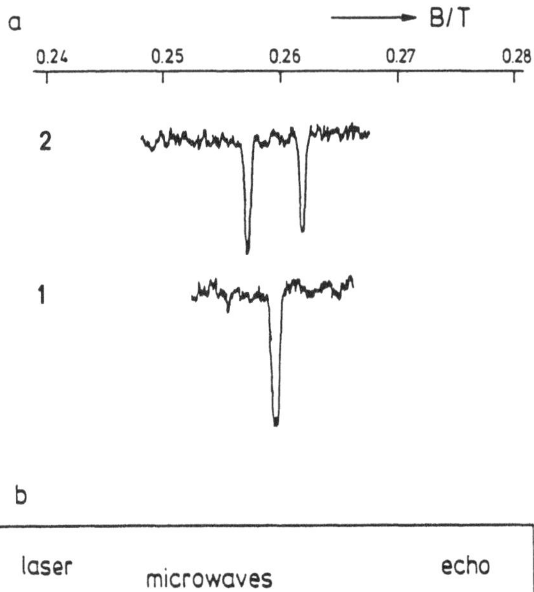

Figure 10. (a) ESE-detected EPR transitions for the metastable triplet state of $CrO_4^{=}$ in a $CaSO_4$ host at 1.2 K [48]. Trace 1 has been recorded with $\vec{B}$ parallel to the crystal $\vec{c}$ axis; all four Jahn-Teller distorted ions then are magnetically equivalent. Trace 2 has been recorded with $\vec{B}$ rotated 0.5° away from $\vec{c}$ towards $\vec{a}$; the signal has split because the distorted ions now are only pair-wise equivalent. For a general orientation of $\vec{B}$ four signals appear. (b) Echo sequence used in the laser-flash ESE experiment. The traces of Fig. 10a were recorded with t_d = 50 ns and τ = 600 ns.

much longer than the spin-lattice relaxation time T_1, and the time needed for the entire experiment, $t_d + 2\tau$, may not substantially exceed the lifetime of the longer living level involved in the transition.

In general the latter two conditions can easily be met, but the first requirement, $\tau \le T_2$, may be difficult or even impossible to meet. For localized triplet states in solids T_2 is usually determined by fluctuations in the local magnetic field caused by "flips" of neighbouring magnetic nuclei. For triplet states of organic molecules carrying protons T_2 tends to be of the order of a few μs [49]; on the replacement of the protons by deuterium with its much weaker magnetic moment, T_2 is lengthened considerably. If, however, one of the sublevels has a sub-microsecond lifetime, a fundamental limitation is encountered because for a transition between two decaying states i and j there is an upper limit to the spin-memory time

$$T_2 \le \left(\frac{1}{\tau_i} + \frac{1}{\tau_j}\right)^{-1} , \qquad (14)$$

where τ_i and τ_j are the lifetimes of the states concerned. Hence, τ can never be chosen much longer than twice the lifetime of a short-lived level involved in the transition. For the metastable triplet state of $CrO_4^=$ in $CaSO_4$ Coremans found sublevel lifetimes $\tau_x = 2.2$, $\tau_y = 0.73$, $\tau_z = 0.023$ υs. Because of the unexpected shortness of τ_z, the interval τ between the two microwave pulses had to be reduced to less than 100 ns for some orientations of the crystal in the ESE experiment. By contrast, echoes could easily be generated in the triplet state of the $Cr_2O_7^=$ ion in a $K_2Cr_2O_7$ crystal where the sublevel lifetimes are of the order of ms and few magnetic nuclei are present.

IV.B. Determination of Kinetics of Populating and Decay

The ESE technique has not only proved its worth for identifying short-lived paramagnetic excited states but, by varying the length of the delay t_d between the laser flash and first mcrowave pulse, it also allows one to determine the kinetics of the excited state generation and decay in the following, elegant manner. The echo height, for a given value of τ, is proportional to the population difference $(N_i - N_j)$ between the two levels i and j connected by the microwave pulses. Thus, on stepwise increasing the delay time t_d between the laser flash and the first microwave pulse, the variation of the echo intensity with t_d should be given by

$$I_{echo}(t_d) \propto N_i(t_d) - N_j(t_d) = p_i e^{-k_i t_d} - p_j e^{-k_j t_d} (\tau = \text{const}). \quad (15)$$

From eq. (15) we again expect a bi-exponential response with exponents equal to the two lifetimes and pre-exponential factors that yield the relative populating rates. A typical example of such a response curve taken on the $m_s \approx 1 \leftrightarrow m_s = 0$ Zeeman transition of the triplet state of CrO_4^{2-} with $\vec{B} \parallel \vec{y}$ is shown in Fig. 11. The fast ingrowth at a rate $k_1 \sim 10.6\ 10^6 s^{-1}$ is barely visible on the displayed timescale; it corresponds to an increase in the population difference due to rapid decay of the upper Zeeman level in which the T_z zero-field component is involved. The subsequent, much slower decrease with t_d reflects the decay of the central Zeeman level at a rate $k_0 = \tau_y^{-1} = 1.37\ 10^6 s^{-1}$.

On the other hand, one can determine the relaxation time T_2 by studying the echo height as a function of the interval τ while keeping the delay t_d constant. For, in the absence of hyperfine coupling, this height will simply decrease as exp $(-2\tau/T_2)$.

IV.C. Electron-Spin-Echo Envelope Modulation

The proviso in the last paragraph about the absence of hyperfine coupling touches on an important discovery made by Mims in the early days of ESE spectroscopy: the nuclear modulation effect [5]. Because of this effect, an oscillatory behaviour is superimposed on the exponential echo decay whenever there are magnetic nuclei which "feel" the local field of the electron spins. This ESE-envelope-modulation (ESE-EM) effect arises because the reorientation of the electron spins during the two brief microwave pulses may be considered a sudden event on the time scale of the nuclear precession. As a result nuclear Zeeman states are coherently excited and beats corresponding to the ENDOR frequencies appear.

In recent years ESE-EM spectroscopy has found wide-spread application in the determination of the structure of excited states [49]. For weakly-coupled nuclei the technique, in general, affords a much higher resolution and greater sensitivity than can be attained with ENDOR. Let me finish by giving an illustration taken from the work by Singel et. al. on the metastable triplet state of free base porphine [50]. The experiments were carried out at 1.4 K on an oriented n-octane single crystal doped with porphine, the peripheral hydrogens of which had been replaced by deuterium in order to increase the spin-memory time T_2. Fig. 12a shows the echo height as a function of the interval τ for a particular orientation of the crystal in the magnetic field, and Fig. 12b represents a Fourier-transform of this spectrum, after the latter had been pre-processed to account for the overall exponential echo decay and limited width of the time window (for all details see [50] and references

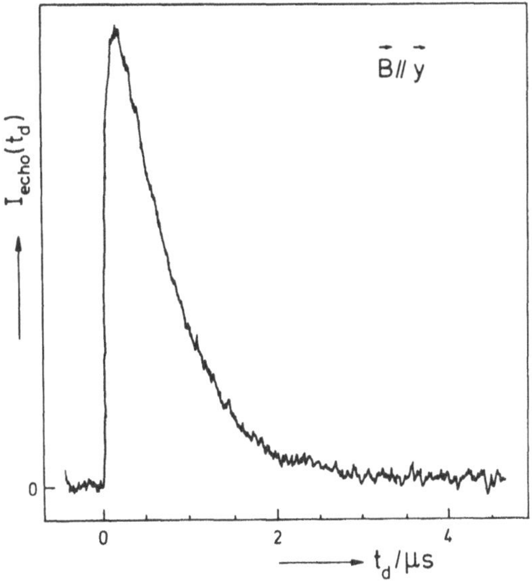

Figure 11. Electron-Spin-Echo intensity as a function of the delay t_d of the first microwave pulse with respect to the laser-flash recorded by Coremans [48] for the metastable triplet state of $CrO_4^{=}$ in $CaSO_4$ with $\vec{B} \parallel \vec{y}$ and τ = 250 ns.

therein). In the Fourier spectrum the frequencies labelled A and B could be assigned to the two distinct pairs of nitrogens in the molecule, those without and with adjacent protons, respectively. The additional peaks marked D arise from coupling to deuterium. (The two protons on the nitrogens are so strongly coupled to the electron spin that their nuclear Zeeman states are not coherently excited by the microwave pulses; their hyperfine coupling was determined in pulsed ENDOR experiments). By

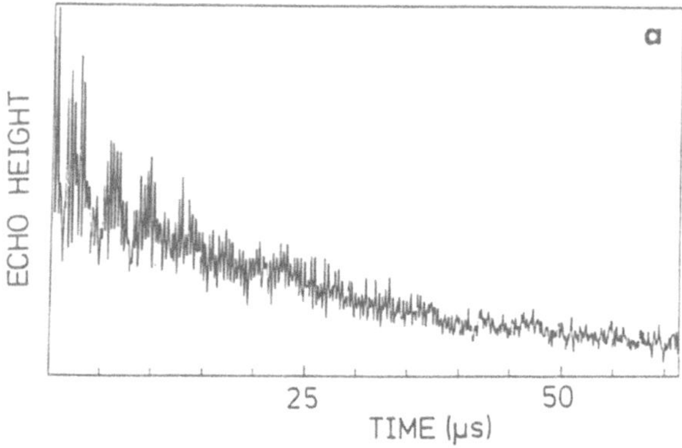

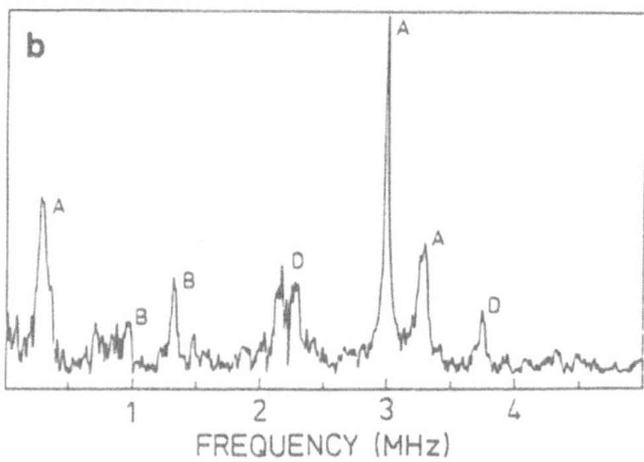

Figure 12. Electron-Spin-Echo-Envelope Modulation experiment on free-base porphine in an n-octane crystal by Singel et.al. [49].
(a) Modulation pattern in the envelope of two-pulse echoes recorded with ν=9.2 GHz and $\vec{B} \parallel 0.997\vec{X} - 0.011\vec{Y} + 0.082\vec{Z}$.
(b) Discrete Fourier modulus transform obtained from the modulation pattern in trace (A).

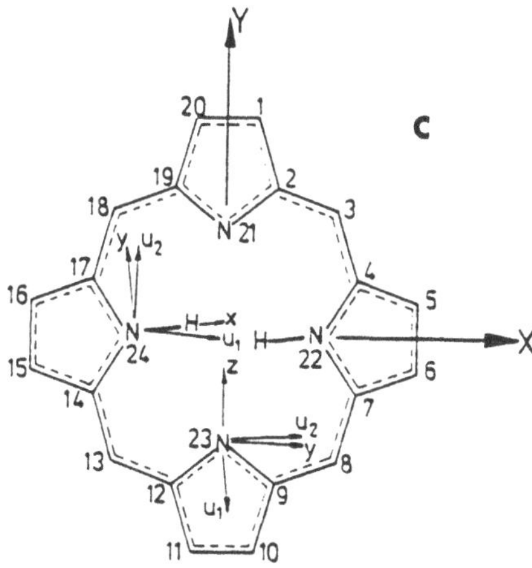

Figure 12 (continued). (c) Structure of free-base porphine with in-plane fine structure ($\vec{X}$ and $\vec{Y}$), ^{14}N hyperfine-structure ($\vec{u}_1,\vec{u}_2$) and quadrupole tensor ($\vec{y}$ and $\vec{x}$ or $\vec{z}$) principal axes. The axes $\vec{u}_1$ and $\vec{Z}$ for the nitrogens 21 and 23 are tilted slightly by 0.028 about $\vec{X}$ out of the idealized molecular plane. For all details see [49].

studying the modulation frequencies as a function of the orientation of the crystal in the field, the ^{14}N hyperfine and quadrupolar coupling tensors could be determined. The directions of their principal axes -- which provide information about the spatial electronic structure -- are sketched in Fig. 12c.

More recently, an intriguing problem was solved by ESE-EM: the identification and structural determination of the elusive triplet states of pyridine (azabenzene) by Buma et.al. [51]. The success of this work holds great promise for the further application of pulsed spin-echo techniques in the study of excited states.

ACKNOWLEDGEMENT

As illustrated by the references, a large part of this review represents a synopsis of ideas developed over a period of thirty years in collaboration with many colleagues. I thank them all. The following figures have been reproduced from previous publications: Figs. 2, 3 and 5 from Acta Physica Polonica A71, 809 (1987), Figs. 6 and 7 from the Ph.D. Thesis by J. van Tol [32], Fig. 8 from "Time Domain Electron Spin Resonance", Ed. L. Kevan and R.N. Schwartz, Wiley Interscience, New York,

1979 [44], Figs. 9 and 11 from the Ph.D. Thesis by C.J.M. Coremans [43], and Figs. 10 and 12 from the Journal of Chemical Physics [48,50].

REFERENCES

[1] J. Brossel and A. Kastler, C.R. Acad. Sci, Paris 229, 1213 (1949); J. Brossel and F. Bitter, Phys. Rev. 86, 308 (1952).

[2] C. Cohen-Tannoudji, Annales de Physique 7, 423, 469 (1962).

[3] C.A. Hutchison and V.W. Mangum J. Chem. Phys. 29, 952 (1958).

[4] S. Geschwind, R.J. Collins and A.L. Schawlow, Phys. Rev. Lett. 3, 544 (1959).

[5] Electron Paramagnetic Resonance, ed. S. Geschwind, Plenum, New York 1972.

[6] A. Jablonski, Nature 131, 839 (1933).

[7] A. Jablonski, Z. Phys. 94, 38 (1935).

[8] G.N. Lewis, D. Lipkin and T.T. Magel, J. Am. Chem. Soc. 63, 3005 (1941).

[9] G.N. Lewis and M. Kasha, J. Am. Chem. Soc. 66, 2100 (1944).

[10] G.N. Lewis and M. Kasha, J. Am. Chem. Soc. 67, 994 (1945).

[11] A. Terenin, Acta Physiochim. SSSR, 18, 210 (1943).

[12] C.A. Hutchison and B.W. Mangum, J. Chem. Phys. 34, 908 (1961).

[13] J.H. van der Waals and M.S. de Groot, Mol. Phys. 2, 333 (1959).

[14] M.S. de Groot and J.H. van der Waals, Mol. Phys. 3, 190 (1960).

[15] W.A. Yager, E. Wasserman and R.M.R. Cramer, J. Chem. Phys. 37, 1148 (1962).

[16] E. Wasserman, L.C. Snijder and W.A. Yager, J. Chem. Phys. 41, 1763 (1964).

[17] S.I. Weissman, J. Chem. Phys. 29, 1189 (1958).

[18] H.M.L. Pryce, Proc. Phys. Soc. A163, 25 (1950).

[19] A. Abragam and H.M.L. Pryce, Proc. Roy. Soc. A205, 135 (1951).

[20] A. Abragam and B. Bleaney, "Electron Paramagnetic Resonance of Transition", Clarendon Press, Oxford (1970).

[21] J.H. van der Waals and M.S. de Groot, in the Triplet State, Ed. A.B. Zahlan, Cambridge University Press (1967).

[22] R.W. Brandon, R.E. Gerkin and C.A. Hutchison, J. Chem. Phys. 41, 3717 (1964).

[23] M.S. de Groot and J.H. van der Waals, Mol. Phys. 6, 545 (1963); M.S. de Groot, I.A.M. Hesselmann and J.H. van der Waals, Mol. Phys. 16, 45 (1969).

[24] M. Sharnoff, J. Chem. Phys. 46, 3263 (1967).

[25] A.L. Kwiram, Chem. Phys. 1, 272 (1967).

[26] J. Schmidt, I.A.M. Hesselmann, M.S. de Groot and J.H. van der Waals, Chem. Phys. Lett. 1, 434 (1967).

[27] J. Schmidt, D.A. Antheunis and J.H. van der Waals, Mol. Phys. 22, 1 (1971).

[28] M. Kinoshita, N. Iwasaki and N. Nishi, Appl. Spectrosc. Rev. 17, 1 (1981).

[29] K.E. Gotberg and D.S. Tinti, Mol. Phys. 47, 97 (1982).

[30] W.A.J.A. van der Poel, J. Herbich and J.H. van der Waals, Chem. Phys. Lett. 103, 253 (1984).

[31] W. Barendswaard and J.H. van der Waals, Mol. PHys. 59, 337 (1986); W. Barendswaard, R.T. Weber and J.H. van der Waals, J. Chem. Phys. 87, 3731 (1987).

[32] J. van Tol, J.A. van Hulst and J.H. van der Waals, Mol. Phys. (1991), to be published; J. van Tol, Ph.D. Thesis, University of Leiden, 1991.

[33] D.S. McClure, J. Chem. Phys. 17, 665 (1949); 20, 682 (1952).

[34] W.S. Veeman and J.H. van der Waals, Mol. Phys. 18, 63 (1970).

[35] W.A.J.A. van der Poel, A.M. Nuys and J.H. van der Waals, J. Phys. Chem. 90, 1537 (1986).

[36] J. Schmidt and J.H. van der Waals, Chem. Phys. Lett. 2, 640 (1968); 3, 546 (1969).

[37] I.Y. Chan in "Triplet State ODMR Spectroscopy", ed. R.H. Clarke, John Wiley & Sons, 1982, p.18.

[38] H.W. van Kesteren, W.Th. Wenckebach and J.A.J.M. Disselhorst, J. Phys. E. 20, 648 (1987).

[39] J. Schmidt, W.S. Veeman and J.H. van der Waals, Chem. Phys. Lett. 4, 341 (1969); D.A. Antheunus, J. Schmidt and J.H. van der Waals, Chem Phys. Lett. 6, 255 (1970).

[40] C.J. Winscom and A.H. Maki, Chem. Phys. Lett. 12, 264 (1971).

[41] D.S. Tinti, M.A. El-Sayed, A.H. Maki and C.B. Harris, Chem. Phys. Lett. 3, 343 (1969).

[42] R.A. Schadee, J. Schmidt and J.H. van der Waals, Chem. Phys. Lett. 41, 435 (1976).

[43] C.J.M. Coremans, Ph.D. Theiss, Leiden 1989; C.J.M. Coremans and J.H. van der Waals, J. Appl. Magn. Res., to be published 1991.

[44] J. Schmidt and J.H. van der Waals in "Time Domain Electron Spin Resonance", Ed. L. Kevan and R.N. Schwarts, Wiley-Interscience, New York, 1979.

[45] R. Vreeker, M. Casalboni and M. Glasbeek, Chem. Phys. Lett. 115, 69 (1985); P.A. van Leeuwen, R. Vreeker and M. Glasbeek, Phys. Rev. B 34, 3483 (1986).

[46] E. van Oort, N.B. Manson and M. Blasbeek, J. PHys. C.: Solid State Phys. 21, 4385 (1988).

[47] Pulsed EPR: A new field of application, ed. C.P. Keyzers, E.J. Reyerse and J. Schmidt, North Holland, Amsterdam 1989.

[48] C.J.M. Coremans, E.J.J. Groenen, J.H. van der Waals, J. Chem. Phys. 93, 3101 (1990).

[49] J. Schmidt and D.J. Singel, Ann. Rev. Phys. Chem. 38, 141 (1987).

[50] D.J. Singel, W.A.J.A. van der Poel, J. Schmidt, J.H. van der Waals and R. de Beer, J. Chem. Phys. 81, 5453 (1984).

[51] W.J. Buma, E.J.J. Gorenen, J. Schmidt and R. de Beer, J. Chem Phys. 91, 6549 (1989).

THE JAHN-TELLER EFFECT IN THE OPTICAL SPECTRA OF IMPURITIES

Gabriele Viliani

Dipartimento di Fisica
Università di Trento
38050 Povo, Trento, Italy

ABSTRACT

The general symmetry properties which lead to the Jahn-Teller instability of electronically degenerate states are presented and discussed. We deal first with the static problem of finding the new stable configurations of the molecule or complex, and subsequently describe the more general situation where dynamical effects are important. Some selected examples are presented.

I. INTRODUCTION

The study of the magnetic and optical properties of impurities in crystals can be carried out, as a first approximation, by considering the host lattice as fixed. However, sometimes the effects of lattice vibrations cannot be neglected. One such case is when one of the electronic levels involved in the optical transition (or the level which determines the magnetic properties of the ion) is degenerate. In this case there is a peculiar coupling between electronic states and lattice vibrations which is called Jahn-Teller (JT) effect. Jahn and Teller (1937) and Jahn (1938) showed that a complex (i.e. impurity ion plus nearest neighbours) or molecule which are in a (orbitally- or spin-) degenerate electronic level and possess a given spatial symmetry, are unstable against distortions which tend to lower the symmetry and to remove the electronic degeneracy. The effect of spin degeneracy alone is quite small (Jahn 1938), and can be neglected in most cases.

There are two exceptions to the validity of the JT theorem: (*i*): Kramers-Wigner degeneracy: the electronic levels of any system with an odd number of electrons are at least doubly degenerate. This is due to the time invariance of the Schroedinger equation and cannot be lifted by any electric field; (*ii*): The degeneracy of linear molecules produced by their rotational symmetry about the molecular axis.

Optical Properties of Excited States in Solids
Edited by B. Di Bartolo, Plenum Press, New York, 1992

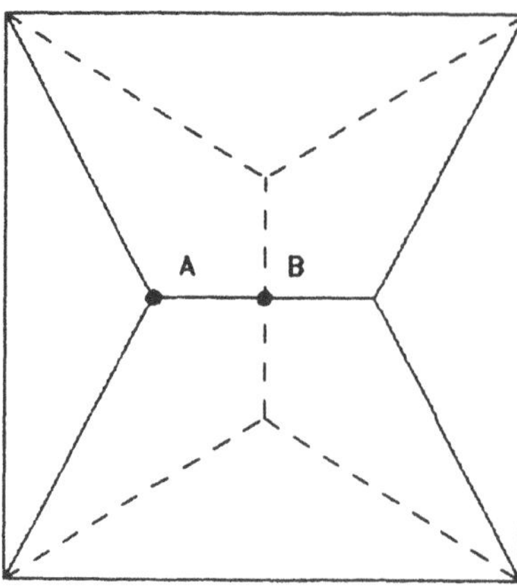

Figure 1. Paths of minimum length connecting the four corners of a square.

This interplay between symmetry and degeneracy is not peculiar to quantum mechanical systems, and the following simple geometrical example may help to clarify most of the features of the JTE. The problem (fig. 1) is to find the shortest path connecting the four corner points of a square of side $2d$. On grounds of symmetry one would think that such path consists of the two diagonals of the square, but a simple calculation shows that the actual solution is the path depicted in the figure, with $\bar{AB} = d(3 - \sqrt{3})/3$. Of course, due to symmetry, the dashed-line path is also a solution.

Thus the square symmetry of the problem produces a degenerate "ground state" (i.e. two paths of the same minimum length), since if we had a rectangle the two paths would clearly have different lengths. On the other hand, the two degenerate solutions, if taken separately, are *less symmetric* than the original problem: each path is invariant under rotations of 180 degrees, while the original problem is invariant under rotations of 90 degrees. This situation is often referred to as a "spontaneously broken symmetry". Note however that the ensemble of the two paths (a "linear combination of degenerate states" in quantum mechanical parlance) recovers the initial symmetry. Much the same qualitative behaviour occurs in molecules (and in any other physical problem): symmetry produces degeneracy, degeneracy tends to destroy symmetry in a sort of perverse competition whose observable result depends on the relative strengths of the symmetry-breaking and symmetry-restoring terms of the hamiltonian. In any case, the ensemble of the degenerate, less symmetric solutions preserves the original symmetry.

In the following we will be forced to use some group-theoretical jargon because the use of the results of group theory enormously simplifies this kind of problems from the mathematical point of view. In order to make the next sections readable to those who are not familiar with this slang, we shall try a brief summary.

The symmetry operations (rotations, reflections, inversion, etc) of a molecule are the geometrical transformations which turn the molecule itself into an identical one;

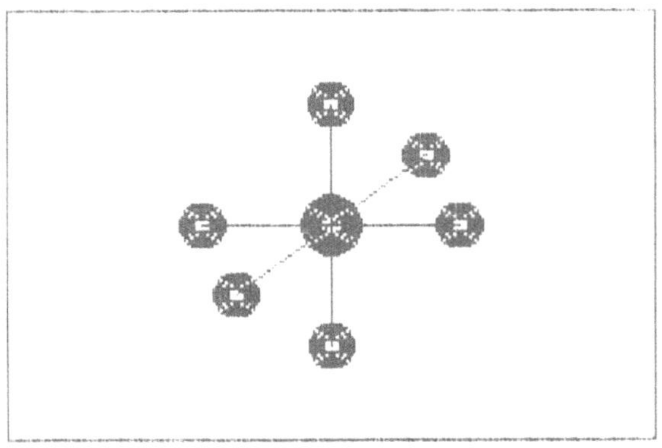

Figure 2. XY_6 octahedral molecule.

in the square that we considered above, seen as a hypothetical square molecule with identical atoms, one such operation would be a rotation of 90 degrees around an axis perpendicular to the molecular plane and passing through the centre.

These operations can be shown to form a group in the mathematical sense; there are 32 possible different combinations of symmetry operations, i.e. any molecule in nature belongs to one of these 32 *point groups.* Since it is very inconvenient to work with geometrical transformations, we must find some mathematical quantities to *represent* them, and matrices turn out to do the job very well. The following example will clarify how this works: consider again the square molecule, and label the atoms 1 to 4; attach an x_i-y_i (i=1-4) reference system to each atom. A clockwise rotation of 90 degrees will produce the following changes: $x_1 \rightarrow -y_2$, $y_1 \rightarrow x_2$, and so on. If you characterise the initial configuration by the vector $\bar{v}_i = |x_1, y_1, x_2; y_2, x_3, y_3, x_4, y_4|$ and the final one by the vector $\bar{v}_f = |-y_2, x_2, -y_3, x_3, -y_4, x_4, -y_1, x_1|$, you see that the 8x8 matrix:

$$M = \begin{vmatrix} 0 & 0 & 0 & -1 & 0 & 0 & 0 & 0 \\ 0 & 0 & 1 & 0 & 0 & 0 & 0 & 0 \\ 0 & 0 & 0 & 0 & 0 & -1 & 0 & 0 \\ 0 & 0 & 0 & 0 & 1 & 0 & 0 & 0 \\ 0 & 0 & 0 & 0 & 0 & 0 & 0 & -1 \\ 0 & 0 & 0 & 0 & 0 & 0 & 1 & 0 \\ 0 & -1 & 0 & 0 & 0 & 0 & 0 & 0 \\ 1 & 0 & 0 & 0 & 0 & 0 & 0 & 0 \end{vmatrix}$$

satisfies $M \cdot \bar{v}_i = \bar{v}_f$; in other words M *represents* the geometrical operation. The set of matrices which represent all the operations of a group is called a *representation* of the group. Of course you can build as many representations as you want, just by changing the initial vector on which the M's operate (the *basis* of the representation). Fortunately, it turns out that only a limited number of representations are significant: these are, in a sense, the representations whose matrices have the smallest possible dimension but still contain all the information. In the example above, to use 8x8 matrices is just pure masochism since the basis that we have chosen (i.e. the 8 components of the vector $\bar{v}_i$) is redundant.

The representations which matter are called *irreducible representations* (IR); it is found that for the groups with a finite number of operations (the ones of interest for us here) the IR can have a maximum dimension of 3 if we neglect spin-orbit interaction (which will suffice for the moment), this value however cannot be realized for all groups but only for the most symmetric ones. The group we shall be concerned with is called O_h, and is the group which pertains, among others, to octahedral molecules of the XY_6 type (see fig. 2); this group is entitled to all possible dimensions of IR: 1 (called A), 2 (called E) and 3 (called T); these letters may be affected by subscripts (*1*,*2*,g,u) which do not interest us for the moment.

The fundamental property of IR is that their dimensions determine the *possible degeneracies* of any physical system with that symmetry. So, for example, in O_h symmetry there may be singlets, doublets and triplets; in lower symmetries triplets (and in some cases doublets) are not allowed. In no case are quartets allowed (neglecting spin-orbit interaction). Moreover, the link among symmetry, degeneracy and wavefunctions is so strong that it turns out that the wavefunctions belonging to a degenerate level form a good basis for the corresponding IR since they have just the right transformation properties under the symmetry operations of the group. Therefore, it is customary to label the levels with the letters of the corresponding IR (for example, we will talk about "the T triplet") and the degenerate wavefunctions with the same name as the basis of the IR (for example, the three states of T will be labelled $|T_x>$, $|T_y>$, $|T_z>$, where the use of the subscripts x, y, z has a meaning that will become clear later).

Consider now the vibrational degrees of freedom of our molecule. There are $3N-6$ of them which correspond to $3N-6$ vibrational frequencies, though some frequencies may appear more than once (vibrational degeneracy). This vibrational degeneracy of course must obey the same rules as electronic degeneracy as regards its relationship with the dimensions of IR. We could chose, for example, the cartesian components of the instantaneous displacement of each atom from its equilibrium position; it turns out, however, that it is far more convenient to use suitable linear combinations of these displacements which are called normal coordinates and which transform as bases for the IR of the group. Some of these normal coordinates are shown for the O_h group in fig. 3 (these are the only ones which matter for the JTE). The so-called totally symmetric coordinate Q_1 does not change the shape of the molecule and it has no relevance in the JTE; unless stated otherwise, in the following sections by "normal coordinates" we will refer *only to the non symmetric ones.* The reason for using normal coordinates is that by group theoretical methods it is simple to evaluate the matrix elements that we will need.

II. THE STATIC JAHN-TELLER EFFECT

II.A. Case of No Spin-Orbit Interaction

Let us label by Q the whole set of nuclear normal *non symmetric* coordinates and by r the electronic ones. The total hamiltonian of the coupled system of electrons and nuclear displacements reads:

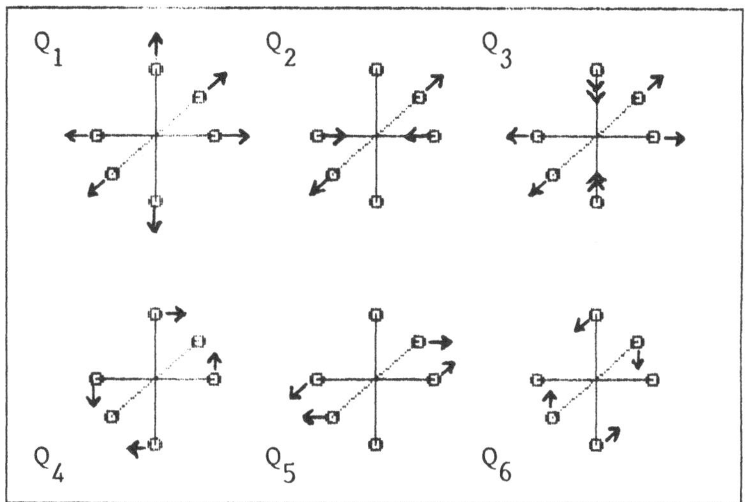

Figure 3. Normal coordinates of the XY_6 molecule; the "breathing mode" Q_1 is not JT active; the E_g modes (Q_2 and Q_3) couple to 2- and 3-fold degenerate electronic states, while the T_{2g} modes (Q_4, Q_5, Q_6) couple only to 3-fold degenerate states.

$$H(r,Q) = H_e(r) + T(Q) + W(r,Q) \tag{1}$$

where H_e is the purely electronic hamiltonian, $T(Q)$ is the nuclear kinetic energy and $W(r,Q)$ is the electron-displacement interaction.

In the spirit of the Born-Oppenheimer approximation we now neglect $T(Q)$ and study the solution of the Schroedinger equation as if the nuclei were fixed; the eigenvalues and eigenvectors will depend on Q since H does. We assume that eigenvalues and eigenvectors of the electronic problem are known for $Q = 0$ (i.e. we assume that we know the solution of the static crystal field problem) and treat $W(r,Q)$ as a perturbation. First let us expand $W(r,Q)$ in power series around $Q = 0$:

$$W(r,Q) = W(r,0) + \sum_i \left(\frac{\partial W}{\partial Q_i}\right)_o Q_i + \sum_{ij} \left(\frac{\partial^2 W}{\partial Q_i \partial Q_j}\right)_o Q_i Q_j + ... \tag{2}$$

Let us now consider an electronic level which is degenerate at $Q = 0$ and let $|\psi_k>$ be the corresponding eigenvectors. For the perturbative calculation we will need all the matrix elements of the type

$$<\psi_k|W(r,Q_i)|\psi_{k'}> = <\psi_k|(\partial W/\partial Q_i)_0|\psi_{k'}> Q_i \tag{3}$$

the equality following from the fact that in first approximation the Q's do not operate on $|\psi_k>$.

At this point it is clear that if any matrix element as in (3) is different from zero the symmetric ($Q = 0$) configuration does not correspond to a minimum of the potential

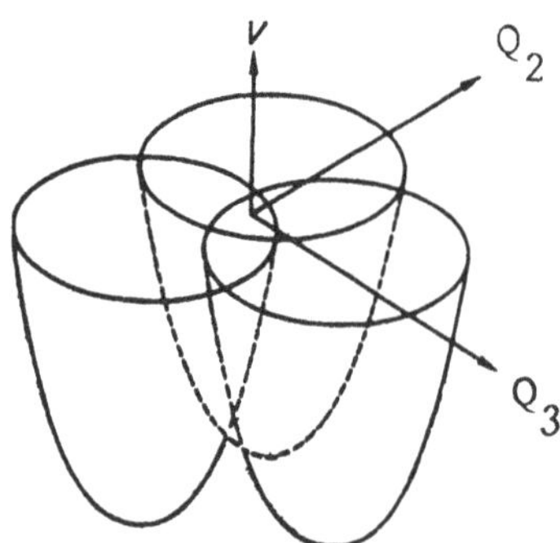

Figure 4. Shape of the potential surface for a T electronic state couplet to E_g modes alone (reprinted from Englman, 1972).

energy because whatever the numerical values of the electronic matrix elements the linear terms will dominate for sufficiently low Q. Jahn and Teller (1937) showed that for degenerate electronic states some finite matrix elements as in (3) *always* exist irrespective of the molecular symmetry. This is the Jahn-Teller theorem.

As for the quadratic terms, let us for the moment consider only those with $i = j$ (we shall see the effect of the others later); these are nothing but the restoring energy in the harmonic approximation and turn out to have only diagonal matrix elements in the electronic basis $|\psi_k>$, whatever the degeneracy of the level; moreover, thanks to the use of normal coordinates and to their symmetry properties the elastic constants relative to Q_i's belonging to the same IR are equal.

We now specialise to the particular case of a threefold degenerate level, T, in O_h symmetry which will be of special interest for us; let us label simply by $|x>$, $|y>$, $|z>$ the three degenerate electronic states at $Q = 0$. As mentioned above, the elastic energy (which we label $V(Q)$) will be diagonal:

$$V(Q) = \left[\frac{1}{2}k_E(Q_2^2 + Q_3^2) + \frac{1}{2}k_T(Q_4^2 + Q_5^2 + Q_6^2)\right] I \tag{4}$$

where the subscripts E and T refer to the IR's spanned by the normal coordinates (see fig. 3), and I is the 3-dimensional identity matrix.

The matrix elements in eq. 3 can be calculated by group theoretical considerations and the secular matrix to be diagonalised (which is often termed the Jahn-Teller hamiltonian) turns out to be:

$$H_{JT} = \begin{vmatrix} b(Q_2 - Q_3/\sqrt{3}) & cQ_6 & cQ_5 \\ cQ_6 & -b(Q_2 - Q_3/\sqrt{3}) & cQ_4 \\ cQ_5 & cQ_4 & 2bQ_3/\sqrt{3} \end{vmatrix} \tag{5}$$

where $b =< x|\partial W/\partial Q_2|x >$ and $c =< x|\partial W/\partial Q_6|y >$.

Eq. 5 has three real solutions $E_1'(Q), E_2'(Q), E_3'(Q)$ which depend on Q; the total potential energy in the Q space will therefore be given by the three hypersurfaces

$$E_i(Q) = E_i'(Q) + V(Q) \tag{6}$$

The secular equation (5) is of third order and can be solved analytically, but it is simpler (and still highly informative) to determine just the positions of the stationary points in the Q space and the conditions for these to be minima, thereby finding the stable distorted configurations. Moreover, the electronic wavefunction in a generic configuration Q will be of the form:

$$|\psi(Q)>= a_1(Q)|x> + a_2(Q)|y> + a_3(Q)|z> \tag{7}$$

where the coefficients a_i depend on Q because the perturbation matrix elements do; the a_i values corresponding to stable configurations can also be determined. The calculation was performed by Opik and Pryce (1957) who found that there are three kinds of stationary points which correspond to distortions of different symmetries: 3 tetragonal distortions along the tetragonal axes x, y, z, which have $(a_1, a_2, a_3) = (0,0,1)$ etc.; 4 trigonal distortions along the diagonals of the cube, which have $(a_1, a_2, a_3) = 1/\sqrt{3}(1,1,1)$ etc.; 6 distortions of intermediate (orthorhombic) symmetry which have $(a_1, a_2, a_3) = 1/\sqrt{2}(1, \pm 1, 0)$ etc. Whether these stationary points are minima (stable) or not depends on the values of the JT and elastic coupling constants; what Opik and Pryce found is that if $b^2/k_E > c^2/k_T$ then the 3 tetragonal distortions are stable, if $b^2/k_E < c^2/k_T$ then the 4 trigonal distortions are stable, while the orthorhombic distortions are never stable. So, as far as linear JT terms and harmonic forces are retained, there is a dichotomy between tetragonal and trigonal minima; this situation changes if higher order terms are included. The full shape of the energy hypersurfaces can easily be visualised if we assume that $c = 0$; in this case the surface in the Q_4-Q_5-Q_6 space is just a hyperparaboloid determined by the harmonic energy and the matrix in eq. 5 is diagonal: the energy surfaces in the $Q_2 - Q_3$ space are three rotation paraboloids displaced from the origin (fig. 4), and the electronic wavefunctions are not mixed (the matrix is diagonal in the $|x, y, z>$ basis): there is no way for the system to pass from one paraboloid to the other moving in the Q_4-Q_5-$Q_6 = 0$ subspace. Such selection rule is of course released if some of the trigonal coordinates have a finite value.

II.B. Effect of the Spin-Orbit Interaction

Let us suppose that the orbital triplet T is also doubly spin degenerate ($s = 1/2$); as mentioned in the introduction in this case there will be Kramers degeneracy, and this cannot be lifted by the JTE. Apart from this, the states will have the form $|x>|+>, |x>|->$ etc., where $|\pm>=|S_z = \pm 1/2>$ and the full perturbation hamiltonian in the Russell-Saunders approximation will read:

$$H = H_{JT} + H_{SO} = H_{JT} + \lambda \bar{L} \cdot \bar{S} \tag{8}$$

whose matrix elements are

$$H_{i+,j-} = \delta_{+-}(H_{JT})_{ij} + \lambda \sum_k <i|L_k|j><+|S_k|->; \; k = x, y, z \tag{9}$$

In our case $<i|L_k|j>= \epsilon_{ikj}$, where ϵ is Ricci's tensor, while the spin matrix elements may be calculated by usual methods. The simplest case, again, is when $c = 0$; this case was treated in detail by Opik and Pryce (1957), and the interesting result that emerges is the following: if $\lambda << b$ the tetragonal minima of the purely orbital problem are still minima, but if λ is sufficiently large the three minima coalesce into the origin. This means that a sufficiently strong spin-orbit interaction quenches the JT effect. The physical reason for this is that a strong spin-orbit interaction splits the electronic levels so that the prerequisite for the JTE (degeneracy or quasi-degeneracy) disappears. Since the JTE operates not only when the levels are exactly degenerate but also when they are sufficiently close in energy ("Pseudo JTE"), there is a continuous shift from a situation of strong distortion ($H_{SO} << H_{JT}$) to a situation of no distortion ($H_{SO} >> H_{JT}$). Of course, in the case $H_{SO} << H_{JT}$ it is the spin-orbit splitting which is quenched by the JTE. We will see that this mutual quenching effect between operators which are not diagonal in the same basis (like H_{JT} and H_{SO}) is probably the most general manifestation of the JTE.

II.C. Higher Order Effects: Quadratic JTE and Anharmonicity

Inclusion of the quadratic terms in equation (3) and/or of anharmonic terms in the restoring forces greatly complicates the problem; we will simply quote the main results and refer the reader to the original papers (Bersuker and Polinger 1973, Ranfagni and Viliani 1974, Bacci et al 1975a, 1975b) for a detailed treatment.

In this case the discussion is made more complicated by the fact that it is necessary to introduce new JT and anharmonic constants in addition to b, c, k_E, k_T, but the qualitative differences with respect to Opik and Pryce's linear JTE results may be stated very simply: (*i*) Orthorhombic distortions may become minima; (*ii*) Minima of different symmetry may coexist on the same energy surface and have different energy; (*iii*) Other stationary points of even lower symmetry appear.

III. THE DYNAMICAL JAHN-TELLER EFFECT

So far we have assumed that the kinetic energy of the nuclei may be neglected; this scheme may be suitable for physical systems where the stabilisation energy due to the JTE (and so the energy barrier between equivalent minima) is large as compared to the vibrational energies, because in this case the tunnelling transitions from one minimum to the others can be neglected. More precisely, we may expect the above scheme to be appropriate when the tunnelling time is long as compared to the characteristic measurement time of the phenomenon we are interested in. As an example, consider the case when the JT active state is the excited state taking part in a luminescent transition; in this case the "characteristic time" of the phenomenon is the radiative lifetime of the excited state: if the tunnelling time is much longer than the radiative lifetime, the system will emit from one of the distorted minima before tunnelling, and the luminescence properties will reflect the distorted nature (lower symmetry) of the system: it can be shown that in this case there is a correlation between the degree of linear polarisation of the exciting and emitted light (Ranfagni et al, 1983). On the contrary, when the tunnelling time is short with respect to the radiative lifetime, the system will undergo many tunnelling transitions before emitting and the distorted character of the different minima will be averaged out in the experimental result.

However, even when the JTE is not strong enough to cause stable distortions it still produces very important observable effects. The full dynamical problem in general

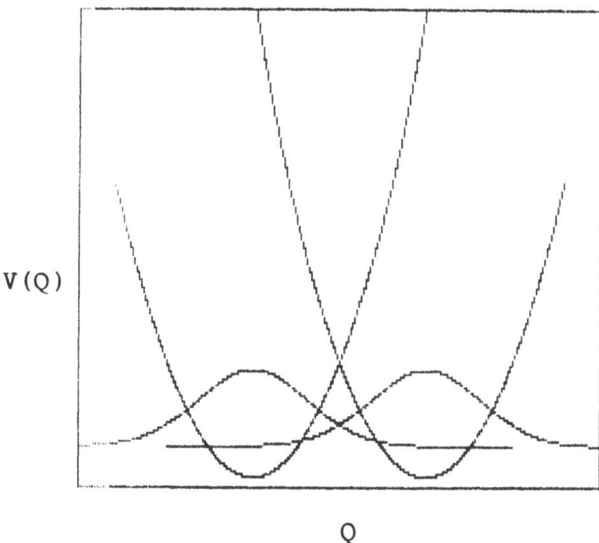

Figure 5. Schematic representation of the Ham effect: the overlap between the vibrational wavefunctions is decreased by the JTE, thereby quenching orbital off-diagonal electronic operators.

requires numerical solution (Englman, 1972), but semiqualitative arguments based on symmetry will give us sufficient insight into the problem.

Let us consider again the triplet T in O_h symmetry and suppose that the coupling constants b and c are very small, practically zero. The JTE is not active and the undistorted configuration is the equilibrium one: in the space $Q_2 - Q_6$ of the normal coordinates the nuclear system is subject only to the restoring potential energy that we assume harmonic. In this situation the solution of the full hamiltonian is simply a 5- dimensional harmonic oscillator. Since there is no electron- vibration coupling the total eigenfunctions will be:

$$|\psi_i^{n_j} >= |i > \Pi_j |\chi_{n_j}(Q_j) > \tag{10}$$

where $i = x, y, z$ labels the electonic states; $j = 2...6$; $n_j = 0 \div \infty$; $|\chi_{n_j} >$ are the vibrational states. Now let us gradually switch on the JTE with the Q_2, Q_3 modes alone; in terms of the static JTE this means that in the Q_2, Q_3 space the three paraboloids start moving away from the origin as in fig. 5. What is important to notice is that the tetragonal JT hamiltonian in the $|x, y, z >$ basis is diagonal (see eq. 5) and so it does not mix the electronic states which retain their paraboloid shape. As a result, in the case of JT coupling to Q_2 and Q_3 alone the complete wavefunctions will have the following form:

$$|\psi_i^{n_j} >= |i > \Pi_j u_i^{n_j}(Q_j - Q_j^0)\phi^m(Q_4, Q_5, Q_6) \tag{11}$$

where $u_i^{n_j}(Q_j - Q_j^0)$ is the eigenfunction of the 2- dimensional oscillator centred at Q_j^0 corresponding to the n- the level and to the electronic state $|i >$, and ϕ^m is the 3-dimensional oscillator wavefunction centred in the origin of the trigonal space and corresponding to the m-th level.

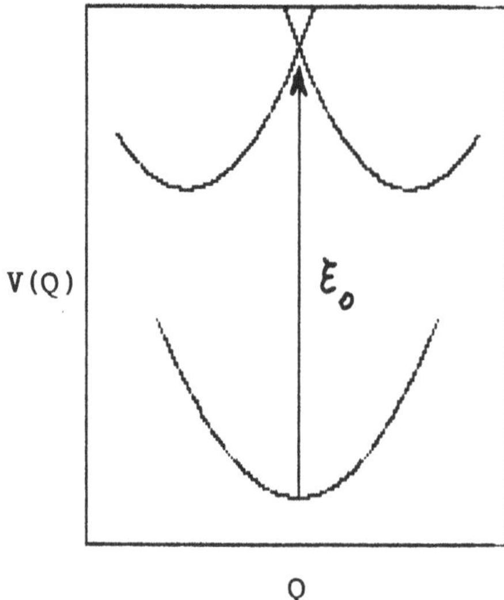

Figure 6. Schematic representation of an absorption transition between a non degenerate ground state and a doubly degenerate excited state.

If $c \neq 0$, the situation is more complicated because the electronic states are mixed; however, if $b^2/k_E > c^2/k_T$ the stable minima are still tetragonal and we may expect that eq. 13 is still a good approximation at least for the ground vibrational state (which, classically, corresponds to small displacements). Therefore, we will use for the total vibronic (i.e. vibrational- electronic) ground level the following three degenerate wavefunctions:

$$\begin{aligned} |\psi_x^o> &= |u_x^o>|\chi_x^o> &= |x>|u_x^o>|\phi^o> \\ |\psi_y^o> &= |u_y^o>|\chi_y^o> &= |y>|u_y^o>|\phi^o> \\ |\psi_z^o> &= |u_z^o>|\chi_z^o> &= |z>|u_z^o>|\phi^o> \end{aligned} \quad (12)$$

By the way, note that the *total* (vibronic) ground level is still degenerate in spite of the JTE: the situation is qualitatively similar to that described in the introductory section.

III.A. Ham Effect: Quenching of Orbital Operators

We will now look for the observable consequences of the dynamical JTE, and for the sake of clarity will do it on a specific example. Assume our system has moderate electron-phonon interaction, so that the zero-phonon line (ZPL) is observable in the absorption spectra from the non-degenerate ground level to a triply degenerate excited level, and assume that the excited level is also spin-degenerate, so that the spin-orbit interaction is active. In this situation, the absorption spectrum will exhibit as many ZPL's as are the levels produced by spin-orbit splitting of the excited level (provided that there are no selection rules which forbid some transitions).

In Russell-Saunders coupling we will write:

$$H_{SO} = \lambda \bar{L} \cdot \bar{S} \tag{13}$$

If $|S>$ is the spin eigenfunction, the ground states will be

$$\begin{aligned} |\psi_x^o> &= |x>|\chi_x^o>|S> \\ |\psi_y^o> &= |y>|\chi_y^o>|S> \\ |\psi_z^o> &= |z>|\chi_z^o>|S> \end{aligned} \tag{14}$$

Since $\bar{L}$ does not operate on $|S>$ and vice versa, and since both do not operate on $|\chi>$, the matrix elements will be:

$$<\psi_i^o|H_{SO}|\psi_j^o> = <i|L_j|k><S_\mu|S_j|S_\nu><u_i^o|u_j^o><\phi^o|\phi^o> \tag{15}$$

where $i, j, k = x, y, z$. For the known properties of the angular momentum operator, in this case we will have $<i|L|k> \propto \epsilon_{ijk}$; therefore H_{SO} has only non- diagonal matrix elements with respect to the orbital electronic states. Then, from eq. 15 it is evident that the whole perturbation matrix is multiplied by the reduction factor $R = <u_i^o|u_j^o>$. R is the overlap integral between two vibrational eigenfunctions centred on different minima; it would be 1 in the absence of JTE since in this case all potential surfaces would coincide, but as schematised in fig. 5 the JTE causes $R < 1$. In the limiting case $E_{JT}/\hbar\omega \to \infty$ we would have $R = 0$, i.e. a complete quenching of the spin-orbit splitting. The quenching of non-diagonal electronic operators that we have schematically described, and which is known as the Ham effect (Ham, 1965), is the most general consequence of the JTE, since it does not require that the effect is as strong as to cause metastable distortions of the system. The spectroscopic result is that the observed spin-orbit splitting may be remarkably lower than expected on the basis of static crystal field theory. Of course, as mentioned in the previous section, the JTE may be quenched by the spin-orbit interaction; quenching of mutually non-diagonal operators is a general occurrence.

In the present case R may be calculated exactly (Ham, 1965):

$$R = exp(-3E_{JT}/2\hbar\omega)$$

If the spin-orbit interaction is strong, eq. 13 is not a good approximation; in this case there are further terms in H_{SO} which in general have diagonal matrix elements also: these second- order contributions are not quenched.

The Ham effect was experimentally observed by Sturge (1970) in the zero-phonon line of the ${}^4A_2 \to {}^4T_2$ absorption transition of $KMgF_3 : V^{2+}$; this transition is splitted into 4 lines by the spin-orbit interaction; the overall splitting expected on the basis of static crystal-field theory is $\approx 100\ cm^{-1}$, while the observed one is $\approx 40\ cm^{-1}$. Taking into account second-order spin-orbit terms which are important in this case, the value $E_{JT}/\hbar\omega \approx 0.8$ was derived (Montagna et al, 1979); since V^{2+} is actually interacting with many crystal modes in this case, $\hbar\omega$ must be interpreted as an effective average frequency.

Other perturbations of spectroscopical relevance which are quenched in octahedral symmetry are: the orbital part of the Zeeman effect, the hyperfine interaction

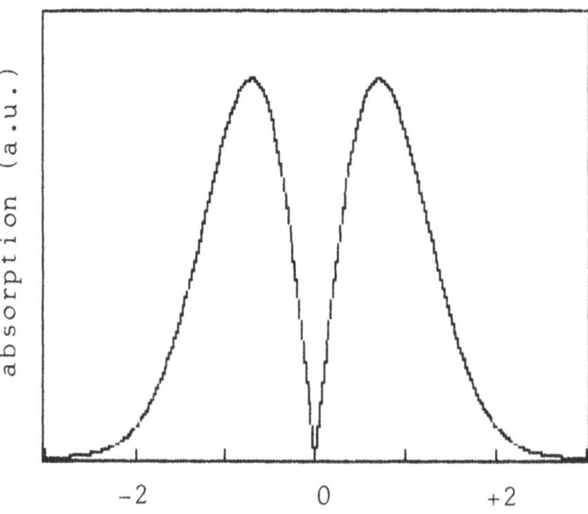

Figure 7. Bandshape of eq. (19) as a function of $\hbar\omega - \varepsilon_o$, computed for $a^2 k_B T = 1$.

between nuclear and electronic spin, uniaxial stress along the [1,1,1] direction, and others (Ham 1972).

If the coupling is to the trigonal Q_4, Q_5, Q_6 coordinates alone ($b = 0$) the results are qualitatively similar to the ones described here, provided we consider orbital perturbations which are non-diagonal in a basis where the JTE to trigonal modes is diagonal. The case of coupling to both tetragonal and trigonal coordinates is more complicated; there are two possible situations: (*i*) if the JTE is weak for both kinds of coordinates, the quenching effect is more or less additive; (*ii*) if the JTE is strong, in practice the quenching effect is produced only by the coordinates which have dominant coupling.

For a detailed discussion and for specific examples see the paper by Ham (1965), from which much of this discussion is derived.

III.B. Absorption Band Shapes

Consider the system schematised in fig. 6 and an optical absorption transition; assume that the JTE in the excited state is strong enough as to mask the vibrational structure in the absorption band. As was shown in 1966 by Toyozawa and Inoue, the JTE produces structures in the spectrum, and it must be stressed that these have nothing to do with the usual vibrational structures which are not peculiar to JT systems.To see this in a simple case, consider a doubly degenerate excited state in octahedral symmetry; it can be shown by group theory that this state couples only to the tetragonal modes Q_2 and Q_3 and the electronic eigenvalues are easily found to be (Englman, 1972):

$$E_{1,2} = \frac{1}{2}k(Q_2^2 + Q_3^2) \pm a\sqrt{Q_2^2 + Q_3^2} \tag{16}$$

where a is the JT coupling constant. In the Condon approximation, and treating the vibrational coordinates as classical quantities, the absorption band shape will be given by (Englman, 1972):

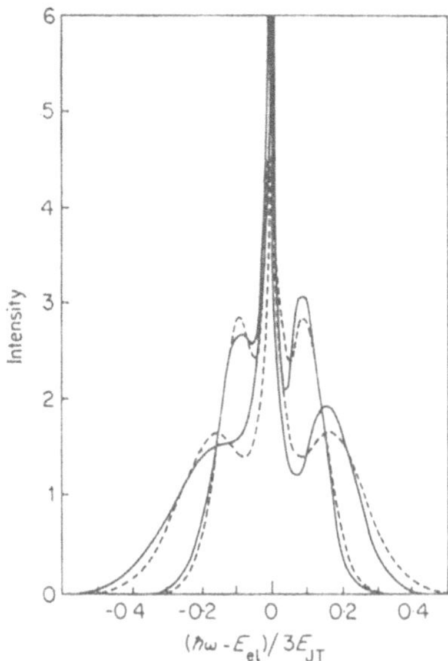

Figure 8. Numerically computed absorption band shapes for the $A \to T$ transition in the presence of JT coupling to T_{2g} modes alone in the excited state. Broader bands correspond to higher temperatures. The (symmetric) broken lines (Toyozawa and Inoue, 1966) include only linear JT coupling, while the (asymmetric) full lines (Honma, 1968) include also a change in curvature between the ground and excited states, which formally results in a quadratic term in the hamiltonian. E_{el} is the same as ε_0 (reprinted from Englman, 1972).

$$I(\omega) = \sum_f \int \dots \int dQ_n \Pi_n P_i(Q_n) \delta \left\{ \hbar\omega - (E_f^{Q_n} - E_i^{Q_n}) \right\} \tag{17}$$

where the summation runs on the excited electronic states, and there are as many integrals as is the number of vibrational coordinates; $\Pi_n P_i(Q_n)$ is the probability that when the system is in the ground electronic state $|i>$ it occupies the volume element $\Pi_n dQ_n$ of the configuration space.

Taking a Boltzmann distribution, $P_i(Q_n) \propto e^{\delta E/k_BT}$, and in the harmonic approximation with $k_E = 2$, we have in our case

$$\Pi_n P_i(Q_n) = \frac{1}{2\hbar k_B T} exp \left\{ -\frac{Q_2^2 + Q_3^2}{k_B T} \right\}$$

Using (16) and considering that for the ground state we have $E_i(Q) \propto (Q_2^2 + Q_3^2)$ we obtain

$$I(\omega) = \frac{1}{2} \sum_{2,3} \int \int dQ_2 dQ_3 \frac{1}{\hbar k_B T} exp \left\{ -\frac{Q_2^2 + Q_3^2}{k_B T} \right\} \times \delta\left(\hbar\omega - E_{1,2}\right) \tag{18}$$

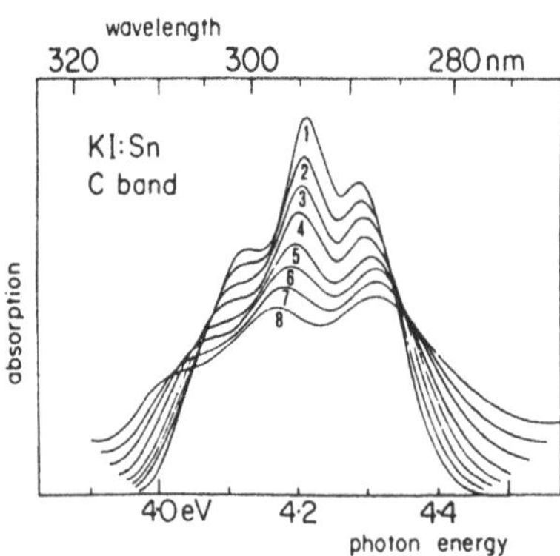

Figure 9. Experimental ${}^1S_0 \to {}^3 P_1$ absorption band shape of Sn^{2+} in KI at various temperatures in the range 10-295 K (Fukuda, 1969; reprinted from Ranfagni et al, 1983)

which is integrated to give

$$I(\omega) = \frac{|\hbar\omega - \varepsilon_o|}{a^2 k_B T} exp\left\{ -\frac{(\hbar\omega - \varepsilon_o)^2}{a^2 k_B T} \right\} \tag{19}$$

where ε_0 is shown in fig. 6. This function has two maxima at different frequencies as shown in fig. 7; this characteristic form is actually observed in JT systems.

The case of threefold electronic degeneracy in the ground state is mathematically more complicated (Toyozawa and Inoue, 1966; Honma, 1968), and leads to a three-peaked band as shown in fig. 8. The asymmetries and temperature-dependences observed in real systems (fig. 9) are qualitatively well reproduced by taking into account spin-orbit interaction (Toyozawa and Inoue, 1966) and/or higher order terms in the hamiltonian (Honma, 1968).

III.C. Selective Intensity Quenching

Going back to the Ham effect in $KMgF_3 : V^{2+}$, though the splittings are well reproduced by quenching the spin-orbit operator the relative intensities of the splitted lines are not. In fact, ${}^4T_2$ is splitted into 4 spin-orbit levels which (in order of increasing energy) are usually labelled $\Gamma_7, \Gamma_8, \Gamma_8^*, \Gamma_6$ and have the following residual degeneracies: 2, 4, 4, 2; since the transitions from the ground ${}^4A_2$ level to these spin-orbit sublevels are all equally allowed, the intensities of the respective transitions should be proportional to the degeneracy. This is not the case; experimentally the low-temperature relative intensity ratios are 1:1.52:1.24:0.57 (instead of 1:2:2:1).

The reason for this is again the dynamical JTE (Montagna et al, 1979). Intuitively, the effect goes as follows. Consider interaction to E_g modes alone; let $b = 0$ (no JTE). In this case the eigenvectors in the excited state are trivial and consist simply of four zero-phonon levels and their respective vibrational replicas, each lying $n\hbar\omega$ higher in energy than the parent level. At this level of approximation (no JTE) and at low temperature all the intensity of the ${}^4A_2 \to {}^4 T_2$ transition is concentrated in the 4 spin-orbit splitted zero- phonon lines. In fact, if M is the dipole moment operator where $|\Gamma_i>$ is one of the 4 spin-orbit electronic states, $|u_A^0>$ is the ground vibrational level relative to $|^4A_2>$, and $|u_T^n>$ is the generic vibrational state relative to $|\Gamma_i>$. If the JTE is absent the vibrational overlap in the above equation is zero unless $n = 0$.

We now switch on the JTE on $|^4T_2>$; this will mix $n = 0$ to $n \neq 0$ states. As a consequence the complete (electronic plus vibrational) Born-Oppenheimer wavefunctions will contain a superposition of different $|u_T^n>$, but only the $|u_T^0>$ components will contribute to the intensity because $< u_A^0|u_T^n> = \delta_{n0}$. It is then clear that the $|\Gamma_7>$ state (which is the lowest in energy) will have more zero-phonon character (and therefore higher relative intensity) than the others, because the energy difference $\Gamma_7(n = 0)$ - $\Gamma_7(n = 1)$ is bigger than the energy differences relative to the other spin- orbit states. It is also evident that this intensity effect will increase by increasing the $H_{SO}/\hbar\omega$ ratio. In some cases the effect may be so strong as to make the Γ_6 transition practically unobserved.

References

Bacci M, Ranfagni A, Fontana MP, Viliani G (1975a) Phys Rev B11, 3052
Bacci M, Ranfagni A, Cetica M, Viliani G (1975b) Phys Rev B12, 5907
Bersuker IB, Polinger VZ (1973) Phys Lett A44, 495
Fukuda A (1969) J Phys Soc Japan 27, 96
Ham FS (1965) Phys Rev 138, A1727
Honma A (1968) J Phys Soc Japan 24, 1082
Jahn HA (1938) Proc Roy Soc A164, 117
Jahn HA, Teller E (1937) Proc Roy Soc A161, 220
Longuet-Higgins HC, Opik U, Pryce MHL, Sack RA (1958) Proc Roy Soc A244, 1
Moffitt W, Thorson W (1957) Phys Rev 108, 1251
Montagna M, Pilla O, Viliani G (1979) J Phys C 12, L699
Opik U, Pryce MHL (1957) Proc Roy Soc A238, 425
Ranfagni A, Mugnai D, Bacci M, Viliani G, Fontana MP (1983) Advances in Physics 32, 823
Ranfagni A, Viliani G (1974) Phys Rev B9, 4448
Sturge MD (1970) Phys Rev B1, 1005
Toyozawa Y, Inoue M (1966) J Phys Soc Japan 21, 1663

Books and articles of general interest on the JTE:
Englman R (1972) *The Jahn-Teller Effect in Molecules and Crystals* (Wiley)
Sturge MD (1967) Solid State Physics 20, 91
Ham FS (1972) in *Electron Paramagnetic Resonance* edited by L Geschwind (Plenum), chap 1

Applications of group theory:
Landau LD, Lifshitz EM (1966) *Mecanique Quantique* (Mir), chap 12 and 13
Griffith JS (1962) *The Irreducible Tensor Method for Molecular Symmetry Groups* (Prentice Hall)
Di Bartolo B (1968) *Optical Interactions in Solids* (Wiley)

SPECTRAL PROPERTIES OF EXCITED STATES IN RESTRICTED GEOMETRIES*

J. Klafter[a] and J.M. Drake[b]

[a]School of Chemistry, Tel Aviv University, Tel Aviv 69978, Israel
[b]Exxon Research and Engineering
Annandale, New Jersey 08801, USA

ABSTRACT

Dynamics of excited states in restricted geometries are known to display nonexponential decays. In this paper we review different approaches to analyzing nonexponential decay functions, with emphasis on the stretched exponential form. We concentrate on a few solvable models for which the stretched exponential law follows in a natural fashion. Although each model describes a different mechanism, it is shown that they have the same underlying reason for the stretched exponential pattern: the existence of scale invariant relaxation times. We bring examples for applications in some real systems.

I. INTRODUCTION

There has been accumulating evidence that nonexponential relaxation patterns in time are common to many disordered condensed matter systems. They appear to be the rule rather than the exception. A few examples for such decays include: fluorescence decay in porous solids [1-3], in polymers [4,5], in self-organized systems such as micelles and vesicles [6], solvation dynamics in polar solvents [7], dielectric relaxation in glasses and in polymers [8], electron scavanging in glasses [9] and birefringence studied in polymers [10].

A number of fitting methods have been proposed in order to analyze nonexponential relaxations and to obtain information about the underlying systems from them. Most of these methods focus on calculating distributions of lifetimes or rate constants in multiexponential expressions [5,11,12].

An interesting empirical observation has been made recently suggesting that the decay of correlation functions for many diverse systems follows the same stretched exponential law,

$$\Phi(t) = \exp\ [-(t/\tau)^{\alpha}],\ o<\alpha<1 \tag{1}$$

The parameters α and τ depend on the material and can be a function of external variables such as temperature. Equation (1) was first proposed by Kohlrausch to describe mechanical creep [13], and was later used by Williams and Watts to describe dielectric relaxation in polymers [14]. The stretched exponential has

* Most of this work was presented at the 29 Yamada Conference in Osaka, Japan, May 1991 and appeared in the conference proceedings.

Optical Properties of Excited States in Solids
Edited by B. Di Bartolo, Plenum Press, New York, 1992

since been employed to fit data in other systems, some of which are mentioned as examples above. Here we show that this ubiquitous decay law can be derived from several models which describe different physical mechanisms. Each of the models generates scale invariant relaxation rates which are responsible for the stretched exponential behavior [15].

Another decay form which has been used to fit relaxation results, although less extensively than the stretched exponential, is the enhanced power law

$$\Phi(t) \sim (t/\tau)^{-B\ln^{\beta-1}(t/\tau)}; \quad \beta\geq 1, \ t>\tau \tag{2}$$

which is essentially the exponential-logarathmic form of Inokuti and Hirayama [16] and reduces to a simple power law for $\beta=1$. Equation (2) has been introduced to explain electron scavanging in glasses and transport properties in amorphous semiconductors [9,17].

In this paper we briefly review three models which lead to stretched exponentials. A more detailed discussion is presented in ref.[15]. We will analyze the Forster direct transfer mechanism [18] which is an example of relaxation via *parallel* channels and relate its mathematical structure to the *serial* hierarchichally constrained dynamics model [19]. The third related theory is that of the target model [20]. In all three cases the relaxation is given by

$$\Phi(t) = \int dR f(R)\, e^{-t/\tau(R)} \tag{3}$$

where f(R) is a weight factor and $\tau(R)$ is a microscopic relaxation time which varies with the relevant variable in the problem (distance or level, vide infra). We show that:

(a) for those cases where $\tau(R)$ scales with R, $\tau(R)\sim R^{\delta}$, a stretched exponential is obtained

(b) for $\tau(R)\sim \exp[\gamma R]$ an enhanced power law is obtained.

Our conclusions concerning the stretched exponential behavior are then extended to spatially restricted structures and applied to direct energy transfer in porous systems and to birefringence experiments in polymer systems.

II. MODELS FOR STRETCHED EXPONENTIALS [15]

The first model we discuss, which esults in a stretched exponential decay, is the direct energy transfer model (DET). This model has been extensively covered in the literature [18,21,22] and is schematically presented in Fig.1. It describes the decay law of an initially prepared static donor due to direct transfer to *randomly* distributed static acceptors. Although DET offers a relatively simple example of spatial disorder, it is still rich enough to obtain the complex relaxations of Eqs. (1) and (2). The donor can simultaneously relax through many competing channels, a process that leads to the non-exponential decay. To each donor there corresponds a *hierarchy* of donor-acceptor distances. The relaxation function in each channel is given by

$$\phi_i = \exp[-tW(R_i)] \tag{4}$$

where $W(R_i)$ is the relaxation rate that depends on the relative locations of a donor-acceptor pair. This defines a relaxation time $\tau(R_i)=[W(R_i)]^{-1}$. One has to include now all the relaxation channels and to configurationally averge (see ref.[21] for details). When a substitutional occupancy of site by acceptors with probability p is assumed,

$$\Phi(t) = \exp[-p \sum_i \{ 1- \exp [-tW(R_i)]\}] \quad ; \qquad p << 1 \tag{5}$$

By introducing a site density function [22]

$$\rho(R) = \sum_i{}' \, \delta (R-R_i)$$

we obtain

$$\Phi(t) = \exp [-p\int dR\rho(R) \{1- \exp[-tW(R)]\}]. \tag{6}$$

Two types of isotopic rates are usually considered [15,22]:

(a) $W(R)=AR^{-s}$, $s \geq 6$, which implies scale invariance $\tau(R) \sim R^s$ (7)

(b)) $W(R)=B\exp[-\gamma R]$, which implies $\tau(R) \sim \exp[\gamma R]$

On regular underlying spatial structures $\rho(R)$=const. and we find for case (a) a stretched exponential

$$\Phi(t) = \exp[-(t/\tau)^{d/s}] \qquad \text{(in d dimensions)} \tag{8}$$

For case (b)

$$\Phi(t) \sim t^{-B} \ln^{d-1}(Bt) \qquad \text{(in d dimensions)} \tag{9}$$

This exhibits an enhanced power law which takes the form of an algebric decay in one dimension.

The nonexponential decays of Eqs. (8) and (9) are both a reuslt of a *parallel* relaxation scheme and a *hierarchy of distances.*

The same types of relaxation patterns can be derived when we restrict the transfer to include only the nearest neighbor acceptors. Then

$$\Phi_{NN} = \int_0^\infty f(R) \exp [-tW(R)] \, dR \tag{10}$$

where $f(R)$ is the probability of having a nearest neighbor at distance R. For randomly placed acceptors in one-dimension

$$f(R) = p\exp(-pR) \tag{11}$$

For $W(R)$ in Eq. (7), case (a),

$$\Phi_{NN} = p\int e^{-pR} \exp (-tAR^{-s}) \, dR \tag{12}$$

which by the method of steepest descent gives

$$\Phi_{NN} = p \exp[-Ct^{\frac{1}{1+s}}] \tag{13}$$

again a stretched exponential but with a smaller exponent than for d=1 in Eq. (8), because the influence of more distant defects has been truncated.

The direct transfer decay laws follow a stretched exponential behavior both for the many parallel channels case and for the fastest decay channel when $W(R) \sim R^{-s}$, namely when the position dependent relaxation time $\tau(R)$ is scale invariant.

Recently Palmer et. al. [19] introduced a model of relaxation which is *serial* rather than parallel. This hierarchical model supposes that relaxation occurs in stages, and the constraint imposed by a faster degree of freedom must relax before a slower degree of freedom can relax. This implies that the time scale of relaxation on one level is subordinated to the relaxation below.

In one possible realization they considered a system with a discrete series of levels n=0,1,2,..., with the degrees of freedom on level n represented by N_n spins which point either up or down as in Fig.1. The spins in level n+1 are only free to change their state when μ_n spins in level n attain one of their 2^{μ_n} possible states. The relaxation time τ_{n+1} of level n+1 is then

$$\tau_{n+1} = 2^{\mu_n} \tau_n \tag{14a}$$

$$= \tau_o \exp\left(\sum_{k=o}^{n} \tilde{\mu}_k \right) \tag{14b}$$

when $\tilde{\mu}_k = \mu_k \ln 2$. The relaxation function $\Phi(t)$ is given by

$$\Phi(t) = \sum_{n=o}^{\infty} \omega_n \exp(-t/\tau_n) \tag{15}$$

where $\omega_n = N_n / \sum_{n=o}^{\infty} N_n$ is a weight factor for level n. Note that the hierarchy of relaxation times generated by Eq. (14b) is similar to the hierarchy of transition rates discussed in the direct transfer model.

One can now choose specific forms for μ_n and ω_n and calculate the corresponding relaxation function. The choice $\mu_n = \mu_o$ implies

$$\tau_n = \tau_o \exp(\tilde{\mu}_o n) \tag{16}$$

which is essentially case (b) of Eq. (7). Choosing

$$\omega_n = \omega_o \lambda^{-n} = \omega_o e^{-n\ln\lambda} \tag{17}$$

which corresponds to Eq. (11), and converting the sum in Eq. (15) to an integral yields

$$\Phi(t) = \omega_o \int_0^{\infty} e^{-n\ln\lambda} \exp[-t \exp(-\tilde{\mu}_o n)/\tau_o] dn \tag{18}$$

which we recognize to be the integral that leads to the algebraic relaxation law that corresponds to d=1 in Eq. (9), namely $\Phi(t) \sim t^{-(\ln\lambda)/\tilde{\mu}_o}$.

The same choice for ω_n, but now coupled with the

$$\mu_n = \mu_o \cdot n^{-1}$$

implies

$$\tau_n = \tau_o \exp (\tilde{\mu}_o \sum_{l=1}^{n} l^{-k})$$

$$\simeq \tau_o n^{\tilde{\mu}_o} \text{ (for k=1)} \tag{19}$$

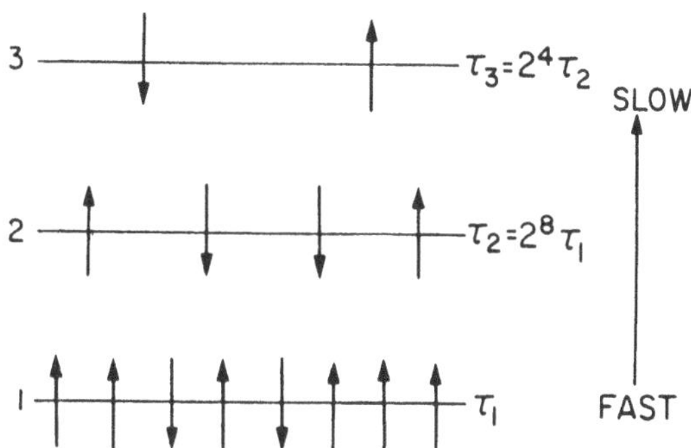

Figure 1. The hierarchically constraint dynamics model.

which corresponds to case (a) of Eq. (7). For k=1 this leads to the relaxation integral

$$\phi(t) = \omega_o \int_0^\infty e^{-n \ln\lambda} \exp [-tn^{-\tilde{\mu}_o}/\tau_o] dn \tag{20}$$

which, as in Eq. (12), produces the stretched exponential law with exponent

$$\beta = \frac{1}{(1+\tilde{\mu}_o)}.$$

Although the physical pictures of relaxing through a serial arrangement of hierarchically constrained levels and relaxing through a direct transfer to a nearest neighbor defect are quite different they both lead to the same relaxation integrals. The weight factor ω_n and the relaxation time τ_n for each level in the hierarchical model correspond to the weight factor f(R) for the defect position, and the transition rate W(R) respectively in the direct transfer model.

The target model offers yet another mechanism for a stretched exponential decay. Unlike the DET case where the donor and acceptors are static, here the acceptors (quenchers) diffuse towards the initially prepared donor (the target) and quench it upon encounter [20]. The target problem can be recast into the same mathematical framework as the two previous models. We assume a one dimensional example where a target is located at the origin and diffusing acceptors are initially randomly distributed around it. Let $f(R_1)$ be the probability of having no acceptors at distance R_1 from the target. As before, for randomly placed, acceptors $f(R_1) = \exp(-pR_1)$. The probability that an acceptor at R_i has not reached the origin by time t is given by $\exp[-t/4R_i^2]$, where the diffusion constant of the acceptors has been set equal to unity. The relaxation law for the target is then

$$\phi(R_1,t) \simeq \exp(-pR_1) \prod_{i=1} \exp[-t/4R_i^2]$$

$$\simeq \exp(-pR_1) \exp\left[-tp\int^{\infty} \frac{dR}{4R^2}\right]$$

$$= \exp[-pR_1 + p\frac{t}{4R_1}] \tag{21}$$

Averaging over R_1 we arrive at

$$\Phi(t) = \int_0^{\infty} \exp(-pR_1) \exp(-pt/4R_1) dR_1 \tag{22}$$

which is again of the form of Eq. (12) in the Förster case and of Eq. (20) in the Palmer et. al. case, now with $\tau(R)=(4/p)R$, and yields the stretched exponential law with $\beta = 1/2$. This result can be generalized to other dimensions (see ref.[23,24]).

We have demonstrated that a common mathematical framework underlies different physical models which lead in a natural way to the common emperically found stretched exponential relaxation law. The unifying feature of the theories is the generation of a scale invariant distribution of relaxation times. One should be able to differentiate among physical mechanisms underlying the relaxation via the variation behaviour of α and τ in Eq. (1) as a function of external variables such as temperautre and pressure.

III. EXTENSIONS OF DET TO RESTRICTED GEOMETRIES [25]

The different models discussed in the previous section have been applied to explain stretched exponential decays in a variety of systems. Here we extend the DET model to restricted geometries. We show that the scale invariant distribution of relaxation times $\tau(R)$ dominates and that even when restrictions are imposed we obtain stretched exponential decays. In order to be able to relate the DET to experimental systems the coefficient A in Eq.(7) case (a) has to be specified:

$$A = \frac{1}{\tau_D} R_0^S \tag{23}$$

where τ_D is the donor lifetime and R_0 is the Förster critical radius, which defines a length in the system. In homogeneous systems the role of R_0 is obvious [18]. However, in the case of DET from a donor to randomly distributed acceptors in restricted geometries a second length enters, which characterizes the spatial restrictions. The model-restricted geometries we discuss here are fractal structures and cylinders. Although fractals introduce the concept of self-similarity, regular shapes such as cylinders and spheres mimic the geometrical properties of pores and molecular assemblies.

In order to apply DET to these model systems we use Eq.(6) which expresses the decay in terms of the site density function $\rho(R)$, which is essentially the two-point correlation function on the structure.

Fractal structures are examples of restricted geometries. They are usually disordered, tenuous but self-similar (such as percolation clusters) [20,22,26]. The self-similar nature of these structures means that there is no typical length that characterizes them and so, when DET is considered, R_0 remains the dominating length. The site density function for fractals is [3]

$$\rho(R) = \frac{F}{V_d}\,\rho_0\left[\frac{\bar{d}}{d}\right] R^{\bar{d}-d} \tag{24}$$

where $\bar{d}$ is the fractal dimension ($1 \leq \bar{d} \leq 3$), d is the Euclidean embedding space, ρ_0 is the density of acceptor sites, and F is an unknown shape factor. The survival of a donor on a fractal is therefore [3]

$$\Phi(t) = \exp[-t/\tau - p\rho_0 F R_0^{\bar{d}}\,\Gamma(1 - \bar{d}/S)t^{\bar{d}/S}] \tag{25}$$

again a stretched-exponential but with anexponent $\bar{d}/S$. The prefactor $p\rho_0 F R_0^{\bar{d}}$ is equal to the number of acceptors with a radius R_0 in $\bar{d}$ dimensions. It has been used to interpret DET experiments on porous Vycor glass [27] on silica gels [3], on zeolites [28] and on Langmuir-Blodgett films [29]. It should be noted, however, that fitting decay curves to the stretched exponential in Eq.(25) may not confirm or even imply that the underlying structure is really a fractal. Behavior similar to that described by Eq.(25) with nonintegral values of $\bar{d}$ can also occur as a result of crossover processes between dimensions. Nevertheless, Eq.(25) is useful in interpreting DET in restricted geometries especially when corroborated by other characterization techniques [2].

More conventional shapes that serve as useful models for restricted geometries are cylinders and spheres. These geometric systems are characterized by their radius R_p. When DET experiments are performed on these geometries, it is the relation between R_0 and R_p that determines the decay patterns of $\Phi(t)$. Cylindrical pores have been used to model local pore characteristics. Many properties of porous materials are being studied in the framework of a cylindrical pore, for example, adsorption, wetting, and diffusion. Spheres are usually chosen to approximate micelle or microemulsion shapes. In order to be able to utilize DET as a tool for structural studies of these geometries, it is essential that one understands the dependence of the energy transfer reaction on the donor-acceptor distributions on or within these geometries. Fig. 2 shows schematically donor and acceptor distributions in cylindrical and spherical shapes.

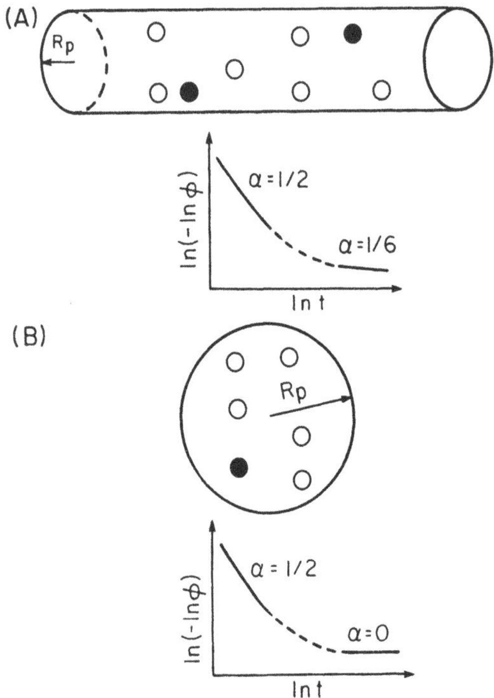

Figure 2. Examples of spatial confinements: (A) cylinder and (B) sphere with the corresponding DET decays.

We now address the case of a cylinder of radius R_p with a donor and randomly distributed acceptors on its surface. Calculating the corresponding site density function and using Eq.(6), one finds that the survival probability can be numerically evaluated. Analytical expressions are easily derived for the short and long time behavior of $\Phi(t)$. The crossover time is estimated from the ratio of the two relevant length scales R_o and R_p:

$$\Phi(t) = \exp\left[-t/\tau - p\rho_o \pi R_o^2\, \Gamma(1 - \tfrac{2}{S})(t/\tau)^{2/S}\right] \tag{26}$$

for $t<\tau(2R_p/R_o)^S$. The short time decay displays a two-dimensional behavior according to both the exponent 2/S and the prefactor $p\rho_o\pi R_o^2$. The donor senses only acceptors on the surface in its immediate vicinity. For long times $t>\tau(2R_p/R_o)^S$,

$$\Phi(t) = \exp[-t/\tau - p\rho_o 4\pi R_p R_o \Gamma(1-1/S)(t/\tau)^{1/S}] \tag{27}$$

This corresponds to a one-dimensional relaxation according to the exponent 1/S. The prefactor $4\pi p\rho_o R_p R_o$ equals, for $R_p< R_o$, the number of acceptors on the cylinder surface within radius R_o. The donor's decay crosses over from a two-dimensional to a one-dimensional form at a crossover time $t_{cross} \sim (2R_p/R_o)^S$. This is a direct result of the finiteness of the systems and a finger print of a length, R_p, that competes with R_o. Both the short-time and the long-time decays are stretched exponentials for which the exponent and prefactor reflect the dimension and molecular arrangement in the system.

One arrives at the same conclusions when DET is performed on a sphere of radius R_p. If we assume that the donor and acceptors are distributed on the surface of the sphere (30), then for short times, $t<\tau(2R_p/R_o)^S$, we recover the same results as for the cylinder, Eq. (16). For long times we obtain

$$\Phi(t) = \exp\left[-t/\tau - p\rho_o 4\pi R_p^2\right] \tag{28}$$

a simple exponential decay. The temporal behavior of $\Phi(t)$ exhibits a crossover that reflects the finiteness of the system. The same approach, based on establishing the site density function $\rho(R)$, can of course be used for other shapes.

What emerges from this analysis is that DET is capable of sensing local dimensionalities as well as crossovers between them. This contributes to a more complete picture of the investigated structures. Both the *exponents* and the scaling of the *prefactors* with R_o and R_p contain desirable information that may lead to an understanding of the morphologies of confinements. When local structures are known, the dependence of the prefactor on R_o can provide insight about the distribution of the acceptors. Crossover times, if measurable, can give clues about the typical size of the restriction, R_p.

The proposed procedure for analyzing experimental decays is to fit relaxation data to a stretched exponential, with the two fitting parameters. The exponent α is given the meaning of $\bar{d}/S$, with $\bar{d}$ as an "effective dimension", and the prefactor is proportional to the number of acceptors within a radius R_o on the structure (and therefore should depend also on R_p). A model can then be proposed for the underlying geometry that is consistent with values of both parameters. If the apparent dimension $\bar{d}$ obtained is not an integer, then the concept of a crossover can be used in the analysis. A modified form can be applied to fit experimental decays:

$$\phi(t) = \exp[-t/\tau - A_o\Gamma(1-\bar{d}/S)(t/\tau)^{\bar{d}/S}] \tag{29}$$

A straightforward connection between the static microstructure of the confinement and DET dynamics is achieved when a site density function $\tau(R)$ can be independently derived. Then, based on Eq.(6), dynamics and statics are directly related. Such an approach is important in establishing the method of using DET to get structural information. In a detailed study on porous Vycor glass, DET measurements and ultrathin transmission-electron-microscopy analysis of $\tau(R)$ were simultaneously carried out. The combined studies confirm the power of DET [30].

DET, using the scheme of a donor transferring energy to randomly distributed acceptors, provides insight into the microstructure of complex restricted geometries. The approach relies on stretched exponential decays and crossover between them and the interplay between two competing lengths: R_p, inherent to the spatial restriction, and R, introduced by the DET method. When DET is applied to study local morphologies, corroboration with other characterization techniques is very useful.

IV. AN EXAMPLE [2]

We now briefly discuss an experimental example where stretched exponential relaxations occur and can be related to our scaling arguments.: direct energy transfer in porous silica gels. For more detailed analyses and discussion we refer the reader to refs.[2,3].

Direct energy transfer experiments have been carried out on the well characterized family of porous silicas, Si-40, Si-60 and Si-100 which were studied and describes extensively [2] with rhodamine 6G and malachite green as the donor-acceptor pair. Each of the silicas has a well defined mean pore size R_p=18 Å (Si-40), R_p=35 Å (Si-60) and R_p=60 Å (Si-100). The experiment has been conducted using a time correlated single photon counting system. As stated, R_o provides an estimate of the length scale probed by DET here $R_o \simeq$ 57 Å. A more realistic measure should take into account the experimental limitations of the detection system. This introduces the length scale R_{max} which is approximately 1.5 R_o. The results on the three porous glasses have been fitted by Eq.(30): $\bar{d}\simeq 2$ (Si-100), $\bar{d}\simeq 2$ (Si-60), $\bar{d}\simeq 3$ (Si-40), the regular Euclidean limits of the equation. When corroborated with the characterization studies on these systems, the following picture emerges, in which the relationship between R_p (from structural characterization) and R_{max} (the optical yardstick) determines what features of the morphology are probed by direct energy transfer:

(1) When $2R_p > R_{max}$, (Si-100) the DET probes scale less than R_p. The process senses therefore only a portion of the local surface as shown in Fig.3A. Here Here $\bar{d}$ is nearly 2, which means a smooth surface. Slight deviations from d=2 occur because of crossovers to higher dimensional configurations.

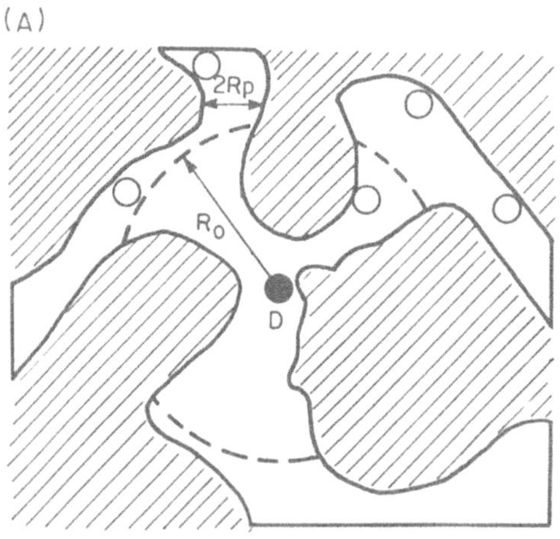

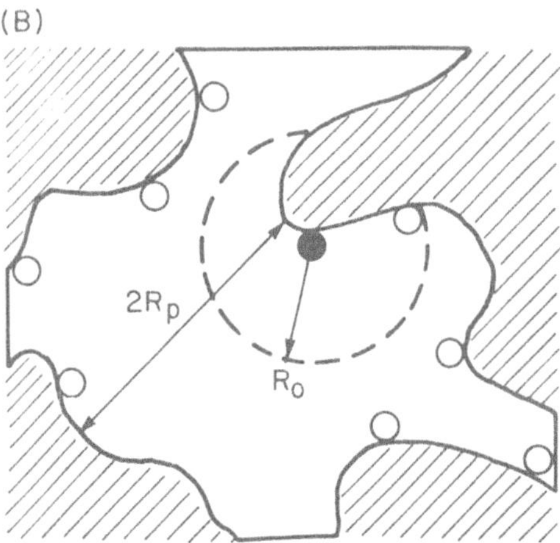

Figure 3. Schematic representation of the two lengths R_0 and $R_{p.}$
(A) $R_0 > R_p$. (B) $R_0 < P_p$.

(2) When $2R_p < R_{max}$, (Si-40) the probing length scale is above R_p. We are then sensitive to the pore network. In other words, R_{max} permits for interpore energy transfer which corresponds to $\bar{d}=3$. This is exemplified in Fig.3B, where the various length involved are presented. A transition to $\bar{d}=2$, for the locally smooth surface, should be possible at higher acceptor concentrations and short times. The latter case is difficult to achieve experimentally.

(3) When $R_p \sim R_{max}$, (Si-60), $\bar{d} \approx 2$. Here again the local vicinity of the primary building blocks is observed and crossover effects are more pronounced.

These conclusions agree with other characterization studies described in previous work [2,3]. Direct energy transfer provides a spectroscopic method to elucidate the local spatial organization of transparent restricted geometries and can serve as an example for a non-diffusive monomolecular reaction in such systems.

REFERENCES

1 J. Klafter and J.M. Drake, Eds Molecular Dynamics in Restricted Geometries (Wiley, New York, 1989).
2 J.M. Drake and J. Klafter, Phys. Today, 43, 46 (1990).
3 P. Levitz, J.M. Drake and J. Klafter, J. Chem. Phys., 89, 5224 (1988).
4 J.D. Byers, M.S. Friedricks, R.A. Friesner and S.E. Webber, in Ref.1.
5 H.F. Kauffmann, G. Landl and H.W. Engl, in Dynamical Processes in Condensed Molecular System, Ed. A. Blumen, J. Klafter and D. Haarer (World Scientific, Singapore, 1991).
6 N. Mataga, in Ref.1.
7 M. Maroncelli and G.R. Fleming, J. Chem. Phys., 86, 6221 (1987).
8 M.F. Shlesinger, Ann. Rev. Phys. Chem., 39, 269 (1988).
9 J.R. Miller, Chem. Phys. Lett., 22, 180 (1973).
10 M. Daoud and J. Klafter, J. Phys., A23, 180 (1973).
11 D.R. James, Y.S. Liu, P. de Mayo and W.R. Ware, Chem. Phys. Lett., 120, 460 (1985).
12 J.M. Beechem, M. Ameloot and L. Brand, Chem. Phys. Lett., 120, 466 (1985).
13 Some history and references are in E.W. Montroll and J.T. Bendler, J. Stat. Phys., 34, 129 (1984).
14 G. Williams and D.C. Watts, Trans Faraday Soc., 66, 80 (1970).
15 J. Klafter and M.F. Shlesinger, Proc. Natl. Acad. Sci. USA, 83, 848 (1986).
16 Inokuti and Hirayama, J. Chem. Phys., 43, 1965 (1978).
17 H. Scher and E.W. Montroll, Phys. Rev., B12, 1434 (1985).
18 T. Förster, Ann.Phys.(Leipzig) 2 (1948) 55; Naturforscher Teil, 4, 321 (1949).
19 R.G. Palmer, D. Stein, E.S. Abrahms and P.W. Anderson, Phys. Rev. Lett. 53, 958 (1984).
20 A. Blumen, J. Klafter and G. Zumofen, in Optical Spectroscopy of Glasses, Ed. I. Zschokke (Reidel, Dordrecht, 1986).
21 A. Blumen, Nuovo Cimento B63, 50 (1981).
22 J. Klafter and A. Blumen, J. Chem. Phys., 80, 875 (1984).
23 S. Redner and K. Kang, J. Phys., A17, L451 (1984).
24 A. Blumen, G. Zumofen and J. Klafter, J. de Physique, Colloque C7, 46 C7-3 (1985).
25 J.M. Drake, J. Klafter and P. Levitz, Science, 251, 1574 (1991).
26 R. Kopelman, Science, 241, 1620 (1988).
27 U. Even, K. Rademann, J. Jortner, N. Manor and R. Reisfeld, Phys. Rev. Lett. 52, 2164 (1984).
28 C.L. Yang, P. Evesque and M.A. El-Sayed, in ref.1.
29 N. Tamai, T. Yamazaki and I. Yamazaki, Chem. Phys. Lett., 147, 25 (1988).

30 P. Levitz and J.M. Drake, in Dynamics in Small Confining Systems, MRS Extended Abstracts, Eds. J.M. Drake, J. Klafter and R. Kopelman (MRS, Pittsburgh, 1990).
31 V.Degiorgio, T. Bellini, R. Piazza, F. Mantegazza and R.E. Goldstein, Phys. Rev. Lett., 64, 1043 (1990).

EXCITED STATE INTERACTIONS IN STABILIZED LASERS

A.M. Buoncristiani

Christopher Newport College
Newport News, VA 23606

S.P. Sandford

NASA Langley Research Center
Hampton, VA 23665

ABSTRACT

Recent developments in the active stabilization of lasers have led to light sources of extremely narrow linewidth and correspondingly long coherence times. Such stability enables several new laser measurement techniques. Stabilized lasers can provide a probe of the interactions between the excited states of laser transitions and the laser cavity itself. This paper will review frequency modulation techniques for stabilizing lasers, with emphasis on the relations between the laser linewidth and random processes within the lasing medium. An experiment to test stabilized lasers in the vibration free and microgravity environment of space and applications of ultra-stable lasers in space will be described.

I. INTRODUCTION

A laser is often described as a monochromatic light source because its bandwidth is much smaller than that of other sources of light. However, any laser has a measurable spread in the frequencies of light that it emits. This spectral width results from several different processes. Like any other oscillator, a laser is influenced by fluctuations or noise in its surroundings or occurring during its operation and these fluctuations broaden the laser lineshape. There are two major sources of such noise. Technical noise is a result of the interactions of the laser with its environment, for example, changes of laser cavity length induced by thermal effects or mechanical vibrations, or by irregularities in operating conditions, such as changes in the pumping intensity. Quantum noise is the result of randomness in the spontaneous emission processes responsible for laser action. Fluctuations which occur much faster than the time it takes to measure the lineshape broaden the line; slower fluctuations cause the line center to change with time or to drift. In this paper we will focus on the more rapid fluctuations contributing to line broadening.

Laser linewidth has been decreasing steadily with improvement in the design of laser systems and in the quality of laser materials. Recently, dramatic reductions in laser linewidth have been achieved by active external stabilization techniques that use feedback to reduce laser phase noise and, consequently, linewidth. As a result of these developments it is now possible to conceive of laser linewidths that are a fraction of a Hertz [1]. For example, the linewidth of a semiconductor laser-diode pumped solid state laser (Nd:YAG), which has an oscillating frequency of 3×10^{14}Hz, has been reduced to 3 Hz; this implies an oscillator stable (in the short term) to one part in 10^{14} [2,3].

Narrow linewidth lasers are of interest in the study of excited state interactions for several reasons. First of all, the laser has become an important tool in the study of the excited state. A principal advantage of a narrow line laser is that it increases the coherence time available for measurements. When the laser linewidth is broad and its correlation time correspondingly short, there can be at most one interaction taking place during a coherence interval. As the linewidth narrows, the correlation time lengthens and more and more interactions occur during the correlation interval; the probe beam begins to resonate with its target. Another reason for the interest in narrow line lasers is that, as the broadening effects of environmental influences on laser linewidth are reduced, intrinsic quantum effects become more important and it becomes possible to study these effects in the laser lineshape. It is possible also to probe the basic interaction responsible for laser emission and interactions between the atoms and the laser cavity.

The plan of this paper is as follows. The next section will give a heuristic discussion of the relationship between noise sources and the linewidth of a free running laser following the arguments developed in [4]. The following section will describe frequency modulation techniques for actively reducing the linewidth. Finally we will discuss an experiment designed to study the operation of ultra-stable lasers in the vibration free and micro-gravity environment of space.

II. LINEWIDTH OF FREE RUNNING LASERS

II.A. Characteristics of the Cold Laser Cavity

The linewidth of a free running laser depends upon the characteristics of the laser cavity supporting its oscillation and upon the power in the coherent radiation emerging from that cavity. The laser cavity can be characterized by looking at the net gain in intensity experienced by a light beam in making one complete circuit of the cavity. A simple Fabry-Perot cavity, shown in Figure 1, consists of two mirrors, separated by a distance L and enclosing an active medium of length ℓ. We assume that one mirror is perfectly reflecting and the other has a reflection coefficient R and transmission (1-R) to allow for some of this circulating energy to escape as a coherent beam. The active medium is characterized by a gain per unit length g and a loss per unit length α. As the beam makes a round trip in the cavity it experiences a net gain given by $G=R \exp(2\ell(g - \alpha))$. We can define the coherence length for a cold cavity (that is, a cavity which is not being pumped so g=0) as the distance a beam must travel before its intensity is reduced by a factor

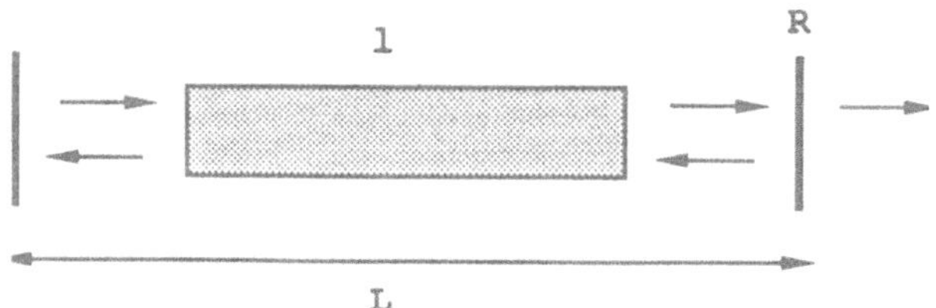

Figure 1. Schematic diagram of a Fabry-Perot laser cavity.

of 1/e. The number of round trips, p, required for this condition to be met is

$$G^{p} = e^{-1} \tag{1}$$

so that the coherence length, L_c, is given by,

$$L_c = 2\ L\ p = \frac{2L}{2\alpha\ell - \ln(R)}. \tag{2}$$

and the cavity coherence time is given by $\tau_c = L_c/c$. Thus, we would expect that the intensity of a beam in a cold cavity decays exponentially with a lifetime τ_c. The loss processes which cause this decay also operate during lasing to produce a loss of coherence in the beam. When the laser is operating continuously, two points along the emergent laser beam, separated by distances much larger than L_c will not be coherent. The coherence time can be used to define a linewidth for the cavity itself,

$$\Delta\omega_c = 2\pi\Delta\nu_c = 1/\tau_c. \tag{3}$$

II.B. The Effect of Noise on Laser Linewidth

The electric field at a point in the laser cavity located by r is given by

$$\vec{E}(\vec{r},t) = \vec{\varepsilon}(E_0+\Delta E(t))e^{i\,\vec{k}\cdot\vec{r} - i\omega t + i\phi(t)}. \tag{4}$$

where $\vec{\varepsilon}$ is the polarization vector, $\vec{k}$ the wave vector defining the wave propagation and $\Delta E(t)$ and $\phi(t)$ are random processes representing amplitude and phase noise on the laser field respectively. In general, the various noise sources may induce fluctuations on both the amplitude and the phase of this field but the response of the laser system to each of these types of noise is quite different.

A fluctuation in the amplitude of the field will be rapidly damped by the strongly non-linear laser dynamics acting to restore the amplitude to its mean value. The field fluctuation may oscillate as it is damped out with characteristic frequencies of relaxation oscillations.

No such damping takes place for fluctuations in the phase. Consequently the phase shift over a time interval τ, $\Delta\phi = \phi(t+\tau)-\phi(t)$, undergoes a random walk, characterized by the time averages, $<\Delta\phi> = 0$ and by $<\Delta\phi^2>$ increasing linearly with τ, that is, by $<\Delta\phi^2> = \gamma\tau$. The correlation function for the field can be shown to depend upon $<\Delta\phi^2>$ exponentially [4] through the expression

$$<E(t)E(t+\tau)> \propto \exp(-<\Delta\phi^2>/2), \tag{5}$$

so that, since $<\Delta\phi^2> = \gamma\tau$, $1/\gamma$ becomes a measure of the laser linewidth. Assuming that the phase shifts occur in discrete jumps, we can represent the variance in the phase shift over time interval τ as the product of the number of phase shifts occurring in that interval $N(\tau)$ and the variance of the single phase shift $<\delta\phi^2>$,

$$\gamma\tau = <\Delta\phi^2> = N(\tau)<\delta\phi^2>. \tag{6}$$

This implies that the width of the line induced by phase variations is given by

$$2\pi\Delta\nu_{laser} = \gamma = \frac{N(\tau)}{\tau} <\delta\phi^2> ; \tag{7}$$

thus, the phase noise induced linewidth is the product of the rate at which phase shifts occur and the variance in a single phase shift. If we assume, further, that the only noise sources operating are those associated with quantum fluctuations we can estimate these two factors separately. A system which has only quantum noise should, on average, experience one phase shift in a coherence time so that phase shifts will occur at a rate given by

$$\frac{N(\tau)}{\tau} = \Delta\omega_c \tag{8}$$

The variance in the phase shifts, $<\delta\phi^2>$, can be estimated by referring to Fig. 2 where the phase shift, $\delta\phi$ of the field is related to the phase angle of the field fluctuation, $\Delta E\sin(\theta) = E_0\delta\phi(\theta)$. We can calculate $<\delta\phi^2>$ from an ensemble average of $\delta\phi(\theta)$ assuming that each angle θ is equally likely,

$$<\delta\phi^2> = \int_0^{2\pi} \frac{1}{2\pi}\,\delta\phi^2(\theta)d\theta = \frac{1}{2}\left(\frac{\Delta E}{E_0}\right)^2. \tag{9}$$

It remains now to calculate the magnitude of the two electric field terms, ΔE and E_0; they can be estimated from the energy density associated with an electric field E, ($\varepsilon_0 E^2/2$, where ε_0 is the permittivity of free space). The field ΔE is that of a single laser photon of energy $h\nu$, so

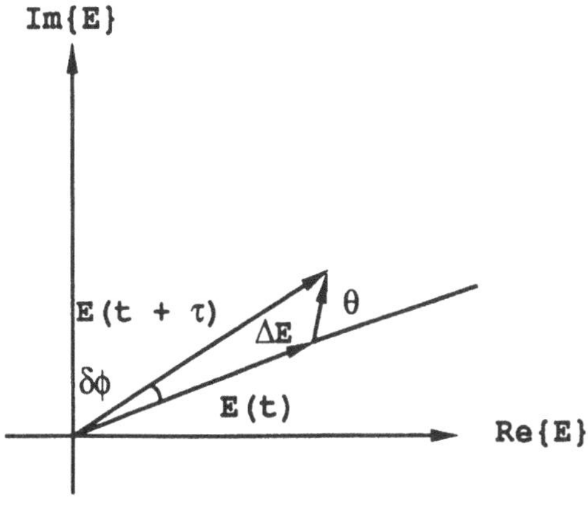

$$E(t)\ \delta\phi \sim \Delta E\ \sin(\theta)$$

Figure 2. Relationship between the phase shift $\delta\phi$ generated by an incremental change in the electric field ΔE and the angle between E and ΔE.

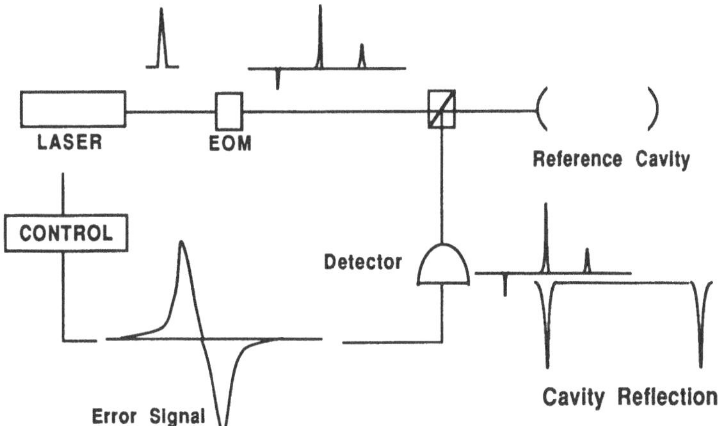

Figure 3. Schematic diagram of the frequency modulation laser stabilization technique [6]. The single laser line (carrier) is modulated by an electro-optic modulator (EOM) to add two out-of-phase sidebands. The sidebands are reflected from the reference cavity while the carrier is averaged within the cavity. The signals are mixed on a detector to generate an electronic error signal which is fed back to the laser.

$$\Delta E^2 = \frac{2h\nu}{\varepsilon_0 V}. \tag{10}$$

V is the volume over which the energy density is calculated. The field E_0 is that of the emergent laser beam and the energy associated with it can be approximated by the product of the laser output power, P_{out} and the cavity coherence time; thus

$$E_0^2 = \frac{2\, P_{out}\, \tau_c}{\varepsilon_0 V}. \tag{11}$$

Combining the equations (9), (10), and (11) we see that the variance of the phase shift due to quantum noise depends upon the laser parameters through

$$<\delta\phi^2> = \frac{1}{2}\left(\frac{h\nu}{P_{out}\tau_c}\right). \tag{12}$$

We can now calculate the laser linewidth from equations (7), (8) and (12)

$$\Delta\nu_{laser} = \pi \frac{h\nu(\Delta\nu_c)^2}{P_{out}} \tag{13}$$

This formula was first derived by Schawlow and Townes (apart from a factor two which has since been corrected in their original work) giving the lower limit to the linewidth of a free running laser [5]. If the only processes responsible for broadening the laser line were those due to intrinsic random phase fluctuations in the laser itself, the laser would have a linewidth given by equation (13).

III. REDUCTION OF LASER LINEWIDTH

In a free running laser, the cavity serves two distinct roles. One is to define the frequencies of axial modes of the laser beam and the other is to maintain some portion of energy circulating within the cavity to convert energy stored in the pumped medium into coherent radiation. A fraction of the circulating radiation is allowed to escape in the emergent beam. These two roles compete with each other in that higher mirror reflectivity (less loss) is needed to define a narrow cavity linewidth but high reflectivity allows less energy for the output beam. In active frequency stabilization these roles are separated. An external reference cavity is used to provide the definition of frequency and this reference cavity can be made with the highest quality mirrors available without influencing the laser output power. The laser cavity can be as lossy as needed to provide for efficiency lasing.

In order to provide for this separation of roles it is necessary that the reference cavity track excursions of the laser frequency from a central reference frequency and convert these excursions into an error signal that can be used to shift that laser frequency back to the reference value. An electro-optic device to generate this error signal was developed by Drever [6]. The idea of this device is simple and elegant.

Light from the laser is modulated electro-optically (at tens of MHz) with a small modulation index so that the modulated output consists of a carrier frequency and two sidebands 180 degrees out of phase with each other. Refer to Fig. 3. The modulated light is passed through a polarizing beam splitter and quarter-wave plate to be incident on a high finesse reference cavity. Light from the carrier passes into the cavity but the sidebands are sufficiently shifted to be reflected from the cavity. The carrier light admitted to the cavity resides there, on average, for a time equal to the cavity round trip time times the finesse. The light emerging from the front of the cavity consists of the reflected sidebands plus the time averaged carrier signal; this light is mixed on a detector where the slowly varying signal at the modulation frequency is extracted, demodulated, and used as an error signal to be fed back to the laser where it controls the frequency by means of a piezoelectric crystal mounted on top of the laser crystal. This servo system reduces the laser noise over the servo bandwidth producing a laser with extremely narrow linewidth.

Research directed by R.L. Byer at the Ginzton Laboratory of Stanford University over the last decade has led to the development of an extremely stable, diode pumped solid state laser [7,8,9]. This laser, now in commercial production, has a free running linewidth of 10 kHz (one part in 10^{10}). During the same period, work at NIST at Boulder, Colorado under J.H. Hall has refined methods for the active stabilization of lasers primarily through the development of the Pound-Drever frequency locking technique [1]. Using the feedback stabilization technique developed at NIST, the Stanford group has recently been able to reduce the linewidth of their Nd:YAG laser to 3 Hz (one part in 10^{14}).

IV. THE SPACE EXPERIMENT

The limits to stability of lasers in terrestrial laboratories is set by environmental and microseismic noise and by gravity induced distortions in the optical devices used as frequency references. Stabilized laser oscillators will operate even more stably in space where vibrational and gravitational effects are significantly reduced. Furthermore, space is the appropriate environment for a variety of important applications of ultra-stable lasers.

In order to examine the possibility of utilizing the space environment to increase the stability of lasers, NASA is preparing for an experiment to test laser stabilization in free flight. This experiment is called SUNLITE, Stanford University NASA Laser Inspace Technology Experiment. The objective of the SUNLITE project is to verify the increased stability of laser oscillators operating in the vibration free and microgravity environment of space by measuring the linewidth variations in the beat note signal from two actively stabilized lasers, and to develop a basic phenomenological understanding and the associated quantitative tools to design ultra-stable lasers for use in space.

The SUNLITE experiment aims at examining the stability characteristics of the most stable oscillator developed to date. In order to do this it must compare the operation of one laser locked to separate high finesse reference cavities will be heterodyned and the beat note signal analyzed. These data will allow the derivation, after the mission, of the standard measures for short term stability and long-term stability. The measurements will also serve to verify understanding of the system when used in conjunction with data obtained on Earth. The experiment design requires the use of Non-Planar Ring Oscillators (NPRO) developed at Stanford. The reference cavity will be built around ultra-high reflectivity mirrors.

An initial conceptual design for this instrument has been completed. Various implementations of this design are now being studied. The current plan calls for a final design by October 1991 and an instrument ready for spaceflight by August 1993. This schedule insures that a stable laser oscillator will be available for several critical technological applications such as, deep-space coherent communications, optical clock technology, global positioning and astrometry, as well as for scientific measurements of relativistic effects and the detection of gravitational waves.

V. APPLICATIONS OF ULTRA-STABLE LASERS IN SPACE

V.A. Gravity Wave Detection

One of the most revolutionary applications of technology emerging from this program is a family of gravitational wave detectors able to probe interactions among the most massive entities in our universe. Using laser gravitational wave interferometers in space to detect gravitational pulses with periods up to a few hours, events involving the motion of large masses, such as the collisions of black holes, can be studied. Ground based systems on a smaller scale (~ 4 km arms) are now being designed, but such devices will have a low frequency cutoff of about 100 Hz. In a proposed experiment titled "Laser Gravitational Wave Observation in Space" (LAGOS), a laser operating at the fundamental linewidth limit will allow the measurement of gravity waves in a one million kilometer interferometer in orbit around the sun. This experiment is enabled specifically by the recent advances in solid state laser technology including the very recent studies on laser stabilization. Several years ago an interferometric gravitational wave antenna was proposed at NASA's Langley Research Center. This antenna would use three geosynchronus satellites connected by laser beams. It would operate as a rotating interferometer and continuously sense the relative separation of the satellites.

V.B. Optical Clock Technology

The laser oscillator used in the SUNLITE experiment has a potential for an accuracy of one part in 10^{17}. This extraordinary stability coupled with recent demonstration of sub-Kelvin temperatures and sub-doppler spectroscopic techniques make it possible to conceive of an optical clock having a short term oscillator or flywheel based on a solid state laser stabilized, in the long term, by interrogating a laser cooled atom or a trapped ion. The flywheel oscillator in the atomic clocks in current use are based on quartz oscillators. Replacing these with an optical frequency oscillator represents a significant advance in clock technology.

V.C. Deep Space Communications

The data link capacity of current deep space communications is approximately 10 kbits/second (Voyager at Uranus). New instruments such as synthetic aperture radar (SAR) and high resolution imaging spectrometer (HIRIS) will require data link capacities on the order of 100 Mbits/second. Future manned missions such as a piloted mission to Mars will require even higher data rates enabling two-way video. Optical communications have several benefits. One is the decreased beam divergence due to Global Positioning System.

Currently, two way, transponded communications between earth based stations and space craft use doppler shift information to determine orbit parameters and/or navigational information. Coherent detection is now used only on the earth station. Plans for coherent detection on spacecraft, either to improve sensitivity of the earth to space system or to enable communication between spacecraft, will require oscillators with significantly more stability. The planned development of the deep space communications network (DSN) requires increased stability both to improve communications from earth to space and to enable coherent communication between spacecraft.

V.D. Tests of Relativity

Several tests of the special and general theory of relativity are possible with sufficiently stable lasers. Recent studies have suggested that a 100-fold increase in the sensitivity of a space-based Kennedy-Thorndike experiment is possible with a stabilized sold state laser. This effort coincides with NASA's long-term interest in contributing to fundamental understanding of space and time.

V.E. Laser Cooled Atoms

Another, equally fundamental study, involves the use of stable laser sources to probe the behavior of laser cooled atoms in the reduced gravity environment of space. Attempts to study the behavior of small ensembles of atoms at temperatures in the tens of microkelvin are hindered by the short observation times available in the 1 g environment. Observation times in space can increase the observation time by a factor from 10 to 100.

ACKNOWLEDGEMENTS

It is a pleasure to acknowledge the many important contributions to this work by the research group of Robert L. Byer of the Ginzton Laboratory at Stanford University, Eric Gustavson, Tom Kane, Allan Nilsson and Tim Day (who made it all work). We have also benefited from the insight of John Hall at NIST and Joe Hafele of Christopher Newport College.

REFERENCES

1. C. Salomon, D. Hils, and J.L. Hall, J. Opt. Soc. Am., 5, 1576 (1988).
2. T. Day, A.C. Nilsson, M.M. Fejer, A.D. Farinas, E.K. Gustafson, C.D. Nabors, and R.L. Byer, Electronics Letters, 25, 810 (1989).

3. T. Day, E.K. Gustafson and R.L. Byer, Opt. Lett, vol. 15, pp. 221-3 (1990).
4. S.F. Jacobs, Am. J. Phys., vol.47, No.7, p.597, (1979).
5. A.L. Schawlow and C.H. Townes, Phys. Rev., vol. 112, pp. 1940-1949, (1958).
6. R.W.P. Drever, J.L. Hall, F.V. Kowalski, J. Hough, G.M. Ford, A.J. Munley, and H. Ward, Appl. Phys. B, 31, 97 (1983).
7. Thomas J. Kane and Robert L. Byer, Opt. Lett., 10, 65 (1985).
8. Thomas J. Kane, Alan C. Nilsson, and Robert L. Byer, Opt. Lett., 12, 175 (1987).
9. A.C. Nilsson, E.K. Gustafson, and R.L. Byer, J. Quant. Electron., 25, 767 (1989).

SEMICONDUCTOR QUANTUM DOTS IN AMORPHOUS MATERIALS

Renata Reisfeld*

Department of Inorganic and Analytical Chemistry
The Hebrew University of Jerusalem, 91904 Jerusalem

ABSTRACT

Quantum dots based on semiconductors have acquired increasing interest recently. Band filling and screening of excitons by free carriers are the most important mechanisms for the large resonant nonlinearity of semiconductors. The information in the literature is mostly concerned with semiconductor crystals or conventional filter glasses doped by semiconductor particles. Due to the spatial confinement of the photogenerated charged carriers, the electronic levels of small size particles rise to higher energies than in bigger particles, In such a case, one can treat the problem in a one-dimensional box; whereas, the energetic band continuum splits into discrete quantized levels which can be detected in the absorption and excitation spectra.

We shall discuss recent techniques of incorporation of semiconductor particles of varying size into glass bulks or films prepared by the sol-gel method, composite materials and polymers. Contrary to the conventional glass filters, the time response of nonlinearity in semiconductor doped glass or polymer films can be much faster because of the higher surface density of the carriers responsible for the nonlinear behavior. It will be shown how excited electronic levels of the quantum dots can be obtained from the inflection points in their absorption spectra and the size and distribution of the particles from luminescence spectra. The semiconductor doped glasses are compared to glasses doped by organic dyes having nonlinear properties.

I. INTRODUCTION

Materials for optical memories, displays, picture processing, optical computing, optical storage, optical signal processing as well as materials

* Enrique Berman Professor of Solar Energy

having nonlinearities for phase conjugation and second harmonic generation are of great scientific and application interest.

Especially, there is a great need for nonlinear optical materials that can be used with low intensity light sources for applications such as image processing and optical switching and phase conjugation.

The phase conjugated waves are able to restore distorted optical beams to their original unaberrated state under reflection from a nonlinear medium [1].

II. THIRD-ORDER SUSCEPTIBILITY

The change of polarization of materials induced by electromagnetic field E is usually expressed as:

$$P = \varepsilon_0 \chi(E) \cdot E \tag{1}$$

where $\chi(E)$ is electric susceptibility and P and E are complex.

To account for effects induced by strong electromagnetic fields it is convenient to expand the susceptibility into a power series:

$$\chi(E) = \chi^{(1)} + \chi^{(2)} \cdot E_1 + \chi^{(3)} \cdot E_1 \cdot E_2 + \ldots \tag{2}$$

where the second- and third-order coefficients are tensors.

In disordered media the second term is zero due to symmetry reasons. The origin of the large nonlinear susceptibility in semiconductor Q particles is different from that of organic molecules having a good singlet triplet transfer. A short summary of nonlinear organic molecules which were discussed previously is given below [2-4].

In organic molecules the optical excitation from the ground single state brings the molecule to the first and second excited singlet states. This is followed by nonradiative relaxation from the second singlet state to the first excited state followed by an intersystem crossing which transfers the excited state population to the lowest lying triplet state. Provided $\Delta E_S - \Delta E_T$, the energy difference between the first excited singlet S_1 and the triplet state T_1 is of appropriate value, delayed fluorescence from the excited siglet state occurs after a back intersystem crossing that brings back the population to the excited singlet states from vibrational excited states of the lowest lying triplet manifold. Many of the relaxation mechanisms that can quench the triplet state do not exist when the dye is held rigidly in a solid matrix, and hence the lowest lying triplet state has a very long lifetime. Because of this longevity of the triplet state, significant population can be trapped in that state even by weak optical fields. The saturation intensity I_s is therefore quite small, less that 100 $mWcm^{-2}$, and since the nonlinear susceptibility $\chi^{(3)}$ varies inversely with I_s it is extremely large, about 0.1 esu as compared with about 10^{-3} esu for Multiple Qantum Wells (MQW), see also Table 2.

The third order nonlinearity of the materials allows four wave and three wave mixing experiments. Long-lived dynamic grating is formed by the interaction of the three beams in the four-wave mixing experiment if such material has a lifetime of tens to hundreds milliseconds.

The third-order susceptibility of such a system may be described by:

$$\chi^{(3)} = (n^2c^2\alpha/24\pi^2I_s)\cdot(\delta + i) \quad (3)$$

where n is the unsaturated refractive index of the material, α is the unsaturated small field attenuation coefficient, I_s is the saturation intensity, defined as the light intensity required to bring half of the population of the saturable absorber to the metastable state and δ is the normalized detuning factor. One must note that the trends of α and δ are opposite, but their frequency dependences differ, therefore the optimum magnitude of $\chi^{(3)}$ can be found at some place off the absorption center.

The value of the saturation intensity can be calculated by fitting the differential equation for intensity, $dI/dz = -(\alpha + \alpha_e)I$, to the experimental saturation curve (transmittance versus applied optical field) where α_e is the absorption of the excited state and α, the absorption coefficient for linearly polarized light, is given by:

$$\alpha = (N\sigma/S)\cdot[1 - (1/\sqrt{3S})\cdot\tan^{-1}\sqrt{3S}] \quad (4)$$

where σ is the unsaturated absorption cross-section and $S = I/I_s$.

For the preliminary estimation of applicability of a saturable dye for phase conjugation, the saturation intensity can be calculated without actually performing the saturation experiment. This can be done by using the well-known expression derived for solutions of saturable absorbers:

$$I_s = h\nu/\sigma\tau \ [mW/cm^2] \quad (5)$$

where τ is the lifetime of the metastable level, which is determined from its decay curve after pulsed laser excitation. This formula ignores both the absorption of the excited state and the fact that in solids the saturable absorber excited by a polarized light is saturated more easily than in solution [5]. Hence the formula overestimates the saturation intensity by a factor of two or three, i.e. one may expect that the actual measurement will reveal saturation intensity lower than predicted by Equation (5). Proceeding further, we can also, by using the predicted value of the saturation intensity and measured values of absorption coefficients, calculate the third-order susceptibility from Equation 3. As a figure of merit we can use only the real part of the formula which, apart from the constants, may be expressed as:

$$Re[\chi^{(3)}] = (N\sigma^2\tau/h\nu^2)\cdot\delta \quad (6)$$

where N is the concentration of the saturable absorber (molecules/cm^3) and σ is the cross-section of absorption.

In order to make these nonlinear materials applicable for technological uses the dyes must be incorporated in stable rigid matrices

Table 1. Lifetimes and saturation intensities of organic dyes in boric acid, heavy glass and in impregnated sol-gel bulks [mW/cm^2].

Compound	Matrix	Lifetime [msec]	Saturation Intensity, [mW/cm^2]
Acridine Yellow	Boric acid	46	493
	Heavy glass	27	200
	Composite	65	300
Acridine Orange	Boric acid	25	652
	Heavy glass	13	273
	Composite	52	310
Fluorescein	Boric acid	1115	2.7
	Heavy glass	25	91
	Composite	175	28

and then, their absorption coefficients and lifetime measured. The well-known technology of dyes in polymer matrices is not sufficient since both plastics and dyes undergo decomposition under prolonged irradiation and under harsh external conditions. On the other hand, one cannot use the current glass technology for incorporation of dyes because of the high temperature of the process at which the dyes decompose. Therefore, these dyes were incorporated in glasses prepared by the sol-gel technique at low temperatures.

First, fluorescein [5] and its derivatives [4, 6] have been studied as materials for nonlinear optics, then other organic dyes, methyl orange and acridine dyes, were studied in composite glasses [7-9]. Lifetimes and saturation intensities of the dyes are presented in Table 1.

III. SOL-GEL GLASSES

Some organic laser dyes have been widely reported as having desirable properties for optical gain, laser applications and other organic molecules for nonlinear materials. Finding suitable host materials for these organics presents a challenge and, in the case of laser dyes, devices have been primarily restructured to liquid state applications. There are a number of devices, however, in which solid state gain media would be quite advantageous. Attempts have been made to fabricate solid state organic dye materials using polymeric hosts such as poly (methylmethacrylate), poly (carbonate), poly (styrene) and poly (vinyl alcohol) [10,11]. These hosts, however, have been shown to be inherently lacking in mechanical and thermal properties, photostability and refractive index uniformity. Inorganic glasses on the other hand, do possess extremely good optical, thermal and chemical stability, however the processing temperatures of conventional glasses promote the rapid decoposition of most organic species. In view of the number of dyes and dopants available the prospect of achieving a wide range of sol-gel based tunable solid state lasers and nonlinear materials are of substantial interest to the optical materials community.

The sol-gel process opens up many possibilities for synthesis of amorphous materials which are either totally inorganic in nature or a composite of inorganics and organics. The process which is based on polymerization, starts from molecular precursors and a macromolecular oxide network is obtained via hydrolysis-condensation reactions which can be controlled by the chemical design of molecular species.
The most widely investigated system involves silica-based precursor glasses which are prepared by polymerization of a silicon alkoxide, $Si(OR)_4$ [2,12-16]. The viscosity properties of sols can be adjusted allowing easy deposition of transparent coatins onto glass, ceramic or polymeric substrates [17]. Sol-gel chemistry is performed in solution at lower temperatures than conventional chemical methods. Homogeneous doping by mixing components at a molecular level, synthesis of metastable or amorphous phases allowing larger concentrations of chromophores and synthesis of mixed organic-inorganic materials can then be performed.

In the case of organosilicate glasses (organically modified ceramics or ormocers) the silicate network may be modified by organic substituents such as alkyl (e.g. methyl) groups or other functional groups (e.g. glycidyloxypropyl) [18,19] which may form organic copolymers that penetrate the silicate structure. In principle, Si may be substituted by Al, B, Ti or Zr to yield ceramics of variable mechanical and optical properties [14]. At present there is no consistent picture of the microstructure of these materials. In the investigation of glassy systems several analytical methods [20-22] have been used among which are neutron scattering and light scattering.

The coating materials based on ormocers can be synthesized from alkoxides of Si, Zr or Ti and from silanes with functional groups (e.g. 3-glycidyloxypropyltrimethoxysilate, abbreviated "GLYMO") by sol-gel processing. During the time of the process a low degree of hydrolysis and condensation may be achieved by the moisture of air. The coating solution can then be poured on PMMA plates or glass plates and dried in air.

Through the impregnation of porous silica gels with organic monomers, such as methylmethacrylate, followed by in situ polymerization, transparent silica gel-PMMA composites can be prepared [21]. Such transparent composites, prepared at relatively low temperatures, have potential applications as optical hosts for organic dye molecules.

The main reason that most polymer-glass composites were opaque until now was due to light scattering. Glass fiber reinforced epoxy, for example, typically uses fibers of 5 to 20 microns in diameter and, often, many metres in length. These large "phase dimensions" result in light scattering, despite the fact that both glass and epoxy are quite transparent in bulk form. In the new composites, however the phase dimension of the order of 10 nm, much smaller than the wavelengths of visible light (390 to 780 nm). Hence, this new class of glass-polymer composites is quite transparent which opens up a wide variety of applications for which composite materials had been previously excluded.

The ability to synthesize inorganic polymers using sol-gel processing with little or no heating makes it possible to dope these gels with a variety of organic and organometallic molecules [23,24] as well as with semiconductor and silver particles. The emission properties of the molecules have been used to optically probe sol-gel chemistry, [2, 25] and structure [2, 26]. Another active area for organic doped sol-gel glasses is to use dopants to induce selected optical properties and to synthesize new optical materials [2, 27]. Among the properties reported to date are solar light concentrators [28], photochromism, nonlinear optical effects [2] and tunable laser action [2, 29]. The latter was achieved by incorporating organic laser dyes in a sol-gel matrix and thus represents a potentially important direction for solid-state laser material in the visible range of the spectrum. Stable solid-state lasers tunable in the visible were first prepared in our group by introduction of perylene based dyes into composite glasses [30]. Slope efficiency and threshold energy of a laser emitting between 510-620 nm were 7.4% and 60μJ. Several other dyes for tunable lasers in the visible are also investigated [31-33].

IV. SEMICONDUCTOR DOPED GLASSES

Absorption and emission spectra of colloidal and microcrystalline particles of CdS [34,35] and silver [36,37] show distinct differences from the respective bulks.

These observations were explained by the quantum size effect [38] resulting from the confinement of the electron and the hole in a small volume, whereby the continuous energy band characteristic of bulk splits into discrete level. The positions of these electronic levels which are shifted to higher energies in respect to the levels of bulk have been calculated by several authors [39,40]. Size quantization effects were observed in the optical spectra of small crystalites of CdS in bulk silica glass, obtained by the sol-gel method [41-44].

The reduction of the dimensionality of the quantum dots was predicted to lead to enhanced nonlinear optical properties [45]. Indeed, the nonlinear behaviour of small particles of CdS incorporated in thin films of polymers was observed [46,47]. The nonlinear properties of CdS incorporated in thin films prepared by the sol-gel method and in composite organic-inorganic polymers was described by us recently [48,49].

The preparation of the glasses doped by CdS was according to the general procedures for preparation of glass films and bulk in references [2,12,13,50]. The thickness of the films is a function of the speed of withdrawal of substrate glass from the precursor solution, pH, concentration of alkoxide in the solution and orientation of the sample in respect to the direction of withdrawal. Control of these conditions is necessary in order to guarantee reproducibility of the coatings.

The films are prepared from precursor solutions having the typical molar ratios: Tetramethoxysilane (TMOS) or Tetraethoxysilate (TEOS): water:alcohol = 1:4(6):11:X, where the number in parentheses refers to TEOS. The molar ratio of Cd varied from 0.0077 to 0.195 (0.12) mol of Cd per 1 mol of TMOS (TEOS). X = CdS

ORMOCER (Organically Modified Ceramics) films are prepared according to the recipes given in [19,50] and modified in our group for incorporation of dyes and colorants [51,52].

The fresh films are then exposed to gaseous hydrogen sulfide for a few minutes, dried and heated to 250°C [48,49]. The thickness of the films is determined by an interface method as described in [51].

The average sizes of the CdS particles in the films are measured by two independent methods. The first is Transmission Electron Microscopy of thin films containing CdS stretched on copper grids.

The second method is based on X-ray diffraction on powders, where Sherer's equation is used to find the correlation between the width (in the units of 2-Theta) of the diffraction peaks with the average size of the particles. Concentration of CdS particles in the films is of the order of 10^{21} molec./cm^3. However, number of molecules in a slab of $cm^2 x 10^{-5}$ cm (which is the order of magnitude of thickness of the films) is about 10^{16} molecules. Such a small number of molecules requires very long scan times of 100-300 seconds/0.05-0.02 deg.

V. NONLINEAR PROPERTIES

Thin sol-gel glass and ORMOCER films doped by CdS nanoparticles produce strong third harmonic generation (THG) signals when irradiated at 1.06 μm and observed at 0.355 μm. The strongest THG signals were about 1500-2000 times stronger than those generated by the glass substrate alone. This can be compared with the recent results of Ohashi et al. [47], where THG signals from CdS in a swollen thin polymer film were about 1.5 times higher than the THG signals of quartz when excited by 1.06 μm laser radiation. They observed also that the ratio increased to about 11 with the change of the fundamental from 1.06 μm to 1.45 μm.

Analysis of THG by thin films can be very involved [53-60], due to multiple reflections in the film, dephasing between fundamental and harmonic waves, and interference betwen THG signals generated in different films layers. In a transmission geometry, these effects can be usually unraveled only by monitoring the THG as a function of the angle of incidence. In our reflection experiments, analogous effects must also be considered. In this case the observed THG may be created in:

a) Nonlinear reflection at the air/film interface.

b) Transmission through the film, followed by linear reflection at the film/substrate interface.

c) Nonlinear reflection at the film substrate interface.

where the term " Nonlinear reflection" refers to the surface layer contributing to THG in reflection at the interface. The depth of this layer is given by

$$d_{eff} = \lambda/[6\Pi\cdot(n_{\omega}\cos\theta_{\omega} + n_{3\omega}\cos\theta_{3\omega})] \qquad (7)$$

which is about 200A in our case. Note that in b) the forward and backward propagating fundamental should be taken into account.

In principle, the observed THG will be given by the coherent addition of THG produced by the above effects. This, however, will be simplified if one of the effects will be dominant. For example, if the film has a $\chi^{(3)}$ much higher than that of the substrate, effect a) or b) should be much stronger than effect c). Whether effect a) or b) is then the most important depends on the film thickness, transparency and the reflection coefficient at the film/substrate interface.

Using s-polarized input, the THG will be also s-polarized arising via χ_{xxxx} term. The THG field generated in reflection via effect a) is proportional to

$$(1+R_{a-f})\cdot(t_{a-f})^3\cdot\chi^{(3)}\cdot d_{eff} \tag{8}$$

and via effect b) to,

$$R_{f-g}t_{f-a}e^{-\alpha}\{[1-e^{-\alpha}]/\alpha\}\cdot(t_{a-f})^3\cdot\chi^{(3)}L/\cos\theta_f \tag{9}$$

Here, **t** and **R** are the Fresnel field factors for reflection and transmission at the interface (**f**=film, **s**=glass substrate, **a**=air), **L** is the film thickness and the terms $e^{-\alpha}$ and $[1-e^{-\alpha}/\alpha]$ refer to self-absorption of the third harmonics field in the film itself. α is related to the optical density of the film, measured at normal incidence, by

$$\alpha = 2.303x(\text{optical density})/2\cos\theta_f \tag{10}$$

Note that in expression (9) for the "bulk" signal we have neglected the backward propagating fundamental wave and dephasing of the fundamental and harmonic waves, due to the close similarity of indices of refraction of sol-gel and regular glass and the small film thickness, respectively.

Straightforward calculation shows that in the sol-gel glass of thickness 0.15μm, refractive index 1.6 and optical density at 355 nm of 0.4 per layer, the THG from reflection at the air/film interface is dominant. In this case the intensity of the reflected signal is proportional to the square of the field hence square root ratio of the measured intensities is,

$$\sqrt{(I_f/I_g)} = e^{-\alpha}\cdot[(1-e^{-\alpha})/\alpha]\chi_f^{(3)}/\chi_g^{(3)} \tag{11}$$

We get here $\chi_f^{(3)} \cong 60\cdot\chi_g^{(3)}$ or $\chi_f^{(3)} \cong 2\cdot10^{-12}$ esu, taking the literature value for $\chi_g^{(3)}$ [60].

On the other hand, for ORMOCER sample (of the kind abbreviated GLYMO), which has thickness of 1 micron, n=1.48 and optical density, OD=0.27 at 355 nm, the larger THG would be expected to come from the interface, although it would be not much larger that the signal from the bulk film. Thus measurement at a single incidence angle is insufficient, as the relative phases of the two contributions are not known. However, we may conclude that to within a factor of two $\chi_f^{(3)} \cong 10^{-12}$ esu.

These values of $\chi^{(3)}$ are much lower than those of pure CdS, but still quite high considering the low volume fraction of CdS particles in the film. At the wavelength of 1.06 μm, $\chi_f^{(3)}$ presumably contains some enhancement due to three photon resonance. According to the work of Ohashi et al. [47] we can expect 10 fold increase of the THG signals by changing the fundamental wavelength from 1.06 μm to 1.45 μm. Table 2 compiles the typical $\chi^{(3)}$ for a number of organic and inorganic materials.

Table 2. Nonlinear optical materials; typical values of third order nonlinear susceptibility.

Material	Matrix	Ref.	$\chi^{(3)}$[esu/cm^3]	Remarks
CdS	glass film	48	10^{-12}	THG, 1-10 w.t%
CdS	bulk	61	10^{-4}	
CdSe	silica	62	$1.3* 10^{-8}$	sputtered film
GaAs	MQW	61	10^{-3}	
InSb	bulk	61	$6*10^{-3}$	
semiconductors	glass	61	10^{-11}-10^{-7}	commercial filters
CuCl	sol-gel glass	63	10^{-8}	microcrystallites
PbI_2	pure	64	10^{-4}	evaporated on quartz
Gold	glass	65	$1.5*10^{-8}$	nanoparticles
Silver	glass	65	$4*10^{-10}$	nanoparticles
polyacetylene	pure	66	$4*10^{-10}$	Oriented film
polydiacetylene	pure	67	$2*10^{-9}$	Film
vinylamines	pure	68	$(1-13)10^{-10}$	Films
PPV	pure	69	$(1-5)*10^{-10}$	Film
Methyl orange	glass	2	$4*10^{-6}$	composite glass
Acridine orange	glass	2	$2*10^{-4}$	composite glass
phtalocyanine	pure	70	10^{-8}	L-B films
Chrysoidine	PVA	71	$2*10^{-7}$	Films
Fluorescein	boric acid	2,5	$(1-10)*10^{-1}$	

PPV: Poly-(p-phenylene vinylene)
PVA: Polyvinyl alcohol.
MWQ: Multiple Quantum Wells.
THG: Third harmonic generation.

Relation between the size of the particles and their absorption spectra is seen from Figures 1-3. Figure 1 presents an example of the absorption spectra and their first derivatives of the samples containing 0.0077 mol of Cd per mol of TMOS (solid curves, left scale) and 0.039 mol of Cd per mol of TMOS (broken curves, right scale). The positions of the inflection points for these and other samples are given in Figure 2, as a function of the molar fraction of Cd (full circle). The data in TEOS samples are also given (open circles). As can be seen from the figure, the wavelengths of the inflection points rise monotonically with the molar fraction of Cd. At molar fractions higher than about 0.15 the wavelengths of the inflection points reach a plateau of about 510 nm. These inflection points can be correlated with the size of CdS particles shown by the measurements by TEM or X-ray diffraction (XRD), Figures 3 and 4.

Figure 3 presents a TEM micrograph of CdS particles in glass film. The majority of particles has the size of 20-50 A. A few particles are larger than 200A.

The inhomogenous distribution of particle sizes in the glass is expected and is analogous to the inhomogenous spectral broadening [72], where there is statistical distribution of crystallographic sites having slightly different energies. However, the dominant behaviour (the maximum

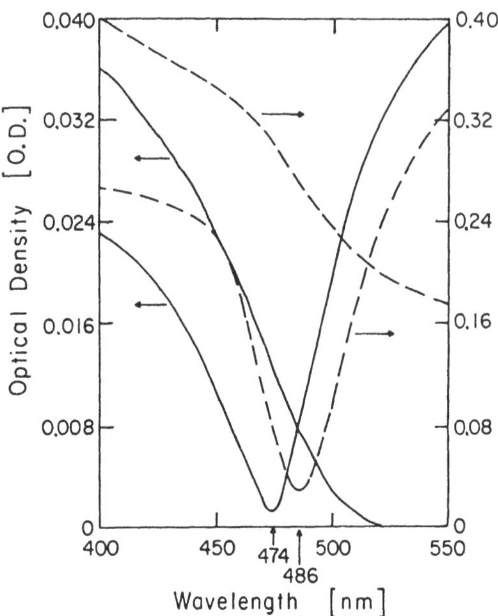

Figure 1. Long-wavelengths part of absorption spectra of CdS in glass films and first derivatives. Solid lines 1: 0.0077 mol CdS per 1 mol of silica, left ordinate. Broken lines: 0.039 mol CdS per 1 mol of silica, right ordinate. Minima of the derivatives denoted by arrows.

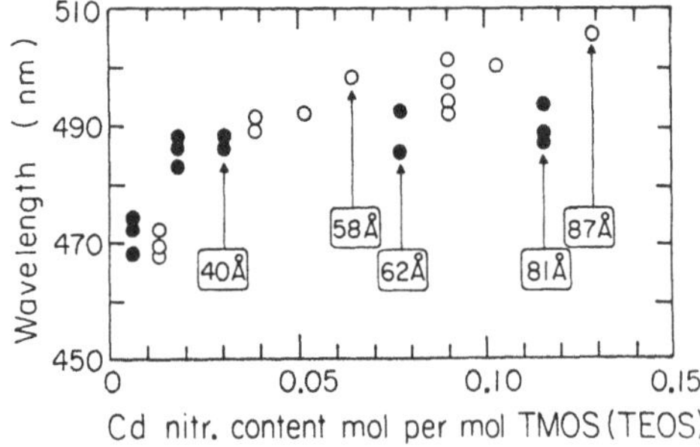

Figure 2. Inflection wavelength as a function of concentration of Cd ions in the precursor solution. Full circles: TMOS films, open circles: TEOS films. Numbers in boxes denote the diameters of CdS particles as determined from X-rays diffraction.

intensities of absorption and emission) in the spectroscopically active sites, the inflection point and nonlinear optical properties are due to a majority of particles having similar sizes.

Figure 4 presents a XRD spectrum from which the structure of CdS is inferred. The lines are recognized as belonging to a hexagonal (wurzite) structure, although some lines belonging to cubic (zinc blende) structure can be observed as well. Especially, the strongest peak at a 26.5 deg. can be assigned equally well to hexagonal as to cubic form. This assignment is in a qualitative agreement with the recent results of Nogami et al. [44] determined for CdS in bulk sol-gel glass. The small peak at about 29.33 deg. is due to a minute amount of unreacted $Cd(NO_3)_2$. The simultaneous existence of zinc blend and wurzite, which differ only in second nearest neighbour arrangement is not surprising since it is known that aqueous precipitation of CdS can yield either wurzite (hexagonal) or zinc blende (cubic) depending on kinetic factors.

The inset presents the XRD spectra of thin films doped by CdS (curves 1,2,3), where the molar fractions of Cd in TMOS are 0.039, 0.078 and 0.117, respectively), the range of 24-29 deg. (2θ), from which the XRD spectrum of undoped film has been substracted.

According to the Sherrer method, the correspondence between the Full Width at Medium Height (FWMH) of the diffraction peak and the average diameter of the diffracting particles can be expressed as:

$$D_{hkl} = \frac{\lambda \cdot 0.9}{\beta_{1/2} \cdot \cos\theta} \tag{12}$$

where $\beta_{1/2}$ is the full width at medium height of the line (FWMH). In our case λ is 1.506 A, the Kα line of Cu and the index (hkl)=(002) for hexagonal form and (hkl)=(001) for cubic form at their strongest peak at about 26.5 deg. The instrumental broadening is negligible here.

Plot of the energy of the inflection point vs.diameter of the CdS particles shows the correct trend, i.e. blue shift with decreasing of the diameter.

Figure 5 presents the room and liquid nitrogen temperature emission and excitation spectra of bulk sol-gel glasses doped by cadmium sulfide prepared from 10^{-3} M/L (curve 1) and 10^{-2} M/L (curve 2) of cadmium nitrate. Curve 3 presents the spectra of the sample 2 at 77°K.

In the room temperature excitation spectra of both concentrations of cadmium sulfide at least three distinct maxima are seen. On the high concentration spectrum an additional bump at about 430 nm can be observed. The emission spectra follow the phenomenon already observed elsewhere [2], whereby at some concentrations range a blue shift of absorption is accompanied by a red shift of emission.

In the low-temperature spectra of the high concentration sample we observe a blue shift of the excitation edge with some sharpening of its structure. This can be ascribed to decreased population of vibrational states due to cooling, which is pronounced for the lowest quantum level at about 430 nm. Its emission maximum does not change upon cooling.

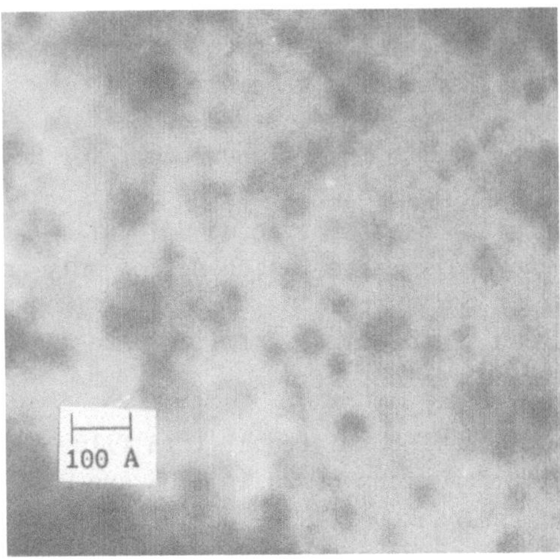

Figure 3. TEM photograph of a thin TEOS film doped by CdS (0.089 mol Cd /1 mol TEOS). range of particle diameters is 20-50 A, although particles as large as 200A can be seen.

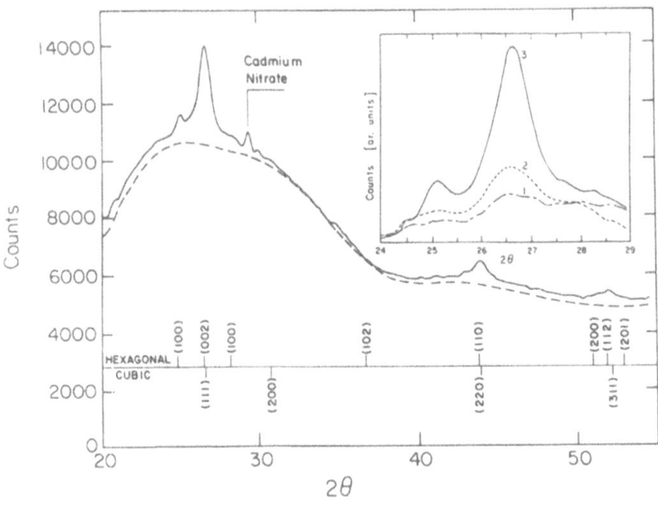

Figure 4. X-Ray diffraction spectrum of a TMOS film containing 0.195 mole CdS/mol TMOS. Lines belonging to both wurzite (hexagonal) and to zinc blende (cubic) structures are recognized. Inset: X-Ray Diffraction spectrum in the range of 24-29 deg. (2θ) of TMOS films containing 0.039 mol (Curve 1) 0.079 mol (curve 2) and 0.117 mol (curve 3) of CdS per mol of TMOS.

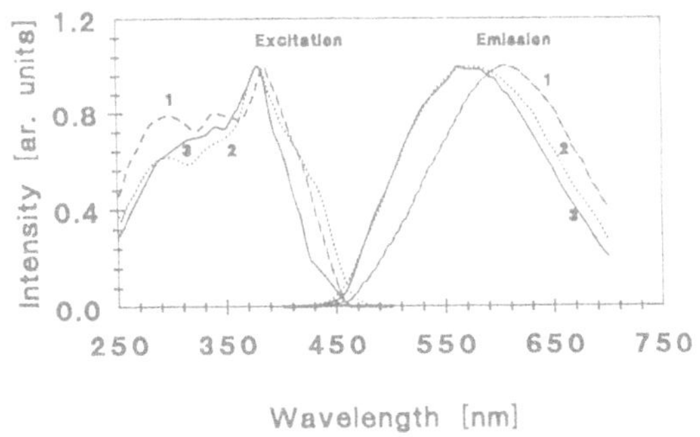

Figure 5. Emission and excitation spectra of CdS in bulk sol-gel glasses. Curves 1: TEOS containing initially 10^{-3} M/L of cadmium nitrate, room temperature. Curves 2: TEOS containing initially 10^{-2} M/L of cadmium nitrate, room temperature. Curve 3: TEOS containing initially 10^{-2} M/L of cadmium nitrate, measured at 77°K.

Thin glass films doped by cadmium sulfide emitted perceptible fluorescence only at 77°K. Figure 6 presents the excitation and the emission spectra of two TEOS films, curve 1, concentration of 0.051 mol of Cd per mol of TEOS and curve 2, concentration of 0.127 mol of Cd per mole of TEOS. Curve 3 presents the emission and excitation spectra of cadmium sulfide (0.117 mol per mol) in GLYMO film. The excitation spectra are well structured and the wavelengths of maxima can be determined. The position and assignment of these levels are compared with theoretical model, a short description of which follows.

Quantum confinement efects in semiconductor microcrystallites have been discussed by Efros and Efros [73] and the first quantitative effort using an idealized model for evaluation of the electronic levels of the quantum dots was performed by Brus [39]. He used wave-mechanical treatment to calculate the lowest eingestate of an exciton confined in a spherical box, having either infinitely high walls or a finite potential barrier.

The calculations gave the correct trend of the blue shift of the lowest exciton level with the decrease of the size of the semiconductor particle.

An improvement of the model was proposed by Fojtik et al.[74]. In their model the electron having the reduced mass $\mu = (1/m_1 + 1/m_2)^{-1}$ (where m_1 and m_2 are the effective masses of electron and hole, 0.19 and 0.8 for bulk CdS respectively) moves around a massless hole placed at the centre of the particle.

The allowed energy levels are found by solving a semiclassical phase integral, as in the models of E.L. Brus [39].

Another approach is proposed by Schmidt and Weller [40]. Here the exciton is allowed to move inside the sphere defined by the radius of the quantum dot, i.e. both electron and hole are allowed to move, whereby the potential term is defined by the Coulomb interaction between the two particles, V(r) is proportional to $e/\epsilon|\mathbf{r}_1 - \mathbf{r}_2|$, where $\mathbf{r}_1$ and $\mathbf{r}_2$ are the coordinates of the electron and the hole respectively. We start from the general Hamiltonian, H

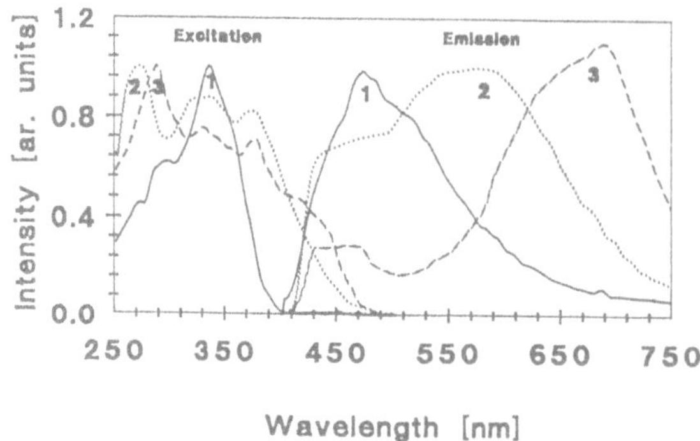

Figure 6. Emission and excitation spectra of thin films containing CdS measured at 77°K. Curves 1: TEOS film containing initially 0.051 mol of cadmium nitrate per mol of TEOS. Curves 2: TEOS film containing initially 0.117 mol of cadmium nitrate per mol of TEOS. Curves 3: GLYMO film containing 0.127 mol of cadmium nitrate per mol of GLYMO.

$$H = -(h^2/2m_1)\nabla_1^2 - (h^2/2m_2)\nabla_2^2 - e/\varepsilon|r_1-r_2| \tag{13}$$

where the first two terms are the kinetic energies of the electron and the hole respectively and the third term represents the Coulomb interaction between the two particles, screened by factor ε, the high frequency dielectric constant of bulk CdS.

The eigenvalues of the system represent the excited energy levels of the system counted from the lower edge of the conduction band. In the infinitely extended material the exciton exhibits hydrogen-like energy levels, proportional to $-(1/2n^2)$ with n=1,2,3,4. As long as the crystal of CdS is much larger than the excitonic radius, these level are bounded slightly below the conduction band edge.

When the electron and the hole are constrained in a small volume of radius R comparable to the radius of the exciton (about 19.3 A) all degeneracies are removed and the excitonic levels become blue-shifted in the respect to the conduction band edge.

Such situation can be treated by the method used for a two-electron atom; The Hamiltonian (13) can be scaled by **r** --> **r**/R to get the form,

$$H = H^\circ + R(-1/r_{1,2}) \tag{14}$$

where R, the radius of the quantum dot, is treated as a perturbation term. The eigenvalues are then approximated by expansion around R;

$$E = 1/R^2\cdot(E^{(0)} + R\cdot E^{(1)} + (R^2/2)\cdot E^{(2)} + \dots) \tag{15}$$

where $E^{(0)}$ is the sum of the kinetic energies of the electron and the hole and $E^{(1)}$ is the corresponding expectation value of $-1/r_{1,2}$.

Eigenfunctions are then a combination of hydrogen like functions, 1s,2s,3s for angular momentum L=0, and 2p,3p... for angular momentum L=1.

For R smaller than the excitonic radius, the lowest state eigenfunction is just a product of the lowest-state functions of the electron and the hole. For larger values of R, however, the function of state should be expanded in terms of the complete orthogonal set of functions,

$$\Psi_a = \sum_{\alpha\beta} c_{\alpha\beta}\Phi_\alpha(r_1)\cdot\Phi_\beta(r_2) \qquad (16)$$

Table 3 presents the values of the parameters $E^{(0)}$ and $E^{(1)}$ for a few lowest eigenfunctions of the quantum dots. In analogy to hydrogen atom the energies calculated from these parameters via eq. 15, are in units of hartree of the exciton; 1 hartree(exc) = $\mu e^2/h^2\varepsilon^2$, which, for μ=0.154 and ε = 5.6 for bulk CdS, is about 0.133 eV. The radii are given in units of bohr of the exciton; 1 bohr(exc) = $h^2\varepsilon/\mu e^2$ = 19.3 A for bulk CdS. Zero energy is the onset of conduction band in bulk CdS, which is about 2.5 eV.

Table 3. Values of parameters for calculation of eigenvalues via eq. 15

L	Configuration	$E^{(0)}$	$E^{(1)}$
0	1s1s	4.935	1.786
0	1s2s	7.776	1.780
0	2p2p	10.093	1.889
0	1s3s	12.511	1.777
0	2p3p	13.883	1.851
1	1s2p	5.926	1.620
1	2p1s	9.104	1.620
1	1s3p	9.715	1.724

Predicted levels are calculated from Table 3 and Eq. 15 in the first-order approximation and drawn in Figure 7. These are compared with the energies calculated from the excitation spectra on Figure 6. The levels fit fairly well the experimental quantum levels of the TEOS samples. The the sol-gel sample shows one level beyond those calculated, while the GLYMO sample shows two such levels. The size dependence of the electronic levels in the glasses is analogous with the behaviour of small colloidal cadmium sulfide particles separated by electrophoresis [75]. Average radii of the CdS particles as indicated by the fit are 17 Å, 22 Å and 26 A, in good agreement with the measurements by X-Ray diffraction.

VI. CONCLUSIONS

The CdS doped thin films show a significant spectral shift of the lowest quantum level as the function of the diameter of the particles. The size of the particles can be controlled by concentrations of the reagents and by environment (viscosity, temperature). Emission and excitation

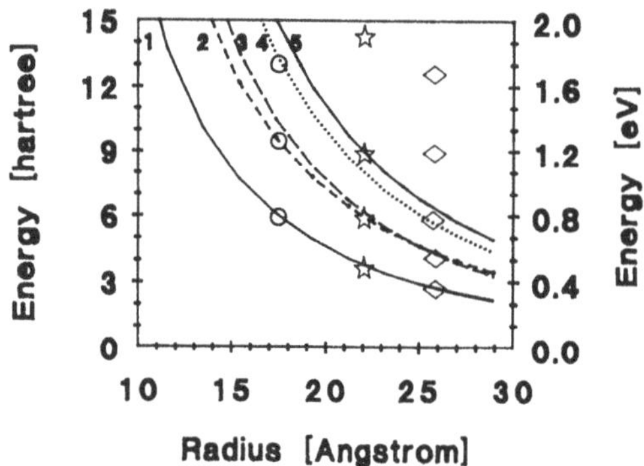

Figure 7. Five lowest quantum levels of quantum dots of CdS obtained by first order approximation of eq. 5 and table I, L=0. Curve 1: 1s1s, curve 2: 1s2s, curve 3: 2p2p, curve 4: 1s3s, curve 5: 2p3p. These are compared with energy levels obtained from excitation spectra of thin films: Circles: low concentration of cadmium sulfide (0.051 mol/mol, sample 1), stars: high concentration of CdS (0.117 mol/mol, sample 2), diamonds: GLYMO (0.127 mol/mol, sample 3).

spectra are also size sensitive. At least three quantum levels were recognized in these spectra.

Relatively strong third order harmonic generation was observed in these materials. The signals are much stronger than those reported in literature for CdS in polymers. Four wave mixing was observed in these samples as well [76]. CdS was discussed extensively in this lecture, however a similar behaviour is also observed in other semiconductor particles embedded in a glass. For example quantum dots of of mixed CdS_xSe_{1-x} in a glass exhibit similar linear and nonlinear properties to those of CdS [77].

VII. ACKNOWLEDGEMENTS

The author is grateful to Professor C.K. Jorgensen and to Dr. M. Eyal for discussion. The work was supported by the Krupp Foundation, by the Volkswagen Foundation and jointly by the European Community and the Israeli Ministry of Science.

REFERENCES

1. V.V.Shkunov, B. Ya. Zeldovich, " Optical phase conjugation",Scientific American 253(12):40 1985.
2. R. Reisfeld and C.K. Jorgensen, "Optical Properties of Colorants or Luminescent Species in Sol-Gel Glasses", Structure & Bonding, 77:(1991).
3. R. Reisfeld, "Luminescence and nonradiative processes in porous glasses" Int'l School of Atomic and Molecular Spectroscopy. 9th Course,Advances in Nonradiative Processes in Solids, Erice, Italy. Plenum , (June 15-19, 1989).

4. R. Reisfeld, M. Eyal, R. Gvishi, "Spectroscopic behaviour of fluorescein and its di(mercury acetate) adduct in glasses", Chem. Phys. Lett., 138:377 (1987).
5. M.A. Kramer, W.R. Tompkin and R.W. Boyd, " Nonlinear optical interactions in fluorescein doped boric acid", Phys. Rev.A, 34:2026 (1986).
6. R. Reisfeld, M. Eyal, R. Gvishi and C.K. Jorgensen, "Photochemical behaviour of luminescent dyes in sol-gel and boric acid glasses", Proc. 7th Int'l Symp. on the Photochemistry and Photophysics of Coordination Compounds, Germany, March 29-April 2, 1987. Springer-Verlag, Heidelberg-New York pp. 313-316 (1987).
7. S. Graham, R. Renner, C. Klingshirn, W. Schrepp, R. Reisfeld, D. Brusilovsky and M. Eyal, "Pump and probe beam measurements in organic materials", Paper presented at Int. Conf. Materials for Nonlinear and Electro-optics, Cambridge, 1989, Inst. Phys. Conf. Ser. No. 103; Section 2.2, Bristol, New York, pp. 157-162 (1989).
8. S. Graham, M. Thoma, M. Eyal, D. Brusilovsky, R. Reisfeld, S.V. Gaponenko, V. Yu Lebed, L.G. Zimin and C. Klingshirn, "Laser Induced Gratings and Nonlinear Optics in Organic Materials", in Organic Materials for Nonlinear Optics II, Edited by R.A. Hann and D. Bloor, Royal Society of Chemistry, Thomas Graham House, Cambridge 1991, p. 142.
9. S. Graham, M. Eyal, M. Thoma, D. Brusilovsky, R. Reisfeld and C. Klingshirn, "Nonlinear absorption and laser induced gratings in glasses doped with acridine orange and methyl orange", J. Luminescence 48&49:325 (1991).
10. U. Itoh, M. Takakusa, T. Moriya, S. Saito, " Optical gain of Coumarine doped thin film laser", Jap. J. Appl. Phys. 16:1059 (1977).
11. R.M. O'Connell, T.T. Saito, " Plastics for high-power laser applications: a review", Opt. Eng.. 22:393 (1983).
12. R. Reisfeld, "Optical behaviour of molecules in glasses prepared by the sol-gel method", Proc. Winter School on Glasses and Ceramics from Gels, SOL-GEL Science and Technology. Brazil, Aegerter, M.A. Jefelici, M. Souza, D.F. Zanotto E.D. eds. World Scientific, Singapore p. 323 (1989) 13. R. Reisfeld, "Spectroscopy and applications of molecules in glasses", J. Non-Cryst. Solids 121:254 (1990).
14. See: J.D. Mackenzie, D.R., Ulrich (ed.) Proc. Third International Conference on Ultrastructure Processing of Ceramics, Glasses and Composites, San Diego, 1987, Wiley, NY (1988).
15. R. Reisfeld, "Theory and applications of spectroscopically active glasses prepared by the sol-gel method", Sol-Gel Optics, SPIE Int'l Symposium on Optical and Optoelectronic Applied Science and Engineering, San Diego, California,July 8-13, 1990, SPIE Proc. 1328: paper 1328-04 (1990).
16. H. Dislich, Thin Films from the sol-gel process, in: "Sol-gel technology for thin films, fibers, preforms, electronics and special shapes", L. Klein, ed., Noyes Publishers, Park Ridge, New Jersey, USA, (1988).
17. C.Sanchez, J. Livage, " Sol-gel density from metal alkoxide precursors ", New J. Chem. 14:513 (1990).
18. J. McKiernan, S.A. Yamanaka, B. Dunn and J.I. Zink, "Spectroscopy and laser action of Rhodamine 6G doped aluminosilicate Xerogels", J. Phys. Chem., 94:5652 (1990).

19. H. Schmidt and H. Wolter," Organically modified ceramics and their applications", J. Non-Cryst. Solids, 121:428 (1990).
20. E.J.A. Pope, J.D. Mackenzie, " Sol-gel processing of silica", J. Non-Cryst. Solids, 87:185 (1986).
21. E.J. A. Pope, J.D. Mackenzie, " Transparent silica gel-PMMA composites", J. Mater. Res., 4(4):1018 (1989).
22. J. Fricke, Phys. Unserer Zeit 17:151 (1986).
23. D. Avnir, D. Levy, R. Reisfeld, "The nature of silica glass cage as reflected by spectral changes and enhanced photostability of trapped Rhodamine 6G", J. Phys. Chem., 88:5956 (1984).
24. D. Avnir, V.R. Kaufman, R. Reisfeld, "Organic fluorescent dyes trapped in silica and silica-titania thin films by the sol-gel method, Photophysical film and cage properties", J. Non-Cryst. Solids, 74:395, (1985).
25. R. Reisfeld, V. Chernyak, M. Eyal, C.K. Jorgensen, "Irreversible spectral changes of cobalt (II) by moderate heating in sol-gel glasses and their ligand field rationalization", Chem. Phys. Lett., 164:307 (1989).
26. J. McKernan, J.C. Pouxviel, B. Dunn, J.I. Zink, "Rigidochromism as a Probe of Gelation and Densification of Silicon and Mixed Aluminum Alkoxide", J. Phys. Chem., 93:2129 (1989).
27. B. Lintner, N. Arfsten, H. Dislich, H. Schmidt, G. Philipp and B. Seiferling, " A first look on the optical properties of ormosils",J. Non Cryst. Solids, 100:378 (1988).
28. R. Reisfeld and C.K. Jorgensen, "Luminescent Solar Concentrators for Energy Conversion", Structure & Bonding, 49:1 (1982).
29. R. Reisfeld and G. Seybold, "Stable Solid-State tunable lasers in the visible", J. Luminescence, Vol. 48&49:98 (1991).
30. R. Reisfeld, D. Brusilovsky, M. Eyal, E. Miron, Z. Burshtein, J. Ivri, "New solid state tunable laser in the visible", Chem. Phys. Lett., 160:43 (1989).
31. Y. Kobayashi, Y. Kurokawa, Y. Imai, "A Transparent Alumina Film Doped with Laser Dye and its Emission Properties", J. Non-Cryst. Solids, 105:198 (1988).
32. R. Reisfeld and G. Seybold, "Solid-state tunable lasers in the visible based on luminescent photoresistant heterocyclic colorants", Chimia, 44:295 (1990).
33. F. Salin, G. LeSaux, P. Georges, A. Brun, C. Bagnall, J. Zarzycki, "Efficient tunable solid-state laser near 630 nm using sulforhodamine 640 doped silica gel", Opt. Lett., 14:785 (1989).
34. A. Henglein, "Small-Particle Research : Physicochemical Properties of Extremely Small Colloidal Metal and Semiconductor Particles", Chem. Rev., 89:1861 (1989).
35. M.G. Bawendi, W.L. Wilson, L. Rothberg, P.J. Carroll, T.M. Jedju, M.L. Steigerwald and L.E. Brus, "Electronic Structure and Photoexcited-Carrier Dynamics in Nanometer-Size CdSe Clusters", Phys. Rev. Lett., 65:1623 (1990).
36. D. Brusilovsky, M. Eyal and R. Reisfeld, "Absorption spectra, energy dispersive analysis of X-rays and transmission electron microscopy of silver particles in sol-gel glass films", Chem. Phys. Lett, 153:203 (1988).

37. H.R. Wilson, "Fluorescent dyes interacting with small silver particles; a system extending the spectral range of fluorescent solar concentrators", Solar Energy Mat., 16:223 (1987).
38. R. Rossetti, J.L. Ellison, J.M. Gibson and L.E. Brus, "Size Effects in the excited electronic states of small colloidal CdS crystallites", Chem. Phys.,80:4464 (1984).
39. L.E. Brus, " Electron-electron and electron-hole interactions in small semiconductor crystallites", J. Chem .Phys., 80:4403 (1984).
40. H.M. Schmidt and H. Weller, "Quantum Size Effects in Semiconductor Crystallites: Calculation of the Energy Spectrum for the Confined Exciton", Chem .Phys. Lett., 129:615 (1986).
41. T. Rajh, M.I. Vucemilovic, N.M. Dimitrijevic, O.I. Micic and A.J. Nozik, "Size Quantization of Colloidal Semiconductor Particles in Silicate Glasses", Chem. Phys. Lett., 143:305 (1988).
42. S. Modes and P. Lianos, "Structural Study of Silicate Glasses by Luminescence Probing: The Nature of Small Semiconductor Particles Formed in Glasses", Chem. Phys. Lett., 153:351 (1988).
43. N. Tohge, M. Asuka and T. Minami, "Doping of CdS Semiconductor Crystallites to SiO_2 Glasses by the sol-gel process", Chem. Express, 5:521 (1990).
44. M. Nogami, K. Nagasaki and M. Takata, "CdS Microcrystal-Doped Silica Glass prepared by the sol-gel process", J. Non Cryst. Solids, 122:101 (1990)
45. S. Schmitt-Rink, D.A.B. Miller and D.S. Chemla," Theory of the linear and nonlinear optical properties of semiconductor microcrystallites", Phys. Rev.B, 35:8113 (1987).
46. Y. Wang, A. Suna and W. Mahler, " Nonlinear optical properties of polymers", Mat. Res. Symp. Proc., 109:18 (1988).
47. Y. Ohashi, M. Ito, T. Hayashi, A. Nitta, H. Matsuda, S. Okada, H. Nakanishi and M. Kato, "CdS Particle Doped Polymer Films for Nonlinear Optics", in: "Nonlinear optics of Organics and Semiconductors", Vol.36 T. Kobayashi (ed.), Springer-Verlag Berlin, Heidelberg p. 81 (1989).
48. H. Minti, M. Eyal and R. Reisfeld, "Quantum dots of CdS in thin glass films prepared by the sol-gel technique", submitted to Chem. Phys. Lett., (1991).
49. R. Reisfeld, H. Minti and M. Eyal, " Active glasses prepared by the sol-gel method including islands of CdS or silver", the Int'l Congress on Optical Sciences and Engineering, 11-10 March 1991 Hague, SPIE Proc., 1513: (1991).
50. H. Schmidt, " Preparation, application and potential of ORMOCERs", Proc. Winter School on Glasses and Ceramics from Gels, Sol-Gel Science and Technology, Brazil, August 1989. Eds. M.A. Aegerter, M. Jafelicci Jr., D. Souza and E.D. Zanotto, World Scientific, Singapore, New-Jersey, London, Hong-Kong, p. 432.
51. R. Reisfeld, V. Chernyak, M. Eyal and A. Weitz, "Laser and spectroscopic characterization of thin films", Proc.Int'l Conf. on Optical Science and Engineering, Optical Materials Technology for Energy Efficiency and Energy Conversion VII, Solar Collecting Devices, SPIE Proceedings, 1016:240, (1988).
52. M. Eyal, R. Reisfeld, V. Chernyak, L. Kaczmarek and A. Grabowska, "Absorption, emission and lifetimes of [2,2-bipyridyl]-3,3'-diol in sol-glass and in polymethylmethacrylate", Chem. Phys. Lett., 176:531 (1991).

53. F. Kajzar and J. Messier, " Third-harmonic generation in liquids", Phys. Rev.A, 32:2352 (1985).
54. G. Berkovic, R. Superfine, P. Guyot-Sionnest. Y.R. Shen and P.N. Prasad, " Study of diacetylene monomer and polymer monolayers using second- and third-harmonic generation"J. Opt. Soc. Am.B, 5(3):668 (1988).
55. F. Kajzar, J. Messier and C. Rosillio,"Third harmonics generation in liquids" J. Appl. Phys., 60:9 (1986).
56. D. Neher, A. Wolf, C.Bubeck and G. Wegner, " Third-harmonic generation in polyphenylacetylene; Exact determination of nonlinear optical susceptibilities in ultrathin films", Chem. Phys. Lett., 163:116 (1989).
57. W.E. Torruellas, R. Zanoni, M.B. Marques, G.I. Stegeman, G.R. Mohlmann, E.W.P Erdhuisen and W.H.G. Horsthuis," Measurements of third-order nonlinearities of side-chain substituted polymers", Chem. Phys. Lett., 175:267 (1990).
58. H.W.K. Tom, T.F. Heinz and Y.R. Shen," Second-harmonics reflection from silicon surfaces and its relation to structural symmetry", Phys. Rev. Lett., 51:1983 (1983).
59. V. Mizrachi and J.E. Sipe, " Phenomenological treatment of suface second-harmonic generation", J. Opt. Soc. Am.B, 5(3):660 (1988).
60. B. Buchalter and G.R. Meredith, " Third order optical susceptibility of glasses determined by third-harmonics generation", Appl. Opt., 21:3221 (1982).
61. J.H. Simmons, E.M. Clausen Jr. and B.G. Potter Jr., " Nonlinear optical composite materials using CdS", in " Ultrastructure processing of advanced ceramics", Editors J.D. Mackenzie, D.R. Ulrich. Proceedings of the Third Int'l Conference on Ultrastructure Processing of Ceramics, Glasses and Composites, 23-27 Feb. 1987, San-Diego, Calif. A Wiley Interscience Publication, John Wiley and Sons, 1988, page 661.
62. J. Yumoto, H. Shinjima and N. Nesugi, K. Tsunetomo, H. Nasu, Y. Osaka, "Optical nonlinearity of CdSe microcrystallites in a sputtered film", Appl. Phys. Lett., 57(23):2393 (1990).
63. M. Nogami, Yi-Qing Zhu, Y. Tohyama and K. Nagasaki, " Preparation and nonlinear optical properties of quantum-sized CuCl doped silica glass by the sol-gel process", J. Am. Ceram. Soc., 74(1):238 (1991).
64. T. Ishihara and T. Goto, " Very large $\chi^{(3)}$ due to coherency of an exciton in PbI_2", in" Nonlinear Optics of Organics and Semiconductors", Editor K. Kobayashi, Springer Proceedings in Physics, Vol.36, page 72, Springer-Verlag 1989.
65. F. Hache, D. Ricard and C. Flytzanis, " Optical nonlinearities of small metal particles: Surface-mediated resonance and quantum effects", J. Opt. Soc. Am. B, 3(12):1647 (1986).
66. M. Sinclair, D. Moses, K. Akagi and A.J. Heeger," Structural relaxation and nonlinear zero-point fluctuations as the origin of the anisotropic third order nonlinear optical susceptibility in trans-$(CH)_x$", in "Nonlinear optical effects in organic polymers", Editors F. Kajzar, P. Prasad and D. Ulrich. NATO ASI Series E: Applied Sciences, vol. 162. Kluwer Academic Publishers, 1989 page. 29.
67. J. Messier, " Third order nonlinear susceptibility in semiconducting polymers", ibid, page 47.
68. L.R. Dalton, " Synthesis of new nonlinear optical ladder polymers", ibid page 123.

69. C. Bubeck, A. Kaltbeitzel, R.W. Lenz, D. Neher, J.R. Stenger-Smith, G. Wegner, " Nonlinear optical properties of poly(p-phenylene vinylene) thin films", ibid, page 143.
70. P.N. Prasad, " Ultrafast third-order nonlinear processes in polymeric films", ibid, page 351.
71. R.A. Lessard, J.J.A. Couture, P. Galarneau, " Application of third order nonlinearities of dyed PVA to real-time holography", ibid, page 343.
72. R. Reisfeld, " Radiative and nonradiative transitions of rare-earth ions in glasses", Structure and Bonding, 22:123 (1975).
73. Al.L. Efros and A.L. Efros, "Interband absorption of light in semiconductor sphere", Sov. Phys. Semicond., 16:772 (1982).
74. A. Fojtik, H. Weller, U. Koch and A. Henglein," Photochemistry of colloidal metal sulfides. 8. Photophysics of extremely small CdS particles: Q-state CdS and magic numbers", Ber. Bunsenges. Physik. Chem., 88:649 (1984).
75. L. Katsikas, A. Eichmuller, M. Gersig and H. Weller, "Discrete excitonic transitions in quantum-sized Cds particles", Chem. Phys. Lett., 172:201 (1990).
76. R. Powell, Private communication.
77. A. Uhrig, L. Banyai, Y.Z. Hu, S.W. Koch, C. Klingshirn and N. Neuroth, High-excitation photoluminescence, studies of $CdS_{1-x}Se_x$ quantum dots", Z. Phys.B, 81:385 (1990).

NEW CRYSTALS FOR LASER APPLICATIONS

A. Kaminskii

Institute of Crystallography
Academy of Sciences of the USSR
Leninsky pr. 59
Moscow 117333, USSR

This talk was divided into two independent parts:

1) New Non-Linear $LaBGeO_5$ Crystals for Self-frequency Doubled Lasers

Intensive spectral investigations have been carried out on trigonal $LaBGeO_5$ crystals. Studies were performed on crystals doped with Nd^{3+} and Pr^{3+}; the pure crystal was also studied. Absorption and luminescence spectra were studied to determine the Stark levels of the activator ions. Luminescence intensity characteristics and pulse stimulated emission parameters were studied for the Nd^{3+} ions at two intermanifold channels ($^4F_{3/2} \rightarrow {}^4I_{11/2}$ and $^4F_{3/2} \rightarrow {}^4I_{13/2}$), and for the Pr^{3+} ions at the visible $^3Po \rightarrow {}^3Hb$ transition. The Raman spectra and non-linear optical properties (including second harmonic generation) were also investigated. A preliminary theoretical analysis of the crystal fields at the activator ion sites has been carried out using a new many-electron semiempirical method, and all the observed transitions have been identified using this technique. From the studies we deduce that $LaBGeO_5$: Nd^{3+} is a promising material for designing a self-frequency doubled laser.

2) Techniques for improving the effectiveness of crystalline lasers.

We present two techniques to improve crystalline lasers. The first: in this lecture we shall discuss spectroscopic laser properties and the peculiarities of crystal containing fluorine and oxygen in a disordered structure. A discussion of the historical background will include the laws of crystal-field disorderness of Ln-activators (static and dynamic structural disorderness), the nature of the multicentricity, and the results of a search for new Ln^{3+} laser crystals. Examples of disordered crystals with Nd^{3+} ions will be given and compared to neodymium laser glasses. Perspectives shall be presented using disordered crystals in lasers, with laser-diode pumping, and in amplifiers of powerful laser machines.

The second: at the present time lamp-pumped stimulated-emission excitation processes of laser crystals containing Ln^{3+} ions rely on the populated Stark levels of the ground state of the activator. In a number of cases this method makes it difficult to obtain effective laser action in the system. A new ground-state lifting method can increase the absorption capability of the lasing crystals, and thus improve the laser action. In this lecture we shall discuss the physical basis of this technique, and possible applications of the technique to fluorine- and oxygen-containing crystals doped with Pr^{3+} ions.

EXCITED STATES AND REORIENTATIONAL PROPERTIES OF COLOR CENTERS WITH AXIAL SYMMETRY

A.Scacco

Dipartimento di Fisica - Università La Sapienza
P.le A. Moro 2, 00185 Roma, Italy

ABSTRACT

Complex color centers, showing lower symmetry than the host ionic crystal, undergo photostimulated reorientation processes when pumped with suitably polarized light. This optical alignment, typical of F_2, F_3, F_A, $(F_2)_A$, F_B, F_H centers in alkali halides, induces dichroic absorption spectra and modifications in the emission features of these defects which can be relevant to their performances in color center lasers.

The reorientation of the complex color centers under optical excitation is a property of their electronic excited states, and then its features (quantum efficiency, activation energy, temperature dependence) strictly reflect the nature of the ionic relaxation and the shape of the electronic wave function after the absorption. Therefore, detailed studies of the photoinduced reorientation can give direct information on the excited states of color centers and also on their geometrical configuration in the lattice.

I. INTRODUCTION

Aggregate color centers in ionic materials constitute a class of point defects particularly interesting because of their peculiar optical properties, which make them suitable for technological applications. The successful use of several kinds of aggregate defects in color center lasers [1-4] fostered investigation on new systems and various emitting centers in order to extend the spectral range of laser tunability and to improve the laser performances.

The most used host materials for these aggregate centers are single crystals of alkali halides, due to their relatively simple preparation and doping and to the cubic symmetry of their lattices. Common features for all aggregate color centers are their formation under proper optical treatment and their configuration, which exhibits a symmetry lower than that of the lattice. This explains why their optical behavior, and then the performances in the laser cavity, are strongly dependent on their microscopic structure and on the sample hystory. Measurements of optical absorption, luminescence, electro-optical and magneto-optical effects are widely used to characterize the fundamental and excited electronic states of the aggregate color centers, giving information on the phonon frequencies coupled to the states and on the energy levels involved in the transitions.

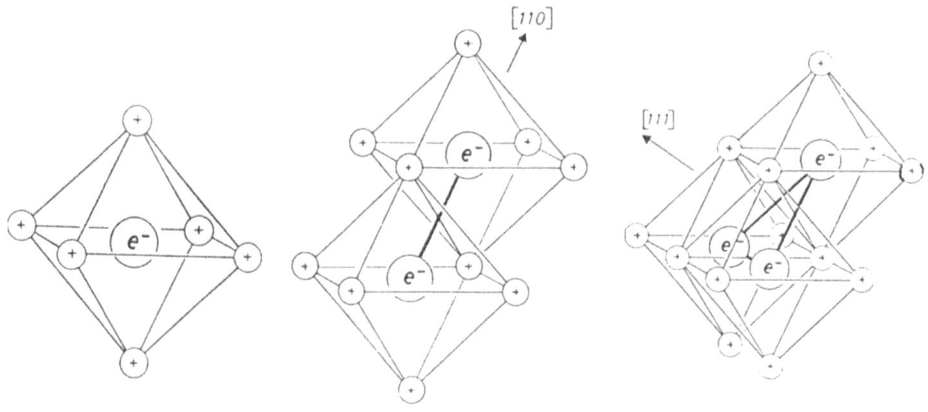

Fig. 1.Ionic configuration of F, F_2, F_3 centers in alkali halides

In this work the main features of the optical behavior of color centers with axial symmetry in alkali halides will be depicted, with particular concern to the nature of their excited states and to their configuration in the crystal lattice. Recent results of experimental and theoretical investigations will be reported and discussed.

II. AGGREGATE COLOR CENTERS

The currently accepted classification of color centers [5] discriminates between intrinsic and extrinsic aggregate defects: the former group, typical of undoped crystals, includes centers formed by aggregation of more than one F center (an anion vacancy trapping an electron), while the latter group contains centers formed by an impurity ion (with positive or negative electric charge) and one (or more than one) F center. Fig. 1 shows the well known structural models of the F, F_2, F_3 centers and the related directions of the optical transition moments [6]. The ionic

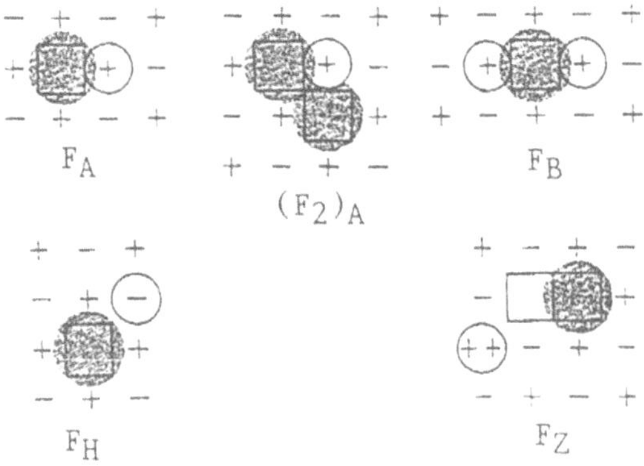

Fig. 2. Ionic configuration of extrinsic aggregate centers in alkali halides.

configurations of extrinsic aggregate defects are displayed in Fig. 2: the F_A and the $(F_2)_A$ centers are given by a monovalent cation (with size considerably smaller than that of the host cation) adjacent to an F center or to an F_2 center, respectively, the F_B center consists of two identical impurity cations bound in two possible non equivalent ways to an F center [7], the F_H center is formed by an impurity anion and an F center [8], the F_Z center is originated by the association of a divalent cation with an F center [9].

Such a list is of course not exhaustive, since more complex defects are possible and have been already observed experimentally, but it includes all centers whose optical properties are at present known in a satisfactory way. In all cases, the O_h symmetry of the F center is reduced in the aggregate defects, and this makes them optically anisotropic centers. Such a feature is related to the nature of the excited state of the defects, in which the degeneracy of the electronic levels (typical of the F center) is removed by the lowered symmetry. The optical anisotropy originates both emission polarization and dichroism in the absorption spectra: the latter is due to the photostimulated reorientation of the centers. This process will be examined in detail for defects showing an axial symmetry, namely for the F_2 centers among the intrinsic and for the F_A centers among the extrinsic defects.

III. EXCITED STATES AND REORIENTATION OF AGGREGATE COLOR CENTERS

The electronic transition responsible for the main absorption of the F center occurs between the 1s and the threefold degenerate 2p energy levels. This means that the excited state is optically isotropic, causing unpolarized emission and no dichroism in the absorption. Transitions to the higher 3p level and to conductive states may originate secondary absorptions (K and L bands, respectively) [10].

In the case of aggregate F_2 and F_A centers, electronic transitions occur from the ground state to excited states which are anisotropic. As a consequence, studies involving the use of linearly polarized light give information on the energy levels and the configuration of the defect after excitation.

A typical feature observed under polarized irradiation is the center reorientation, caused by the possible jumping (during the relaxation process in the excited state) of a vacancy from the initial lattice site to a new position situated in a different crystallographic direction with respect to the other vacancy or to the impurity ion. Since the reoriented center can no longer be excited with unchanged efficiency by the same polarized light, the uniform initial distribution of centers along equivalent directions is changed in favour of one orientation. Then, in the dichroic absorption the bands due to different transitions show relative heights which depend on the orientation of the polarization vector of the incident light.

The quantum efficiency of the reorientation η, defined as the number of reoriented centers per number of observed light quanta, is generally given by

$$\eta = \eta_0 \exp(-E_R/KT) \qquad (1)$$

where η_0 is the efficiency at high values of the temperature T and E_R indicates the activation energy needed for the vacancy jump.

IV. F_2 CENTER

The F_2 center, consisting of two nearest-neighbor F centers along a <110> lattice direction, exhibits a D_2 symmetry [6]. Its normal ground state is given by a singlet state with antiparallel spins of the two electrons. The primary transition of the center is excited by light polarized parallel to the <110> direction (the defect axis), and the excited state consists of a linear combination of the product of a 1s state on one vacancy and a $2p_{<110>}$ state on the other vacancy. Other transitions are excited by light polarized perpendicular to the defect axis, and further absorptions are caused by transitions to higher excited states (analogously to the F center) or by transitions involving triplet states and spin-orbit mixing [10]. The F_2 emission is given by one band, corresponding to the transition from the lowest excited state to the ground state and polarized parallel to the center axis.

The reorientation of F_2 centers has been studied in several alkali halides [11-15], but different hypotheses about its mechanism were proposed. The common interpretation postulated that the reorientation occurs during deexcitation from a state of the F_2 center [16], but experimental evidence was given of an indirect reorientation in KCl at low temperature [13]: the F_2 center would be first ionized to F_2^+ by light absorption, then it would absorb and rotate, and finally the reoriented F_2^+ center would capture an electron originating a reoriented F_2 center. Such a model is interesting because it involves ionized F_2^+ centers, which are laser active [17].

The temperature dependence of the reorientation efficiency in KCl, shown in Fig. 3, is in agreement with eq. (1). Calculation of the dichroic spectrum can be performed by considering the F_2 centers (Fig. 4) lying along the six <110> directions of the cubic cell, and studying the steady-state situation reached, for example, under irradiation with <011> polarized light at wavelength coincident with that of the primary F_2 absorption band and propagating along the <100> direction. In these conditions, the center are preferentially aligned and their populations (initially randomly distributed in all equivalent directions) are no more equal and become

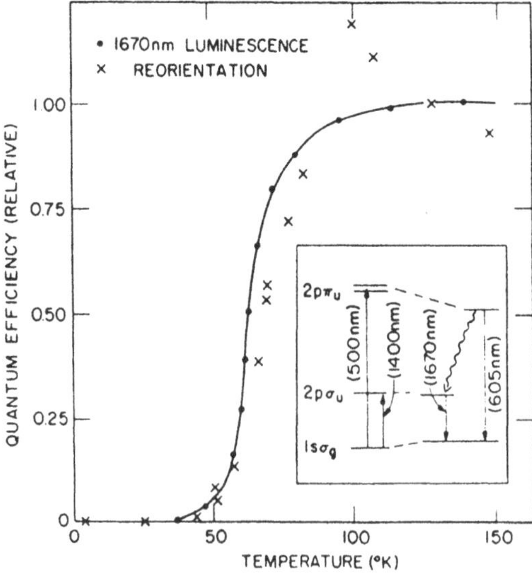

Fig. 3. Temperature dependence of the reorientation efficiency of F_2 centers in KCl (Ref. 13).

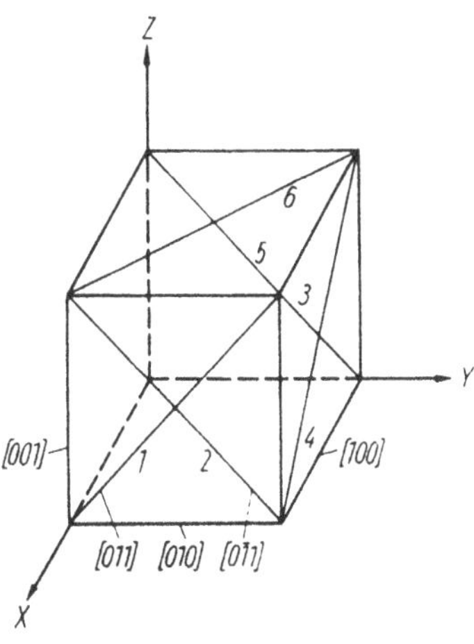

Fig. 4. Schematic model of F_2 center orientations in alkali halide crystals

$$N_2 > N_1 \neq N_3 = N_4 = N_5 = N_6 \qquad (2)$$

where N_i is the number of F_2 centers along the i direction (of course, $\Sigma_i N_i = 100\%$). The dichroic spectrum is given by

$$\Delta K_{[011]} = K_{[011]} - K_{[0\bar{1}1]} = \beta\ (N_2 - N_1) - \alpha\ (N_2 - N_1) \qquad (3)$$

in which $K_{[011]}$ and $K_{[0\bar{1}1]}$ are the absorption coefficients measured with [011] and [0̄11] polarized light, respectively, and α, β depend on wavelength and are typical of transitions having [011] and [011] optical dipole moments, respectively.

Another dichroic spectrum can be calculated by measuring the absorption spectra, after alignment of the F_2 centers with the same light of the previous experiment but [010] polarized, so that

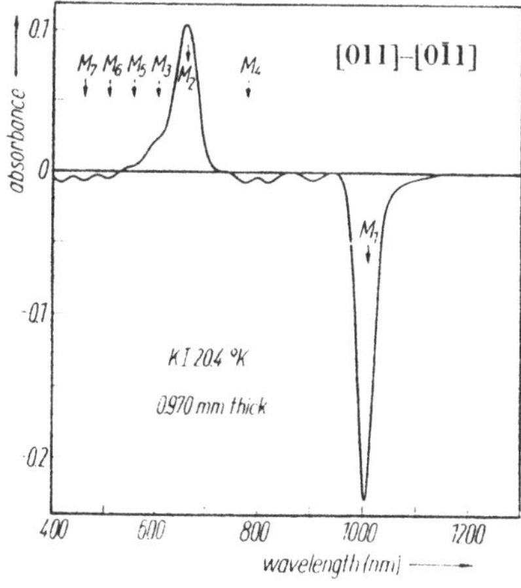

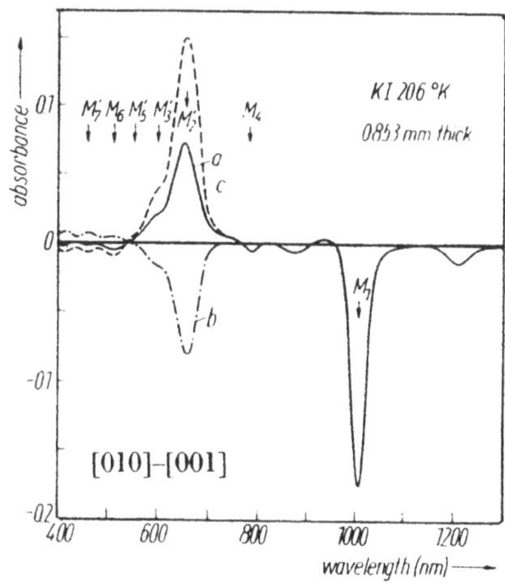

Fig. 5. Dichroic absorption spectra of F_2 centers in KI (Ref. 15).

$$N_3' = N_4' > N_5' = N_6' = N_1' = N_2' \tag{4}$$

with light polarized along the [010] and the [001] directions, and taking the difference which is given by

$$\Delta K_{[010]} = K_{[010]} - K_{[001]} = 2\gamma(N_3' - N_5') - \beta(N_3' - N_5') - \alpha(N_3' - N_5') \tag{5}$$

where the wavelength-dependent γ is typical of transitions having <100> optical dipole moments, and the primes indicate the new steady-state center populations. In this way, the experimental determination of four polarized spectra in two different experiments allows the identification and characterization of all absorptions of F_2 centers due to the various possible optical transitions (Fig. 5). The alignment of the F_2 centers can be almost quantitative: in KCl polarized irradiation at - 150 °C during several hours gives N_2 = 89.0%, N_1=N_3=N_4=N_5=N_6=2.2% [14].

V. F_A CENTER

The F center and the impurity cation, forming the F_A center, are aligned along one of the <100> directions, so that the resulting dipole exhibit a C_{4V} symmetry [18]. This causes a splitting of the threefold degenerate F transition into the components F_{A1}, which is a transition polarized parallel to the center axis, and F_{A2}, which is a twofold degenerate transition polarized perpendicular to it. As a consequence, the absorption spectrum consists of two bands (Fig. 6) and shows an evident dichroism when measured with linearly polarized light.

The reorientation quantum efficiency is temperature dependent up to 100 K, in agreement with eq. (1), and then saturates to 2/3 for all F_A

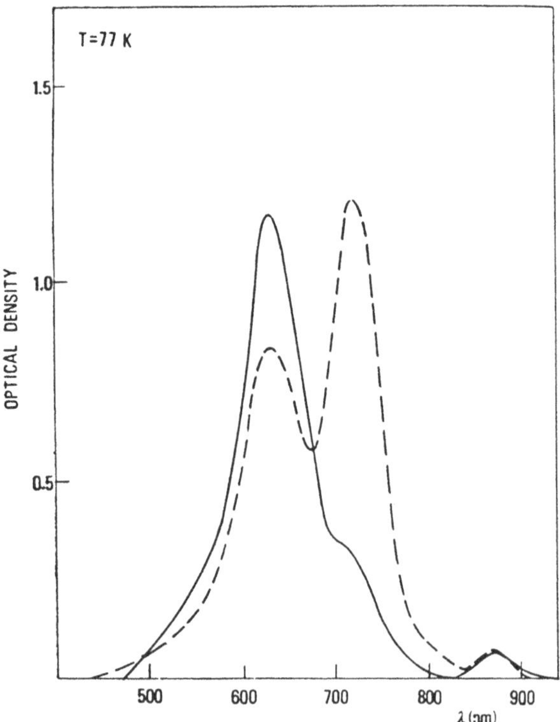

Fig. 6. Dichroic absorption spectra of F_A centers in $RbCl:Li^+$ (Ref. 34).

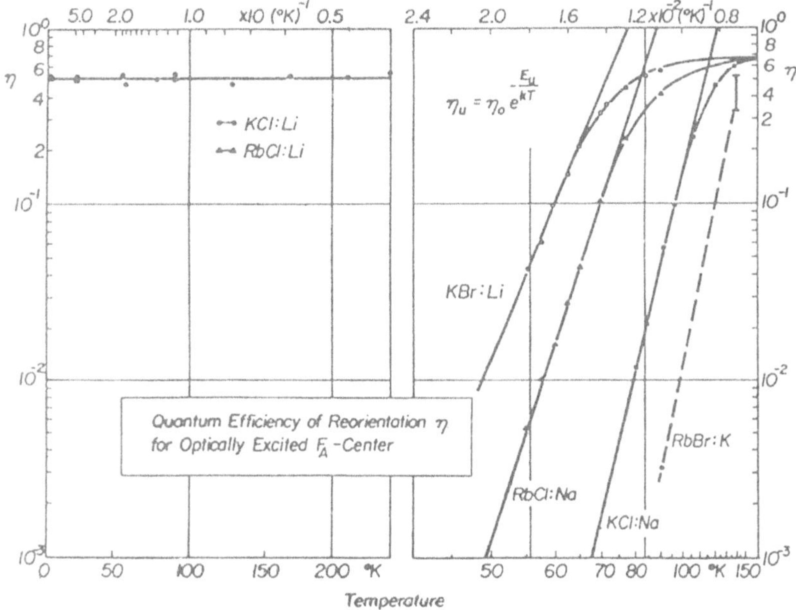

Fig. 7. Temperature dependence of the reorientation efficiency of F_A(I) (right) and F_A(II) centers (left) in alkali halides (Ref. 18).

centers containing K^+ or Na^+ impurities (with some behavior peculiarity in the case of KF:Na^+ [19]), and for those known containing Li^+, with the exception of KCl, KF, RbBr and RbCl. In these systems η = 1/2 (Fig. 7) in the whole temperature range (for RbCl:Li^+ at least above 20 K). Hence the classification of F_A centers into type I (reorientation thermally activated) and type II (no activation energy for reorientation), which takes into account the difference in the relaxed excited states of the defects.

The F_A emission is given by only one band, due to the existence of only one relaxed excited state, which is reached during the relaxation from

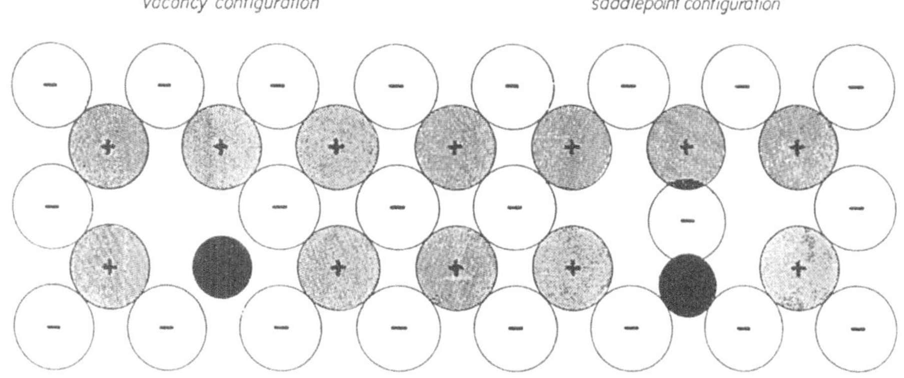

Fig. 8. Configurations of the relaxed excited state for F_A(I) (left) and F_A(II) (right) centers (Ref. 18).

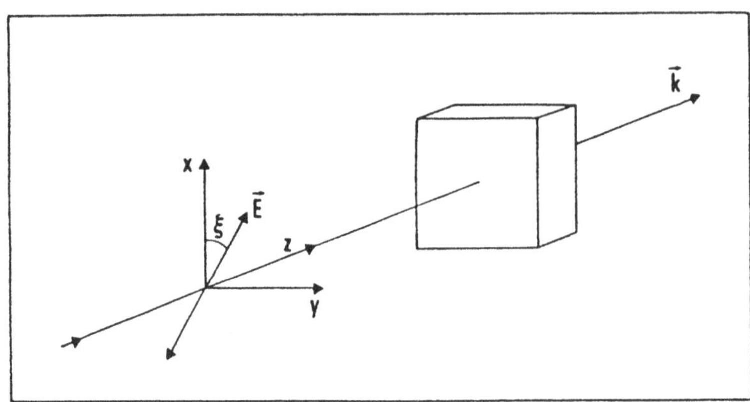

Fig. 9. Typical experimental arrangement for measuring the reorientation of color centers (the angle ξ is formed by the x crystal axis and the direction of the light polarization vector E) (Ref. 26).

different configurations in the two classes of F_A centers. In type I defects it is very similar to that of the F center (Fig. 8), with a diffuse vacancy-centered wavefunction (vacancy configuration): this originates spectral position and width of the F_A(I) emission very close to those of the F luminescence. Type II defects exhibit a luminescence with large Stokes shift and narrow bandwidth: these features are due to the so-called saddle-point configuration of the relaxed excited state, formed by two vacancies separated by an interstitial anion and sharing the electron (Fig. 8). The optical behavior of the F_A center in a system, and then its classification, are then determined by the geometrical parameters (impurity - host cation distance and anionic size) [18] or by the effective charge parameters [20,21] which make or not possible the saddle-point configuration. Recent results [22] show that there are not clear cut criteria for the F_A(II) center existence, since in the same system ($RbCl:Li^+$) the F_A defect exhibits both type I and type II behaviors in different temperature regions (which can perhaps explain the anomalies of the F_A(Na) center in KF).

The reorientational properties of the F_A centers, strongly dependent on the nature of the relaxed excited states, have received in the last few years particular attention because of the relevance of the alignment process to the color center laser performances [23]. The reorientation kinetics can be described (referring to Fig. 9) by the general rate equations [24]

$$\frac{1}{\eta I_o}\frac{dN_x}{dt} = -\sigma_x N_x + \frac{\sigma_y N_y}{2} + \frac{\sigma_z N_z}{2}$$

$$\frac{1}{\eta I_o}\frac{dN_y}{dt} = \frac{\sigma_x N_x}{2} - \sigma_y N_y + \frac{\sigma_z N_z}{2} \qquad (6)$$

$$\frac{1}{\eta I_o}\frac{dN_z}{dt} = \frac{\sigma_x N_x}{2} + \frac{\sigma_y N_y}{2} - \sigma_z N_z$$

where I_o is the intensity of the polarized pumping light, σ_i (i=x,y,z) is the absorption cross section of the dipoles lying along the i direction, whose number is N_i, and the total number of centers N is

$$N = N_x + N_y + N_z \qquad (7)$$

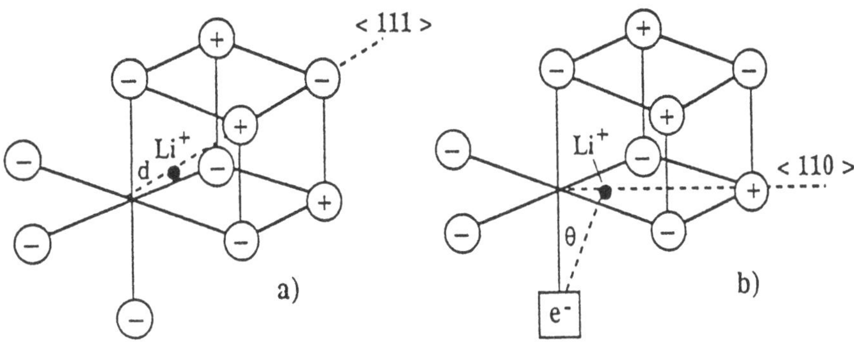

Fig. 10. Off-center displacements of the Li^+ ion when: a) isolated impurity b) coupled to an F center (Ref. 43).

The photostimulated reorientation can be monitored by measuring the changes in the absorption coefficient a of the crystal, which is

$$\alpha = N_x\sigma_x + N_y\sigma_y + N_z\sigma_z \tag{8}$$

or in the luminescence L, emitted from a crystal of thickness d, given by

$$L = c\, I_o\, [1 - \exp(-\alpha d)] \tag{9}$$

where c is a proportionality coefficient depending on the experimental conditions.

The calculation of the absorption cross sections must take into account the overlap between the F_{A1} and F_{A2} absorption bands, which causes both types of transitions under optical pumping:

$$\sigma_i = \sigma_{1i} + \sigma_{2i} \tag{10}$$

Since there is a large difference in the low temperature values of η for $F_A(I)$ and $F_A(II)$ centers, the two kinetics must be treated separately. From the results of such calculations for $F_A(I)$ centers [25,26], the real absorption of F_{A1} and F_{A2} transitions and then the lineshape of the $F_A(I)$ absorption bands can be evaluated from luminescence measurements, removing problems of overlap with other absorptions and background subtraction.

In the case of $F_A(II)$ centers, the mathematical treatment gives general solutions for the rate equations (6) which have already been reported [24]. The calculation of the absorption cross sections includes for these defects the effect of their off-axis configuration, caused by the known off-center displacement from the normal lattice site of the Li^+ ion in some alkali halides [27]. The isolated Li^+ impurity occupies, at least in KCl, an off-center position (Fig. 10a) shifted along one of the eight equivalent <111> directions [28]. When associated with the F center, the Li^+ ion is displaced (Fig. 10b) along one of the four equivalent <110> directions [29,30], and a fast tunneling motion among them takes place.

By taking into account the off-axis angle θ formed by the F_A dipole and the crystallographic axis (Fig. 11), the values (10) for σ_i can be determined and eq. (6) solved for $F_A(II)$ centers [31,32]. From the comparison of the experimental absorption and luminescence data with the theoretical equations (Fig. 12), an evaluation of the tilt of the F_A dipole

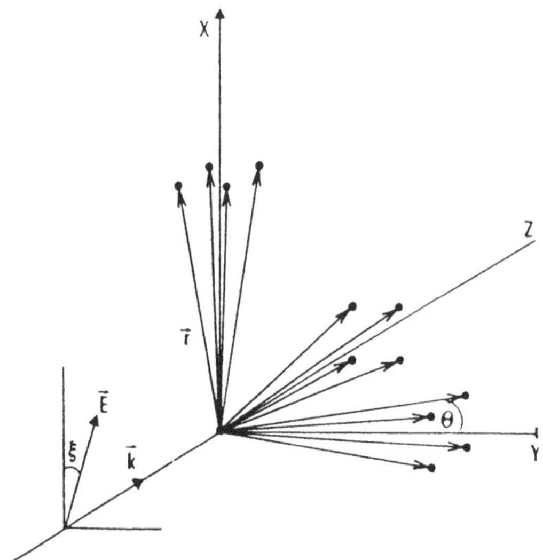

Fig. 11. Off-axis configuration of the $F_A(Li)$ dipoles in a cubic lattice (Ref. 24).

Table 1. Experimental and theoretical values for the off-axis tilt of the F centre-Li^+ ion dipoles in alkali halides containing $F_A(II)$ centres, for a <110> displacement of the impurity.

OFF-AXIS ANGLE θ (Degrees)

System	Theory			Experiment		
	a	b	c	d	e	f
$KCl:Li^+$	5.4	11.9	14.6	5 ± 1	16 ± 2	2.5
$RbCl:Li^+$	9.7	13.5	-	7 ± 1	21	-
$KF:Li^+$	12.3	14.0	11.9	14 ± 2	-	-

a - Ref.37; b - Ref.38; c - Ref.39; d - Ref.31,34,36; e - Ref.40,41; f - Ref.42

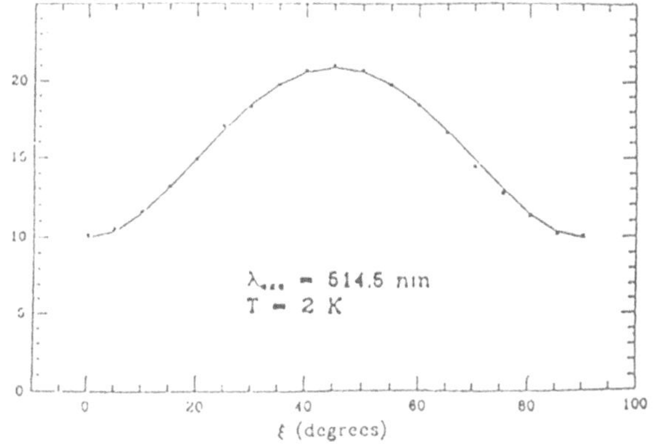

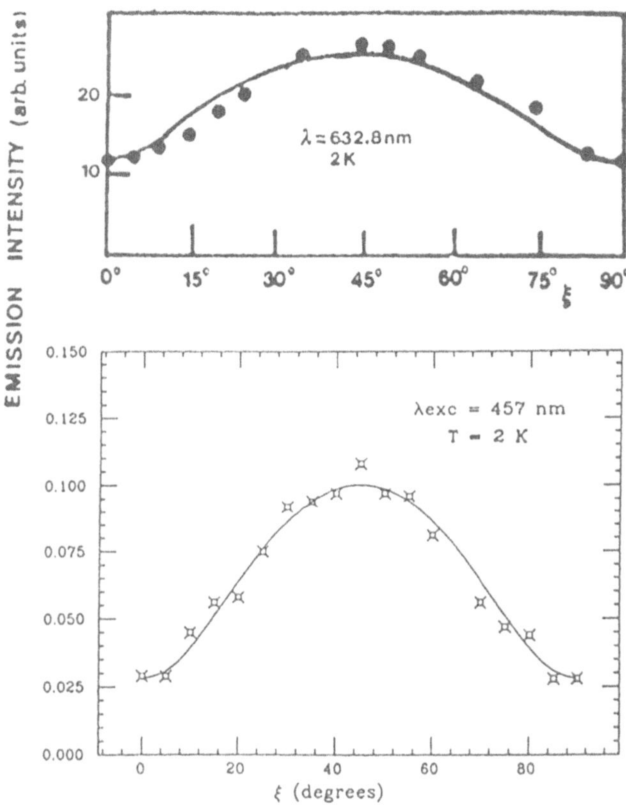

$$L = cI_0(1 - e^{-\alpha d})$$

$$\alpha = 3N\sigma_{02}\left[\frac{1}{R(\cos^2\theta\cos^2\xi + \frac{1}{2}\sin^2\theta\sin^2\xi) + \frac{1+\cos^2\theta}{2}\sin^2\xi + \sin^2\theta\cos^2\xi} + \frac{1}{R(\cos^2\theta\sin^2\xi + \frac{1}{2}\sin^2\theta\cos^2\xi) + \frac{1+\cos^2\theta}{2}\cos^2\xi + \sin^2\theta\sin^2\xi} + \frac{1}{R\frac{\sin^2\theta}{2} + \frac{1+\cos^2\theta}{2}}\right]^{-1}$$

Fig. 12. Azimuthal dependence of the F_A(II) luminescence in KCl:Li^+ (above), RbCl:Li^+ (middle), KF:Li^+ (below), fitted with equations taken from Ref. 26.

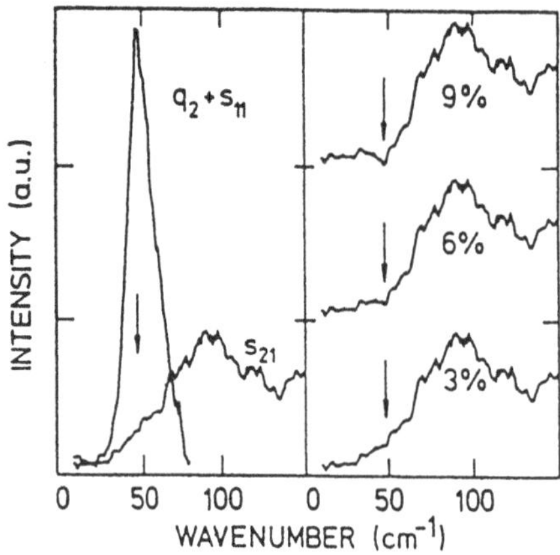

Fig. 13. Raman scattering spectra for F_A(Li) centers in KCl (Ref. 45).

was obtained so far for KCl:Li^+ [31,32], RbCl:Li^+ [33,34] and KF:Li^+ [35,36] and is planned for RbBr:Li^+. The results (Table I) indicate qualitatively that all investigated F_A(II) centers exhibit an off-axis configuration, and that the dipole tilt increases with increasing cation size and much more with decreasing anion size in the host lattice. More quantitatively, the values of the off-axis angle θ are quite large, if compared with previous findings in electro-optical experiments [42] but considerably smaller than those derived from optical dichroic measurements neglecting the F_A band overlap [40,41], reported as well in Table I.

A possible test of the reliability of the results is the comparison with the theoretical calculations performed on the off-center displacements of the Li^+ ion in various alkali halides, under the rough assumption that the F center replacing a host anion does not modify appreciably the displacement of the impurity ion in the <110> direction. The conclusion of such test can be derived from Table I, where all the calculated values for the off-axis angle (obtained with simple geometrical considerations from the off-center shifts) are also shown. The agreement between the data determined by photostimulated reorientation measurements and those calculated with shell model potentials adapted to the crystal elastic behavior is excellent and cannot be achieved for all three studied systems with any of the other models nor any of the other experimental results [43]. Then, such a simple successful correlation between the off-center and the off-axis configurations can be used to predict off-axis behavior of F_A centers in systems where the off-center shift of the Li^+ ions occurs.

It has to be pointed out that the problem of the off-axis involves considerations under several points of view, which have been formulated by different researchers and demonstrate the stimulating interest of the question.

Raman scattering experiments (Fig. 13) fully confirmed the off-axis geometry of F_A(Li) centers in KCl, RbCl, KF [44]. Moreover, even if not able to quantitatively determine the tilting angle, they supplied for it an upper limit in KCl which rules out the high values obtained taking not in account the overlap of the F_A bands [45].

On the other hand, effects of self-induced change of the light polarization for F_A(Li) centers have been measured in KCl [46,47] and would in some extent modify the optical reorientation of the centers under polarized pumping (Fig. 14).

From the theoretical point of view, a new criterion for forecasting off-center displacements of impurity ions in ionic crystals and for the classification of F_A centers was recently proposed [48,49]: it takes into account that in non-molecular crystals all bonds are polar covalent, and then atoms contain partial effective charges which modify their radii. The derived model is in generally good agreement with the known data, but again some uncertainties concerning a limited number of systems are found and then push for further studies.

Very recent developments [50] considered the isolated Li^+ ion in alkali halides electronically coupled to an anharmonic oscillator, composed of three halogen ion pairs vibrating in the T_{1u} mode and dragging the impurity into rotation over <111> sites (Fig. 15a). The capital effect of an F center, nearest neighbor in a <001> site, is to reduce the vibration dimensionality and to force the impurity to rotate in the (001) plane (Fig. 15b), shrinking the off-center displacement. In such a vibronic model, which is alternative to the polarization approach, the off-center can then be regarded as a pseudo-Jahn-Teller phenomenon, in which the coupling strength increases along the <100>, <110>, <111> displacement sequence and also in the order KCl-RbCl-KF.

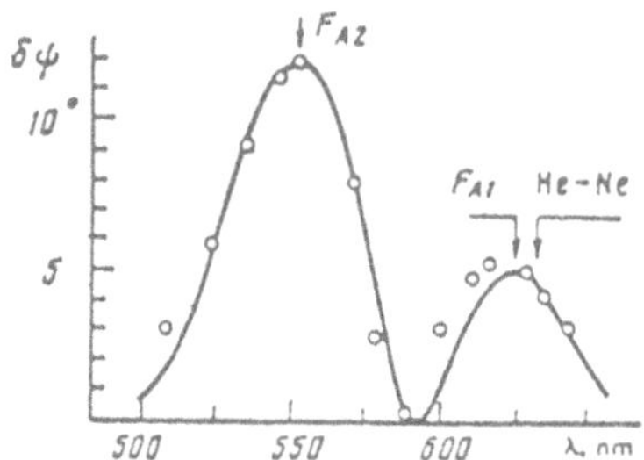

Fig. 14. Self-induced change of the light polarization for F_A(Li) centers in KCl (Ref. 46).

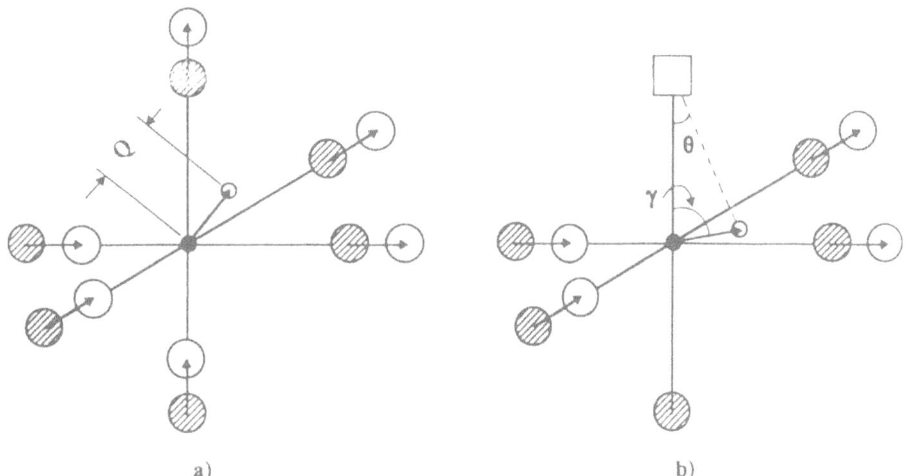

Fig. 15. Vibronic model for the off-center displacement of: a) an isolated Li^+ ion, b) a Li^+ ion coupled to an F center.

VI. CONCLUSIONS

F_2 and F_A centers represent typical cases of defects with axial symmetry: simple optical investigations on their photoinduced reorientation give, as shown above, detailed information on the excited states and on the geometrical configuration of the centers. Reorientational properties are also particularly interesting in the case of other aggregate (intrinsic or extrinsic) centers, which can be used in color center lasers: the optimization of laser emission is always related to the maximization of the number of active centers, which depends on the relative orientation of the defects with respect to the polarization of the pumping light. Then, studies on the symmetry and the electronic transitions of aggregate centers are of primary importance not only for fundamental knowledge but also for technological applications.

REFERENCES

[1] L. F. Mollenauer and D. H. Olson J. Appl. Phys. 46, 3109 (1975).
[2] W. Gellermann, K. P. Koch and F. Lüty, Laser Focus, April 1982.
[3] G. Baldacchini, Cryst. Latt. Def. Amorph. Mat. 18, 43 (1989).
[4] W. Gellermann, J. Phys. Chem. Solids 52, 249 (1991).
[5] F. Lüty, Proc. Symp. on Optical Properties of Solids and Physics of Surfaces, Mexico, 1979.
[6] H. Pick, in Optical Properties of Solids, Ed. F. Abelés, North-Holland, Amsterdam (1972), Chap. 9.
[7] N. Nishimaki, Y. Matsusaka and Y. Doi, J. Phys. Soc. Japan 33, 424 (1972).
[8] D. Schoemaker, Phys. Rev. 149, 693 (1966).
[9] H. Härtel and F. Lüty, Z. Phys. 182, 111 (1964).
[10] W. B. Fowler, in Physics of Color Centers, Ed. W.B. Fowler, Academic, New York (1968), Chap. 2.
[11] M. Ueta, J. Phys. Soc. Japan 7, 107 (1952).
[12] T. J. Neubert and S. Susman, J. Chem. Phys. 43, 2819 (1965).
[13] I. Schneider, Phys. Rev. Lett. 24, 1296 (1970).
[14] M. A. Aegerter and F. Lüty, phys. stat. sol. (b) 43, 227 (1971).
[15] A. E. Gudat and T. J. Neubert, phys. stat. sol. (b) 56, 145 (1973).
[16] C. Z. Van Doorn and Y. Haven, Phys. Rev. 100, 753 (1955).
[17] L. F. Mollenauer, Opt. Lett. 1, 164 (1977).
[18] F. Lüty, in Physics of Color Centers, Ed. W.B. Fowler, Academic, New York (1968), Chap. 3.
[19] W. C. Collins and I. Schneider, J. Phys. Chem. Solids 37, 917 (1976).
[20] A. Y. S. Kung and J. M. Vail, phys. stat. sol. (b) 79, 663 (1977).
[21] A. Y. S. Kung, J. Lagowski and J. M. Vail, phys. stat. sol. (b) 100, 621 (1980).
[22] G. Baldacchini, E. Giovenale, F. De Matteis, A. Scacco, F. Somma, M. Casalboni and U.M. Grassano, Europhys. Lett. 7, 647 (1988).
[23] M. Meucci, M. Tonelli, G. Baldacchini, U.M. Grassano, A. Scacco, F. Somma, Opt. Commun. 51, 33 (1984).
[24] G. Baldacchini, G. P. Gallerano, U. M. Grassano, A. Lanciano, A. Scacco, F. Somma, M. Meucci and M. Tonelli, Phys. Rev. B 33, 4273 (1986).
[25] G. Baldacchini, G. Cellai, U.M. Grassano, M. Meucci, A. Scacco, K. Somaiah, F. Somma and M. Tonelli, Cryst. Latt. Def. Amorph. Mat. 15, 177 (1987).
[26] G. Baldacchini, E. Giovenale, F. De Matteis, A. Scacco, F. Somma, U.M. Grassano, Nuovo Cim. 10 D, 693 (1988).
[27] A. M. Stoneham, Theory of Defects in Solids, Clarendon, Oxford (1975), Chap. 21.
[28] V. Narayanamurti and R. O. Pohl, Rev. Mod. Phys. 42, 201 (1970).
[29] F. Rosenberger and F. Lüty, Phys. Rev. Lett. 21, 25 (1968).
[30] F. Rosenberger and F. Lüty, Sol. St. Commun. 7, 983 (1969).

[31] G. Baldacchini, K. Somaiah, U.M. Grassano, A. Scacco and F. Somma, Nuovo Cim. 9 D, 1105 (1987).
[32] G. Baldacchini, U.M. Grassano, A. Scacco and F. Somma, Phys. Scripta 37, 381 (1988).
[33] G. Baldacchini, F. De Matteis, U.M. Grassano, A. Scacco and F. Somma, J. Lumin. 40-41, 351 (1988).
[34] G. Baldacchini, F. De Matteis, E. Giovenale, U.M. Grassano, A. Scacco and F. Somma, Phys. Rev. B 37, 7014 (1988).
[35] A. Scacco, F. Somma, M. Rossi, F. De Matteis, G. Baldacchini and U. M. Grassano, Helv. Phys. Acta 62, 746 (1989).
[36] F. De Matteis, M. Rossi, A. Scacco, F. Somma, G. Baldacchini, M. Cremona and U. M. Grassano, J. Phys. Chem. Solids 51, 1053 (1990).
[37] P. B. Fitzsimons, J. Corish and P. W. M. Jacobs, Cryst. Latt. Def. Amorph. Mat. 15, 7 (1987).
[38] M. J. L. Sangster, J. Phys. C 13, 5279 (1980).
[39] J. A. Van Winsum, T. Lee, H. W. Den Hartog and A. J. Dekker, J. Phys. Chem. Solids 39, 1217 (1978).
[40] M. May, E. Rzepka, S. Debrus and J. P. Hong, Opt. Commun. 61, 325 (1987).
[41] M. May, S. Debrus, J. P. Hong, B. Quernet, E. Rzepka and A. Bouazi, Proc. Intern. Conf. on Defects in Insulat. Cryst., Parma, Italy, p. 133.
[42] F. Lüty, Surf. Sci. 37, 120 (1973).
[43] F. De Matteis, M. Rossi, A. Scacco, F. Somma, G. Baldacchini, M. Cremona, U. M. Grassano, Radiat. Eff. Def. Solids, in press.
[44] M. Leblans, W. Joosen and D. Schoemaker, Phys. Rev. B 42, 7220 (1990).
[45] M. Leblans, Ph. D. Thesis, University of Antwerp, Belgium (1990), p. 63-64.
[46] S. A. Boiko, M. Ya. Valakh, M. I. Dykman, M. P. Lisitsa, G. Yu. Rud'ko and G. G. Tarasov, Sov. Phys. Solid State 26, 1747 (1984).
[47] S. A. Boiko, M. Ya. Valakh, M. I. Dykman, M. P. Lisitsa, G. G. Tarasov and A. M. Shpak, Sov. Phys. Solid State 28, 1550 (1986).
[48] L. Bosi and M. Nimis, Nuovo Cim. 12 D, 851 (1990).
[49] L. Bosi and M. Nimis, phys. stat. sol. (b) 156, K5, (1989).
[50] G. Baldacchini, U. M. Grassano, A. Scacco, F. Somma, M. Staikova and M. Georgiev, to be published.

DE-EXCITATION PROCESSES OF THE OPTICALLY EXCITED STATES OF THE F-CENTERS (ABSTRACT ONLY)

H. Okhura

Department of Electronics
Okayama University of Science
1-1 Ridai-Cho
Okayama, 700
JAPAN

We explain the de-excitation processes of the optically excited states of the F centers in alkali halides through the observation of the resonant secondary emission (RSE). Here, the RSE occurs successively as resonant Raman scattering, hot luminescence, and ordinary luminescence. By studying carefully the RSE and the transient effects of ultra-fast emission in de-excitation, we explain the dynamical processes of lattice relaxation within a strong interaction scheme of lattice and electron excited optically. We also show specific evidences of de-excitation caused by two-photon absorption of F centers in KCl with a Nd:YAG laser excitation. Finally, magneto-optical processes in the optical pumping process of the F center are discussed.

TWO-PHOTON SPECTROSCOPY IN INSULATING CRYSTALS

U.M. Grassano

Dipartimento di Fisica
Università di Roma — Tor Vergata
00173 Roma - Italy

ABSTRACT

The optical spectroscopy has been a valuable tool in the study of the energy levels of defects and impurities absorbing in the transparent region of the wide gap materials such as the insulating crystals.

Two-photon spectroscopy is a technique complementary to the ordinary, one-photon spectroscopy because the selection rules, different from those of the one-photon spectroscopy, allow the detection of transitions towards states with the same parity of the ground state.

Examples of the problems studied with this nonlinear technique are the symmetry and the energy levels of color centers, of impurities such as rare earths, silver, or copper ions and of excitons in alkali halides.

I. INTRODUCTION

The study of radiation-matter interaction yield a large amount of information on the properties of solids. Measurements of light absorption, emission, polarization dependence, lifetimes, have provided most of the experimental data to which the calculated crystal band structures have been compared.

A further broadening of the knowledge of the energy spectra of the matter can be obtained by means of "two-photon spectroscopy" or in general of "nonlinear spectroscopy". This kind of spectroscopy can be thought as deriving from the simultaneous absorption of two photons and indeed it was first theoretically calculated as second order transition probability by M. Goeppert-Mayer in 1931 [1].

An alternative description of this effect considers how the optical properties of a medium are changed by the presence of a perturbation. In our case the perturbation is the electric field of a second optical source such as an intense laser beam. The electric field associated to a laser whose intensity is 10 MW/cm^2, is of the order of 10^5V/cm, very close to the dielectric breakdown of several media.

The main advantage of two-photon spectroscopy derives from the possibility of studying transitions towards final states more numerous and in general different from those reached by one-photon allowed transitions. This fact is due to the different selection rules valid for one-photon or two-photon transitions. From these considerations one can imagine the complementary characteristics of the two spectroscopical techniques.

A second property of the nonlinear spectroscopy is the possibility of studying true bulk effects and not surface effects in strongly absorbing samples. At energies above the absorption edge, the linear absorption coefficient can reach easily values of the order of $10^5 \div 10^6$ cm^{-1}, and consequently the penetration depth is about $10^2 \div 10^3$ Å. In these cases, both the transmition measurements in thin films, and the reflectivity measurements, yield values of the optical constants that might not be those appropriate to bulk specimens. The much smaller values of two-photon absorption coefficients prevent this source of error.

II. DEFINITIONS

As mentioned above, in two-photon spectroscopy one measures (as a function of energy, $\hbar\omega$) the change of intensity $I(\omega)$ of the probe beam when it crosses a sample of thickness d in presence of a second light beam of intensity $I_L(\omega_L)$. The second beam is a strong laser beam usually at fixed frequency, ω_L. The intensity change of the probe beam is obtained by integration of the following equation:

$$\frac{\mathrm{d}I}{\mathrm{d}x} = -\alpha^{(1)} I - \beta^{(2)} I_L I \tag{1}$$

where $\alpha^{(1)}$ and $\beta^{(2)}$ are the coefficient of a series expansion of the absorption coefficient $\alpha(\omega)$ in powers of I_L:

$$\alpha(\omega) = \alpha^{(1)}(\omega) + \beta^{(2)}(\omega) I_L + \gamma^{(3)}(\omega) I_L^2 + \dots \; . \tag{2}$$

If $\alpha^{(1)} = 0$ at the frequency of the probe beam and assuming that the intensity changes of the probe beam are small, after integration one has:

$$\frac{\Delta I}{I} \simeq \beta^{(2)} I_L d = \alpha^{(2)} d \; . \tag{3}$$

The second order absorption coefficient $\alpha^{(2)}$, normalized to the laser intensity I_L usually expressed in MW/cm^2, is given by $\beta^{(2)}$. The dimensions of $\beta^{(2)}$ are no longer [cm^{-1}] but are expressed in a hybrid system as $\left[\frac{\mathrm{cm}}{\mathrm{MW}}\right]$.

Because the thickness of the samples is generally $d <$ 1cm and because it is difficult to measure accurately $\frac{\Delta I}{I} < 10^{-3}$ one has to look for intense laser sources in order to have

$$\alpha^{(2)} = \beta^{(2)} I_L \geq 10^{-3}\mathrm{cm}^{-1} \; . \tag{4}$$

These values can be obtained with focused laser beams whose intensity is $10 \div 50$ MW/cm^2. Even in this case the two-photon absorption coefficients are at least six order of magnitude smaller than the allowed one-photon absorption. For impurities in solids or atoms in liquids and gases the two-photon absorption coefficient depends linearly on their concentration N_0. Usually one defines the two-photon absorption cross section $\sigma^{(2)}$:

$$\sigma^{(2)} = \frac{\alpha^{(2)}}{N_0 F_L} \left[\frac{\mathrm{cm}^{-1}}{\left(\frac{\mathrm{atoms}}{\mathrm{cm}^3}\right) \cdot \left(\frac{\mathrm{photons}}{\mathrm{cm}^2 \mathrm{s}}\right)} \right] \; . \tag{5}$$

In this definition, the cross-section is normalized to the photon flux F_L and not to the energy flux I_L as for $\beta^{(2)}$. Typical values of the two-photon absorption cross section are

$$\sigma^{(2)} \simeq 10^{-50} \mathrm{cm}^4 \cdot s \cdot \mathrm{atoms}^{-1} \cdot \ \mathrm{photons}^{-1} \ . \tag{6}$$

The smallness of the two-photon effects makes impossible their observation whenever linear effects are present at the same time. Therefore the photons of the probe beam, in the case of crystals, must have energy smaller than the band gap of the material. Because the energy of the laser photons (Nd-YAG, Excimer, dye) is at most few eV, it is not possible to explore, with this technique, an energy interval larger than few eV above the conduction band.

Two-photon absorption should not be mistaken with two-step absorption in which the first photon populates an intermediate excited state and the second is responsible of the second transition to the final state. As will be clarified below, in two-photon processes no energy conservation rule is present for the virtual transition to the intermediate states.

Before outlining briefly the theory and the experimental techniques used in two-photon spectroscopy, I would like to mention the very useful review article by H. Mahr [2] to which these lecture notes are largely indebted.

III. THEORETICAL FRAMEWORK

III.A. Macroscopic theory

Nonlinear effects due to intense electro-magnetic fields can be described by appropriate coefficients in the relation between the induced polarization and the fields. In the simpler approximation one considers only electric dipole polarization P, induced by electric field E:

$$\underline{P} = \underline{P}_L + \underline{P}_{NL} = \chi_L \cdot \underline{E} + \chi^{(2)}_{NL} : \underline{EE} + \chi^{(3)}_{NL} \vdots \underline{EEE} \tag{7}$$

we have labelled the linear (L) and the nonlinear (NL) contributions. The χ's are the dielectric susceptibilities of the various orders.

Let us recall the relation between the absorption coefficient and the dielectric susceptibility. The absorption coefficient, α is defined as:

$$\alpha = \langle \frac{d}{dt} \left(\frac{\text{absorbed energy}}{\text{volume}} \right) \rangle_{\text{time}} / \text{energy flux} \ . \tag{8}$$

The time average of the rate of energy absorption is given in general as:

$$\langle \frac{d}{dt} \left(\frac{\text{absorbed energy}}{\text{volume}} \right) \rangle_{\text{time}} = \langle \underline{J} \cdot \underline{E} \rangle \tag{9}$$

when $\underline{J}$ is the current induced by the electric field $\underline{E}$ in a medium without free charges or currents.

Neglecting magnetic dipoles or electric multipoles one has

$$\underline{J} = \frac{\partial \underline{P}}{\partial t} \ . \tag{10}$$

Let us consider separately the susceptibilities to the various orders.

1. Linear Dipole Susceptibility. The polarization due to a periodic field $Ee^{-i\omega t}$ is

$$P(t) = \chi_L(\omega)Ee^{-i\omega t} \tag{11}$$

and therefore

$$J = \frac{dP}{dt} = -i\omega\chi_L Ee^{-i\omega t} , \tag{12}$$

$$\langle \underline{J} \cdot \underline{E} \rangle = \langle \frac{1}{2} Re[\tilde{J} \cdot \tilde{E}^*] \rangle = \frac{1}{2} Re[-i\omega\chi_L(\omega)EE^*] . \tag{13}$$

Recalling that the energy flux in a medium of refractive index n, is:

$$I = \frac{EE^*}{8\pi} nc \tag{14}$$

where c is the light velocity, and writing explicitly the real and imaginary part of $\chi_L = \chi_L' + i\chi_L''$, one obtains:

$$\alpha = \alpha^{(1)} = \frac{4\pi\omega}{nc}\chi_L'' = \frac{\omega\varepsilon_2}{nc} . \tag{15}$$

The last relation derives from the relations between the dielectric constant $\tilde{\varepsilon}$ and the susceptibility χ

$$\varepsilon = \varepsilon_1 + i\varepsilon_2 = 1 + 4\pi\chi_L = 1 + 4\pi\chi_L' + i4\pi\chi_L'' . \tag{16}$$

2. Second-Order Electric Dipole Susceptibility. From the definition

$$P^{(2)}t = \chi_{NL}^{(2)}(\omega_1\omega_2)E_1 e^{-i\omega_1 t}E_2 e^{-i\omega_2 t} \tag{17}$$

is evident that no absorption can occur at the frequencies ω_1 or ω_2. The polarization in the medium will have the frequency $\omega_1 \pm \omega_2$. In the case of a simple electric field due to a laser of frequency ω, the induced polarization will appears at the frequencies 2ω or 0. The second order coefficient $\chi_{NL}^{(2)}$ will be responsible of frequency doubling (i.e. second harmonic generation) and of optical rectification (i.e. production of a constant polarization).

One can formally calculate the absorption due to $\chi_{NL}^{(2)}$ and verify that it is zero for example at the probe frequency ω_1

$$\begin{aligned} \langle \underline{J} \cdot \underline{E} \rangle = \langle \frac{\partial \tilde{P}}{\partial t} \cdot \tilde{E}^* \rangle &= \text{const}\langle E_1 E_2 e^{-i(\omega_1+\omega_2)t} \cdot E_1^* e^{i\omega_1 t} \rangle \\ &= \text{const}\langle e^{-i\omega_2 t} \rangle = 0 . \end{aligned} \tag{18}$$

3. Third-Order Electric Dipole Susceptibility. In general one can define

$$P_i^{(3)}(\omega_1 \pm \omega_2 \pm \omega_3) = \chi_{ijkl}^3(\omega_1\omega_2\omega_3)E_j(\omega_1)E_k(\omega_2)E_l(\omega_3) . \tag{19}$$

Because we want to study the polarization at the frequency ω_1 produced by an intense electric field at frequency ω_2, we chose $\omega_3 = -\omega_2$. We obtain

$$Pe^{-i\omega_i t} = \chi_{NL}^{(3)}(\omega_1\omega_2, -\omega_2)E_1 e^{-i\omega_1 t}E_2 e^{-i\omega_2 t}E_2^* e^{i\omega_2 t} . \tag{20}$$

As before we have:

$$J = \frac{\partial P}{\partial t} = -i\omega_1 \chi^{(3)}_{NL}(\omega_1, \omega_2, -\omega_2) E_1 E_2 E_2^* e^{-i\omega_1 t} \ . \tag{21}$$

The energy absorbed at the frequency ω_1 is:

$$\langle \underline{\tilde{J}} \cdot \underline{\tilde{E}}^* \rangle = \frac{1}{2} Re[-i\omega_1 \chi^{(3)}_{NL} E_1 E_2 E_2^* E_1^*] \ . \tag{22}$$

Introducing the real and the imaginary part of $\chi^{(3)}_{NL} = \chi''^{(3)}_{NL} + i\chi''^{(3)}_{NL}$ one obtains:

$$\alpha^{(2)}(\omega_1) = \beta^{(2)} I_2 = \frac{4\pi\omega_1}{n_1 c} \chi''^{(3)}_{NL}(\omega_1 \omega_2 - \omega_2)|E_2|^2 \tag{23}$$

and with eq. (14)

$$\beta^{(2)} = \frac{32\pi^2 \omega_1}{n_1 n_2 c} \chi''^{(3)}_{NL}(\omega_1 \omega_2 - \omega_2) \tag{24}$$

where n_1 and n_2 are the refractive indexes at the frequencies ω_1 and ω_2 respectively.

The fourth rank tenor $\chi^{(3)}_{ijkl}$ has 81 elements. Some of them may be equal or zero in the different symmetries. For example in a cubic crystal only 21 elements are different from zero and only 4 linearly independent χ_{xxxx}, χ_{yyzz} χ_{yzyz} and χ_{yzzy}. The physical properties of the system under study will be derived from $\chi''^{(3)}_{NL}$ that (as χ''_L) will be very large close to resonance.

III.B. Microscopic theory

A different description of two-photon processes, and often more directly connected to the electronic properties of matter, derives from the second order transition probability calculated in the framework of the perturbation theory.

The relation with the previous description is straightforward. In linear processes with photons of energy $\hbar\omega$, one has:

$$\langle \frac{d}{dt} \left(\frac{\text{energy absorbed}}{\text{volume}} \right) \rangle = W_{gf} \hbar\omega N_0 \tag{25}$$

where N_0 is the density of the absorbity species and W_{gf} is the transition probability from a ground state g toward a final state f.

The absorption coefficient of eq. (8) can be written as

$$\alpha^{(1)} = W_{gf} \frac{\hbar\omega N_0}{I} \ . \tag{26}$$

The energy flux $I = N\hbar\omega c/n$ (N is the photon density and n as above the refraction index).

By comparison of eq. (26) with eq. (15) one gets the relation

$$N_0 W_{gf} = \frac{4\pi I}{c\hbar} \chi''_L \ . \tag{27}$$

The analogous extension to the nonlinear case is

$$N_0 W_{gf}^{(2)} = \frac{32\pi^2}{c^2\hbar} I_1 I_2 \chi''^{(3)}_{NL} \ . \tag{28}$$

The second order transition probability $W_{gf}^{(2)}$ due to a perturbation expressed for one electron in dipole approximation as $-e\underline{E}\cdot\underline{r}$, is given by [3]

$$W_{gf}^{(2)} = \frac{16\pi^3\hbar e^4 N_1 N_2 \omega_1 \omega_2}{n_1^2 n_2^2} \left| (1+P_{12}) \sum_i \frac{\langle f|\underline{r}\cdot\underline{\varepsilon}_1|i\rangle\langle i|\underline{r}\cdot\underline{\varepsilon}_2|g\rangle}{\omega_{ig}-\omega_1} \right|^2 \cdot$$
$$\cdot\, \delta(E_f - E_g - \hbar\omega_1 - \hbar\omega_2) \tag{29}$$

where $\underline{\varepsilon}_i$ are the polarization vectors of the two beams, P_{12} is a permutation operator of the index 1 and 2 equivalent to the exchange of the two photons, and $\hbar\omega_{ig}$ is the energy difference between states $|i>$ and $|g>$.

Fig. 1 shows a schematic diagram of energy levels with the virtual transitions from the ground state $|g>$ to the intermediate states $|i>$ and from these to the final state $|g>$. The energy conservation, expressed by the δ function in eq. 29, holds only for the initial and final states. The right side of Fig. 1 shows a case of near resonance with an intermediate state between $|g>$ and $|f>$. In this case $\omega_{ig} \simeq \omega_1$ and $W_{gf}^{(2)}$ might become very large.

The transition probability and therefore the absorption coefficient depends on the dipole matrix elements and on the direction of the vectors ε_i relative to each other and to the crystal axes. Calculations have been performed for the various symmetries [4-7].

The selection rules for the cubic symmetry (space group O_h) can be easily derived.

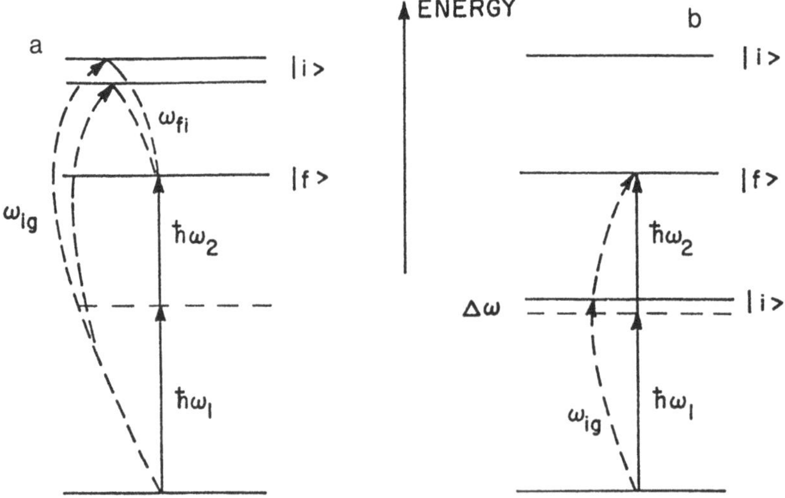

Fig. 1. Schematic representation of two-photon transitions from a ground state $|g>$ to a final state $|f>$ via intermediate states $|i>$. In (a) no resonance is possible between $\hbar\omega_1$, $\hbar\omega_2$ and $\hbar\omega_{ig}$. In (b) a near-resonance case ($\omega_1 \approx \omega_{ig}$) is shown with a particular intermediate state $|i>$.

Table I. Polarization Dependence of Two-Photon Transitions for the Space Group O_h [(a)].

Transition	Polarization dependence for linearly polarized beams[(b)]
$\Gamma_1^+ \to \Gamma_1^+$	$(l_1 \cdot l_2 + m_1 \cdot m_2 + n_1 \cdot n_2)^2 = (\hat{\boldsymbol{\varepsilon}}_1 \cdot \hat{\boldsymbol{\varepsilon}}_2)^2$
$\Gamma_1^+ \to \Gamma_3^+$	$(l_1^2 \cdot l_2^2 + m_1^2 \cdot m_2^2 + n_1^2 \cdot n_2^2)$ $-(l_1 \cdot l_2 \cdot m_1 \cdot m_2 + m_1 \cdot m_2 \cdot n_1 \cdot n_2 + n_1 \cdot n_2 \cdot l_1 \cdot l_2 \cdot)$
$\Gamma_1^+ \to \Gamma_4^+$	$1 - (l_1 \cdot l_2 + m_1 \cdot m_2 + n_1 \cdot n_2)^2 = (\hat{\boldsymbol{\varepsilon}}_1 \times \hat{\boldsymbol{\varepsilon}}_2)^2$
$\Gamma_1^+ \to \Gamma_5^+$	$1 - (l_1^2 \cdot l_2^2 + m_1^2 \cdot m_2^2 + n_1^2 \cdot n_2^2)$ $+2(l_1 \cdot l_2 \cdot m_1 \cdot m_2 + m_1 \cdot m_2 \cdot n_1 \cdot n_2 + n_1 \cdot n_2 \cdot l_1 \cdot l_2)$
	Polarization dependence for circularly polarized beams[(c)]
$\Gamma_1^+ \to \Gamma_1^+$	$\frac{1}{4}[l \mp (\hat{\mathbf{s}}_1 \cdot \hat{\mathbf{s}}_2)]^2;$
$\Gamma_1^+ \to \Gamma_3^+$	$\frac{1}{4}\left\{\frac{3}{4} + \left[\frac{1}{2} \pm (\hat{\mathbf{s}}_1 \cdot \hat{\mathbf{s}}_2)\right]^2 - 3(s_{1x}s_{2x}s_{1y}s_{2y} + \right.$ $\left. + s_{1x}s_{2x}s_{1z}s_{2z} + s_{1y}s_{2y}s_{1z}s_{2z}\right\};$
$\Gamma_1^+ \to \Gamma_4^+$	$1 - \frac{1}{4}[1 \pm (\hat{\mathbf{s}}_1 \cdot \hat{\mathbf{s}}_2)]^2;$
$\Gamma_1^+ \to \Gamma_5^+$	$1 - \frac{1}{4}[1 \mp (\hat{\mathbf{s}}_1 \cdot \hat{\mathbf{s}}_2)]^2 + (s_{1x}s_{2x}s_{1y}s_{2y} +$ $+ s_{1x}s_{2x}s_{1z}s_{2z} + s_{1y}s_{2y}s_{1z}s_{2z})$

[(a)] From ref. 5 and 7;
[(b)] $\hat{\boldsymbol{\varepsilon}}_1$ and $\hat{\boldsymbol{\varepsilon}}_2$ are the polarization vector of beams 1 and 2 respectively; l, m, n are their direction cosines;
[(c)] $\hat{\mathbf{s}}_1$ and $\hat{\mathbf{s}}_2$ are the unit propagation vectors. The upper sign holds when both beams have the same polarization (both right or both left) the lower sign holds for mixed polarization.

Neglecting the spin-orbit interaction, the levels can be classified into 10 irreducible representations: $\Gamma_1^+, \Gamma_2^+, \ldots \Gamma_5^+, \Gamma_1^-, \ldots \Gamma_5^-$ or with equivalent notation: $A_{1g}, A_{2g}, E_g, T_{1g}, T_{2g}, A_{1u}, \ldots T_{2u}$.

An allowed-allowed transition connects two states of the same parity trough an intermediate state of the opposite parity. From a ground state of symmetry Γ_1^+ are allowed transitions towards all the states whose representation is included into the two-photon transition operator:

$$\Gamma_4^- \oplus \Gamma_4^- = \Gamma_1^+ + \Gamma_3^+ + \Gamma_4^+ + \Gamma_5^+ \tag{30}$$

This multiplicity of transitions has to be compared with the single one-photon allowed transition $\Gamma_1^+ \to \Gamma_4^-$.

With linearly or circularly polarized light one can distinguish the symmetries of the various excited states. Indicating by $(l_1 m_1 n_1)$ and $(l_2 m_2 n_2)$ the direction

Table II. Angular Dependence of the Intensity of Two-Photon Transition[a].

Transition		Linear polarization		Circular polarization
		$\hat{\varepsilon}_1 \parallel \hat{\varepsilon}_2$	$\hat{\varepsilon}_1 \perp \hat{\varepsilon}_2$	
$\Gamma^+ \to \Gamma_1^+$	$A_{1g} \to A_{1g}$	c_0	0	0
$\Gamma_1^+ \to \Gamma_3^+$	$A_{1g} \to E_g$	$c_1\left[1 - \frac{3}{4}\sin^2(2\theta)\right]$	$c_1\left[\frac{3}{4}\sin^2(2\theta)\right]$	$\frac{3}{4}c_1$
$\Gamma_1^+ \to \Gamma_4^+$	$A_{1g} \to T_{1g}$	0	c_2	0
$\Gamma_1^+ \to \Gamma_5^+$	$A_{1g} \to T_{2g}$	$c_3 \sin^2(2\theta)$	$c_3 \cos^2(2\theta)$	c_3

[a] Incident direction is the [001] direction. θ is the angle between the direction of polarization of the incident light and the [100] crystal axis. The $\hat{\varepsilon}_i$ are the unit polarization vectors of the two beams. The c_i are the relative weights of the contributions of the different irreducible representations.

cosines of the linear polarization vector ε_1 and ε_2, and with s_1 and s_2 the propagation vectors for circularly polarized light, one can calculate the polarization dependence of the different two-photon transitions. These dependence for the cubic group are reported in Table I.

For the circularly polarized light the top sign holds for beams with the same elicity, the bottom one for beams with opposite elicity. It is noteworthy that the transitions $\Gamma_1^+ \to \Gamma_1^+$, and $\Gamma_1^+ \to \Gamma_4^-$ depends only upon the mutual orientation of the polarization or propagation vectors and not on their orientation with respect to the crystal axes. In particular using a two photons from a single beam, $\hat{\varepsilon}_1 = \hat{\varepsilon}_2$ or $\hat{s}_1 = \hat{s}_2$ and therefore the transition $\Gamma_1^+ \to \Gamma_4^+$ is always forbidden.

The results of Table I can be rewritten for the case of two identical photons of a laser beam propagating along the [001] direction. By introducing the polarization angle θ of the beam with respect to the [100] axis one obtains the results reported in Table II.

IV. EXPERIMENTAL TECHNIQUES

IV.A. Direct absorption measurements

In these measurements, one observes the attenuation of a probe beam in presence of an intense pump (laser) beam. The two-photon absorption spectrum is of course obtained by changing the wavelength of the probe beam. This direct technique yield more straighford results and allows more easily the study of the polarization dependence of the nonlinear effect. The difficulties of these measurements are very severe in the study of impurities and defects in solids whose density ($10^{16} \div 10^{18}$ cm^{-3}) is at least one thousand time smaller than that of valence band electrons. In the case of impurities and defects the size of the two-photon effect is unmeasurably small ($\Delta I/I \leq 10^{-6}$).

In principle the measurements of two-photon absorption are performed as indicated in Fig. 2. In the sample the probe beam of intensity $I(\omega)$ overlaps the laser beam $I_L(\omega_L)$ and the simultaneous absorption of two photons is detected through the change of the transmitted intensity of the probe beam $\Delta I(\omega)$. As already mentioned only with laser of high intensity one obtains measurable effects (see eq. 3). In Table III are reported typical values of photon energy, peak power and flux for several lasers used in these experiments. These values of power are available in pulsed lasers and we have supposed pulsewidth of 10 ns and area of the beam focus $\simeq 0.25$ cm^2.

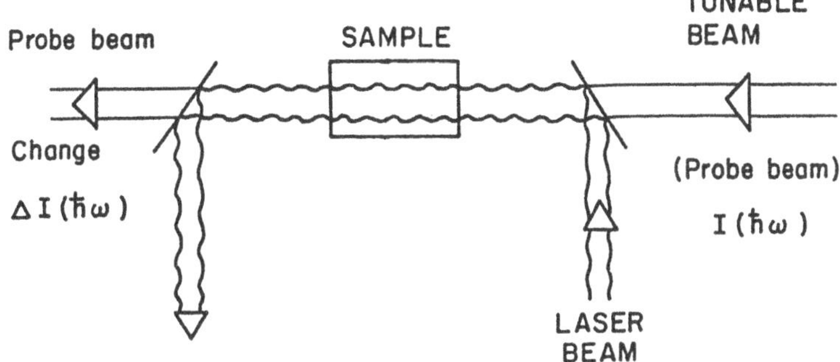

Fig. 2. Principle of two-photon absorption spectroscopy. Two light beams, a fixed frequency beam (laser beam) of photon energy $\hbar\omega_L$ and a frequency-tunable beam (probe beam) overlap inside a sample. If two photons $\hbar\omega$ and $\hbar\omega_L$ are simultaneously absorbed, a change ΔI in the intensity of the probe beam will be registered in presence of the laser beam.

In presence of intermediate states close to the final state (as usually happens in band-to-band transitions in solids) the energy denominator in eq. 29 decreases (and the transition probability increases) if one comes close to resonance. This is obtained with low energy lasers such as CO_2 laser or using Raman-Stokes shifted Nd-YAG. In this case however the available spectral range is markedly decreased.

A block diagram of an experimental apparatus used by Fröhlich and co-workers [8] is shown in Fig. 3. The high intensity source is a Nd-YAG laser Raman shifted

Table III. Characteristics of laser sources used in two-photon experiments.

	Photon energy (eV)	Peak power (MW)	Flux (photons/cm^2·s)
Nd-YAG	1.17	30	6.10^{26}
2nd harmonic of Nd-YAG	2.34	5	5.10^{25}
1st order Stokes shift of Nd-YAG	0.65	5	2.10^{26}
Ruby	1.78	100	1.10^{27}
CO_2	0.117	2	4.10^{26}
Excimer XeCl	4.03	8	5.10^{25}

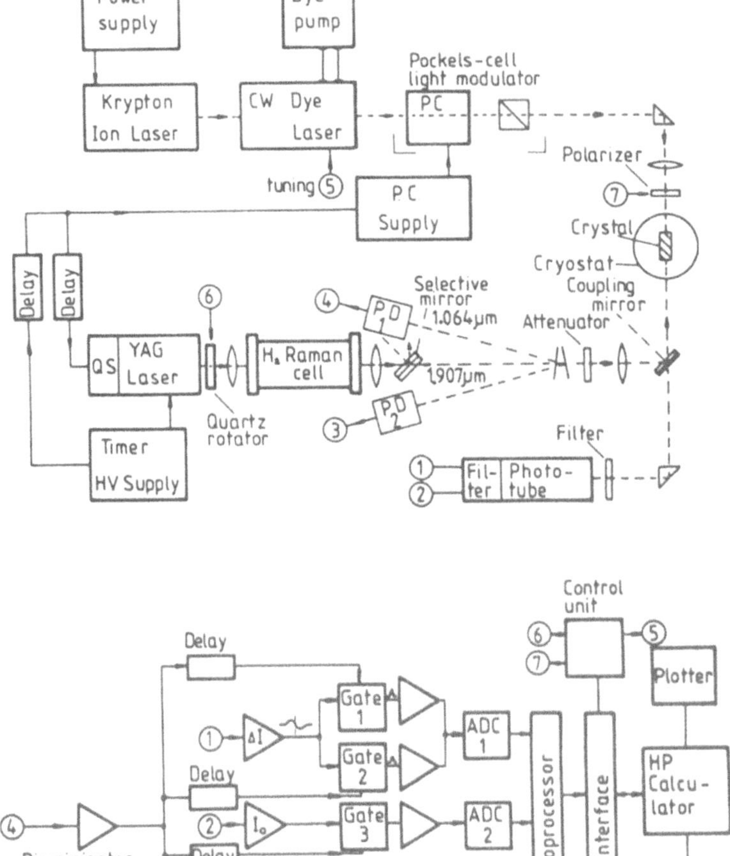

Fig. 3. Block diagram for two-photon absorption setup: PC, Pockels cell; PD, Photodiode; QS, Pockels cell Q switch; ADC, analog-to-digital converter [from Ref. 8].

to lower energy by a H_2 cell. The tunable probe beam is a Kripton pumped dye laser. The two beams propagate antiparallel in the sample, their intensity can be changed by means of attenuators. Polarization direction can be also changed rotating the polarizers. Both lasers were pulsed but with different repetition rate and pulse width so that proper synchronization between pulses had to be obtained. The pulsewidth of the probe beam pulse is longer than that of the pump so that the intensity change due to pump beam is detected as a fast transient on the relatively flat background of the longer pulse. The same phototube records this fast change ΔI and the average value I. The pump laser intensity I_L is also monitored to obtain the values of $\beta^{(2)}$. The automatic data acquisition system allow the recording of two-photon absorption spectra with a resolution (due to the dye laser) of 0.1 meV

With this or analogous set-up two-photon absorption spectra have been measured in various wavelength intervals depending on the dye laser tunability. While the energy scale and therefore peaks positions, structures in the absorption bands are measured with great accuracy, the intensities of the two-photon absorption are generally reported in arbitrary units because of the difficulty of an absolute calibration of a signal critically dependent on the temporal and spatial overlapping of the two beams, on the accurate evaluation of the beamwaist of a supposedly Gaussian beam, on the pulse-to-pulse reproducibility of the laser output.

The absolute measurements of the two-photon absorption cross section are generally performed at fixed wavelength with suitable normalization techniques [9,10]. The review paper of Nathan et al. [11] reports extensively the works on nonlinear absorption cross sections.

IV.B. Indirect measurements

The indirect methods measure effects that follow a two-photon absorption. The most used of these methods is the detection of the luminescence and indeed with this method a two-photon effect was discovered for the first time by Kaiser and Garret in 1961 [12]. Both photons in that experiment came from a single ruby laser $\hbar\omega = 1.78$ eV and were absorbed by a CaF_2 crystals containing Eu^{2+}. The luminescence following the two-photon absorption had an energy of 2.9 eV higher than that of excitation and its intensity increased as the square of the laser intensity (see Fig. 4). The quadratic dependence of the emission on the laser intensity is the clearer indication that a two-photon absorption has occured in the material; it is almost a fingerprint of the two photon effect.

The detection of the luminescence is a technique much more sensitive that the direct measure of the absorption and has been universally employed in the measurements of nonlinear effects of defects in solids. Other effects produced by two-photon absorptions are photoelectric emission, photodecomposition, defects formation... All these phenomena, implying excitation of the samples with photons of energy above gap, have been observed with laser sources of below-gap energy: of course they are due to two-photon absorption.

The main shortcoming of these indirect measurements is the greater difficulty of going back to the optical transitions that have taken place. For example in the

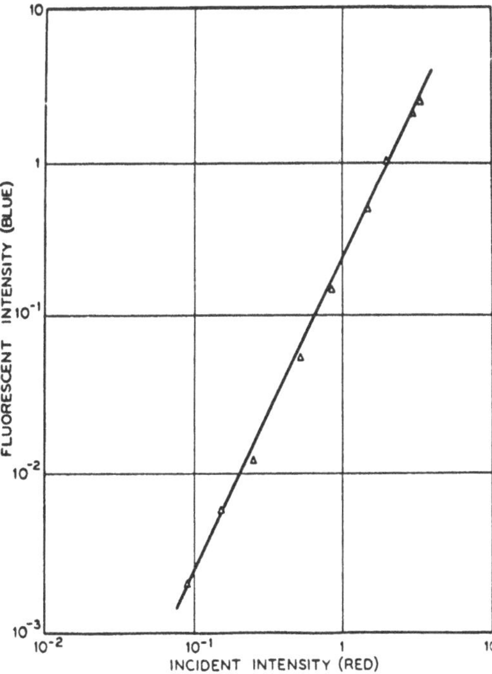

Fig. 4. Quadratic dependence of the blue fluorescence of CaF_2: Eu^{2+} upon the intensity of the extensity of the exciting laser beam (from ref. 12).

case of the luminescence, the existence of nonradiative decays or the possibility of photoinduced reactions between excited species make uncertain most of the quantitative determinations of the cross sections.

A simplified experimental apparatus for two photon luminescence excitation is shown in Fig. 5. [13] The excitation is provided by two photons of the same or of two different dye lasers with variable polarizations. The emission is collected at right angle with respect to the excitation. Particular care has to be used to reject the scattered pumping light from the detection system. The emission in solids is usually Stokes shifted with respect to the excitation and often its lifetime is longer than the laser pulsewidth. Rejection of the scattered light is therefore possible with filters, monochromators and time discriminators.

It is generally observed that after the one-photon or the two photon excitation the system relaxes toward the same relaxed excited states and indeed the emission bands are the same in both cases. On assuming therefore the same quantum efficiency η in both cases one can derive absolute values of the absorption coefficients if the apparatus is calibrated for the measure of the absolute number of photons emitted I_{em} for laser pulse.

ΔI_{ass}, the number of absorbed photons, can be derived from the emitted pho-

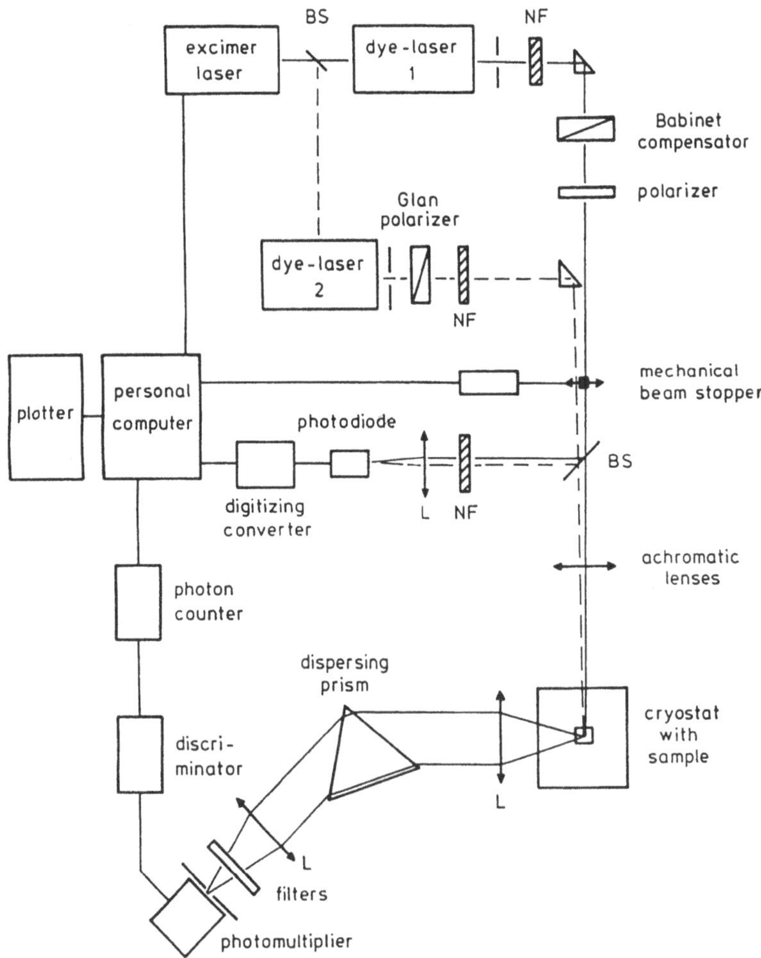

Fig. 5. Apparatus for two-photon excitation studies. The extensions for the double-beam setup are drawn with a dashed line (from ref. 13). BS: beam splitter; L: lens; NF: neutral filter.

tons I_{em} because:

$$\Delta I_{ass} = \frac{2I_{em}}{\eta} \tag{31}$$

and the absorption cross section through the definitions (3) and (5) is given by:

$$\sigma^{(2)} = \frac{\Delta I_{ass}}{IdN_0F_L} \ . \tag{32}$$

As in the direct measurements also in the emission experiments the relative polarizations of the laser beams (when a two beam system is used) and their relation to the direction of the crystal axes are very useful to determine the symmetries of the energy levels.

V. EXAMPLES

Results of two-photon spectroscopy in insulating crystals are too numerous to discuss them all. As already mentioned review papers have been published by H. Mahr in 1975 [2] and more recently by V. Nathan, A.H. Guenter and S.S. Mitra in 1984 [11]. As typical examples I will briefly present the results of investigations of different localized defects: color centers, impurity ions and excitons in alkali halides.

V.A. Color centers

The first color center to be investigated through a nonlinear technique was of course the F center (an electron bound to a negative ion vacancy). The lucky coincidence that the F-band in KCl occurs at twice the energy of the Nd-YAG laser is counterbalanced by the fact that the Stokes shifted F-center emission (whose detection, as we have seen, is essential to reveal a two-photon absorption) overlaps the Nd-Yag laser line. The problems of suppression of the scattered laser light are therefore very severe. Conflicting evidences have been reported on the lifetimes of the one-photon and two-photon excited luminescences [14,15]. New emissions and different from that of the F center have been observed by our group by nonlinear excitation of colored KCl [16].

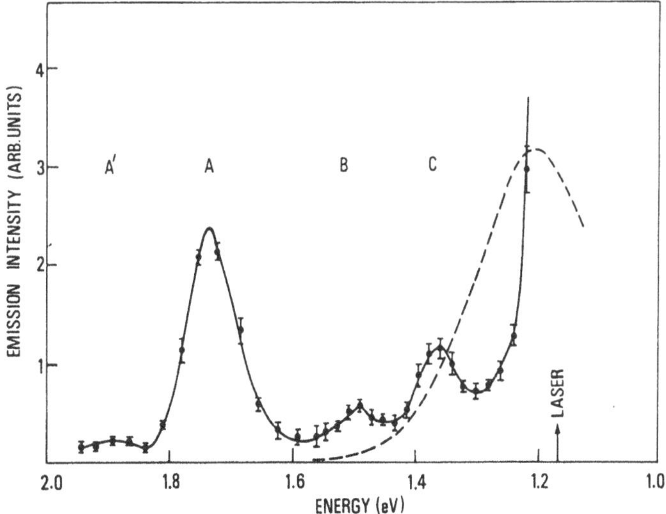

Fig. 6. Emission spectrum of a KCl crystal containing F centers excited by a Nd-YAG laser at 1.17 eV (full curve) and by the second harmonic of the same laser (dashed curve) (from ref. 16).

This result is reported in Fig. 6. The one-photon excitation at 2.34 eV (dashed line) shows the normal F-center emission. The nonlinear excitation with photons of 1.17 eV produces new bands (full line) even if the sample contains essentially only F-centers. The increase of the new emissions intensity with the aggregation of F centers into complexes (F_2 and F_3 centers) lead us to the attribution of the nonlinear luminescence to these centers. More recent investigations have assigned these emissions to loose couples of F-centers [17]. Two-photon excitation of complex color centers was also observed in NaF and an estimate of the absorption cross section was given [18].

V.B. Impurity ions

The importance of two-photon spectroscopy in the study of energy levels and the intensities of the one-photon forbidden transitions within the $4f^n$ multiplet of the rare-earth ions has been already stressed in the lectures of Prof. Imbush at his school [19].

I will therefore confine myself to the study of an ion with $(ns)^2$ electronic configuration: the Ag^- ion [13,20,21]. Alkali halides doped with the corresponding silver halides grown by the Kyropulos technique contains substitutional Ag^+ impurities. The conversion from Ag^+ to Ag^- is obtained via an additive coloration of the crystal with potassium vapor. The excess of vacancies introduced by the coloration is removed from the sample by a subsequent electrolitic discoloration [22]. The Ag^- ion, now substitutional of a Cl^- ion, has, in the ground state, an electronic configuration $A_{1g}(5s)^2$. This electronic level and those belonging to the excited configurations are reported in Fig. 7. Two emission bands (indicated by A' and C') are one-photon excited after the symmetry allowed absorption into the A, B or C band. These A' and C' emissions are used to measure the two-photon excitation spectrum. This spectrum reveals the two-photon allowed transitions toward the even parity excited states A_{1g} E_g and T_{2g}. The excitation spectra are very dif-

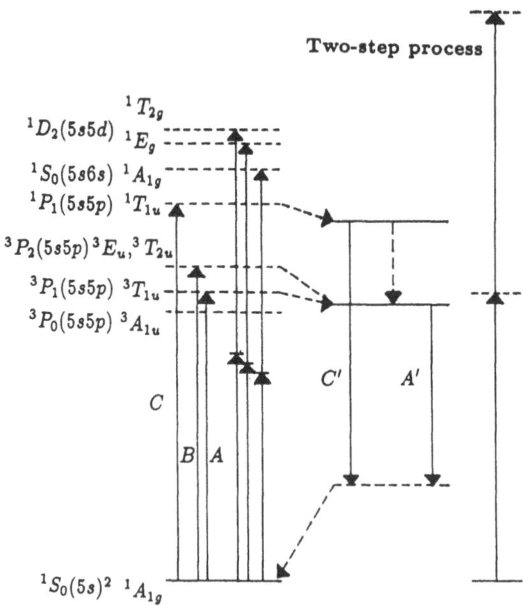

Fig. 7. Energy-level scheme of a Ag^- ion: free-ion on the left, ion in a cubic environment at the center, and two-step process on the right. Radiative transitions (solid arrows), nonradiative transitions (dashed arrows), relaxed states (solid lines), and unrelaxed states (dashed lines).

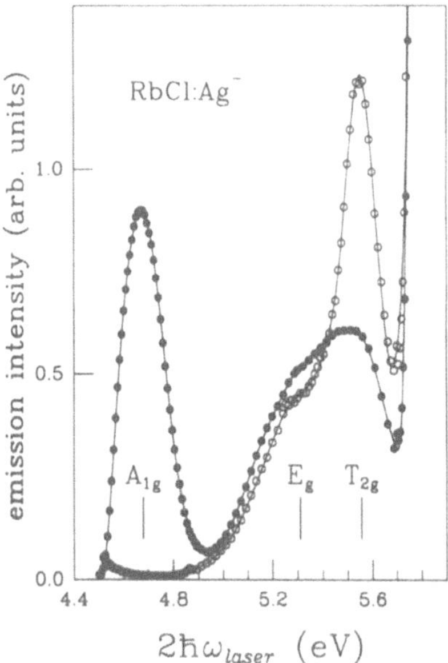

Fig. 8. Two-photon excitation spectra of RbCl:Ag$^-$ for two different polarizations of the exciting laser beam. Linear polarization parallel to a [100] crystal direction (solid circles), circular polarization (open circles).

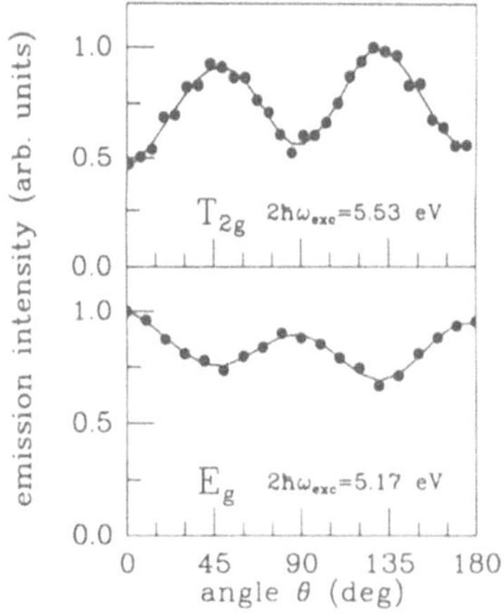

Fig. 9. Angular dependence of the two-photon excitation of RbCl:Ag$^-$ at two different excitation energies.

ferent using linear and circularly polarized exciting light as shown in Fig. 8 and in Fig. 9 for RbCl:Ag^-. In the latter figure the linear polarization of the exciting beam (propagating along the [001] direction) is rotated by an angle θ ($0 \leq \theta \leq \pi$) with respect to the [100] crystal axis. By checking the angular dependences reported in Table II one can assign the two-photons transitions to the appropriate excited state.

Also the ions with d^{10} closed shells (Cu^+ and Ag^+) have been throughly studied with analogous nonlinear techniques [23,24].

V.C. Excitons

The exciton is a very peculiar type of defects formed by a bound electron-hole pair. Its absorption is very close to the band-to-band transition. The symmetry of the valence and the conduction bands in alkali halides makes one photon allowed exciton with *s*-like evelope function [3]. The lack of this absorption in two photon spectroscopy has allowed an easier localization of the band gap energy [25].

The excitons in the alkali chlorides and fluorides occurs in the vacuum ultraviolet spectral region and have been studied with two-photon techniques by using as a tunable source, the synchrotron radiation light of the Adone storage ring. Two-photon absorption spectra have been obtained after solving the difficult problems of time coincidence and detection of very narrow light pulses (1 ns the VUV and 10 ns the Nd YAG pulses) [26,27].

An example of these measurements is reported in Fig. 10. The dashed line shows the one-photon absorption with the 1s exciton structure. The two-photon absorption (full line) shows structures that we have attributed to 2p and 3p excitons. With an hydrogenic model we have determined the energy gap and the exciton effective mass [28].

Before closing I would like also to mention the important contributions given to the field of nonlinear spectroscopy in insulating crystals by A. Cingolani and his school, with the measurements of nonlinear absorption cross section of several band-to-band transitions [30,31]. More recently the same group has investigated the spectral dependence of the two-photon and three-photon absorption spectra revealing the different weights of the possible processes involving combination of allowed and forbidden virtual transition depending on the symmetry of the intermediate levels [32,33].

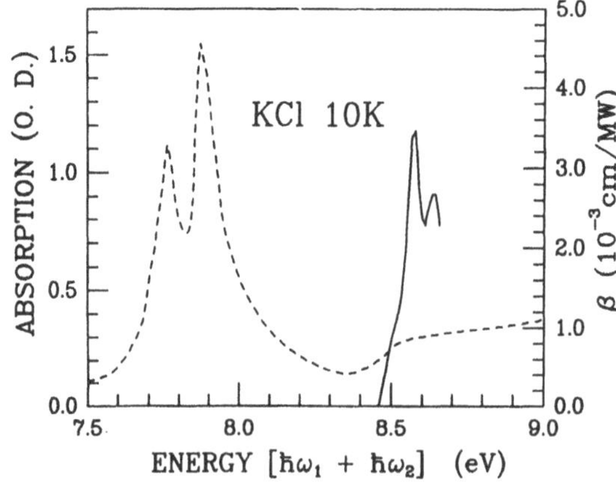

Fig. 10. Two-photon absorption spectrum (continuous line) of KCl at 10 K, compared with the one-photon absorption spectrum (dashed line) from ref. 29.

ACKNOWLEDGEMENTS

The author gratefully acknowledges the stimulating discussions and the co-operation with Drs. M. Casalboni, C. Cianci, R. Francini, F. Lomheier, R. Pizzoferrato, N. Zema that over the years made possible the development of the Tor Vergata research group on nonlinear spectroscopy of solids.

REFERENCES

1. M. Goeppert-Mayer, Ann. Phys. 9, 273 (1931).
2. H. Mahr, Quantum Electronics, (Eds. H. Rabin and C.L. Tang) (Academic Press, New York, 1975), vol. 1, p. 285.
3. F. Bassani and G. Pastori Parravicini, in: "Electronic States and Optical Transitions in Solids", (Pergamon Press Oxford 1975).
4. M. Inoue and Y. Toyozawa, J. Phys. Soc. Japan 20, 363 (1965).
5. T.R. Bader and A. Gold, Phys. Rev. 171, 997 (1968).
6. E. Doni, R. Girlanda and G. Pastori-Parravicini, Phys. Stat. Sol. (b) 65, 203 (1974).
7. D. Frohlich, B. Staginnus and S. Thurm, Phys. Stat. Sol. 40, 287 (1970).
8. Ch. Nihlein, D. Fröhlich and R. Kenklies, Phys. Rev. B23, 2731 (1981).
9. H. Lotem and C.B. de Araujo, Phys. Rev. B16, 1711 (1977).
10. C.B. da Araujo and H. Lotem, Phys. Rev. B18, 30 (1978).
11. V. Nathan, A.H. Guenther and S.S. Mitra, J. Opt. Soc. America B2, 294 (1985).
12. W. Kaiser and C.G.B. Garrett, Phys. Rev. Lett. 7, 229 (1961).
13. F.J. Lohmeier, R. Francini and U.M. Grassano, Phys. Rev. B38, 1468 (1988).
14. F. De Martini, G. Giuliani and P. Mataloni, Phys. Rev. Lett. 35, 1464 (1975).
15. K. Vogler, Phys. Stat. Sol. b107, 195 (1981).
16. M. Casalboni, R. Francini, U.M. Grassano and R. Pizzoferrato, Phys. Stat. Sol. b117, 493 (1983).
17. H. Hanzawa, Y. Mori and H. Ohkura, Europhysical Conference Lattice Defects in Ionic Materials, (Groningen 1990) Abstract p. 313.
18. M. Casalboni, U.M. Grassano and A. Tanga, Phys. Rev. B19, 3306 (1979).
19. G. Imbush, Lectures at this school.
20. F.J. Lohmeier and K. Schmitt, Optics Commun. 63, 49 (1987).
21. M. Bellatreccia, M. Casalboni, R. Francini and U.M. Grassano, Phys. Rev. B43, 2334 (1991).
22. H. Reimer and F. Fisher, Phys. Stat. Sol. b124, 61 (1984).
23. S.A. Payne, A.B. Goldberg and D.S. McClure, J. Chem, Phys. 81, 1529 (1984).
24. B. Moine and C. Pedrini, J. Physique 45, 1491 (1984).
25. D. Frohlich, in: "Festkorperprobleme", (Vieweg Verlag Braunshweig, 1970), vol. X, p. 227.
26. R. Pizzoferrato, M. Casalboni, R. Francini, U.M. Grassano, F. Antonangeli, M. Piacentini, N. Zema and F. Bassani, Europhys. Lett. 2, 571 (1986).
27. R. Pizzoferrato and M. Casalboni, J. Phys. E.: Sci. Instrum. 20, 897 (1987).
28. M. Casalboni, C. Cianci, R. Francini, U.M. Grassano, M. Piacentini and N. Zema, Phys. Rev. to be published.
29. K.J. Teegarden and G. Baldini, Phys. Rev. 155, 986 (1967).
30. I.M. Catalano, A. Cingolani and A. Minafra, Phys. Rev. B5, 1629 (1972).
31. I.M. Catalano, A. Cingolani and A. Minafra, Opt. Commun. 5, 212 (1972).
32. I.M. Catalano, A. Cingolani and M. Lepore, Phys. Rev. B33, 7270 (1986).
33. I.M. Catalano, A. Cingolani, R. Cingolani and M. Lepore, Phys. Rev. B38, 3438 (1988).

SPECIAL TOPICS

PARTICLES AND ELEMENTARY EXCITATIONS

G. Costa

Department of Physics, University of Padova, Italy
Istituto Nazionale di Fisica Nucleare, Sezione di Padova, Italy

ABSTRACT

Analogies in the formulation of quantum field theory (QFT) for elementary particles and of many-body theory for condensed matter systems are briefly discussed. One one side, photons, gravitons, etc. are identified with the quanta of radiation fields; on the other side, phonons, plasmons, etc. appear from quantization of collective waves in condensed matter.

Both in elementary particle and in solid state physics, a fundamental role is played by symmetry. Specifically, we consider the mechanism of spontaneous symmetry breaking, which takes place when the ground state of a system possesses a lower symmetry than the equations of motion. In the case of continuous symmetry, the breaking is accompanied by the appearance of Nambu-Goldstone modes, which are massless bosons in QFT and elementary excitations without energy-gap in non-relativistic many-body theory. A relevant example is the case of superfluidity.

When long-range interactions are present, such as those mediated by massless vector fields, the would-be Goldstone mode acquires a non-vanishing mass, and it could be interpreted as the longitudinal component of the vector field which becomes massive. These considerations are applied to the case of superconductivity.

I. INTRODUCTION

The formulation of the quantum properties of very different physical systems, such as a bulk of condensed matter or a high energy particle reaction, often reveals surprisingly strong analogies. This is true in spite of the fact that each system is usually described in terms of its appropriate basic elements: in fact, the concept of "elementary constituent" is relative not only to the specific system but also to the energy scale of the phenomena which are involved.

Physical properties of matter could be described, in principle, in terms of what are considered its microscopic constituents: atoms, ions, electrons. However, it usually appears more convenient to adopt other elements as the basic ingredients of the description.

At the level of nuclear reactions, one has to deal explicitly with nucleons (protons and neutrons) and their interactions. At higher energy scales new forms of matter appear, which are not observable in ordinary conditions and require the introduction of more fundamental objects. In fact, nucleons and their excited states appear to be composite and it is appropriate to describe them in terms of "quarks". The energies which now can be reached in the laboratories are of the order of a few hundred GeV (1 GeV = 10^9 eV). It might be that, at still higher energy scales, even quarks reveal a composite structure.

The present picture considers each nucleon made up of three quarks. Many excited states with half-integer spin have been found: they are called baryons. Other states have integer spin; they are called mesons and each of them is described in terms of a quark-antiquark pair.

On the other hand, at the available energy scales, electrons appear still elementary; other particles similar to but heavier than electrons have been found: μ and τ, and also the massless $\frac{1}{2}$ -spin neutrinos. All together, they are called leptons.

Thus we can say that the elementary constituents of matter are quarks and leptons.* Their mutual interactions are mediated by fields, i.e. by "radiation". There are four kind of fundamental interactions: going from the classical to the quantum description, each field appears as an assembly of quanta, or in other words of "particles", with integer spin (i.e. bosons). The situation is summarized in Table 1.

Table 1 - Fundamental Interactions

Type of interaction	Quanta	Spin	Mass
EM	photons	1	0
Weak	intermediate bosons	1	very big
Strong	gluons	1	0
Gravitation	gravitons	2	0

We note that, from the point of view of quantum field theory, the distinction between matter and radiation becomes less evident, since also matter is described in terms of fields: then all elementary particles correspond to the quanta of these fields.

When one considers a bulk of matter, it is not at all pratical to describe its properties in terms of the fundamental particles and interactions, but it is convenient to describe collective motions of (macroscopic) groups of particles in the system.

Therefore, one makes use of the concept of elementary excitations, which are nothing else that the quanta associated with the collective motions, similarly to the case of photons which are the quanta of the EM radiation field.

Well known examples of elementary excitations are the following:

phonons, which are the quanta of sound waves, corresponding to collective motions in a crystal; phonons are also present in a quantum liquid, like a superfluid;

plasmons, which are the quanta of wave-like fluctuations of the electron density in the electron gas in a metal;

magnons, which are the quanta of the "spin-waves", i.e. the density fluctuatios of spin in a ferromagnet.

In the following, I shall discuss some rather general theoretical schemes, which describe in similar ways phenomena of condensed state physics and elementary particle physics.

II. QUANTUM MECHANICS OF SYSTEMS WITH INFINITE DEGREES OF FREEDOM

Quantum formulation of systems with infinite degrees of freedom [2,3] goes under various names: second quantization, quantum field theory, many-body theory, etc.

---

* For a brief elementary introduction see ref. [1].

On one side, the appropriate description of elementary particles and their interactions is provided by relativistic quantum field theory.

On the other side, the description of macroscopic systems in terms of forces between their constituents involve also infinite degrees of freedom. The essential feature is that a very large number of particles involve intensive or thermodynamical properties (like particle density, mean energy, etc.) the effects of which are of order $1/N$ and $1/V$, while the density $n = N/V$ is kept finite. Then it is appropriate to consider the limiting situation

$$N \to \infty, \; V \to \infty \quad \text{with} \quad N/V \text{ finite.}$$

In general, it is possible to perform a transformation from a system of strongly interacting constituents (particles, atoms, ions, etc.) to a system consisting of approximately independent elementary excitations, for which a new ground state can be defined. As a matter of fact, when the number of elementary excitations is sufficiently small, one can neglect their mutual interactions, and their assembly can be regarded as an ideal gas. We recall that the elementary excitations represent the quanta associated with collective motions of macroscopic groups of constituents in the system.

The transformations can be represent symbolically in the form:

$$\begin{aligned} H &= \sum_i H(\vec{p}_i, \vec{r}_i) + \frac{1}{2} V(\vec{r}_i, \vec{r}_j; \vec{p}_i, \vec{p}_j) \\ &\to E_0 + \sum_k \omega_k a_k a_k^+ + f(a_k, a_k^+), \end{aligned} \tag{1}$$

where E_0 is the ground state energy in the new description, ω_k represents the excitation energy spectrum and f contains small corrections; the meaning of a_k, a_k^+ will become clear in the following.

The transformed system is described in terms of a set of canonical variables and conjugate momenta, which satisfy appropriate commutation relations. The states of the system are analyzed by making reference to the ground state and by specifying the number of each type of elementary excitations.

These states are represented by basic vectors, which are denoted by

$$|n_1, n_2, \cdots n_i, \cdots\rangle \quad \text{with} \quad \sum_i n_i = n, \tag{2}$$

and which are transformed into each other by application of the annihilation and creation operators a_i and a_i^+:

$$\begin{aligned} a_i |n_1, n_2, \cdots n_i, \cdots\rangle &= \sqrt{n_i} |n_1, n_2, \cdots n_i - 1, \cdots\rangle \\ a_i^+ |n_1, n_2, \cdots n_i, \cdots\rangle &= \sqrt{n_i + 1} |n_1, n_2, \cdots n_i + 1, \cdots\rangle \end{aligned} \tag{3}$$

The ground state or "vacuum" is defined by the vanishing of all occupation numbers: $n_i = 0$.

There are two different possibilities for the operators a_i, a_i^+.

a) In one case they satisfy commutation relations $[A, B] = AB - BA$:

$$\begin{aligned} [a_i, a_j] &= 0 \\ [a_i^+, a_j^+] &= 0 \\ [a_i, a_j^+] &= \delta_{ij} \end{aligned} \tag{4}$$

The occupation number operator

$$N_i = a_i^+ a_i \tag{5}$$

has all integer eigenvalues

$$n_i = 0, 1, 2, \cdots$$

This choice describes correctly the case of bosonic excitations, i.e. quanta obeying Bose-Einstein statistics (completely symmetrical states).

b) In the other case they satisfy anticommuting relations $\{A, B\} = AB + BA$:

$$\begin{aligned} \{a_i, a_j\} &= 0 \\ \{a_i^+, a_j^+\} &= 0 \\ \{a_i, a_j^+\} &= \delta_{ij} \end{aligned} \tag{6}$$

In this case the occupation number operator N_i satisfies the condition

$$(N_i)^2 = N_i, \tag{7}$$

so that the only allowed eigenvalues are

$$n_i = 0, 1.$$

The present choice corresponds to fermionic excitations, i.e. quanta satisfying Fermi-Dirac statistics (completely antisymmetrical states).

For both choices one can define the following field operators (in the non-relativistic case and in the Schödinger picture):

$$\begin{aligned} \varphi(\vec{x}) &= \sum_i f_i(\vec{x}) a_i \\ \varphi^+(\vec{x}) &= \sum_i f_i^*(\vec{x}) a_i^+ \end{aligned} \tag{8}$$

where

$$f_i(\vec{x}) = \langle \vec{x} | \alpha_i \rangle \tag{9}$$

stands for the wave function of one-particle states, and α_i specifies momentum, spin, etc. For plane-wave case one has explicitly

$$f_i(\vec{x}) = \frac{1}{\sqrt{V}} e^{i\vec{k}\vec{x}}. \tag{10}$$

In the bosonic case, the field operators satisfy the commutation relations

$$\begin{aligned} [\varphi(\vec{x}), \varphi(\vec{x}\,')] &= 0 \\ [\varphi^+(\vec{x}), \varphi^+(\vec{x}\,')] &= 0 \\ [\varphi(\vec{x}), \varphi^+(\vec{x}\,')] &= \delta^3(\vec{x} - \vec{x}\,'). \end{aligned} \tag{11}$$

The Hamiltonian of the system, by taking into account only the term corresponding to non-interacting elementary excitations, is given by

$$H = \sum_i \omega_i a_i^+ a_i = \sum_i \omega_i N_i \tag{12}$$

or, in terms of the field operators $\varphi(\vec{x}), \varphi^+(\vec{x})$, by:

$$H = \int d^3x \varphi^+(\vec{x}) \left(-\frac{1}{2m} \nabla^2 \right) \varphi(\vec{x}), \tag{13}$$

where we refer to the specific case $\omega = k^2/2m$.

III. SYMMETRY AND BREAKING: GOLDSTONE MODES

In general, the analysis of the symmetry properties of a system is very important because one can gain information from considerations of symmetry alone, without having to solve the equations of motion.

Schematically one can make the following links:

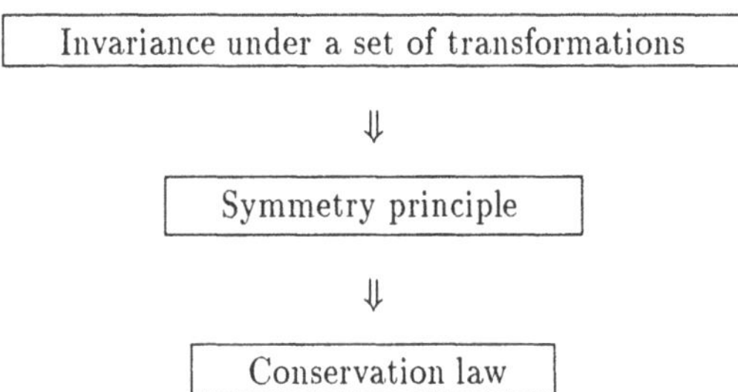

One can remind the well-known examples:

i) invariance under translations corresponds to homogeneity of space, and to conservation of linear momentum;

ii) invariance under rotations corresponds to isotropy, and then to conservation of angular momentum;

iii) invariance under phase transformations corresponds to internal symmetries and to the conservation of internal quantum number, such as the electric charge.

Some symmetries appear to be exact: they are related to strictly conserved quantities. The Lagrangian of the system and its ground state are invariant under the corresponding set of transformations.

Other symmetries appear to be only approximate: the situation can be described by adding, to the main symmetric term of the Lagrangian, a small breaking term, which corresponds to the presence of non-invariant interactions.

An alternative possibility is realized in the following way: the Lagrangian and the equations of motion of the system possess the full symmetry, but the ground state corresponds to a lower symmetry. In this case one usually says that the symmetry is spontaneously broken; as a matter of fact, the symmetry is not destroyed, but is it somewhat hidden.

In order to illustrate this important mechanism, we consider a toy model taken from field theory. It is based on the use of a complex scalar field $\varphi(x)$; the corresponding quanta have zero spin.

The Lagrangian density of the system has the form

$$\mathcal{L} = \frac{1}{2}\partial_\mu \varphi^+(x) \partial^\mu \varphi(x) - \frac{1}{2} m^2 \varphi^+(x)\varphi(x) - \frac{1}{4} g(\varphi^+(x)\varphi(x))^2, \tag{14}$$

which is invariant under the phase transformations

$$\begin{aligned} \varphi(x) &\rightarrow e^{i\alpha}\varphi(x) \\ \varphi^{+}(x) &\rightarrow e^{-i\alpha}\varphi^{+}(x). \end{aligned} \tag{15}$$

In order to analyze the properties of the ground state of the system, it is useful to split $\mathcal{L}$ into two parts:

$$\mathcal{L} = \mathcal{L}_{kin} - V(\varphi), \tag{16}$$

where the first is the kinetic term and $V(\varphi)$ represents the "potential" term:

$$V(\varphi) = \frac{1}{2}m^2|\varphi|^2 + \frac{1}{4}g|\varphi|^4 \tag{17}$$

The ground state corresponds to the minimum of the potential:

$$\frac{dV}{d|\varphi|} = m^2|\varphi| + g|\varphi|^3 = 0 \tag{18}$$

i.e.

$$|\varphi|(m^2 + g|\varphi|^2) = 0 \tag{19}$$

There are two possible kinds of solutions, which are represented in Fig. 1:

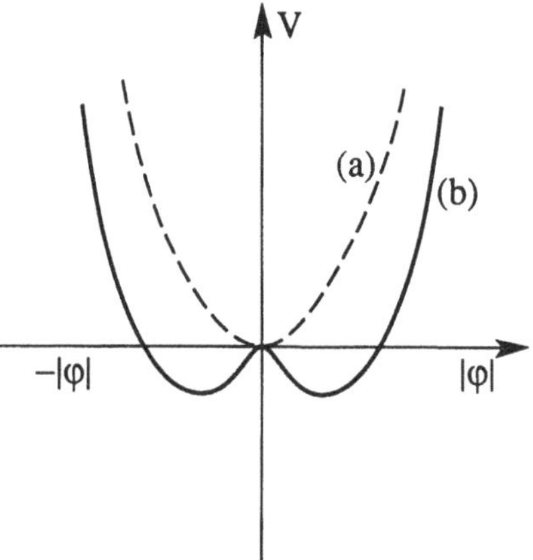

Fig. 1. The potential $V(\varphi)$ for the two solutions (a) and (b).

a) the usual solution (for $m^2 > 0$)

$$|\varphi|_0 = 0, \tag{20}$$

b) the alternative solution (if $m^2 < 0$)

$$|\varphi|_0^2 = -\frac{m^2}{g} \tag{21}$$

We note that the quantity m^2 is a function of the energy scale - or, in other words, of the temperature - and it can change its sign.

The solution (b) represents, in fact, a set of solutions, since the phase is arbitrary:

$$\varphi_0 = |\varphi|_0 e^{i\gamma} \tag{22}$$

One can choose the particular solution corresponding to vanishing phase; writing the field in terms of its real and imaginary parts

$$\varphi(x) = \varphi_1(x) + i\varphi_2(x), \tag{23}$$

one gets

$$|\varphi_1|_0 = \sqrt{-\frac{m^2}{g}}, \quad |\varphi_2|_0 = 0 \tag{24}$$

We point out that the solution (a) is invariant under the phase transformations (15), while the solution (24) is no longer invariant.

It is convenient to define

$$v \equiv |\varphi_1|_0 = \sqrt{-\frac{m^2}{g}} \tag{25}$$

from which it follows

$$g\, v^2 + m^2 = 0, \tag{26}$$

and re-define the real component of the field $\varphi(x)$:

$$\tilde{\varphi}_1(x) = \varphi_1(x) - v \tag{27}$$

In terms of $\tilde{\varphi}_1(x)$ and $\varphi_2(x)$ the Lagrangian density (14) reads

$$\begin{aligned}\mathcal{L} = \frac{1}{2}\partial_\mu\tilde{\varphi}_1\partial^\mu\tilde{\varphi}_1 + \frac{1}{2}\partial_\mu\varphi_2\partial^\mu\varphi_2 - \frac{1}{2}\tilde{m}^{\,2}\tilde{\varphi}_1^2 \\ - \frac{1}{4}g(\tilde{\varphi}_1^{\,2} + \varphi_2^2)^2 - gv\tilde{\varphi}_1(\tilde{\varphi}_1^2 + \varphi_2^2)\end{aligned} \tag{28}$$

where $\tilde{m}^2 = -2m^2$.

It is useful to compare the expression of the Hamiltonian densities in the two cases (a) and (b):

$$\mathcal{H} = \frac{1}{2}|\nabla\varphi|^2 + \frac{1}{2}|\dot{\varphi}|^2 + \frac{1}{2}m^2|\varphi|^2 + \frac{1}{4}g|\varphi|^4 \tag{29}$$

for case (a), with $m^2 > 0$, and

$$\begin{aligned}\mathcal{H} = \frac{1}{2}\{(\nabla\tilde{\varphi}_1)^2 + (\nabla\varphi_2)^2\} + \frac{1}{2}(\dot{\tilde{\varphi}}_1^2 + \dot{\varphi}_2^2) + \frac{1}{2}\tilde{m}^2\tilde{\varphi}_1^2 + \\ + \frac{1}{4}g(\tilde{\varphi}_1^2 + \varphi_2^2)^2 + gv\tilde{\varphi}_1(\tilde{\varphi}_1^2 + \varphi_2^2)\end{aligned} \tag{30}$$

for case (b), with $\tilde{m}^2 = -2m^2 > 0$.

We point out that, while in the case (a) the two components of the field $\varphi(x)$ correspond to quanta of equal mass m, in the case (b) one component remains massive (massive mode with mass $\tilde{m}$), and the other results to be a massless mode of oscillation.

The situation of case (b) is illustrated in Fig. 2.

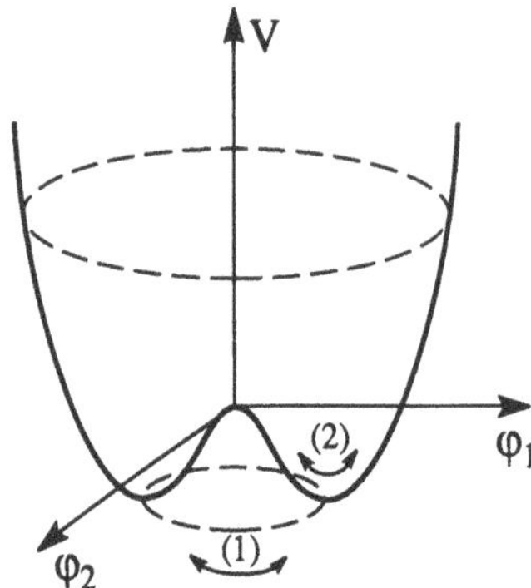

Fig. 2. Representation of the potential V for case (b): (1) massless mode; (2) massive mode.

The above result emerging in a simple way in the toy model, which has been employed for the sake of simplicity, are generalized by the famous Goldstone's theorem [4].

The theorem can be stated in the following form:

i) in relativistic quantum field theory, when a continuous symmetry is spontaneosuly broken, there appear massless bosons (the so-called Nambu-Goldstone bosons)[4,5];

ii) in the case of non-relativistic quantum systems, spontaneous breaking of a continuous symmetry leads to elementary excitations with no-energy gap ($\omega(k) \to 0$ as $\vec{k} \to 0$).

IV. A PHYSICAL EXAMPLE: SUPERFLUIDITY

An interesting physical example of the mechanism predicted by the Goldstone's theorem for a condensed state system, i.e. the appearence of oscillation modes with no gap in the energy spectrum, is provided by the phenomenon of superfluidity, which occurs in liquid helium at low enough temperatures ($T < T_c = 2.18K$).

This phenomenon, which is related to the Bose character of helium atoms, can be described schematically by considering a body of mass M moving inside liquid helium. If its velocity v is lower than a critical value v_c, than the viscosity appears to be zero and the body is no longer slowed down.

At microscopic level, a non-vanishing viscosity can be interpreted in terms of excitations of elementary modes in the fluid, at the expenses of the energy and momentum of the moving body [3,6]. Denoting by $\omega(k)$ the energy of an elementary excitation, conservation of energy and momentum gives the relations

$$\begin{aligned} \frac{1}{2}Mv^2 &= \frac{1}{2}Mv'^2 + \omega(k) \\ M\vec{v} &= M\vec{v}\,' + \vec{k} \end{aligned} \tag{31}$$

from which it follows:

$$\vec{v}\cdot\vec{k} = \frac{k^2}{2M} + \omega(k) \tag{32}$$

The above equation is satisfied provided

$$v \geq \frac{k}{2M} + \frac{\omega(k)}{k}, \tag{33}$$

or in other words provided v is greater or equal to a minimum value v_c defined by

$$v_c = \min\left(\frac{k}{2M} + \frac{\omega(k)}{k}\right). \tag{34}$$

If the excitation energy $\omega(k)$ is given by*

$$\omega(k) = \frac{k^2}{2m}, \tag{35}$$

as in the case of a free gas of bosons of mass m, one gets $v_c = 0$, and elementary excitations can appear for any value of v.

On the other hand, if the excitation energy is given by

$$\omega(k) = uk, \tag{36}$$

where u is a constant, the critical velocity is given by $v_c = u$ and excitations can occur only for $v \geq v_c$, but not for $v < v_c$.

The energy spectrum (36) is typical of the phonon behaviour. In fact, liquid helium at very low temperatures can have phonon-like excitations as in a crystal; the difference being that only longitudinal modes are present. Superfluidity is thus related to the phonon-like energy spectrum: $\omega(k) \to 0$ as $k \to 0$.

The phenemenon of superfluidity can be related to the field model considered in the previous section, if one adopts the language of fields. In terms of the "order field" $\varphi(x)$, which represents the fluctuations around the "order parameter" $|\varphi|_0$, the Hamiltonian density, with an appropriate normalization, has the same expressions given by (29), (30). A phase transition occurs at a critical temperature T_c.

For $T > T_c$ only the normal phase is present in the system, corresponding to the situation $m^2 > 0, |\varphi|_0 = 0$.

For $T < T_c$, $m^2 < 0$, the minimum occurs at $\varphi_0 = |\varphi|_0 e^{i\gamma}$, and there are "size" fluctuations around $|\varphi|_0$ as well as "phase" fluctuations; the latter correspond to Nambu-Goldstone massless bosons, i.e. to phonons, or in other words to the superfluid phase. For $T \ll T_c$, size fluctuations can be neglected and only phase fluctuations are relevant. The resulting energy density corresponds to the hydrodynamic energy of the superfluid.**

* We remind that this expression corresponds to the non relativistic case described by eq. (13).

** More details on these points can be found in ref. [7].

V. LONG-RANGE INTERACTIONS: HIGGS MECHANISM

The situation examined so far corresponds to the cases in which only short-range interactions are effective. The previous results are modified if also long-range interactions are present.

Going back to the field theory toy model described by eq. (14), the inclusion of long-range interaction can be represented by adding to the scalar field $\varphi(x)$ a massless vector field $A^\mu(x)$, such as the electromagnetic (EM) four-vector potential. The EM fields are described by the antisymmetric tensor

$$F^{\mu\nu} = \partial^\mu A^\nu - \partial^\nu A^\mu \tag{37}$$

and the Lagrangian (14) is replaced by

$$\begin{aligned}\mathcal{L}(\varphi, A_\mu) =& \frac{1}{2}(\partial_\mu \varphi^+ - ie\ A_\mu \varphi^+)(\partial^\mu \varphi + ie\ A^\mu \varphi) - V(\varphi) \\ & - \frac{1}{4} F_{\mu\nu} F^{\mu\nu},\end{aligned} \tag{38}$$

or equivalently by

$$\begin{aligned}\mathcal{L}(\varphi, A_\mu) =& \frac{1}{2}\partial_\mu \varphi^+ \partial^\mu \varphi - V(\varphi) - \frac{1}{4} F_{\mu\nu} F^{\mu\nu} \\ & + \frac{1}{2} ie\ A_\mu (\varphi^+ \partial^\mu \varphi - \varphi \partial^\mu \varphi^+) + \frac{1}{2} e^2 A_\mu A^\mu \varphi^+ \varphi\end{aligned} \tag{39}$$

This modified Lagrangian is not only invariant under the phase transformations (15), but also under the <u>local</u> gauge transformations:

$$\begin{aligned}\varphi(x) &\to e^{ie\alpha(x)} \varphi(x) \\ \varphi^+(x) &\to e^{-ie\alpha(x)} \varphi^+(x)\end{aligned} \tag{40}$$

and

$$A_\mu(x) \to A_\mu(x) + \partial_\mu \alpha(x) \tag{41}$$

The SSB (spontaneous symmetry breaking) solution is again given by eq. (21). In terms of

$$\tilde{\varphi}_1(x) \equiv \varphi_1(x) - v$$

the Lagrangian can be re-written in the form

$$\begin{aligned}\mathcal{L} =& \frac{1}{2}\partial_\mu \tilde{\varphi}_1 \partial^\mu \tilde{\varphi}_1 - \frac{1}{2}\tilde{m}^2 \tilde{\varphi}_1^2 + \frac{1}{2}\partial_\mu \varphi_2 \partial^\mu \varphi_2 \\ & - \frac{1}{4} F_{\mu\nu} F^{\mu\nu} + \frac{1}{2} e^2 v^2 A_\mu A^\mu + (\text{interaction terms})\end{aligned} \tag{42}$$

The situation concerning the fields components $\tilde{\varphi}_1$ and φ_2 appears to be the same as in eq. (28). However, it is possible to show that the Goldstone mode φ_2 gets mixed with the longitudinal components of A_μ (see e.g. ref. [8]), and that the physical particles are the following:

i) a <u>scalar</u> boson of mass $\tilde{m}$ ($\tilde{m}^2 = -2m^2$);

ii) a <u>vector</u> boson of mass $M = ev$.

We notice that by this mechanism, which is the well-known Higgs mechanism [9], the massless vector field A_μ, which has only two physical (transverse) components, acquires a third (longitudinal) component and become massive. The would-be Nambu-Goldstone mode is absorbed by this longitudinal component.

VI. A PHYSICAL EXAMPLE: SUPERCONDUCTIVITY

An interesting physical example of the mechanism described in the previous section is provided by superconductivity. It is well known that at very low temperatures a phase transition occurs in some specific materials (e.g. metals) and the electric resistence drops to zero.

In a superconductor, a flow of electric current is equivalent to a common shift of the velocity of the electrons (or electron pairs). Denoting by M the total mass of the electron system, the ground energy will be shifted by $1/2\ Mv^2$. One expects that, if the source of the electric current is switched off, the current flow will decrease; however, this happen only if elementary excitations can be emitted, thus leading to an attenuation of the current. On the other hand, if these modes cannot be excited, the electric current is not reduced and it continues to flow [10,11,3].

Denoting by $E(p)$ the energy of the elementary excitations, the energy-momentum conservation requires

$$\begin{aligned} \frac{1}{2}Mv^2 &= \frac{1}{2}Mv'^2 + E(p) \\ M\vec{v} &= M\vec{v}\,' + \vec{p}, \end{aligned} \tag{43}$$

from which it follows

$$\vec{v}\cdot\vec{p} = \frac{p^2}{2M} + E(p) \tag{44}$$

The above condition is satisfied provided

$$v \geq v_c, \tag{45}$$

with

$$v_c = \min\left(\frac{p}{2m} + \frac{E(p)}{p}\right) = \min\frac{E(p)}{p}; \tag{46}$$

while it cannot be satisfied if $v_c \neq 0$ and $v < v_c$ (this situation would correspond to superconductive regime).

In the case of a free electron gas, $E(p)$ represents the energy above the Fermi sphere:

$$E(p) = \frac{p^2}{2m} - \frac{p_F^2}{2m}. \tag{47}$$

In the limit $p \to p_F$ one gets $E(p) \to 0$ and $v_c = 0$ (no superconductivity).

On the other hand, an energy spectrum of the form

$$E(p) = \frac{1}{2m}\sqrt{(p^2 - p_F^2)^2 + 4m^2\Delta^2} \tag{48}$$

leads to a critical velocity

$$v_c = \Delta/p_F, \tag{49}$$

and superconductivity can occur. We see that a necessary condition is the presence of an energy gap in the spectrum:

$$E(p) \to \Delta \quad \text{as} \quad p \to p_F. \tag{50}$$

Collective effects associated with the condensation of electron pairs (the so-called Cooper pairs) [12] give rise to an energy spectrum of the form (48), and therefore to a critical velocity $v_c \neq 0$ related to the energy gap Δ.

In the language of order fields, the situation looks at first sight similar to that of superfluidity. One can start from a Hamiltonian density as in eq. (29): at $T > T_c$ there are two independent fluctuation modes corresponding to density fluctuations of the electron gas, i.e. to ordinary plasmons. At $T < T_c$, the ground state corresponds to a SSB solution; there are two different kind of fluctuations: size and phase fluctuations, and only the latter remain effective at very low temperature.

However, the situation is somewhat different, since we are not dealing with a neutral gas, but with a plasma. What happens is that electron pairs are formed, due to attractive forces between the electrons coupled to the lattice ions, which are overwhelming the Coulomb repulsion. Below $T < T_c$, we are still dealing with a Bose gas, made of Cooper electron pairs which are carrying electric charge.

Therefore we have to include the interaction of the charged field φ with the EM vector potential $\vec{A}$ in the Hamiltonian (29), obtaining in the static case

$$\begin{aligned}\mathcal{H} = \frac{1}{2}|(\nabla - iq\vec{A})\varphi|^2 + \frac{1}{2}m^2|\varphi|^2 + \frac{1}{4}g|\varphi|^4 \\ + \frac{1}{2}(\nabla \times \vec{A})^2\end{aligned} \tag{51}$$

The above expression can be re-written in the form

$$\mathcal{H} = \mathcal{H}_0 + \frac{1}{2}(\nabla \times \vec{A})^2 - \vec{j}\cdot\vec{A} + \frac{1}{2}q^2|\varphi|^2\vec{A}^2, \tag{52}$$

where $\mathcal{H}_0$ is the unmodified Hamiltonian of the scalar field and

$$\vec{j} = \frac{1}{2i}(\varphi^+\nabla\varphi - \varphi\nabla\varphi^+). \tag{53}$$

The equation of motion for the vector potential $\vec{A}$ is then:

$$\nabla \times \nabla \times \vec{A} = -\vec{j} + q^2|\varphi|^2\vec{A}. \tag{54}$$

In the case of SSB, at the minimum $|\varphi|_0 = v$, eq. (54) reduces to

$$\nabla \times \nabla \times \vec{A} = M^2\vec{A} \tag{55}$$

where

$$M = qv \tag{56}$$

Since the magnetic field is given by $\vec{B} = \nabla \times \vec{A}$, and it satisfies the condition $\nabla \cdot \vec{B} = 0$, one gets from (55):

$$\nabla^2\vec{B} = M\vec{B}. \tag{57}$$

This equation has a physical solution of the type

$$\vec{B} = \vec{B}_0 e^{-M|\vec{x}|}, \tag{58}$$

corresponding to vanishing magnetic field B at the distance λ (the so-called screening length):

$$\lambda = 1/M = 1/gv. \tag{59}$$

For physical systems, λ is the order of 10^2 Å, so one can say that, in the superconducting phase, the conductor becomes a perfect diamagnetic material since it cannot be penetrated by a magnetic field: this is the famous Meissner effect.

In conclusion, due to the coupling with the long-range Coulomb field, the plasmons present an energy gap in the dispersion law. In fact, they could be interpreted as the longitudinal components of the vector field A_μ: the corresponding quanta acquire now a mass $M = qv$.

VII. CONCLUSION

In these lecture we have concentrated mainly on the mechanism of spontaneous symmetry breaking (SSB), which provides a unifying frame for a set of phenomena occurring in different branches of physics.

Two relevant situations of broken symmetries have been considered: one refers to many-body systems with finite range interactions and to the appearance of Nambu-Goldstone excitation modes; the other occurs in systems where infinite-range interactions are present and the Higgs mechanism becomes effective.

As an illustration of the first case, we have briefly examined the well-known phenomenon of superfluidity, pointing out the interpretation of phonons in the superfluid phase as Nambu-Goldstone modes. But other examples of SSB could be considered in non-relativistic many body systems, such as the Heisenberg ferromagnet, where the rotational symmetry is spontaneously broken and the zero-frequency modes are spin waves [13]. Another interesting example is related to the Jahn-Teller effect which causes a distorsion in the symmetry of a complex of ions in a solid [14].

The Goldstone's theorem has given rise to several applications also in relativistic quantum field theory: we limit ourself to mention the interpretation of the π-meson (which can be extended to the octet of pesudoscalar mesons) in terms of the Nambu-Goldstone boson related to the breaking of "chiral" symmetry [15].

As an illustration of the second case, in which the Goldstone's theorem is evaded due to the presence of long-rang interactions, we have considered the phenomenon of superconductivity. In this framework, the Meissner effect provides a striking example of the Higgs mechanism.

In particle physics it is well known that this mechanism is a key ingredient in the unifying theory of electro-weak interactions [16], where the short-range character of the weak interactions at low energy emerges from the mass generation of the intermediate heavy bosons, which are the counterparts of the massless EM vector field A_μ.

In conclusion, theoretical methods provide a link between different topics of physics: thus a comparison between apparently unrelated phenomena may reveal deep conceptual analogies.

REFERENCES

1. G. Costa, in Disordered Solids, edit. by B. Di Bartolo, G. Ozen and J.M. Collins, Plenum Publishing Corporation (1989), p. 397.

2. A.L. Fetter and J.D. Walecka, Quantum Theory of Many-Particle Systems, McGraw-Hill Book Co. (1971).
3. F. Strocchi. Elements of Quantum Mechanics of Infinite Systems, World Scientific Publishing Co. (1985).
4. J. Goldstone, Nuovo Cimento 19, 154 (1961).
5. Y. Nambu, Phys. Rev. Letters 4, 380 (1960).
6. L.D. Landau and E.M. Lifshitz, Statistical Physics, Pergamon Press (1958).
7. H. Kleinert, Gauge Field in Condensed Matter, World Scientific Publishing Co. (1989).
8. I.J.R. Aitchison and A.J.G. Hey, Gauge Theories in Particle Physics, Adam Hilger (1989).
9. P.W. Higgs, Phys. Rev. Letters 13, 508 (1964); Phys. Rev. 145, 1156 (1966).
10. E.A. Lynton, Superconductivity, John Wiley and Sons, Inc. (1962).
11. P.W. Anderson, Phys. Rev. 130, 439 (1963).
12. L.N. Cooper, Phys. Rev. 104, 1189 (1956).
13. R.V. Lange, Phys. Rev. 146, 301 (1966).
14. J. Sarfatt and A.M. Storeham, Proc. Phys. Soc. 91, 214 (1967).
15. Y. Nambu and G. Jona-Lasinio, Phys. Rev. 122, 345 (1961); 124, 246 (1961).
16. S.L. Glashow, Nucl. Phys. 22, 579 (1961); S. Weinberg, Phys. Rev. Letters 19, 1269 (1967); A. Salam, Proceed. of the VIII Nobel Symposium, ed. N. Svartholm; Almquist and Wiksells (1968), p. 367.

SUPERCONDUCTIVITY

M.J. Graf*

Department of Physics
Boston College
Chestnut Hill, MA 02167

and

J.D. Hettinger**
Argonne National Laboratory
Argonne, IL 60439

ABSTRACT

A general review of superconductivity is presented, including discussion of the fundamental parameters of superconductivity, their physical interpretation, and the basic theory of superconductivity, the BCS model. Current trends in superconductivity research are discussed, including current work on the possible 'unconventional' superconductor UBe_{13}, as well as a discussion on high-temperature superconductors.

I. INTRODUCTION

Since its discovery by Kammerling-Onnes in 1911, the phenomenon of superconductivity has been an area of intense theoretical and experimental research. Below a critical temperature T_c, the resistance of certain metals is found to drop abruptly to zero (precisely), and the material expels any small applied magnetic field (the Meissner effect) and exhibits perfect diamagnetism. These effects were not understood theoretically until 1957 when Bardeen, Cooper, and Schreiffer published their historic work [1], now called BCS theory. Here they showed that at low enough temperatures a metal can consist of bound pairs of electrons ('Cooper pairs') which in turn form into a collective ground state which has all the experimentally observed properties of superconductors. In recent years this field has become even more active as new discoveries have challenged some long-held ideas concerning the nature and uses of superconducting materials. Strong experimental evidence now exists for 'unconventional superconductors', and, of course, the newly-discovered high temperature superconductors [2] have the potential to make the widespread application of superconductors economically feasible.

---

* Visiting Scientist, Francis Bitter National Magnet Laboratory, MIT, Cambridge, MA 02139 USA.

** JDH acknowledges the support of the US Department of Energy under contract No. W-31-109-EN6-38.

In this paper we will first define and discuss the basic physical parameters and the phenomenological classifications of superconductors. Next, we will briefly review the theoretical model which describes most 'conventional' superconductors, the BCS theory. Finally, we will describe some recent trends in superconductivity research, and will discuss one possible candidate for unconventional superconductivity, the heavy-fermion system UBe_{13}. For a good general overview of superconductivity (and superfluidity), see Tilley and Tilley [3], or for a more rigorous discussion, see Tinkham [4].

II. EXPERIMENTAL PARAMETERS OF SUPERCONDUCTING MATERIALS

A phenomenological approach will now be used to describe the basic physical parameters which characterize superconductors. The most obvious is the temperature at which the material enters the superconducting state, T_c. Studies show that at T_c not only does the resistance drop to zero, but a jump occurs in the specific heat which is consistent with a nearly ideal second-order phase transition, indicating an abrupt transition from a disordered to an ordered state as the temperature is lowered [5]. The nature of this ordering will be discussed in Section III. Below T_c the specific heat falls off exponentially, following an Arrhenius-type behavior. This suggests that an energy gap has opened up between the superconducting ground state and its excited states, unlike the normal metallic state, where the ground state (all states below the Fermi energy occupied) is not separated from the excited states (the empty states above the Fermi energy). The energy gap (which is a function of temperature) is usually written as $2\Delta(T)$. The energy gap vanishes at $T = T_c$, and it has been experimentally determined that for many superconductors $2\Delta(T=0)$ is approximately equal to $3.5\ k_BT_c$, where k_B is Boltzmann's constant. The typical behavior of the resistance and specific heat versus temperature is illustrated in Figure 1.

The energy gap is of fundamental importance in understanding the behavior of superconductors. Because an energy gap exists between the ground and excited states, the superconducting carriers cannot transfer any energy of less than 2Δ through collisons, and so current flow through the material is dissipationless (i.e., zero-resistance). At large enough currents, however, the kinetic energy of flow is sufficient to exceed the gap 2Δ, and a non-zero resistance is restored. The value of current per cross-sectional area of flow where this destruction of the zero-resistance state occurs is called the critical current density $J_c(T)$, and, like the energy gap, is a function of temperature and becomes zero at T_c.

The response of the superconductor to applied magnetic fields is also of interest. For small applied fields, the magnetic flux is expelled from the bulk of the superconductor. Superconducting charge carriers flow through a thin layer of thickness λ near the surface in such a way as to cancel the applied magnetic field inside the material. This length λ, which is also the distance over which a magnetic field may penetrate into the material, is called the penetration depth. As the temperature of a superconductor is increased towards T_c, the magnetic field penetrates the entire sample, and so $\lambda(T)$ diverges at $T=T_c$.

The response of the superconductor to strong magnetic fields can be of two types. For 'type-I' superconductors, the surface currents

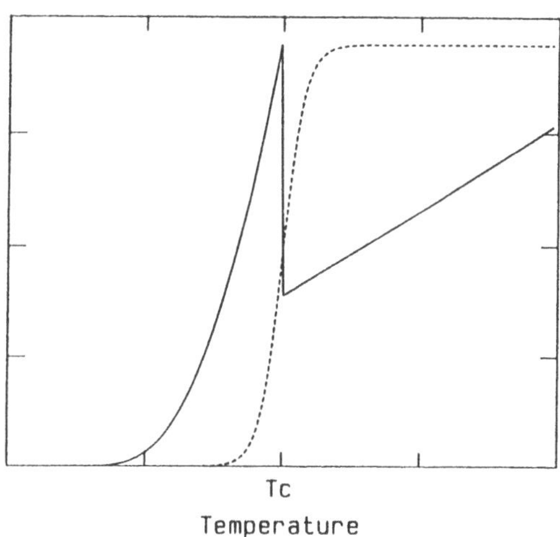

Figure 1. Characteristic change in the specific heat (solid line) and resistance (dashed line) for a superconductor near temperature T_c.

continue to completely shield the bulk of the material from applied fields. However, above some critical field H_c, the required current densities near the surface exceed the critical current density, superconductivity breaks down, and the magnetic field permeates throughout the sample; above Hc the material has re-entered the 'normal state'. Typically, pure elemental metals are type-I superconductors. The magnetization versus applied field for a type-I superconductor is shown in Figure 2a: below H_c the magnetization is that of a perfect diamagnet, while above H_c the magnetization assumes the (small) value of the metal being studied.

A more complex behavior occurs for 'type-II' superconductors. At low field the response is the same as for type-I, until the perfect diamagnetic response is lost above a field H_{c1}, called the lower critical field. Above H_{c1}, however, the field does not permeate throughout the entire sample, but only in thin filaments evenly spaced throughout the material. Thus the material has normal filaments, called vortices, surrounded by regions which are in the superconducting state; the material is said to be in the mixed state. The vortices are so named because supercurrents flow in a circular path outside the core to create a non-zero magnetic field inside the core region only. As the field is further increased, the density of the vortices becomes greater, until they finally fill up the bulk of the sample. The field at which the sample becomes completely normal is called the upper critical field H_{c2}. The magnetization versus applied field for a type-II superconductor is shown in Figure 2b. Note that even though the response to a magnetic field is no longer perfectly diamagnetic for $H_{c1} < H < H_{c2}$, superconducting paths exist for current flow and the resistance is still zero. Most type-II superconductors are alloys, with a notable exception being pure Nb, which is also type-II.

The normal-state core of vortices have a characteristic size ζ, the coherence length, which contains the smallest unit of magnetic clux

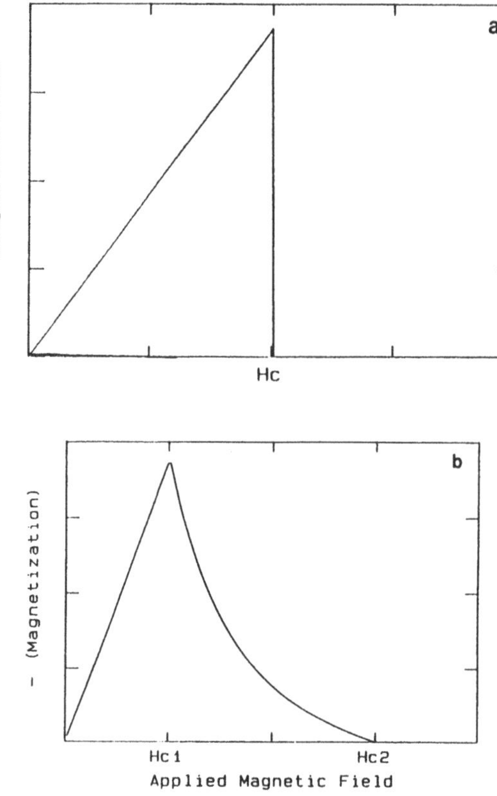

Figure 2. The negative value of magentization plotted versus applied magnetization field for (a) type-I and (b) type-II superconductors.

possible, the magnetic clux quantum Φ_0. For superconductors with carriers of charge 2e (that is, the Cooper pair), $\Phi_0 = h/2e$, where h is Planck's constant. The upper critical field H_{c2} is the field strength as which the sample plane (normal to the applied field) is completely filled with vortex cores, each containing one quantum of flux. If the sample area is A_s, then the total flux is $H_{c2}A_s$. This must equal $\Phi_0\ A_s/A_{core}$, where A_{core} is the vortex core area, roughly equal to $\pi\zeta^2$. This rough estimate yields H_{c2} in terms of ζ; using a mean-field theoretical approch ('Ginzburg-Landau theory' [6], also see the article by Costa in this volume), one gets the more exact relationship

$$H_{c2}\ (T) = \Phi_0/2\pi\zeta(T)^2 \quad . \tag{1}$$

$\zeta(T)$, like $\lambda(T)$, diverges as the temperature approaches T_c, resulting in $H_{c2}(T_c)=0$.

On a more fundamental level, ζ can be thought of as the minimum distance over which the superconducting wavefunction can be reduced to zero (at a normal-superconducting interface), and as such is a parameter which describes both type-I and type-II superconductors. For relatively pure metals, this intrinsic uncertainty in position (the spatial extent of a Cooper pair) is given by the uncertainty principle as $\delta x = \zeta_0 = \hbar/\delta p$. Pippard first introduced this phenomenological parameter to describe the non-locality of electrodynamics in superconductors [7], and ζ_0 is called the Pippard coherence length. δp is roughly the gap energy $\Delta(0)$ divided by the Fermi velocity v_F, and we obtain

Table I. Superconducting parameters of two different materials. For all temperature-dependent quantities the T = 0 K value has been given. For the type-II material $YBa_2Cu_3O_{7-\delta}$, the 'Hc' value is the upper critical field.

	In [8]	$YBa_2Cu_3O_{7-\delta}$ [9]
T_c(K)	3.4	92
H_c(T)	0.03	>100
J_c(A/cm^2)	5×10^7	10^7
λ(Å)	430	1400
ζ_0(Å)	4400	14

$$\zeta_0 = \alpha\ \hbar v_F/\Delta(0) \quad , \qquad (2)$$

where α is a constant of order unity. Unlike the $\zeta(T)$ defined from the core size, ζ_0 is a constant. For alloys, granular materials, etc. (that is, type-II materials), this intrinsic length will be modified by the mean-free-path L

$$1/\zeta_{eff} = 1/\zeta_0 + 1/L \quad . \qquad (3)$$

ζ_{eff} and $\zeta(T)$ are different parameters, but are roughly equal at T = 0 K.

Due to their ability to accommodate strong applied magnetic fields, type-II superconductors have upper critical fields which are typically much larger than the critical fields of type-I superconductors. In Table I, we have listed the various superconducting parameters for indium (type-I), and $YBa_2Cu_3O_{7-\delta}$(type-II). Note that for type-I superconductors, $\lambda < \zeta$, and for type-II, $\lambda > \zeta$. This condition can be derived from considering the energy associated with the creation of a superconducting-normal interface [6].

III. BCS THEORY

The dissipationless current flow that occurs in superconductors is clearly a quantum mechanical effect, and London [10] most clearly elucidated this by comparing supercurrent flow to the electron orbital motion in a hydrogen atom. Here the electron is 'flowing' without energy dissipation, and will remain in the ground state unless given sufficient energy to enter the first excited state. By analogy, London reasoned that the current-carrying electrons in a superconductor must also be described by a single, macroscopic phase-coherent wavefunction, separated from the excited states by an energy gap.

An important advance [11] was made by Cooper [12], who showed that two electrons added to the Fermi sea could form a bound pair (with oppositely directed spin and momenta) if any weak attractive potential existed between the two electrons. The mechanism considered was phonon coupling -- an electron moving through the lattice affects a positive ion in the lattice, which in turn has an effect on a second electron. Because this is a retarded interaction (the response of the relatively heavy ion is much slower than the motion of the electrons), this attractive interaction circumvents the Coulomb repulsion between the electrons. The model for the interaction is simple: the two electrons have a constant attractive potential of magnitude V if the electrons have energies within $\hbar\omega_D$ of the spherical Fermi surface, and no interaction otherwise (where ω_D is the Debye frequency, the maximum phonon frequency in the Debye model of a solid).

A more formidable problem is to show that the Fermi sea itself is unstable to the creation of Cooper pairs. BCS constructed a many-body wavefunction from the individual Cooper pair states, and by performing a variational calculation showed that a ground state of the paired electrons had a negative condensation energy. Also, this state was shown to be separated from the excited states by an energy gap, which vanished at a particular temperature T_c.

The transition temperature T_c can be evaluated rather easily in the case of weak electron-phonon coupling, $N(E_F)V \ll 1$, where $N(E_F)$ is the electron density of states evaluated at the Fermi energy. Using this, one finds

$$T_c = (\hbar\omega_D/k_B) \exp[-1/N(E_F)V] \ , \tag{4}$$

This result is in excellent agreement with the observed 'isotope effect' [13]. Upon replacement of the ions in a superconductor with a different mass isotope, the transition temperature is experimentally observed to change according to

$$T_{c1}/T_{c2} = (M_2/M_1)^{\alpha} \ , \tag{5}$$

where M_1 and M_2 are the isotope masses, T_{c1} and T_{c2} are the corresponding transition temperatures, and α is a constant, typically around 0.5.

Using Equation (4), and realizing that ω_D is proportional to $M^{-1/2}$, one can explain the experimental results. Another striking prediction in the weak-coupling limit is that

$$2\Delta(0)/k_BT_c = 3.53 , \tag{6}$$

again in excellent agreement with observations. Specific heat data, tunnelling measurements, and a host of other experimental observations are all well-described by the results of BCS theory. Of the materials which have properties which deviate from BCS behavior, most can be explained by a breakdown of the weak-coupling approximation, as in the case of mercury (interestingly, the first known superconductor). Good agreement can then be obtained in the 'strong-coupling' approximation [14].

IV. CURRENT RESEARCH TRENDS

In this section we will discuss some of the more recent trends in superconductivity research. One major area of research concerns the modification of the properties of superconductors by modifying their physical structure. For example, using modern materials processing techniques, it is possible to fabricate artificially "layered" superconductors- thin superconducting films separated from one another by layers of insulators or normal metals [15]. First, the fundamental parameters are found to change due to the fact that each individual layer is thinner: T_c is modified (it can be either decreased or increased) because the phonon spectrum of the superconductor is altered, and ζ is changed because the layer thickness can be the same size as, or smaller than, the bulk mean free path L. Second, the properties of the layered material are different from those of the individual films because, for layers separated by distances of the order of the coherence length, there is wavefunction overlap between layers. The degree of coupling affects the ability to carry high currents, the structure of the vortices in an applied magnetic field, and other properties. There is particular interest in these nearly-ideally layered structures because many of the properties of high-T_c superconductors are believed to be due to the anisotropic layered structure of the crystal- for example, conduction occurs easily along the Cu-O planes, but coupling between the layers is weak. Another example of a structurally modified conventional superconductor occurs when indium metal is injected into a porous glass host material [16], where the pore size can be much smaller than the bulk coherence length. This material is similar to other granular superconductors [17], with the added twist of having the disorder of the host glass imposed on the superconducting indium network.

In the cases just described interesting new properties of superconductors arose from physically modifying BCS superconductors (with the possible exception of high-T_c superconductors, which will be discussed more fully in the next section). Another interesting area of research is to find examples of superconductors which are non-BCS; this is typically taken to mean that the mechanism of electron pairing is not via phonon exchange, but by some other mechanism. Such superconductors could have properties which deviate from BCS-like behavior. Now we will discuss one possible candidate for such a material, the heavy-fermion superconductor UBe_{13}.

Like other members of the class of heavy fermion materials [17], such as UPt_3, $CeCu_2Si_2$, or $CeAl_3$), UBe_{13} (and the thorium-doped

$U_{1-x}Th_xBe_{13}$) is an intermetallic compound (cubic crystalline structure) with conduction electrons which strongly interact with localized U f-electron magnetic moments. At high temperatures, this material has properties similar to other magnetic materials, including a Curie-Weiss-like magnetic susceptibility. At lower temperatures, however, rather than entering a simple magnetically-ordered state, a state develops characterized by heavy charge carriers with effective mass over 300 times larger than the bare electron mass (hence the name "heavy fermion"). This large effective mass manifests itself in an enhanced specific heat and magnetic susceptibility. While the heavy fermion state is not well understood, it is clear that it results from the conduction electron-local moment interaction. For example, the observed low temperature magnetoresistance can be described by a single-ion Kondo model where the conduction electrons are scattered by dilute magnetic impurities in the host material [18]. This is quite unexpected as the local moments are not dilute, but are regularly and "densely" placed in the host crystal lattice (these systems are often called "Kondo lattices").

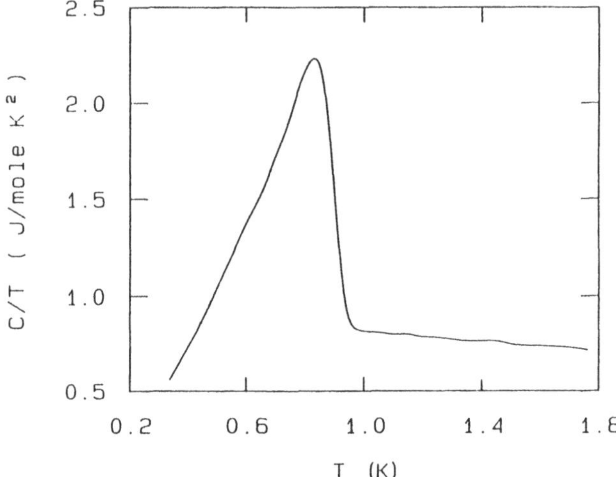

Figure 3. The specific heat divided by temperature versus temperature near the superconducting transition for UBe_{13} [21].

Even more interesting is that a few of these heavy fermion systems, including both UBe_{13} and $U_{1-x}Th_xBe_{13}$($x<0.06$), undergo superconducting transitions [19], with paired heavy electrons; observation of an enhanced jump in the specific heat shows that the heavy electrons are the ones which undergo a superconducting transition. The coexistence of magnetism and superconductivity is extremely interesting, as the BCS model predicts paired electrons with opposite spins; the presence of local magnetic moments tends to align electron spins, thereby breaking up the Cooper pair. Conventional BCS superconductivity is hard to explain in a system such as this where the electrons and magnetic moments obviously have strong coupling. This unconventional behavior also manifests itself in the superconducting properties. The jump in the specific heat at T_c is much larger than the value predicted by BCS theory in the weak-coupling approximation; below T_c the temperature dependence of the specific heat is not exponential, but appears to have a power law dependence [20]. Typical data [21] is shown in Figure 3. Other details of the behavior of superconducting properties -- such as the temperature dependence of the penetration depth $\lambda(T)$ [22] -- also deviate from the BCS prediction. At present it is not clear whether the results can be explained by "strong-coupling" effects within the BCS model, but alternative coupling mechanisms are being considered.

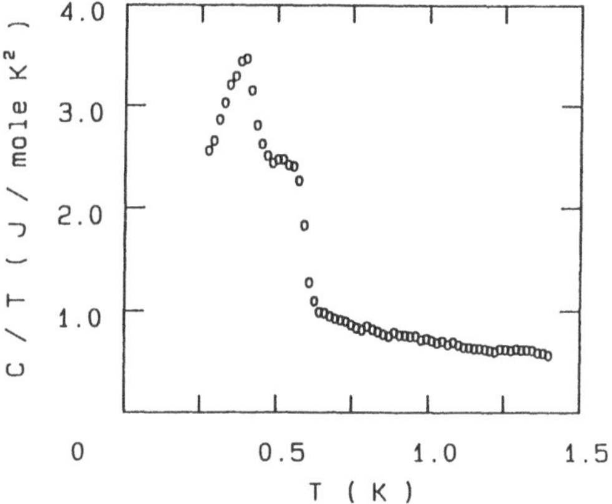

Figure 4. The double transition in the specific heat divided by temperature versus temperature for $U_{0.967}Th_{0.033}Be_{13}$ at the superconducting transition [21].

One possibility is that the electrons can exchange spin-fluctuations to result in a paired state. This type of mechanism is appealing given the strong magnetic interactions present in this system. Further, there is a precedent for this type of behavior: it is now accepted that superfluid ^{3}He has pairs linked by spin fluctuations [3]. There is a large body of theoretical work on this interesting system, and a good deal of success in describing the experimental results. Two transitions are observed in the superfluid state (superfluid ^{3}He-A and ^{3}He-B states), consistent with the two possible ground states predicted for this system [23]. The complexity of this system results from the electron pairing mechanism. For the case of phonon-mediated pairing, paired electrons have opposite spin and momenta (the 'singlet' state). This is not the case for ^{3}He atoms paired by magnetic fluctuations, where spin-aligned electrons are also possible (the 'triplet' state). The similarity of the temperature dependence of the specific heat of liquid ^{3}He in the normal state to the specific heat of normal state UBe_{13} is striking, indicating both have the same type interactions present before formation of their superconducting/ superfluid states. More intriguing -- and controversial -- is that <u>two</u> transitions are observed for $U_{1-x}Th_xBe_{13}$ $(0.017 < x < 0.040)$ [24] (and also for UPt_3 [25]). Typical data is shown for $U_{0.967}Th_{0.033}Be_{13}$ [21] in Figure 4. It is tempting to assign the two transitions to the analogous phases in superfluid ^{3}He, but at present there is contradictory evidence as to whether the lower temperature transition is into a second superconducting phase, or if it is a magnetic ordering transition [26]. Clearly, this is still a very active area of research, with many questions left unanswered.

In the next section, we will point out some interesting aspects of another very topical area of superconductivity research, investigations into high-temperature superconductivity.

V. HIGH-TEMPERATURE SUPERCONDUCTIVITY

Before 1986 the critical temperatures of superconductors were below 25 K, and with the exception of a few complex systems, such as organic and heavy fermion superconductors, superconductivity was thought to be well understood. However, in September of 1986, Bednorz and Müller discovered the first 'high-temperature superconductor' (HTS), LaBaCuO [27]. Following this, several discoveries led to a maximum critical temperature of 125 K for $Tl_2Ba_2Ca_2Cu_3O_x$ [28]. Since this time, the desire for superconducting device applications at temperatures above 77 K has led to intense research directed toward the understanding of the pairing mechanism, and the nature and dynamics of the superconducting mixed state in these type-II materials. In this section we will describe results for one of the fundamental physical parameters, the energy gap, as well as some inconsistencies of these results with the BCS weak-coupling prediction. In addition, the configuration of the vortices in these extreme type-II HTS is an important area of current research, and will also be discussed.

The superconducting energy gap $2\Delta(0)$ is predicted by the BCS model in the weak-coupling limit to be roughly $3.5\ k_BT_c$, as stated previously. For HTS, measurements of the gap energy via several spectroscopic methods have yielded values of the ratio $2\Delta(0)/k_BT_c$ greater than 8 [29]. Generally all measurements of the gap yield a ratio of roughly 5.3, and this is clearly much larger than the predicted value. These anomalously large values for $\Delta(0)$ are in contrast to the weak-coupling behavior one would expect based on the small density of states at the Fermi energy measured for temperatures just above T_c. Also, the jump in the specific heat at the critical temperature is more than a factor of two larger than the expected BCS value. These results have generated a great deal of theoretical research into possible alternative electron pairing mechanisms [30] (it has been experimentally established that the superconducting carriers are pairs of charge 2e [31]). Accurate and consistent results for both the energy gap and specific heat jump at T_c are necessary to test these theories.

The interpretation of tunneling and specific heat measurements of HTS is unfortunately an area of some controversy. Tunnelling spectroscopy, the technique traditionally used for accurate determination of $\Delta(0)$, only samples a layer of the superconductor of depth comparable to the coherence length. This is only 10 Å for HTS, and so the results are very sensitive to the condition of the surface of the sample being studied. This is particularly problematic for HTS since they are very sensitive to oxygen content. Specific heat measurements are also difficult to interpret due to the relatively small size of the electronic heat capacity as compared to the lattice contribution, as well as other unwanted contributions such as magnetic fluctuations, ionic tunnelling, and Schottky levels. The overall size of the electronic contribution is roughly 1-3%, and is difficult to extract accurately from the data [32]. As a result of these complications, it is difficult to test the many different models for superconductivity proposed for HTS.

Practical applications, as well as the extraction of basic parameters, rely on an understanding of the HTS mixed state -- superconductors below TC and in magnetic fields $H_{c1} < H < H_{c2}$. It is well known that in this state vortices will form a triangular lattice in the absence of crystalline defects. This effect has been understood since Abrikosov's theoretical study of the mixed state in 1957 [33]. In this ideal case the current flow will be dissipationless up to the critical current, as described previously. With the addition of defects however, the long range correlations in the vortex lattice can be destroyed due to vortex pinning at defect sites [34]. The implications are important because if the vortex lattice is destroyed, the dynamics of the vortices are 'pinning dominated', and, as will be mentioned below, may lead to different transport properties. Interest in the vortex structure has been renewed with the realization that the observed broad resistive transitions in HTS may be associated with the mixed state vortex dynamics [35]. The HTS crystal anisotropy may strongly modify the way vortices interact and move through the superconductor; this is apparently due to the strength of coupling between the Cu-O planes, which determine the 'stiffness' of the vortex lines [36].

For more anisotropic HTS, the resistive transitions become broader, indicating less stiff vortices which can move more freely through the material. This behavior has great importance in device applications, where it is often desirable to carry large currents in the presence of magnetic fields.

Beyond the practical implications of this behavior, there are more fundamental questions concerning the vortex structure in the mixed state. In 1962 Anderson proposed a model [37] where the vortices experience a random potential resulting from crystal lattice defects and vortex-vortex interactions. At any non-zero temperature the vortices may be thermally excited from one local potential minimum to another. In the presence of a superconducting current, vortices will experience a Lorentz force, the thermally excited vortices can flow and a dissipation will result. Thus this process, known as thermally activated flux creep (TAFC), will result in a non-zero resistance in the presence of a magnetic field for all non-zero temperatures. This model has been somewhat successful at predicting certain properties of the mixed state of HTS, but breaks down at low temperatures and magnetic fields. Recent calculations [38], however, have indicated that within the mixed state a vortex phase transition may occur as the temperature is lowered. The freely moving vortices may undergo a true phase transition to a 'vortex glass' phase. Even though the system of vortices has no true long range order (that is, the vortices do not form a lattice), the system may 'solidify' as do many liquids into a glass-like phase (this transition is similar to a spin-glass transition). In contrast to the TAFC

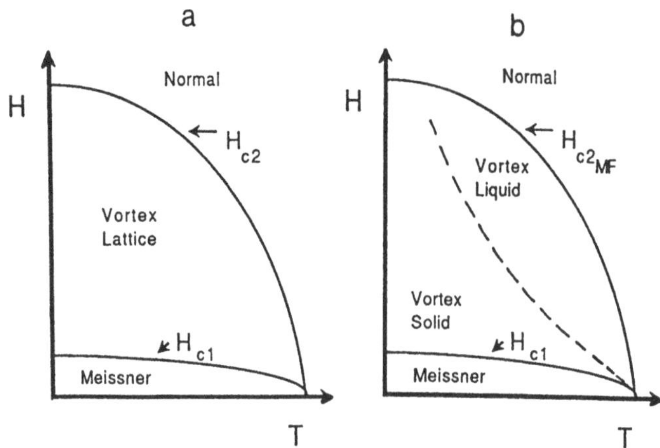

Figure 5. The H-T phase diagram for a type-II superconductor (a) as usually presented, and (b) as proposed in the 'vortex-glass' model.

prediction, the vortex glass phase would be characterized by zero resistance because the vortices are not free to move. There is now some experimental evidence for the existence of this phase in HTS [39]. In Figure 5a the conventional temperature-magnetic field phase diagram for type-II superconductors is shown, while in Figure 5b we show the proposed phase diagram which includes the vortex glass phase.

In summary, there is a considerable body of evidence indicating that HTS are not traditional BCS superconductors; however, there are still many problems to address, both theoretical and experimental, before this conclusion can be confirmed. In addition, studies of the mixed state are not only working towards improving the device characteristics but also addressing some fundamental questions on the structure and ordering of vortices.

VI. SUMMARY

We have reviewed the fundamental experimental parameters used to describe the superconducting state, and tried to give some simple physical interpretations of these parameters. The BCS model was then described, some predictions of this model were given, and shown to describe the experimental results for most superconductors quite well. Some general trends in current research were described. Advances in materials processing techniques have allowed researchers to physically alter 'typical' superconductors and construct interesting new systems which test our current understanding of superconductivity. Also, some systems exist which appear to not be well-described by the BCS model; two such systems were described here, the heavy fermion superconductor UBe_{13} and the recently discovered high-temperature superconductors. While our understanding of both these systems has improved greatly in recent years, their complexity has made it difficult to create simple models to describe all the observed properties.

VII. REFERENCES

[1] J. Bardeen, L.N. Cooper, and J.R. Schrieffer, Phys. Rev. 106, 162 (1957); Phys. Rev. 108, 1175 (1957).

[2] See the June issue of Physics Today issue 44 (1991) for several review articles on the state of high-temperature superconductivity.

[3] Superfluidity and Superconductivity- 2nd edition, D.R. Tilley and J. Tilley (Adam Hilger, Bristol, 1986).

[4] Introduction to Superconductivity, M. Tinkham (Krieger, Malabar, 1980).

[5] Landau and Lifshitz Course of Theoretical Physics, Volume 5: Statistical PHysics, Part I - 3rd edition, E.M. Lifshitz and L.P. Pitaevskii (Pergamon, Oxford, 1980).

[6] V.L. Ginzburg and L.D. Landau, Zh. Eksp. Teor. Fiz. 20, 1064 (1950).

[7] A.B. Pippard, Proc. R. Soc. London, Ser. A. 216, 547 (1953).

[8] R.W. Shaw, D.E. Mapother, and D.C. Hopkins, Phys. Rev. 120, 88 (1960); P.N. Dheer, Proc. Roy. Soc. London A260, 333 (1961); the critical current value is estimated by assuming that at the critical field H_c, the critical current density J_c is flowing through a region of thickness λ along the surface of a wire: thus $J_c = H_c/\lambda$ (note that H_c must be in units ampere/m).

[9] B. Batlogg, Physica B 169, 7 (1991); J.D. Hettinger et al., Phys. Rev. Lett. 62, 2044 (1989); also see Reference [2]. Values for ζ_0 and λ are in the 'ab' plane of single-crystal $YBa_2Cu_3O_{7-\delta}$, and the J_c results are on a thin-film sample.

[10] Superfluids, Volume 1, F. London (Wiley, New York, 1950).

[11] For an extremely interesting historical summary of the development of BCS theory, see J.R. Schrieffer in the April issue of Physics Today 45, 46 (1992).

[12] L.N. Cooper, Phys. Rev. 104, 1189 (1956).

[13] E. Maxwell, Phys. Rev. 78, 477 (1950); C.A. Reynolds, B. Serin, W.H. Wright, and L. B. Nesbitt, Phys. Rev. 78, 487 (1950).

[14] G.M. Eliashberg, Zh. Eksp. Teor. Fiz 38, 966 (1960) [Sov. Phys. JETP 11, 696 (1960)].

[15] S.T. Ruggiero, T.W. Barbee, and M.R. Beasley, Phys. Rev. Lett. 45, 1299 (1980).

[16]. M.J. Graf, T.E. Huber, and C.A. Huber, Phys. Rev. B 45, 3133 (1992) and references therein.

[17]. See G.R. Stewart, Rev. Mod. Phys. 56, 755 (1984), F. Steglich, J. Phys. Chem. Solids 50, 225 (1988), or J.M. Lawrence and D.L. Mills, Comments Cond. Mat. Phys. 15, 163 (1991).

[18] U. Rauchschwalbe, F. Steglich, and H. Rietschel, Physica 148B, 33 (1987) and references therein, or B.A. Allor, M.J. Graf, J.L. Smith, and Z. Fisk, Physica B 165 and 166, 359 (1990).

[19] H.R. Ott, Physica 126B, 100 (1984).

[20] H.R. Ott et al., Phys. Rev. Lett. 52, 1915 (1984).

[21] M.J. Graf et al., Phys. Rev. B 40, 9358 (1989).

[22] D. Einzel et al., Phys. Rev. Lett 56, 2513 (1986); F. Gross et al., Z. Phys. B - Condensed Matter 64, 175 (1986).

[23] P.W. Anderson and P. Morel, Phys. Rev. 123, 1911 (1961) and P.W. Anderson and W. F. Brinkman, Phys. Rev. Lett 30, 1108 (1973); R. Balian and N.R. Werthamer, Phys. Rev. 131, 1553 (1963).

[24] H.R. Ott, H. Rudigier, Z. Fisk, and J.L. Smith, Phys. Rev. B 31, 1651 (1985).

[25] R.A. Fisher et al., Phys. Rev. Lett. 62, 1411 (1989).

[26] R.H. Heffner et al., Phys. Rev. B 39, 11345 (1989).

[27] J.G. Bednorz and K.A. Müller, Z. Phys. B64, 189 (1986).

[28] Z.Z. Sheng et al., Phys. Rev. Lett. 60, 937 (1988).

[29] B. Batlogg, Physica B 169, 7 (1991) and references therein.

[30] For a review of recent theories, see P.B. Allen in High Temperature Superconductivity, J. W. Lynn, ed. (Springer-Verlag, New York, 1990).

[31] P.L. Gammel et al., Phys. Rev. Lett 59, 2592 (1987).

[32] J.E. Crow and N.P. Ong in reference 31.

[33] A.A. Abrikosov, Zh. Teor. Eksp. Fiz. 32, 1442 (1957).

[34] A.I. Larkin, Zh. Eksp. Teor. Fiz. 58, 1466 (1970).

[35] T.T.M. Palstra, B. Batlogg, L.F. Schneemeyer, and J.V. Waszczak, Phys. Rev. Lett. 61, 1662 (1988); also J.D. Hettinger et al., reference 9, and reference 40.

[36] D.H. Kim et al., Physica C 177, 431 (1991).

[37] P.W. Anderson, Phys. Rev. Lett 9, 309 (1962).

[38] M.P.A. Fisher, Phys. Rev. Lett 62, 1415 (1989); D.S. Fisher, M.P.A. Fisher, and D.A. Huse, phys. Rev. B 43, 130 (1990).

[39] R.H. Koch et al., Phys. Rev. Lett 63, 1511 (1989).

SHORT SEMINARS

THE LUMINESCENT EXCITED STATE OF THE VANADATE ION STUDIED BY OPTICALLY-DETECTED MAGNETIC RESONANCE

J. Hans van Tol

Rijks Universiteit Leiden
Huygens Laboratory (MAT)
Postbus 9504
2300 RA Leiden
THE NETHERLANDS

The Vanadate ion, VO_4^{3-}, shows a bright luminescence upon UV excitation. The structureless optical spectra, however, do not provide many clues concerning the electron configuration of the excited state. The long lifetimes of the radiative decay point towards a spin-triplet character, which has been confirmed by optically-detected EPR. The EPR experiments reveal that in the excited state a static Jahn-Teller deformation of the nuclear frame occurs. They further show that the generally accepted description for the lower excited states in terms of a $t_1 \rightarrow e$ excitation of the tetrahedral ion is not adequate, but rather that the excitation is of a more local nature and can be described by the excitation of a 2p electron from only one of the oxygens to a molecular orbital consisting mainly of a 3d orbital on the vanadium.

RARE EARTH SPECTROSCOPY IN GLASSES, A FRACTION

J. Lincoln

Physics Department
Southhampton University
Highfield Road
Southhampton SO9 5NH
UNITED KINGDOM

This seminar dealt with the work done at Southhampton University on Rare Earths doped into optical glass fibres. The influence of the glass host on the optical properties of the Rare Earth was discussed with particular reference to multiphonon decay in glass and its deviations from theories developed for well defined crystal systems.

THE INFLUENCE OF IMPURITIES ON THE QUANTUM YIELD OF Y_2O_3:3%Eu^{3+}

W. van Schaik

Debye Research Institute
University of Utrecht
P.O. Box 80 000
3508 TA Utrecht
THE NETHERLANDS

The decrease of the quantum yield of Y_2O_3:3%Eu^{3+}, a commercial phosphor, upon 254 nm excitation due to codoping with impurities (Fe, Cr, Mn, Ce, and Tb) is caused by competitive absorption. Codoping with Zr^{4+} or firing in air induces interstitial oxygens in the host lattice. Firing in air or in oxygen can also induce interstitial oxygens. These oxygen defects give rise to competitive absorption at 254 nm and hence to a decrease of the quantum yield upon 254 nm excitation. The oxygen defects decrease the quantum yield also upon host lattice excitation (185 nm) since the oxygen defects hinder energy transfer from the host lattice to the Eu^{3+} ion.

FLUORESCENCE MECHANISMS OF MIXED CRYSTALS $Sr_{1-x}Ba_xF_2:Eu^{2+}$

C. Dujardin

Université de Lyon I, Bat 205
Lab. de Physico-Chimie des Mat.-Lum.
43, Bd. du 11 Novembre 1918
69622 Villeurbanne
FRANCE

The photoionization phenomenon of impurity ions in insulating materials may occur at relatively low energy and therefore compete with the fluorescence processes. In some cases, it can even completely change the optical properties of the systems.

While the $CaF_2:Eu^{2+}$ and $SrF_2:Eu^{2+}$ compounds give rise to a normal blue fluorescence, $BaF_2:Eu^{2+}$ presents an "anomalous" yellow fluorescence. This phenomenon is correlated with the photoionization process of impurity ion and in fact is due to an impurity-trapped exciton.

The mixed crystals $Sr_{1-x}Ba_xF_2:Eu^{2+}$ are expected to exhibit both normal blue and anomalous yellow fluorescence. We are able, in this way, to follow the evolution of those two extreme cases (SrF_2,BaF_2).

When x increases, the exciton energy decreases with the energy gap and the interaction between the $4f^6 5d$ configuration and the exciton level of the conduction band also increases. We obtain measurements of this interaction from the half-width of the zero-phonon line for several values of x. This broadening is interpreted in terms of a Fano effect. By convoluting the $SrF_2:Eu^{2+}$ absorption spectra by a Fano lineshpae with different widths, we can reproduce the mixed crystal spectras. We find a good agreement between those two methods.

Finally, we have to set aside the hypothesis of an inhomogeneous broadening by comparing the zero-phonon line of $Sr_{1-x}Ca_xF_2:Eu^{2+}$ which gives rise to an inhomogeneous effect only. We would like to point out the extreme difference between the $Sr_{1-x}Ba_xF_2:Eu^{2+}$ and $Sr_{1-x}Ca_xF_2:Eu^{2+}$ crystals, where the half-width of the zero-phonon line is much narrower for the $Sr_{1-x}Ca_xF_2:Eu^{2+}$ mixed crystals.

The author wishes to acknowledge co-workers B. Moine, C. Pedrin.

SPECTROSCOPY OF Er^{3+}:GGG AND CALCULATION OF THE JUDD-OFELT LIFETIME PARAMETERS

B. Dinerman

Schwartz Electro Optics
45 Winthrop Street
Concord, MA 01742
U.S.A.

The spectroscopy of Er^{3+}:GGG was studied to find the optimum pumping transition necessary for an efficient diode-pumped cw 2.8μm laser. Results indicate that the 970 nm $^4I_{15/2} \rightarrow {}^4I_{13/2}$ transition is more favorable to efficient emission than the 790 nm $^4I_{15/2} \rightarrow {}^4I_{9/2}$ transition. A preliminary Judd-Ofelt analysis was also conducted to calculate laser level lifetimes. The calculated lifetimes appear to be in general agreement with the observed lifetimes, suggesting that this is a valid method for calculating lifetimes when experimental data is not available.

LASER SPECTROSCOPIC STUDIES OF SOLID STATE DEFECT CHEMISTRY IN PEROVSKITES

E.M. Standifer

University of Wisconsin-Madison
Department of Chemistry
Madison, WI 53706
U.S.A.

Mixed-metal oxides of formula ABO_3 crystallize in the perovskite ($CaTiO_3$) structure when the A and B cations differ drastically in radius. Many oxide perovskites display interesting and technologically useful properties including photoconductivity, ferroelectricity and very high dielectric constants, piezoelectricity, high nonlinear optical coefficients, water photoelectrolysis, and superconductivity. These properties show sensitivity to preparative parameters which influence defect chemistry and grain growth, e.g. stoichiometry, impurity concentrations, and annealing history. To characterize fully the defects in ABO_3 perovskites, luminescent probe ions must be separately substituted for both the A and B cation sites. Several such systems are currently under study in our laboratories.

New spectroscopic results from our group on the doped perovskitic materials $Sr_{n+1}Ti_nO_{3n+1}:Eu^{3+}$ (with $n=1,2,3,\infty$) will be discussed in terms of point and layered defects in $SrTiO_3$.

The author wishes to acknowledge co-workers N.J. Cockroft, S.H. Lee, and J.C. Wright.

PICOSECOND TIME-RESOLVED CARS: APPLICATION TO VIBRONS IN MOLECULAR CRYSTALS

J. De Kinder

Universitaire Instelling Antwerpen
Physics Department
Universiteitsplein 1
B-2610 Antwerpen (Wilrijk)
BELGIUM

We shall discuss the technique of time-resolved coherent anti-Stokes Raman scattering (TR-CARS) and the construction of a setup with a picosecond time-resolution. The analysis of the obtained decay spectra for the $k \approx 0$ vibron states in a α-N_2, α-$(^{15}N_2)_x(^{14}N_2)_{1-x}$, and α-$Ar_x(N_2)_{1-x}$ mixed crystals [1,2,3] enabled us to separate homogeneous and inhomogeneous contributions to the linewidth and to determine the crystal field splitting between the A_g and T_g mode with an accuracy up to 0.007 cm^{-1}.

REFERENCES

[1] J. De Kinder, E. Goovaerts, A. Bouwen, and D. Schoemaker, J. Lumin. 45, 423 (1990)
[2] J. De Kinder, E. Goovaerts, A. Bouwen, and D. Schoemaker, Phys. Rev. 42, 5953 (1990)
[3] J. De Kinder, E. Goovaerts, A. Bouwen, and D. Schoemaker, accepted for publication in J. Chem Phys.

The author wishes to acknowledge co-workers E. Goovaerts, A. Bouwen, and D. Schoemaker.

EXPERIMENTAL STUDIES OF UPCONVERSION LASER MATERIALS AND UPCONVERSION LASERS

R.A. McFarlane

Hughes Research Laboratories
3011 Malibu Canyon Road
Malibu, CA 90265
U.S.A.

Results will be presented of our spectrocopic measurements on Er:YLi:F_4 and Er:BaY_2F_8. These materials have now been employed to make red, green and blue lasers that are pumped in the infrared. An output power 0.5 W has been measured at 551 nm with a 14% slope efficiency. Measurements as another class of host crystal suggest that laser operation at room temperature might be achievable.

LUMINESCENCE OF THE Eu^{3+} ION IN CALCIUM COMPOUNDS

D. van der Voort

Debye Research Institute
University of Utrecht
P.O. Box 80 000
3508 TA Utrecht
THE NETHERLANDS

The luminescence of the Eu^{3+} ion in calcium compounds is reported. The effectively positive charge of the Eu^{3+} ion gives rise to high rates of radiationless processes in the excited charge-transfer state. This can be understood from a model which is based on the Configurational Coordinate Model. It is shown that an adapting of structural parameters may increase the luminescence output, but even in the ideal configuration about half of the input of light quanta is lost nonradiatively.

THEORETICAL STUDY OF ULTRA-FAST DEPHASING BY FOUR-WAVE MIXING

C. Hoerner

I.P.C.M.S.
Groupe d'Optique Nonlinéaire et d'Optoélectronique
5, Rue de l'Université
67084 Strasbourg Cedex
FRANCE

Microscopic dynamics of materials can be studied by time delayed four-wave mixing spectroscopy. Dephasing constants can be measured by such experiments. Experimental results are easily analyzed when the correlation time τ_c of the exciting field is much shorter than the dephasing time, T_2. However, when these quantities are on the same time scale, numerical fits are necessary. In the present work, we analyzed theoretically a time delayed four-wave mixing experiment. The material is described by an ensemble of two-level systems. Using the third order term of the density matrix, we establish a relation between T_2, τ_c, and τ_m the time delay for which the maximum of the diffracted light intensity is observed. The condition to obtain an estimate of T_2 from the measurement of τ_m are discussed.

A NEW WAY TO THE RELAXED EXCITED STATE IN LOCALIZED CENTERS: THE PULSE MODEL

M. Dominoni

Dipartimento di Fisica
Sezione Fisica dello Stato Solido e Nuovi Materiali
Via Celoria 16
20133 Milano
ITALY

We present a new approach to the study of the relaxation in the excited state of localized centers, the Pulse Model [1,2].

We considered a highly localized initial pulse, as that generated by exciting a bound electron of an impurity in the Condon approximation.

We have studied the pulse dynamics in the ballistic regime, where the contribution coming from inelastic scattering are probably irrelevant and can be safely neglected.

The ionic relaxation around the impurity has been calculated classically through the time-evolution of the average displacements, of the host lattice ions around the excited center (from equilibrium).

In particular, we have considered some color centers in alkali halides. We have evaluated the relaxation both in strong (TI^+ in KI,KBr and KCl) and in intermediate ($LiF:F_2^+$) electron-phonon coupling. In the former case we have found a relaxation time of about 1 ps. In the latter case we have found a relaxation time of about 105 fs, which is in a good agreement with the experimental one [3].

REFERENCES

[1] M. Dominoni, N. Terzi, How to Play with Springs and Pulses in a Classical Harmonic Crystal, in Advances in Non-radiative Processes, ed. by B. Di Bartolo, (Plenum Press, New York) 1991.
[2] M. Dominoni, N. Terzi, accepted for publication in Euro. Phys. Lett.
[3] W.H. Knox, L.F. Mollenauer, R.L. Fork, Chem. Phys. Series 46, 277, (Springer) 1986.

The author wishes to acknowledge co-worker, N. Terzi.

PROPOSITION OF EFFECTIVE WAVEFUNCTION FOR 2DEG WITHIN MODFET HETEROSTRUCTURES

E.A. Anagnostakis

Section of Solid State Physics
Department of Physics
University of Athens
22 Kalamakiou Ave.
174 55 Alimos
Athens, GREECE

In this seminar we propose a form of effective wavefunctions for the 2-Dimensional Electron Gas (2DEG) confined within the channel of a Modulation-Doped Field Effect Transistor (MODFET). The wavefunction $\Phi_n(x)$ of the nth conduction sub-level is taken as a linear superposition $\sum_j \{c_{nj}\, \varphi_{nj}(x)\}$ of harmonic oscillator eigenfunctions $\varphi_{nj}(x)$ with a frequency $\omega_j(\xi)$ dependent upon the sheet density ξ of the 2DEG. For the ξ-dependence of $w_j(\xi)$, we propose a rational function $[j\alpha/(\beta+\xi)]$ consistent with the requirements for low ξ (quantum limit) and high ξ (3D bulk limit) behaviour.

EPITAXY OF CdS-THIN FILMS BY PULSED LASER EVAPORATION (PLE)

M. Müller

Universität Kaiserslautern
FB Physik
Erwin-Schrödingerstr.
D-6750 Kaiserslautern
GERMANY

We have constructed a new vacuum chamber for PLE. This new technique is very promising. A typical deposition rate is 0.1 - 1Å/Laserpulse. As a result, it should be possible to grow heterostructures of two materials with well-defined single layer thickness.

The energy distribution of the ablated atoms and ions is non-thermal. The high energy leads to a high surface mobility of the atoms. By applying an electronic field, it is also possible to influence this energy distribution. Since the evaporation process is far away from thermodynamic equilibrium, it is also possible to vaporize incongruently evaporating compound materials.

The first evaporated CdS-films on BaF_2 are of good crystallinity and can be seen by Laue-diffraction and luminescence spectroscopy. The surface is shiny and no grains or inhomogeneities can be observed. More detailed information about the optical properties will be given in the short seminar of Harald Giessen.

GROWTH AND OPTICAL PROPERTIES OF THIN CdS FILMS

H. Giessen

Universität Kaiserslautern
Fachbereich Physik
Erwin-Schrödingerstr.
D-6750 Kaiserslautern
GERMANY

We grow thin CdS films on different substrates, e.g. $BaF_2(111)$, $SrF_2(111)$, and GaAs(111) by Hot Wall Epitaxy and Pulsed Laser Evaporation.

We get highly oriented ($c \perp$ substrate) crystalline films and investigate them at T = 4.2K by transmission and reflection spectroscopy, and study the luminescence under low (HBO-lamp) and high (N_2-laser) excitation. The samples show strong A and B excitonic resonances, bound excitons, D-A-pair-transitions and an electron-hole-plasma.

Conclusions about sample quality can be made. Strain effects shift the excitonic resonances energetically, the A exciton down and the B exciton up to higher energies. A theoretical model helps to understand the experimental results. We split up the biaxial strain into hydrostatic pressure and uniaxial strain. Strain effects at the interface between the substrate and the film are different from strain effects at the film surface.

LUMINESCENCE OF NEW STORAGE PHOSPHORS: ALKALINE EARTH FLUORO-HALIDES DOPED WITH DIVALENT YTTERBIUM

W.J. Schipper

Debye Research Institute
University of Utrecht
P.O. Box 80 000
3508 TA Utrecht
THE NETHERLANDS

A storage phosphor is a material that is able to store energy from incident radiation. When a crystalline insulator is irradiated with X-rays, electrons may be promoted to the conduction band. These mobile electrons may be trapped at e.g. anion vacancies. An analog process takes place for the hole which is left in the valence band. When an impurity, such as Eu^{2+}, is present, the hole may be trapped at or near it. If the energy difference between the trap and the band is large enough the electron and hole will remain separated at room temperature for a considerable time. When the electron is released from its trap by either thermal stimulation or irradiation with an appropriate wavelength, recomination on the hole trap occurs, giving rise to characteristic emission in the case of Eu^{2+}. The role of Eu^{2+} in the trapping process is still under discussion; in the present study a comparison is made between Eu^{2+} and Yb^{2+} who both may act as a hole trap in alkaline earth fluorohalides (MFX; M=Ca,Sr,Ba; X=Cl,Br). It is shown that Yb^{2+} has a 5d → 4f emission in the Ca and Sr compounds whereas there is no luminescence in BaFX:Yb^{2+}. The thermoluminescence behavior of the Yb-doped Ca and Sr fluorohalides is the same as that of the Eu-doped fluorohalides; hence it is concluded that the role of the Yb^{2+} and Eu^{2+} in the trapping process is the same.

APPLICATION OF PHOSPHORS IN X-RAY COMPUTED TOMOGRAPHY

W. Rossner

Siemens AG
Otto-Hahn Ring 6
P.O. Box 83 09 53
D-8000 München 83
GERMANY

One of the most attractive applications of phosphors is their use as X-ray conversion detectors in X-ray computed tomography (CT), the dominating method of medical radiology. For the last few years a new class of phosphors based on ceramic materials has been introduced by several groups for CT use. For example, Europium-doped Yttriagadolinia ceramics, Y_2O_3-Gd_2O_3:Eu are chosen to demonstrate the afterglow problem. In many cases the afterglow behvior of such phosphors is affected by excitation conditions. By high energy excitation with photons up to 150 keV defects are generated which may act as trapping centers increasing afterglow. For the next generation of high performance CT systems an improvement of afterglow is of special interest.

DISSOCIATION OF POLYATOMIC MOLECULES BY INFRARED LASERS

B. Bowlby

Physics Department
Boston College
Chestnut Hill, MA 02167
U.S.A.

In recent years, dissociation of polyatomic molecules through the use of infrared lasers has become a subject of much interest. Since the dissociation energy is much greater than the energy of a single photon, this is obviously a multiphoton process. Also, it was shown that this reaction is isotopically selective. Finally, the size of the molecule plays an important role in the reaction rate. In this talk I shall present a simple model that incorporates current theory to explain this process.

A QUANTITATIVE ANALYTIC THEORY OF THE SPECTRA OF DIATOMIC MOLECULES

J.F. Ogilvie

Academia Sinica
Institute of Atomic and Molecular Sciences
P.O. Box 23-166
Taipei 10764
TAIWAN

To treat the spectral properties of free diatomic molecules within any particular electronic state, we have developed a quantitative analytic theory to enable the reduction of radial functions that are independent of nuclear mass. From the frequencies of spectral lines for transitions of any type involving a change of rotational angular momentum, the measured data can lead to radial functions for potential energy and for adiabatic and nonadiabatic effects of either nucleus. For electronic states with net electronic spin or orbital angular momentum, the interactions between the various angular momenta produce further radial functions; so far $^2\Sigma$, $^1\Pi$, $^3\Sigma$ and $^2\Pi$ states have been treated in specific applications of the theory. Likewise for intrinsic nuclear angular momenta, several interactions have been treated so as to determine the corresponding radial functions. From either the effects of applied electric and magnetic fields or the intensities of spectral lines due to vibration-rotational transitions, the radial functions for the electric dipole moment, magnetic dipole moment, and electric quadrupole moment have been determined for absorption and emission spectra, and prospectively the electric (dipole) polarizability and (nonlinear) susceptibilities from scattering spectra. The resulting radial functions reflect the accuracy of the spectral data; these data can be reproduced within the uncertainty of the measurements. The analytic form of the expressions produced by the application of the theory makes its use simple and rapid but accurate.

THERMAL BEHAVIOR OF SPECTRAL LINE POSITIONS AND WIDTHS OF Nd^{3+} IN GSGG

X. Chen

Department of Physics
Boston College
Chestnut Hill, MA 02167
U.S.A.

The thermal effects on luminescence line positions and widths of Nd^{3+} in $Gd_3Sc_2Ga_3O_{12}$:Cr^{3+}, Nd^{3+} were studied from 78 to 600 K. The temperature dependence of linewidths and lineshifts is explained in terms of an ion-phonon interaction as a perturbation using a Debye phonon distribution. The Debye temperature is treated as an adjustable parameter to get the best fit to the experimental data. The linewidths are related to Raman phonon scattering, direct phonon processes, and random microscopic strains of the crystal. The lineshifts are due to the stationary effect of the ion-phonon interaction.

The author wishes to acknowledge co-worker B. Di Bartolo.

COMPARISON OF Er^{3+} SPECTROSCOPY IN DOPED GLASS FIBERS AND IN GLASS BULK SAMPLES

D. Meichenin

Centre National D'Études
des Télécommunications
196, Av. H. Ravera
92220 Bagneux
FRANCE

Because of their shape (diameter of few microns, length of few meters) optical fiber geometry may strongly influence the spectroscopy of the doping ions.

Due to the strong confinement, a large power density is easily obtained even at moderate excitation level, giving rise to absorption saturation.

Fluorescence spectra are also modified by such saturation which gives amplification by stimulated emission (ASE). Also, dynamical properties may be modified by a strong optical trapping which gives enlarged values for the emission lifetime.

Examples of such phenomena for Erbium-doped fluoride and silica fibers will be presented.

The author wishes to acknowledge co-worker F. Auzel.

PASSIVE INTRACAVITY STABILIZATION OF WIDE GAIN LASER BY Er^{3+}-DOPED MATERIALS

Zhou, B.W.

Centre National d'Études
des Télécommunications
196, Av. H. Ravera
92220 Bagneux
FRANCE

We show that intracavity Er^{3+}-doped materials can be used to stabilize wide gain lasers against temperature and pumping variations.

The frequency is determined by the intensity of the crystal stark lines for Er^{3+}. We present oscillator strength calculations beyond the usual Judd-Ofelt theory for the stark structure of the $^4I_{15/2} \rightarrow {}^4I_{13/2}$ transition extending from 1.4 μm to 1.6 μm.

The first experiments on the frequency stabilization of $NaCl:OH^-$ laser and of an InGaAsP semiconductor laser emitting in the 1.5 μm region are presented.

OPTICAL PROPERTIES OF F_3^+ CENTER IN LiF (TRIPLET STATE)

M. Cremona

ENEA, INN-SVIL
Via E. Fermi, 27
00044 Frascati (Rome)
ITALY

It is well known that the excitation in the M band of a colored LiF crystal produces a "green" emission band with a peak at about 530 nm and a "red" band around 670 nm. Detailed optical investigations including the polarization measurements have shown that the emitting centers are F_3^+ and F_2 respectively. Moreover a quenching of the emission intensity of the F_3^+ centers was observed at high pump power. This behavior is strongly dependent upon the presence of a triplet absorption of this center. We have observed change in the absorption and in the luminescence of the F_3^+ centers while exciting the LiF crystal with the 457 nm line of the Ar^+ laser. To our knowledge, no previous measurements have shown in such direct way the existence of a triplet state for the F_3^+ color center in LiF whose kinetic has also been studied. Our investigation is useful for the application of the F_3^+ emission in color center lasers.

The author wishes to acknowledge co-workers R.M. Montereali and V. Kalinov.

TWO-PHOTON SPECTROSCOPY IN THE F-SHELL

G. Vandenberghe

Katholieke Universiteit Leuven
Departement Scheikunde
Afdeling Anorganische
en Analytische Scheikunde
Celestijnenlaan 200 F
B-3001 Leuven
BELGIUM

The formalism of second quantization provides a powerful method to describe the various intermediate coupling mechanisms of two-photon transitions in lanthanide ions. A complete analysis has been carried out by means of a fourth-order mechanism, which combines the two-photon interaction with spin-orbit coupling and crystal field elements. The resulting selection rules are discussed, in relationship with the recently obtained TP-spectra of Tb^{3+} in the octahedral elpasolite lattices.

TWO-PHOTON TRANSITION INTENSITIES WITHIN SYMMETRY-ADAPTED EIGENVECTOR APPROACH: Ni^{2+} IN O_h SYMMETRY

J. Sztucki

Institut für Festkörperphysik
Fachgebiet Technische Physik
Fachbereich 5
Hochschulstrasse 8
W-6100 Darmstadt
GERMANY

A short overview of the advantages of two-photon spectroscopy in solids is presented. A symmetry-adapted wavefunction formulation of the theory of two-photon transition intensities is given and the particular example of Ni^{3+} ion in O_h symmetry host lattice is discussed.

The theoretical results are compared with recent experimental report by C. Compachiaro and D.S. McClure, P. Rabinowitz and S. Dougal (Phys. Rev., B43, 14 (1991)).

CHARACTERISTIC ELECTROLUMINESCENCE AT THE SEMICONDUCTOR/ELECTROLYTE INTERFACE

E. A. Meulenkamp

Debye Research Institute
University of Utrecht
P.O. Box 80 000
3508 TA Utrecht
THE NETHERLANDS

Electroluminescence (EL) of an n-type semiconductor can be observed when a hole, which is injected into the valence band by a strong oxidizing agent, and an electron recombine radiatively. The study of this EL is a valuable tool in the characterization of processes occurring at the semiconductor/electrolyte interface. The use of dopant ions, whose spectral properties are known, can provide extra information. An example is the characteristic Tb^{3+}-emission in thin Ta_2O_5 films. Photoluminescence measurements indicate that the Tb^{3+}-ion acts as a recombination center. The spatial origin of the EL can then be deduced from the decay time of the Tb^{3+}-emission and from the relative intensity of the various 4f-4f transitions.

LOCALIZED $^3\pi\pi^*$ EXCITATIONS OF $[Rh(phpy)_2bipy]PF_6$ (phpy = 2-phenylpyridine, bipy = 2,2'-bipyridine)

G. Frei

Institut für Anorganische Chemie
Universität Bern
Freiestrasse 3
3009 Bern
SWITZERLAND

Polarized single-crystal absorption spectra of $[Rh(phpy)_2bipy]PF_6$ at 10K show three origin lines C,D, and N with very sharp vibrational sidebands. From a comparative study of the highly resolved low-temperature absorption spectra of $[Rh(phpy)_2bipy]PF_6$, $[Rh(phpy)_2bipy]B\Phi_4$, $[Rh(phpy)_2en]PF_6$, and $[Rh(phpy)_2en]B\Phi_4$ (en-enthylenediamine, $B\Phi_4^-$=tetraphenylborate anion) it can be concluded that the lines C and D correspond to ligand-centered $^3\pi\pi^*$ excitations localized on the two crystallographically inequivalent phpy- ligands and that line N can be assigned to an analogous transition on the bipy ligand.

From polarized absorption spectra of $[Rh(phpy)_2bipy]PF_6$ and its crystal structure the orientation of the transition moments relative to the ligands can be determined: Line N corresponds to a short-axis polarized transition on bipy, whereas C and D are in-plane polarized transitions on the two phpy- ligands with transition moments oriented approximately parallel to the respective Rh-N bonds.

The author wishes to acknowledge co-workers A. Zilian and H.U. Güdel.

CROSS RELAXATION OF EXCITED STATES IN A ONE-DIMENSIONAL COMPOUND

M.P. Hehlen

Institut für Anorganische Chemie
Universität Bern
Freiestrasse 3
3009 Bern
SWITZERLAND

If a transition from an excited state to a lower state is in resonance with an absorption from the ground state of a nearby ion, efficient excited state relaxation may occur by cross relaxation. This is shown for the one-dimensional compound $Pr(CH_3COO)_3 \cdot H_2O$.

The Pr^{3+} electronic excited state $^1D_2$ may relax to $^1G_4$ by transferring this energy nonradiatively to a neighboring Pr^{3+} ion which is excited from its $^3H_4$ electronic ground state to the $^3F_4$ state. This cross relaxation, which efficiently quenches the luminescence from $^1D_2$, is strongly suppressed in the diluted sample $Ce(CH_3COO)_3 \cdot H_2O{:}1\%Pr^{3+}$. There, Pr^{3+} ions mostly have Ce^{3+} neighbors which cannot act as acceptors for the energy of any transition from $^1D_2$. That is why an enhancement of the $^1D_2$ luminescence intensity of two orders of magnitude can be observed with dilution to 1%.

The author wishes to acknowledge co-worker H.U. Güdel.

LUMINESCENCE OF THE V=O $\bar{\underline{V}}$ COMPLEX

M.F. Hazenkamp

Debye Research Institute
University of Utrecht
P.O. Box 80 000
3508 TA Utrecht
THE NETHERLANDS

The luminescence properties of silica supported Vanadium Oxide catalysts and of the crystalline compound $KVOF_4$ are rather similar. The luminescence strongly differs from tetrahedral oxo vanadate complexes. In the present cases there is vibrational structure observed in the emission spectra, and the lifetimes of the emitting states are unusually long. These observations suggest that the luminescent center in the catalyst is a monomeric Vanadate entity which has one very short V-O bond. $KVOF_4$ serves as a model compound to try to undersand the unusual luminescence properties.

RADIATIONLESS VIBRONIC RELAXATION AND ELECTRON TRANSFER OF THE F-CENTER IN NaBr

M. Leblans

Universitaire Instelling Antwerpen
Physics Department
Universiteitsplein 1
B-2610 Antwerpen (Wilrijk)
BELGIUM

By means of a pump-probe technique for induced transparency the relaxation of the optically excited F-center in NaBr is studied. At 10K a relaxation time of 6±1 ns is established. The decay process is thermally activated above 70K, yielding a lifetime of 38 picoseconds at 150K. These results are discussed in connection with the Dexter-Klick-Russell criterion, which predicts quenching of the F-center emission in NaBr as a result of a much faster non-radiative process. At temperatures above 100K a second decay component is observed. It is related to the transfer of the optically excited F-center electron to another F-center (resulting in an empty vacancy and a F-center, a vacancy containing two electrons), and the reverse process. The mechanism of this electron transfer is not yet completely understood.

The author wishes to acknowledge coworkers F. De Matteis, E. Gustin, and D. Schoemaker.

LUMINESCENCE OF BROAD BANDS IN Mn-DOPED n-TYPE GaP

T. Monteiro

Universidade de Aveiro
Departamento de Fisica
3800 Aveiro
PORTUGAL

Several broad unstructured bands are found in the luminescence spectrum of n-type GaP.

Before annealing, a broad band emission [1], which occurs above 70K with a maximum at 1.64 eV, shifts towards higher energies upon increasing temperatures and is shown to be due to emission from two thermalizing levels which are 78 meV apart.

After annealing, it is known that Mn n-type GaP undergoes complexation. The main feature of the spectra after annealing is the appearance of a new broad band, peaking at 2.03 eV at low temperatures[2].

A comparison between the spectra before and after annealing was made and the origin of these bands and its relationship with Mn is discussed and compared with centers of similar characteristics in GaP.

REFERENCES

[1] T. Monteiro, E. Pereira, Journal of Lum. 48 & 49 (1991) 671.
[2] T. Monteiro, E. Pereira, Appl. Surface Science (1991) (to be published).

The author wishes to acknowledge co-worker E. Pereira.

OPTICAL SPECTROSCOPY OF THE MATRIX-ISOLATED NH RADICAL

C. Blindauer

Institut für Physikalische Chemie
Wegelerstr. 12
D-5300 Bonn 1
GERMANY

Matrix-isolation is a wonderful technique to chemically stabilize radicals and to investigate molecules which are effectively quenched in the gas phase. This is valid especially for metastable excited states. On the other hand, a matrix guest is a powerful indicator of solid state properties like site-symmetry, phonon spectrum, etc. Moreover, matrix data are often useful in understanding van der Waals-complexes.

All these items will be discussed taking as an example NH in a rare gas solid.

INTERACTION BETWEEN COLOR CENTER AND DISLOCATION IN ALKALI HALIDES

R.B. Pode

Department of Physics
Nagpur University
Nagpur 440010
INDIA

Contribution of individual defects such as point defects and dislocations to the physical properties of solids have been studied extensively. Not much is known about the contribution of their interaction to physical properties. Attempts have been made to study the interaction between color centers and dislocations in alkali halides. Color center-dislocation interaction has been studied by measuring diffused reflectance spectra and stability of color center in microcrystalline powders of alkali halides.

SUMMARY OF THE MEETING

(G.F. Imbusch)

The subject of this school - Optical Properties of Excited States in Solids - covers a very wide range of material and phenomena, and this was reflected in the variety of topics discussed at the school. The introductory lectures by the director, Professor Baldassare Di Bartolo, aimed at setting a broad theoretical framework which would encompass many of the concepts to be discussed by the later speakers. In these lectures, Professor Di Bartolo analyzed the properties of a two level electronic system interacting with the radiation field, deriving the optical Bloch equations, Rabi oscillations, and introducing phenomenological T_1 and T_2 relaxation parameters. In his treatment he made use of a density matrix formulation. In the later lectures, stress was laid on coherence effects, and concepts such as photon echoes, optical nutation, and free induction decay were introduced. These concepts arose again in the presentations of later lecturers.

Dr. Francois Auzel examined the quantum mechanical processes giving rise to the various coherent relaxation times introduced phenomelogoically by Professor Di Bartolo. A particular feature was the use of the second-quantized formulation of the electromagnetic field, but the treatment was given in a manner which used the minimum of mathematical formalism, merely enabling the basic physics to be kept in focus at all time. He showed how spontaneous emission merges into stimulated and coherent emission, and he drew a careful distinction between superradiance and superfluoresence. The efficiency of the APTE upconversion process was emphasize by him and illustrated by reference to his experimental studies of the Er^{3+} ion, showing the conversion of up to five infrared photons into visible light. As Dr. Ross McFarlane's later seminar showed, upconversion processes in Er^{3+}-doped materials can lead to multicolour laser emission.

We were reminded by Professor J.H. van der Waals that many of the concepts described by Di Bartolo in his treatment of the two-level system have their origins in the fields of nuclear magnetic resonance (NMR) and electron paramagnetic resonance (EPR). After a short introduction in which the classical optical pumping experiments on mercury vapour by Kastler, Brossel, and Bitter were reviewed, Professor van der Waals' lectures centered on optically-detected magnetic resonance (ODMR) in optically- excited metastable spin triplet states. The ODMR technique allows the excites state to be probed with microwave resolution. This technique can be used to study the dynamics of the generation and decay of the excited state. In a later seminar Hans van Tol, who worked under Professor van der Waals, described his ODMR experiments on the excited triplet state of the VO_4^{3-}, MoO_4^{2-} and Cu^+ ions in organic hosts.

The availability of tunable stable laser beams with extremely narrow bandwidths (order of 1 MHz or less) now allows one to probe optical transitions (typically inhomogeneous widths in trivalent rare earth systems are around 1 cm^{-1} or 30 GHz) with exceptional resolution, as Dr. Roger Macfarlane showed in his lectures. The technique employed is that of holeburning. The narrow hole burned in the broad inhomogeneous line is regarded as the spectroscopic signal and, because of its narrow width, one can easily study Stark and Zeeman splittings, obscure isotopic shifts as well as such small effects as nuclear Zeeman splittings. A number of coherence effects, such as were described by previous speakers were demonstrated in these experiments. Despite the many gains already made, Dr. Macfarlane felt that this form of high resolution spectroscopy is still a young field and he felt that there is still much to be done.

The use of very narrow band laser beams to probe the defect sites in perovskites was the topic of Eugene Standifer's later short seminar.

Professor J. Reuss, in his lectures pointed out that solid state spectroscopists have much to learn from molecular spectroscopists. He described his experiments on the interaction of infrared laser light with a beam of SF_6 molecules, whereby various vibrational modes of the molecule could be excited. A sensitive bolometer detection technique was used. By increasing the laser power, multiphoton transitions could be generated. In his experiments he was able to demonstrate a number of the coherence concepts introduced in Professor Di Bartolo's introductory lectures.

Dr. John Ogilvie, in his short seminar, showed that the bound states of simple diatomic molecules can now be very accurately parameterized with the result that the large number of observed transitions can be predicted with exceptionally high accuracy. The problems of explaining the photodissociation of polyatomic molecules was addressed by Brian Bowlby. He outlined current theories on the subject.

Staying on the topic of molecules, we were reminded by Professor N. Schwentner of the difficulty, which still exists, of providing an adequate description of the dissociation of molecules in the condensed liquid phase. Professor Schwentner presented his results on the photodissociation of an analogous but much simpler system - that of a simple small molecule in a rare gas matrix. The surrounding rare gas lattice acts as a cage to hinder the movement of the components of the molecule and of the dissociation products. Extensive molecular dynamics calculations of the molecular components and of their lattice atoms have been made and predictions compared with experimental results obtained on carefully prepared samples. One of the interesting spectroscopic species discussed by him was the exciton state consisting of a hole in the Xe lattice bound to the H^- atom. In the case of the F_2 molecule in a solid Xe matrix, the long range migration of the F fragments after photodissociation and the ensuing production of XeF molecules, which can be used to generate excimer laser action on the solid state crystal, were described. In respect to molecular dynamics calculations, Professor Schwentner did not share Marshall Stoneham's view about the accuracy of available interatomic potentials. However, in the case of low energies and bound states, the regions of Stoneham's interest, a more positive view may be warranted.

The study of isolated spectroscopic entities embedded in rare gas matrices was continued in a number of short seminars. Carsten Blindauer pointed to the advantage of this technique when dealing with chemically unstable molecules or molecules which condense at low temperatures. He described his experiments with the NH molecule in a rare gas solid.

Dr. Marshall Stoneham addressed the question of the behaviour of the bound electronic states of centres in a solid immediately after optical excitation. In introducing the ubiquitous configurational coordinate model, he emphasized its many approximations but felt that it was a convenient model for basing a simple description of the relaxation process. He referred to the general rule proposed by Dexter, Klick and Russel for predicting whether or not a colour centre would emit luminescence; this model is based on the position of the crossover point between ground state and excited state parabolae, and it has some qualitative validity. Making accurate quantitative estimates of nonradiative relaxation phenomena, he stated, is a very difficult problem for which the simpler models based on the harmonic approximation and the Condon approximation should no longer be regarded as adequate. Dr. Stoneham was of the opinion that sufficiently accurate potentials are now available so that, given adequate computing facilities, accurate relaxation rates can be calculated.

There was an interesting short seminar by Matteo Dominoni, who described the classical analysis of ionic relaxation about an excited impurity ion immediately after excitation. When applied to some specific systems, his calculated relaxation rates are close to measured values.

The general field of colour centres was surveyed by Dr. Giuseppe Baldacchini in his introductory lecture before concentrating on the nature of the excited state. He pointed out the approximate validity of the Dexter-Klick-Russell rule, but he explained how the existence of hot luminescence, i.e. the release of photons by the colour centre before it reaches its lowest relaxed excited state, shows that the relaxation process in the excited state is really quite complicated. He described the pioneering experiments and analysis carried out by Mollenauer and Baldacchini which demonstrated the extended nature of the relaxed excited state, whose wavefunction stretches well beyond the firs few atomic neighbours.

We learned from Dr. Baldacchini about the very puzzling question of the nature of the higher relaxed excited states. There is still an absence of agreement on the nature of these higher states as well as a disturbing conflict between different sets of experimental data. Dr. Baldacchini favours a model with weak coupling between electronic and vibrational states. On the other hand, Professor Hiroshi Ohkura, who presented a seminar on the topic, prefers a model which visualized a strong coupling between electronic and vibrational states, and he shared an appropriate set of experimental data to support his viewpoints.

The experiment described by Dr. Marc Leblans in his short seminar - using a pulse-probe technique to observe time-resolved induced transparency - would seem to be the technique to throw light on this contentious issue of the return of relaxation of the excited F-centre. Dr. Marco Cremona described his experiments on the F_3^+ centre in LiF. The F_3^+ centre is an important colour centre for laser action. And in a further seminar in the field of colour centres, Dr. Augusto Scacco described his experiments on the relaxation of F_2 centres in ionic crystals.

Dr. Charles Struck in his seminar made much use of the single configurational coordinate model. He stressed the influence of the charge transfer states on the spectroscopy of Eu^{3+} -doped Yttrium and lanthanum oxysulphides. He showed how a good quantitative description of the spectra can be obtained in terms of a configurational coordinate model which includes a small change in the force constant for the charge transfer state. He pointed out that an accurate description of the nature of the charge transfer states in these material has yet to be obtained, and he posed a number of puzzling questions, still awaiting clarification, about these charge transfer states. An interesting long time energy storage, in the form of trapped electron-hole pairs, was shown to occur in these materials.

Dick van der Voort, in his short seminar, spoke on the luminescence of Eu^{3+} in calcium compounds. The excess charge on the Eu^{3+} ion pulls down the charge transfer states, leading to a lower cross-over with the Eu^{3+} configurational coordinate parabolae and increasing the nonradiative decay rate.

Professor G.F. Imbusch, in his lectures presented a fairly broad survey of spectroscopic studies of the excited luminescent states of rare earth and transition metal ions in solids. In many cases the optical transition to the excited state involves a change in the coupling of the ion with the surrounding lattice, leading to a distortion of the ionic surroundings of the excited ions, and this has an influence on the bandshape and on the luminescence efficiency of the transition. Again, much use was made of the configurational coordinate model. Recent photocalorimetric techniques for accurate estimation of quantum efficiency were described. The complementary role of one-photon and two-photon absorption spectroscopy was illustrated by some examples. Professor Umberto Grassano in his seminar also emphasized the complementary nature of one- and two-photon spectroscopy with reference to his recent study of $KCl:Eu^{2+}$. In his short seminar, Geert Vandenberghe delved more deeply into the selection rules for two-photon absorption in rare earth ion systems. He compared his calculations with observed spectra. And, in his short seminar, Jaroslaw Sztucki discussed two-photon selection rules for Ni^{2+} in MgO, for which experimental data are available.

Strong coupling to specific vibrational modes can show up as multiphonon peaks on the vibronic transition. An interesting example of this was shown by Menno Hazenkamp in the case of the luminescence of the vanadate centre in silica-supported vanadium oxide catalysts, where the appearance of many peaks separated by 1020 cm^{-1} indicated the existence of a short V-O band in this centre. And on the subject of vibronic peaks, Gabriela Frei showed very sharp electronic and vibronic absorption transitions from a single crystal of an organic material. She showed how these vibronic spectra could be used as a characteristic "fingerprint" to determine the ligand on which the absorption transition occurred. Marcus Hehlen described the strong cross-relaxation which can occur between Pr^{3+} ions in the one dimensional compound $Pr(CH_3COO)_3 \cdot H_2O$ resulting in a low luminescence efficiency of the Pr^{3+} ions. And in a short seminar Christopher Dujardin reported on the complicated nature of the luminescence from the mixed crystal $Sr_{1-x}Ba_xF_2:Eu^{2+}$. This exhibits both normal Eu^{2+} emission, such as is observed from $SrF_2:Eu^{2+}$, and luminescence from an impurity-trapped exciton, such as is observed from $BaF_2:Eu^{2+}$. He described the variation of the spectrum with changing x.

The topic of the Jahn-Teller effect was treated in an elegant didactic manner by Professor Gabriele Viliani, making use of symmetry arguments and simple examples. He distinguished between the static and dynamic Jahn-Teller effects, and he gave examples of how these effects manifest themselves in the optical spectra of solids.

Much of the justification for our research is the development of technically useful materials - new laser materials, new and efficient phosphors, etc. Such practical topics were addressed by a number of speakers. Professor Georges Boulon reported on his studies of solid state laser materials based on garnet hosts doped with transition metal ions or codoped with transition metal ions and rare earth ions. He showed examples of efficient excitation transfer from transition metal to rare earth ions in these materials. Professor Martin Buoncristiani discussed sources of noise in stabilized lasers, the ultimate objective being to produce laser systems for fundamental experiments to be carried out in space.

Professor Alexander Kaminskii announced the recent results from the major laser development program being carried out in his laboratory in Moscow. He described the new material $LaBGeO_5:Nd^{3+}$ whose molecular properties allow it to be employed as a self-frequency-doubling laser crystal. He pointed out the advantages of using crystalline hosts with disordered structures, and demonstrated the improvements in optical pumping of lanthanide ions by using a double pumping scheme.

The Er^{3+} ion and its employment in laser systems was the topic of a number of lectures and seminars. In addition to the reports by Auzel and McFarlane mentioned above, short seminars on the topic were presented by Brad Dinerman, who was seeking to optimize the pumping of Er^{3+} in GGG so as to achieve 2.8μm laser action, Daniel Meichenin, who pointed out the interesting differences in spectroscopic behaviour between Er^{3+} ions in glass fibers and in bulk glass samples, and Bei-Wen Zhou who achieved stabilization of wide gain lasers against temperature and pumping variations by incorporating Er^{3+}-doped material in the laser cavity.

We were reminded by Dr. Bruno Smets of the important industrial use of phosphors in X-ray intensifying screens, cathode ray tubes, and in fluorescent lamps. He gave a number of illustrative examples of how energy transfer processes and nonradiative relaxation mechanisms can strongly affect the efficiency of phosphors. There is an increasing use of phosphor lamps for suntanning and the treatment of skin conditions, and the design of such lamps poses a number of special problems. He recounted the steps involved in bringing the new tri-colour and special deluxe lamps to the market place.

Willem Schipper, in his short seminar, described his search for an X-ray storage phosphor in which trapped electrons are released at a later time by laser excitation, and the released electron energy is emitted as Eu^{3+} luminescence. Dr. Wolfgang Rossner spoke on the development of new phosphors for the new generation of high performance X-ray Computer Tomography systems. Willem van Schaik reported on the factors which limit the quantum yield of the important phosphor $Y_2O_3:Eu^{3+}$, and John Lincoln reported on some peculiarities of the excited state relaxation of rare earth ions in glass fibres.

The excited luminescence states of semiconductors were treated by Professor Claus Klingshirn. In his lectures he gave a brief introduction on semiconductors. He described how the main elementary excitations of semiconductors - phonon, plasmons, excitons - arise, how these interact with the radiation field, starting with the case of weak coupling, then moving to the case of strong coupling, when the polariton concept arises. The basic concepts of the non-linear optics of semiconductors were outlined by him, he discussed how various coherence effects arise and supported these concepts with experimental observations. He was able to relate these effects to the general treatment of Professor Di Bartolo.

In a short seminar Martin Müller described an apparatus which uses pulsed laser evaporation to lay down a very controlled thin film of CdS. This was followed by Harald Giessen who described the spectroscopic parametrization of thin films. The luminescence properties of Mn^{2+} in GaP were described by Teresa Monteiro.

An important semiconductor topic today is the production and analysis of quantum dots. Professor Renata Reisfeld described her experiments on incorporating CdS particles of varying small size into glasses prepared by the sol-gel method. Because of the tiny size of these particles one expects an increase in the energy levels of the electronic states compared with that found in bulk CdS. She presents some evidence indicating such energy variations in the spectra in these materials.

Before concentrating on the spectral properties of excited states in restricted geometries, Dr. Joseph Klafter described how the stretched exponential formula describes the relaxation behaviour of many different disordered systems. He used different energy transfer models, with which experimentalists are familiar, to illustrate his theoretical analysis. The variety of theoretical formulae which describe relaxation processes in materials with restricted geometries and different dimensionalities were discussed by him, and he commented on the problems of fitting experimental data to these theoretical expressions.

There were a number of other interesting short seminars. John Hamilton described the sophisticated apparatus he is building for pulsed four-wave mixing experiments using three separate lasers. Claudine Hoerner presented a theoretical analysis of ultrafast dephasing during pulsed four-wave mixing. Professor Emmanuel Anagnostakis described his proposed wave functions for the two-dimensional degenerate electron gas within the channel of a MODFET device. Xuesheng Chen explained her attempts to fit measured temperature-dependent linewidths and energies of Nd^{3+} transitions in GSGG to standard theoretical models. Eric Muelenkamp described his use of Tb^{3+} ions to help to unravel the complex electroluminescence behaviour at a semicondutor electrolyte interface. Jan de Kinder discussed time-resolved CARS measurements of the k=0 vibron states of the α-N crystals.

It is becoming traditional to have a small number of lectures on topics outside of the main theme of the school. In the first such lecture Professor Giovanni Costa described the phenomena of solid state physics using the language of particle physics, using terms such as "dressed" particles, anti-particles, and Goldstone modes to describe our familiar solid state excitations. Professor Michael Graf reminded us of the phenomenon of low temperature superconductivity and the successful understanding of this phenomenon through the accumulated work of London, Bardeen, Schrieffer, and Cooper. An adequate explanation for high temperature superconductivity, however, continues to elude us.

In addition to the formal lectures and seminars, which were periods of intellectual concentration, all the participants at the school enjoyed the excursions, marveled at many monuments to Sicily's glorious past, and were refreshed by the informality and camaraderie which are distinguished features of the school of spectroscopy.

The participants of the NATO Advanced Study Institute on "Optical Properties of Excited States in Solids", held June 16-30, 1991 in Erice, Italy.

PARTICIPANTS

	Names	Fields of Work
1)	Anagnostakis, Emmanuel A. Section of Solid State Physics Department of Physics University of Athens 22 Kalamakiou Ave., 17455 Alimos Athens, GREECE Tel: (30) 1 9817810 FAX: (30) 1 7234100	Quantum Electronics; Quantum Optics
2)	Auzel, Francois Centre National d'Etudes des Telecommunications 196, Ave. H. Ravera 92220 Bagneaux FRANCE Tel: (33) 1 45295202 FAX: (33) 1 45295405	RE Spectroscopy; Laser Materials; Energy Transfers; Upconversion; Nonradiative Processes; Superfluorescence; Optical Doped Fibers
3)	Baldacchini, Giuseppe ENEA Centro Richerche Energia Frascati Via E. Fermi, 27 00044 Frascati (Roma) ITALY Tel: (39) 06 94005365 FAX: (39) 06 94005400	Color Centers Physics R & D; High-resolution Molecular Spectroscopy
4)	Balda de LaCruz, Rolindes Departamento de Fisica Aplicada I Escuela Tecnica Superior de Ingenieros Industriales Y de Telecomunicacion Alda, Urquijo, S/N 48013, Bilbao, SPAIN Tel: (34) 4 4416400 FAX: (34) 4 4414041	Optical Spectroscopy; Time-resolved Spectroscopy; Photoacoustic Spectroscopy
5)	Beckwith, Clyfe* Department of Physics Boston College Chestnut Hill, MA 02167 USA Tel: (1) 617 5523575 FAX: (1) 617 5528478 E-MAIL: bitnet:beckwicl@bcvms	Absorption and Dispersion of Sound in a Gas in a Resonant Cavity via Photo- acoustic Spectroscopy

* Scientific Secretary of the Course

6) Blindauer, Carsten
Institut für Physikalische Chemie
Wegelerstr. 12
D-5300 Bonn 1
W. GERMANY
Tel: (49) 228 732631
FAX: (49) 228 732551

Optical Spectroscopy and Relaxation Behavior of Matrix-isolated Radicals

7) Borel, Corinne
CENG/LETI
Departement D'Optronique
Service des Matérieux et
Dispositifs Optiques, BP 85X
38041 Grenoble Cedex
FRANCE
Tel: (33) 76884214
FAX: (33) 76885157

Crystal Growth and Optical Characterization of RE-doped Oxide Crystals, for Solid State IR Laser Application

8) Boulon, Georges
Université Claude Bernard-Lyon
Laboratoire de Physico-Chimie
des Matériaux Luminescents
69622 Villeurbanne Cedex
FRANCE
Tel: (33) 72448271 or (33) 72448321
FAX: (33) 78894415

Spectroscopy of Doped-inorganic Luminescent Materials; Laser Crystals

9) Bowlby, Brian
Physics Department
Boston College
Chestnut Hill, MA 02167
USA
Tel: (1) 617 5524661
FAX: (1) 617 5528478

Spectroscopy of Laser Crystals

10) Buoncristiani, Martin
NASA Langely Research Center MS468
Hampton, VA 23665-5225
USA
Tel: (1) 804 5947192
FAX: (1) 804 5947772
E-MAIL: martinb@pcs.cnc.edu

Physics of Laser Crystals

11) Chen, Xuesheng
Department of Physics
Boston College
Chestnut Hill, MA 02167
USA
Tel: (1) 617 5524661
FAX: (1) 617 5528478
E-MAIL: chenxu@bcvms.bitnet

Spectroscopy of Laser Crystals

12) Costa, Giovanni
Università degli Studi
Instituto di Fisica Galileo Galilei
Via F. Marzolo 8
35100 Padova, ITALY

Elementary Particles

13) Cremona, Marco
ENEA, INN-SVIL
Via E. Fermi, 27
00044 Frascati (Rome)
ITALY
Tel: (39) 6 94005668
FAX: (39) 6 94005400
VAX Address (DECNET): efr419::cremona

Color Center Lasers and Spectroscopy

14) Croci, Mauro
Laboratorium für Physikalische Chemie
Universitätsstrasse 22
Eidgenössische Technische Hochschule
CH - 8092 Zürich
SWITZERLAND
Tel: (41) 01 2564384
FAX: (41) 01 2523402
E-MAIL: macr@ppc.lpc.ethz.ch

Hole-burning and Holographic Detection of Spectral Holes in Dye-doped Polymer-films (or Color Centers or RE-doped Inorganic Materials)

15) Daemen, Carine
Philips Lighting B.V.
Lamps I
P.O. Box 80020 Building EDW 633
5600 JM Eindhoven
THE NETHERLANDS
Tel: (31) 40756223
FAX: (31) 40755861
Telex: 35000phtcnl

Lamp Phosphors

16) De Kinder, Jan
Universitaire Instelling Antwerpen
Physics Department
Universiteitsplein 1
B-2610 Antwerpen (Wilrijk)
BELGIUM
Tel: (32) 3 8202451
FAX: (32) 3 8202245
E-MAIL: dekinder@phs.uia.ac.be

Time-resolved Spectroscopy; Molecular Crystals; Color Centers; Nonlinear Optics

17) De Matteis, Fabio
Universitaire Instelling Antwerpen
Physics Department
Universiteitsplein 1
B-2610 Antwerpen (Wilrijk)
BELGIUM
Tel: (32) 3 8202452
FAX: (32) 3 8202245

Time-resolved Spectroscopy of Defect Centeres in Alkali Halides

18) Di Bartolo, Baldassare[†]
Department of Physics
Boston College
Chestnut Hill, MA 02167, USA
Tel: (1) 617 5523601
FAX: (1) 617 5528478

Luminescence and Molecular Spectroscopy; Flash Photolysis; Photoacoustic Spectroscopy

---

[†]Director of the Course

19) Di Bartolo, John‡ — Theoretical Physics
Department of Physics
Boston College
Chestnut Hill, MA 02167, USA
Tel: (1) 617 5523575
FAX: (1) 617 5528478

20) Dinerman, Brad — RE Spectroscopy; Solid-state Infrared Lasers; Volleyball; Chinese Food
Schwartz Electro Optics
45 Winthrop Street
Concord, MA 01742, USA
Tel: (1) 508 3712299
FAX: (1) 508 3711265

21) Djeu, Nicholas — Solid State Lasers; Energy Transfer Crystal Fiber Devices
Department of Physics
University of South Florida
Tampa, FL 33620-5700
USA
Tel: (1) 813 9712121

22) Dominoni, Matteo — Energy Transfer Mechanisms and Photoacoustics
Dipartimento di Fisica
Sezione Fisica dello Stato Solido
e Nuovi Materiali
Via Celoria 16
20133 Milano, ITALY
Tel: (39) 2 2392410
FAX: (39) 2 2392357

23) Dujardin, Christophe — Photoionization and Fano Interactions in the Fluoride Systems Doped with Eu^{2+}
Université de Lyon I, Bat 205
Lab. de Physico-Chimmie des Mat.-Lum.
43, Bd. du 11 Novembre 1918
69622 Villeurbanne
FRANCE
Tel: (33) 72443371
FAX: (33) 78894415

24) Edvardsson, Sverker — Molecular Dynamics Applied to Judd/Ofelt Theory
Department of Inorganic Chemistry
Uppsala University
Box 531
S-751 21 Uppsala
SWEDEN
Tel: (46) 18183729
FAX: (46) 18108542

25) Fernandez, Joaquin — Optical Spectroscopy; Photoacoustic Spectroscopy
Departamento de Fisica Aplicada I
Escuela Tecnica Superior de
Ingenieros Industriales Y
de Telecomunicacion
Alda, Urquijo, S/N
48013, Bilbao, SPAIN
Tel: (34) 4 4416400
FAX: (34) 4 4414041

‡Administrative Secretary of the Course

26) Frei, Gabriela
Institut für Anorganische Chemie
Universität Bern
Freiestrasse 3
3009 Bern
SWITZERLAND
Tel: (41) 031 654254
FAX: (41) 031 654499

Optical Spectroscopy of d^6-chelate Complexes

27) Giessen, Harald
Universität Kaiserslautern
Fachbereich Physik
Erwin-Schrödingerstr.
D-6750 Kaiserslautern
GERMANY
Tel: (49) 6312053157
FAX: (49) 6312053300
E-MAIL: kphy0301@dkluni01.bitnet

HWE & PLE - Growth of II-VI Semiconductor-Films (CdS,CdSe) and Linear and Nonlinear Optical Properties of these Films
Nonlinear Optics;
Optical Computers

28) Graf, Michael
Physics Department
Boston College
Chestnut Hill, MA 02167
USA
Tel: (1) 617 5524128
FAX: (1) 617 5528478
E-MAIL: bitnet%"grafm@bcmvs"

Low-temperature Physics;
Strong Magnetic Fields;
Quantum Solids & Liquids

29) Grassano, Umberto M.
Dipartimento di Fisica
II Università degli Studi di Roma
Via O. Raimondo
00173 Roma
ITALY
Tel: (39) 679794521
FAX: (39) 62023507
E-MAIL: vaxtov::grassano

Two-photon Spectroscopy;
Laser Materials;
Picosecond Spectroscopy

30) Hamilton, James
Department of Chemistry
University of Wisconsin-Madison
1101 University Avenue
Madison, WI 53706
USA
Tel: (1) 608 2631082
FAX: (1) 608 2620381
E-MAIL: Hamilton@bert.chem.wisc.edu

Nonlinear Spectroscopy
Coherent & Incoherent Laser Spectroscopy

31) Hazenkamp, Menno F.
Debye Research Institute
University of Utrecht
P.O. Box 80 000
3508 TA Utrecht
THE NETHERLANDS
Tel: (31) 30532214
FAX: (31) 30517629

Luminescence of Inorganic Solids;
Luminescent Centers at Solid Surfaces;
Heterogeneous Catalysts

32) Hehlen, Markus P.
Institut für Anorganische Chemie
Universität Bern
Freiestrasse 3
3009 Bern
SWITZERLAND
Tel: (41) 031 654254
FAX: (41) 031 654499

Upconversion of light by Lanthanide Ions in Solids

33) Heindl, Rudolf
CNRS, Laboratoires de Bellevue, LPCM
1, Place A. Briand
92195 Meudon Principal Cedex
FRANCE
Tel: (33) 1 45075048

Radiothermoluminescence; Scintillators; OSL (Optical Stimulated Luminescence, X-rays Phosphors)

34) Hoerner, Claudine
I.P.C.M.S.
Groupe d'Optique Nonlinéaire
et d'Optoélectronique
5, Rue de l'Université
67084 Strasbourg Cedex
FRANCE
Tel: (33) 88358142

Nonlinear Optics (Organic Materials) Four Wave Mixing

35) Hölsä, Jorma P.K.
Helsinki Unviersity of Technology
Department of Chemical Engineering
Kemistintie 1A
SF-02150 ESPOO
FINLAND
Tel: (358) 0 4512598
FAX: (358) 0 462373
E-MAIL: jholsa@sorvi.hut.fi

RE Spectroscopy in Solids; Crystal Field Analysis

36) Huguenin, Denis
Rhône-Poulenc Recherches
52, Rue de la Haie Coq
93308 Aubervilliers Cedex
FRANCE
Tel: (33) 1 49376262
FAX: (33) 1 49376100

RE for Optical Applications

37) Imbusch, George F.
Department of Physics
University College
Galway, IRELAND
Tel: (353) 91 24411
FAX: (353) 91 25700

Optical Spectroscopy of Inorganic Insulating Materials

38) Jiang, Jien-Ping
Chemistry Department
University of California
Irvine, CA 92717
USA
Tel: (1) 714 8566162
FAX: (1) 714 8568571
E-MAIL: jjiang@vmsa.oac.uci.edu

Picosecond Spectroscopy of Third Order Optical Nonlinearity in Ladder Polymers; Optical Storage

39) Kaminskii, Alexander
Institute of Crystallography
Academy of Sciences of the USSR
Leninsky pr. 59
Moscow 117333, USSR
Tel: (95) 1352210
FAX: (95) 1351011

Physics and Spectroscopy of Laser-insulating Crystals

40) Kirkby, Scott
Department of Chemistry
University of Toronto
80 St. George Street
Toronto, Ontario
CANADA M5S 1A1
Tel: (1) 416 9784735

Linear and Nonlinear Optical Properties of Small Semiconductor Clusters in Zeolite Matrices

41) Klafter, Joseph
School of Chemistry
Tel Aviv University
Tel Aviv 69978
ISRAEL
Tel: (972) 3 5450254
FAX: (972) 3 6426212

Theories of Transport and Energy Transfer in Disordered Systems and Restricted Geometries

42) Klingshirn, Claus
Fachbereich Physik
Universität Kaiserslautern
Erwin-Schrödingerstr. 1
6750 Kaiserslautern
GERMANY
Tel: (49) 631 2052319
FAX: (49) 631 2053300

Linear and Nonlinear Semiconductor Optics; Growth of Semiconductor Layers and Dots; Nonlinear Dynamics

43) Kunttu, Henrik
Department of Physical Chemistry
University of Helsinki
Meritullinkatu 1C
SF-00170, Helsinki
FINLAND
FAX: (358) 0 1913528

Photochemistry; Charge Transfer and Laser Applications of Small Guests in Rare Gas Solids

44) Leblans, Marc
Universitaire Instelling Antwerpen
Physics Department
Universiteitsplein 1
B-2610 Antwerpen (Wilrijk)
BELGIUM
Tel: (32) 3 8202452
FAX: (32) 3 8202245

Raman Scattering and Time-resolved Spectroscopy on Point Defects in Alkali Halides

45) Lincoln, John
Physics Department
Southhampton University
Highfield Road
Southhampton SO9 5NH
UNITED KINGDOM
Tel: (44) 703 592194
FAX: (44) 703 585813

RE Doped Glasses and Crystals;
Effects of Glass Host and Host Geometry on RE

46) Litzenburger, Bernd
Technische Universität Berlin
Institut für Festkörperphysik
Sekr. PN 4-1
Hardenbergstr. 36
D-1000 Berlin 112
GERMANY
Tel: (49) 30 31424698

Optical Spectroscopy: Luminescence of Impurities (Fe, Mn) in II-VI-Compounds

47) Martin, Philippe
DRECAM
SRSIM
Bat 624, CEN Saclay
91191 Gif sur Yvette
FRANCE
FAX: (33) 1 69089063

Laser Surface Interactions

48) de Matos Gomes, Maria de Jesus
Department of Physics
University of Minho, Largo do Paço
4719 Braga Codex
PORTUGAL
Tel: (351) 53 612234 or
(351) 53 612252
FAX: (351) 53 612367

Linear and Nonlinear Optical Spectroscopy; Semiconductor (II-VI) Thin Films: Growth and Optical Properties; Optical Doped Glasses

49) Macfarlane, Roger
IBM Almaden Research Center K32/802
650 Harry Road
San Jose, CA 95120, USA
Tel: (1) 408 9272428
FAX: (1) 408 9272100

Laser Spectroscopy of Solids; Upconversion Lasers

50) McFarlane, Ross
Hughes Research Laboratories
3011 Malibu Canyon Road
Malibu, CA 90265, USA
Tel: (1) 213 3175445
FAX: (1) 213 3175483

Solid State Lasers and Spectroscopy; Semiconductor Quantum Confined Structures

51) Meichenin, Daniel
Centre National d'Études
des Télécommunications
196, Av. H. Ravera
92220 Bagneux
FRANCE
Tel: (33) 1 45295334
FAX: (33) 1 45295405

RE Spectroscopy

52) Mervic, Alberto
Universitá degli Studi di Milano
Dipartimento di Fisica
Via Celoria 16
20133 Milano
ITALY
Tel: (39) 02 2392341

Photoluminescence in SiO_2

53) Meulenkamp, Eric A.
Debye Research Institute
University of Utrecht
P.O. Box 80 000
3508 TA Utrecht
THE NETHERLANDS
Tel: (31) 30 532408
FAX: (31) 30 517629

Electroluminescence in Semiconductors;
RE-, s^2-, d^{10}-ions Luminescence

54) Monteiro, Teresa
Universidade de Aveiro
Departamento de Fisica
3800 Aveiro
PORTUGAL
Tel: (351) 34 25085

Luminescence of Defects in III-V Semiconductors Doped with T.M. Ions

55) Müller, Martin
Universität Kaiserslautern
FB Physik
Erwin-Schrödingerstr.
D-6750 Kaiserslautern
GERMANY
Tel: (49) 631 2053168
FAX: (49) 631 2053300

Epitaxy of II-VI Semiconductor Films by Pulsed Laser Evaporation;
Optical and Structural Characterization

56) Ogilvie, J.F.
Academia Sinica
Institute of Atomic and Molecular Sciences
P.O. Box 23-166
Taipei 10764, TAIWAN
FAX: (886) 2 3620200
E-MAIL: ogilvie@twniams.bitnet

Physical Chemistry and Molecular Physics

57) Ohkura, Hiroshi
Department of Electronics
Okayama University of Science
1-1 Ridai-Cho
Okayama, 700
JAPAN
Tel: (0862) 52 3161
FAX: (0862) 55 3611

Dynamical Processes in Resonant Secondary Radiation Emitted from Strong Coupling Systems

58) Pekcan, Önder
Technical University of Istanbul
Department of Physics
Maslak-Istanbul 80626
TURKEY
Tel: (90) 11763213

Fluorescence Studies in Restricted Geometries and Fractal Structure in Polymer Blends;
Polymer Diffusion Using Fluorescence Technique

59) Pode, Ramchandra B.
Department of Physics
Nagpur University
Nagpur 440010
INDIA
Tel: (91) 712 531946

Color Center Lasers

60) Reisfeld, Renata
Department of Inorganic
and Analytical Chemistry
The Hebrew University of Jerusalem
Jerusalem, ISRAEL
Tel: (972) 2 585323
FAX: (972) 2 585319

Optical Properties of Ions and Molecules in Glasses; Radiative and Nonradiative Transitions; Lasers; Energy Transfer; Nonlinear Optics; Quantum Dots; Restricted Geometry

61) Reuss, J.
Faculty of Sciences
Catholic University of Nÿmegen
6525ED Nÿmegen
THE NETHERLANDS
Tel: (31) 80 652101
E-MAIL: u630005@hnykun11

Molecular Spectroscopy of Molecules and Clusters; Relaxation Phenomena; Nonlinear Optics

62) Rossner, Wolfgang
Siemens AG
Otto-Hahn Ring 6
P.O. Box 83 09 53
D-8000 München 83
GERMANY
Tel: (49) 89 6362685
FAX: (49) 89 63648131

Development of Phosphors and Scintillators for Application in Medical Radiology and Nuclear Medicine

63) Saleh, Mona
Universität Kaiserslautern
Fachbereich Physik
Erwin-Schrödingerstr.
D-6750 Kaiserslautern
GERMANY
Tel: (49) 631 2053112

Preparation of Inorganic Silicate Glasses Containing Semiconductor Microcrystals of CdSSe and Studying Them Spectroscopically

64) Savasta, Salvatore
c/o Prof. F. Bogani
Dipartimento di Fisica
Largo Enrico Fermi 2
50125 FIRENZE (Arcetri)
ITALY
Tel: (39) 055 2298141
FAX: (39) 055 224032

Two photon excitation of Exciton-polaritons in II-VI Compounds (Spectroscopy in K Space)

65) Scacco, Augusto
Dipartimento di Fisica
Universitá di Roma - La Sapienza
Piazza A. Moro, 2
00185 Roma
ITALY
Tel: (39) 6 49914394
FAX: (39) 6 4957697

Crystal Growth; Optical Spectroscopy of Point Defects in Ionic Crystals

66) Schäffner, Mario
Universität Regensburg
Fakultät der Physik
Universitätstrasse 31
8400 Regensburg
F.R. of GERMANY
Tel: (49) 941 9434202

Nonlinear Optics; Two-photon-absorption; Second Harmonic Generation; Amplified Spontaneous Emission in CdSe

67) Schipper, Willem J.
Debye Research Institute
University of Utrecht
P.O. Box 80 000
3508 TA Utrecht
THE NETHERLANDS
Tel: (31) 30532408
FAX: (31) 30517629

Photostimulable X-ray Storage Phosphors

68) Schwentner, N.
Freie Universität Berlin
Institut für Experimentalphysik (WE1)
1000 Berlin 33
GERMANY
Tel: (49) 30 8386035
FAX: (49) 30 8386560

Matrix Isolation Spectroscopy; Photochemistry; Nonradiative Processes; VUV-stimulated Emission from Crystals

69) Smets, Bruno M.J.
Philips Lighting B.V.
Lamps I
P.O. Box 80020 - Building EDW-6
5600 TM, Eindhoven
THE NETHERLANDS
Tel: (31) 40756971
Telex: 35000phtcnl
FAX: (31) 40755861

Lamp Phosphors

70) Standifer, Eugene M.
University of Wisconsin-Madison
Department of Chemistry
Madison, WI 53706, USA
Tel: (1) 608 2631082
FAX: (1) 608 2620381
E-Mail: standifer@bert.chem.wisc.edu

Site-selective Laser Spectroscopy of Probe Ions in Perovskites

71) Stoneham, A. Marshall
A.E.R.E. Harwell
Didcot, Oxon
OX 11 ORA
UNITED KINGDOM
Tel: (44) 0235 432288

Realistic Theory of Defects; Defect Processes in Solids and at Surfaces

72) Struck, Charles
GTE Laboratories
40 Sylvan Road
Waltham, MA 02254, USA
Tel: (1) 617 4662326
FAX: (1) 617 8909320

Luminescence; Thermodynamics

73) Sztucki, Jaroslaw
Institut für Festkörperphysik
Fachgebiet Technische Physik
Fachbereich 5
Hochschulstrasse 8
W-6100 Darmstadt
GERMANY
Tel: (49) 6151 163484

Two-photon Spectroscopy of Transition Metal and RE Ions (theory); Transition Metal Pairs in Crystals (strongly coupled)

74) Ture, Iskender Engin
Gaziantep University
Department of Physics
Gaziantep 27310
TURKEY
Tel: (90) 85 160137
FAX: (90) 85 181749

Optoelectronic Devices; Electrical and Optical Characterization; Growth of II-VI & III-V Compounds, Defects and Characterization

75) Vacca, Roberto
Via Oddone di Cluny, 3
00153 ROMA
ITALY
Tel: (39) 6 5741264

Computer Science; Electric Control Systems; Writing

76) Vandenberghe, Geert
Katholieke Universiteit Leuven
Departement Scheikunde
Afdeling Anorganische
en Analytische Scheikunde
Celestijnenlaan 200 F
B-3001 LEUVEN
BELGIUM
Tel: (32) 16 200656 (ext. 3393)
FAX: (32) 16 223375
E-MAIL: fgcba16@blekul11

Two-photon Transitions in Lanthanide Ions (theoretical work - second quantization)

77) van der Voort, Dick
Debye Research Institute
University of Utrecht
P.O. Box 80 000
3508 TA Utrecht
THE NETHERLANDS
Tel: (31) 30532321
FAX: (31) 30517629

Luminescence of RE Ions with an Effective Charge

78) van der Waals, J.H.
Huygens Laboratorium
University of Leiden
P.O. Box 9504
2300 RA Leiden
THE NETHERLANDS
Tel: (31) 71275913 or
(31) 71275910
FAX: (31) 71275819

Excited States of Molecules and Ions and their Electronic Structure

79) van Schaik, Willem
Debye Research Institute
University of Utrecht
P.O. Box 80 000
3508 TA Utrecht
THE NETHERLANDS
Tel: (31) 30532214
FAX: (31) 30517629

Influence of Impurities on the Quantum Yield of Lamp Phosphors

80) van Tol, J. Hans
Rijks Universiteit Leiden
Huygens Laboratory (MAT)
Postbus 9504
2300 RA Leiden
THE NETHERLANDS
Tel: (31) 71275908

Study of Excited States by Magnetic Resonance Techniques

81) Viliani, Gabriele
Dipartimento di Fisica
Universita degli Studi di Trento
38050 Povo (Trento)
ITALY
Tel: (39) 0461 881555

Dynamical Properties of Disordered Solids

82) Wolf, Mats
Department of Inorganic Chemistry
Uppsala University
Box 531
SWEDEN
Tel: (46) 18183729
FAX: (46) 18108542

Optical Properties of β'' Aluminas (RE-doped); Structural Work: Molecular Dynamics; Judd/Ofelt Analysis

83) Wrachtrup, Jörg
FU Berlin
FB Physik
Arnimallee 14
W-1000 Berlin 33
GERMANY
Tel: (49) 30 8386074
FAX: (49) 30 8386560

Study of Coherent Process by ODMR; Spectroscopy on Molecular Aggregates in Solids

84) Wright, Andrew
University of Wisconsin
Department of Chemistry
1101 University Avenue
Madison, WI 53706, USA
Tel: (1) 608 2631082
FAX: (1) 608 2620381

Study of Defect Systems in Solids by Means of Site Selective Laser Spectroscopy

85) Zhou, Bei-Wen
Centre National d'Études
des Télécommunications
196, Av. H. Ravera
92220 Bagneux
FRANCE
Tel: (33) 1 45295334
FAX: (33) 1 45295405

RE Spectroscopy

INDEX

The manufacturer's authorised representative in the EU is Springer Nature Customer Service Centre GmbH, Europaplatz 3, 69115 Heidelberg, Germany. If you have any concerns regarding our products, please contact ProductSafety@springernature.com

Printed and bound by CPI Group (UK) Ltd, Croydon, CR0 4YY
16/03/2026
02072181-0013